INDUSTRIAL ORGANIC CHEMICALS

Starting Materials
and Intermediates

VOLUME 7

Weinheim · New York · Chichester · Brisbane · Singapore · Toronto

INDUSTRIAL ORGANIC CHEMICALS

VOLUME 1
Acetaldehyde to **Aniline**

VOLUME 2
Anthracene to **Cellulose Ethers**

VOLUME 3
Chlorinated Hydrocarbons to **Dicarboxylic Acids, Aliphatic**

VOLUME 4
Dimethyl Ether to **Fatty Acids**

VOLUME 5
Fatty Alcohols to **Melamine and Guanamines**

VOLUME 6
Mercaptoacetic Acid and Derivatives to **Phosphorus Compounds, Organic**

VOLUME 7
Phthalic Acid and Derivatives to **Sulfones and Sulfoxides**

VOLUME 8
Sulfonic Acids, Aliphatic to **Xylidines**

Index

INDUSTRIAL ORGANIC CHEMICALS

Starting Materials and Intermediates

VOLUME 7
Phthalic Acid and Derivatives
to **Sulfones and Sulfoxides**

Weinheim · New York · Chichester · Brisbane · Singapore · Toronto

This book was carefully produced. Nevertheless, authors and publisher do not warrant the information contained therein to be free of errors. Readers are advised to keep in mind that statements, data, illustrations, procedural details or other items may inadvertently be inaccurate.

Library of Congress Card No.: Applied for.
British Library Cataloguing-in-Publication Data: A catalogue record for this book is available from the British Library.

Die Deutsche Bibliothek – CIP-Einheitsaufnahme
Industrial organic chemicals : starting materials and intermediates ;
an Ullmann's encyclopedia. – Weinheim ; New York ;
Chichester ; Brisbane ; Singapore ; Toronto : Wiley-VCH
   ISBN 3-527-29645-X
Vol. 7. Phtalic Acid and Derivatives to Sulfones and Sulfoxides. – 1. Aufl. – 1999.

© WILEY-VCH Verlag GmbH, D-69469 Weinheim (Federal Republic of Germany), 1999
Printed on acid-free and chlorine-free paper.
All rights reserved (including those of translation in other languages). No part of this book may be reproduced in any form – by photoprinting, microfilm, or any other means – nor transmitted or translated into machine language without written permission from the publishers. Registered names, trademarks, etc. used in this book, even when not specifically marked as such, are not to be considered unprotected by law.

Composition and Printing: Rombach GmbH, Druck- und Verlagshaus, D-79115 Freiburg
Bookbinding: Wilhelm Osswald & Co., D-67433 Neustadt (Weinstraße)
Cover design: mmad, Michel Meyer, D-69469 Weinheim
Printed in the Federal Republic of Germany

Contents

1 Phthalic Acid and Derivatives

1. Phthalic Acid................ 3866
2. Phthalic Anhydride........... 3866
3. Phthalimide................. 3880
4. Phthalonitrile 3883
5. Phthalates.................. 3886
6. Toxicology.................. 3894
7. References.................. 3904

2 Phthalocyanines

1. Introduction 3920
2. Physical Properties........... 3923
3. Chemical Properties 3928
4. Production 3933
5. Phthalocyanine Derivatives...... 3944
6. Legal Aspects and Environmental Protection................. 3952
7. Quality Specifications and Analysis . 3953
8. Uses...................... 3953
9. Economic Aspects............. 3956
10. Toxicology and Occupational Health . 3957
11. References.................. 3959

3 Polysaccharides

1. Introduction 3972
2. Analysis and Characterization 3973
3. Pectin..................... 3975
4. Alginates 3993
5. Carrageenan 4003
6. Agar 4009
7. Gum Arabic 4012
8. Gum Tragacanth 4015
9. Gum Karaya 4016
10. Gum Ghatti................. 4017
11. Xanthan Gum 4018
12. Gellan Gum 4023
13. Galactomannans.............. 4023
14. References.................. 4028

4 Propanal

1. Introduction 4037
2. Physical Properties........... 4037
3. Chemical Properties and Uses..... 4038
4. Production 4041
5. Quality Specifications 4041
6. Economic Aspects............. 4042
7. Storage and Transportation 4042
8. Toxicology and Occupational Health 4042
9. References.................. 4043

5 Propanediols

1. 1,2-Propanediol and Higher Propylene Glycols................. 4047
2. 1,3-Propanediol 4055
3. Toxicology.................. 4057
4. References.................. 4059

V

6 Propanols

1. Introduction 4061
2. Physical Properties 4062
3. Chemical Properties 4063
4. Production 4065
5. Uses 4070
6. Specifications 4072
7. Economic Aspects 4073
8. Storage and Transportation ... 4074
9. Toxicology and Occupational Health 4075
10. References 4076

7 Propene

1. Introduction 4081
2. Physical Properties 4082
3. Chemical Properties 4082
4. Production 4082
5. Storage, Transportation, Quality Requirements 4093
6. Uses 4094
7. Economic Aspects 4098
8. References 4098

8 Propionic Acid and Derivatives

1. Introduction 4102
2. Physical Properties 4102
3. Chemical Properties 4104
4. Production 4105
5. Storage and Transportation ... 4110
6. Quality Specifications 4110
7. Environmental Protection 4111
8. Uses 4111
9. Economic Aspects 4114
10. Toxicology 4115
11. Derivatives 4116
12. References 4124

9 Propylene Oxide

1. Introduction 4129
2. Physical Properties 4130
3. Chemical Properties 4132
4. Production and Raw Materials . 4134
5. Environmental Protection and Ecotoxicology 4149
6. Quality Specifications and Analysis . 4151
7. Handling, Storage, and Transportation 4152
8. Uses 4154
9. Economic Aspects 4156
10. Toxicology and Occupational Health 4156
11. References 4159

10 Purine Derivatives

1. Introduction 4165
2. Properties 4166
3. Occurrence and Production ... 4167
4. Quality Specifications and Analysis . 4170
5. Uses 4170
6. Economic Aspects 4172
7. References 4172

11 Pyridine and Pyridine Derivatives

1. Introduction 4176
2. Pyridine and Alkylpyridines 4176
3. Pyridine Derivatives 4189
4. Toxicology 4213
5. References 4221

12 Pyrimidine and Pyrimidine Derivatives

1. Introduction 4227
2. Chemical Properties 4228
3. Production 4229
4. Important Naturally Occurring Derivatives of Pyrimidine 4230
5. Industrially Important Pyrimidines . 4232
6. References 4234

13 Pyrrole

1. Pyrrole 4235
2. Pyrrole Derivatives 4237
3. Toxicology and Occupational Health 4238
4. References 4239

14 2-Pyrrolidine

1. Introduction 4241
2. 2-Pyrrolidine 4241
3. *N*-Methyl-2-pyrrolidine 4243
4. *N*-Vinyl-2-pyrrolidine 4246
5. Other Pyrrolidone Derivatives 4247
6. Toxicology and Occupational Health 4248
7. References 4250

15 Quinoline and Isoquinoline

1. Quinoline 4253
2. Isoquinoline 4256
3. Toxicology 4257
4. References 4258

16 Resorcinol

1. Introduction 4261
2. Physical Properties 4261
3. Chemical Properties 4263
4. Production 4263
5. Environmental Protection 4267
6. Quality Specifications and Analysis . 4267
7. Uses 4268
8. Transport Classifications and Occupational Health 4268
9. References 4269

17 Salicylic Acid

1. Introduction 4271
2. Physical and Chemical Properties .. 4272
3. Production 4273
4. Quality Specifications, Storage and Transportation, and Environmental Protection 4275
5. Uses and Economic Aspects 4276
6. Salicylic Acid Derivatives 4277
7. Toxicology 4279
8. References 4280

18 Saponins

1. Introduction 4284
2. Pharmacology 4287
3. Plant Saponins 4288
4. Animal Saponins 4299
5. References 4301

19 Silicon Compounds, Organic

1. Introduction 4306
2. Fundamental Synthetic Routes for Organosilicon Compounds 4310
3. Silicon-Functional Organosilicon Compounds 4319
4. Organofunctional Organosilicon Compounds 4330
5. Other Organosilanes 4343
6. Uses 4345
7. Toxicology and Environmental Aspects 4355
8. Economic Aspects 4356
9. References 4357

20 Sorbic Acid

1. Introduction 4365
2. Physical Properties 4366
3. Chemical Properties and Derivatives 4366
4. Production 4368
5. Environmental Protection 4369
6. Quality Specifications 4369
7. Analysis 4370
8. Storage and Transportation 4370
9. Legal Aspects 4370
10. Mode of Action and Uses 4371
11. Economic Aspects 4373
12. Toxicology 4373
13. References 4374

21 Starch

1. Starch 4379
2. Modified Starches 4396
3. Economic Aspects 4411
4. References 4412

22 Styrene

1. Introduction 4417
2. Physical Properties 4418
3. Chemical Properties 4419
4. Production 4422
5. Quality and Testing 4431
6. Storage and Transportation 4431
7. Uses and Economic Aspects 4432
8. Related Monomers 4434
9. Toxicology and Occupational Health 4439
10. References 4440

23 Sulfamic Acid

1. Introduction 4443
2. Physical Properties 4443
3. Chemical Properties 4444
4. Production 4445
5. Uses 4446
6. Toxicology 4446
7. Sulfamates 4446
8. References 4447

24 Sulfinic Acids and Derivatives

1. Introduction 4449
2. Properties 4450
3. Preparation of Sulfinic Acids and their Salts 4452
4. Reactions of Sulfinic Acids and their Derivatives 4457
5. Sulfinic Acid Derivatives 4458
6. 1-Hydroxyalkanesulfinates 4465
7. Formamidinesulfinic Acid 4468
8. Industrial Uses of Sulfinic Acids and their Derivatives 4469
9. Toxicology 4470
10. References 4471

25 Sulfones and Sulfoxides

1. Sulfones 4477
2. Sulfoxides 4486
3. Toxicology 4496
4. References 4497

IX

Phthalic Acid and Derivatives

Individual keywords: Isophthalic acid→Terephthalic Acid and Dimethyl Terephthalate

PETER M. LORZ, BASF AG, Ludwigshafen, Federal Republic of Germany (Chaps. 1–4)
FRIEDRICH K. TOWAE, BASF AG, Ludwigshafen, Federal Republic of Germany (Chaps. 1–4)
WALTER ENKE, BASF AG, Ludwigshafen, Federal Republic of Germany (Chap. 5)
RUDOLF JÄCKH, BASF AG, Ludwigshafen, Federal Republic of Germany (Chap. 6)
NARESH BHARGAVA, BASF Canada Inc., Cornwall, Ontario, Canada (Chaps. 1–5)

1.	**Phthalic Acid**........ 3866	3.3.	Uses................ 3883
2.	**Phthalic Anhydride**...... 3866	4.	**Phthalonitrile**......... 3883
2.1.	**Physical Properties**...... 3867	4.1.	Properties............ 3883
2.2.	**Chemical Properties**...... 3868	4.2.	Production........... 3883
2.3.	**Resources and Raw Materials**. 3869	4.2.1.	Production from *o*-Xylene..... 3884
2.4.	**Production**............ 3869	4.2.2.	Production from Phthalic Acid Derivatives............ 3885
2.4.1.	Gas-Phase Oxidation........ 3869	4.3.	Uses................ 3885
2.4.1.1.	Catalyst and Reaction Mechanism 3871	5.	**Phthalates**............ 3886
2.4.1.2.	Apparatus and Important Process Steps in the Gas-Phase Oxidation of *o*-Xylene............. 3873	5.1.	**Physical and Chemical Properties**............ 3887
2.4.2.	Fluidized-Bed Oxidation...... 3876	5.2.	**Raw Materials**......... 3887
2.4.3.	Liquid-Phase Oxidation of *o*-Xylene............... 3877	5.3.	**Production**........... 3888
		5.4.	**Environmental Protection**... 3890
2.5.	**Environmental Protection**... 3877	5.5.	**Quality Specifications**...... 3891
2.6.	**Quality Specifications and Analysis**.............. 3878	5.6.	**Storage and Transportation**.. 3892
2.7.	**Economic Aspects**........ 3879	5.7.	**Uses**................ 3892
2.8.	**Storage and Transportation**.. 3879	5.8.	**Economic Aspects**........ 3893
2.9.	**Uses**................ 3880	6.	**Toxicology**............ 3894
3.	**Phthalimide**............ 3880	6.1.	**Phthalic Acid**........... 3894
3.1.	**Properties**............. 3880	6.2.	**Phthalimide**............ 3894
3.2.	**Production**............ 3881	6.3.	**Phthalic Anhydride**....... 3894
3.2.1.	Production from Phthalic Anhydride and Ammonia..... 3881	6.4.	**Phthalonitrile**.......... 3895
3.2.2.	Production from Phthalic Anhydride and Urea......... 3882	6.5.	**Phthalate Esters**......... 3896
		6.5.1.	General Profile........... 3896
3.2.3.	Production from *o*-Xylene..... 3882	6.5.2.	Acute Toxicity and Irritating Potential................ 3898

6.5.3.	Liver Effects; Peroxisome Proliferation 3898	6.5.6.	Species Differences and Relevance to Humans 3901
6.5.4.	Carcinogenicity 3899	6.5.7.	Reproductive Toxicity 3901
		6.5.8.	Metabolism and Toxicokinetics. . 3903
6.5.5.	Genotoxicity and Mutagenicity . . 3900	7.	**References**. 3904

1. Phthalic Acid

Phthalic acid [*88-99-3*], 1,2-benzenedicarboxylic acid, has remained unimportant industrially. It is formed as a byproduct in the manufacture of phthalic anhydride.

Physical Properties. Phthalic acid, $C_8H_6O_4$, M_r 166.14, forms colorless, monoclinic crystals which melt at 191 °C (sealed tube) and are converted into phthalic anhydride with the elimination of water at 210 – 211 °C. Some physical properties of phthalic acid are as follows [1] – [3]:

Density (15 °C)	1.593 g/cm^3
Heat of fusion	315.3 J/g
Specific heat of solid (0 – 99 °C)	1.214 J g^{-1} K^{-1}
Heat of combustion	19 657.03 J/g
Heat of formation	43 714.34 J/g
Heat of solution at 25 °C	123.55 J/g
Flash point	168 °C

2. Phthalic Anhydride

Phthalic anhydride [*85-44-9*], isobenzofuran-1,3-dione, has been commercially produced continuously since 1872 when BASF developed the naphthalene oxidation process. It was the first anhydride of a dicarboxylic acid to be used commercially and is comparable in its importance to acetic acid.

The most important derivatives of phthalic anhydride are plasticizers and, to a lesser degree, polyester resins and dyes.

About 60 years after its discovery in 1836 by A. LAURENT, a more effective commercial process for its production was introduced, which was based on mercury-catalyzed liquid-phase oxidation of naphthalene.

The breakthrough that led to commercial production of a quality product was the development of the gas-phase oxidation of naphthalene or *o*-xylene in an air stream with vanadium oxide as catalyst [4].

Table 1. Solubility of phthalic anhydride

Solvent	Temperature, °C	Solubility, g/100 g
Water	20	1.64
Water	50	1.74
Water	100	19.0
Carbon disulfide	20	0.7
Formic acid	20	4.7
Pyridine	20	80
Benzene		soluble
Ethanol	20	soluble
Diethyl ether	20	slightly soluble

2.1. Physical Properties

Phthalic anhydride, $C_8H_4O_3$, M_r 148.12, forms colorless needles or platelets, with a monoclinic or rhombic crystalline form. Some important physical properties of phthalic anhydride are as follows [1], [5], [6]:

Density of solid (4 °C)	1.527 g/cm^3
Specific vapor density (1013 mbar)	6.61 kg/m^3
Melting point	131.6 °C
Boiling point (1013 mbar)	295.1 °C
Heat of fusion	159.1 J/g
Heat of combustion	22 160.7 J/g
Heat of formation from naphthalene	12 058 J/g
Heat of formation from o-xylene	8625 J/g
Heat of sublimation	601 J/g
Heat of evaporation	441.7 J/g
Flash point	152 °C
Ignition temperature	580 °C
Upper limit of flammability (1013 mbar)	10.5 vol%
Lower limit of flammability (1013 mbar)	1.7 vol%
Lower dust explosion limit	25 g/m^3

The density of liquid phthalic anhydride in the range 140 – 240 °C can be calculated by using the following equation [7]:

$$\varrho/\text{kg m}^{-3} = 1321.55 - 0.6697\,(t/°C) - 0.000905\,(t/°C)^2$$

Table 1 lists the solubility of phthalic anhydride in various solvents.

The reported explosion hazard data of phthalic anhydride in air vary significantly [4], [5]. Explosions can occur at concentrations below 100 g/m^3, depending on the impurities present. Recent incidents in production plants indicate that phthalic anhydride concentrations exceeding 35 g/m^3 in the reaction product gas are capable of ignition if heat-transfer salt enters the reactor due to broken reactor tubes.

2.2. Chemical Properties

As a cyclic anhydride, phthalic anhydride is a reactive compound, but in addition, the otherwise very stable aromatic ring is capable of reaction. Phthalic anhydride reactions which have achieved commercial importance are summarized below. The most important is the reaction with alcohols or diols to give esters or polyesters.

Unsaturated polyester resins are obtained by polycondensation in the presence of maleic anhydride or fumaric acid.

One or both of the carboxy groups can react with ammonia to give phthalic monoamide and phthalimide [85-41-6] or phthalonitrile [91-15-6].

Phthalein and rhodamine dyes, some of which have been in production for over 100 years and have not yet lost their importance, are obtained by reaction of phthalic anhydride with phenols, aminophenols or quinaldine derivatives. The Friedel–Crafts reaction of phthalic anhydride with benzene derivatives followed by ring closure to form anthraquinone derivatives is of importance as a route to Indanthrene dyes.

Much attention has been devoted to the rearrangement of dipotassium phthalate to produce terephthalic acid [100-21-0] [8], but due to technical problems the process is no longer used.

3,5-Dihydrophthalic acid can be produced by electrochemical hydrogenation of phthalic anhydride [4]. Hydrogenation with a nickel catalyst produces phthalide [529-20-4].

2.3. Resources and Raw Materials

In the 1960s a fundamental shift took place in the raw material base for phthalic anhydride. From 1960 to 1975 the production was switched from 100% coal-tar naphthalene to about 75% o-xylene. In 1991, 85% of phthalic anhydride was produced from o-xylene [9]. Naphthalene derived from petroleum, which is produced mainly in the United States, has not gained importance as a raw material for phthalic anhydride.

The changeover to o-xylene as the feedstock was inevitable because the quantities of naphthalene derived from coal tar depend on the production of coke and these were unable to keep pace with the increasing demand for naphthalene. On the other hand, naphthalene is an inevitable byproduct of coke production which should be used commercially; therefore, the phthalic anhydride process will continue to use some naphthalene as feedstock in the future. o-Xylene, which can readily be separated from the mixture of xylenes containing roughly one third o-xylene and two thirds p-xylene, is nowadays available in adequate quantities from cracking plants and refineries. However, in the past, variations in the demand for p-xylene have often affected the availability and price of o-xylene, as has the need to increase the amount of alkyl aromatics in unleaded gasoline.

These facts and the need to be able to react to varying raw material prices have resulted in plants being planned which are capable of processing o-xylene or naphthalene or mixtures of the two with tailor-made catalysts.

2.4. Production

2.4.1. Gas-Phase Oxidation

Phthalic anhydride is predominantly produced on an industrial scale by gas-phase oxidation of o-xylene or naphthalene.

General Features. Preheated o-xylene is introduced into a stream of hot air. The o-xylene – air mixture is passed through a tubular reactor where the exothermic oxidation takes place on a highly selective catalyst. The heat of reaction is used to produce steam, only part of which is utilized in the plant itself. Any excess steam can be used elsewhere. The gases emerging from the reactor are precooled. At high loadings of phthalic anhydride in the product gases, some liquid phthalic anhydride can be won in a liquid condenser. The product gases are then fed to a switch condenser system where the phthalic anhydride is condensed on the finned tubes as a solid. The switch condensers are cooled by a heat-transfer oil in an automated switching cycle. During the heating cycle the deposited phthalic anhydride is melted and collected in a storage tank.

After the phthalic anhydride has been separated, the exhaust gases still contain byproducts and small quantities of phtalic anhydride and must be cleaned by scrubbing with water, or catalytically or thermally incinerated. If scrubbing with water is employed, it is possible to concentrate the main byproduct, maleic acid, and the scrubbing solution can then be processed further to yield fumaric acid or maleic anhydride [10], [11]. If the scrubbing of the exhaust gases is combined with production of maleic anhydride, the discharge of any polluted wastewater from the plant can be avoided.

The crude phthalic anhydride is transferred to a continuous thermal/chemical treatment system, which converts the phthalic acid formed as a byproduct into the anhydride. The crude product is then purified in a continuous two-stage distillation system [12].

The BASF process can be operated with a wide range of o-xylene loadings up to 105 g/m^3 (STP). The reactor outlet gases are further treated catalytically in a finishing reactor to decrease the amount of byproducts and to improve product quality. Flexible operation of the reactor system enables the phthalic anhydride yield to be optimized and decreases the amount of residues and volatile organic compounds. Simultaneously a smoother operation of the catalyst in the tubular reactor prolongs its life. Addition of SO_2 to activate the catalyst is not required, and there is no need for a liquid condenser before collecting the crude phthalic anhydride in the switch condensers. Because of the low byproduct content, chemical treatment of the crude phthalic anhydride is unnecessary. The plant design is optimized for low energy consumption and a high net export of steam and electrictity [13].

The Wacker process makes it possible to use *o*-xylene or naphthalene or mixtures of the two. With *o*-xylene, loadings of 90–100 g/m^3 (STP) are possible [14], [15]. Modifications of the process are aimed at saving energy [16].

The Nippon Shokubai VGR Process. In patent literature, the Nippon Shokubai VGR (vent gas recycling) process is described. Its characteristic feature is, that exhaust gas is recycled and added to the mixture of *o*-xylene and process air to reduce the oxygen concentration to less than 10 vol%. This makes it possible to work outside the limits of flammability despite a high *o*-xylene loading (up to 85 g/m^3 STP). Despite the low O_2 concentration, yields of up to 116 g PA/100 g xylene (given as 116% by PA producers) have been reported. This is attributed to a specially developed catalyst system. It is also possible to use naphthalene instead of *o*-xylene in this process [17]–[21].

The Alusuisse–Ftalital LAR Process. In the LAR (low air ratio) process, *o*-xylene loadings in the process air of up to 134 g/m^3 (STP) are suggested. This could make it possible to achieve an appreciable reduction in energy and the size of the equipment [22], [23]. In commercial application *o*-xylene loadings up to 80 g/m^3 (STP) are reached [107]. Catalysts in the form of rings or half rings are used. The catalyst is capable of processing both *o*-xylene and naphthalene or mixtures of the two [24], [25].

The Rhône-Poulenc process employs *o*-xylene as raw material. The crude product is subjected to a chemical posttreatment before being purified by two-stage distillation. The waste gas is incinerated [26], [27].

The ELF Atochem/Nippon Shokubai process uses a Nippon Shokubai catalyst for reacting *o*-xylene or naphthalene. With *o*-xylene, feed loadings of up to 75 g/m^3 (STP) are used industrially [107].

2.4.1.1. Catalyst and Reaction Mechanism

The oxidation of *o*-xylene and naphthalene is nowadays carried out almost exclusively in tubular reactors cooled by a molten salt. The spherical catalysts previously employed are only used in isolated cases in old plants. In modern plants ring-type catalysts have become established for energy saving reasons [28], but catalysts on half-shell supports are also available [24], [25]. Commercial catalysts are normally composed of an inert ring-type support, made of silicate, silicon carbide, porcelain, alumina, or quartz measuring 5 – 10 mm in diameter, on which a thin active layer of a mixture of finely divided titanium dioxide and vanadium oxide is deposited. Compounds of antimony, rubidium, cesium, niobium, and phosphorus are added to improve the selectivity [29] – [35]. The use of a two-zone catalyst, i.e., a combination of catalysts with low activity with catalysts of higher activity, has become generally established [36]. For high loadings even three-zone catalysts are known. Some catalysts require addition of SO$_2$ for activation and longer service life [37], [38]. The reaction is driven to an almost quantitative turnover of *o*-xylene, to ensure low concentrations of disturbing byproducts (e.g., phthalide). It is summarized by the following equation:

$$\text{o-xylene} \xrightarrow[360-390\,°C]{3\,O_2} \text{phthalic anhydride} + 3\,H_2O$$

The heat of formation is 1108.7 kJ/mol. Total combustion of *o*-xylene yields 4380 kJ/mol, i.e., almost four times the heat of formation of phthalic anhydride. Oxidation in tubular reactors produces a heat of reaction between 1300 and 1800 kJ/mol of *o*-xylene.

Scheme shows the important byproducts found in the reaction product. Although the intermediate alcohol has so far not been detected, it is probable that the formation of phthalic anhydride passes through it. A standard interpretation of the course of the reaction is not available in the literature. All the conclusions are very dependent on the method of measurement used and on the experimental setup. So far no commercially used catalyst has been investigated in depth. There are also no literature data relating to aging phenomena. It is generally accepted that a redox mechanism, in which the selective oxidation proceeds at oxygen atoms in the lattice, is involved. Attempts are made in [39]–[59] to describe the course of the reaction.

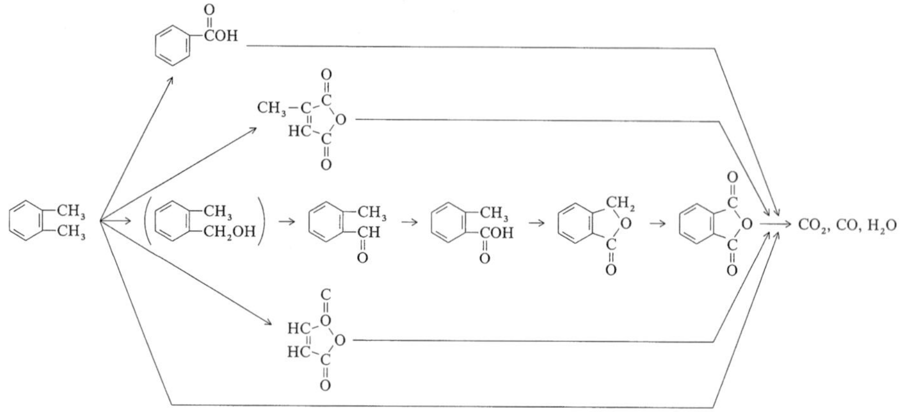

Scheme 1.

According to literature data, it is now possible to achieve yields of 120 kg PA /100 kg o-xylene [35], but it is generally accepted that reactor yields exceeding 112–114 kg PA/ 100 kg o-xylene (calc. 100 %) are improbable on an industrial scale. This is equivalent to 80–82 % of stoichiometric yield. In o-xylene based processes, after allowing for the losses during condensation, dehydration, and distillation, pure phthalic anhydride yields of 110–112 kg PA/100 kg o-xylene may be expected.

The oxidation of naphthalene proceeds in accordance with the following equation:

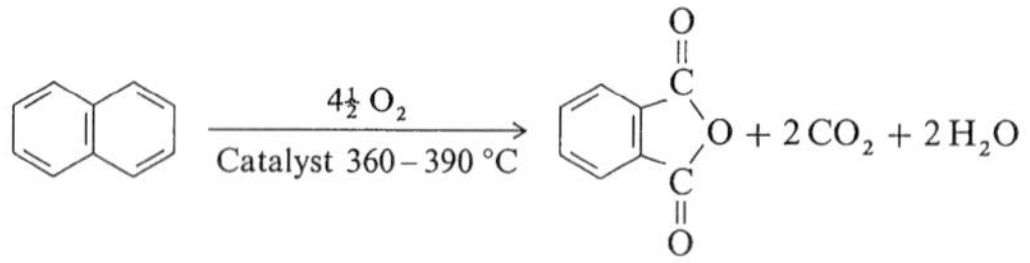

$$naphthalene \xrightarrow[\text{Catalyst 360–390 °C}]{4\tfrac{1}{2} O_2} \text{phthalic anhydride} + 2\,CO_2 + 2\,H_2O$$

The heat of formation is 1788 kJ/mol; total combustion of naphthalene yields 5050 kJ/mol. Depending on the yield achieved and the byproducts formed, a heat of reaction of 2100–2500 kJ/mol is expected. Possible side reactions are shown in Scheme 1.

Besides maleic anhydride, naphthoquinone is also formed as a byproduct, which requires a high performance of the purification stage. Data on the course of the reaction can be found in [60]–[63].

According to the literature, yields of up to 102 kg/100kg naphthalene are achievable in commercial plants, but the final product yields generally do not exceed 98 kg/100kg naphthalene, i.e., 85% of theory.

2.4.1.2. Apparatus and Important Process Steps in the Gas-Phase Oxidation of o-Xylene

Procedure Employing an Ignitable Xylene–Air Mixture. The current objective of phthalic anhydride technology is essentially to limit energy consumption as far as possible. This is supplemented by efforts to construct as complete a system as possible for utilizing the steam produced in the plant, or alternatively by considering driving the air blower with a steam turbine [64], [65]. It is possible to save energy costs primarily by increasing the concentration of o-xylene in the process air, thus reducing the quantity of process air for the same capacity and increasing the steam yield [66].

It is possible to operate existing older plants, which were originally designed for loadings of 40 g/m^3 (STP), with o-xylene loadings of up to 80 g/m^3 (STP) by employing modern catalysts and modifying the reactor cooling system. The attainable increase in loading depends primarily on the feasibility of removing safely the increased heat of reaction from the reactor.

The lower limit of flammability is exceeded with a loading of ca. 44 g/m^3. In order to exceed the limit of flammability, it is essential to estimate the risks to the safety of the personnel and of the plants. It is impossible to completely eliminate the risk of ignition of an explosive mixture, but the equipment may be designed to be inherently safe, i.e., to withstand the highest possible pressure. Alternatively the equipment may be safeguarded by pressure relief devices so that pressure surges are relieved safely. Probability

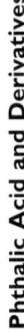
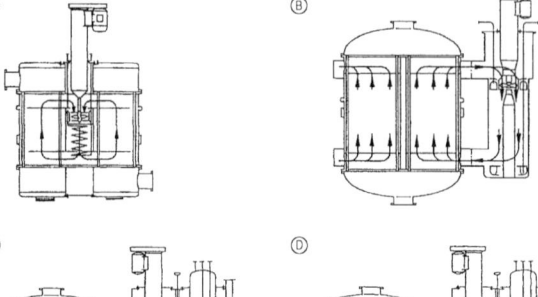
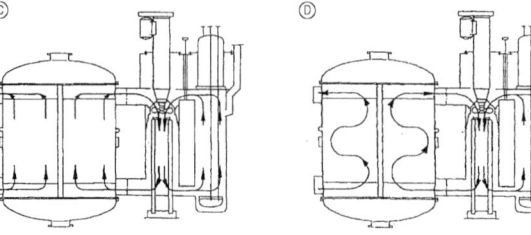

Figure 1. Reactor types used for phthalic anhydride production
A) Reactor designed for the von Heyden catalyst; B) Reactor with an external cooling system; C) Reactor with an external cooling system in which saturated steam is produced; D) Reactor with radial salt flow (DWE-type)

of ignition inside the catalyst is minimal as the autoignition temperature for phthalic anhydride is 580 °C, which is far above the temperature of the reaction.

Reactors. Figure 1 shows the development of the reactor types used to produce most of the phthalic anhydride worldwide. The reactor (A) was originally designed for the von Heyden catalyst and has a salt bath circulation pump and cooler fitted internally. This reactor gave way to reactor (B), which has an external cooling system [67]; the latter makes it possible to achieve a more advantageous temperature distribution for larger diameters. Superheated steam at pressures up to ca. 70 bar is produced in both reactors. The reactor (C), with an external cooling system in which saturated steam is produced, was developed for processes with higher loadings [68]. Reactor (D) is the common type for high-load processes. It uses radial salt flow for improved heat transfer. Reactors can be machined in one piece up to a diameter of 8 m, but can also be fabricated in segments and assembled on site [69]. Such reactors may incorporate more than 20 000 catalyst tubes and the construction of single-line plants having phthalic anhydride capacities exceeding 60 000 t/a is possible.

Switch Condenser. A decisive factor for the efficiency of a phthalic anhydride plant is the performance of the condenser system in which the phthalic anhydride (which is contained in the process gas at a concentration of less than 1 vol%) must be condensed as completely as possible. Several separating methods are discussed in the literature. Processes in which the phthalic anhydride reaction gas is extracted by absorption into solvents or liquid phthalic anhydride [70]–[72] have not been exploited commercially.

Generally, switch condensers equipped with finned tubes which are alternately heated or cooled with a hot or cold heat-transfer medium are used. Solid phthalic anhydride is desublimated on the surface of finned tube bundles and subsequently melted out and collected in a storage tank.

A milestone in the improvement of switch condenser technology was the changing of the design from the upflow-type to the downflow-type. In the old switch condensers the reaction gas enters from the bottom and passes through the internal tube bundles from the bottom to the top. Most of the phthalic anhydride is deposited on the lowest bundle.

In the downflow type, the reaction gas enters at the top and leaves the condenser at the bottom. Most of the solid product deposits on the uppermost bundle, and the lower tube bundles are washed by molten phthalic anhydride during each melting cycle. In this way maleic acid and phthalic acid, which are byproducts of the process, are removed from the tubes. Phthalic acid and especially maleic acid are highly corrosive, and their iron salts, being pyrophoric, can cause fires by self-ignition.

Cooling and heating are usually carried out by a heat-transfer oil, fed through the finned tubes. Cooling with water and heating with steam has been proposed but has no obvious advantage.

Modern condensers can separate more than 99.5% of the phthalic anhydride from the reaction gas. Improvements in the arrangement of the finned tubes resulted in reduced oil throughput and hence lower energy consumption compared with older condensers. The mechanical flexibility has been improved by changing the design, leading to a reduction of thermal stresses when switching from hot oil to cold oil and vice versa.

The increase of the capacities in the recent years have also affected the size of equipment, and switch condensers with a surface area of up to 4000 m^2 are being used today. At the beginning of the melting cycle, when the cold oil is replaced by hot oil, the system requires the highest amount of steam to maintain the oil temperature due to the heating-up of the swith condenser and to the heat of fusion of the phthalic anhydride. By installation of an additional heat reservoir connected to the hot oil system, it is possible to eliminate peaks and make the steam demand more uniform [73]–[76]. This effect is especially pronounced when the air blower is driven by a turbine using the steam from the same plant.

Treatment of Crude Phthalic Anhydride. To obtain high-purity phthalic anhydride, the crude product is subjected to thermal treatment before purification by distillation. Treatment with chemicals is generally necessary in the case of the crude product obtained from naphthalene in order to remove the byproduct, naphthoquinone, but is also used for products manufactured from *o*-xylene. Apart from sulfuric acid, which has been used for a long time, the following substances are used to destroy naphthoquinone: sodium hydroxide [77]; boric acid/sulfuric acid [78], [79]; potassium disulfide [80]; sodium carbonate [81]–[83]; potassium maleate [84]; and aliphatic polydienes [85], [86]. The most widely used method is probably simple heat treatment, but this is only suitable for crude products with low contents of impurities. It involves heating the crude product to 100–400 °C in a cascade of vessels. Normally the temperature is limited to 230–300 °C, and the retention times are 10–24 h. In the case of phthalic anhydride produced from *o*-xylene, the treatment is intended to remove water and low-

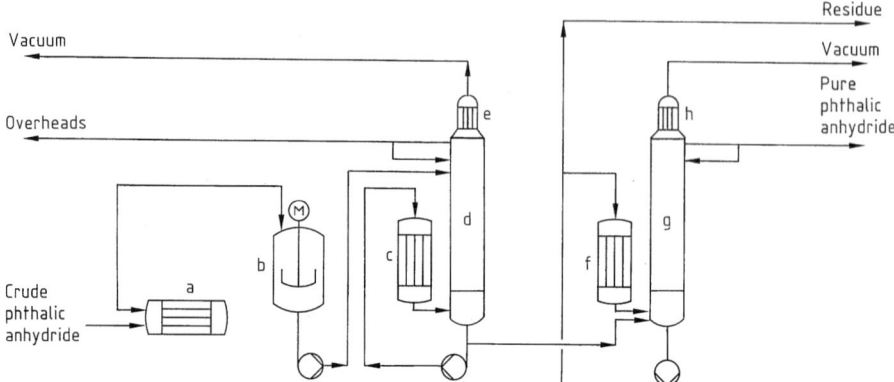

Figure 2. Purification of crude phthalic anhydride by distillation
a) Crude phthalic anhydride heater; b) Predecomposer; c) Reboiler; d) Stripper column; e) Condenser; f) Reboiler; g) Distillation column; h) Condenser

boiling contaminants such as maleic anhydride and o-tolualdehyde. The destruction of phthalide is also achieved in pretreatments [87]–[90].

Distillation. The final purification of the crude phthalic anhydride is normally carried out in a continuous distillation system. Figure 2 shows schematically a typical distillation plant. The crude phthalic anhydride is first pretreated in a predecomposer where byproduct phthalic acid is converted to phthalic anhydride. The product is then introduced into a column in which the low-boiling constituents such as maleic anhydride and benzoic acid are concentrated at the head and removed. The bottoms are introduced into a distillation column, from which pure phthalic anhydride is removed from the head while the residue is discharged from the bottom.

2.4.2. Fluidized-Bed Oxidation

Of the many fluidized-bed processes for the oxidation of naphthalene, the process developed by Badger may still be in use today [91]–[93]. The capacity of such plants has been increased by injecting oxygen and air into the fluidized bed, thus enabling the naphthalene concentration to be increased.

Liquid naphthalene is injected directly at the bottom of the catalyst bed and evaporates immediately, distributing itself over the entire bed. Here it comes in contact with the catalyst and reacts with the atmospheric oxygen fed in via a distributor plate.

The vigorous agitation and mixing in the fluidized bed results in a uniform temperature being maintained throughout the bed, with the temperature being in the range 345–385 °C. The heat is removed by cooling tubes installed in the bed and is used to produce high-pressure steam. Any entrained catalyst is separated by cyclones and, after being retained in specially constructed ceramic filters, is blown back into the reactor.

Several filter units are used, and one of them is always blown back with a stream of air in order to remove catalyst from the filter surfaces. Up to 60% of the phthalic anhydride can be recovered in a liquid condenser, the remainder being desublimated in switch condensers. The crude product is then purified in a distillation column under vacuum. A proposal for regenerating a deactivated catalyst with SO_2 has been published [94].

Kawasaki Steel Corporation is working on the fluidized-bed oxidation of o-xylene. According to patent literature a stoichiometric yield of up to 84% at an o-xylene conversion of 94% is reached [95]–[99].

2.4.3. Liquid-Phase Oxidation of o-Xylene

In the liquid-phase oxidation of o-xylene, a mixture of acetic acid, o-xylene, and catalyst, which consists mainly of Co, Mn, and Br, is fed to the first vessel in a cascade. The reaction, which is completed in the subsequent vessels, is initiated under pressure while the air is injected. The water produced by the reaction is removed in the first vessel by azeotropic distillation with o-xylene. The isomers of phthalic acid are removed from the reaction mixture, and the phthalic anhydride is obtained by crystallization. The crude product is subjected to a pretreatment which imposes special requirements because of the bromine content. Purification can then be carried out by distillation. The o-xylene must have an ortho isomer content > 99% if the process is to achieve high yields of 130 kg of phthalic anhydride from 100 kg of o-xylene. Companies such as Hüls have investigated the process [100] and, in addition, further work has been done on it by Standard Oil of Indiana [101].

Sisas has described a two-step process: in the first liquid-phase oxidation step crude o-toluic acid is formed with a yield of 1.19 g per gram of o-xylene at partial o-xylene turnover; unreacted o-xylene is recycled. The second step is the gas-phase oxidation of the o-toluic acid to phthalic anhydride. The overall molar yield based on o-xylene is up to 88% [102].

2.5. Environmental Protection

Exhaust Gas. The exhaust gas from the switch condensers still contains an appreciable quantity of organic substances and therefore must be purified before it is discharged from the plant. Scrubbing with water in a scrubbing tower is suitable for extracting maleic anhydride, the main byproduct in the exhaust gas. In a recovery process, the organic acids from the scrubbed gas are concentrated in several scrubbing towers in which the scrubbing water is circulated until the concentration is ca. 30%. The scrubbing solution is then worked up to recover maleic anhydride or fumaric acid. It is also possible to produce sodium maleate by adding sodium hydroxide or sodium carbonate [103].

The efficiency of the scrubbing depends on the sophistication of the equipment and the number of scrubbing stages. The concentration of all organic compounds can be reduced to < 150 mg per m^3 exhaust air discharged from the plant.

The main substance emitted from phthalic anhydride plants is carbon monoxide, which cannot be removed by exhaust-gas scrubbers. A plant emits about 1100 t CO per 10 000 t of phthalic anhydride produced, depending on the operating conditions.

Apart from scrubbing towers, two processes are at present available for purifying exhaust gas:

1) Thermal combustion at 650–850 °C which involves burning the organics and carbon monoxide in the presence of an oil-fed or gas-fed reverberatory flame.
2) Catalytic combustion at 250–450 °C, which normally does not require additional fuel. This process employs noble-metal or mixed-oxide catalysts.

Such plants are supplied, for example, by Haldør Topsoe in Denmark who employ their Catox process or by Engelhard who market the Torvex process.

A comparison of thermal and catalytic exhaust-gas combustion for phthalic anhydride plants is discussed in [104].

Catalyst. Depending on the operating procedure, the catalyst normally has a service life of 2–8 years, after which it must be replaced because its activity is too low. Used catalysts may be deposited, for example, at a secure landfill site; this depends on local regulations owing to the toxicity of the additives. As a result of various problems, washing and reusing the active material has so far been uneconomical, even though proposals are repeatedly being made in this connection [105], [106].

2.6. Quality Specifications and Analysis

The specifications of technically pure phthalic anhydride are listed in Table 2. A suitable method of determining the acids by gas chromatography is to form derivatives, for example, with diazomethane, but direct determination on packed columns or capillary columns is also possible.

Table 2. Specifications of technical-grade phthalic anhydride

Property	Average value	Method	Specification limit
Purity, %	99.9	GC	99.8 min
Solidification point, °C	131.0	ASTM 1493-67	130.8 min
Maleic anhydride, %	0.03	GC	0.05 max
Benzoic acid, %	0.03	GC	0.1 max
Color index of melt (APHA)	5–10	ASTM 3366-74	20 max
Heat stability (APHA)	10–20	ASTM 3366-74	40 max

2.7. Economic Aspects

In 1996 world production capacity for phthalic anhydride was $> 3.7 \times 10^6$ t/a and this figure is expected to exceed 4.6×10^6 t/a in the year 2000. A remarkable increase in capacity is occurring in south-east Asia.

The world capacities for phthalic anhydride (10^3 t) in 1996 can be broken down as follows:

Western Europe	869
Eastern Europe	641
North America	568
South America	256
Japan	320
South Korea	260
India	153
Taiwan	95
Remainder of Asia	155
Africa and Near East	124

2.8. Storage and Transportation

Pure phthalic anhydride is a very stable compound and may be stored for a prolonged period in the molten state without any change in its properties. To prevent fires, it is recommended that storage containers and transportation are blanketed with nitrogen.

Molten phthalic anhydride is subject to the following safety and transport regulations:

EC directive no.	607-009-00-4
Symbols	X_i
R phrases	36/37/38
UN no.	3256
Packaging group	III
Transportation regulations	
GGVE/RID	3.61 c
GGVS/ADR	3.61 c
ADN/ADNR	3.61 c
GGV See/IMDG code	3.3, UN-No. 3256-PG III
IATA Dangerous Goods Regulation	3, UN-No. 3256

2.9. Uses

The most important application of phthalic anhydride is the production of plasticizers. The main uses of phthalic anhydride are [13]:

Plasticizers	55 %
Unsaturated polyester resins	14 %
Alkyd resins	15 %
Other	16 %

3. Phthalimide

3.1. Properties

Phthalimide [85-41-6], 1,3-dioxoisoindoline, $C_8H_5O_2N$, M_r 147.14, crystallizes from solution as white needles or prisms, while sublimation produces platelets. The compound is sparingly soluble in water (0.3 g at 20 °C, 0.9 g at 50 °C, 2.2 g at 100 °C in 100 g of water) and readily soluble in acetic acid, sodium hydroxide solution, or potassium hydroxide solution. Some physical properties of phthalimide are as follows:

Melting point	238 °C
Heat of combustion	3560 kJ/mol
Heat of fusion	187.6 J/g
Specific heat at 100 °C	1.21 J g^{-1} K^{-1}
Vapor pressure at	
120 °C	0.10 mbar
150 °C	0.95 mbar
180 °C	5.93 mbar
220 °C	30.7 mbar
254 °C	187.6 mbar
Flash point	214 °C
Ignition temperature	530 °C

With bases, phthalimide forms water-soluble salts, which react with halogens to form the corresponding N-chloro, N-bromo, or N-iodo compounds. These N-halo compounds are also obtained if alkali-metal phthalimides are reacted with hypochlorous or hypobromous acid. When the N-halo compounds are heated, isatoic anhydride [118-48-9] or anthranilic acid (o-aminobenzoic acid) [118-92-3] is formed in a Hofmann degradation [108], [109]. The reaction of alkali-metal phthalimides with alkyl halides to give N-alkyl phthalimides, and subsequent hydrolysis or hydrazinolysis affords primary amines (Gabriel synthesis) [110]:

$$\text{Phth-NK} + RX \xrightarrow{-KX} \text{Phth-NR} \xrightarrow{N_2H_4/H_2O, H^+} \text{Phth(NH)(NH)} + RNH_2$$

3.2. Production

Phthalimide is produced almost exclusively from phthalic anhydride and ammonia, but processes based on phthalic anhydride and urea or oxidative ammonolysis of *o*-xylene are also known.

3.2.1. Production from Phthalic Anhydride and Ammonia

The continuous processes for producing phthalimide are particularly important industrially. One of the continuous processes is carried out in an externally heated, vertical reaction tube filled with packing material [111]. A gas-tight connection links the bottom of the reaction tube to a sublimation chamber from which a discharge arm discharges the phthalimide through a slot by means of a screw conveyor. The exhaust gases are discharged from the sublimation chamber via built-in baffles.

Molten phthalic anhydride and excess ammonia are continuously fed into the reaction tube at the top and are reacted at 250–280 °C. The reaction gases are cooled to 170–180 °C in the sublimation chamber, and solid phthalimide is deposited and discharged. Water and excess ammonia are removed via an exhaust-gas pipe. A 98 % yield of phthalimide is obtained with a purity of 99 %.

Another continuous process for producing phthalimide from phthalic anhydride is a countercurrent process [112]. Molten phthalic anhydride is continuously fed to the head of a reactor while ammonia is continuously fed to the bottom. The temperature increases from ≥ 150 °C at the top of the reactor to a maximum of 270 °C at the bottom. The molten phthalimide emerging from the bottom of the reactor with a purity of 99 % is cooled and flaked or dissolved in aqueous alkali, and supplied as a starting material for further syntheses, e.g., to produce anthranilic acid.

The exhaust gas from the head of the reactor, which is composed of water vapor, sublimed phthalimide and phthalic anhydride, and ammonia, is fed to the bottom of a scrubbing column where it is scrubbed in countercurrent with molten phthalimide from the bottom of the reactor. The melt taken from the bottom of the scrubbing column, which contains unreacted phthalic anhydride, is returned to the reactor, while the exhaust gas from the scrubbing column, which contains predominantly water vapor and phthalimide, is dissolved in aqueous alkali, and the solution is used to produce anthranilic acid.

3.2.2. Production from Phthalic Anhydride and Urea

Solvent-Free Process. A mixture of phthalic anhydride and urea is introduced into a heatable, tiltable reaction vessel. Mounted on the cover of the vessel is an extraction tube for the reaction gases, CO_2 and water vapor. The reactor is heated to 130–140 °C until the contents are molten, and as soon as the elimination of CO_2 and water starts, heating is ceased. The heat of reaction causes the temperature to rise to ca. 160 °C. The reaction is complete when the contents of the reactor have swelled to several times their original volume as a result of foaming and have solidified.

The cover of the reactor is then removed. The pure phthalimide is discharged into a collecting vessel by tilting the reactor, is cooled and ground, and used without further treatment. The yield is > 90 % [113].

Production in a Solvent. Phthalic acid or phthalic anhydride can be reacted with urea in the presence of a solvent to form phthalimide. Suitable solvents are substituted and unsubstituted hydrocarbons and aromatics and heteroaromatics such as *n*-propylbenzene, cumene, 1,2-dichlorbenzene, and picoline. Urea must be insoluble in the reaction medium, while phthalic acid or phthalic anhydride must be only slightly soluble. The reaction is carried out below the boiling point of the solvent, normally at 160–170 °C. When the reaction is complete, the pure phthalimide is filtered off and washed with water. The yield is 95–100 % [114].

3.2.3. Production from o-Xylene

o-Xylene is reacted in the gas phase with ammonia in the presence of a metal-oxide catalyst as oxygen donor. Phthalimide, phthalamide, or phthalonitrile can be selectively produced, depending on how the reaction is controlled [115].

3.3. Uses

Phthalimide is of industrial importance as the starting material for producing anthranilic acid by Hofmann degradation, and a large number of primary amines can be produced by the Gabriel synthesis. Phthalimide is an intermediate in the production of agricultural pesticides and wood preservatives, pigments, and pharmaceuticals.

4. Phthalonitrile

4.1. Properties

Phthalonitrile [91-15-6], 1,2-dicyanobenzene, 1,2-benzenedicarbonitrile, $C_8H_4N_2$, M_r 128.14, is a crystalline powder having a faint grayish yellow color and a slighty aromatic odor, similar to benzonitrile. Phthalonitrile was first described in 1896 when it was isolated during the diazotization of 2-aminobenzonitrile.

The compound is sparingly soluble in water (ca. 1 g/L) and soluble in acetone, nitrobenzene, and benzonitrile. It cannot be distilled and polymerizes if heated above the melting point. It is not explosive and is difficult to ignite, but the dust can explode. Some physical properties of phthalonitrile are as follows:

Melting point	141 °C
Heat of combustion	4013 kJ/mol
Heat of evaporation	67 kJ/mol
Specific heat at 30 °C	1.30 J g^{-1} K^{-1}
Vapor pressure at 20 °C	0.05 mbar
Density	1.238 g/cm^3
Apparent density	ca. 0.5 g/cm^3
Flash point	162 °C

4.2. Production

Phthalonitrile is produced commercially from *o*-xylene, phthalic acid, phthalic anhydride, phthalamide (**1**), or phthalimide (**2**).

4.2.1. Production from o-Xylene

Ammoxidation. In a single-stage continuous process, o-xylene is converted to phthalonitrile by reaction with ammonia and oxygen in the gas phase in a fluidized-bed reactor. Generally, metal oxide mixtures containing vanadium, antimony, chromium, and molybdenum, with further active components such as iron, tungsten, and alkali-metal oxides, on an alumina or silica support are used as catalysts [116]–[120], [121]–[123].

$$\text{o-C}_6\text{H}_4(\text{CH}_3)_2 + 2\,\text{NH}_3 + 3\,\text{O}_2 \longrightarrow \text{o-C}_6\text{H}_4(\text{C}\equiv\text{N})_2 + 6\,\text{H}_2\text{O}$$

The course of the reaction and the industrial importance of ammoxidation have been described in depth in [124]–[126]. The commercial process of producing phthalonitrile from o-xylene by oxidative ammonolysis is as follows [127], [128]: A gaseous mixture of o-xylene, NH_3, and O_2 is passed through a distributor plate into a fluidized-bed reactor. Either a vanadium oxide–antimony oxide catalyst or a vanadium oxide–chromium oxide catalyst on an alumina support is used. The optimum temperature for pressureless ammoxidation is 480 °C. Temperatures below 480 °C lead to the formation of up to 10 % of the undesirable byproducts phthalamide and phthalimide. Above 500 °C, ammonia begins to burn. The cooling elements incorporated in the fluidized bed make it possible to keep the temperature constant despite the large heat of reaction. The hot reaction gases from the reactor are quenched in a product settler with an aqueous suspension of phthalonitrile. After settling and cooling, the phthalonitrile is separated in decanters and dried.

The phthalonitrile produced in this way, which contains $< 0.1\,\%$ of acid and phthalimide and $< 0.1\,\%$ water, can be processed directly without further purification to yield phthalocyanine pigments. The yield is 80–85 %.

Ammonia is removed from the gas mixture emerging from the product settler (the latter contains NH_3, CO_2, CO, N_2, and traces of HCN) in an ammonia recovery plant, and the exhaust gas is burnt. The ammonia is returned to the process. Unreacted o-xylene and the intermediate o-toluonitrile can be worked up to isolate o-toluonitrile or can be fed back into the fluidized-bed reactor for complete conversion to phthalonitrile.

Oxidative Ammonolysis. The continuous production of phthalonitrile from o-xylene by oxidative ammonolysis has been studied in a pilot plant [115], [129]. In this process o-xylene is reacted with ammonia in the presence of a metal-oxide catalyst in the gas phase at 350–450 °C using a vanadium oxide–molybdenum oxide catalyst on an alumina or silica support. The oxidation medium is the metal-oxide catalyst, part of which is continuously removed, reoxidized, and then fed back into the reactor. The phthalonitrile is quenched from the hot reaction gases in the same way as in the ammoxidation process, while byproducts such as o-toluonitrile, phthalamide, and phthalimide are fed back into the phthalonitrile reactor or can be converted into phthalonitrile in a second reactor at 400 °C in the presence of NH_3 on a boron phosphate–alumina catalyst.

Ammonia is recovered from the residual gas mixture, which contains NH_3, CO_2, CO, N_2, and traces of HCN, and fed back into the process. Noxious gases such as CO and HCN are burnt in the regenerator and N_2 and CO_2 are eliminated from the process.

4.2.2. Production from Phthalic Acid Derivatives

Phthalonitrile can be produced from phthalic acid, phthalic anhydride, phthalamide, or phthalimide by reaction with ammonia and elimination of water at 300–500 °C in the gas phase in the presence of a catalyst.

Catalysts mentioned in the patent literature are the oxides of thorium, copper, beryllium, zirconium, or tungsten on a silica, alumina, or phosphate support, or on silicates, borates, and basic alumina.

Molten phthalic anhydride heated to ca. 160 °C is evaporated in a heating apparatus into which preheated circulating gas containing NH_3 in a concentration of ca. 90 % is also fed. The gas mixture is reacted under virtually unpressurized conditions at 350–400 °C in a fixed-bed reactor downstream of the heating apparatus using an Al_2O_3 catalyst. The reaction gas emerging from the reactor is quenched with water, and the phthalonitrile is separated from the aqueous suspension in decanters and dried. The concentration of NH_3 in the circulating gas is kept at ca. 90 % by adding NH_3.

Numerous other production methods have been described [130], [131], but have no commercial significance. For example, phthalonitrile can be prepared by eliminating water from phthalamide in the presence of acid halides such as phosgene or thionyl chloride. To prevent hydrolysis of the phthalamide by the acid formed, diluents such as benzene or chlorobenzene are added or the acids are bound by tertiary amines such as *N,N*-diethyl-*o*-toluidine or pyridine. It is also possible to use acylated secondary amines such as *N*-methylformamide.

4.3. Uses

Phthalonitrile is used as a starting material for producing phthalocyanine pigments (→ Phthalocyanines), fluorescent brightners, and photographic sensitizers.

Table 3. Commercially important phthalates

Chemical denomination	Code * DIN 7723	CAS registry no.	M_r	Viscosity at 20 °C, mPa·s	Density at 20 °C, g/m³	Refractive index at 20 °C
Dimethyl phthalate	DMP	[131-11-3]	194.19	18	1.191	1.516
Diethyl phthalate	DEP	[84-66-2]	222.24	13	1.117	1.502
Dibutyl phthalate	DBP	[84-74-2]	278.35	21	1.047	1.493
Diisobutyl phthalate	DIBP	[84-69-5]	278.35	41	1.039	1.490
Butyl benzyl phthalate	BBP	[85-68-7]	312.36	59	1.121	1.538
Diisopentyl phthalate	DIPP	[84777-06-0]	306.40	30	1.024	1.490
Diheptyl phthalate	DHP	[68515-50-4]	362.51	49	0.993	1.487
Di-2-ethylhexyl phthalate	DOP(DEHP)	[117-81-7]	390.56	80	0.983	1.487
Diisooctyl phthalate	DIOP	[27554-26-3]	390.56	75–84	0.984	1.488
Di-n-octyl phthalate	DNOP	[117-84-0]	390.56	39	0.980	1.485
Di(hexyl-octyl-decyl) phthalate	HXODP(610P)	[68515-51-5]		44	0.974	1.484
Di(octyl-nonyl-decyl) phthalate	ONDP(810P)	[71662-46-9]		49	0.969	1.483
Di(heptyl-nonyl) phthalate	HNP(79P)	[68515-41-3]		44	0.986	1.486
Di(heptyl-nonyl-undecyl) phthalate	HNUP(711P)	[68515-42-4]		50	0.972	1.484
Di(nonyl-decyl-undecyl) phthalate	NDUP(911P)	[68515-43-5]		73	0.962	1.484
Diisononyl phthalate	DINP **	[28553-12-0]	418.62	80	0.975	1.486
Diisononyl phthalate	DINP **	[68515-48-0]	418.62	105	0.974	1.486
Di(3,5,5-trimethylhexyl) phthalate	DINP **	[14103-61-8]	418.62	110	0.968	1.486
Diisononyl phthalate	DINP **	[28553-12-0]	418.62	165	0.978	1.488
Diisodecyl phthalate	DIDP	[26761-40-0]	446.67	130	0.966	1.486
Diundecyl phthalate	DUP	[3648-20-2]	474.72	70	0.955	1.482
Diisoundecyl phthalate	DIUP	[85507-79-5]	474.72	200	0.964	1.487
Diisotridecyl phthalate	DTDP	[68515-47-9]	530.83	310	0.948	1.482
Di(methoxyethyl) phthalate	(DMEP)	[117-82-8]	282.29	55	1.170	1.502
Di(butoxyethyl) phthalate	(DBEP)	[117-83-9]	366.45	42	1.061	1.486

* Codes in parentheses are used in English.
** DINP types are based on different branched isononyl alcohols.

5. Phthalates

A wide variety of dialkyl phthalates are produced and marketed (Table 3).

In the early 1900s, the availability of monohydric alcohols was limited to those with a chain length of up to four carbon atoms. This resulted in the development and manufacture of four esters; e.g., diethyl phthalate (DEP) was used as a heat-transfer oil, and dibutyl phthalate (DBP) was used to reduce the hygroscopicity of explosives. Both DEP and DBP were also used as carriers in the perfume industry [132].

Also at this time camphor was being used as a plasticizer for cellulose nitrate and cellulose acetate. In the early 1920s it was found that phthalates could replace the expensive plasticizer camphor. This increased the demand for phthalates [132]. In 1937, the "Chemiker Taschenbuch" [133] listed seven phthalates for use as plasticizers. The late 1920s saw the start of the development of poly(vinyl chloride), PVC. PVC is rigid and brittle at ambient temperature. However, when plasticized it was found to be suitable for replacing natural rubber [132]. In 1929 a U.S. patent for plasticized PVC was granted to OSTROMISLENSKI [134], and KYRIDES applied for U.S. patent for the use of di-2-ethylhexyl phthalate (DEHP) as plasticizer for PVC, which was granted in 1933 [135].

The most important phthalates are DEHP, the standard general purpose plasticizer for PVC, followed by DINP and DIDP [136]. In trade, DEHP is also referred to as DOP. Besides the phthalates listed in Table 3, phthalate blends and coesters are used for special applications [137].

5.1. Physical and Chemical Properties

All phthalates listed in Table 3 are clear, oily liquids at room temperature. They are soluble in common organic solvents and are miscible with other PVC plasticizers. When added to plastics and resins, they improve the workability during fabrication, modify the properties, or give rise to new improved properties not exhibited by the original material [138].

Some of the important properties of plasticizers are listed below. These properties are influenced by the manufacturing process and can vary significantly. The values given are typical ranges in which commercial products are available. The analytical methods are listed in Table 4.

APHA color	5–60
Acid number, mg KOH/g	0.02–0.1
Water content, wt%	0.05–0.1
Purity, area% or wt%	99.0–99.5
Odor	odorless or low

5.2. Raw Materials

Phthalic anhydride (PA) and monohydric alcohols are generally the raw materials for phthalates of commercial significance. Phthalic anhy-dride is mainly used in the molten form. The alcohols are based on the C_2-C_4 olefins, which are produced in either the steam cracking process or in petrochemical plants. These alcohols can be produced by (1) oligomerization or Ziegler synthesis followed by hydroformylation and hydrogenation, or (2) by aldol condensation with hydrogenation (e.g., the production of 2-ethylhexanol from propylene). The linear or slightly branched C_6 to C_{11} alcohols are based on ethylene. Isononanols are produced from the C_4 fraction of the steam cracker output or from C_8-C_9 olefins from a poly gas unit (olefin oligomerization) [133].

5.3. Production

The formation of phthalates is as follows:

$$\text{PA} \xrightarrow[\text{fast}]{\text{ROH}} \text{monoester (COOR, COOH)} \underset{\text{slow}}{\overset{\text{ROH}}{\rightleftharpoons}} \text{diester (COOR, COOR)} + H_2O$$

The first step, alcoholysis of PA to give the monoester, is rapid and goes to completion. The reaction generally starts at elevated temperatures and proceeds exothermically.

The second step is the conversion of the monoester to a diester with the formation of water. This is a reversible reaction and proceeds more slowly than the first, thus determining the overall rate of reaction. To shift the equilibrium towards the diester, the water of reaction is removed by distillation. The rate of reaction can be influenced by the choice of catalyst and the reaction temperature. For fast conversion rates, high reaction temperatures are generally used. However, these are influenced by the boiling point of the alcohol and/or the type of catalyst.

Sulfuric acid [132], methanesulfonic acid [139], [140] and p-toluenesulfonic acid [132] are effective esterification catalysts [140]. Sodium hydrogensulfate [132] or the monoester itself [132], [141] can also be used. Acid catalysts can be used for esterification reactions at temperatures up to 140–165 °C [141], [132]. Their use above these temperatures results in undesirable side reactions such as alcohol degradation with the formation of olefins, ethers, and colored products [141]. Acid catalysts are the preferred catalysts for the manufacture of phthalates from low-boiling alcohols up to C_4. Acid catalysts have also been used for the production of phthalates from higher alcohols in the presence of a water entrainer, e.g., aromatics or chlorinated hydrocarbons. Alternatively, the esterification can be carried out at reduced pressure, which allows the distillation of a water–alcohol azeotrope at low temperature; however, longer reaction times are required.

The autocatalytic esterification [132] can be carried out at temperatures above 200 °C, but the conversion to diester is not complete, and the recovery of unreacted monoester is necessary.

Currently, nearly all major phthalate producers use amphoteric catalysts for the esterification of high boiling alcohols. The first claim for the use of titanates as esterification catalysts was made in 1955 in the United States [142]. Since then titanates and tin(II) compounds have been widely used as esterification catalysts [140], [143]–[145].

The reaction temperatures for the amphoteric catalysts are about 200 °C [143]. At this temperature side reactions are minimized, and the alcohol can be recycled without purification. By using this type of catalyst, over 99.5 % conversion to diester can be achieved [143]–[145].

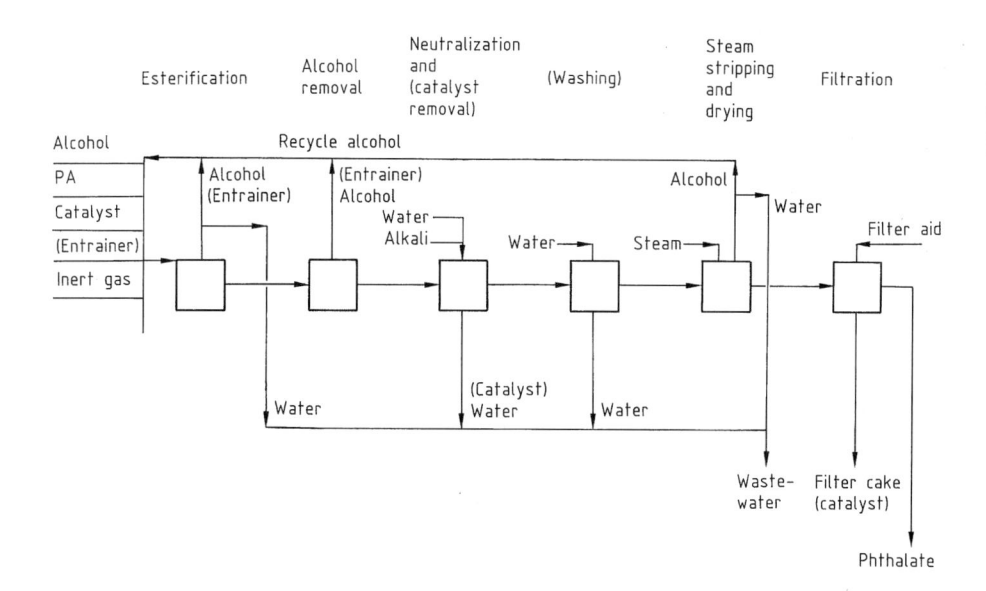

Figure 3. Simplified flow sheet for the production of phthalates. Terms in parentheses are optional.

Phthalates can be produced continuously [132], [141] or batchwise. The process steps outlined in Figure 3 are applicable to both methods. Variations are possible depending on the catalyst used, the reactant alcohol, and the process type.

For batch processes the reactor is usually a stirred vessel made of stainless steel, which can be rapidly heated to maintain a high distillation rate during diester formation for removing the water of reaction.

For the esterification of the water-soluble C_1-C_4 alcohols, the overhead vapors of the reactor are fed into a rectification column where the evaporated alcohol is dried before it is recycled to the reactor. If higher alcohols of very low water solubility are used, the vapors from the reactor can be fed directly into a condenser which drains into a alcohol–water separator. The water is separated here by settling and the alcohol is recycled to the reactor.

Usually a 20–25% excess of alcohol is used [140]. The catalyst can be added with the alcohol. Amphoteric catalysts are preferably added after monoester formation has taken place at 160 °C [140]. PA can be charged as flakes or molten. The use of molten PA reduces the cycle time of a batch.

After charging the raw materials heating is continued, e.g., for DOP to a maximum of 224 °C [145]. In order to remove the water of reaction, a good boil up rate is maintained. This can be achieved by adjusting the pressure with a vacuum source. The water of reaction can also be stripped off by an inert gas sparge. When the required acid number is achieved, the excess alcohol is removed under reduced pressure [145].

The purification procedure required after the esterification is determined by the catalyst type. For acid catalysts, neutralization with aqueous caustic soda is necessary.

However, traces of alkali remain in the organic phase, and therefore a water wash after the neutralization step is advisable.

Titanates are generally removed by alkaline hydrolysis. The product is cooled to ca. 100 °C [140], and lime or soda ash or sodium hydroxide are added with water. The mixture is stirred and the water is then removed by vacuum distillation.

After neutralization, residual alcohol is removed by steam distillation [140]. The temperature, pressure, and steam for injection are adjusted so that the product after this treatment is nearly alcohol free and dry. The final filtration is performed with a rotary vacuum filter or plate and frame filters. If the hydrolyzed catalyst was not removed with the water phase in the neutralization step, the catalyst can be removed in the filtration step by adding a filter aid.

The continuous BASF process described in [132], [141] can be carried out autocatalytically or preferably by using an amphoteric catalyst. The esterification is carried out in a multistage cascade of agitated vessels. The reaction mixture leaving the last reactor is pumped to a vacuum flashing chamber to recover the excess alcohol [132]. To remove the catalyst, the crude ester is treated with a mild aqueous alkali solution in a mixer/settler system. To remove residual volatiles such as water, alcohol, and other low-boiling compounds a steam distillation is carried out under reduced pressure. For the final treatment a filter aid is added. The product is subjected first to coarse and then to final filtration.

5.4. Environmental Protection

The wastewater from phthalate processes is contaminated with phthalates, alcohols, and other organic compounds. Therefore, the process wastewater must be treated before discharging to a municipal system or a water body. Local regulations dictate the treatment and disposal of this water. Generally, biological treatment is required. Studies on phthalates have shown that they can be biodegraded by many species of bacteria [146]–[148]. In Germany only three phtha-lates (diethyl phthalate, diallyl phthalate, benzyl butyl phthalate) are listed in the catalog for water-endangering materials [149]. They are classified as materials with moderate hazard potential (WGK 2).

In the United States and Canada there are no current regulations limiting the concentration of phthalates in wastewater.

In the TA Luft [150], only DOP is mentioned as material of class II (intermediate danger potential). The other phthalates have been classified according to the hazard potential for the environment by the rules given in [150]. Exxon [146] lists the C_4–C_{11} phthalates, with the exception of DOP, in class III. Because of their low vapor pressure, the phthalates easily meet the concentration limits for phthalates specified by TA Luft.

Owing to the high working temperatures, air emissions from compounding and extrusion processes for flexible PVC generally have high phthalate concentrations.

Therefore emission controls may be required for these discharges. Exxon [146] gives the following data for the environmetal levels of DBP and DEHP:

- Fresh and estuarine/sea waters: 10 and 0.7 ppb, respectively
- Sediments: a few ppb up to 100 ppm
- Air: 3 ng/m^3 over the sea and up to 130 ng/m^3 in cities.

Detailed information on DBP and DEHP distribution in the environment are given in [147], [148].

In North America, waste containing DEHP and DBP is regarded as hazardous waste and must be disposed off in a secured landfill site.

5.5. Quality Specifications

For manufacturing flexible plastics the following properties are important for the selection of a plasticizer [134]: (1) compatibility with plastics, (2) solvating temperature (a measure of the solvating capability on plastics), (3) efficiency for plasticizing, (4) volatility in plastics, (5) extractability and diffusion losses in plastics, and (6) electrical properties. The properties required in the final products determine the type and the quality specification of the plasticizer.

For manufacturing flexible PVC for blood bags or other goods in medical application the plasticizer used (e.g., DEHP) should have a purity of min. 99.5%, a low alcohol content, a low acid number (max. 0.02 mg KOH/g), and no odor [151].

Flexible vinyl sheets used for automotive interior trim should be plasticized with a phthalate that ensures low fogging of the interior windshield. Also low-temperature flexibility, light and thermal stability, and processability are required. BASF [152] recommends a low-branched DINP or a HNUP for this application.

Phthalates used for plasticizing PVC for wire, cable, and conductor sheathing need high electrical volume resistivity. Products with volume resistivities $> 1.0 \times 10^{11}$ $\Omega \cdot$ cm at 20 °C are required. Typical resistivity of DOP is 1.0×10^{11}, and of DIDP 1.0×10^{12} $\Omega \cdot$ cm.

Quality tests are carried out both on the phthalate itself and on the plasticized polymers. Table 4 lists standards for testing phthalates as PVC plasticizers.

Nowadays, almost all demands of plasticizers for use in flexible plastics can be met by phthalates.

Table 4. Standards for testing of phthalates

Property	Methods	
	DIN	ASTM
Phthalates		
Viscosity at 20 °C	51 561	D 445
Density at 20 °C	51 757	D 1045
Refractive index at 20 °C	51 423	D 1218-61
APHA color	ISO 6271	D 1209
Acid value	53 402	D 1045
Pour point	ISO 3016	D 97-57
Water content	51 777	D 1364
Purity	gas chromatography	
Plasticized PVC		
Solvating temperature	53 408	
Tensile strength	53 455	D 412
Elongation at break	53 455	D 412
Brittleness temperature	53 372	D 746
Clash & Berg torsion stiffness at 310 N/mm^2	53 447	D 1043
Shore B hardness	53 505	D 2240
Volume resistivity at 25 °C	53 482	D 1169

5.6. Storage and Transportation

Because phthalates are neutral liquids they can be stored in carbon steel tanks. The more expensive stainless steel or aluminum alloy tanks are also used. All phthalates have a flash point above 100 °C. Thus bulk storage of phthalates does not require any special fire protection equipment [153]. German regulations [154] classify phthalates as having an endangering potential to water. Therefore storage tanks have to be installed in a diked area which has a containment capacity of the largest tank. Detailed proposals for storing phthalates are given in [155]. Phthalates can be stored indefinitely and can withstand tropical conditions.

Phthalates are transported in tank cars, tank trucks, ships, or mild steel drums. Canada and California have special labeling requirements for DBP and DEHP.

5.7. Uses

The C_1–C_4 phthalates including DMEP and DBEP are mainly used as plasticizers for cellulosic resins and some vinyl ester resins. DBP and BBP are fast-fusing plasticizers for PVC and are mostly used in combination with DEHP. C_4 phthalates are also appropriate plasticizers for nitrocellulose lacquers.

DEHP, DINP, DIDP are the general purpose plasticizers for PVC in most applications. For wire and cable, DIDP is preferred. The linear phthalates, e.g., 711P or 911P, are used for PVC in various applications. They are less volatile and confer better low-

Table 5. Major phthalate producers and their trade names

Company	Location	Trade name
Aristech Chemical Corp.	United States	PX
ATOCHEM	Europe	Garbeflex
BASF	Europe, United States	Palatinol
BP Chemicals	Europe	Bisoflex
Chemische Werke Hüls	Europe, United States	Vestinol, Nuoplaz
Eastman Chemical	United States	Kodaflex
Enichimica	Europe	Sicol
Exxon Chemical	Europe, United States	Jayflex
Hoechst	Europe	Genomoll
ICI	Europe	Hexaplas
Kyowa Hakko	Japan	Kyowacizer
Mitsubishi Kasei Vinyl	Japan	
Neste Chemicals	Europe	
New Japan Chem.	Japan	

temperature properties than general-purpose phthalates. DUP and DTDP are recommended for use in wire insulation for use at ca. 100 °C.

87% of the phthalates produced are used for formulating flexible PVC, which is consumed for manufacturing the following goods [156]:

Wire and cable	25%
Film and sheeting	23%
Flooring	15%
Profiles and tubing	10%
Plastisol spread coatings	11%
Other plastisols	8%
Miscellaneous (shoe soles, blood bags, gloves)	8%

5.8. Economic Aspects

The major phthalate producers in Europe, the United States, and Japan are listed in Table 5. The capacities of these companies range from 70 000 to 350 000 t/a.

In 1988 world consumption of plasticizers, including nonphthalates, was 3.85×10^6 t [157]. Based on these figures and uses in other applications, world consumption for phthalates is estimated at 3.25×10^6 t, of which DEHP accounts for ca. 2.1×10^6 t. The estimated total consumption of all phthalates and consumption of DEHP (in 10^3 t) by geographic region is:

	Total	DEHP
Western Europe	900	465
North America	730	155
Eastern Asia	530	490
Japan	320	245
Others	720	765
World total	3250	2120

Prognoses for the world market show an annual increasing demand of ca. 2.5%.

6. Toxicology

6.1. Phthalic Acid

Only few investigations exist; however, the toxicity appears to be low. Acute toxicity (LD_{50}, mouse, i.p.): 550 mg/kg [158].

Metabolism. Most of the material is excreted in the urine, either directly (as in dogs) or partially conjugated (as in rats and rabbits). A minor part may be decarboxylated and excreted as benzoic acid [159].

6.2. Phthalimide

Few data exist; however, a low toxicity may be assumed. The material is hydrolyzed to phthalic acid and ammonia [160]. Acute toxicity (LD_{50}, mouse, oral): 5000 mg/kg [161].

6.3. Phthalic Anhydride

Phthalic anhydride is an inhalation allergen and may cause occupational asthma in exposed persons [162], [163], [164].

Acute toxicity data indicate minor toxicity. Acute toxicity: LD_{50}, rat, oral: 1500–4000 mg/kg [165]; LC_{50}, rat inhalation: > 210 mg/m^3/h (dust) [166]; LD_{50}, rabbit, dermal: > 10 000 mg/kg [166]. Dermal irritation: nonirritating or slightly irritating in rabbit skin [165]. Eye irritation (rabbit): irritating [165], [166]. Dermal sensitization: sensitizing in the mouse ear swelling test [164].

Subacute Toxicity. Feeding experiments with 250, 1000, and 3800 ppm in the diet (17, 67, and 253 mg kg^{-1} d^{-1}) were tolerated for 28 d without significant effects [166].

Gavage experiments in female rats, beginning with 20 mg kg^{-1} d^{-1} and subsequently increasing doses up to 4800 mg kg^{-1} d^{-1} over a period of 8 weeks, caused toxic effects on the kidneys and the gastric mucosa at doses exceeding 1200 mg kg^{-1} d^{-1} [167].

Inhalation experiments with sublimed material in rats, mice, rabbits, and cats showed some irritation and impaired respiration at 10 000 mg/m^3 for 4 h per day for two weeks [165].

In female guinea pigs, vapor concentrations of 615 mg/m^3 for 30 min per day over 4 and 8 d caused irritative effects on the eyes and respiratory tract [167].

Repeated inhalation of pure phthalic anhydride dust in concentrations of 8.5 mg/m^3 for 3 h per day for several periods of 4 consecutive days, alternating with 10-d recovery periods for a total experimental period of 8 months, caused irritation of the respiratory tract and frequent pneumonias [167].

Genotoxicity. No mutagenic effects were detected in the Ames test. No sister chromatid exchanges or chromosome aberrations were found in CHO cells. All experiments were conducted with and without metabolic activation [168], [169].

Chronic Toxicity and Carcinogenicity. Long-term bioassays with 7500 and 15 000 ppm in the diet of rats (500 and 1000 mg kg^{-1} d^{-1}) did not lead to an increased tumor incidence compared to control animals. Similarly, mice dosed with 3500 and 7100 mg kg^{-1} d^{-1} for 32 weeks and with 900 and 1800 (females) or 1800 and 3600 (males) mg kg^{-1} d^{-1} for a further 72 weeks did not respond with an increase of tumors [170], [171].

Reproductive Toxicity. Only screening experiments with intraperitoneal injection in CD-1 mice have been reported. 80 mg/kg injections were carried out from gestation days 8 to 10 or 11 to 13. Malformations (cleft palates, malformations of ribs and vertebrae) were detected at doses $\geq$ 55.5 mg kg^{-1} d^{-1} [172] – [174]. The experiments are inconclusive for the quantitative assessment of a teratogenic risk, since the exposure route was artificial and mice are prone to respond to experimental stress with these types of malformations.

6.4. Phthalonitrile

Acute toxicity (LD$_{50}$, rat, oral): 35 – 125 mg/kg [175].

Epileptiformic convulsions were observed after occupational exposure with a latency period of several hours to days [176]. Also dermal exposure may be relevant [175]. Since the material does not appear to be metabolized to cyanide, the typical measures taken against cyanide intoxications are not promising.

The material does not cause mutagenic effects in the Ames test [177] and the CHO/HGPRT gene mutation assay [178].

After oral and subcutaneous administration to rats and mice, and after percutaneous administration to mice, an induction of leukemias was reported. The study description is, however, vague and does not allow a conclusive assessment [179].

6.5. Phthalate Esters

6.5.1. General Profile

Most phthalates show very low acute toxicity.

Subacute and chronic effects as well as reproductive effects depend to some extent on chain length and structure of the alcohol moiety. Thus, a differentiation between short-, medium- and long-chain esters appears to be justified. However, the risks involved during handling and workplace exposure are generally low. Only under special circumstances such as parenteral exposure (e.g., eluates from medical devices), may the systemic intake be considerable and of potential toxicological relevance [180], [181], [182]. On the other hand, the plasticizer most commonly employed for medical use, di-2-ethylhexyl phthalate (DEHP), and its corresponding monoester, mono-2-ethylhexyl phthalate (MEHP), appear to stabilize the membranes of stored erythrocytes [183]. Dermal resorption has been measured in vitro [184] and in vivo [185] for several phthalate esters and appears to decrease with increasing chain length. Inhalation of vapors, mists, and aerosols are regulated in the United States by TLVs (TWA 5 mg/m^3, STEL 10 mg/m^3 for DEHP, dibutyl phthalate and dimethyl phthalate). In Germany the MAK value is 10 mg/m^3. These values may also be regarded as adequate safety margins for other phthalates.

Having entered the organism, one of the two ester bonds is cleaved and the alcohol is released [180], [186]–[189]. Toxicological effects may result from the phthalic acid monoester and its sequel products and/or from the alcohol. In some cases (e.g., dimethoxyethyl phthalate) the impact of the alcohol is of greater importance [190]–[192].

Short-Chain Phthalate Esters (C_1–C_3). Dermal exposure to *dimethyl phthalate* in humans occurs widely due to its use in insect repellants; so far, no significant toxicological side effects were observed [193]. *Dimethyl* and *diethyl phthalate* are used in cosmetics at concentrations of < 10% and are dermally resorbed. They are considered to be nonirritants, nonsensitizers, and nonphototoxic agents [193]. Feeding studies in rats and mice with diethyl phthalate caused liver weight increases at high doses (≥ 0.5% in the diet), but no testicular toxicity [194]–[197]. Peroxisome proliferation is only marginal and requires high doses [198]. Developmental toxicity is regarded as insignificant [199]–[201].

Dimethoxyethyl phthalate has a teratogenic potential [190]–[192] due to the released methoxyethanol.

Diallyl phthalate is acutely relatively toxic (LD$_{50}$, rat, oral: 800–1700 mg/kg) and irritating [202], [203]. In a long-term bioassay with gavage administration there was no clear carcinogenic activity, but allyl alcohol-related toxicity was observed (see Section 6.5.4).

Medium-Chain Phthalate Esters (C_4–C_7) exhibit degenerative testicular effects (see Section 6.5.7, testicular toxicity) upon repeated oral administration in all species investigated [195]–[197], [204]–[213]. Furthermore, in rodents, but not in other species, hepatic alteration (peroxisome proliferation and hepatomegaly [210], [214]–[216]) is observed. The peroxisome proliferation is somewhat more pronounced in the group of longer chain phthalates, especially of the branched-chain type (see below). The testicular effects are also obtained with DEHP to a similar degree (see below) but are not observed with the linear longer-chain di-*n*-octyl phthalate or with short-chain phthalates. 1% of dibutyl phthalate (DBP) in the diet of ICR/ICL mice also caused embryotoxicity and teratogenicity [208], [217], [218]. After a 6-month inhalation exposure of rats to 50 mg DBP/m^3 for 6 h per day, an accumulation of DBP in the brain and increase of brain weight were found; at 5 mg/m^3 these effects were marginal [219].

Butyl benzyl phthalate at 2.5 and 5.0% in the feed caused testicular effects, enlargement of liver and kidneys and decrease of bone marrow cellularity and thyroid function [220].

Longer-Chain Phthalate Esters (C_8–C_{12}). Diethylhexyl phthalate (DEHP) is the substance examined most extensively, and hundreds of publications on toxicological profile, biochemistry, and metabolism exist. Far fewer data have been generated on other phthalates of this group.

Branched-chain phthalates commonly share a potential to induce peroxisomes and hepatomegaly in livers of rats and mice at high doses (see Sections 6.5.3 and 6.5.6). This was demonstrated with diethylhexyl phthalate (DEHP), diisononyl phthalate and diisodecyl phthalate. These materials may cause peroxisome proliferation and hepatomegaly in rodents (DEHP at doses exceeding 50 mg kg^{-1} d^{-1}) [198], [214]–[216], [221]–[229]. In long-term feeding experiments, DEHP [227] and related phthalates [215] also led to an increase of liver tumors in rodents. Diisodecyl phthalate hitherto was not investigated for carcinogenicity, but high no-adverse-effect levels were achieved for potential causative factors (liver enlargement and peroxisome proliferation). Non-rodents and primates do not show these liver effects [225], [230]–[233] and hence no carcinogenicity is expected for humans (see Section 6.5.6).

DEHP [195], [196], [210]–[212], [229], [234] also exhibited testicular effects after oral intake. Oral uptake of >50 mg kg^{-1} d^{-1} of DEHP also caused embryotoxic and teratogenic effects in mice [194], [217], [218], [235]–[238]. In rats, some embryotoxicity was observed in feeding studies at maternally toxic levels [239], but not upon inhalation [240]. DINP and DIDP, are expected to be examined for prenatal toxicity in the near future.

Di-*n*-octyl phthalate did not exert testicular toxicity in vivo [194], [238] nor significant peroxisome proliferation in the liver [227]. It was shown to be a liver tumor promotor in a rat feeding experiment [241], probably as a sequel of fat accumulation and hepatomegaly [226].

Chronic administration of $\geq 0.5\%$ of DEHP in the diet also lead to kidney toxicity in rats and mice with organ weight increase, cystic changes, dysfunctions [242]–[245], and alterations in thyroids [243], [246], [247], with a fall in plasma thyroxine levels associated with hyperactivity and changes of the thyroidal colloid spaces. This is also observed with other peroxisome proliferators [246].

The straight-chain analogues di-*n*-hexyl and di-*n*-octyl phthalate also show thyroid effects although they differ in terms of hepatic toxicity [243], [249]. Di-*n*-octyl phthalate causes a different type of renal toxicity than DEHP [250].

6.5.2. Acute Toxicity and Irritating Potential

Acute toxicity is as follows: Dimethyl phthalate (LD_{50}, rat, oral): 2400–8200 mg/kg; diethyl phthalate (LD_{50}, rat, oral): 4270–9200 mg/kg; DBP (LD_{50}, rat, oral): 8000–23 000 mg/kg; DEHP (LD_{50}, rat, oral): 30 600–34 000 mg/kg; and diallyl phthalate (LD_{50}, rat, oral): 1500 mg/kg.

As a rule, most phthalates do not cause significant irritation [215], [251], [252] or sensitization of the skin [193], [253]. Diallyl phthalate [202], [203], however, may exert a weak irritating potential. Earlier reports of respiratory sensitizations appear to be related to contamination of the esters with phthalic anhydride and maleic anhydride.

6.5.3. Liver Effects; Peroxisome Proliferation

DEHP and several other phthalates act on the liver lipid metabolism in a qualitatively similar manner to a group of structurally unrelated compounds which are used or investigated as hypolipidemic drugs to reduce serum cholesterol and triglycerides. Increased enzyme activities involved in mitochondrial and peroxisomal fatty acid β- and ω-oxidation appear to be early effects of DEHP [254]–[256]. Initially, short transient inhibition of β-oxidation and lipid accumulation may occur, but subsequently can be overcome [257]–[259]. In rodents, but not in primates, the numbers and volumes of peroxisomes increase, and the liver undergoes a short period of DNA replication and cellular proliferation with a maximal mitogenic response after 24 h [198], [222], [223], [226]. This results in a dose-related increase in liver weight [214], [216], [221]–[224]. Plasma cholesterol values decrease in the same way as is observed with hypolipidemic drugs.

One study in Sprague–Dawley rats reports a continuous increase of peroxisomal enzyme activities upon prolonged treatment over 2 years [228]. In Fischer rats, increased peroxisomal enzyme activity was still present after 2 years, but not hepatomegaly or cellular proliferation [274]. 25 mg/kg of DEHP appears to be the no-effect level for the increase of peroxisome-related enzyme activity [216]. The no-effect levels for DINP and DIDP are somewhat higher; DBP is much less active [229].

Peroxisome proliferators have been recognized as a group of epigenetic rodent liver carcinogens which widely differ in their carcinogenic activity [230], [260]. Several hypotheses exist on the mechanism of action. One is that increased intracellular

H$_2$O$_2$ and reactive O$_2$ species released from the peroxisomes [230], [261]–[263] may cause continuous accumulation of 8-hydroxyguanine in the DNA [264]. Leakage of H$_2$O$_2$ from peroxisomes is also reflected by increases of oxidized glutathione in the bile [260]. Conjugated dienes and accumulation of lipofuscin pigments in lysosomes were also observed and indicate oxidative stress to the cell [222], [260], [265], [266]. Other authors place greater emphasis on the liver weight increase and the replicative DNA synthesis [223] or otherwise on a reduced expression of growth factor receptors [267]–[270]. No direct genetic activity is observed with phthalate esters (see also Section 6.5.5), and the same also applies to hypolipidemic drugs, even if they stimulate peroxisomes at very low dose levels and are strongly carcinogenic in rats and mice [260]. These rodent liver tumors are therefore regarded as epigenetic and a sequel of the liver effects.

The present knowledge allows the assumption that threshold levels for these early appearing liver effects also reflect a threshold for the appearance of liver tumors in rodents. In addition, phthalate esters do not cause peroxisome proliferation in non-rodents and primates.

6.5.4. Carcinogenicity

DEHP is the phthalate ester most thoroughly investigated. Earlier chronic toxicity studies of DEHP showed no tumors in Wistar or Sherman rats over two-year periods at dietary levels of 4000 and 5000 ppm [271], [272]. In one of two more recent studies, Sprague–Dawley rats did not respond with liver tumors or preneoplastic lesions when fed with doses of 200, 2000, and 20 000 ppm; a continuous and sustained dose- and time-dependent peroxisomal proliferation was observed [228]. Due to the low numbers of animals employed, all these studies are of limited value in assessing the carcinogenicity risk.

Feeding studies in Fischer-344 rats with 50 animals per dose group and sex, exposed to 6000 and 12 000 ppm DEHP in the diet, and in B6C3F1 mice with 3000 and 6000 ppm, showed an increased incidence of hepatocellular carcinoma, adenoma, and preneoplastic foci [228]. The actual uptake from the feed was estimated to be 320–770 mg kg^{-1} d^{-1} in rats, depending on dose, sex, and food consumption, and 600–1800 mg kg^{-1} d^{-1} in mice [265].

Similar results at comparable dose ranges (1.2–2.0% in the diet) were obtained in consecutive studies with DEHP in smaller collectives of Fischer rats [224], [273]–[275] and in bioassays with two different forms of diisononyl phthalate and with a phthalate of a mixture of heptanols, nonanols, and undecanols (C$_7$–C$_{11}$ fraction) [215], [276]. Other investigators [277] employing lower dose levels (up to 6000 ppm) did not find a statistically significant increase of liver tumors with a diisononyl phthalate. Under these conditions (6000 ppm over two years) the material did not exhibit peroxisome proliferation at the end of the 2-year observation period. Different forms of diisononyl phthalates and related materials may exhibit slightly different profiles in relation to their structure and the type of branching and molecular shape.

In hamsters, inhalation of atmospheres saturated with DEHP vapor (ca. 15 mg/m^3; ca. 7–10 mg/kg) for 2 years or intraperitoneal injection up to a total dose of 54 g/kg, administered by weekly injections, did not increase the numbers of tumors [278]. Similarly, feeding of comparably low levels (300 and 1000 ppm) of DEHP did not lead to liver tumors [273].

Butyl benzyl phthalate, which is not a significant peroxisome proliferator, caused some mononuclear cell leukemia in Fischer rats at 1.2% in the diet, but no liver tumors [215], [279].

Diallyl phthalate, similarly, did not cause liver tumors, but necrotic and fibrotic lesions of the liver and bile duct hyperplasia in rats, and gastric inflammation and hyperplasia in mice were observed. These effects were probably mediated via release of allyl alcohol. The doses in this study were 100 and 300 mg/kg in rats, and 50 and 150 mg/kg in mice [171].

Initiation/Promotion Experiments. No initiating activity was found with DEHP after single oral administration of 10 g/kg or after 12 weeks feeding of 1.2%, followed by 0.05% feeding of phenobarbital as promotor [280]. Tumor promoting activity of DEHP was investigated with several experimental procedures. The results were either negative or weak [269], [281]–[284]; in some regimens antipromoting activities were also recorded [283], [285], [286]. There is evidence that peroxisome proliferators exert significant promoting effects only late in the rodent's life [287]. Di-*n*-octyl phthalate, which is not a peroxisome proliferator, was shown to be a liver tumor promotor [240], probably due to lipid accumulation and hepatomegaly [225].

6.5.5. Genotoxicity and Mutagenicity

DEHP and its metabolites have shown little if any genotoxic activity in a wide variety of short-term tests [278], [288]–[293]. A few in vitro experiments on DEHP and MEHP were weakly positive and probably related to the cytotoxicity of MEHP [294]–[297]. Two dominant lethal studies [298], [299] using very high doses and exposure routes of little relevance (subcutaneous, intraperitoneal) showed positive results. However, the effects were of borderline significance and appear to reflect testicular toxicity rather than true genotoxicity. After oral administration of DEHP and MEHP, no dominant lethal effects were observed [300], [301].

Other phthalates so far investigated showed similarly negative genotoxicity test results in most of the experiments [193], [214], [296].

DEHP has been shown not to react covalently with rat liver DNA after gavage administration following a 3-week prefeeding period [302]. When high doses of DEHP, known to cause peroxisomal proliferation, were repeatedly administered, an increase of 8-hydroxyguanine in the DNA was found in the rodent liver. This obviously reflects an oxidative DNA alteration [264], which, e.g., is also observed in rats fed on a choline-free diet [303], and other conditions of oxidative stress [304]. However, it is far from clear whether this is a major driving force for the liver carcinogenicity.

6.5.6. Species Differences and Relevance to Humans

Hepatomegaly, peroxisome proliferation and related liver effects are described in Section 6.5.3.

In hamsters, DEHP exerted a much lower extent of peroxisome proliferation [221]. Furthermore, hepatocytes of guinea pigs, ferrets, marmosets, and humans did not respond with peroxisome proliferation [231], [232]. Marmosets did not show peroxisome proliferation after oral administration of DEHP (2000 $mg\,kg^{-1}\,d^{-1}$ for 14 d), whereas in a parallel rat experiment marked peroxisome proliferation was observed [225]. Peroxisome proliferation was also not observed in cynomolgus monkeys which were dosed with 500 $mg\,kg^{-1}\,d^{-1}$ of DEHP for 3 weeks [305].

Several reasons for this species difference are assumed: The primary metabolite, MEHP (see Section 6.5.8), undergoes much less glucuronidation in rats. Therefore, renal excretion is much slower in rats, and tissue levels are higher. This has been demonstrated in experiments with rats and monkeys [225], [305]. Furthermore, it was shown that in rats, due to saturation of a final β-oxidation step in the ω-oxidation process, a secondary metabolic pathway (ω-1 oxidation) is induced to a far greater extent, leading to the 5-keto derivate of MEHP, which was shown to be the ultimate peroxisome proliferator [305]. In addition to these kinetic differences, there may also be intrinsic toxicodynamic differences among the species. The overall evidence indicates that human hepatocytes either do not respond to peroxisome proliferators or do so only to a much lesser extent [260]. The conclusion is that, in the case of phthalates, rodent experiments overestimate the risks of chronic toxicity to humans.

6.5.7. Reproductive Toxicity

Prenatal Toxicity; Embryotoxicity; Teratogenicity. Embryotoxic and teratogenic effects of DEHP and DBP depend on dose, species, and route of administration. Mice are more sensitive than rats. Feeding appears to be more effective than gavage or inhalation.

In *rats*, only embryotoxic effects were observed, with little indication of teratogenicity, even at dose levels that led to some maternal toxicity [239]. Inhalation of DEHP aerosols up to 300 mg/m^3 did not cause embryotoxic or teratogenic effects [240].

In *mice*, DEHP and DBP caused teratogenicity after oral administration [194], [217], [218], [235]–[238]: In CD-1 mice [236], [261] 0.1 and 0.15% of DEHP in the diet produced both maternal and fetal toxicity. At 0.05%, malformations (eye and tail defects, exencephaly, aortic and pulmonary arch defects, and skeletal defects) were observed in the absence of maternal or fetal toxicity. No effects were reported at the lowest dietary level in this study (0.025% DEHP, equivalent to ca. 35 mg kg^{-1} d^{-1}).

DBP caused reduction of the numbers of litters and of live pups per litter at the 1.0% level, but not at 0.3% in a continuous breeding study [194]. In another feeding study in mice, with 1% DBP in the diet, a slight increase in malformations (exencephaly and spina bifida) was observed [217], [218].

Butyl benzyl phthalate at 2% in the diet of rats caused complete resorption of all litters (embryonic or fetal death), whereas 1% and 0.5% were without effect [306]. In another rat feeding study [248] 2% of butyl benzyl phthalate in the diet did not lead to complete resorption of the litters; the number of malformations was greatly increased. In mice, malformations occurred, with a dose-dependent increase at 0.5 and 1.25%. The maternal and developmental no-adverse-effect level was 0.1% [248].

No teratogenicity or embryotoxicity was found with diethyl [201] and dimethyl phthalate [200] in rat feeding studies (0.25–5.0% in the diet) or with dimethyl phthalate in rats after occlusive epicutaneous administration (0.5, 1.0 and 2.0 mL kg^{-1} d^{-1}; 2 h/d; gestation days 6–15) [199].

Since DEHP in the diet might interfere with zinc resorption and metabolism (see below "Testicular Toxicity") the prenatal toxicity may partly be due to a zinc depletion, which is known to cause teratogenic effects [307].

Di(methoxyethyl) phthalate is teratogenic due to the release of highly teratogenic methoxyethanol [190]–[192].

Testicular Toxicity. Phthalate esters of medium and longer chain length (C_3–C_6, DEHP) cause testicular damage [195]–[197], [204], [205]–[213], [230], [234], [308]. The lesion is characterized by an initial effect on Sertoli cells [309], [310] leading to the exfoliation of spermatocytes and spermatids with atrophic degeneration of seminiferous tubules containing cytoplasmic remnants of Sertoli cells and spermatogonia [309], [311]. Germ cell detachment may also be observed in vitro with MEHP [309], [312] but not with 2-ethylhexanol [313]. Further hormonal sequels of the testicular atrophy are decreases of serum testosterone and increases of serum FSH (follicle stimulating hormones) and LH (luteinizing hormones) [314].

After administration of 2 g kg^{-1} d^{-1} DEHP for 2 weeks to rats, only slight reversal of the testicular lesions occurred after 45 d [211]. An incomplete reversal has also been observed after 60-d feeding of 20 000 ppm of DEHP in the diet of rats and a 70-d postobservation period [314]. The effects are age- and, to some extent, species-related [234]. Young rats (4 weeks old) appear to be more susceptible (lower doses, shorter duration required) than adult rats. This age susceptibility has been successfully modeled with MEHP in primary testicular cultures from rats of varying age [312].

DEHP can also induce testicular atrophy in mice [208], [238], guinea pigs [311], ferrets [315], and hamsters [221]. Marmosets appear to be more resistant, since oral doses of DEHP (2 g/kg for 2 weeks) caused no testicular toxicity [225]; the reasons may be less intestinal resorption and higher rate of excretion [255].

Testicular zinc content is decreased while urinary zinc increases [195], [206], [207], [212]. Detailed studies with di-*n*-pentyl phthalate [316] have not been able to indicate whether the changes in testicular zinc content are solely a secondary phenomenon. A zinc-supplemented diet partially protected rats from DEHP-induced testicular damage [317], whereas rats on a zinc-deficient diet were more sensitive [317]. Hypolipidemia and hepatomegaly appeared to occur independently of zinc supplementation.

In acute studies, relatively high (> 250 mg/kg) oral doses are required to induce testicular damage. In experiments of longer duration the effective levels are considerably lower. Serum testosterone [314] and male fertility in continuous breeding protocols (see below) appear to be sensitive parameters. The no-effect levels are in the range 25 – 50 mg kg^{-1} d^{-1}. 1250 ppm (70 mg kg^{-1} d^{-1}) in the diet of rats decreased serum testosterone within 60 d [314]. This is a dose range which caused malformations in a mouse teratology study [236], [261].

Effects on Fertility. In CD-1 mice exposed to DEHP in the feed in a continuous breeding protocol, complete suppression of fertility at the highest dose (0.3% = ca. 430 mg kg^{-1} d^{-1}) was observed. At 0.1% a significant reduction in fertility was induced. No effects were seen at 0.01% ($\approx$ 20 mg kg^{-1} d^{-1}). A cross-over mating employing treated males and females from the 0.3% groups indicated significant effects on both the male and female reproductive function [194], [261].

DBP exposure in mice resulted in a reduction in the number of litters at 1.0% in the diet, but not at 0.3%. Females were more affected than males [194]. Di-*n*-hexyl phthalate affected both sexes of mice in a dose-related fashion [194] at 0.3, 0.6, and 1.2%. Di-*n*-pentyl phthalate completely inhibited fertility at 1.25% and 2.5%. Di-*n*-propyl phthalate caused complete inhibition at 5.0% and reduced fertility at 0.5% and 2.5%. Di-*n*-pentyl phthalate was more toxic to males, and di-*n*-propyl phthalate more toxic to females [238]. Diethyl and di-*n*-octyl phthalate showed no adverse effects on fertility and the testes [194], [238].

6.5.8. Metabolism and Toxicokinetics

Phthalate esters in general are rapidly hydrolyzed to the corresponding monoesters in the intestine [186]. These are readily absorbed from the gut [188], with some species difference in absorption and excretion. The pattern of DEHP metabolites in the urine has been elucidated [187]. MEHP in native or — depending on the species — in glucuronidated form (at the carboxylate moiety) appears in the urine along with degradation products from ω and ω-1 oxidation in the alcohol side chain. ω-Oxidation leads to a carboxylic moiety at the C-terminal site of the alcohol chain. ω-1 Oxidation leads to the 5-keto derivative of MEHP, which was shown to be the metabolite ultimately responsible for peroxisome proliferation [231]. The ω-1 oxidative pathway is more pronounced in the rat than in the cynomolgus monkey [306].

Rats dosed with 2000 mg kg^{-1} d^{-1} of DEHP for 14 d by gavage eliminated (on day 13) 56.2% of the administered daily dose in the urine, and 58.7% in the feces. The peak blood level on day 14 was 368 µg equivalents of DEHP per gram blood. In a parallel marmoset study the same dose regimen led to only 13 µg equivalents of DEHP per gram blood. Tissue levels were also much lower in marmosets, and no peroxisome proliferation or testicular effects could be detected [225].

Urinary excretion of phthalate ester metabolites (such as monoesters in conjugated or unconjugated form) may be taken as an indirect measure for intestinal absorption. After in vitro hydrolysis of all metabolites phthalic acid may be taken as summaric parameter. Two volunteers, each ingesting 30 mg of DEHP, excreted 11 and 15% of the dose in the urine within 24 h. Four daily doses of 10 mg lead to a urinary recovery rate of 15 and 25% [189].

Short-chain phthalates such as dimethyl phthalate are either excreted unchanged in the urine or completely hydrolyzed to phthalic acid and excreted [318].

After intravenous administration of DEHP to the African green monkey, 80% of the urinary metabolites were excreted as glucuronide conjugates, in contrast to rats, which did not excrete conjugates in the urine [319]. This indicates that species differences in the rate of glucuronidation occur. They may cause considerable variations in biological half-life times and tissue levels. As a consequence, lower absorption of DEHP and/or its metabolites from the gastrointestinal tract and the higher glucuronidation (detoxification) rate of the metabolites in primates may contribute to the higher resistance to DEHP-induced toxicity.

7. References

[1] R. C. Weast: *Handbook of Chemistry and Physics*, 60th ed., CRC Press, Boca Raton, Fla., 1979/1980, C 436.
[2] *Handbuch des Chemikers*, VEB-Verlag Technik, Berlin 1956.
[3] *Kirk-Othmer*, 2nd ed., **15**, 445.
[4] H. Suter: *Phthalsäureanhydrid und seine Verwendung*, Dr. Dietrich Steinkopf-Verlag, Darmstadt 1972.
[5] K. Nabert, G. Schön: *Sicherheitstechnische Kennzahlen brennbarer Gase und Dämpfe*, 2nd ed., Deutscher Eichverlag, 1970.
[6] R. H. Perry: *Chemical Engineer's Handbook*, 6th ed., McGraw-Hill, New York 1973.
[7] BASF, unpublished results, Ludwigshafen 1989.
[8] B. Raeke, *Angew. Chem.* **70** (1958) 1.
[9] K. Weissermel, H.-J. Arpe: *Industrielle organische Chemie*, 4th ed., VCH Verlagsgesellschaft, Weinheim 1994, p. 415.
[10] F. Wirth, *Hydrocarbon Process.* **54** (1975) 107.
[11] BASF, DE 2 356 049, 1973 (F. Wirth et al.).
[12] BASF, DE 1 266 748, 1966 (G. Schaefer et al.).
[13] BASF company brochure: "Phthalic Anhydride Advanced Catalytic PA-Technology," 1995.
[14] *Hydrocarbon Process.* **76** (1997) no. 3, 146.

[15] *Europa Chemie* (1997) no. 4, 12.
[16] O. Wiedemann et al., *Chem. Eng. (N.Y.)* **86** (1979) 62.
[17] Nippon Shokubai Kagaku Kogyo, DE 2 948 163, 1978 (T. Sato et al.).
[18] Nippon Shokubai Kagaku Kogyo, US 4 284 571, 1978 (T. Sato et al.).
[19] T. Sato et al., *Hydrocarbon Process.* **62** (1983) no. 10, 107.
[20] Nippon Shokubai Kagaku Kogyo, US 4 046 780, 1975 (Y. Nakanishi et al.).
[21] Nippon Shokubai Kagaku Kogyo, US 4 356 112, 1976 (Y. Nakanishi et al.).
[22] L. Verde, A. Neri, *Hydrocarbon Process.* **63** (1984) no. 11, 83.
[23] *Hydrocarbon Process.* **64** (1985) no. 11, 156.
[24] Alusuisse Italia S.P.A., EP 37 492, 1980 (A. Neri et al.).
[25] Alusuisse Italia S.P.A., US 4 489 204, 1980 (A. Neri et al.).
[26] *Hydrocarbon Process.* **59** (1979) no. 11, 208.
[27] Rhône Progil, DE 2 416 457, 1973 (R. Dumont et al.).
[28] BASF, DE 2 510 994, 1975 (K. Blechschmitt et al.).
[29] BASF, DE 2 547 624, 1975 (K. Blechschmitt et al.).
[30] BASF, DE 2 914 683, 1979 (P. Reuter et al.).
[31] Nippon Shokubai Kagaku Kogyo, DE 3 045 624, 1979 (Y. Nakanishi et al.).
[32] Nippon Shokubai Kagaku Kogyo, JP 56 73 543, 1981.
[33] Nippon Shokubai Kagaku Kogyo, JP 57 105 241, 1982.
[34] Nippon Shokubai Kagaku Kogyo, JP 57 180 430, 1982.
[35] Mitsui Toatsu Chemicals Inc., JP 54 5 891, 1979 (K. Oshima).
[36] BASF, DE 2 546 268, 1975 (K. Blechschmitt et al.).
[37] N. G. Glukhovskii et al., SU 978 910, 1981;*Chem. Abstr.* **98** (21): 178 962 b.
[38] D. W. B. Westerman et al., *Appl. Catal.* **3** (1982) 151.
[39] M. S. Wainwright et al., *Can. J. Chem. Eng.* **55** (1977) 557.
[40] W. Fiebig et al., *Chem. Tech. (Leipzig)* **28** (1976) 673.
[41] D. Vankove et al., *J. Catal.* **36** (1975) 6.
[42] J. Herten, G. F. Froment, *Ind. Eng. Chem. Process Des. Dev.* **7** (1968) 516.
[43] P. H. Calderbank, *Chem. Eng. Sci.* **32** (1977) 1435.
[44] M. S. Wainwright et al., *Catal. Rev. Sci. Eng.* **19** (1979) 211.
[45] D. D. McLean et al., *Can. J. Chem. Eng.* **58** (1980) 608.
[46] A. G. Lyubarksii et al., *Kinet. Katal.* **14** (1973) 956.
[47] K. Hertwig et al., *Chem. Tech. (Leipzig)* **23** (1971) 584.
[48] J. Skrzypek et al., *Chem. Eng. Sci.* **40** (1985) 611.
[49] M. Blanchard, D. Vanhove, *Bull. Soc. Chim. Fr.* 1971, 3291.
[50] R. Römer et al., *Chem. Ing. Tech.* **46** (1974) 653.
[51] G. Emig et al., *Chem. Ing. Tech.* **47** (1975) 717, 888, 1012.
[52] I. E. Wachs et al., *Stnd. Surf. Sci. Catal.* **19** (1984) 275.
[53] I. E. Wachs et al., *Appl. Catal.* **15** (1985) 339.
[54] R. Saleh et al., *Appl. Catal.* **31** (1987) 87.
[55] V. Nikolov et al., *Stnd. Surf. Sci. Catal.* **34** (1987) 173.
[56] I. E. Wachs et al., *J. Catal.* **91** (1985) 366.
[57] G. F. Froment et al., *Appl. Catal. A* **120** (1994) 17.
[58] G. F. Froment et al., *Chem. Eng. Scie.* **51** (1996) 2091.
[59] G. C. Bond et al., *J. Catal.* **164** (1996) 276.
[60] N. L. Franklin et al., *Trans. Inst. Chem. Eng.* **34** (1956) 280.
[61] F. de Maria et al., *Ind. Eng. Chem.* **53** (1961) 259.

[62] F. Bernardini et al., *Chim. Ind. (Milan)* **48** (1966) 9.
[63] A. Farkas, *Hydrocarbon Process.* **49** (1970) 121.
[64] Davy Powergas, DE 2 602 895, 1976 (L. Hellmer et al.).
[65] R. M. Dow et al., *Hydrocarbon Process.* **56** (1977) 167.
[66] *Chem. Eng.* **84** (1977) 65.
[67] BASF, DE 1 601 162, 1967 (F. Lorenz et al.).
[68] Deggendorfer Werft und Eisenbau, DE 2 207 166, 1972 (O. Wanka et al.).
[69] Deggendorfer Werft und Eisenbau, DE 2 543 758, 1975 (R. Vogl).
[70] Davy McKee AG, DE 2 855 629, 1978 (G. Keunecke et al.).
[71] Davy McKee AG, DE 2 855 630, 1978 (G. Keunecke et al.).
[72] BASF, DE 3 207 208, 1982 (E. Danz et al.).
[73] BASF, DE 3 007 627, 1980 (W. Muff).
[74] Davy McKee AG, DE 3 339 392, 1983 (H. Saffran).
[75] Metallgesellschaft AG, DE 3 411 732, 1984 (V. Franz et al.).
[76] GEA Luftkühler GmbH, DE 3 512 903, 1985 (W. Rudowski).
[77] Metallgesellschaft AG, DE 1 227 443, 1964 (W. Scheiber).
[78] Pittsburgh Coke and Chemical Company, US 2 850 440, 1955 (M. O. Shrader et al.).
[79] Mitsui Toatsu Chem., DE 1 948 374, 1968 (K. Hayakawa et al.).
[80] Japan Gas Chem., JP 45/10 333, 1970.
[81] Rhône-Progil, DE 2 452 171, 1976.
[82] Nippon Shokubai Kagaku Kogyo, JP-Kokai 58 8074, 1981.
[83] Nippon Steel Chem., JP-Kokai 58 45 434, 1977.
[84] Nippon Shokubai Kagaku Kogyo, DE 3 225 079, 1981 (Y. Kita et al.).
[85] Chemische Werke Hüls AG, EP 120 199, 1983 (F. Gude et al.).
[86] Chemische Werke Hüls AG, DE 3 309 310, 1983 (F. Gude et al.).
[87] BASF, US 3 507 886, 1966 (H. Suter et al.).
[88] Chemiebau Dr. A. Zieren GmbH, US 3 655 521, 1968 (H. Gehrken et al.).
[89] Nippon Steel Chem. Co, JP-Kokai 57 49 549, 1977.
[90] Nippon Steel Chem. Co, JP-Kokai 58 45 435, 1977.
[91] *Hydrocarbon Process.* **46** (1967) 215.
[92] The Badger Company Inc., US 4 435 580, 1982 (C. D. Miserlis).
[93] The Badger Company Inc., US 4 435 581, 1982 (C. D. Miserlis).
[94] Kawasaki Steel Corp., JP-Kokai 63 141 645, 1986.
[95] Kawasaki Steel Corp., US 5 252 752, 1991 (T. Aona et al.).
[96] Kawasaki Steel Corp., JP 61 90 281, 1994 (Y. Asami et al.).
[97] Kawasaki Steel Corp., EP 451 614, 1991 (T. Aona et al.).
[98] Shokubai Kasei Kogyo, JP 80 99 040, 1994.
[99] Kawasaki Steel Corp., JP 57 32 096, 1994.
[100] Chemische Werke Hüls AG, DE 1 643 827, 1967 (F. List et al.).
[101] Standard Oil Company, US 4 215 051, 1979 (H. Schroeder et al.); US 4 215 052, 1979 (H. Schroeder et al.); US 4 215 053, 1979 (D. A. Palmer et al.); US 4 215 054, 1979 (H. Schroeder et al.); US 4 215 055, 1979 (D. A. Palmer et al.); US 4 215 056, 1979 (H. Schroeder et al.).
[102] Sisas, EP 256 352, 1987 (F. Celeste et al.).
[103] Mitsubishi Gas Chemical Co, JP 53/40 711, 1978.
[104] V. N. Orlik, *Khim. Tekhnol.* **6** (1981) 51.
[105] S. Slavov, *Khim. Ind. (Sofia)* **56** (1984) 252.
[106] S. Slavov, *Geterog. Katal.* **1** (1983) 81.

[107] Chem. Sys. Inc., Phthalic Anydride, 93–3, 05. 1995.
[108] BASF, DE 2 902 978, 1979 (G. Kilpper et al.).
[109] BASF, EP 4635, 1978 (G. Kilpper et al.).
[110] *Beilstein* **E II 21** (1953) 348.
[111] LENTIA GmbH, DE 2 056 891, 1970 (J. Schweighofer).
[112] BASF, DE 2 334 379, 1973 (E. Hetzel et al.); DE 2 334 916, 1973 (E. Hetzel et al.); DE 2 911 245, 1979 (G. Kilpper et al.).
[113] O. Joklik, AT 286 271, 1969.
[114] Dawe's Lab Inc., US 3 819 648, 1971 (W. R. Boehme et al.).
[115] M. C. Sze et al., *Hydrocarbon Process.* **55** (1976) no. 2, 103.
[116] Sun Res. + Develop. Co, DE 2 256 596, 1972 (R. D. Bushick et al.).
[117] Degussa, DE 1 948 714, 1969 (T. Lüssling et al.).
[118] Allied Chem. Corp., US 2 833 807, 1956 (A. Farkas et al.).
[119] Mitsubishi Gas. Chem. Co, DE 2 427 191, 1979 (M. Saito et al.).
[120] Institut Nefttecnimitscheskich USSR, DE 2 627 068, 1979 (S. D. Mechtjev et al.).
[121] BASF, EP 699 476, 1996 (L. Karrer et al.).
[122] Takeda Yakuhin Kogyo, JP 55 17 360, 1980 (I. Onishi et al.).
[123] BASF, EP 766 985, 1997 (L. Karrer et al.).
[124] G. Stefani, *Chim. Ind.* **54** (1972) 984.
[125] K. K. Moll et al., *Chem. Tech. (Leipzig)* **20** (1968) 600.
[126] T. Kudo, *CEER Chem. Econ. Eng. Rev.* **2** (1970) 43.
[127] BASF, DE 1 172 253, 1962. (H. Kröper et al.).
[128] BASF, DE 1 643 630, 1967 (M. Decker et al.).
[129] Lummus Corp., US 3 776 937, 1972 (A. P. Gelbein); DE 2 008 648, 1970 (A. P. Gelbein).
[130] *Ullmann*, 3rd ed., **13,** 733–735.
[131] *Beilstein*, E III 9, 4199–4200.
[132] H. Suter: *Phthalsäureanhydrid und seine Verwendung,* D. Steinkopff Verlag, Darmstadt 1972.
[133] H. Rein, *Chemiker Taschenbuch,* 58th ed.,Part 2, Springer-Verlag, Berlin 1937.
[134] L. Meier: "Weichmacher für PVC," in G. W. Becker, D. Braun (eds.): *Kunststoff-Handbuch,* vol. 2/1: Polyvinylchlorid, Hanser-Verlag, München 1986.
[135] Monsanto US 1 923 938, 1933 (L. P. Kyrides).
[136] G. Lützel, J. Holzmann: "Ein neuer Weichmacher für PVC auf Basis DINP," *Kunststoffe* **74** (1984) no. 4, 1088–1091.
[137] *Modern Plastics Encyclopedia,* McGraw-Hill, New York (1988).
[138] US. International Trade Commission, Synthetic Organic Chemicals, US. Production and Sales,1987. US. Government Printing Office, Washington, 1987.
[139] Bulletin S-107 B, Methane Sulfonic Acid, Pennwalt Corporation, Pa 19 102, USA.
[140] TIL, Tioxide UK Ltd., Booklet TIL 12, Tilcom Catalysts For Ester Manufacture, Billingham 1982.
[141] *Ullmann,* 4th ed. **18,** 534–538.
[142] F. X. Werber, B. F. Goodrich, US 3 056 818, 1962.
[143] Du Pont Company, Chemical & Pigment Dept., Booklet E-69 996-1 4/87 (2 M), Du Pont Performance Products, Wilmington, DE, 1987.
[144] Dynamit Nobel, Technical Information, ZN 666/2.80/500/2439, Titanates as Catalysts for the manufacture of plasticizers, Troisdorf, 1980.

[145] M & T, Chemicals Inc., How to make DOP and other Phthalate Plasticizers easier with FASCAT 2001, and M & T Technical Data Sheet No. 341, Rev. 9/78, and Material Safety Data Sheet 3/82 for Stannous Oxide, Rahway, NJ 1978.

[146] Exxon Chemical International. A practical guide to protecting man and the environment, Kraainem, Belgium 1988.

[147] BUA Report 22 on Di-butyl phthalate, VCH-Verlagsgesellschaft, Weinheim, Dec. 1987.

[148] BUA Report 4 on Di-2-ethylhexyl phthalate, VCH-Verlagsgesellschaft, Weinheim, Jan. 1986.

[149] Gemeinsames Ministerialblatt (GMBl), Bonn, Germany 1990.

[150] Gemeinsames Ministerialblatt (GMBl), G 3191 A, Bonn, W. Germany, 28. Feb. 1986.

[151] BASF, Technical Leaflet, Palatinol AH-L (med), Ludwigshafen, 1989.

[152] B. L. Wadey, J. Holzmann, SAE Technical Paper Series, 870 318, Plasticizer for Automotive Interior Trim, SAE, 400 Commonwealth Drive Warrendale, PA 15 096, USA 1987.

[153] Verordnung über Anlagen zur Lagerung, Abfüllung und Beförderung brennbarer Flüssigkeiten – VbF, Bonn, v. 27. Feb. 1980.

[154] Regulations for water (Wasserhaushaltsgesetz-WHG) Bonn, last edition Sept. 23., 1986 (BGBl. I, S. 1529).

[155] BASF, Hinweise zur Lagerung von Weichmachern für PVC, TI-CIW/ES 003 d, Ludwigshafen, Dec. 1987.

[156] Verband Kunststofferzeugende Industrie, Argumente 3, Zum Thema Weichmacher, Frankfurt, 1988.

[157] BASF, Marketing CIW, Ludwigshafen, 1988.

[158] Registry of Toxic Effects of Chemical Substances (NIOSH; 1990), *Proc. Soc. Exp. Biol. Med.* **49** (1942) 471.

[159] E. Sandmeyer, C. Kirwin, jr., in G. D. Clayton, F. E. Clayton (eds.): *Patty's Industrial Hygiene and Toxicology,* vol. **II A,** Wiley-Interscience, New York (1982) p. 2350.

[160] *Patty,* 3rd ed., **IIA,** 2710.

[161] Registry of Toxic Effects of Chemical Substances (NIOSH; 1990), *Farmaco Ed. Sci.* **20** (1965) 3.

[162] M. Wernfors, J. Nielsen, A. Schütz, S. Skerfving: "Phthalic Anhydride-Induced Occupational Asthma," *Int. Arch. Allergy Appl. Immunol.* **79** (1986) 77–82.

[163] J. Nielsen, H. Welinder, A. Schütz, S. Skerfving: "Specific Serum Antibodies Against Phthalic Anhydride in Occupationally Exposed Subjects," *J. Allergy Clin. Immunol.* **82** (1988) 126–133. Alternative Dermal Sensitization Test: The Mouse Ear Swelling Test (MEST)," *Toxicol. Appl. Pharmacol.* **84** (1986) 93–114.

[164] S. C. Gad et al.: "Development and Validation of an Alternative Dermal Sensitization Test: The Mouse Ear Swelling Test (MEST)," *Toxicol. Appl. Pharmacol.* **84** (1986) 93–114.

[165] BAYER AG, Institut für Toxikologie, unpublished observations, 1978.

[166] Biofax Industrial Bio-Test Laboratories, Inc., Northbrook, Ill. 60 062 Data Sheets 13-4/70.

[167] E. Gross, H. Friebel, Prüfung der Toxizität von reinem Phthalsäureanhydrid und Rohprodukten aus der industriellen Phthalsäuresynthese, unpublished results, Bayer, 1955.

[168] E. Zeiger, S. Haworth, K. Mortelmans, W. Speck, "Mutagenicity Testing of Di(2-ethylhexyl) Phthalate and Related Chemicals in Salmonella," *Environm. Mutagen.* **7** (1985) 213–232.

[169] I. Florin, L. Rutberg, M. Curvall, C. R. Enzell, "Screening of Tobacco Smoke Constituents for Mutagenicity Using the Ames Test," *Toxicol.* **15** (1980) 219–232.

[170] NTP Bioassay of Phthalic Anhydride for Possible Carcinogenicity, Techn. Report 159, NCI, Bethesda, Md. 1979.

[171] "NTP Technical Report on the Carcinogenesis Bioassays of Diallylphthalate in F 344/N Rats and B 6 C 3 F 1/N Mice," NTP-TR Report no. 284, NTP-81-083, US Department of Health and Human Services, 1985. NIH Publ. 85–2540 and 82–1798.

[172] R. L. Dixon, G. E. Shull, S. Fabro, "Relative Teratogenicity of Selected Anhydrides,"*Arch. Toxicol.* **39** Suppl. 1 (1978) 327 A.

[173] N. A. Brown, G. E. Shull, R. L. Dixon, S. E. Fabro, "The Relationship between Acylating Ability and Teratogenicity of Selected Anhydrides and Imides," *Toxicol. Appl. Pharmacol.* **45** (1978) no. 361, 333 A.

[174] S. Fabro, G. Shull, N. A. Brown, "The Relative Teratogenic Index and Teratogenic Potency: Proposed Components of Developmental Toxicity Risk Assessment," *Teratogen. Carcinogen. Mutagen.* **2** (1982) 61–76.

[175] H. Zeller, H. T. Hofmann, A. M. Thiess, W. Hey, "Zur Toxizität der Nitrile," *Zentrallbl. Arbeitsmed. Arbeitsschutz* **19** (1969) 225.

[176] A. M. Thiess, "Beobachtungen von Gesundheitsschädigungen und Vergiftungen durch Einwirkung von o-Phthalodinitril" *Zentrallbl. Arbeitsmed. Arbeitsschutz* **18** (1968) 303.

[177] F. Oesch, Ames-Test for o-Phthalodinitrile, unpublished investigations BASF, Ludwigshafen, 1978.

[178] H. G. Miltenburger: "Detection of Gene Mutations in Somatic Mammalian Cells in Culture: HGPRT-test with V 79 Cells; ortho-Phthalodinitrile," in *BUA-Ber.*, Benzoldicarbonitrile, VCH Verlagsgesellschaft, Weinheim, 1988.

[179] G. B. Pliss, N. J. Wolfson: ber die leukämogene Wirkung des Phthalodinitrils (Russian). Vorposy Onkologii XVIII (1), 81–86, 1972 (cited in *BUA-Ber.*, Benzoldicarbonitrile, Dez. 1988, VCH, Weinheim, 1989).

[180] G. M. Pollack et al.: "Circulating Concentrations of Di(2-ethylhexyl) Phthalates and its De-esterified Phthalic Acid Products Following Plasticizer Exposure in Patients Receiving Hemodialysis," *Toxicol. Appl. Pharmacol.* **79** (1985) 257–267.

[181] P. Sjöberg, U. Bondesson, G. Sedin, J. Gustafsson: "Dispositions of Di- and Mono-(2-ethylhexyl) Phthalate in Newborn Infants Subjected to Exchange Transfusions,"*Eur. J. Clin. Invest.* **15** (1985) 426–430.

[182] Y. A. Barry, R. S. Labow, W. J. Keon, M. Tocchi: "Atropine Inhibition of the Cardiodepressive Effect of Mono(2-ethylhexyl) Phthalate on Human Myocardium," *Toxicol. Appl. Pharmacol.* **106** (1990) 48–52.

[183] G. Rock, R. S. Labow, M. Tocchi: "Distribution of Di(2-ethylhexyl) Phthalate and Products in Blood and Blood Components," *EPH Environ. Health Perspect.* **65** (1986) 309–316.

[184] R. C. Scott, P. H. Dugard, J. D. Ramsey, C. Rhodes: "In Vitro Absorption of some o-Phthalate Diesters through Human and Rat Skin,"*EHP Environ. Health Perspect.* **74** (1987) 223–227.

[185] A. E. Elsisi, D. E. Carter, I. G. Sipes: "Dermal Absorption of Phthalate Diesters in Rats," *Fundam. Appl. Toxicol.* **12,** 1989, 70–77.

[186] P. W. Albro, R. O. Thomas: "Enzymatic Hydrolysis of Di(2-ethylhexyl) Phthalate by Lipases," *Biochim. Biophys. Acta* **360** (1973) 380–390.

[187] P. W. Albro, R. Thomas, L. Fishbein: "Metabolism of Diethylhexyl Phthalate by Rats. Isolation and Characterization of the Urinary Metabolites," *J. Chromatogr.* **76** (1973) 321–330.

[188] J. W. Daniel, H. Bratt: "The Absorption, Metabolism and Tissue Distribution of Di(2-ethylhexyl) Phthalate in Rats," *Toxicology* **2** (1974) 51–65.

[189] P. Schmidt, C. Schlatter: "Excretion and Metabolism of Di(2-ethylhexyl) Phthalate in Man," *Xenobiotica* **15** (1985) 251–256.

[190] A. R. Singh, W. H. Lawrence, J. Autian: "Teratogenicity of Phthalate Esters in Rats," *J. Pharm. Sci.* **61** (1972) 51–55.

[191] J. Yonemoto, N. A. Brown, M. Webb: "Effects of Dimethoxyethyl Phthalate, Monomethoxyethyl Phthalate, 2-Methoxyethanol and Methoxyacetic Acid on Post Implantation Rat Embryos in Culture," *Toxicol. Lett.* **21** (1984) 97–102.

[192] E. J. Ritter, W. J. Scott, J. L. Randall: "Mechanistic Studies of the Teratogenic Action of Di(2-methoxyethyl) phthalate (DMEP) and 2-Methoxyethanol in Wistar Rats," *Teratology* **29** (1984) 54 A Suppl.

[193] "Final Report on the Safety Assessment of Dibutyl Phthalate, Dimethyl Phthalate, and Diethyl Phthalate," Am. Coll. Toxicol. **4** (1985) 267–303.

[194] J. C. Lamp IV. et al.: "Reproductive Effects of Four Phthalic Acid Esters in the Mouse," *Toxicol. Appl. Pharmacol.* **88** (1987) 255–269.

[195] S. Oishi, K. Hiraga: "Testicular Atrophy Induced by Phthalic Esters: Effect on Testosterone and Zinc Concentrations," *Toxicol. Appl. Pharmacol.* **53** (1980) 35–41.

[196] S. Oishi, K. Hiraga: "Effect of Phthalic Acid Esters on Gonadal Function in Male Rats," *Bull. Environ.Contam. Toxicol.* **21** (1979) 65–67.

[197] P. M. D. Foster, L. V. Thomas, M. W. Cook, S. D. Gangolli: "Study of the Testicular Effects and Changes in Zinc Excretion Produced by some n-Alkyl Phthalates in the Rat," *Toxicol. Appl. Pharmacol.* **54** (1980) 392–398.

[198] D. E. Moody, J. K. Reddy: "Hepatic Peroxisome (Microbody) Proliferation in Rats Fed Plasticizers and Related Compounds," *Toxicol. Appl. Pharmacol.* **45** (1978) 497–504.

[199] E. Hansen, O. Meyer: "No Embryotoxic or Teratogenic Effect of Dimethyl Phthalate in Rats after Epicutaneous Application,"*Pharmacol. Toxicol.* (Amsterdam) 64 (1989) 237–238.

[200] E. A. Field et al.: "Developmental Toxicity Evaluation of Dimethyl Phthalate (DMP) in CD Rats," *Teratology* **39** (1989) 452, P 76 [A].

[201] C. J. Price et al.: "Developmental Toxicity Evaluation of Diethyl Phthalate (DEP) in CD Rats," *Teratology* **39** (1989) 473, P 66 [A].

[202] J. Autian: "Toxicity and Health Threats of Phthalate Esters: Review of Literature," *EHP Environ. Health Perspect.* **4** (1973) 3.

[203] D. E. Carter, B. Feldmann, J. G. Sipes: "Liver and Lung Toxicity of Diallyl Phthalate,"*Toxicol. Appl. Pharmacol.* **45** Abstr. 81 (1978) 254.

[204] P. M. D. Foster et al.: "Differences in Urinary Metabolic Profile from Di-n-butyl Phthalate-Treated Rats and Hamster. A Possible Explanation for Species Differences in Susceptibility to Testicular Atrophy," *Drug. Metab. Dispos.* **11** (1983) 59–61.

[205] J. A. Ward, H. Zenick, W. W. Wilfinger, R. J. Niewenhuis: "Effects of Dibutyl Phthalate (DBP) on the Reproductive System of Two Strains of Young Male Rats," *The Toxicologist* **4** (1984) 28 A.

[206] S. Oishi, K. Higara: "Testicular Atrophy Induced by Phthalic Acid Esters: Effect on Testosterone and Zinc Concentrations," *Toxicol. Appl. Pharmacol.* **53** (1980) 35–41.

[207] B. R. Cater, M. W. Cook, S. D. Gangolli, P. Grasso: "Studies on Di-butyl Phthalate-Induced Testicular Atrophy in the Rat: Effect on Zinc Metabolism," *Toxicol. Appl. Pharmacol.* **41** (1977) 609–618.

[208] J. R. Reel, A. D. Lawton, D. B. Feldmann, J. C. Lamb: "Di(n-butyl) Phthalate: Reproduction and Fertility Assessment in CD-1 Mice, when Administered in the Feed," Report 1984, NTP-84-411, *Gov. Rep. Announce. Index (U.S.)* **85** (1985) no. 7, 35.

[209] T. J. B. Gray, K. R. Butterworth: "Testicular Atrophy Produced by Phthalate Esters," *Arch. Toxicol. Suppl.* **4** (1980) 452–455.

[210] T. J. B. Gray et al.: "Short-Term Toxicity Study of Di(2-ethylhexyl) Phthalate in Rats," *Food. Cosmet. Toxicol.* **15** (1977) 389–399.

[211] S. Oishi: "Reversibility of Testicular Atrophy Induced by Di(2-ethylhexyl) Phthalate in Rats," *Environ. Res.* **36** (1984) 160–169.

[212] S. Oishi, K. Hiraga: "Testicular Atrophy Induced by Di-2-ethylhexyl Phthalate: Effect of Zinc Supplement," *Toxicol. Appl. Pharmacol.* **70** (1983) 43–48.

[213] S. P. Srivastava et al.: "Testicular Effects of Di-n-butyl Phthalate (DBP): Biochemical and Histopathological Alterations," *Arch. Toxicol.* **64** (1990) 148–152.

[214] K. N. Woodward, A. M. Smith, S. P. Mariscotti, N. J. Tomlinson: "Review of the Toxicity of the Esters of o-Phthalic Acid (Phthalate Esters),"*Toxic. Rev.* **14** (1986).

[215] J. W. Hirzy: "Carcinogenicity of General-Purpose Phthalates: Structure-Activity Relationships," *Drug Metab. Rev.* **21** (1989) 55–63.

[216] B. G. Lake, T. J. B. Gray, S. D. Gangolli: "Hepatic Effects of Phthalate Esters and Related Compounds – In Vivo and In Vitro Correlations,"*EHP Environ. Health Perspect.* **67** (1986)283–290.

[217] K. Shiota, M. J. Chou, H. Nishimura: "Embryotoxic Effects of Di-2-ethylhexyl Phthalate (DEHP) and Di-n-butyl Phthalate (DBP) in Mice," *Environ. Res.* **22** (1980) 245–253.

[218] K. Shiota, H. Nishimura: "Teratogenicity of Di(2-ethylhexyl) Phthalate (DEHP) and Di-n-butyl Phthalate (DBP) in Mice," *EHP Environ. Health Perspect.* **45** (1982) 65–70.

[219] M. Kawano: "Toxicological Studies on Phthalate Esters: 1. Inhalation of Effects of Dibutyl Phthalate (DBP) on Rats. 2. Metabolism, Accumulation and Excretion of Phthalate Esters in Rats," *Nippon Eiseigaku Zaschi* **35** (1980/81) 693–701; *Nippon Eiseigaku Zaschi* **35** (1981) 687–701.

[220] D. K. Agarwal, R. R. Maronpot, J. C. Lamb, W. M. Kluwe: "Adverse Effects of Butyl Benzyl Phthalate on the Reproductive and Hematopoietic Systems of Male Rats," *Toxicology* **35** (1985) 189–206.

[221] B. G. Lake et al.: "Comparative Studies on Di(2-ethylhexyl) Phthalate-Induced Hepatic Peroxisome Proliferation in the Rat and Hamster," *Toxicol. Appl. Pharmacol.* **72** (1984) 46–60.

[222] F. E. Mitchell et al.: "Time and Dose-Response Study of the Effects on Rats of the Plasticizers Di(2-ethylhexyl) Phthalate," *Toxicol. Appl. Pharmacol.* **81** (1985) 371–392.

[223] B. E. Butterworth, D. J. Loury, T. Smith-Oliver, R. C. Cattley: "The Potential Role of Chemically Induced Hyperplasia in the Carcinogenic Activity of the Hypolipidemic Carcinogens," *Toxicol. Ind. Health* **3** (1987) 129–148.

[224] J. A. Popp, L. K. Garvey, R. C. Cattley: "In Vivo Studies on the Mechanism of di(2-Ethylhexyl) Phthalate Carcinogenesis," *Toxicol. Ind. Health* **3** (1987) 151–163.

[225] C. T. C. Rhodes et al.: "Comparative Pharmacokinetics and Subacute Toxicity of di(2-Ethylhexyl) Phthalate (DEHP) in Rats and Marmosets: Extrapolation of Effects in Rodents to Man," *EHP Environ. Health Perspect.* **65** (1986) 299–308.

[226] A. H. Mann et al.: "Comparison of the Short-Term Effects of Di(2-ethylhexyl) Phthalate, Di(n-hexyl) Phthalate, and Di(n-octyl) Phthalate in Rats," *Toxicol. Appl. Pharmacol.* **77** (1985) 116–132.

[227] W. Kluwe, J. K. Haseman, J. F. Douglas, J. E. Huff: "The Carcinogenicity of Dietary di(2-Ethylhexyl) Phthalate (DEHP) in Fisher-344 Rats and B6C3F1 Mice," *J. Toxicol. Environ. Health* **10** (1982) 797–815.

[228] A. E. Ganning, J. J. Olsson, U. Brunk, G. Dallner: "Effects of Prolonged Treatment with Phthalate Ester on Rat Liver," *Pharmacol. Toxicol.* (Amsterdam) **68** (1991) 392–401.

[229] *An Investigation of the Effect of Di-2-Ethylhexyl Phthalate (DEHP), Dibutylphthalate (DBP), Di-iso-decylphthalate (DIDP) on Rat Hepatic Peroxisomes,* British Industrial Biological Research Association (BIBRA), Surrey 1990.

[230] J. J. Reddy, N. D. Lalwani: "Carcinogenesis by Hepatic Peroxisome Proliferators: Evaluation of the Risk of Hypolipidemic Drugs and Industrial Plasticizers to Humans," *CRC Crit. Rev. Toxicol.* **12** (1983) 1–58.

[231] A. M. Mitchell, J. C. Lhuguenot, J. W. Bridges, C. R. Elcombe: "Identification of the Proximate Peroxisome Proliferator(s) Derived from Di(2-ethylhexyl) Phthalate," *Toxicol. Appl. Pharmacol.* **80** (1985) 23–32.

[232] D. J. Benford et al.: "Species Differences in Response of Cultured Hepatocytes to Phthalate Esters," *Food Chem. Toxicol.* **24** (1986) 799–800.

[233] J. V. Rodricks, D. Turnbull: "Interspecies Differences in Peroxisomes and Peroxisome Proliferation," *Toxicol. Ind. Health* **3** (1987) 197.

[234] P. Sjöberg, N. G. Lindquist, L. Plöen: "Age-Dependent Response of the Rat Testes to Di(2-ethylhexyl) Phthalate," *EHP Environ. Health Perspect.* **65** (1986) 234–242.

[235] K. Shiota, S. Mima: "Assessment of the Teratogenicity of Di(2-ethylhexyl) Phthalate and Mono-(2-ethylhexyl) Phthalate in Mice," *Arch. Toxicol.* **56** (1985) 263–266.

[236] R. Wolowsky-Tyl, C. Jones-Price, M. C. Marr, C. A. Kimmel: "Teratologic Evaluation of Diethylhexyl Phthalate (DEHP) in CD-1 Mice," *Teratology* **27** (1983) 84A–85A.

[237] Y. Nakamura, Y. Yagi, I. Tomita, K. Tsuchikawa: "Teratogenicity of Di(2-ethylhexyl) Phthalate in Mice," *Toxicol. Lett.* **4** (1979) 113–117.

[238] J. J. Heindel et al.: "Reproductive Toxicity of Three Phthalic Acid Esters in a Continuous Breeding Protocol," *Fundam. Appl. Toxicol.* **12** (1989) 508–518.

[239] R. Wolowsky-Tyl, C. Jones-Price, M. C. Marr, C. A. Kimmel: "Teratologic Evaluation of Diethylhexyl Phthalate (DEHP) in Fischer-344 Rats," *Teratology* **27** (1983) 85A.

[240] J. Merkle, H. J. Klimisch, R. Jäckh: "Developmental Toxictiy in Rats after Inhalation Exposure of Di-2-ethylhexyl Phthalate (DEHP)," *Toxicol. Lett.* **42** (1988) 215–223.

[241] A. B. DeAngelo, C. T. Garrett, L. A. Manolukas, T. Yario: "Di-n-octyl Phthalate (DOP), a Relatively Ineffective Peroxisome Inducing Straight Chain Isomer of the Environmental Contaminant Di(2-ethylhexyl) Phthalate (DEHP), Enhances the Development of Putative Preneoplastic Lesions in Rat Liver," *Toxicology* **41** (1986) 279–288.

[242] J. M. Ward et al.: "The Chronic Hepatic or Renal Toxicity of Di(2-ethylhexyl) Phthalate, Acetaminophen, Sodium Barbital, and Phenobarbital in Male B 6 C 3 F 1 Mice: Autoradiographic, Immunohistochemical, and Biochemical Evidence for Levels of DNA Synthesis not Associated with Carcinogenesis or Tumor Promotion," *Toxicol. Appl. Pharmacol.* **96** (1988) 494–506.

[243] R. H. Hinton et al.: "Effects of Phthalic Acid Esters on the Liver and Thyroid," *EPH Environ. Health Perspect.* **70** (1986) 195–210.

[244] K. N. Woodward: "Phthalate Esters, Cystic Kidney Disease in Animals and Possible Effects on Human Health: A Review," *Hum. Exp. Toxicol.* **9** (1990) 397–401.

[245] J. F. S. Crocker, S. H. Safe, P. Acott: "Effects of Chronic Phthalate Exposure on the Kidney," *J. Toxicol. Environ. Health* **23** (1988) 433–444.

[246] S. C. Price et al.: "Alterations in the Thyroids of Rats Treated for Long Periods with Di(2-ethylhexyl) Phthalate or with Hypolipidaemic Agents," *Toxicol. Lett.* **40** (1988) 37–46.

[247] S. C. Price et al.: "Alterations in the Thyroids of Rats Treated for Long Periods with Di(2-ethylhexyl) Phthalate or with Hypolipidaemic Agents," *Toxicol. Lett.* **40** (1988) 37–46.

[248]   C. J. Price et al.: "Developmental Toxicity of Butyl Benzyl Phthalate (BBP) in Mice and Rats," *Teratology* **41** (1990) 586; *Chem. Abstr.* P 51.

[249]   A. H. Mann et al.: "Comparison of the Short-Term Effects of Di(2-ethylhexyl) Phthalate, Di(n-hexyl) Phthalate, and Di(n-octyl) Phthalate in Rats," *Toxicol. Appl. Pharmacol.* **77** (1985) 116–132.

[250]   S. Khanna et al., *J. Environ. Biol.* **11** (1990) 22–34.

[251]   D. Calley, J. Autina, W. L. Guess: "Toxicology of a Series of Phthalate Esters," *J. Pharm. Sci.* **55** (1966) 158–162.

[252]   W. H. Lawrence et al.: " A Toxicological Investigation of Some Acute, Short-Term, and Chronic Effects of Administering Di-2-Ethylhexyl Phthalate (DEHP) and Other Phthalate Esters," *Environ. Res.* **9** (1975) 1–11.

[253]   G. Klecak, H. Geleik, J. R. Frey: "Screening of Fragrance Materials for Allergenicity in the Guinea Pig," *J. Soc. Cosmet. Chem.* **28** (1977) 53–64.

[254]   R. Z. Christiansen, H. Osmundsen, B. Borrebaek, J. Bremer: "The effects of Clofibrate Feeding on the Metabolism of Palmitate and Erucate in Isolated Hepatocytes," *Lipids* **13** (1978) 487–419.

[255]   C. R. Elcombe, A. M. Mitchell: "Peroxisome Proliferation due to Di-(2-ethylhexyl) Phthalate: Species Differences and Possible Mechanisms," *EPH Environ. Health Perspect.* **70** (1986) 211–219.

[256]   G. P. Mannaerts, L. J. Debeer, J. Thomas, P. J. Schepper: "Mitochondrial and Peroxisomal Fatty Acid Oxidation in Liver Homogenates and Isolated Hepatocytes from Control and Clofibrate-Treated Rats," *J. Biol. Chem.* **254** (1979) 4585–4595.

[257]   E. A. Lock, A. M. Mitchell, C. R. Elcombe: "Biochemical Mechanisms of Induction of Hepatic Peroxisome Proliferation," *Annu. Rev. Pharmacol. Toxicol.* **29** (1989) 145–163.

[258]   P. I. Eacho et al.: "Hepatic Peroxisomal Changes Induced by a Tetrazole-Substituted Alkoxyacetophenone in Rats and Comparison with other Species," *Toxicol. Appl. Pharmacol.* **83** (1986) 430–437.

[259]   S. P. Foxworthy, D. N. Perry, D. M. Hoover, P. I. Eacho: "Changes in Hepatic Lipid Metabolism Associated with Lipid Accumulation and its Reversal in Rats given the Peroxisome Proliferator LY 171 883," *Toxicol. Appl. Pharmacol.* **106** (1990) 375–383.

[260]   J. G. Conway, R. C. Cattley, J. A. Popp, B. E. Butterworth: "Possible Mechanisms in Hepatocarcinogenesis by the Peroxisome Proliferator Di-(2-ethylhexyl) Phthalate," *Drug Metab. Rev.* **21** (1989) no. 1, 65–102.

[261]   R. L. Melnick. R. E. Morrissey, K. E. Tomaszewski: "Studies by the National Toxicology Program on Di(2-ethylhexyl) Phthalate," *Toxicol. Ind. Health* **3** (1987) 99–118.

[262]   K. E. Tomaszewski, D. A. Agarwal, R. L. Melnick: "In Vitro Steady-State Levels of Hydrogen Peroxide after Exposure of Male F 344 Rats and Female B 6 C 3 F 1 Mice to Hepatic Peroxisome Proliferators," *Carcinogenesis* (London) **7** (1986) 1871–1876.

[263]   H. Tamura, T. Iida, T. Watanabe, T. Suga:"Long-Term Effects of Hypolipidemic Peroxisome Proliferator Administration on Hepatic Hydrogen Peroxide Metabolism in Rats,"*Carcinogenesis (London)* **11** (1990) 445–450.

[264]   A. Takagi et al.: "Relationship between Hepatic Peroxisome Proliferation and 8-Hydroxyoxyguanosine Formation in Liver DNA of Rats following Long-Term Exposure to Three Peroxisome Proliferators; Di(2-ethylhexyl) Phthalate, Aluminium Clofibrate and Simfibrate," *Cancer Lett.* (Shannon Irel.) **53** (1990) 33–38.

[265] J. M. Makowska, F. W. Bonner, G. G. Gibson: "Hepatic Induction Potency of Hypolipidaemic Drugs in the Rat Following Long-Term Administration: Influence of Different Dosing Regiments," *Xenobiotica* **20** (1990) 1112–1128.

[266] J. G. Conway et al.: "Relationship of Oxidative Damage to the Hepatocarcinogenicity of the Peroxisome Proliferators Di(2-ethylhexyl) Phthalate and Wy-14,643," *Carcinogenesis (London)* **10** (1989) 513–519.

[267] R. R. Shukla, P. W. Albro: "In Vitro Modulation of Protein Kinase C Activity by Environmental Chemical Pollutants," *Biochem. Biophys. Res. Commun.* **142** (1987) 567–572.

[268] C. Gupta, A. Hattori, H. Shinozuka: "Short Communication; Suppression of EFG binding in Rat Liver by the Hypolipidemic Peroxisome Proliferators, 4-chloro-6-(2,3-xylidino)-2-pyrimidinylthio(N-β-hydroxyethyl)acetamide and Di(2-ethylhexyl) Phthalate," *Carcinogenesis (London)* **9** (1988) 167–169.

[269] T. Yario et al.: "Effect of Di-(2-ethylhexyl) Phthalate (DEHP) on DNA Synthesis and on Promotion of Glucose-6-phosphatase Negative Preneoplastic Lesions in Rat Liver," *Carcinogenesis (London)* **27** 148, Abstr. 585 (1986).

[270] C. Gupta, A. Hattori, H. Shinozuka: "Suppression of EGF Binding in Rat Liver by the Hypolipidemic Peroxisome Proliferators, 4-chloro-6(2,3-xylidino)-2-pyrimidinylthio(N-β-hydroxyl)acetamide and Di-(2-ethylhexyl) Phthalate," *Carcinogenesis (London)* **9** (1988) 167–169.

[271] C. P. Carpenter, C. S. Weil, H. F. Smyth: "Chronic Oral Toxicity of Di(2-ethylhexyl) Phthalate for Rats, Guinea Pigs, and Dogs," *Arch. Ind. Hyg. Occup. Med.* **8** (1953) 219–226.

[272] R. S. Harris, H. C. Hodge, E. A. Maynard, H. J. Blanchet jr.: "Chronic Oral Toxicity of 2-Ethylhexyl Phthalate in Rats and Dogs,"*AMA Arch. Ind. Health* **13** (1956) 259–264.

[273] G. Lake et al.: "Effect of Prolonged Administration of Clofibric Acid and Di-(2-ethylhexyl) Phthalate on Hepatic Enzyme Activities and Lipid Peroxidation in the Rat," *Toxicology* **44** (1987) 213–228.

[274] C. Cattley, J. G. Conway, J. A. Popp: "Association of Persistent Peroxisome Proliferation and Oxidative Injury with Hepatocarcinogenicity in Female F-344 Rats Fed Di-(2-ethylhexyl) Phthalate for 2 years," *Cancer Lett. (Shannon Irel.)* **38** (1987) 15–22.

[275] M. S. Rao, N. Usuda, V. Subbarao, J. K. Reddy: "Absence of γ-Glutamyl Transpeptidase Activity in Neoplastic Lesions induced in the Liver of Male F-344 Rats by Di-(2-ethylhexyl) Phthalate, a Peroxisome Proliferator," *Carcinogenesis* **8** (1987) 1347–1350.

[276] Biodynamics Inc., A Chronic Toxicity Carcinogenicity Feeding Study in Rats with Santicizer 900, Proj. no. 81-2572, Monsarto Company, 1986.

[277] A. Lington, M. Bird, R. Blutnick, J. Quance: *The Toxicologist* **7** Abstr. (1987) 405.

[278] P. Schmezer et al.: "Various Short-Term Assays and Two Long-Term Studies with the Plasticizer Di-(2-ethylhexyl) Phthalate in the Syrian Golden Hamster," *Carcinogenesis (London)* **9** (1988) 37–43.

[279] Carcinogenesis Bioassay of Butyl Benzyl Phthalate (CAS No. 85-68-7) in F 344/N Rats and B 6 C 3 F 1 Mice (Feed Study). NTP-Rep. 80-25, NIH Publication 82-1769, US Dpt. Health and Human Services, Washington, DC, 1982.

[280] L. K. Garvey, J. A. Swenberg, T. E. Hamm, J. A. Popp: "Di(2-ethylhexyl) Phthalate: Lack of Initiating Activity in the Liver of Female F 334 Rats," *Carcinogenesis (London)* **8** (1987) 285–290.

[281] A. B. DeAngelo, C. T. Garrett, A. E. Queral: "Inhibition of Phenobarbital and Dietary Choline Deficiency Promoted Preneoplastic Lesions in Rat Liver by Environmental Contaminant Di(2-ethylhexyl) Phthalate," *Cancer Lett. (Shannon Irel.)* **23** (1984) 323–330.

[282] J. M. Ward et al.: "Tumor-Initiating and Promoting Activities of Di(2-ethylhexyl) Phthalate in Vivo and in Vitro," *EHP Environ. Health Perspect.* **65** (1986) 279–291.

[283] M. I. R. Perera, H. Shinozuka: "Accelerated Regression of Carcinogen-Induced Preneoplastic Hepatocyte Foci by Peroxisome Proliferators, BR 931, 4-Chloro-6-(2,3-xylidino)-2-pyrimidinylthio (N-B-hydroxyethyl)acetamide, and Di(2-ethylhexyl) Phthalate," *Carcinogenesis (London)* **5** (1984) 1193–1198.

[284] D. Oesterle, E. Deml: "Promoting Activity of Di (2-ethylhexyl) Phthalate in Rat Liver Foci Bioassay," *J. Cancer Res. Clin. Oncol.* **114** (1988) 133–136.

[285] A. B. DeAngelo, A. E. Queral, C. T. Garrett: "Concentration-Dependent Inhibition of Development of GGT Positive Foci in Rat Liver by the Environmental Contaminant Di(2-ethylhexyl) Phthalate," *EPH Environ. Health Perspect.* **60** (1985) 381–385.

[286] A. B. DeAngelo, C. T. Garrett: "Inhibition of Development of Preneoplastic Lesions in the Livers of Rats Fed a Weakly Carcinogenic Environmental Contaminant," *Cancer Lett. (Shannon Irel.)* **20** (1983) 199–205.

[287] B. Kraupp-Grasi, W. Huber, H. Taper, R. Schulte-Hermann: "Increased Susceptibility of Aged Rats to Hepatocarcinogenesis by the Peroxisome Proliferator Nafenopin and the Possible Involvement of Altered Liver Foci Occurring Spontaneously," *Cancer Res.* **51** (1991) in press.

[288] K. Yoshikawa, A. Tanaka, T. Yamaha, H. Kurata: "Mutagenicity Study of Nine Monoalkyl Phthalates and a Dialkyl Phthalate using Salmonella Typhimurium and Escherichia Coli," *Food Chem. Toxicol.* **21** (1983) no. 2, 221–223.

[289] R. J. Rubin, K. Kozumbo, R. Kroll: "Ames Mutagenic Assay of a Series of Phthalic Acid Esters: Positive response of the Dimethyl and Diethylesters in TA 100," *Toxicol. Appl. Pharmacol.* **48** (1979) A 133.

[290] D. K. Agarwal, W. H. Lawrence, L. J. Nunez, J. Autian: "Mutagenicity Evaluation of Phthalic Acid Esters and Metabolites in Salmonella Typhimurium Cultures," *J. Toxicol. Environ. Health* **16** (1985) 61–69.

[291] B. E. Butterworth et al.: "Lack of Genotoxic Activity of Di(2-ethylhexyl) Phthalate (DEHP) in Rat and Human Hepatocytes," *Carcinogenesis (London)* **5** (1984) 1329–1335.

[292] D. J. Kornbrust, T. R. Barfknecht, P. Ingram, J. D. Shelburne: "Effect of Di(2-ethylhexyl) Phthalate on DNA Repair and Lipid Peroxidation in Rat Hepatocytes and on Metabolic Cooperation in Chinese Hamster V-79 Cells," *J. Toxicol. Environ. Health* **13** (1984) 99–116.

[293] G. Zeiger, S. Haworth, K. Mortelmans, W. Speck: "Mutagenicity Testing of Di(2-ethylhexyl) Phthalate and Related Chemicals in Salmonella," *Environ. Mutagen.* **7** (1985) 213–232.

[294] J. M. Parry, N. Danford, E. M. Parry: "In Vitro Techniques for the Detection of Chemicals Capable of Inducing Mitotic Chromosome Aneuploidy," *ATLA Altern. Lab. Anim.* **11** (1984) 117–128.

[295] B. J. Phillips, D. Anderson, S. D. Gangolli: "Studies on the Genetic Effects of Phthalic Acid Esters on Cells in Culture," *EHP Environ. Health Perspect.* **263** (1986) 263–266.

[296] N. N. Woodward: *Phthalate Esters: Toxicity and Metabolism*, vol. **II**, CRC Press, Boca Raton, Florida 1988, pp. 47–75.

[297] B. J. Phillips, T. T. B. James, S. D. Gangolli: "Genotoxicity Studies of DEHP and its Metabolites in CHO Cells," *Mutat. Res.* **102** (1982) 297–304.

[298] A. R. Singh, W. H. Lawrence, J. Autian: "Mutagenic and Antifertility Sensitivities of Mice to Di-2-ethylhexyl Phthalate (DEHP) and Dimethoxyethyl Phthalate (DMEP)," *Toxicol. Appl. Pharmacol.* **29** (1974) 35–46.

[299] D. K. Agarwal, W. H. Lawrence, L. J. Nunez, J. Autian: "Antifertility and Mutagenic Effects in Mice from Parenteral Administration of Di-2-ethylhexyl Phthalate (DEHP)," *J. Toxicol. Environ. Health* **16** (1985) 61–69.

[300] Y. Hamano et al.: "Dominant lethal test of DEHP in mice," *Food Hygiene Series* **10** (1980) 1–10 (Japan);" Chem. Abstr. Sect. Toxicol. 92 2 099 960, 145.

[301] C. J. Rushbrook, T. A. Jorgenson, J. R. Hodgson: "Dominant Lethal Study of Di-2-(ethylhexyl) Phthalate and its Major Metabolites in ICR/SIM Mice," *Environ. Mutagen.* **4** (1982) 387 A.

[302] A. von Däniken, W. K. Lutz, R. Jäckh, C. Schlatter: "Investigation of the Potential for Binding of Di(2-ethylhexyl) Phthalate (DEHP) and Di(2-ethylhexyl) Adipate (DEHA) to Liver DNA in Vivo," *Toxicol. Appl. Pharmacol.* **73** (1984) 373–387.

[303] L. I. Hinrichsen, R. A. Floyd, O. Sudilovsky: "Is 8-Hydroxydeoxyguanosine a Mediator of Carcinogenesis by a Choline-Devoid Diet in the Rat Liver?" *Carcinogenesis (London)* **11** (1990) 1879–1881.

[304] R. A. Floyd: "The role of 8-Hydroxyguanine in Carcinogenesis," *Carcinogenesis (London)* **11** (1990) 1447–1450.

[305] R. D. Short, E. C. Robinson, A. W. Lington, A. E. Chin: "Metabolic and Peroxisome Proliferation Studies with Di(2-ethylhexyl) Phthalate in Rats and Monkeys," *Toxicol. Ind. Health* **3** (1987) 185–196.

[306] M. Ema, T. Murai, T. Itami, H. Kawasaki: "Evaluation of the Teratogenic Potential of the Plasticizer Butyl Benzyl Phthalate in Rats," *JAT, J. Apll. Toxicol.* **10** (1990) 339–343.

[307] H. Swenerton, L. S. Hurley: "Teratogenic Effects of a Chelating Agent and their Prevention by Zinc," *Science* (Washington D.C.) **173** (1971) 62–63.

[308] M. Fukuoka, Y. Zou, A. Tanaka: "Mechanism of Testicular Atrophy induced by Di-n-butyl Phthalate in Rats, part 2. The Effects on Some Testicular Enzymes," *JAT, J. Appl. Toxicol.* **10** (1990) 285–293.

[309] T. J. B. Gray, S. D. Gangolli: "Aspects of the Testicular Toxicity of Phthalate Esters," *EPH Environ. Health Perspect.* **65** (1986) 229–235.

[310] L. A. Dostal et al.: "Testicular Toxicity and Reduced Sertoli Cell Numbers in Neonatal Rats by Di(2-ethylhexyl) Phthalate and the Recovery of Fertility as Adults," *Toxicol. Appl. Pharmacol.* **95** (1988) 104–121.

[311] S. D. Gangolli: "Testicular Effects of Phthalate Esters," *EPH Environ. Health Perspect.* **45** (1982) 77–84.

[312] T. J. B. Gray, J. A. Beamand: "Effect of some Phthalate Esters and other Testicular Toxins on Primary Cultures of Testicular Cells," *Food Chem. Toxicol.* **22** (1983) 123.

[313] C. Rhodes et al.: "The Absence of Testicular Atrophy in Vivo and in Vitro Effects on Hepatocyte Morphology and Peroxisomal Enzyme Activities in Male Rats Following the Administration of Several Alkanols," *Toxicol. Lett.* **21** (1984) 103–109.

[314] D. K. Agarwal et al.: "Effects of Di(2-ethylhexyl) Phthalate on the Gondal Pathophysiology, Sperm Morphology, and Reproductive Performance of Male Rats," *EPH Environ. Health Perspect.* **65** (1986) 343–350.

[315] B. G. Lake et al.: "Studies on the Effects of Orally Administered Di-(2-ethylhexyl) Phthalate in the Ferret," *Toxicology* **6** (1976) 341.

[316] P. M. D. Foster et al.: "Changes in Ultrastructure and Cytochemical Localization of Zinc in Rat Testis Following the Administration of Di-n-pentyl Phthalate," *Toxicol. Appl. Pharmacol.* **63** (1982) 120–132.

[317] D. K. Agarwal et al.: "Influence of Dietary Zinc on Di(2-ethylhexyl) Phthalate-Induced Testicular Atrophy and Zinc Depletion in Adult Rats," *Toxicol. Appl. Pharmacol.* **84** (1986) 12–24.

[318] W. M. Kluwe: "Overview of Phthalate Ester Pharmacokinetics in Mammalian Species," *EPH Environ. Health Perspect.* **45** (1982) 3–10.
[319] P. W. Albro et al.: "Identification of the Metabolites of Di-(2-ethylhexyl) Phthalate in Urine from African Green Monkey," *Drug, Metab. Dispos.* **9** (1981) 23.

Phthalocyanines

GERD LÖBBERT, BASF Aktiengesellschaft, Ludwigshafen, Federal Republic of Germany

1.	Introduction	3920	4.5.	Stabilization of
2.	Physical Properties	3923		Phthalocyanines. 3943
2.1.	General Data.	3923	5.	Phthalocyanine Derivatives. . . 3944
2.2.	Particle Size	3924	5.1.	Halogenated Phthalocyanines . 3945
2.3.	Color Properties	3924	5.2.	Phthalocyanine Sulfonic Acids
2.4.	Crystal Structure and			and Sulfonyl Chlorides. 3947
	Polymorphism.	3925	5.3.	Chloromethylphthalocyanines . 3948
2.5.	Surface and Interfacial		5.4.	Phthalocyanines with Other
	Properties.	3927		Substituents. 3948
2.6.	Other Physical Investigations .	3928	5.5.	Polymeric Phthalocyanines. . . 3949
3.	Chemical Properties.	3928	5.6.	Phthalocyanine Analogues . . . 3949
3.1.	Oxidation	3928	6.	Legal Aspects and
3.2.	Reduction	3931		Environmental Protection . . . 3952
3.3.	Catalytic Properties	3932	7.	Quality Specifications and
4.	Production	3933		Analysis 3953
4.1.	General Synthetic Methods. . .	3933	8.	Uses 3953
4.2.	Reaction Mechanism	3934	8.1.	Phthalocyanine Pigments 3954
4.3.	Industrial Production	3936	8.2.	Phthalocyanine Dyes 3955
4.3.1.	Copper Phthalocyanine	3936	9.	Economic Aspects 3956
4.3.2.	Phthalocyanine	3939	10.	Toxicology and Occupational
4.4.	Pigment Finishing	3939		Health. 3957
4.4.1.	Finishing by Treatment with Acid	3940		
4.4.2.	Finishing by Grinding.	3942	11.	References. 3959

I. Introduction

The term phthalocyanine was first used by R. P. LINSTEAD in 1933 [4] to describe a class of organic dyes, whose colors range from reddish blue to yellowish green. The name phthalocyanine originates from the Greek terms *naphtha* for mineral oil and *cyanine* for dark blue. In 1930–1940, LINSTEAD et al. elucidated the structure of phthalocyanine (H_2Pc) and its metal complexes [4]–[14]. The basic structure is represented by phthalocyanine (**1**) itself:

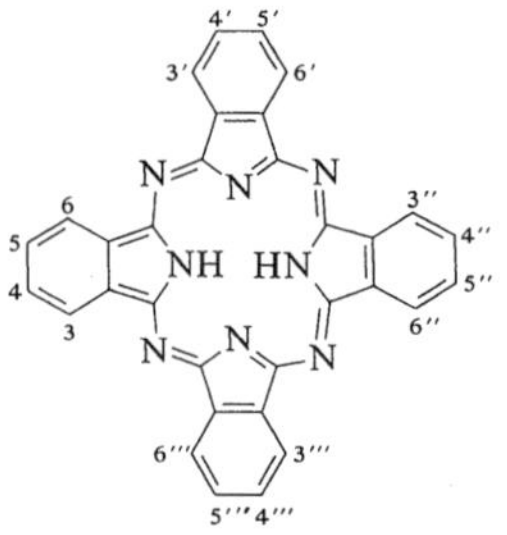

1

Phthalocyanine forms complexes with numerous metals of the periodic table. Today, 66 complexes with various elements are known [15]–[21]. Metal phthalocyanines MPc (**2**) and compounds with metalloids such as B, Si, Ge, and As or nonmetals such as P display a wide variety in their coordination chemistry.

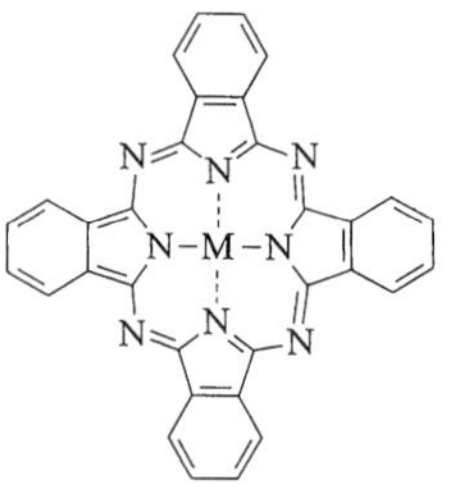

2

The coordination number of the square planar complexes of Cu, Ni, or Pt is 4. Higher coordination numbers of 5 or 6 with one or two additional ligands such as water or ammonia result in square-based pyramidal, tetrahedral, or octahedral structures (**3**) [22]–[25]. Complexes of lanthanides and actinides (e.g., $Nd^{III}HPc_2$ or $U^{IV}Pc_2$) have an unusual sandwich structure (**4**) with eight coordinated N atoms [26]–[28]. In contrast, reaction of 1,2-dicyanobenzene with uranium chloride in DMF gives a phthalocyanine ring with five isoindole units (**5**; super phthalocyanine) [29], [30]. A boron compound (**6**) with three former dinitrile units can be synthesized from boron halides and 1,2-dicyanobenzene [31].

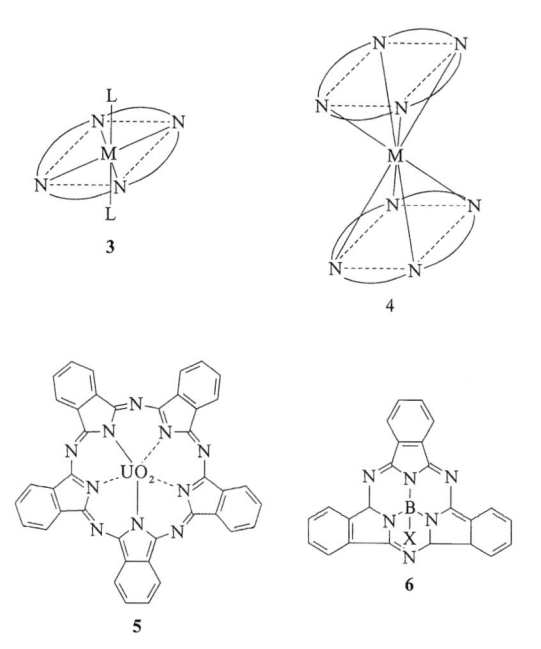

Phthalocyanines form polymers in which the central atom is a part of the polymeric chain. Elements such as Si^{IV}, Ge^{IV}, or Sn^{IV} yield covalently bonded polymers (**7**) [32]–[37]. The degree of polymerization n ranges from 65 to 140 for $(SiO_2Pc)_n$.

The phthalocyanines are structurally related to the macrocyclic ring system porphyrin (**8**). Formally, phthalocyanine can be regarded as tetrabenzotetraazaporphyrin and as the condensation product of four isoindole units.

The phthalocyanines are structurally similar to naturally occurring porphyrins such as hemoglobin (**9**), chlorophyll a, vitamin B_{12} and turacin (a red dye from the wings of an African bird) [39]. Phthalocyanines themselves do not occur in nature.

Compounds with naphthalene or anthracene rings in place of the benzene nucleus also belong to the phthalocyanine family. The hydrogen atoms of benzene rings can be substituted in many ways (e.g., by halogens or by sulfonic acid, sulfonyl chlorides, or chloromethyl groups) (Chap. 5). Various polymers with interesting properties can be synthesized.

History. BRAUN and TSCHERNAK [40] obtained phthalocyanine for the first time in 1907 as a byproduct of the preparation of *o*-cyanobenzamide from phthalimide and acetic anhydride. However, this discovery was of no special interest at the time. In 1927, DE DIESBACH and VON DER WEID prepared CuPc in 23% yield by reacting *o*-dibromobenzene with copper cyanide in pyridine [41]. Instead of the colorless dinitriles, they obtained deep-blue CuPc and observed the exceptional stability of their product to sulfuric acid, alkalies, and heat. The third observation of a phthalocyanine was made at Scottish Dyes, in 1929 [42]. During the preparation of phthalimide from phthalic anhydride and ammonia in an enamel vessel, a greenish blue impurity appeared. DUNSWORTH and DRESCHER carried out a preliminary examination of the compound, which was analyzed as an iron complex. It was formed in a chipped region of the enamel with iron from the vessel. Further experiments yielded FePc, CuPc, and NiPc. It was soon realized that these products could be used as pigments or textile colorants. LINSTEAD et al. at the University of London discovered the structure of phthalocyanines and developed improved synthetic methods for several metal phthalocyanines from 1929 to 1934 [4]–[14]. For example, the reaction of dicyano aromatics and metal salts was patented in 1939 [43]. The important CuPc could not be protected by a patent, because it had been described earlier in the literature [41]. Based on LINSTEAD's work the structure of phthalocyanines was confirmed by several physicochemical measurements [44]–[50]. Methods such as X-ray or electron microscopy verified the planarity of this macrocyclic system. Properties such as polymorphism, absorption spectra, magnetic and catalytic characteristics, oxidation and reduction, photoconductivity and semiconductivity, solubility, and photochemical and dielectric properties were investigated from the 1930s to the 1950s.

Copper phthalocyanine was first manufactured by ICI in 1935, where its production from phthalic anhydride, urea, and metal salts was developed [51]. Use of catalysts such as ammonium molybdate improved the method substantially [52]. In 1936, I.G. Farbenindustrie began production of CuPc at Ludwigshafen, and in 1937 Du Pont followed in the United States. The most important of the phthalocyanines, CuPc, is now produced worldwide. The first phthalocyanine dye was a phthalocyanine polysulfonate [42]. Other derivatives, such as sulfonyl chlorides, ammonium salts of pyridyl phthalocyanine derivatives, sulfur and azo dyes, and chrome and triazine dyes, have been patented since 1930. At that time, the use of phthalocyanines as colorants for printing ink, paint, plastics, and textiles began. Of the industrial uses, the application of CuPc in printing inks is its most important use. The greenish blue CuPc shade is suitable for color printing. Other favorable properties such as light, heat, and solvent resistance led to the use of this blue pigment for paints and plastics. The chloro and bromo derivatives are important green organic pigments. Other derivatives are used in textile dyeing and printing or for the manufacture of high-quality inks (pastes for ballpoint pens, ink jets, etc.).

2. Physical Properties

2.1. General Data

The molecular mass of phthalocyanine, $C_{32}H_{18}N_8$, is 514.55. The densities of Pc compounds depend strongly on chemical composition:

β-H$_2$Pc	ϱ=1.43 g/cm^3
β-CuPc	1.61 g/cm^3
Polychloro CuPc	2.14 g/cm^3
β-CoPc	1.50 g/cm^3
β-NiPc	1.59 g/cm^3

The color of most Pc's ranges from blueblack to metallic bronze, depending on the manufacturing process. Ground powders exhibit colors from green to blue. Most compounds do not melt but sublime above 200 °C, which can be used for purification. An exception is Si(OC$_{18}$H$_{37}$)$_2$Pc, which melts at 152 °C [53].

Solubility and Thermal Stability. H$_2$Pc, CuPc, and halogenated phthalocyanines have very poor solubility in organic solvents. Only in some high-boiling solvents such as quinoline, trichlorobenzene, and benzophenone is recrystallization possible at higher temperature. However, the solubilities have a maximum of several milligrams per liter. In common solvents such as alcohols, ethers, or ketones the solubility is considerable lower. One liter of benzene dissolves 0.15 mg of α- or γ-CuPc and 0.046 mg of β-CuPc at 40 °C [54]. The heats of solution are 34.3 kJ/mol for α-CuPc and 41.4 kJ/mol for β-CuPc [54]. The solubilities of other Pc complexes depend on the central atom.

Phthalocyanine and its unsubstituted metal derivatives dissolve in highly acidic media such as concentrated sulfuric acid, chlorosulfuric acid, or anhydrous hydrofluoric acid, presumably due to protonation of the bridging nitrogen atoms [55]–[57]. In the presence of strong bases, reversible deprotonation of the central imino groups occurs [57]. The solubility in sulfuric acid depends on temperature and concentration [58]. In particular, the removal of copper from solid phthalocyanine by 25–65% H$_2$SO$_4$ at 25–70 °C ranges from 0.6 to 9.2×10^{-6} mol/L after 5 to 100 h. The rate of decomposition of CuPc increases with increasing H$_2$SO$_4$ concentration, reaching a maximum at about 80% H$_2$SO$_4$. The stability of metal phthalocyanines increases in the order: ZnPc < CuPc < CoPc < NiPc < CuPcCl$_{15}$ [59]. The color of phthalocyanine solutions in sulfuric acid depends on the degree of protonation (the N atoms of the ring systems are protonated by H$_2$SO$_4$; metals, e.g., Cu influence this protonation): H$_2$Pc gives a brownish yellow color; CuPc, greenish yellow to olive. The phthalocyanines can be precipitated from these solutions by addition of water. CuPc dissolves to the extent of 20 mg in 100 mL of liquid ammonia at −33.5 °C [60]. Solubility can be improved in some cases by reversible oxidation with organic peroxides or hypochlorites; the Pc's are oxidized to substances soluble in organic solvents from which they can be regenerated by reduction [61].

Both H$_2$Pc and its derivatives exhibit high thermal stability. For example, CuPc can be sublimed without decomposition at 500–580 °C under inert gas and normal pressure [8]. In vacuum, stability up to 900 °C has been reported [62]. Polychloro CuPc is thermally stable up to 600 °C in vacuum. At higher temperature it decomposes without sublimation. CuPc decomposes vigorously at 405–420 °C in air. In nitrogen, sublimation and decomposition occur simultaneously at 460–630 °C [63], [64]. Generally all metal Pc's are more stable thermally in N$_2$ than in O$_2$. CuPc changes from the α- to the β-form at 250–430 °C [65]. Several MPc polyimides exhibit good thermal and thermooxidative stability at > 500 °C in air and nitrogen, and are therefore of interest for heat-stable films, coatings, and fibers [66].

2.2. Particle Size

Particle size and particle-size distribution are important criteria for Pc's used as pigments; they can be determined by the Debye–Scherrer-method [67], [68]; by adsorption measurements [67], [69], [70]; or by electron microscopy [67], [70]–[75]. The particle size calculated from adsorption data is ca. 0.044 μm for α-CuPc, 0.054 μm for β-CuPc, and 0.037 μm for polychloro CuPc [73].

The particle size of pigments dispersed, for example, in paint or printing ink binders can be determined by sedimentation analysis in centrifuges [69], [70], [75]–[79]. The particle diameters measured by this method are larger than those obtained by electron microscopy [69], [70]. This is due to incomplete dispersion of agglomerates from the pigment powder. These agglomerates can be dispersed totally only by prolonged grinding [69]. Lattice dimensions and crystal defects can be determined from the diffraction pattern that appears on high magnification under the electron microscope [80]. The particle size and particle form of Pc pigments are discussed in [70], [71], [73], [81].

2.3. Color Properties

Phthalocyanines absorb strongly in the visible range of 600 to 700 nm. They are thus blue to blue-green colorants. Their absorption spectra in solution are very different from those of the solids. The color of solid pigments is also influenced by the crystal modification [54], [82]. Color properties may depend to some extent on pleochroistic effects [70], [79] and on the nature of the central metal atom. In the series metal-free H$_2$Pc, NiPc, CuPc, CoPc with the same crystal modification, the color shifts from strongly greenish blue to strongly reddish blue, and the color purity decreases. The only metal phthalocyanine with a brilliant color suited for the cyan tone in color

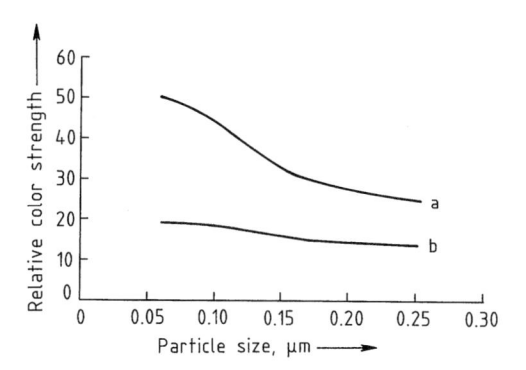

Figure 1. Relative color strength of β copper phthalocyanine (a) and polychloro copper phthalocyanine (b) as a function of particle size

printing is CuPc. Color strength increases with decreasing particle size of the crystals, aggregates, and agglomerates. Figure 1 illustrates this for β-CuPc and polychloro CuPc [75].

Substitution of the benzene rings in phthalocyanine results in a color shift to greenish blue or green. Polychloro CuPc is bluish green. Partial substitution of the Cl atoms by Br atoms results in a yellowish green. Alkyl, aryl or other functional groups have a similar effect (Chap. 5.). A red shift can be attained by replacement of benzene rings by pyridine rings [83]. This compound has a more turbid color shade than CuPc, but it exhibits similar pigment properties.

2.4. Crystal Structure and Polymorphism

The crystal structure and polymorphism of phthalocyanines are reviewed in [84]–[87]. These are among the first organic solids whose crystal structures were determined by X-ray analysis of single crystals [45], [47], [48]. Phthalocyanine and its important metal derivatives (Fe, Co, Ni, Cu, Pt) have monoclinic lattices. Lattice constants and other X-ray data are tabulated in [88], [89]. The Pc ring system is always planar. Bond lengths and angles are influenced only slightly by the nature of the central metal atom [89]. Phthalocyanine and its metal derivatives occur in different crystal modifications [90], which has been shown by X-ray analysis and by IR and ESR spectroscopy. The modifications of H_2Pc, CuPc, and other derivatives differ in regard to color and physical properties. Five polymorphous modifications of CuPc are known; they are designated α, β, γ, δ, and ε [54]. Two more modifications have been described in the literature [27], [91]–[96].

According to their solubilities in benzene (good solubility implies instability), thermodynamic stability increases in the sequence $\alpha = \gamma < \delta < \varepsilon < \beta$ [54]. The enthalpy difference between α- and β-CuPC is 10.75 kJ/mol [97]. Three modifications of H_2Pc — α, β [98], [99], and X [100] — have been found. Polychloro CuPc has only an α-modification [93], which is related to the α-modification of CuPc. All modifica-

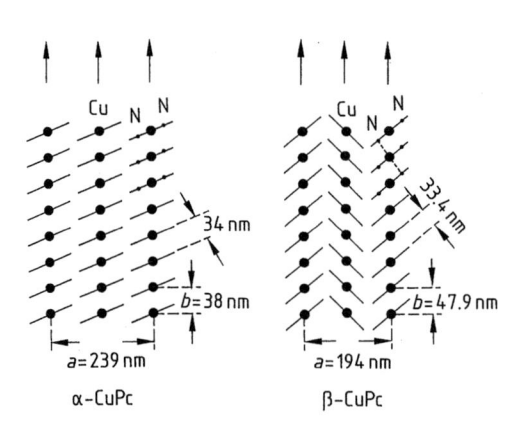

Figure 2. Arrangement of copper phthalocyanine molecules in the α- and β-modifications
The planar molecules appear as lines in this projection.

α-CuPc: a = 239 nm, b = 38 nm, 34 nm
β-CuPc: a = 194 nm, b = 47.9 nm, 33.4 nm

tions exhibit one-dimensional stacking of the molecules (Fig. 2). However, the arrangement of the stacks with respect to one another differs, as illustrated by the structures of α- and β-CuPc (Fig. 2).

Phase transformations in crystals are of interest, impelled by the fact that the most desirable pigment in terms of color properties in a given medium may not be the most resistant to changes in crystal modification [2,][3], [54], [101]. During synthesis the most stable modification is formed; in the case of CuPc and H_2Pc this is the β-modification. Dissolving β-CuPc in concentrated sulfuric acid or treatment with 40–90% sulfuric acid yields the finely divided α-form, after hydrolysis of the greenish yellow sulfate [2], [3], [99]. The β-modification is often converted to the α-modification by dry grinding in the presence of additives (e.g., salts). Figure 3 shows examples of the temperature-dependent β → α transition by dry grinding. The rate of transformation decreases with increasing temperature.

Preparation of the γ- [102], δ- [103], and ε-modifications [104], [105] is described in the literature, and further modifications (π, x) are believed to exist [27], [91]–[96], [106]. All modifications can be transformed into the most stable form by heating in a high-boiling, inert solvent. In the absence of solvent, the transformation occurs at ca. 200 °C [2][3]. Crystal growth has been studied by X-ray diffraction and electron microscopy [107], [108]. The transformation of CuPc in various organic media has been studied by electron microscopy and X-ray diffraction [109]. The preliminary growth of metastable α crystals precedes the transition to stable β crystals. Other studies of the α → β transition by visible spectroscopy, X-ray diffraction, electron microscopy, and BET determination of the specific surface area [110] indicate a three-stage process in terms of the classical theories of nucleation and crystal growth:

1) Ostwald ripening of the α-phase and nucleation of the β-phase
2) Disappearence of the α-phase and production of the β-phase
3) Final crystallization and ripening of the β-phase

The process is determined by nucleation of the stable phase [54], [111]. Treatment of β-CuPc with boiling solvents causes the β crystals to grow; treatment of α-CuPc with

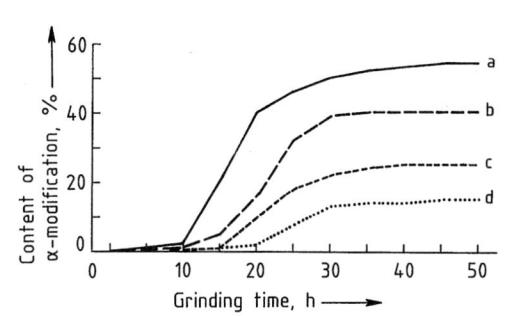

Figure 3. The β → α transformation by dry grinding as a function of grinding time and temperature
a) 70 °C; b) 90 °C; c) 110 °C; d) 130 °C

solvents causes the crystals to grow before transition to the β-phase occurs, and the β crystals grow [112]. The behavior of metal phthalocyanines depends on the central metal ion [113]. Suspending α-CuPc with metal oxides such as alumina, magnesia, or zinc oxide in organic media results in an acceleration of the α → β transformation [114]. Solvent-stable, flocculation-resistant α-, β-, and γ-modifications of CuPc and NiPc are prepared by combining phenyl-, chloro-, and sulfo-substituted tetraazaporphyrins with the MPc [115].

2.5. Surface and Interfacial Properties

A variety of surface and interfacial properties have been studied [2], [3]. They include the effects of size reduction or milling, surface ionization energies [116], orientation overgrowth (epitaxial growth of crystals on a substrate surface) [117], the change of surface properties on crystal transformation [118], electric and photoelectric effects, interphase surface activity [119], and measurement of surface free energy and polarity [120]. Interfacial properties such as heats of adsorption and adsorption isotherms with organic liquids and water demonstrate the hydrophobic character of phthalocyanine pigments, which is in contrast to many organic and inorganic pigments [70], [73], [121]. In their properties, Pc's are very similar to graphite or diamond powder. Flocculation stability and rheological properties are influenced disadvantageously. This effect, however, can be counteracted by addition of hydrophilic phthalocyanine derivatives [122]–[124]. In dispersions in organic media such as paints and binders, CuPc and polychloro CuPc are electrostatically charged [125], [126]. However, in contrast to many aqueous systems, this does not lead to stabilization of the dispersions. Adsorption of polymers improves flocculation stability. Both positive and negative charges are present on the same CuPc particles, which promotes flocculation, especially for rodlike structures [126]. Theoretical and experimental studies of the adsorption of a variety of organic compounds on phthalocyanines support the theory of privileged sites for adsorption on phthalocyanine at the central metal atom and along certain axes [127]. The stacking of phthalocyanine molecules results in two different surfaces. The surface parallel to the longitudinal axis of the stack is characterized primarily by benzene ring substituents (nonpolar hydrogen atoms). At the base of the stack, the dominant structural units are π-systems (nitrogen atoms and the central metal atom). Only the bases of the acicular crystals consist of easily polarizable atoms [128].

The specific surface area s of CuPc decreases on heating in vacuum at 300 °C from ca. 90 to 45 m^2/g, and its coloring power decreases by 10% [2], [3], [121]. In vacuum α-CuPc, has s=30, 11, and 4 m^2/g at 20, 200, and 300 °C, respectively [121].

2.6. Other Physical Investigations

The physical properties of phthalocyanines have been investigated theoretically and in regard to application as semiconductors, catalysts, etc., by a variety of physical methods. Absorption, fluorescence and phosphorescence, NMR, IR, Raman, and ESR spectra, as well as electric, photoelectric, magnetic, and semiconductor properties, are reviewed in [2], [3], [15], [129] – [133]. Some of these properties depend on the purity of the compounds [134]. The relationship to the catalytic properties of some Pc's is given in [135]. Mössbauer, ESCA, and inelastic tunneling spectroscopy, along with field ion and field electron microscopy, are reviewed in [2], [3], [136], [137] – [140]. Although many of these physical properties suggest possible applications of phthalocyanines in semiconductors, catalysts, lubricants, lasers, or photography, they have not yet been exploited industrially.

3. Chemical Properties

The chemical properties of phthalocyanines are determined by the nature of the central atom, with each complex undergoing specific reactions. Nevertheless, some properties are universal.

3.1. Oxidation

The behavior of phthalocyanines upon electrochemical or chemical oxidation and reduction is important for determination of the redox potentials [141], [142] by cyclic voltametry or polarography, for the use of Pc's in preparative chemistry, and for application in photoredox processes [20]. Compared to porphyrins, phthalocyanines are easy to oxidize or reduce [20]. Oxidation can be reversible or irreversible, depending on the conditions. Phthalocyanines are stable to atmospheric oxygen up to ca. 100 °C. Stability depends on the central metal atom [143]. In aqueous solutions of strong oxidants, the phthalocyanine ring system is completely destroyed. Oxidation of one molecule of CuPc consumes one oxygen atom, which was confirmed by oxidation with cerium(IV) sulfate in sulfuric acid [10]:

$[C_8H_4N_2]_4Cu + 3H_2SO_4 + 7H_2O + [O] \longrightarrow 4\ C_8H_5O_2N + CuSO_4 + 2(NH_4)_2SO_4$

The phthalocyanine ring is oxidized to phthalimide. This reaction can be used for quantitative determination of CuPc content. Other possibilities are oxidation with sodium vanadate [144], nitric acid [145], or potassium dichromate in sulfuric acid [146]. During oxidation of CuPc, intermediates with an intense violet color are formed [8], [147]–[150]. Addition of a solution of CuPc in concentrated sulfuric acid to ice water containing 3% nitric acid gives a violet precipitate, which can be converted back to the pigment by drying [151]. The stability of CuPc toward ozone indicates that this oxidant has little or no effect on its structure [152]. Adducts of $Fe^{II}Pc$, $Cr^{IV}Pc$, $Mn^{II}Pc$, and $Co^{II}Pc$ with nitrogen oxides are also known [151]. Well-defined oxidation products of phthalocyanines can be obtained in nonaqueous solution. Either oxidation occurs in the ring system, or the oxidation state of the central metal atom is changed. Oxidation of the ring system yield derivatives of tetracycloindolenines **10** and **11** [153], [154].

Oxidation with nitric acid in toluene, acetic acid, methanol, or pyridine yields **10** and **11**, with R=OH. Bromine in methanol or methanol–pyridine gives alkoxy or bromine derivatives (R=OCH_3 or Br) [61], [153], [154]. Acyl peroxides lead to acyl derivatives (R=OCOR) [154].

Oxidation products **10** and **11** are formed by metal-free phthalocyanine and MPc (M=Mg, Zn, Fe, Co, Ni, and Cu) [154]; polychloro CuPc does not react in this way. These oxidation products are much lighter than the blue dyes. For example, the nitrate of CuPc is yellow, and that of CoPc, yellow-orange [61], [153], [154]. Many of these complexes are crystalline and stable in the absence of reducing agents. They can be reduced to the corresponding Pc's in good yield [153], [154]. The same effect is observed on heating to 140–220 °C [61]. The good solubility of the oxidation products in organic solvents such as chloroform, pyridine, DMF, and methanol is noteworthy. These phthalocyanine derivatives are of considerable industrial interest for textile dyeing. Their synthesis and applications are described in many patents (see Chap. 8).

Oxidation can also occur at the central metal atom of the phthalocyanine system. Transition metals, which occur in several oxidation states, exhibit this behavior. They usually form six-coordinate complexes on oxidation:

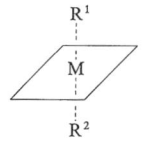

The chemistry of these metal phthalocyanines is reviewed in [15]. Manganese phthalocyanine, which occurs in five oxidation states from 0 to +4, is unique in this context [15]. Synthesis from manganese and phthalodinitrile yields $Mn^{II}Pc$ (**12**) [11]. Oxidation in organic solvents such as alcohol or chloroform or in the presence of a trace of acid yields $Mn^{III}Pc$ (**13**). In pyridine, dimethyl sulfoxide, diethylamine, or alkaline alcohol, this is converted to $Mn^{IV}Pc$ (**14**) [155]–[157]. The $Mn^{II}Pc$ reversibly chemisorbs oxygen in pyridine, which was first demonstrated by ELVIDGE and LEVER [156]. The oxygen absorbed at room temperature in pyridine solution can be removed by heating to boiling temperature. This process can be repeated several times, whereby small amounts of the phthalocyanine complex are decomposed. The mechanism of the process follows [157], [158]:

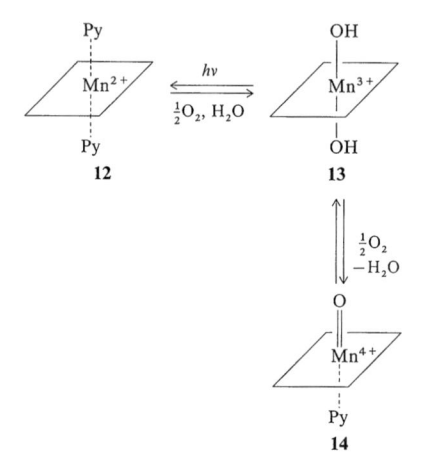

$$[Py\text{-}Mn^{2+}\text{-}Py] \underset{\frac{1}{2}O_2, H_2O}{\overset{h\nu}{\rightleftarrows}} [HO\text{-}Mn^{3+}\text{-}OH]$$

$$\updownarrow\; \tfrac{1}{2}O_2,\; -H_2O$$

$$[O=Mn^{4+}\text{-}Py]$$

12   **13**   **14**

Complex **13** is not very stable; irradiation with light causes reduction to **12**. In the dark, **13** disproportionates to **12** and **14**. In the presence of oxygen, the oxidation of **13** continues to give **14**. In **14**, the manganese atoms are bridged by oxygen atoms [159]. This reversible uptake of oxygen seems to be analogous to oxygen transfer by heme, in which oxygen is fixed by a π-bond in the first step [160]. The catalytic properties of other phthalocyanines (e.g., FePc) can be interpreted in the same way [161].

Reduction of MnPc with lithium benzophenone in the presence of tetrahydrofuran results in formation of the lithium salt of $Mn^{I}Pc$ and the dilithium salt of $Mn^{0}Pc$ [162], [163].

The compound $Co^{I}Pc$ can be converted to $RCo^{III}Pc$ by reaction with alkyl halides [20], [164], [165]; $Co^{I}Pc$ can also be used as a selective reducing agent for aliphatic or aromatic nitro compounds [166] and as a reagent for cleaving β-halo alkyl groups, which are widely used to protect amino, carboxyl or phosphoryl groups in nucleotide or peptide chemistry [167]. Applications of MPc in ammonia synthesis [20], [168]–[170], Fischer–Tropsch synthesis [171], [172], or hydroformylation [173] are described in the literature.

3.2. Reduction

Either the central metal atoms or the carbon atoms of the phthalocyanine ring system can be reduced. The totally reduced form of the ring system, hexadecahydrophthalocyanine (**15**), can be produced from 3,4,5,6-tetrahydrophthalodinitrile [174].

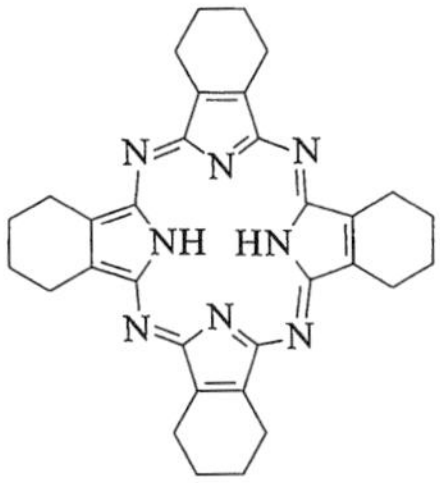

15

In reversible reduction, electrons are removed from the inner ring system or the central metal atom. In alkaline sodium dithionite solution, for example, conversion to a leuco form is possible [42], [55], [175], [176]. The reduction products are weakly colored: greenish for CoPc and brownish yellow for FePc. With reducing agents such as alkali metals, multiple reductions (MPc + n e$^-$ → MPc^{n-}; $n = 1-5$ for Mn, $n=1-4$ for Ni) should be possible [20], [177]. For example, CuPc can be reduced with potassium in liquid ammonia [178], [179]. The solubility of the leuco form is usually too low for industrial purposes (e.g., CoPc). It can be increased by partial sulfonation. Indanthrene brilliant blue 4G (C.I. Vat Blue 29, 74 140) is used for spin dyeing [180]. Although the structure of the leuco form has not been determined totally, a structure of type **16** can be assumed [181].

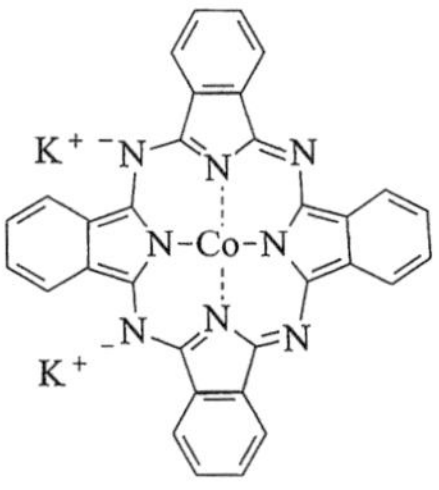

16

Reduction products with similar structures are known for other metal phthalocyanines [1, pp. 278–279]. Depending on the metal, reduction can take place either only in the ligand (Ni, Cu, Zn, Li) or first in the metal and then in the ligand (Cr, Mn, Fe, Co) [182].

3.3. Catalytic Properties

Many metal phthalocyanines, particularly the transition-metal complexes, can bind further ligands to form octahedral complexes. In some cases, this is accompanied by a change in oxidation state. This phenomenon is related to the catalytic properties exhibited by many members of this class of compounds [2, p. 61], [134], [135].

Two review articles deal with heterogeneous catalysis in dehydrogenation, oxidation, and electrocatalysis [183] and heterogeneous catalysis in gas-phase reactions by metal Pc's [184]. Catalysis by metal phthalocyanines was first discovered in 1936 in the exchange reaction between molecular hydrogen and water and the catalysis of water formation from oxygen and hydrogen [185]. Oxygen transfer reactions and catalysis of the decomposition of hydrogen peroxide were described later [186]. This suggested the similarity of metal phthalocyanines to heme and to enzymes such as hydrogenase [187]. Oxygen transfer reactions, particularly of iron phthalocyanines, were later observed in other reactions. Alkanes, olefins and aromatics, alcohols, aldehydes, alkyl aromatics, phenols, amines, thiols, cumenes, polymers, and sugars are oxidized by molecular oxygen in the presence of iron, copper, or cobalt phthalocyanine. In addition, hydrogenation, dehydrogenation, degradation, polymerization, isomerization, hydrogen-exchange reactions, reductive dehalogenation, hydrogenative thermal cracking, autooxidation, epoxidation, decarboxylation, and Fischer–Tropsch synthesis are catalyzed by phthalocyanines [2, p. 61], [3, I, p. 79–102], [15, p. 92], [134], [135], [183], [184].

A major interest in the reduction of molecular oxygen by Pc catalysts is the development of fuel cells [188]. The recognition of phthalocyanines as oxygen reduction catalysts is due to JASINSKI, JAHNKE, and SCHOENBORN, who studied catalysts for electrochemical reactions in fuel cells [189]–[192]. The phthalocyanines were the first platinum-free catalysts suitable for this purpose. The electrocatalytic activity of metal Pc's depends considerably on the central metal ion [161], [186], [193]. The order of decreasing electrocatalytic activity is Fe > Co > Ni > Cu > H, with H_2Pc having very low activity [3, I, p. 81]. The high activity and stability of FePc electrodes are reported in [194], [195]. Light is involved in the electrocatalytic reduction of oxygen by FePc, CuPc, and H_2Pc on a graphite carrier [2, I, p. 81]. Phthalocyanines also catalyze the reduction on carbon cathodes of substances other than oxygen [196].

The mechanism of these heterogeneous catalysts is basically similar to that of other solid catalysts. An advantage is the homogeneity of the active centers. In contrast to the usual catalysts, the metal atoms here are always within the same electronic surrounding of the macromolecule in the surface of the crystals. Depending on the structure of the electron shells of the metal atoms, adducts can be formed with the reactants [161], [197], [198]. The catalytic activity is therefore closely related to the interactions between substrate and catalyst [197]. Activity increases with increasing electrical conductivity of the solid phthalocyanines [135] and is also influenced by the carrier material [199]. The stability of metal complexes in the presence of other chemicals is still a problem, and in fact the most active compounds are often the least stable [161]. Polymeric phthalocyanines exhibit higher stability [134].

Polymeric CoPc, for example, is an effective catalyst in secondary fuel cells. The oxygen formed in electrolysis is stored in the polymer matrix and reduced during electrochemical discharge [200]. Sulfonic acids of CoPc and CuPc are used in the oxidation of thiols in crude oil [2, p. 300], [201].

4. Production

4.1. General Synthetic Methods

Phthalocyanine complexes have been synthesized with nearly all the metals of the periodic table [1], [15], [21]. Despite the apparently complex structure of the Pc system, it is formed in a single-step reaction from readily available starting materials. The reaction is strongly exothermic. For example, the synthesis of CuPc from phthalodinitrile (4 $C_8H_4N_2$ + Cu → $C_{32}H_{16}N_8Cu$) has a reaction enthalpy of −829.9 kJ/mol. The low energy of the final product can be accounted for by resonance stabilization; this explains at least partially the relatively facile formation of the complex. The most important metal phthalocyanines are derived from phthalodinitrile, phthalic anhydride, Pc derivatives, or alkali-metal Pc salts.

From o-Phthalonitrile.

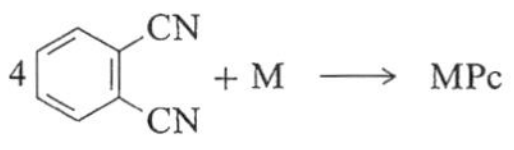

$$4\,C_6H_4(CN)_2 + M \longrightarrow MPc$$

where M is a metal, a metal halide (MX_2), or a metal alkoxide [$M(OR)_2$]. The reaction is carried out in a solvent at ca. 180 °C or by heating a mixture of solid reactants to ca. 300 °C.

From Phthalic Anhydride.

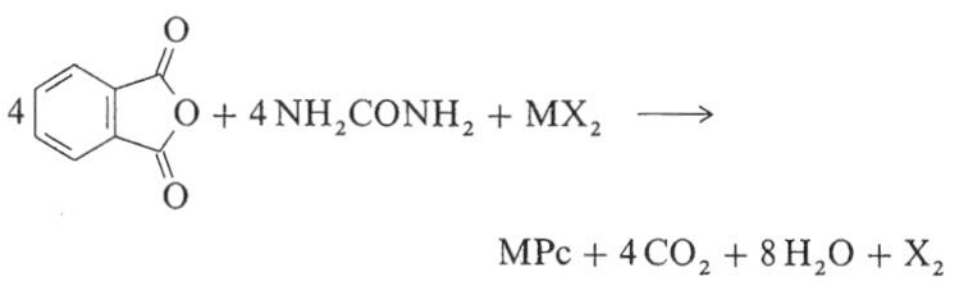

$$4\,C_6H_4(CO)_2O + 4\,NH_2CONH_2 + MX_2 \longrightarrow MPc + 4\,CO_2 + 8\,H_2O + X_2$$

This synthesis is carried out either in a solvent at 200 °C or without solvent at 300 °C.

From phthalimide derivatives, e.g., diimidophthalimide.

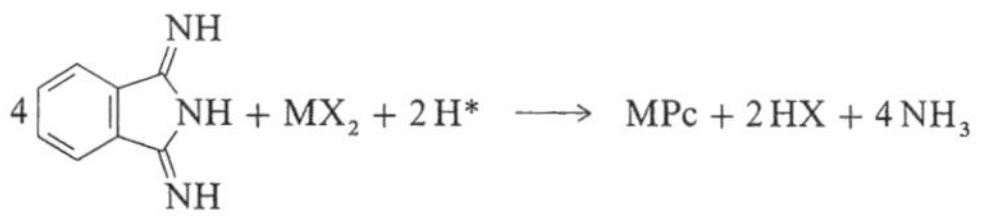

$$4\,C_6H_4(CNH)_2NH + MX_2 + 2\,H^* \longrightarrow MPc + 2\,HX + 4\,NH_3$$

*from reducing agent

This synthesis is carried out in a solvent (e.g., formamide).

From Alkali-Metal Pc Salts. Dilithium phthalocyanine can be used as starting material for the production of other metal Pc's. The reaction is carried out in a solvent.

$Li_2Pc + MX_2 \longrightarrow 2LiX + MPc$

Metal-free phthalocyanine is obtained by the following procedures [1]–[3], [15]:

1) Decomposition of an unstable MPc with alcohol or acid

$PcNa_2 + 2H_3O^+ \longrightarrow PcH_2 + 2Na^+ + 2H_2O$

2) Direct synthesis (e.g., from phthalodinitrile)

Syntheses of MPc from phthalodinitrile or phthalic anhydride in the presence of urea are the two most important laboratory and industrial methods. They were also used originally by LINSTEAD and coworkers [11], [12]. This procedure allows the production of many phthalocyanine compounds [1]–[3]. Catalysts such as boric acid, molybdenum oxide, zirconium and titanium tetrachloride, or ammonium molybdate are used to accelerate the reaction and improve the yield [2], [3]. Ammonium molybdate is especially effective. Reaction is carried out either in a solvent or by heating the solid components. When metal chlorides and phthalodinitrile are used as starting materials, the reaction products are partially chlorinated:

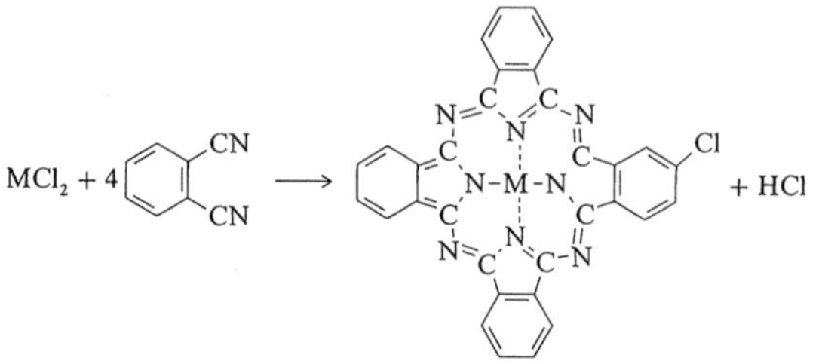

Lowering the reaction temperature or adding urea or basic solvents decreases the extent of chlorination. Solvents such as nitrobenzene, trichlorobenzene, alcohols, glycols, pyridine, and aliphatic hydrocarbons are employed. By using substituted phthalic acid compounds such as 4-chlorophthalic acid anhydride, 4-sulfophthalic acid anhydride or 4-nitrophthalimide, phthalocyanines with inner substitution can be produced. The products can often be purified by sublimation in vacuo at 300–400 °C. Soluble Pc's can be purified by recrystallization.

4.2. Reaction Mechanism

In the synthesis of phthalocyanines from phthalodinitrile or from pthalic anhydride and urea, intermediate isoindolenines (**17**) and polyisoindolenines are formed. According to [153], isoindolenines (**17**) are formed in the first step from dinitrile and polar compounds as follows:

Metal salts, ammonia, primary or secondary amines, hydrogen sulfide, hydrogen bromide, or alkoxides are used as polar compounds. The isoindolenines condense to form the ring system around a central metal atom.

The intermediates cannot be isolated under normal synthesis conditions of ca. 200 °C. Some phthalocyanine developing dyes however use these intermediates, which react with metal salts on the fiber to give metal phthalocyanines (Chap. 8) [2], [3], [153], [202].

The synthesis of phthalocyanine from phthalic anhydride or similar phthalic acid derivatives also results in the formation of an intermediate compound [1]–[3]:

The 1,3-diiminoisoindolenine can be isolated as the nitrate [153].

In a urea melt, the reaction involves formation of polyisoindolenines and their reaction with copper salts to form CuPc. Synthesis with ^{14}C-labeled urea or phthalimide showed that the urea carbon atom is not incorporated into the phthalocyanine framework but is released as ^{14}CO [203], [204].

4.3. Industrial Production

4.3.1. Copper Phthalocyanine

Two processes are commonly used for the production of copper phthalocyanine: the phthalic anhydride–urea process patented by ICI [51], [52] and the I.G. Farben dinitrile process [205]. Both can be carried out continuously or batchwise in a solvent or by melting the starting materials together (bake process). The type and amount of catalyst used are crucial for the yield. Especially effective as catalysts are molybdenum(IV) oxide and ammonium molybdate. Copper salts or copper powder is used as the copper source [1]–[3]; use of copper(I) chloride results in a very smooth synthesis. Use of copper(I) chloride as starting material leads to the formation of small amounts of chloro CuPc. In the absence of base, especially in the bake process, up to 0.5 mol of chlorine can be introduced per mole of CuPc with CuCl, and up to 1 mol with $CuCl_2$. During pigment finishing (section 4.4) these products can be processed to the stable α-modification, but not to the β-modification, of copper phthalocyanine.

The patent literature gives details of modifications and refinements of the original processes. A review of older processes is given in [1], [2], whereas examples of more modern production methods are described in [3], [206].

As apparatus for the batch process, an enamel or steel reactor with an agitator and pressure steam or oil heating suffices. Apparatuses used in the continuous synthesis in the presence of solvents and in the bake process are described in [207] and [208], [209], respectively. The choice of process depends on the availability and cost of the starting materials phthalodinitrile or phthalic anhydride. Although the phthalodinitrile process has certain advantages over the phthalic anhydride process, the latter is preferred worldwide because of the ready accessibility of phthalic anhydride. In this process the molar ratio of phthalic anhydride, urea, and copper(I) chloride is 4:16:1, with ammonium molybdate as catalyst. The mixture is heated in a high-boiling solvent such as trichlorobenzene, nitrobenzene, or kerosene. The solvent is removed after the formation of copper phthalocyanine. Frequently a purification step follows. A few typical examples are described below.

Phthalic Anhydride–Urea Process [52], [141, pp. 274–278]. Phthalic anhydride (100 kg), urea (135 kg), and trichlorobenzene (300 kg) (technical isomer mixture) are mixed in a 1000-L vessel, equipped with an oil bath and stirrer; 24 kg of copper(II) chloride and 0.5 kg of ammonium molybdate are then added. The mixture is heated up slowly within 1 h to 200 °C. Gas evolution (mostly CO_2 and some NH_3) with formation of phthalimide begins at 130 °C. Formation of CuPc begins at 160–170 °C, with simultaneous release of CO and NH_3. After stirring at 200–205 °C for 1 h, formation of the pigment is complete. The phthalocyanine is filtered off; washed with hot trichlorobenzene, trichlorobenzene at 50 °C, methanol, and hot water; and dried. The yield of CuPc is 87 kg (90%).

Carrying out the reaction under pressure gives a high-purity CuPc pigment [210].

Several *dry processes* have also been described [3, II, p. 32].

In one of these [211], urea, copper(II) chloride, and ammonium molybdate are ground for 4 h in a ball mill, phthalic anhydride is added, and grinding is continued for 80 min. A small amount (10 to 60% of the weight of phthalic anhydride) of solvent (nitrobenzene, naphthalene, or dodecylbenzene) is added, and the slurry is spread in stainless steel pans. These are heated in an oven at 195 °C for 8 h. The product is ground and slurried in dilute hydrochloric acid, filtered hot, washed with water, and dried. The yield is 84.5 – 86.6% or 88 – 90% when the reaction is carried out in a paddle drier [3].

The solvent can be replaced by ammonium chloride [212], a fourfold excess of phthalic anhydride, sodium chloride [213], or a 1:1 NaCl–MgCl$_2$ mixture [3], [214]. In the dry reaction, the ammonium molybdate catalyst can be replaced by a molybdenum or molybdenum alloy agitator [2][3].

Another dry process is run continuously [215]. The dry, powdered reaction mixture is fed into a rotary furnace kept at 180 °C, and the dry product is discharged into a drum at a yield of 96% [2][3]. A vacuum method for the preparation of relatively pure CuPc is described in [216].

Phthalodinitrile Process. In the bake process [143, pp. 343 – 344], 100 kg of phthalodinitrile (ca. 95% pure) is mixed with 24 kg of dry CuCl (1.25 mol) and 400 kg of anhydrous Na$_2$SO$_4$. The mixture is then slowly passed through a tunnel dryer within 1 h at 200 °C. The baked mixture is washed with hot water and 1% hydrochloric acid, filtered off, and dried. The process can be carried out continuously.

One improvement of the process consists of grinding 100 kg of phthalodinitrile, 19 kg of anhydrous CuCl, and 20 kg of urea; mixing the powder thoroughly (or grinding in a ball mill); and heating it to 150 °C. The temperature increases to 310 °C due to the heat of reaction, thus completing the reaction within a few minutes. After purification the yield is 97%, and the product contains 0.3% Cl [3, II, p. 30], [217]. Carrying out the reaction in the presence of a salt that decomposes at 30 – 200 °C to form ammonia improves the yield [218].

The reaction has also been carried out in solvents such as trichlorobenzene in the presence of pyridine [143]. Pyridine converts the insoluble copper(I) chloride into a soluble complex, which reacts more quickly.

A mixture of 300 kg of trichlorobenzene, 64 kg of dry CuCl, and 25 kg of pure pyridine is stirred at 80 °C for 30 min; 750 kg of trichlorobenzene and 150 kg of phthalodinitrile (95% pure) are then added. The mixture is heated quickly to 200 – 205 °C and stirred for 6 h. The resulting melt can be purified as described above.

The reaction can be accelerated by the use of sodium hydroxide or sulfonic or carboxylic acids [219] instead of pyridine. Other high-boiling solvents such as nitrobenzene, benzophenone, or naphthalene can be used instead of trichlorobenzene [2], [3], [220] – [223].

The phthalodinitrile–solvent (pyridine) process can be used generally for the production of various metal phthalocyanines and is often suitable when other methods fail. A series of patents cover the low-temperature formation of copper phthalocyanine and continuous processes [3, II, pp. 31–32].

Dichlorobenzene Process. Until now, this synthesis [41] has found no use in industry. Heating *o*-dichlorobenzene (50 g) with quinoline (130 g), copper(I) bromide (49 g), and copper(I) cyanide (61 g) at 205–220 °C for 4.5 h gives 34.8 g of crude, well-crystallized CuPc (71% yield) [143, pp. 274–278].

Diiminoisoindolenine Process. An alternative route is the formation of the isoindolenines, which are then treated with copper(II) salts [224]–[226]. 1,3-Diiminoisoindolenine is prepared by reaction of phthalonitrile with ammonia. The isoindolenine is then reacted with copper acetate in ethylene glycol and 2-chlorobenzonitrile at 60–70 °C for 1 h. The presence of benzyl cyanide can improve the solvent stability of the pigment [2][3], [227], [228].

Copper phthalocyanine can also further be made by milling 145 g of diiminoisoindolenine, 24.8 g of copper(I) chloride, 700 g of anhydrous sodium sulfate, and 145 g of ethylene glycol for 3 h at 100–110 °C. After slurrying in water, filtration, reslurrying in 1% hydrochloric acid, and dilution with sodium hydroxide, the product is suitable as a pigment [2], [3], [229]. Other solvents and additives are also used for the reaction [230]–[233].

Production of γ-, δ-, ε-, π-, and X-CuPc. γ-CuPc is obtained by treating the α-modification with 30% sulfuric acid and glycolmonobutylether at 110 °C [2][3], [234]–[236].

δ-CuPc is formed when 25 g of CuPc is dissolved in 250 g of 98% sulfuric acid and the resulting solution is added to 440 g of benzene, 800 mL of water, and 5 g of turkey red oil at 20 °C over 30 min. After emulsion separation and suspension of CuPc in the benzene layer for 48 h, it is washed with alcohol and water until acid free and dried at 70 °C [237]. Other examples are given in the literature [2], [3], [238].

ε-CuPc is made by mixing 350 g of urea and 295 g of 92% sulfuric acid in a heavy-duty mixer at 80 °C; adding 400 g of 85% CuPc made by the phthalic anhydride–urea process; and then adding 1500 g of sodium chloride and 100 g of water. After being mixed for 18 h at 50–60 °C the mixture is poured into 17 L of water, boiled, filtered, washed until acid and chloride free, and dried at 60 °C [2], [3]. Other variations are given in [182], [239]–[243].

π-CuPc is made by refluxing *o*-phthalodinitrile and copper(I) cyanide in 2-(dimethylamino)ethanol and ammonia [2], [3], [91], [92].

X-CuPc is prepared by synthesis of π-CuPc in the presence of a catalytic amount of X-CuPc [93], [94], by ball-milling of the α- and β-forms [95], or by vaporization methods [96].

For toxicological and ecological reasons, some of the solvents used in phthalocyanine synthesis, especially trichlorobenzene, must be replaced. High-boiling hydrocarbons [244]–[246] or esters [247] have been suggested as replacements. Other processes aim to produce the finished phthalocyanine pigment in order to avoid the pigment finishing step (Section 4.4). In these processes, syntheses are carried out mostly in polar media (e.g., glycols) at low temperature and often lead to formation of the α-modification [248]–[253].

4.3.2. Phthalocyanine

Metal-free phthalocyanine has limited production for speciality uses. It is manufactured according to the original ICI process by reaction of phthalodinitrile with sodium amylate and alcoholysis of the resulting disodium phthalocyanine.

Under nitrogen, 20 kg of sodium is dissolved in 800 kg of amyl alcohol; 208 kg of phthalodinitrile is added, and the mixture is heated for 6 h to 125 °C. The Na_2Pc is filtered off, washed with amyl alcohol, and dried under vacuum. The dried and ground Na_2Pc is added to 850 kg of methanol at 5–10 °C, whereby metal-free Pc is formed. Stirring is continued until no variation of the sample is visible under the microscope. After dilution with water and filtration through a pressure funnel, washing with methanol and water, and drying under vacuum, H_2Pc is obtained as a powder in 80 % yield.

Phthalocyanine can also be obtained by melting phthalic anhydride and urea in the presence of catalytic amounts of ammonium molybdate [254], [255].

Anhydrous sodium polysulfide can be used instead of sodium alkoxides [256]. An electro-chemical process that uses phthalonitrile as starting material in the presence of solvents is described in [257]. Other methods are given in [3, II, p. 29], [258]–[263], which also deal with production of the crystal modifications.

4.4. Pigment Finishing

Crude phthalocyanine or CuPc is obtained during production in the β-modification. The crystals thus obtained are too large to develop optimal pigment characteristics. They must therefore be subjected to a finishing or conditioning step in which the required modification, having optimal particle size with respect to properties such as color strength, color tone, color purity, grinding, flocculation, and rheology, is formed. Finishing methods are described in [2, p. 152], [264].

The particle size of crude phthalocyanine can be reduced by chemical or mechanical methods. The former involves dissolving phthalocyanine in concentrated sulfuric acid and precipitating it by addition to water. Phthalocyanines with a high halogen content must be treated with oleum or chlorosulfonic acid for purification. Phosphoric acid or sulfonic acids have been also proposed as solvents [2], [3]. The use of dilute sulfuric acid results in formation of a sulfate but leaves the pigment partially undissolved (soaking process). Depending on the sulfuric acid concentration, various sulfates are formed. Hydrolysis of the sulfate formed with 68% sulfuric acid gives a very finely divided pigment [2], [3], [143]. Sulfates formed at other sulfuric acid concentrations must be ground before hydrolysis. The α-modification of the pigment is obtained by these methods. A variety of patents cover these and other processes [3, II, pp. 35–36].

The second finishing alternative involves grinding the dry or humid form, cold or warm, with or without addition of grinding agents such as sodium chloride. Organic solvents, especially aromatic hydrocarbons and their derivatives such as xylene, nitrobenzene or chlorobenzene, alcohols, ketones, or esters can also be used [2], [3]. Grinding is carried out in ball mills or kneaders.

In the absence of solvent, the unrefined pigment in the β-modification is converted to the α-form [2], [3], [264]–[266]. Figure 4 illustrates the change of particle-size distribution during milling (agglomeration of the primary crystals) and after finishing (deagglomeration and crystal ripening) of the crude pigment in an organic solvent.

Grinding additives such as sodium chloride are removed by boiling with water, and the pigment is isolated by filtration and drying. Little is known about the mechanism by which β-CuPc is ground and simultaneously transformed to the α-modification [264]. In the presence of solvent, this phase transformation is hindered, and a finely divided pigment in the β-modification is obtained.

4.4.1. Finishing by Treatment with Acid

Dissolving in Acid. Crude product is dissolved in concentrated sulfuric acid and precipitated with water. The degree of agitation and a correct rate of addition are important factors in precipitation [267]. The fineness of the suspension can be improved by addition of emulsifying agents [268], [2, p. 152] or simultaneous use of water-immiscible solvents [269], [2, p. 152], [270].

Hydrophilic agglomeration of pigment particles during drying must be avoided. Otherwise, the pigment may be difficult to disperse later. The tendency to agglomerate increases with decreasing particle size. Agglomeration can be reduced by various additives [2, p. 152] (e.g., ammonium naphthenate) [271], polyalkanolamines of fatty acids [272], nonaqueous solvents [273], or resins [274], [275]. Additives, particularly resins, coat the pigment particles, thereby impeding their agglomeration [276]. Pigments that are easy to disperse are obtained when they are mixed in water and the resin is deposited on the surface of the particles [277], [278].

Production of α-CuPc from Crude CuPc [143, p. 295], [198]. Crude CuPc (700 kg) is added quickly to concentrated sulfuric acid (96%, 6300 kg), and the temperature is allowed to rise to 50 °C. The suspension is stirred overnight. When the phthalocyanine is completely dissolved, the suspension is sprayed into 35 000 L of water at 70 °C, and the temperature is allowed to rise to 85–90 °C. The

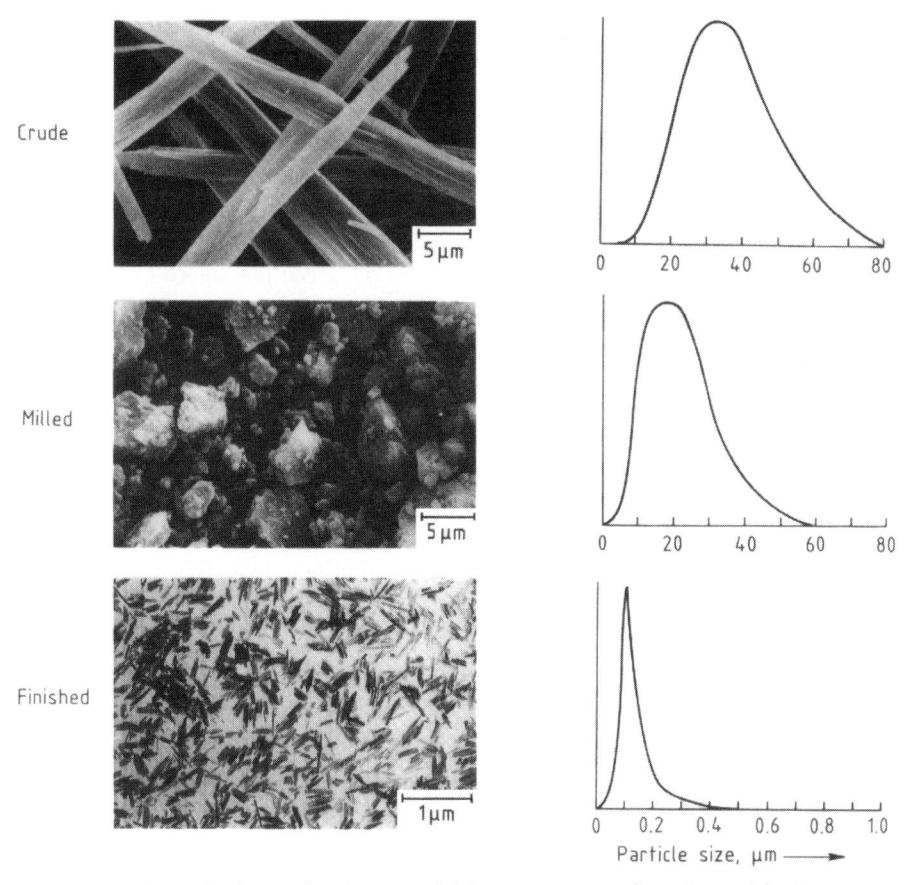

Figure 4. Particle-size distribution of crude copper phthalocyanine pigment after milling and finishing

precipitated pigment is filtered off in a filter press and washed until acid free. The press cake is then mixed with 25 000 L of water, and the remaining acid is neutralized with potassium carbonate and ammonia. After addition of 35 kg of lorol sulfonate the suspension is heated for 1 h until a clear filtrate is obtained. After filtration of the suspension, the press cake is washed with hot water and homogenized with sufficient water to give a 20% paste, which is passed three to five times through a disintegrator (3000 rpm).

Pigment formation by soaking has several advantages compared to dissolution: less sulfuric acid is required, the unrefined phthalocyanine can be processed as a moist press cake, and the resulting pigment has improved color and dispersion properties. To obtain good results, the sulfuric acid concentration must be adjusted within exact limits. An industrial process in which the crude product is allowed to soak in 62–72% sulfuric acid and then hydrolyzed with water [143, p. 294, 304] is carried out as follows:

At 35 °C, 200 kg of CuPc press cake (50% water) is added to a mixture of 565 kg of 96% sulfuric acid and 125 kg of ice. The relative density of a filtrate sample should correspond to a sulfuric acid concentration of 68%. The suspension is stirred for 24 h. Conversion of the β-form to the α-form is monitored by microscopy or X-ray diffraction. Then 6000 L of water is pumped into the suspension. The acid-free washed pigment is again mixed with 2000 L of water. The mixture is made slightly alkaline with ammonia and boiled for 10 min, and the pigment is filtered off and dried under vacuum (14 h at 38–40 °C).

A similar soaking process can be used to form the green polychloro CuPc. Because the product is only sparingly soluble in sulfuric acid, it is stirred for 18–20 h at 35–40 °C in 95% sulfuric acid, and the sulfate is decomposed at 70–93 °C with water.

Conditioning can also be carried out with polyphosphoric acid.

Process Example [279]. One part of crude CuPc is added to seven parts of polyphosphoric acid at 80–90 °C. Within 16 h the temperature rises to 195 °C; 30 parts of water (20–30 °C) is added, and the mixture is stirred for 2 h. The pigment is isolated by filtration.

For some uses, even a simple hot-water purification may be sufficient [280].

4.4.2. Finishing by Grinding

Finishing by grinding was first described in 1936 [281], [282]. In dry grinding, with salts such as sodium chloride or anhydrous sodium sulfate as grinding agents, the β-modification is converted to the α-modification. The coarse crystals of unrefined material are simultaneously ground to smaller size. Fine particles, high color strength, and good dispersive properties can be obtained when grinding agents are used [283]. Washing out the salts with water gives a color-intensive α-pigment. This procedure has been modified often [3, II, p. 37]. For example, pigment quality is improved by addition of surfactants, dispersants, or resins [2, p. 157], [3, II, p. 37].

The grinding process for β-CuPc is of greater importance than that for α-CuPc production [2, p. 157]. The simplest process involves grinding crude CuPc at 70 °C with a water-soluble salt and small amounts of aromatic or halogenated hydrocarbons or alcohols [2, p. 157], [3, II, p. 37], [284]–[288].

Process Example [288]. Iron rods (454 kg) of ca. 12.7-mm diameter and 24.4-mm length, crude CuPc (3.6 kg) (chlorine free), aluminum sulfate (22.7 kg), tetrachloroethylene (0.5 kg), and 0.15 kg of a surfactant containing cetyltrimethylammonium chloride are added to a 227-L ball mill. The mill is rotated for 12 h at 70–75% of its critical velocity, and the temperature is allowed to increase to 60–70 °C. The charge is then removed and digested with hot, dilute sulfuric acid. A green CuPc pigment with high color intensity and good crystal stability is obtained. X-Ray analysis shows that the pigment is predominantly in the β-form.

Crude CuPc pigment can also be ground in the dry state without grinding aids. A surfactant is often added. This milled product exhibits satisfactory pigment properties only after further treatment in water immiscible organic solvents, for example, which are then removed by distillation after the finishing process is complete [289]. The β-modification of CuPc can also be produced by wet milling in the presence of surfactants [2, p. 157], [290]. Another widely used finishing process is the wet milling of crude pigments in mixers for highly viscous pastes (e.g., kneaders). Dry, crude pigment is kneaded with a water-soluble salt in an organic medium (polyalcohols or polyether alcohols).

Process Example [291]. In a kneader, 283 g of crude CuPc (chlorine free) is ground for 2 h with 1131 g of sodium chloride and 140 g of poly(ethylene glycol) ($M_r = 400$). Then, 30 g of this paste is boiled in 1 L of water and filtered to remove water-soluble substances. Washing is continued until no trace of water-soluble substance can be found. The moist paste is dried at 70 – 75 °C. A color-intensive β-CuPc is obtained.

A similar finishing process is used for partially or fully halogenated crude pigments. Another process for making an aqueous pigment dispersion involves activating the crude pigment in the presence of a phase-directing solvent (i.e., a hydrophilic agent that activates the pigment surface, facilitating its transfer to the aqueous phase), followed by fine milling in an emulsion containing a second phase-directing solvent [292].

Process Example [292]. A ball mill is charged with 906 g of chlorine-free crude CuPc and 45 g of turpentine and then rotated for 24 h at 70 % of the critical speed. An emulsion is produced by stirring 160 g of water, 12.5 g of turpentine, and 2.8 g of Aerosol OT (75 %) in a laboratory attritor for 2 – 3 min. At 375 rpm, 40 g of milled crude CuPc is added to the emulsion, and the water and turpentine are separated from the pigment slurry by filtration through a screen. The slurry is then added to 1 L of water containing sufficient hydrochloric acid to yield a pH of 1 – 2. After being heated to 75 – 90 °C for 30 min, the slurry is filtered, washed acid free, and dried.

4.5. Stabilization of Phthalocyanines

Unsubstituted and halogenated phthalocyanine and copper phthalocyanine possess a few principal disadvantages for industrial applications [2, p. 344], [125], [264]. In pure form they are extremely hydrophobic solids (Chap. 2.5.) and therefore tend to flocculate in binders and paints [70], [73], which results in reduction of the color strength and poor flow properties of the dispersions. This effect is particularly apparent in combination with titanium dioxide [125]. The process can be observed by optical and electron microscopy [293]. A review with emphasis on CuPc and titanium dioxide in nonpolar media is given in [294]. A further disadvantage, especially of unchlorinated α-CuPc, is its instability in organic solvents, where conversion into the β-modification and formation of coarse crystals can occur [2, p. 165], [264].

These deficiencies are eliminated by various means. Unstable modifications, mainly the α-form of CuPc, can be stabilized by partial chlorination [2, p. 165]. Chlorine atoms in the 4-position are much more effective than those in the 3-position [170]. Partial sulfonation of CuPc increases flocculation stability. A combination of both methods leads to a high-quality pigment, particularly for solvent-containing coatings [2, p. 165]. Another method is to add phthalocyanines of tin, vanadium, aluminum, or magnesium [2, p. 165]. The addition of CuPc derivatives has also become important. These additives introduce new adsorption centers on the pigment surface. Monomeric and polymeric derivatives with basic substituents are common today (**18** [295] and **19** [296], [297]).

$$\text{CuPc}\left[\text{CH}_2\text{N}\begin{array}{c}R\\ \diagdown\\ R'\end{array}\right]_n \qquad R-\text{NH}\left[\text{CH}_2-\underset{(\text{CH}_2-\text{NHR})_{0-6}}{\text{CuPc}}-\text{CH}_2-\text{N}\begin{array}{c}R\\|\end{array}\text{H}\right]_x$$

18, $n = 1-8$ **19**, $x \geq 2$

These molecules are very strongly adsorbed on the surface of the pigment, thus fixing basic centers that promote the adsorption of binder molecules. The colloid stability is due to the adsorbed layer of binder [122], [124], [276]. The addition of CuPc derivatives with long aliphatic chains gives pigments that also show high flocculation stability in the absence of polymers in aliphatic hydrocarbons [298]. Other additives are derived from the amides or salts of CuPc sulfonic acids [299]–[302].

These additives also strongly hinder recrystallization of the pigments and can therefore be used to stabilize unstable modifications such as the reddish, chlorine-free α-CuPc [303]–[306]. Other methods involve the use of surfactants [307]–[310] or adsorbed resin layers [311].

5. Phthalocyanine Derivatives

The copper phthalocyanine derivatives are of major industrial importance as green organic pigments (halogenated products) and dyes. The first pigments of this class were commercialized in 1938. The first phthalocyanine dye was polysulfonated CuPc [42]. Since then, many patents describing various phthalocyanine compounds have been registered [1]. Substituted phthalocyanines are either accessible through synthesis from phthalocyanine derivatives (with the advantage of defined products) or through substitution of phthalocyanines [1, p. 255], [2, pp. 171, 192]. The latter method is favored in industry for economic reasons.

Synthesis. Usually 4 mol of a substituted phthalodinitrile or a substituted phthalic acid is used as starting material. A mixture of an unsubstituted and substituted starting material in approximate ratios, respectively, of 1:3, 2:2, or 3:1 can also be used. When reactivities of the two starting materials are approximately equal, Pc derivatives whose degree of substitution closely corresponds to the ratio of the starting materials are obtained. More often, however, a mixture of products results.

With the exception of tetrachlorophthalic acid, substituted phthalic acids, phthalimides, or phthalonitriles are industrially not readily accessible in pure form. Therefore, synthesis from phthalic acid derivatives is used only when the substitution reaction fails. For example, tetranitrophthalocyanines can be synthesized from 4-nitrophthalimide, whereas direct nitration of phthalocyanines gives only the oxidation products. Other substituted copper phthalocyanines such as tetra-4-phenyl copper phthalocyanine cannot be produced by direct introduction of the desired substituents.

Substitution. Copper phthalocyanine is preferred as starting material. Very little is known about the position of substitution. With the exception of hexadecachloro CuPc, all commercial Pc substitution products, as well as the tetrasubstituted derivatives synthesized from monosubstituted phthalic acids, are mixtures of isomers. Despite the 16 hydrogen atoms that can be substituted, only two different monosubstituted Pc's are possible. The number of disubstituted isomers is higher. Mono- to heptasubstituted Pc derivatives have not yet been isolated in isomerically pure form. In addition, only a limited number of isomers are accessible in pure form by synthesis. Only symmetrically substituted phthalic acids, phthalimides, or phthalodinitriles (3,6-di-, 4,5-di-, or 3,4,5,6-tetrasubstituted derivatives) yield pure isomers of octa- or hexadecasubstituted phthalocyanine derivatives. All other substituted phthalic acids give mixtures of isomers.

5.1. Halogenated Phthalocyanines

Perchloro and perchlorobromo copper phthalocyanines are of major industrial importance as green organic pigments with high lightfastness. Commercial development of these pigments began in 1935 [312], [313]. They are obtained by direct chlorination of copper phthalocyanine. The color changes only slightly from blue to green with the introduction of the first five to seven chlorine atoms, but a strong change in color occurs when the tenth to fourteenth chlorine atoms are introduced. Introduction of the last two chlorine atoms does not influence the shade. Replacement of chlorine by bromine displaces the color from blue to yellowish green [314]. These products were first marketed in 1959 [315]. Because the introduction of 15 to 16 bromine atoms is very difficult, commercial yellowish CuPc pigments thus contain 11 to 12 bromine atoms and 4 to 5 chlorine atoms per molecule. The halogenated derivatives are prepared by direct chlorination of copper phthalocyanine in a eutectic melt of aluminum and sodium chloride [143], [205], [313].

Hexadecachloro Copper Phthalocyanine [316]. In a 2.4-m^3 vessel equipped with a stirrer, 1800 kg of aluminum chloride and 420 kg of sodium chloride are heated to give a homogeneous melt; the temperature rises to 160 °C. The temperature is lowered to 100 – 110 °C in 1 – 2 h by external cooling; then 600 kg of CuPc is added over a period of 1 h. Chlorine is introduced initially at a rate of

30 kg/h, which is increased to 50 kg/h after 30 min and then to an average rate of 100 kg/h when addition of CuPc is complete. The temperature is kept below 150 °C by external cooling and regulation of the chlorine flow rate between 80 and 120 kg/h. The chlorine flow is stopped when the required degree of chlorination is attained (normally after 11 – 13 h). The melt is then heated to 180 °C, held at this temperature for 30 min, and discharged into water. The polychlorophthalocyanine is isolated from the aqueous suspension. The yield is 1100 kg, with a chlorine content of 48 – 49 %. The aluminum content is < 0.05 %.

When CuPc is halogenated with bromine instead of chlorine in an $AlCl_3$ – NaCl melt in the presence of copper(II) chloride, a polybromo-CuPc is obtained that contains eleven bromine atoms and three chlorine atoms. A further development of the procedure includes the addition of sulfur trioxide or chlorosulfonic acid [317] and sulfur dichloride to the melt [318].

Low-melting complexes of aluminum trichloride with SO_3, $SOCl_2$, or SO_2Cl_2 can also be used as reaction media for the perchlorination of phthalocyanines, particularly CuPc, at 60 – 200 °C. Especially significant is that on decomposition of this chlorination melt, perchlorinated CuPc precipitates in a form that does not require recrystallization from sulfuric acid [319].

Inorganic acids such as sulfuric or phosphoric acid are unsuitable solvents for chlorination because they lead to partial oxidative degradation of CuPc.

Organic solvents such as tri- or tetrachlorobenzene, acid chlorides, carbon tetrachloride, or nitrobenzene are also less suitable for halogenation of phthalocyanines [2, pp. 171, 192]. Only perchlorocyclopentene and perchloroindane have proved suitable [320]. Chlorination in molten phthalic anhydride or its chlorinated derivatives leads to complete chlorination of phthalocyanines. Chlorination of metal-free phthalocyanine (e.g., in phthalic anhydride) up to a content of about 12 chlorine atoms gives a yellowish green pigment [143]. Chlorination of CuPc in the absence of solvents is possible at 200 – 250 °C; the crude CuPc can be diluted with salts [321], [322].

Synthesis from chlorinated phthalic anhydride gives the α-modification of copper phthalocyanine derivatives directly. Either 4-chlorophthalic anhydride and urea or phthalodinitrile is used as starting material under conditions that lead to partial chlorination [323], [324]. The crude product often contains some of the β-modification, which is converted to the stable α-modification during pigment finishing.

Fluorinated phthalocyanines are also known [325], [326]. Hexadecafluorophthalocyanine can be prepared from tetrafluorophthalocyanine [203]. These products have surprising properties; for example, they are soluble in many organic solvents.

Halogenated phthalocyanines must be converted to useful pigments by finishing (Section 4.4). Solvents such as methanol, butanol, ethylene glycol, triethanolamine, esters of aliphatic and aromatic carboxylic acids [327], mixtures of aliphatic ketones and water, [328] or benzoic acid [329] are used as finishing agents.

5.2. Phthalocyanine Sulfonic Acids and Sulfonyl Chlorides

The sulfonic acids and sulfonyl chlorides, especially those of CuPc, are readily accessible. The sulfonic acids, $CuPc(SO_3H)_n$ with $n = 2$, 3, or 4, were significant direct cotton dyes (C.I. Direct Blue 86, 74 186 and 87, 74 200). The sulfonyl chlorides are intermediates in the production of various copper phthalocyanine colorants [1, p. 259], [2, p. 192]. The watersoluble sulfonic acids are produced by heating copper phthalocyanines in oleum. By varying concentration, reaction temperature, and time, one to four sulfo groups can be introduced in the 4-position [330]. The products synthesized from 4-sulfophthalic acid exhibit slightly different properties. This is due to the different isomer distribution with respect to the 4- and 5-positions. Only one sulfo group is introduced in each benzene ring, as shown by the fact that only 4-sulfophthalimide is obtained on oxidative degradation. The most important dyes have two sulfonic acid groups per molecule and are prepared as follows:

Copper Phthalocyanine Disulfonic Acid [205]. Within 1 h, 150 kg of CuPc (97 – 98 % pure) is added to 1200 kg of 26 % oleum, with the temperature kept below 45 °C. The mixture is heated for ca. 12 h at 60 – 61 °C. The reaction is complete when a sample is fully soluble in dilute soda solution and the addition of sodium sulfate does not lead to any precipitation [198, p. 59]. The solution is then stirred into a mixture of 5000 L of saturated sodium chloride solution and 4000 kg of ice. The suspension is brought to 30 000 L with cold water and filtered through a filter press. The press cake is dried inside an air-circulating oven at 60 – 70 °C and then ground with soda.

Partially sulfonated cobalt phthalocyanine, which is an important dye (C.I. Vat Blue 29, 74 140), is prepared in a similar manner [331]. The corresponding disulfonic acid is used as an oxidation catalyst (Section 3.3). The cobalt [332], nickel [333], iron [334], chromium [335], manganese [336], and zinc [336] phthalocyanine sulfonic acids are synthesized by the urea process from 4-sulfophthalic acid.

The sulfonyl chlorides are prepared directly from chlorosulfonic acid and phthalocyanines [337]. Some of the sulfonyl chloride groups are hydrolyzed by the water formed in the reaction. Nearly pure phthalocyanine sulfonyl chlorides are obtained when phosphorus tri- or pentachloride or thionyl chloride is added. The positions of the sulfonyl chloride groups are the same as those of the sulfonic acids [330].

Copper Phthalocyanine Tetrasulfonyl Tetrachloride [338]. To 1500 parts of chlorosulfonic acid is added 193 parts of CuPc within 1 h. The mixture is heated 1 h to 70 – 75 °C, in 1.5 h to 75 – 130 °C, and for 4 h at 130 – 135 °C. The brown color of the starting material changes to green. After being cooled to 80 °C, 600 parts of thionyl chloride is slowly added and the mixture is stirred for 1 h at 75 – 80 °C. The solution is stirred into ice; the sulfonyl chloride is filtered off and washed with water. The tetrakis(3-sulfonyl chloride) is free of sulfonic acid groups.

A broad range of sulfonamide dyes are accessible by substitution of CuPc sulfonyl chlorides with amines. Dyes thus obtained can be used in nearly all fields of dyeing [1, pp. 171, 192], [2, pp. 171, 192].

5.3. Chloromethylphthalocyanines

After sulfonic acid chlorides, the chloromethyl derivatives constitute the second most important group of phthalocyanine derivatives. Like acid chlorides, they can be converted to a wide range of colorants by reaction with amines or phenols. In the original procedure [339] introduced by HADDOCK and WOOD, the chloromethyl compounds are formed by substitution of phthalocyanines with bis(chloromethyl) ether in the presence of aluminum chloride and tertiary bases such as triethylamine or pyridine. Up to eight chloromethyl groups can be introduced. Another procedure involves the substitution of phthalocyanines with paraformaldehyde in sulfuric acid or chlorosulfonic acid [340]. Since bis-(chloromethyl) ether is carcinogenic [341], the reaction must be carried out in a hermetically sealed apparatus. Dichloromethyl ether is also produced in synthesis from paraformaldehyde in the presence of substances such as chlorosulfonic acid that produce hydrochloric acid.

5.4. Phthalocyanines with Other Substituents

Other phthalocyanine derivatives with various substituents, represented by formula **20**, are known for R=*tert*-butyl [342]–[346]; NR_2 [346]–[348]; OR [343], [344], [349]–[354]; SR [355], [356]; SiR_3 [343], [344]; CF_3 [357]; SCF_3 [358]; SO_2NR_2 [359]–[361]; COOH [362]–[366]; COOR [367], [368]; CN [347], [369]–[371]; NO_3 [372], [373]; and OH [374]. These derivatives are greener and more turbid than the copper phthalocyanines. More details and further references can be found in [1, pp. 192, 308], [2, pp. 192, 308].

20

5.5. Polymeric Phthalocyanines

Combination of a phthalocyanine with a polymer or incorporation of a phthalocyanine into a polymer matrix is a powerful tool for designing new materials with special properties. Polymer phthalocyanines are of interest as catalysts, semiconductors, and lubricants because they are more stable than the corresponding monomers.

The following principles are used to form phthalocyanine polymers:

1) Linking through the central metal atom as part of the polymeric chain [30]–[35]
2) Linking of neighboring phenyl rings (diphenyl structure)
3) Linking through substitution (e.g., ether bridges, ethylene bridges)
4) Linking through phenyl rings that belong to two phthalocyanine systems
5) Linking of the phthalocyanines to a polymer chain
6) Linking the metal of a phthalocyanine with a polymer donor ligand, or electrostatic bonding of a charged phthalocyanine with a charged polymer chain

Details of the polymeric phthalocyanines are given in [1, pp. 274–275], [2, p. 328], [3, II, p. 91], [375, pp. 59–125].

5.6. Phthalocyanine Analogues

Phthalocyanine analogues include compounds such as porphyrins (**8**, Chap. 1), in which the four pyrrole rings are linked through one to four aza bridges and up to eight β-hydrogen atoms of the pyrrole are substituted [2, p. 308], [3, I, p. 195]. Like the phthalocyanines, many of these compounds exhibit very low solubility and are therefore potential pigments. Many such derivatives have been synthesized and studied. Because of their high cost and various application problems, none of these products have found use in industry.

Replacement of Benzo Rings by Other Ring Systems. The four benzo rings of phthalocyanines can be replaced by other aromatic ring systems such as naphthalene or aromatic heterocycles, for example, **21–27**.

21

22

23

24

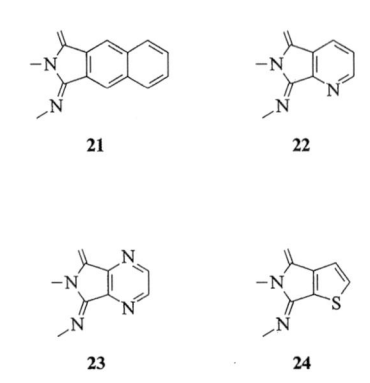

These compounds are derived from the corresponding o-dicarboxylic acids, their anhydrides, or o-dinitriles. The naphthalophthalocyanines (**21**) [376], their derivatives, [377], [378], and their metal complexes show reactions similar to phthalocyanines [13], [379]. The analogue **22** is redder than phthalocyanine [143, pp. 334–336], [380], and **23** is redder still but also more turbid [143, pp. 334–336], [381]. These are the reddest phthalocyanine analogues known. The tetrathiophenotetraazaporphyrin **24** has a greenish color [382]. Crown ether derivatives (**25**) [383]–[385] and cumyl phenoxy derivatives [386] have also been reported. Octaphenyl Pc's (**26**) and analogues **27** [387] have been prepared from the corresponding diphenyls and dicyanotriphenylenes. An anthracene analogue of Pc is prepared from 2,3-dicyano-9,10-diphenylanthracene [388].

Derivatives of Tetraazaporphyrin. Alkyl and aryl derivatives of tetraazaporphyrin (**28**) such as tetramethyltetraazaporphyrin (**28**, R=CH$_3$, R'=H) [389], octamethyltetraazaporphyrin (**28**, R = R' = CH$_3$) [390]–[392], and octaphenyltetraazaporphyrin (**27**, R = R' = C$_6$H$_5$) [393]–[395] belong to this group.

Hexadecahydrophthalocyanine can also be included in this group [177], [396]. The copper complexes of hexadecahydrophthalocyanine are among the reddest phthalocyanine analogues known. Starting materials for the latter two derivatives (the dialkylmaleic anhydride or dinitrile) [397] are not readily accessible; therefore these compounds are not in industrial use. Tetra- and octaphenyltetraazaporphyrins and their alkyl derivatives are potential oil-soluble dyes and siccatives [398]–[400].

Other Aza- and Thiaporphyrins. Metal complexes derived from mono-, di-, or triazatetrabenzoporphyrin or thiatetrabenzoporphyrins (**29**) are greener and more turbid than CuPc. They have no industrial importance. Tetrathio Pc's (**30**) are more strongly colored than the corresponding Pc's themselves [401]–[403].

6. Legal Aspects and Environmental Protection

Although CuPc pigments exhibit very low toxicity, producers and users are currently faced with various regulatory changes, which are related to the presence of copper in these substances [404], [405].

Since the inhalation of dusts and mists of copper salts can result in irritation of nasal mucous membranes, OSHA has established PEL of 0.1 mg/m^3 for copper fumes, and 1 mg/m^3 for dusts and mists of copper salts [405]. The German TA Luft calls for 5 mg/m^3 Cu in off-gases. The MAK value for copper is 1 mg/m^3 [341]. The issue of copper in wastewater is more complex because concerns are based on the toxicity of soluble copper to various aquatic species. The EPA ambient water criteria for copper call for a 4-day average of 12 µg/L (1 day max. 18 µg/L), whereas the maximum contamination level of copper for public drinking water is 1 mg/L [406]. Thus, the 12–18-ppb copper in wastewater from CuPc production may be a problem and is handled by authorities on a case-by-case basis. Regulations in Germany call for 0.2–0.3 mg/L Cu in wastewater.

A method for removing copper(II) ions from wastewater in CuPc production is treatment with hydrosulfite to give a copper concentration of < 1 mg/L [407]. Another possibility is the addition of MoO$_3$, then the precipitation of molybdenum with a high molecular mass amine, and finally copper flotation with alkylxanthogenate [408]. The Resource Conservation and Recovery Act (RCRA; USA) of 1976 and the Hazardous and Solid Waste Amendments (HSWA; USA) of 1984 regulate the generation, transportation, treatment, storage, and disposal of hazardous and nonhazardous waste. Copper-containing waste is classified as nonhazardous under RCRA [405]; other regulations apply only to soluble or bioavailable copper, not to CuPc pigments. For the production of other metal phthalocyanine pigments, similar treatment of wastewater, off-gas, solid, and liquid waste; transportation; storage; and disposal involving the regulations must be taken into account, especially with respect to bioavailable metals.

Polychlorinated biphenyls (PCBs) have been detected in pigments manufactured in trichlorobenzene but not in those made with nonchlorinated solvents [409].

7. Quality Specifications and Analysis

Several procedures are applied to the testing of phthalocyanines. Tests for crystallization, flocculation, and applications in paints, plastics, and printing inks have been standardized [3, II, pp. 109–119].

Crystallization tests are carried out in xylene or in various paints [410], [411]. The flocculation test follows the ASTM standard specification for copper phthalocyanine [412]. This specification covers CuPc blue pigment in dry powder form for use in paint, printing ink, and related products. Tests include pigment composition, color, character of tint, oil absorption, reactions in identification tests, and dispersion and storage stability of the end product. Application tests for printing inks, paints, and plastics are described in [3, II, pp. 112–115], [412].

Identification methods consist of the trichloroacetic acid test [413], the N.P.I.R.I. standard test for pigment identification [414], and tests for various applications.

Quantitative determinations of the pigment are possible with cerium(IV) sulfate [10] or sodium vanadate [415]. Copper and chlorine determinations are often important [416].

8. Uses

Phthalocyanines are used predominantly (ca. 90%) as pigments. These pigments are almost always copper phthalocyanines or their halogenated derivatives. This group of pigments accounts for 25% of the total organic pigment capacity. Worldwide consumption of organic pigments in 1989 was estimated to be 250×10^3 t. Modification of the nearly insoluble phthalocyanines by introduction of various substituents has led to the use of Pc's as dyes for textiles, paper, and leather or as specialty dyes (e.g., for ballpoint pen inks). The colors of these dyes range from brilliant blue through turquoise to green. Phthalocyanine dyes are therefore used in almost all cotton dyes, synthetic-fiber dyes, sulfur and azo developing dyes, and reactive dyes.

8.1. Phthalocyanine Pigments

Because of their excellent color properties and high fastness to light, weathering, and various solvents, phthalocyanine pigments are used in nearly all fields. Furthermore, they are less expensive than other organic pigments. Phthalocyanine pigments are particularly important for gravure, offset, and flexo printing. The β-form of copper phthalocyanines is used predominantly in this field. Other important applications are in the paint, coating, plastics, textile, and spin-dyeing industries. Through special procedures during finishing of the crude pigment, optimal pigments are produced for each application. The use of additives leads to improved rheological properties and reduced flocculation. Apart from pure pigment powders, preparations are manufactured in which the pigment is already dispersed in a suitable binder system (e.g., polyethylene, nitrocellulose, polyamide, or liquid systems). The range of phthalocyanine products is now very large. For example, the Heliogen trademarks (powder pigments) of BASF include 48 products, of which 13 are the α-modification, 22 the β-modification, 2 the ε-modification, 1 is metal free, and 10 contain halogen.

Phthalocyanine pigment producers and trade names are as follows:

American Cyanamid, USA	Cyan Blue, Cyan Green
Aziende Colori Nazionali Affini, Italy	Turchese Segnale Luce, Verde Segnale Luce
BASF, Germany	Heliogen-Blau, Heliogen-Grün
Bayer, Germany	Helioecht-Blau, Helioecht-Grün
Ciba-Geigy, Switzerland	Irgalith-Blau, Irgalith-Grün
Dainichi Seika Colors, Japan	Chromofine Blue, Chromofine Green
Dainippon Ink and Chem. Co., Japan	Cyanine Blue, Cyanine Green, Fastogen Blue, Fastogen Green
Du Pont, USA	Cinquasia Blue, Cinquasia Green
Harmon Colors-National Aniline Division, USA	Phthalo-Blue, Phthalo-Green
Hilton Davis Chemical Co., USA	Peacoline Blue
Hoechst, Germany	Hostapermblau, Hostapermgrün
ICI, UK	Monastral Blue, Monastral Green
Kemisk Vaerk Koge, Denmark	Isophthalblau, Isophthalgrün
Produits Chimiques Ugine Kuhlmann, France	Cyanine Lutetia, Emerande Lutetia
Sandoz, Switzerland	Sandorin-Blau, Graphtol-Blau, Graphtol-Grün
Sumitomo Chemical Co., Japan	Sumitone Cyanine Blue, Sumitone Cyanine Green
Sun Chemical Co., USA	Fastolux Blue, Fastolux Green

Further information about manufacturers and trade names can be found under Pigment Blue 15, 74 160, in the Colour Index.

15	unstable α-CuPc
15:1	crystalline, stable α-CuPc
15:2	crystalline, stable, nonflocculating α-CuPc
15:3	β-CuPc
15:4	nonflocculating β-CuPc
15:6	ε-CuPc

C.I. Pigment Blue 16, 74 100 is metal free.
C.I. Pigment Green 7 is 74 260; C.I. Pigment Green 36, 74 265.

8.2. Phthalocyanine Dyes

The first phthalocyanine dye was a sulfonic acid derivative, synthesized in 1929 [42]. These dyes are still in use today to dye cotton. The sodium salts of copper phthalocyanine disulfonic acid (C.I. Direct Blue 86) and trisulfonic acid (C.I. Direct Blue 87) and their derivatives are also used. C.I. Acid Blue 185 is a wool dye. Finishing additives such as Solidogen (Hoechst) or Levogen (Bayer) improve the wetfastness of these CuPc sulfonic acids. Further applications are leather and paper dyeing.

Brilliant dyes that are readily soluble in alcohols and glycol ether can be obtained by reaction of Pc sulfonyl chlorides with amines. In addition to sulfonamide groups, these dyes contain ammonium sulfonate substituents. They can be used to color transparent paint, flexo and gravure printing ink, wood stains, plastics, and ballpoint ink. Commercial products include C.I. Solvent Blue 24, 74 380 and C.I. Solvent Blue 42, 44, 46, and 52. Substitution of copper phthalocyanine sulfonyl chlorides with primary amines that also contain a tertiary or secondary amino group leads to dyes that are soluble in acidic media. They are used mostly for printing ink, ballpoint ink, ink ribbons, copying ink, and carbon paper. Commercial products include Astrablau 6 GLL (Bayer) and Blaubase KG (BASF).

In addition to these widely applicable phthalocyanine dyes, a number of specialty dyes are used in the textile industry. In one case, functional groups are introduced into the phthalocyanine molecule, which solubilize the compound and are later cleaved on the fiber by further treatment. For example, the Inthion dyes (Hoechst) have thiosulfate esters in the side chains that are cleaved during the dyeing process [417].

The reaction of trichloromethyl copper phthalocyanine with tetramethylthiourea gives an isothiuronium salt, whose hydrophilic groups can be cleaved by weak bases [418]. Trade names can be found under C.I. Ingrain Blue 1, 74 240 (e.g., Alican Blue 8 G, ICI). Another possibility is linking a copper or nickel phthalocyanine derivative with a reactive group on the fiber (e.g., via sulfo groups) [419]. Commercial products of this type are listed in [420]. Examples are the copper phthalocyanine derivatives C.I. Reactive Blue 7, 74 460 (e.g., Cibacron Turquoise Blue, Ciba Geigy; Procion Turquoise MX-G, ICI; Remazol Turquoise Blue, Hoechst; Levafix Turquoise Blue E-G, Bayer; Primazin Turquoise Blue, BASF and nickel phthalocyanine derivatives such as Remazol Brilliant Green 6 B (Hoechst) and Cibacron Turquoise Blue (Ciba Geigy).

In contrast to the above-mentioned dyes, phthalocyanine developing dyes do not contain any substituents. They thus give purer, more genuine colors. Details of manufacturing, composition, and applications are given in [202], [421].

The phthalocyanine is generally synthesized from an isoindolenine or its derivative and a metal compound by drying and subsequent heating to 120 °C. The starting materials are impregnated in the fibers with water-soluble solvents such as glycol and diglycol monoethyl ether or diethanolamine. The solvents and additives act as reductants [421]. Important products are C.I. Ingrain Blue 2:1, 74 160 (e.g., Phthalogen

Brilliant Blue IF 3 GM, Bayer and Sumilogen Brilliant Blue, Sumitomo). Other products include Phthalogen Turquoise 3 FBM and Phthalogen Marine Blue IRRM (Bayer).

Phthalocyanine dye producers and trade names are as follows:

American Cyanamid, USA	Calcomine Turquoise, Calcotone Green
Aziende Colori Nazionali Affini, Italy	Segnale Light Green, Eliama Light Turquoise
BASF, Germany	Lurantin Light Blue, Blaubase KG, Primazintürkisblau
Bayer, Germany	Astrablau, Sirius Supra Turquoise
Ciba Geigy, Switzerland	Chlorantine Fast Turquoise, Pergasol Turqoise Blue
Dainippon Ink and Chem. Co., Japan	Daivougen Blue, Symulon Turquoise, Fastogen Blue, Fastogen Green
Du Pont, USA	Pontamine Fast Turquoise, Brilliant Bond Blue
Hilton Davis Chemical Co., USA	Hiltamine Turquoise Blue
Hoechst, Germany	Remazol Turquoise Blue, Hostaperm Green
ICI, UK	Durazol Blue, Alican Blue, Monosol Blue
Sandoz, Switzerland	Pyrazol Turquoise, Cuprofix Blue, Finisol Blue Green
Sumitomo Chemical Co., Japan	Sumilight Supra Turquoise, Sumilogen Brilliant Blue

Further information can be found in the Colour Index:

C.I.	Ingrain Blue 2:1, 74 160	developing dyes
C.I.	Direct Blue 86, 74 180	CuPc disulfonate salt
C.I.	Direct Blue 87, 74 200	CuPc trisulfonic acid
C.I.	Ingrain Blue 3, 74 280	diiminoisoindolenine precursor of copper tetraphenylphthalocyanine
C.I.	Solvent Blue 24, 74 380	CuPc, sulfonated
C.I.	Reactive Blue 7, 74 460	CuPc tetrasulfonic acid containing 1 amide and 1 sulfophenylamide group

9. Economic Aspects

World capacity for crude phthalocyanine in 1987 was estimated at ca. 56 500 t [422]. Capacity in tonnes by geographic area is

North America	
United States	4 700
Mexico	750
South America	
Brazil	700
Argentina	250
Venezuela	250
Western Europe	
Germany	4 500
United Kingdom	3 100
France	2 000
Italy	3 700
Switzerland	200
Eastern Europe	3 000
Far East, Oceania	
Japan	18 000
India	3 400

South Korea	6 500
Taiwan	3 700
Thailand	300
China	1 000
Australia	400
Total	56 450

Table 1. Major producers of crude phthalocyanine

Company	Location	Estimated 1989 capacity, t
Toyo/Cosmos	Japan, Taiwan	9000
DIC/Sudarshan	Japan, India	8000
Ciba Geigy	United Kingdom, Switzerland, Korea	4500
Phthalchem	United States	4500
BASF	Germany, Brazil	4500
Acna	Italy	3500
Woo Song	South Korea	3000
DNS	Japan	3000
ICI	United Kingdom, France	2000
Sigma	Taiwan	1500
Maintop	Taiwan	1000
Sanyo	Japan	1000

The major world producers are listed in Table 1. The producers were operating at ca. 80 % of capacity in 1989.

Over the past ten years, only one new company in the United States has begun producing intermediates while many have dropped out. The greatest growth in intermediate production today is taking place in India, Korea, and Taiwan, and these countries are likely to become even more significant producers [422].

Figure 5 shows the capacity distribution according to geographical regions; 59 % of the capacity is concentrated in the Far East. World capacity in the year 2000 is expected to reach 75 000 t.

Consumption data for phthalocyanines in paint, printing ink, plastics, textiles, dyes, and specialty uses (leather, paper, ballpoint ink) are listed in Table 2.

10. Toxicology and Occupational Health

Copper phthalocyanine pigments have found widespread acceptance because of their durability, excellent color, and good working properties. Adding to their attractiveness is the fact that Pc pigments are insoluble in water and lipids. No substantial scientific evidence exists to indicate that Pc pigments pose a significant risk to human health in the workplace or to the environment [404].

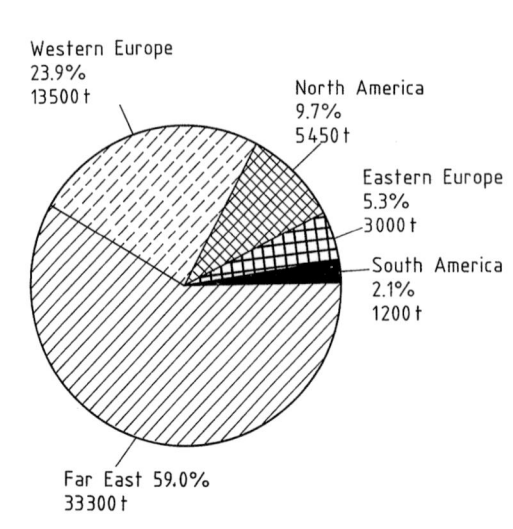

Figure 5. Production capacity for crude phthalocyanine according to geographical regions

Western Europe 23.9% 13500 t
North America 9.7% 5450 t
Eastern Europe 5.3% 3000 t
South America 2.1% 1200 t
Far East 59.0% 33300 t

Table 2. Consumption data for phthalocyanines (1989)

Application	Estimated consumption, t
Printing ink	19 500
Paint	9 500
Plastics	8 000
Textiles	3 000
Dyes	2 000
Specialty uses	2 000

Acute Toxicity Tests. The CuPc pigments have LD_{50} values of > 10 g/kg in rats, which is more than 40 times less toxic than CuO or 70 times less toxic than $CuCl_2$ [404], [405]. The LD_{50} values were confirmed for the following CuPc pigments: phthalocyanine Blue 15 and Green 7 [423]. Doses of CuPc sulfonate up to 100 mg/kg in mice, guinea pigs, rabbits, cats, and dogs did not induce any toxic effects [424]. At concentrations of 5000 ppm, two commercial CuPc sulfonates were nontoxic to protozoans, small crustaceans, small fish, and, when injected, to rabbits, rats, and guinea pigs [425]. Skin and eye irritation studies [426] on CuPc and the Ames tests for mutagenicity were negative [427].

Chronic Toxicity Tests. No carcinogenic risk to humans was indicated in a study of laboratory mice, in which tumors failed to develop over an eight-month period [428]. A subchronic feeding study using doses up to 5000 mg/kg of body weight did not reveal any toxicity or pathological changes related to the Pc pigments administered [405]. The National Toxicity Program in the United States canceled a long-term bioassay on Pc's, because a 13-week feeding study of mice and rats did not indicate any toxicity [429]. Based on the very low solubility of the pigments in aqueous and nonaqueous media and the conclusion of nonbioavailability to humans, testing of a chemical fate and ecological effects of Pc green pigment was suspended by the TSCA Interagency Testing Committee. The FDA has approved the use of copper Pc in general and opthalmic surgery (e.g., for coloring polypropylene sutures, foils, and packaging), for contact lenses, and for food packaging [406]. The MAK value of copper in fumes is 0.1 mg/m^3 and in dust 1 mg/m^3. The TLV value for copper dust is 1 mg/m^3, which corresponds to 10 mg/m^3 CuPc [430] because the copper content of CuPc is usually <12%. A Russian study recommends a maximum permissible concentration of 5 mg/m^3 in the workplace [431]. Workplace exposure can readily be reduced to meet these limits through engineering controls and protective equipment [404].

Studies of the uptake of sulfonated phthalocyanines (Ni, Zn, Cu, Co, Al) by cultured mammalian cells [432] and chick embryos [433] point to possible teratogenic effects. However, control experiments with copper(II) chloride suggest that at high concentrations copper is responsible for these findings rather than sulfonated phthalocyanine.

11. References

General References

[1] G. Booth: "Phthalocyanines," in K. Venkataraman (ed.): *The Chemistry of Synthetic Dyes*, vol. **V**, Academic Press, New York 1971, p. 241.
[2] F. H. Moser, A. L. Thomas: *Phthalocyanine Compounds*. Reinhold Publ. Co., New York, Chapman and Hall, London 1963.
[3] F. H. Moser, A. L. Thomas: *The Phthalocyanines*, vol. **I** and **II**, CRC Press, Boca Raton, Fla. 33431, 1983.

Specific References

[4] R. P. Linstead, *Br. Assoc. Adv. Sci. Rep.* 1933, 465.
[5] R. P. Linstead, *J. Chem. Soc.* 1934, 1016.
[6] G. T. Byrne, R. P. Linstead, A. R. Lowe, *J. Chem. Soc.* 1934, 1017.
[7] R. P. Linstead, A. R. Lowe, *J. Chem. Soc.* 1934, 1022.
[8] C. E. Dent, R. P. Linstead, *J. Chem. Soc.* 1934, 1027.
[9] R. P. Linstead, A. R. Lowe, *J. Chem. Soc.* 1934, 1031.
[10] C. E. Dent, R. P. Linstead, A. P. Lowe, *J. Chem. Soc.* 1934, 1033.
[11] P. A. Barret, C. E. Dent, R. P. Linstead, *J. Chem. Soc.* 1936, 1719.
[12] P. A. Barret, D. A. Frye, R. P. Linstead, *J. Chem. Soc.* 1938, 1157.
[13] J. S. Anderson, E. F. Bradbrook, A. H. Cook, R. P. Linstead, *J. Chem. Soc.* 1938, 1151.
[14] R. P. Linstead, *Ber. Dtsch. Chem. Ges.* **A 72** (1939) 93.
[15] A. B. P. Lever, *Adv. Inorg. Radiochem.* **7** (1965) 27.
[16] L. J. Boucher in G. A. Melson (ed.): *Coordination Chemistry of Macrocyclic Compounds*. Plenum Press, New York 1979, chap. 7.
[17] K. Kasuga, M. Tsutsui, *Coord. Chem. Rev.* **32** (1980) 67.
[18] P. Sayer, M. Goutermann, C. R. Connell, *Acc. Chem. Res.* **15** (1982) 73.

[19] C. Hamann et al. in P. Goerlich (ed.): *Organische Festkoerper und duenne Schichten*, Akademische Verlagsgesellschaft, Leipzig 1978, chap. 2.
[20] D. Woehrle, G. Meyer: "Phthalocyanine – ein System ungewoehnlicher Struktur und Eigenschaften," *Kontakte* **3** (1985) 38; **1** (1986) 24.
[21] F. Lux in C. J. Kevane, T. Moeller, *Proc. Rare Earth Res. Conf. 10th* 1973, 871.
[22] M. S. Fischer, D. H. Templeton, A. Zalkin, M. Calvin, *J. Am. Chem. Soc.* **93** (1971) 2622.
[23] T. Kobayashi et al., *Bull. Chem. Soc. Jpn.* **44** (1971) 2095.
[24] F. Cariati, F. Morazzoni, M. Zocchi, *Inorg. Chim. Acta* **14** (1975) L 31.
[25] J. R. Mooney, C. K. Choy, K. Knox, M. Kenney, *J. Am. Chem. Soc.* **97** (1975) 3033.
[26] W. E. Bennet, D. E. Broberg, N. C. Baenziger, *Inorg. Chem.* **12** (1973) 930.
[27] A. Gieren, W. Hoppe, *J. Chem. Soc., Chem. Commun.* 1971, 413.
[28] K. Kasuga et al., *J. Am. Chem. Soc.* **102** (1980) 4835.
[29] F. Lux et al., *Radiochim. Acta* **14** (1970) 57.
[30] T. J. Marks, *J. Coat. Technol.* **48** (1976) 53.
[31] A. Meller, A. Ossko, *Monatsh. Chem.* **103** (1972) 150.
[32] D. Wöhrle, *Adv. Poly. Sci.* **50** (1983) 45.
[33] K. Fischer, M. Hanack, *Chem. Ber.* **116** (1983) 1860.
[34] J. Metz, G. Pawlowski, M. Hanack, *Z. Naturforsch.* **38 b** (1983) 378.
[35] E. A. Orthmann, V. Enkelmann, G. Wegner, *Makromol. Chem. Rapid Commun.* **4** (1983) 687.
[36] C. W. Dirk, T. Inabe, K. Schoch, T. J. Marks, *J. Am. Chem. Soc.* **105** (1983) 1539.
[37] T. Inabe, J. W. Lyding, T. J. Marks, *J. Am. Chem. Soc., Chem. Commun.* 1983, 1084.
[38] B. L. Wheeler et al., *J. Am. Chem. Soc.* **106** (1984) 7404.
[39] O. Völker, *Naturwiss. Rundsch.* **8** (1955) 268.
[40] A. Braun, J. Tscherniak, *Ber. Dtsch. Chem. Ges.* **40** (1907) 2711.
[41] H. de Diesbach, E. von der Weid, *Helv. Chim. Acta* **10** (1927) 886.
[42] Scottish Dyes, GB 322 169, 1929; DE 586 906, 1929.
[43] ICI, GB 410 814, 1934; US 2 116 602, 1939 (I. M. Heilbron, F. Irving, R. P. Linstead).
[44] R. P. Linstead, A. R. Lowe, *J. Chem. Soc.* **1934**, 1031.
[45] I. M. Robertson, *J. Chem. Soc.* **1935**, 615; 1936, 1195.
[46] I. M. Robertson, R. P. Linstead, C. E. Dent, *Nature (London)* **135** (1935) 506.
[47] I. M. Robertson, I. Woodword, *J. Chem. Soc.* **1937**, 219; 1940, 36.
[48] R. P. Linstead, I. M. Robertson, *J. Chem. Soc.* **1936**, 1195; 1936, 1736.
[49] E. W. Mueller, *Naturwissenschaften* **37** (1950) 933.
[50] A. J. Melmed, E. W. Mueller, *J. Chem. Phys.* **29** (1958) 1037.
[51] ICI, GB 464 126, 1935 (M. Wyler).
[52] ICI, GB 476 243, 1936 (A. Riley).
[53] P. C. Krueger, M. E. Kennedy, *J. Org. Chem.* **28** (1963) 3379.
[54] D. Horn, B. Honigmann, *XII. Fatipec Kongreßbuch*, Verlag Chemie, Weinheim 1974, p. 181.
[55] M. A. Dahlen, *Ind. Eng. Chem.* **31** (1939) 839.
[56] D. L. Ledson, M. V. Twigg, *Chem. Ind.* **3** (1975) 129.
[57] D. L. Ledson, M. V. Twigg, *Inorg. Chim. Acta* **13** (1975) no. 1, 43.
[58] B. D. Berezin, *Izv. Vyssh. Uchebn. Zaved. Khim. Khim. Technol.* **2** (1959) no. 10, 169.
[59] B. D. Berezin, *Izv. Vyssh. Uchebn. Zaved. Khim. Khim. Technol.* **6** (1963) no. 5, 841; **7** (1964) no. 1, 111.
[60] G. W. Watt, J. W. Dawes, *J. Inorg. Nucl. Chem.* **14** (1960) 32.
[61] C. J. Petersen, *J. Org. Chem.* **22** (1957) 127.
[62] E. A. Lawton, *J. Phys. Chem.* **62** (1958) 384.

[63]   Y. Taru, K. Takoaka, *Shikizai Kyokaishi* **55** (1982) no. 1, 2.
[64]   M. Radulescu, R. Vîlceanu, *J. Therm. Analysis* **7** (1975) no. 1, 209.
[65]   M. Zhan, B. Zheng, X. Gu, *Xiamen Daxue Xuebao Ziran Kexueban* **25** (1986) no. 2, 192.
[66]   B. N. Archor, G. M. Fohlen, J. A. Parker, *J. Polym. Sci., Polym. Chem. Ed.* **21** (1983) no. 4, 1025; **21** (1983) no. 11, 3063; **22** (1984) no. 2, 319.
[67]   B. Honigmann, J. Stabenow, *VI. Fatipec Kongreßbuch*, Verlag Chemie, Weinheim 1962, p. 89.
[68]   F. M. Smith, *Br. Ink Maker* **7** (1964) 21.
[69]   M. Herrmann, B. Honigmann, *Farbe+Lack* **75** (1969) 337.
[70]   R. Sappok, B. Honigmann: "Organic Pigments," in G. D. Parfitt, K. S. W. Sing (eds): *Characterization of Powder Surfaces.* Academic Press, New York 1976, p. 231.
[71]   B. Honigmann, *Ber. Bunsen-Ges. Phys. Chem.* **71** (1967) 239.
[72]   W. Jettmar, *VII. Fatipec Kongreßbuch*, Verlag Chemie, Weinheim 1964, p. 343.
[73]   R. Sappok, *J. Oil Colour Chem. Assoc.* **61** (1978) 299.
[74]   W. Jettmar in J. Grehn (ed.): *Handbuch der Mikroskopie in der Technik*, vol. **7**, Umschau Verlag, Frankfurt 1975, p. 369.
[75]   P. Hauser, M. Herrmann, B. Honigmann, *Farbe + Lack* **76** (1970) 545; **77** (1971) 1097.
[76]   J. Beresford, *J. Oil Colour Chem. Assoc.* **50** (1967) 594.
[77]   W. Carr, *J. Paint Technol.* **42** (1970) 696.
[78]   W. Carr, *J. Oil Colour Chem. Assoc.* **54** (1971) 155.
[79]   P. Hauser, D. Horn, R. Sappok, *XII. Fatipec Kongreßbuch*, Verlag Chemie, Weinheim 1974, p. 191.
[80]   J. Stabenow, *Ber. Bunsen-Ges. Phys. Chem.* **72** (1968) 374.
[81]   B. Honigmann, *Farbe+Lack* **82** (1976) 815.
[82]   F. M. Smith, J. D. Easton, *J. Oil Colour Chem. Assoc.* **49** (1966) 614.
[83]   IG Farbenind., DE 696 590, 1937 (B. Bienert); FIAT 1313, 3, 324 (1948).
[84]   M. Suito, N. Uyeda, M. Ashida, *Senryo to Yakuhin* **12** (1967) no. 2, 41.
[85]   M. Tomita, *Senshoku Kogyo* **21** (1973) no. 12, 725.
[86]   E. Suito, *Kotai Butsuri* **1** (1976) no. 3, 151.
[87]   E. Suito, *J. Electronmicrosc.* **18** (1969) no. 4, 341.
[88]   B. Honigmann, H. U. Lenne, R. Schroedel, *Z. Kristallogr. Kristallgeom. Kristallphys. Kristallchem.* **122** (1965) 185.
[89]   C. J. Brown, *J. Chem. Soc.* **A 1968**, 2488; **1968**, 2494.
[90]   G. V. Susisch, *Anal. Chem.* **22** (1950) 425; FIAT 1313, III, p. 412 (1948).
[91]   P. J. Brach, M. A. Lardon, DE-OS 2 218 788, 1972; GB 1 395 769, 1972; US 3 708 293, 1972; FR 2 138 865, 1972; CA 995 211, 1972.
[92]   P. J. Brach, H. A. Six, DE-OS 2 218 767, 1972; GB 1 395 615, 1972; US 3 708 292, 1972; FR 2 138 730, 1972; CA 996 931, 1972.
[93]   Xerox Corp., DE-AS 2 026 057, 1970 (P. J. Brach, H. A. Six).
[94]   P. J. Brach, H. A. Six, DE-OS 2 026 057, 1970; GB 1 312 946, 1975; US 3 927 026, 1975.
[95]   J. F. Byrne, P. F. Kurz, US 3 357 989, 1967; GB 1 169 901, 1967; DE 1 619 654, 1967.
[96]   J. H. Sharp, R. L. Miller, M. A. Lardon, DE-OS 1 944 021, 1970; GB 1 268 422, 1972; US 3 862 127, 1972; FR 2 016 641, 1972.
[97]   J. H. Beynon, A. R. Humphreys, *Trans. Faraday Soc.* **51** (1955) 1065.
[98]   M. Shigemitsu, *Bull. Chem. Soc. Jpn.* **32** (1959) 607.
[99]   B. Honigmann, *Farbe+Lack* **70** (1964) 787.
[100] Xerox Corp., US 3 357 989, 1966 (J. F. Byrne, P. F. Kurz).
[101] T. Silina, B. G. Aristov, *Zh. Vses. Khim. Ova. im. D. I. Mendeleeva* **19** (1974) no. 1, 77.

[102] American Cyanamid Comp., US 2 770 629, 1952 (J. W. Eastes).
[103] ICI, GB 912 526, 1962 (B. P. Brand).
[104] Kemisk Vaerk Koge, GB 981 364, 1961; US 3 160 635, 1960.
[105] BASF, DE-OS 2 210 072, 1972 (B. Honigmann, I. Kram).
[106] Nippon Shokubai Kagaku Kogyo, DE-OS 2 659 211, 1977 (A. Komai et al.).
[107] C. Hamann, H. Wagner, *Krist. Tech.* **6** (1971) 307.
[108] M. Ashida, N. Uyeda, E. Suito, *J. Cryst. Growth* **8** (1971) 45.
[109] E. Suito, N. Uyeda, *Kolloid Z., Z. Polym.* **19** (1974) no. 1, 77; **193** (1963) no. 2, 97.
[110] B. Honigmann, D. Horn, *SCI Monogr.* **38** (1973) 283.
[111] B. Honigmann, D. Horn: *Particle Growth in Suspensions*, Academic Press, New York 1973.
[112] B. G. Aristov et al., *Congr. FATIPEC* **13** (1976) 112.
[113] N. Uyeda, *Prog. Org. Coat.* **2** (1973) no. 2, 131.
[114] N. Kawashima, T. Suzuki, K. Meguro, *Bull. Chem. Soc. Jpn.* **49** (1976) no. 8, 2029.
[115] Bayer, DE-OS 2 104 200, 1972 (H. Leister, K. Wolf).
[116] M. Pope, *J. Chem. Phys.* **36** (1962) 2810.
[117] M. Ashida, *Bull. Chem. Soc. Jpn.* **39** (1966) no. 12, 2625; **39** (1966) no. 12, 2632.
[118] K. Meguro et al., *Kogyo Kagaku Zasshi* **69** (1966) no. 6, 1724.
[119] O. J. Schmitz, P. J. Sell, K. Hamann, *Farbe+Lack* **79** (1973) no. 11, 1049.
[120] S. Wu, K. J. Brzozowski, *J. Colloid Interface Sci.* **33** (1971) no. 3, 686.
[121] V. Y. Davidov, A. V. Kiselev, T. V. Silina, *Kolloidn. Zh.* **34** (1972) no. 1, 30; **36** (1974) no. 2, 359; **36** (1974) no. 4, 762; **36** (1974) no. 5, 945.
[122] W. Black, F. T. Hesselink, A. Topham, *Kolloid Z. Z. Polym.* **213** (1966) 150.
[123] K. Merkle, H. Schaefer, in T. C. Patton (ed.): *Pigment Handbook*, vol. **3,** J. Wiley & Sons, New York 1973, p. 157.
[124] A. Topham, *Prog. Org. Coat.* **5** (1977) 237.
[125] V. T. Crowl, *J. Oil Colour Chem. Assoc.* **50** (1967) 1023.
[126] W. Ditter, D. Horn, *Proc. Int. Conf. Org. Coat. Sci. Technol. 4th* 1978.
[127] A. V. Kuznetov, C. Vidal-Madjar, G. Guiochon, *Bull. Soc. Chim. Fr.* **5** (1960) 1440.
[128] U. Kaluza in: *Physical/Chemical Fundamentals of Pigment Processing for Paints and Printing Inks*, E. Moeller, Filderstadt 1981.
[129] J. M. Assour, S. E. Harrison, *J. Am. Chem. Soc.* **87** (1965) no. 3, 651.
[130] N. M. Ksenofontana et al., *Zh. Prikl. Spektrosk.* **205** (1974) no. 5, 834.
[131] G. B. Birell, C. Burke, P. Dehlinger, O. H. Griffith, *Biophys. J.* **13** (1973) no. 5, 462.
[132] H. F. Shurvell, L. Pinzuti, *Can. J. Chem.* **44** (1965) no. 2, 125.
[133] M. Starke, H. Wagner, *Z. Chem.* **9** (1969) no. 5, 193.
[134] K. Beales, D. D. Eley, D. J. Hazeldine, T. F. Palmer in H. Kropf, F. Steinbach (eds.): *Katalyse an Phthalocyaninen*, Thieme, Stuttgart 1973, p. 1.
[135] H. Meier: *Organic Semiconductors*, Verlag Chemie, Weinheim 1974.
[136] H. Fluck in [134], p. 37.
[137] H. Rupp, U. Weser, *Biochim. Biophys. Acta* **446** (1976) no. 1, 151.
[138] C. K. Joergensen, H. Berthov, *Mat.-Fys. Medd.-K. Dan. Vidensk. Selsk.* **38** (1972) no. 15, 15.
[139] C. H. Kleint, K. Moeckel, *Surf. Sci.* **40** (1973) no. 2, 343.
[140] S. de Cheveigne, A. Leger, J. Klein, *Proc. Int. Conf. Low Temp. Phys. 14th*, vol. **3,** 1975, p. 491.
[141] A. B. P. Lever, P. C. Minor, *Inorg. Chem.* **20** (1981) 4015.
[142] A. B. P. Lever et al., *Adv. Chem. Ser.* **201** (1982) 237.
[143] FIAT 1313, vol. **III** (1948).
[144] G. G. Rao, T. P. Sastri, *Fresenius Z. Anal. Chem.* **169** (1959) 11.

[145] N. P. Kanyaev, A. A. Spryskov, *Zh. Prikl. Khim. (Leningrad)* **25** (1952) 1220.
[146] A. M. Islam, A. Waser, A. A. El-Mariah, A. A. Salman, *J. Oil Colour Chem. Assoc.* **57** (1974) no. 4, 134.
[147] P. A. Barett, R. P. Linstead, G. A. P. Tuey, *J. Chem. Soc.* 1939, 1809.
[148] R. P. Linstead, F. T. Weis, *J. Chem. Soc.* 1950, 2981.
[149] A. E. Cahill, H. Taube, *J. Am. Chem. Soc.* **73** (1951) 2487.
[150] P. George, D. J. E. Ingram, J. E. Bennett, *J. Am. Chem. Soc.* **79** (1957) 1870.
[151] C. Ercolani, C. Neri, *J. Chem. Soc.* **A 1967**, 1715; **1968**, 2123.
[152] C. W. Frank et al., *Electrophotogr. Int. Conf. 2nd*, 1974, p. 52.
[153] F. Baumann et al. *Angew. Chem.* **68** (1956) 133.
[154] Du Pont, US 26 628 957, 1951 (C. I. Pedersen).
[155] H. A. Rutter, J. D. McQueen, *J. Inorg. Nucl. Chem.* **12** (1960) 361.
[156] J. A. E. Elvidge, A. B. P. Lever, *Proc. Chem. Soc., London* **1959**, 195.
[157] G. Engelsma, A. Yamamtuoto, E. Markham, M. Calvin, *J. Phys. Chem.* **66** (1962) 2517.
[158] A. Yamatuoto, L. K. Philips, M. Calvin, *Inorg. Chem.* **7** (1968) 847.
[159] L. H. Vogt, A. Zalkin, D. H. Templeton, *Inorg. Chem.* **6** (1967) 1725.
[160] J. S. Griffiths, *Proc. R. Soc. London, A*, **235** (1956) 23.
[161] F. Beck et al., *Z. Naturforsch. A: Phys. Phys. Chem. Kosmophys.* **28 A** (1973) 1009.
[162] R. Taube, H. Munke, *Angew. Chem.* **75** (1963) 639; *Angew. Chem. Int. Ed. Engl.* **2** (1963) 477.
[163] R. Taube, M. Zach, K. A. Stanske, T. Heidrich, *Z. Chem.* **3** (1963) 392.
[164] R. Taube, *Pure Appl. Chem.* **38** (1974) 427.
[165] R. Taube, H. Drevs, D. Steinborn, *Z. Chem.* **6** (1978) 8.
[166] H. Eckert, *Angew. Chem.* **93** (1981) 216; *Angew. Chem. Int. Ed. Engl.* **20** (1981) 208.
[167] H. Eckert, Y. Kiesel, *Synthesis* 1980, 947.
[168] S. Naito, K. Tamuro, *Z. Phys. Chem.* **94** (1975) 150.
[169] M. Ichikawa, M. Soma, T. Onishi, K. Tamaru, *Bull. Chem. Soc. Jpn.* **41** (1968) 1739.
[170] S. Mizuo et al., *J. Phys. Chem.* **73** (1969) 1174.
[171] M. Ichikawa et al., *J. Am. Chem. Soc.* **91** (1969) 1538.
[172] K. Watanabe et al., *Proc. R. Soc. London, A:* **333** (1971) 51.
[173] E. H. Homeier, US 3 984 478, 1975.
[174] G. E. Ficken, R. P. Linstead, *J. Chem. Soc.* **1952**, 4847.
[175] M. Shigemitsu, *Bull. Chem. Soc. Jpn.* **32** (1959) 544.
[176] M. Shigemitsu, *Bull. Chem. Soc. Jpn.* **32** (1959) 502.
[177] N. M. Bigelow, M. A. Perkins in H. A. Lubs (ed): *The Chemistry of Synthetic Dyes and Pigments*, Rheinhold Publ., New York 1955, p. 577.
[178] G. W. Watt, J. W. Daws, *J. Inorg. Nucl. Chem.* **14** (1960) 32.
[179] W. A. Alexander, P. L. Pauson, *J. Inorg. Nucl. Chem.* **17** (1961) 186.
[180] Bayer, US 2 613 128, 1949 (B. Bienert, F. Baumann); GB 704 310, 1949. M. Kunz, *Text. Rundsch.* **6** (1951) 546.
[181] F. Grund, *J. Soc. Dyers Colour.* **69** (1953) 671.
[182] B. Honigmann, J. Kranz, DE-OS 2 210 072, 1973; GB 1 411 880, 1973; FR 2 174 089, 1973.
[183] S. Manassen, *Catal. Rev.* **9** (1974) no. 2, 223.
[184] F. Steinbach, H. H. Schmidt, M. Zobel, *Catal. Proc. Int. Symp.*, 1975, 417.
[185] M. Calvin, E. G. Cockbain, M. Polanyi, *Trans. Far. Soc.* **32** (1936) 1436. M. Polanyi, *Trans. Far. Soc.* **34** (1938) 1191.
[186] A. Cook, *J. Chem. Soc.* 1938, 1761, 1768, 1774, 1845.
[187] D. Rittenberg, A. I. Krasna, *Discuss. Faraday Soc.* **20** (1955) 185.

[188] J. H. Wood, C. W. Keenan, W. E. Bull in *Fundamentals of College Chemistry*, 3rd ed., Harper and Row, New York 1972.
[189] R. Jasinski, *J. Electrochem. Soc.* **112** (1965) 526.
[190] H. Jahnke, *Ber. Bunsen-Ges. Phys. Chem.* **72** (1968) 1053.
[191] H. Jahnke, M. Schoenborn, *19th CITCE Meeting,* Detroit 1968; *Journ. Int. Etude Piles Combust. C. R. 3rd,* 1969, 60.
[192] H. Jahnke, M. Schoenborn, G. Zimmermann, *Proc. Symp. Electrocatal.* 1974, 303.
[193] H. Kropf, *Angew. Chem.* **84** (1972) 219; *Angew. Chem. Int. Ed. Engl.* **11** (1972) 239.
[194] J. Yamaki, *J. Appl. Electrochem.* **15** (1985) 441.
[195] J. Yamaki, *J. Electrochem. Soc.* **132** (1985) 2125.
[196] S. Meshitsuka, M. Ichikawa, K. Tamaru, *J. Chem. Soc., Chem. Commun.* **5** (1974) 168.
[197] H. J. Joswig, H. H. Schmidt, F. Steinbach, R. Stritzel, *Proc. Int. Congr. Catal. 6th,* 1976, 9583.
[198] F. Steinbach, K. Mittner in [134] p. 122.
[199] H. Meier, E. Zimmerhackl, W. Albrecht, M. Tschirwitz in [134] p. 104.
[200] O. Hirabaru et al., *J. Chem. Soc., Chem. Commun.* **1983**, 481.
[201] Universal Oil Products, GB 849 998, 1959.
[202] H. Vollmann: "Phthalogen Dyestuffs," in K. Venkatamaran (ed.): *The Chemistry of Synthetic Dyes,* vol. V., Academic Press, New York 1971, p. 283.
[203] S. N. Brumfield, V. W. Folz, C. M. McGhee, A. L. Thomas, *J. Org. Chem.* **27** (1962) 2266.
[204] S. N. Brumfield, B. C. Mays, A. L. Thomas, *J. Org. Chem.* **29** (1964) 2484.
[205] BIOS Final Report 960.
[206] Y. L. Meltzer: "Phthalocyanine Technology," *Chem. Process. Rev.* **42** (1970).
[207] BASF, US 3 412 102, 1968 (G. Schulz, R. Polster).
[208] Hoechst, DE 2 256 170, 1972 (W. Deucker, E. Spietschka, D. Steide).
[209] E. F. Klenke Jr., US 2 964 532, 1960.
[210] Dainichiseika Color Chem., JP 02 9851, 1981 (Y. Abe).
[211] Ciba Geigy, GB 909 375, 1962; DE 1 202 419, 1967; CH 428 046, 1967 (D. Razavi, W. Fioroni, V. Meister).
[212] Allied Chemical Corp., GB 991 419, 1965.
[213] Y. Abe, M. Muto, JP 70 07661, 1970.
[214] S. Ohira, M. Muto, JP 70 07662, 1970.
[215] D. E. Mack, US 3 188 318, 1965.
[216] T. P. Prassad, *Res. Ind.* **33** (1988) no. 2, 144.
[217] E. Spieltschka, W. Deucker, DE-OS 1 644 679, 1970; GB 1 228 997, 1970.
[218] J. C. R. Nicaise, L. A. Cabut, US 3 985 767, 1977.
[219] H. Hiller, W. Kirshenlohr, DE-OS 2 045 908, 1972; GB 1 353 049, 1972; FR 2 107 595, 1972.
[220] Y. Kuwabara, JP 65 27186, 1965.
[221] Y. Abe, S. Horiguchi, JP-Kokai 77 36130, 1977.
[222] R. Polster, R. Schroedel, D. v. Pigenot, GB 1 073 348, 1967; FR 1 410 814, 1967.
[223] K. Hoelzle, CH 352 436, 1961.
[224] S. Susuki, Y. Bansho, Y. Sakashita, K. Ohara, *Nippon Kagaku Kaishi* **51** (1976) 1460.
[225] BASF, DE 2 250 938, 1972 (M. Gaeng).
[226] Dainichiseika Colour Chem., JP 48 022 117, 1973.
[227] M. Gaeng, DE-OS 2 250 938, 1974.
[228] H. Leister, K. H. Wolf, R. Hoernle, DE-OS 2 012 507, 1971; GB 1 273 968, 1971.
[229] T. Akamatsu, Y. Tezuka, DE-OS 2 136 767, 1972; GB 1 306 055, 1972; FR 2 103 195, 1972.
[230] T. Akamatsu, Y. Tezuka, JP-Kokai 73 10128, 1973.

[231]	A. Komai, N. Shirane, JP-Kokai 75 127 928, 1975.
[232]	A. Komai, M. Ninomiya, N. Shirane, JP-Kokai 75 128 720, 1975.
[233]	T. Kasiwagi, Y. Kuwahara, JP-Kokai 76 22720, 1976.
[234]	Dainichiseika Colour Chem., JA 67 0970, 1967 (Y. Abe).
[235]	S. Suzuki, Y. Bansho, Y. Tanabe, *Kogyo Kagaku Zasshi* **72** (1969) no. 3, 720.
[236]	S. Sekiguchi, Y. Bansho, O. Kaneko, *Kogyo Kagaku Zasshi* **70** (1967) no. 4, 499.
[237]	B. P. Brand, US 3 150 150, 1964; GB 912 526, 1964.
[238]	Dainichiseika Colour Chem., JP 67 0870, 1967 (Y. Abe).
[239]	Ciba Geigy, DE 2 742 066, 1977 (I. R. Wheeler).
[240]	M. Miyatake, S. Tamura, S. Ishizuka, JP-Kokai 73 76 925, 1973.
[241]	I. Kumano, M. Miyatake, S. Tamura, S. Ishizuka, JP-Kokai 74 59 136, 1974.
[242]	Dainippon Ink Chem., JP 71 68950, 1981 (N. Fukada).
[243]	Dainippon Ink Chem., JP 71 49358, 1981 (N. Fukada).
[244]	Kawasaki Chem. Ind. Chem., JP 49 116 121, 1974 (R. Matsuura).
[245]	Dainichiseika Colour Chem., JP-Kokai 52 036130, 1977 (Y. Abe).
[246]	Toyo Ink MFG KK, DE-OS 2 642 416, 1976 (A. Sato et al.).
[247]	Toyo Ink MFG KK, JP-Kokai 52 010 326, 1977 (T. Funazo).
[248]	Ten Horn Pigmentchem., DE-OS 2 618 429, 1976 (J. J. Einerhand et al.).
[249]	Ciba-Geigy, DE-OS 2 637 202, 1976 (R. Barraclough, R. Laugley).
[250]	ICI, GB 1 374 793, 1972 (F. Hauxwell, H. R. Murton).
[251]	Japan Cataly. Chem. Ind. KK, JP-Kokai 50 124 927, 1975 (S. Kitano).
[252]	ICI, DE-OS 2 224 117, 1972 (F. Hauxwell, H. R. Murton).
[253]	Dainichiseika Colour Chem., JP 73 38332, 1971; JP 49 131 224, 1974 (Y. Abe).
[254]	Du Pont, US 2 820 796, 1954 (F. F. Ehrich).
[255]	ICI, US 200 005 152, 1932 (J. F. Thorpe, J. Thomas).
[256]	Ciba-Geigy, CH 579 131, 1976 (Z. Seha).
[257]	BASF, DE 2 711 005, 1977 (P. Thoma, W. Habermann, J. Kranz).
[258]	E. Stocker, A. Pugin, US 3 060 189, 1962; BE 609 166, 1962; GB 952 775, 1962; DE 1 218 639, 1962.
[259]	Z. Seha, DE-OS 2 364 689, 1974; GB 1 410 310, 1974; FR 2 212 399, 1974; JP-Kokai 75 157 417, 1974; CH 579 131, 1974; US 3 998 839, 1976.
[260]	BASF, GB 998 255, 1965; NL 6 405 836, 1965 (H. Mueller).
[261]	L. Weinberger, P. J. Brach, S. J. Grammatica, US 3 509 146, 1970; GB 1 232 241, 1970; DE 1 770 779, 1970.
[262]	Toyo Ink MFG KK., JP-Kokai 57 164 158, 1982 (R. Ohshima).
[263]	American Cyanamid Corp., US 4 320 059, 1980 (B. Casas).
[264]	F. M. Smith, J. D. Easton, *J. Oil Colour Chem. Assoc.* **49** (1966) 614.
[265]	Ciba Geigy, US 2 791 589, 1957 (A. Pugin).
[266]	Bayer, FR 1 265 104, 1960 (H. Raab, R. Hoernle).
[267]	Du Pont, US 2 334 812, 1940 (S. R. Detrick, C. R. Brandt).
[268]	Interchem. Corp., US 2 262 229, 1941 (V. Giambaivo).
[269]	American Cyanamid Corp., US 2 359 737, 1941 (H. T. Lacey, H. Z. Lecher).
[270]	BASF, EP 63 321, 1981; DE-OS 114 928, 1981 (T. Leary).
[271]	Du Pont, US 2 294 381, 1940 (H. E. Burdick).
[272]	Du Pont, US 2 305 379, 1941 (S. R. Detrick, J. W. Lang).
[273]	Du Pont, US 2 540 775, 1948 (S. R. Detrick).
[274]	Du Pont, US 2 173 699, 1936 (A. Siegel).

[275] GAF Corp., DE-AS 1 003 885, 1955 (R. Brouillard, T. D. Mutaffis).
[276] J. Moilliet, D. A. Plant, *J. Oil Colour Chem. Assoc.* **52** (1969) 289.
[277] ICI, GB 978 242, 1964 (A. E. Ambler, R. W. Tomlinson).
[278] ICI, GB 957 440, 957 984, 1964 (A. E. Ambler, R. W. Tomlinson).
[279] Mobay Chem. Corp., US 304 127, 1981.
[280] BASF Corp., CA 1 211 431 A, 1982; US 425 784, 1986.
[281] IG Farbenind., GB 470 079, 1937 (G. W. Johnson).
[282] General Aniline Works Inc., GB 457 426, 1936 (J. Y. Johnson).
[283] Du Pont, US 24 021 670, 1942 (F. W. Lane).
[284] American Cyanamid Corp., US 2 486 304, 1946 (S. A. Loukomsky).
[285] Du Pont, US 255 672 630, 1949 (F. W. Lane).
[286] Ciba, DE 846 757, 1950; 913 216, 1951 (W. Wettstein).
[287] Geigy, FR 1 113 323, 1954.
[288] Du Pont, DE-AS 1 161 533, 1960 (J. Jackson).
[289] Du Pont, US 3 017 414, 1959 (J. W. Minnick, R. L. Sweet).
[290] Bayer, DE-OS 2 160 208, 1971 (H. J. Moells et al.).
[291] California Ink Co., US 2 982 666, 1958 (M. Chun, A. M. Enskine).
[292] BASF Corp., US 123 670, 1980 (T. E. Donegan, J. H. Bantjes, T. G. Leary).
[293] W. Jettmar, *Dtsch. Farben Z.* **24** (1970) 55.
[294] R. B. McKay, *Surfactant Sci. Ser.* **21** (1987) 361.
[295] BASF, GB 949 739, 1964; GB 985 620, 1965 (A. Schoellig, R. Schroedel, H. J. Sasse).
[296] ICI, GB 1 082 945, 1964; GB 972 805, 1964 (W. Black, J. Mitchell, A. Topham).
[297] B. G. Aristov, R. G. Feizulova, *Zh. Fiz. Khim.* **53** (1979) no. 7, 1859.
[298] ICI, GB 1 343 606, 1971 (H. P. Dryhurst-Paget, L. R. Rogers, A. Topham, J. K. D. Royle, J. F. Stansfield).
[299] Ciba-Geigy, DE-OS 2 720 464, 1977 (I. R. Wheeler, G. H. Robertson).
[300] Ciba-Geigy, DE-OS 2 800 181, 1977 (I. R. Wheeler, G. H. Robertson); DE-OS 2 753 008, 1977.
[301] ICI, DE-OS 2 461 567, 1974 (G. Hull, A. Topham).
[302] V. A. Smrchek, S. G. Vilner, V. P. Gorin, G. A. Burgrov, *Lakokras. Mater. Ikh. Primen.* **2** (1984) 11.
[303] BASF, DE 2 519 753, 1975; DE 2 516 054, 1975 (W. Jettmar, B. Honigmann, R. Sappok).
[304] Produits Chimiques Ugine Kuhlmann, US 3 985 570, 1975; FR 7 041 758, 1970; FR 7 041 759, 1970 (L. A. Cabut, J. C. H. R. Hardouin, M. E. K. Huille, D. F. X. Pigasse).
[305] Bayer, GB 1 422 834, 1973 (H. Leister).
[306] L. Cabuf, J. C. Hardouin, H. Ciceron, A. Chapelle, *Double Liaison Chim. Peint.* **21** (1974) no. 224; **159** (1974) no. 225, 212.
[307] A. Lorinc, C. Szterjopulosz, *Abh. Akad. Wiss. DDR* 1976, 529.
[308] A. Lorinc, C. Szterjopulosz, T. Mezhdunar, *Kongr. Poverkhin.-Akt. Veshchestvam, 7th,* **3** (1976) 669.
[309] B. G. Aristov, R. K. Feizulova, Y. G. Frolov, *Izv. Vyssh. Uchebn. Zaved., Khim. Khim. Tekhnol.* **23** (1980) no. 10, 1269.
[310] V. A. Moroz, V. L. Tikhonov, *Lakrokas. Mater. Ikh. Primen.* **5** (1981) 17.
[311] B. G. Aristov et al., *Lakrokas. Mater. Ikh. Primen.* **4** (1974) 10.
[312] ICI, GB 461 268, 1937 (R. P. Linstead, C. E. Dent).
[313] IG Farbenind., DE 717 164, 1935 (H. Tietjens).
[314] Du Pont, US 2 247 752, 1937 (A. L. Fox).
[315] E. A. Wich, *Am. Ink Maker* **37** (1959) 26.

[316] BASF, DE 2 504 150, 1975 (H. Geeren, W. Fabian).
[317] Ciba, US 2 862 929, 1958 (A. Caliezi, W. Kern, T. Holbro).
[318] Du Pont, US 2 833 784, 1958 (F. F. Ehrich).
[319] GAF Corp., US 2 873 279, 1959 (D. I. Randall, J. Taras).
[320] Bayer, DE 928 344, 1952 (H. Vollmann).
[321] Du Pont, US 2 586 598, 1952 (G. Barnhart, R. W. Grimble).
[322] IG Farbenind., FR 838 009, 1939.
[323] Du Pont, US 2 933 505, 1957 (J. Jackson).
[324] BASF, GB 1 073 348, 1964 (R. Polster, R. Schroedel, D. V. Pigenot).
[325] Du Pont, US 3 006 921, 1957 (V. Weinmayr).
[326] Imperial Smelting Corp., GB 1 037 657, 1964 (D. E. M. Wotton).
[327] Mobay Corp., US 236 955, 1981; US 275 488, 1981; EP-A 58 888, 1986 (J. F. Santimauro).
[328] BASF, DE 3 636 428, 1986 (J. Kranz, K. Schmeidl).
[329] BASF, EP-A 0 182 207, 1986 (J. Kranz).
[330] R. P. Linstead, F. T. Weiss, *J. Chem. Soc.* 1950, 2977.
[331] Bayer, US 2 613 128, 1949 (F. Baumann, B. Bienert).
[332] N. Fukada, *Nippon Kagaku Zasshi* **75** (1954) 1141; *Chem. Abstr.* **51** (1957) 12 729.
[333] F. Fukada, *Nippon Kagaku Zasshi* **75** (1954) 586; *Chem. Abstr.* **51** (1957) 11 154.
[334] F. Fukada, *Nippon Kagaku Zasshi* **76** (1955) 1378; *Chem. Abstr.* **51** (1957) 17 945.
[335] F. Fukada, *Nippon Kagaku Zasshi* **77** (1956) 1421.
[336] F. Fukada, *Nippon Kagaku Zasshi* **79** (1958) 396, 980; *Chem. Abstr.* **54** (1960) 4612.
[337] IG Farbenind., DE 891 121, 1937 (F. Nadler, H. Hoyer, O. Bayer).
[338] Ciba, BE 565 782, 1957 (E. Parette, F. Maes).
[339] ICI, GB 586 340, 1944 (N. H. Haddock, C. Wood).
[340] BASF, DE 843 726, 1950 (A. Tartter).
[341] Berufsgenossenschaft der chemischen Industrie: *MAK-Werte 1989,* Nov. 30, 1989.
[342] M. J. Camenzind, C. L. Hill, *J. Heterocycl. Chem.* **22** (1985) 2.
[343] M. Hanack, J. Metz, G. Pawloski, *Chem. Ber.* **115** (1982) 2836.
[344] J. Metz, O. Schneider, M. Hanack, *Inorg. Chem.* **23** (1984) 1065.
[345] A. B. P. Lever et al., *J. Am. Chem. Soc.* **103** (1981) 6800.
[346] S. A. Mikhalenko, V. M. Derkacheva, E. A. Luk'-yanets, *Zh. Obshch. Khim.* **51** (1981) 1650; *Chem. Abstr.* **95** (1981) 196 503 v.
[347] D. Woehrle, B. Schulte, *Makromol. Chem.* **186** (1985) no. 11, 2229.
[348] S. A. Mikhalenko, E. A. Luk'yanets, *Zh. Obshch. Khim.* **46** (1976) no. 108, 9, 2156.
[349] N. O. Sigl, *J. Heterocycl. Chem.* **18** (1981) 1613.
[350] T. M. Keller, J. R. Griffith, *J. Flourine Chem.* **12** (1978) 73.
[351] V. M. Derkacheva, S. S. Jodko, O. L. Kaliya, E. A. Luk'yanets, *Zh. Obshch. Khim.* **51** (1981) 2319; *Chem. Abstr.* **96** (1982) 21 207 k.
[352] C. L. Leznoff, T. W. Hall, *Tetrahedron Lett.* **1982,** 3023.
[353] G. Pawlowski, M. Hanack, *Synthesis* **1980,** 287.
[354] C. Piechocki et al., *J. Am. Chem. Soc.* **104** (1982) 5245.
[355] A. W. Snow, N. L. Jarvis, *J. Am. Chem. Soc.* **106** (1984) 4707.
[356] A. W. Snow, J. R. Griffith, N. P. Marullo, *Macromolecules* **17** (1984) 1614.
[357] G. Pawlowski, M. Hanack, *Synth. Commun.* **11** (1981) 351.
[358] I. G. Oksengendler, N. V. Kondratenko, E. A. Luk'-yanets, L. M. Yagupolskii, *Zh. Org. Khim.* **14** (1978) no. 5, 1046.

[359] L. I. Solovèva, S. Mikhailenko, E. V. Chernykh, E. A. Luk'yanets, *Zh. Obshch. Khim.* **52** (1982) 90; *Chem. Abstr.* **96** (1982) 219 256 d.

[360] BASF, DE 3 622 363, 1986 (M. J. Degen, H. Kraus).

[361] K. H. Yahya, A. A. Dawood, *J. Chem. Eng. Data.* **33** (1988) no. 4, 529.

[362] H. Shirai, A. Marayuma, K. Kobayashi, N. Hoyo, *Macromol. Chem.* **181** (1980) 575.

[363] S. Higaki, K. Hanabusa, H. Shirai, N. Hojo, *Macromol. Chem.* **184** (1983) 691.

[364] J. H. Schutten, J. Zwart, *J. Mol. Catal.* **5** (1979) 109.

[365] B. N. Achar, G. M. Fohlen, J. A. Parker, J. Keshavayya, *Indian J. Chem. Sect. A.***27 A** (1988) no. 5, 411.

[366] L. I. Solov'eva, S. A. Luk'yanets, *Zh. Obshch. Khim.* **50** (1980) no. 112, 1122.

[367] H. Shirai et al., *Makromol. Chem.* **185** (1984) no. 12, 2537.

[368] Nippon Teleg. & Teleph., JP 02 3788, 1985 (A. I. Korakawa).

[369] D. Woehrle, G. Meyer, B. Wahl, *Macromol. Chem.* **181** (1980) 2127.

[370] D. Woehrle et al., *Ber. Bunsen-Ges. Phys. Chem.* **91** (1987) no. 9, 975.

[371] D. Woehrle, B. Wahl, *Tetrahedron Lett.* **3** (1979) 227.

[372] V. N. Klyuev, L. S. Shiryaeva, SU 260 044, 1960.

[373] S. Horiguchi, *Shikizai Kyokaishi* **38** (1965) no. 3, 99.

[374] V. F. Borodkin, R. D. Komarov, SU 419 539, 1974.

[375] D. Woehrle in C. C. Leznoff, A. B. P. Lever (eds.): *Phthalocyanines: Properties and Applications*, VCH Verlagsgesellschaft, Weinheim 1989, pp. 59 ff.

[376] S. A. Mikhalenko, E. A. Luk'yanets, *Zh. Obshch. Khim.* **39** (1969) no. 11, 2554.

[377] T. A. Shatskaya, M. G. Gal'pern, V. R. Skvarchenko, E. A. Luk'yanets, *Zh. Obshch. Khim.* **57** (1987) no. 10, 2364.

[378] S. A. Mikhalenko, L. A. Yagodina, E. A. Luk'yanets, *Zh. Obshch. Khim.* **46** (1976) no. 7, 1598.

[379] IG Farbenind., GB 457 526, 1936 (A. V. Aerden).

[380] IG Farbenind., DE 696 590, 1940 (R. J. Recidon).

[381] IG Farbenind., GB 471 418, 1937 (W. W. Groves).

[382] R. P. Linstead, E. G. Noble, J. M. Wright, *J. Chem. Soc.* **1937**, 911.

[383] A. R. Koray, V. Ahsen, O. Bekaroglu, *J. Chem. Soc., Chem. Commun.* **12** (1986) 932.

[384] R. Hendriks, Ot E. Sielcken, W. Drenth, R. J. M. Roeland, *J. Chem. Soc., Chem. Commun.* **19** (1986) 1464.

[385] N. Kobayashi, Y. Nishiyama, *J. Chem. Soc., Chem. Commun.* **19** (1986) 1462.

[386] A. W. Snow, N. L. Jarvis, *J. Am. Chem. Soc.* **106** (1984) 4706.

[387] S. A. Mikhalenko, L. A. Yagodina, E. A. Luk'yanets, *Zh. Obshch. Khim.* **46** (1976) no. 7, 1598.

[388] V. N. Kopranenkov, E. A. Luk'yanets, *Zh. Obshch. Khim.* **41** (1971) no. 10, 2341.

[389] M. Whalley, P. M. Brown, D. B. Spiers, *J. Chem. Soc.* **1957**, 2882.

[390] M. E. Baguley, H. France, R. P. Linstead, M. Whalley, *J. Chem. Soc.* **1955**, 3521.

[391] ICI, GB 750 240, 1954; US 2 744 913, 1951; US 2 681 341, 1951 (W. F. Beech et al.).

[392] Sumitomo Chem. Co., JA 633, 1959.

[393] H. P. Kaufmann, DE 920 666, 1951.

[394] IG Farbenind., DE 663 552, 1935.

[395] A. H. Cook, R. P. Linstead, *J. Chem. Soc.* 1937, 929.

[396] ICI, US 2 681 344, 1954; US 2 681 345, 1954 (H. France).

[397] Bayer, DE 964 324, 1954; DE 1 001 785, 1954 (G. Roesch, H. Klappert, W. Wolff).

[398] Hoechst, DE-AS 1 017 720, 1954 (F. Erbe).

[399] American Cyanamid, US 2 850 505, 1954 (D. W. Hein).

[400] H. P. Kaufmann, DE-AS 1 034 794, 1954.

[401] V. N. Klyuev, F. P. Snegireva, *Izv. Vyssh. Ucheb. Zaved. Khim. Khim. Tekhnol.* **14** (1971) no. 2, 258.
[402] V. N. Klyuev, F. P. Snegireva, SU 196 856, 1967.
[403] V. N. Klyuev, F. P. Snegireva, SU 293 024, 1971.
[404] P. G. Webb, *Am. Ink Maker* **65** (1987) no. 2, 11.
[405] R. E. Gosselin, H. C. Hodge, R. P. Smith, M. N. Gleason: *Chemical Toxicology of Commercial Products*, 4th ed., Williams and Wilkins, Baltimore 1976.
[406] USEPA: "Ambient Water Quality Criteria for Copper," PB 85 227 023 (1984).
[407] E. Kwiatkowska, E. Zawadzka, A. Stach, *Biul. Inf. Barwniki Srodki Pomocniczc* **29** (1985) no. 1, 17.
[408] J. Formanek, H. Kutna, CS 218 295, 1980.
[409] R. C. Buchta et al., *J. Chromatography* **325** (1985) no. 2, 456.
[410] E. C. Botti, *Off. Dig. Fed. Paint Varn. Prod. Clubs* **305** (1950) 408.
[411] M. J. H. Turk, *Verfkroniek* **32** (1959) 494.
[412] A.S.T.M., *Standards*, part 28, American Society of Testing Materials, Philadelphia, Pa., 1978, p. 211.
[413] R. L. P. R. Hepsworth, *Chem. Ind.* **1952**, 272.
[414] C. E. Moore, *Paint Manuf.* **27** (1957) 377.
[415] G. G. Rao, T. P. Sastri, *Z. Anal. Chem.* 1959, 11.
[416] D. Kruh, *Anal. Chem.* **42** (1970) no. 14, 1849.
[417] Hoechst, BE 590 125, 1960 (J. Gevers).
[418] ICI, GB 576 270, 1944; GB 587 636, 1944 (N. H. Haddock, C. Wood); GB 633 160, 1947.
[419] K. H. Schuendehuette: "Chromophore Systems," in K. Venkataraman (ed.): *The Chemistry of Synthetic Dyes*, vol. **VI**, Academic Press, New York 1974, p. 312.
[420] P. Rys, H. Zollinger, *Text. Chem. Color.* **6** (1974) 62.
[421] J. Eibl, *Melliand Textilber. Int.* **56** (1975) 398.
[422] A. P. Hopmeier, *Am. Ink Maker* **67** (1989) no. 9, 58.
[423] *Kirk-Othmer*, **17**, 787.
[424] F. R. Wrenn Jr., M. L. Good, P. Handler, *Science* **113** (1951) 525.
[425] E. Neuzil, J. Ballenger, *Compt. Rend. Soc. Biol.* **146** (1952) 1108.
[426] *NPIRI Raw Materials Data Handbook*, vol. **4**, 1984.
[427] P. Miloy, K. Kay, *J. Toxicol. Environ. Health* **4** (1978) 31.
[428] A. Haddow, E. Hornung, *J. Nat. Cancer Inst.* **24** (1960) 109.
[429] *NTP Technical Bulletin* **1** (1981) no. 5, 8.
[430] American Conference of Governmental Industrial Hygienists: *Documentation of the Threshold Limit Values and Biological Exposure Indices*, 5th ed., Cincinnati 1986.
[431] B. A. Kurlyandski et al., *Gig. Sanit.* **1** (1975) 92.
[432] E. Ben-Hur, A. Newman, S. W. Crane, I. Rosenthal, *Cancer Lett.* **38** (1987) no. 1/2, 215.
[433] S. Sandor, O. Prelipceanu, I. Checiu, *Rev. Roum. Morphol. Embryol. Physiol. Morphol. Embryol.* **31** (1985) no. 3, 173.

Polysaccharides

ALPHONS C. J. VORAGEN, Wageningen, Agricultural University, Department of Food Science, Wageningen, The Netherlands

WALTER PILNIK, Wageningen, Agricultural University, Department of Food Science, Wageningen, The Netherlands

CLAUS ROLIN, The Copenhagen Pectin Factory Ltd., Lille Skensved, Denmark (Chap. 3)

BEINTA U. MARR, The Copenhagen Pectin Factory Ltd., Lille Skensved, Denmark (Chap. 3)

1.	Introduction	3972
2.	Analysis and Characterization	3973
3.	Pectin	3975
3.1.	Occurrence and Structure	3975
3.2.	Pectolytic Enzymes	3978
3.3.	Production	3979
3.4.	Properties	3982
3.4.1.	Physical Properties	3983
3.4.2.	Gel Properties	3984
3.4.3.	Stability and Chemical Reactions	3988
3.5.	Analysis	3990
3.5.1.	Measurement and Standardization of Gel-Forming Capacity	3990
3.5.2.	Chemical Analysis	3990
3.6.	Pharmaceutical and Nutritional Characteristics	3992
3.7.	Application in the Food Industry	3993
3.8.	Market	3993
4.	Alginates	3993
4.1.	Occurrence and Structure	3994
4.2.	Production	3996
4.3.	Properties	3998
4.4.	Analysis	4001
4.5.	Applications	4002
4.6.	Market	4002
5.	Carrageenan	4003
5.1.	Occurrence and Structure	4003
5.2.	Production	4004
5.3.	Properties	4006
5.4.	Analysis	4008
5.5.	Applications	4009
5.6.	Market	4009
6.	Agar	4009
6.1.	Occurrence and Structure	4010
6.2.	Production	4010
6.3.	Properties and Analysis	4011
6.4.	Applications	4012
6.5.	Market	4012
7.	Gum Arabic	4012
8.	Gum Tragacanth	4015
9.	Gum Karaya	4016
10.	Gum Ghatti	4017
11.	Xanthan Gum	4018
11.1.	Production	4019
11.2.	Structure and Properties	4020
11.3.	Analysis	4021
11.4.	Applications, Market	4022
12.	Gellan Gum	4023
13.	Galactomannans	4023
13.1.	Structure	4024
13.2.	Production	4025
13.3.	Properties	4025

13.4.	Analysis and Composition of Commercial Preparations...	4026	13.6.	Applications........... 4027
			13.7.	Market.............. 4028
13.5.	Derivatives............	4027	14.	References........... 4028

1. Introduction [1]–[12]

For a general definition of the term polysaccharides, see → Carbohydrates.

Polysaccharides made up of only one type of neutral monosaccharide structural unit and with only one type of glycosidic linkage — as in cellulose or amylose — are denoted as *perfectly linear* polysaccharides. In branched polysaccharides the frequency of branching sites and the length of the side chains can vary greatly. Molecules with a long "backbone" chain and many short side chains are called *linearly branched* polysaccharides.

Polysaccharides are water soluble or swell in water, giving colloidal, highly viscous solutions or dispersions with plastic or pseudoplastic flow properties. Functional properties such as thickening, water holding and binding, stabilization of suspensions and emulsions, and gelling, are based on this behavior. Therefore, polysaccharides are often referred to as gelling or thickening agents, stabilizers, water binders, or fillers. A more generic name is gums. The terms hydrocolloids or, in the food industry, food colloids are also in use, but they include proteins with similar functional properties.

In this article the economically important polysaccharides, with the exception of starch (see → Starch) and cellulose are described. Table 1 lists the polysaccharides, and gives an overview of their sources and composition. Their most important and various applications are summarized in Table 2. In only a few cases is their use specific. Examples are the use of pectin in commercial jam and marmalade production, where its heat stability coincides with its ability to gel in the typical pH range of these products, or the stabilization of drinking chocolate with carrageenan, which has a specific interaction with milk proteins and chocolate particles. In many cases, the manufacturer has various options and price plays an important role. For papermaking choice may be between, for example, alginates, galactomannans, and starch derivatives; for oil drilling, xanthan, starch derivatives, cellulose derivatives, galactomannans, or alginates can be used. Milk gels can be prepared with pectin, starch, alginates, or carrageenans. Often mixtures of polysaccharides are preferred to single polysaccharides because they combine various desirable properties (e.g., in ice cream or salad dressing). In a number of cases a synergistic interaction on the molecular level between polysaccharides can be exploited (e.g., improvement of the structure of carrageenan gels by

Table 1. Origin and main sugar moieties of polysaccharides

Name	Origin	Main sugar moieties
Pectin	cell walls and middle lamella of higher land plants	D-galacturonic acid D-galacturonic acid methylester
Alginate	cells walls of brown seaweeds exopolysacchrides of *Azetobacter vinelandii*	D-mannuronic acid L-guluronic acid and their acetyl derivatives
Carrageenan	cell walls of red seaweeds	D-galactose 3,6-anhydro-D-galactose both sugars sulfated to higher or lower degree
Agar	cell walls of red seaweeds	D-galactose 3,6-anhydro-L-galactose few sulfate groups
Gum arabic	exudate of acacia species	L-arabinose, D-galactose L-rhamnose, (4-O-methyl) D-glucuronic acid
Gum tragacanth	exudate of astragalus species	L-arabinose, D-galactose D-galacturonic acid methylester D-xylose, L-rhamnose, L-fucose
Gum karaya	exudate of sterculia species	D-galacturonic acid, L-rhamnose, D-galactose D-glucuronic acid
Gum ghatti	exudate of *Anageissus latifolia*	L-arabinose, D-galactose D-mannose, D-xylose, D-glucuronic acid, L-rhamnose
Guar gum	endosperm of seeds of the guar plant (*Xyampsis tetragonolobus* L. Taub)	D-mannose D-galactose
Locust bean gum	endosperm of the seeds of the carob tree (*Ceratnia siliqua* L.)	D-mannose D-galactose
Tara gum	endosperm of the seeds of the tara tree (*Caesalpinia spinoza*)	D-mannose D-galactose
Xanthan gum	exopolysaccharide of *Xanthomonas campestris*	D-glucose, D-glucoronic acid, D-mannose substituted with acetate or pyruvate groups
Gellan gum	exopolysacchride *Pseudomonas elodea*	D-glucose, D-glucuronic acid, L-rhamnose

the addition of guar gum or the preparation of gels by mixing xanthan and locust bean gum).

2. Analysis and Characterization

The analysis and characterization of polysaccharides can be aimed at the following:

1) Estimation of their content in raw materials
2) Qualitative and quantitative identification of one or more polysaccharides in a preparation
3) Qualitative and quantitative identification of one or more polysaccharides in a (food) product
4) Estimation of specific properties such as gelling strength or viscosity
5) Estimation of specifications for purity (e.g., color, content of heavy metals or cations, moisture, ash, or microbial count)

The manufacturer is interested primarily in item 1); items 2) and 3) are of concern to the user and 3) for state inspection services; item 4) is of interest to the manufacturer in terms of standardization, and the user in terms of applications; item 5) is of particular

Table 2. Areas of application of polysaccharides *

Foods
1. Thickening and gelatinizing of fruit with a higher or lower content of sugar: preserves, marmalades, jellies, fruit for yogurt and ice cream, marmalade for cooking.
2. Gelatinizing of heat-reversible jelly and sugar glaze. Jellied sugar products, jellied sweets, dessert jellies (powder form).
3. Prevention of starch breakdown in bread and baked products, and in fruit-filled products based on starch.
4. Cloud stabilization in fruit juice. Providing body in refreshing powdered drinks (→ Beverages, Nonalcoholic).
5. Stabilization of powdered flavor emulsions. Microencapsulation of flavorings (→ Microencapsulation).
6. Gelatinization and thickening of milk: puddings and cream for hot and cold preparations; also as powdered instant products.
7. Stabilization of chocolate milk and evaporated milk: prevention of sedimentation or separation of the cream fraction.
8. Stabilization of ice cream: prevention of formation of ice crystals and dripping. Stabilization of the fat emulsion, imparting favorable slip properties for mold release.
9. Water binding in cream cheese-type spreads, soft cheese, and cheese preparations.
10. Stabilization of fat emulsion in powdered coffee whiteners and low-fat margarine.
11. Stabilization of egg-white froth, whipped cream, and imitation whipped cream, meringues, and beer foam.
12. Stabilization and thickening of soups, sauces, dips, salad cream, mayonnaise, catsup, etc. Prevention of syneresis. Imparting freeze – thaw stability, and body in fat and starch-reduced products.
13. Binding agents in ground meat products (e.g., corned beef, sausages) and in canned dog and cat foods.
14. Gelatinized (jellied meat) or thickened stock from meat, fish, and vegetables.
15. Binding agents for (potato flour) snacks.

Cosmetics
16. Thickening agents in creams, ointments, lotions, and hair gels.

Household Goods and Industry
17. Solid air fresheners.
18. Stabilization of emulsified and suspended active agents in insecticides and herbicide sprays.
19. Stabilization of suspended abrasion materials in polishes and cleaning agents.
20. Glue.

Pharmaceuticals
21. Thickening agents in creams, lotions, ointments, and syrups.
22. Toothpaste (water-binding, consistency.
23. Denture adhesives.
24. Tablet-binding agents.
25. Dental impression materials.

Industry
26. Binding agents in explosives.
27. Binding agents for layered coating of welding electrodes.
28. Different functions of the rinsing liquid used in deep drilling technology.
29. Paper binding agents. Paper coating agents (fat-resistant properties, improved printability, uniform distribution of pigment despite extremely high speeds of modern papermaking machines; additives to improve absorbency.
30. Sizing and finishing agents used in the textile industry.
31. Thickening of textile printing pastes and coloring agents to ease application and prevent color bleeding.
32. Flocculation agents for water processing.
33. Pigment stabilization and paint consistency (thixotropy in emulsion paints).

* The numbering has been introduced merely to allow reference in the text and does not rank the applications in order of importance.

interest to the consumer and to state inspection services in pharmaceutical and food applications.

For determination of items 2) and 3), several possiblities exist [13], [14]. In most cases, solutions or extracts must be prepared in which the polysaccharides are estimated qualitatively by precipitation reactions [15], or a special analytical procedure is used to separate anionic polysaccharides from neutral polysaccharides by precipitation with cetylpyridinium chloride [16], [17]. When the necessary equipment is available, quantitative analysis, based on the gas chromatographic determination of the constituent sugar moieties after hydrolysis and derivatization of the purified polysaccharide is often quicker and more reliable [29]. Recently, HPLC has been used more often in the analysis of constituent sugar moieties. Another new development is the use of immunological methods based on the development of antibodies against the polysaccharides. The requirements indicated under item 5) and the corresponding analytical methods are defined in pharmacopeias, food legislation, and specifications of national and international organizations [41], [62], [63]. The latter also describe identification reactions. Additional information on the analysis of individual polysaccharides is given in the corresponding sections.

3. Pectin

Pectin or pectic substances, also called galacturonans or rhamnogalacturonans in scientific literature, is a collective name for heteropolysaccharides that consist predominantly of partially methylated galacturonic acid residues [74], [85], [96], [104], [18], [105], [106]–[111]. The name pectin (Greek: *Pectos* = gelled) was coined by BRACONNOT [112] who first described this compound in detail in 1825 and indicated its primary use as a gelling agent. Native pectin plays an important role in the consistency of fruits and vegetables, and in textural changes during ripening and storage [85], [113]. The often desired cloud stability of fruit juices and fruit drinks is lost when enzymes endogenous to the fruit degrade pectin (tomatoes) or cause it to precipitate as calcium pectate after demethylation. On the other hand, enzymes may be added as processing aids to degrade native pectin, for example, to apple juice to facilitate clarification or to the pulp of berries to improve press and color yield [114]–[116]. Pectin is extracted on an industrial scale from the press residues in apple and citrus juice manufacture and used mainly as gelling and stabilizing agents in the food industry [117].

3.1. Occurrence and Structure

Occurrence. Pectins occur in virtually all higher plants, Zosteraceae seaweed, and certain freshwater algae. Pectins are major structural components of the primary cell wall and the middle lamella of young growing plant tissues (meristimatic and parenchymatic) but do not occur in more mature tissue. The composition of the cell wall therefore is of major importance in the texture of fruit and vegetables [19], [20]. The primary cell wall consists of 90% polysaccharide and 10% glycoprotein on a dry matter basis [21]. The polysaccharide composition is 30–60% cellulose, 15–45% pectic substances, and 15–25% hemicellulose [22]. A schematic structure of the plant cell wall is illustrated in Figure 1.

The biosynthesis of pectin takes place in the cell plate during cell division. Pectin is formed as polygalacturonic acid with UDP-D-galacturonic acid—arising from UDP-D-glucose by an epimerase-catalyzed reaction—as the most active glycosyl donor. Immediately after the galacturonan chain has been formed, methoxyl groups are formed with S-adenosylmethionine as the methyl group donor [120].

Pectic substances can be partially solubilized from plant tissues without degradation by using weakly acidic, aqueous solvents with or without calcium chelating agents. The pectin fraction that is not extractable with these extractants because of its attachment to other cell wall components by chemical, physical, or mechanical (enmeshment) bonds, is often designated as *protopectin*. Commercial pectin extraction must break down protopectin to a soluble, high molar mass pectin. This is achieved by acid hydrolysis in which, on the one hand, the molar mass of pectin molecules is lowered and, on the other hand, connections to the hemicellulose fractions are split. This transition from protopectin to soluble pectin also occurs during ripening of fruit or cooking of vegetables, resulting in textural changes [85], [113]. Pectin technology is therefore interested in the nature of protopectin and its fixation in the cell walls in an attempt to develop more efficient methods of extraction. FRY [119] has discussed cross-links in cell walls and agents used to cleave the individual bonds. Pure enzymes may also be used for extraction of firmly bound pectin [24], [25].

Structure. Pectin is composed of 1,4-linked α-D-galactopyranosyluronic acid units in the $^4C_1$ conformation, with the glycosidic linkages arranged diaxially (Fig. 2). A proportion of the carboxyl groups is esterified with methanol. Commercial pectins are divided into low-ester and high-ester pectin. In *low-ester pectins (LM-pectins)* less than 50% of the carboxyl groups are methylated (typical range is from 20 to 40%) whereas in *high-ester pectins (HM-pectins)* more than 50% are methylated (typical range is from 55 to 75%). If less than 10% of the carboxyl groups are methylated the polysaccharide is called *pectate* or *pectic acid*. Pectins prepared from pears, potatoes, sugar beets, and sunflower heads are acetylated to varying degrees at the secondary hydroxyl groups of the galacturonic acid residues [118].

The heteropolysaccharide nature of pectin derives from the fact that other sugars are incorporated in the pectin molecule. The most common ones being L-rhamnose (Fig. 3), occasionally inserted by α-1,2-linkages in the galacturonan backbone, providing "kinks" in the molecular chain. Other sugars are β-D-xylose, attached as single-unit side chains mainly to O-3 of the galactopyranosyluronic acid residues in the backbone; and D-galactose and L-arabinose, which occur in long side chains, only attached to rhamnopyranosyl residues (for projection formulae, see → Carbohydrates).

Pectin

Figure 1. Schematic structure of the plant cell wall [23]

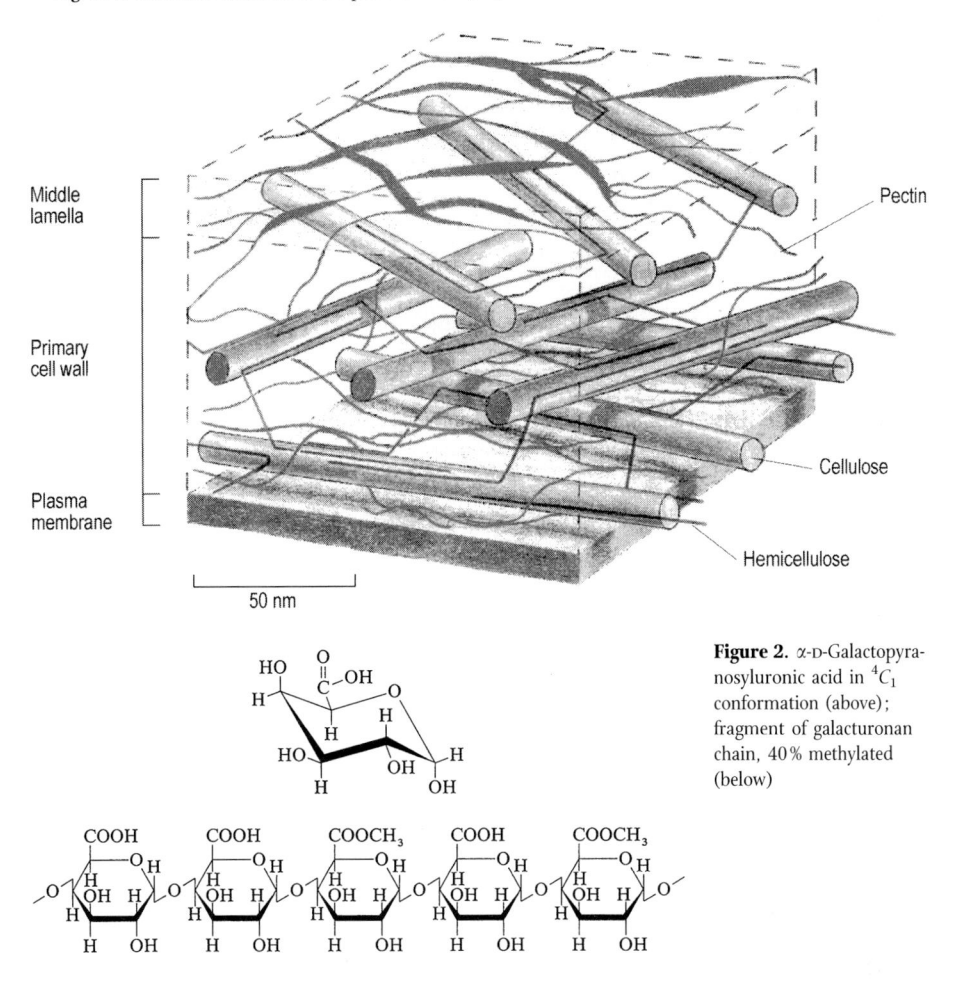

Figure 2. α-D-Galactopyranosyluronic acid in $^4C_1$ conformation (above); fragment of galacturonan chain, 40% methylated (below)

Figure 3. 1,2-Linked L-rhamnopyranosyl unit in $^1C_4$ conformation

The frequency of rhamnose occurrence remains to be established, however it has been suggested that α-rhamnosyl units are concentrated in rhamnose-rich areas. In other words, the soluble pectin is built up of homogalacturonan-dominated areas, so-called *smooth regions* linked to rhamnogalacturonan areas rich in neutral sugars, so-called *hairy regions* (Fig. 4). The neutral sugars account for 10–15% of the weight of the

3977

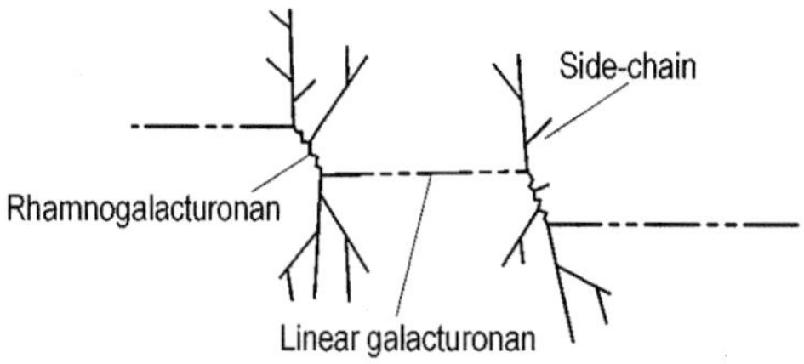

Figure 4. Schematic representation of pectin backbone, showing the "hairy" regions (rhamnogalacturonan and side-chains) and the "smooth" regions (linear galacturonan) [31]

pectin. In the hairy regions, the neutral sugar chain length may be in the range 8 to 20 residues [26], [27]. By degradation with chemical β-elimination and endo-polygalacturonase, it was found that 90% of the rhamnose units are found in the hairy region [28], [30].

By acid hydrolysis, i.e., splitting the acid labile glycosidic bonds between rhamnose and galacturonic acid, nearly pure homogalacturonic acids with molecular masses in the range 20 000 to 25 000 have been obtained [32], [33]. This corresponds to a chain length of 75 to 100 galacturonic acid residues. For comparison, a molecular mass of 90 000 has been quoted for intact pectin [34], but the molecular mass of pectin is somewhat uncertain (see Section 3.5.2).

Commercial pectins may also contain neutral polysaccharides that are not covalently attached to the pectic backbone, such as galactans, arabinans, arabinogalactans, and starch, often referred to as "ballast" compounds. Purified pectins prepared from apple pomace or citrus peel may contain 75–90% anhydrogalacturonic acid on an ash- and moisture-free basis.

3.2. Pectolytic Enzymes

Pectins can be attacked by various enzymes [114], [115], [121]. The significance of native pectic substances in food technology can be evaluated properly only when the activity of these enzymes is taken into account.

Pectinesterase (PE, pectin methylesterase, pectase, pectin demethoxylase, pectin pectylhydrolase, EC 3.1.1.11) splits off the methoxyl groups and converts high-methoxyl pectins to low-methoxyl pectins. The latter are extremely sensitive to complex formation and precipitation with Ca^{2+} ions, particularly when a pectinesterase of plant origin is used. This type of enzyme does not saponify methyl esters in a random fashion as microbial PEs do, but acts along the galacturonan chain, creating blocks of free carboxyl groups. Pectinesterase occurs in many higher plants, particularly tomatoes, citrus, and other fruits; it is also produced by many fungi and bacteria.

Polygalacturonase (PG, pectinase, pectate hydrolase, poly-α-1,4-D-galacturonide glycanohydrolase, EC 3.2.1.15 and 3.2.1.67) preferentially hydrolyzes low-methoxyl pectins or pectic acid because these enzymes can cleave glycosidic linkages only next to free carboxyl groups. PGs can be divided into enzymes that degrade their substrate by an endo attack (splitting randomly in the backbone, *endoPG*), and enzymes that act from the nonreducing end removing mono- or digalacturonic acid (*exoPG*). PGs are produced by fungi and certain bacteria, and also occur in higher plants (tomatoes). The endoPGs, with their strong depolymerizing action, are of particular technological importance.

Pectate lyase (PAL = pectic acid lyase, PATE = pectic acid transeliminase, LMPL = low-methoxyl pectin lyase, poly-α-1,4-D-galacturonide lyase, EC 4.2.2.2 and 4.2.2.9) also splits glycosidic linkages next to free carboxyl groups. In this group of enzymes, endo and exo enzymes also exist. The preferential substrates for *endoPAL* are LM-pectins rather than pectic acid. Pectate lyases have an absolute requirement for Ca^{2+} ions. The glycosidic linkages are split by a *trans*-elimination reaction. PALs are produced predominantly by bacteria and are not important in fruit and vegetable processing because of their high optimum pH (8.5 – 9.5).

Pectin lyase (PL, PTE = pectin transeliminase, pectinase, poly-α-1,4-D-methoxygalacturonide lyase, EC 4.2.2.10) splits glycosidic linkages between methoxylated galacturonide residues by a *trans*-elimination reaction. These enzymes therefore have a preference for HM-pectins. Pectin lyases are produced only by fungi.

Recently a new pectin-degrading enzyme was described that, in cooperation with a pectin acetylesterase acts only on highly branched regions of pectin, to release oligosaccharides consisting of alternating sequences of α-1,2-linked L-rhamnosyl residues and α-1,4-linked D-galacturonosyl residues, with galactosyl residues β-1,4-linked to part of the rhamnosyl units. The nonreducing end is always a rhamnose unit [122], [123].

Commercially available pectinases, used on an industrial scale in fruit and vegetable processing are of fungal origin and generally contain in addition to PE, PG, and PL, proteases and various hemicellulases and cellulases.

3.3. Production [106], [109], [117], [122], [124]

The production of pectin is summarized in Figure 5.

Raw materials of importance to pectin manufacturing are currently various kinds of citrus peel, and apple pomace. Lemon and lime are the preferred citrus sources, and more pectin is produced from these than from apple or the less preferred citrus materials, orange and grapefruit. Some of the pectin producing companies which historically developed in connection to apple production now partly or wholly base their production on imported citrus peel. Sugar beet was once used as a pectin source to some extent [35], but the pectin is inferior as a gelling agent compared to citrus or apple pectin. It has recently been reintroduced, and is currently being marketed as a stabi-

Figure 5. Flow chart of pectin production

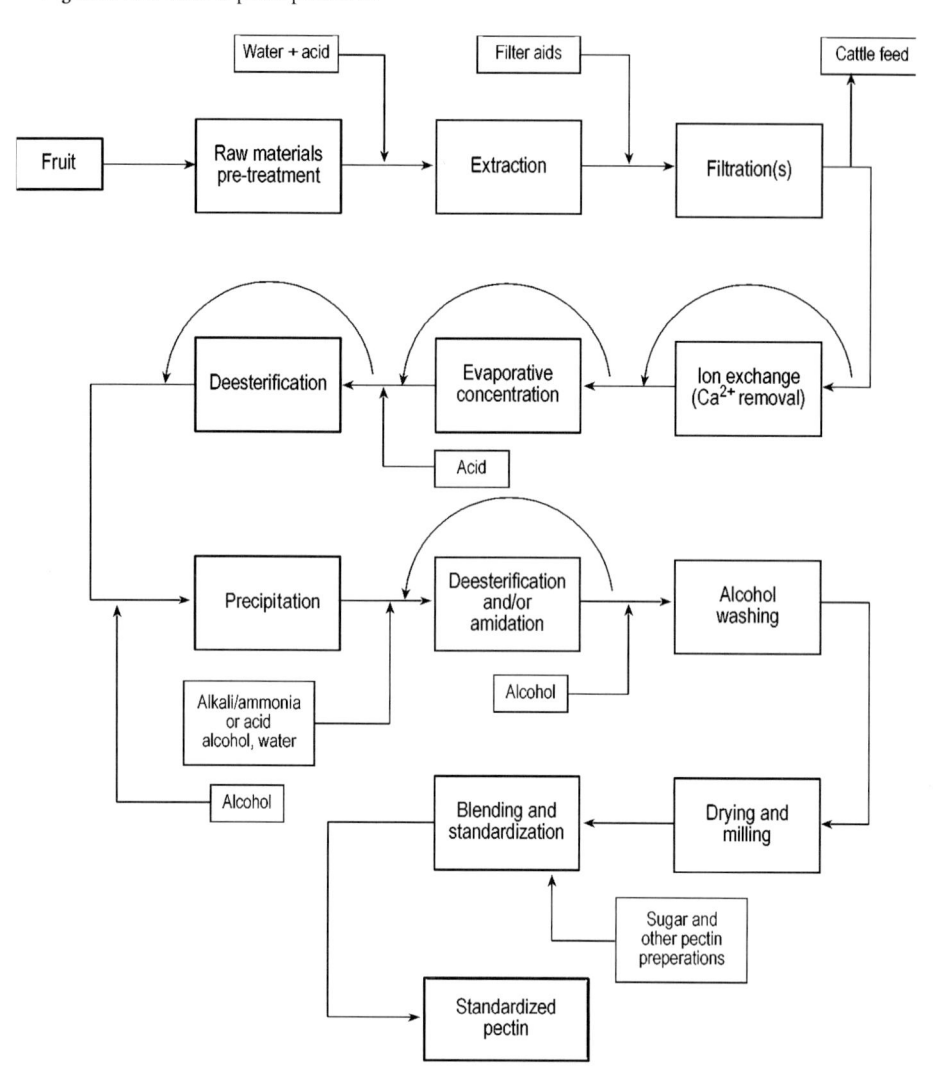

lizer. Numerous other sources like mango [36], pea hulls [37], sunflower heads [38], [39], and pumpkin [40] have been suggested.

Citrus peel and apple pomace are available as byproducts from juice manufacturing. They are usually washed in water and dried before being used for pectin manufacturing, but some citrus material is used in pectin plants neighboring the juice production without previous drying. In either case processing of the raw materials has to commence immediately after juice production in order to prevent microbial degradation. The washing leaches out organic acids, sugar, and pigments and is thus one of the separation processes in the purification of pectin. Most importantly, it prevents discoloration either from browning of pigments or from caramelization during the raw materials drying.

Extraction. The pretreated raw material is extracted in water which has been acidified with e.g., hydrochloric or nitric acid. Typical conditions are: pH 1 to 3, temperature 50 to 90 °C, duration 3 to 12 h. During the extraction, limited depolymerization of the pectin and possibly of other connecting biopolymers takes place, and the pectin dissolves. The low pH further dissociates ionic linkages which hold the pectin in the plant tissue. In addition to hydrolyzing glycosidic bonds, the extraction conditions also hydrolyze ester linkages, more specifically the methyl ester at C-6, and the acetate to which pectin may be esterified by its hydroxyl groups. The extraction process thus causes a reduction in degree of polymerization as well as in degree of methyl and acetate esterification. The pectin yield increases with the acidity, the temperature, and the duration, but the product will lose too much in degree of polymerization if all these parameters are at their maximum. The combination of low pH and low temperature favors hydrolysis of ester linkages over hydrolysis of glycosidic bonds, and it is thus preferred for production of pectin with a relatively low degree of esterification.

Filtration in one or more stages separates the extract containing the solubilized pectin from the insoluble, but at this stage very soft and fragile, plant tissue. The rather difficult filtration requires reasonably low viscosity, and as a consequence the pectin concentration must be less than 0.6 to 1%, depending upon the pectin type. Further, the solids must not have been comminuted by excessive mechanical treatment such as vigorous agitation. Water-insoluble materials like wood cellulose or diatomaceous earth may be added in order to improve the porosity and mechanical strength of filter cakes. Amylase may be added to remove starch from apple pectin extracts.

The spent plant raw material is typically used for cattle feed.

Isolation. Following filtration, the extract may optionally be passed through a column with cation-exchange resin and concentrated by evaporation. The pectin is then precipitated by mixing the extract with an appropriate alcohol, e.g., 2-propanol. Finally, the precipitate is separated from the spent alcohol, washed in more alcohol, pressed to drain as much liquid as possible, and then dried and milled. The powder is now ready for standardization, i.e., mixing with other pectin batches and/or sucrose in order to ensure uniformity. The alcohol is recovered by distillation. An alternative to alcohol precipitation is precipitation by adding appropriate metal salts to the extract. Pectin forms insoluble salts with, e.g., Cu^{2+} and Al^{3+}. The Al^{3+} precipitation [125] was previously used industrially. Removal of the metal ions from the precipitated pectin is done by washing in acidified aqueous alcohol.

Modification. Pectin derived from citrus or apple raw material as described above will normally have a degree of esterification between 55 and 75. A lower degree of esterification can be achieved by acidifying the extract and leaving it for some time before precipitation, or by treating precipitated (but not dried) pectin with acid or alkali during suspension in aqueous alcohol. Ammonia may convert methyl-esterified carboxylate groups of a pectin to primary amides [42], [43]. This is done industrially by suspending precipitated pectin in a mixture of alcohol and water with dissolved ammonia [127], [128]. Deesterification takes place concurrently. By choosing proper conditions with respect to ammonia concentration, water activity, and temperature, pectins with various proportions of amidated, methyl esterified and free carboxylate groups can be produced.

Standardization. The properties of botanical raw materials like those used for pectin fluctuate due to, e.g., weather conditions or sorts variation. Pectin as it appears directly from milling contains this variation. In order to maintain a constant quality, the pectin manufacturer may mix different batches. Further, pectin intended for food is typically diluted with sucrose in order to achieve a uniform grade (i.e., "strength," for a definition, see Section 3.5.1). Pectin without admixed sugar, e.g., for pharmaceutical purposes, is also available from the major manufacturers.

Due to the multitude of ways in which pectins may vary, it is not possible to ensure batch to batch consistency with respect to all possible attributes at the same time. Pectin is normally standardized with respect to a few properties which are measured in defined chemical systems which simulate the applications, e.g., breaking strength of a gel, gelation temperature, etc. The major pectin manufacturers have developed a great number of specialty types which are tailor-made for individual applications and which have each their set of standardization criteria (control methods). In fact, a pectin type is defined by its set of standardization criteria. When using pectin, it is obviously important to choose a type which has been standardized in a way that corresponds reasonably to the intended use.

3.4. Properties

A range of parameters — intrinsic and extrinsic — are important for the performance of pectin which is in most cases used to impart certain rheological properties, e.g., by forming a gel. The *intrinsic parameters* determine the nature of the gel and may include molecular mass, degree of esterification (DE), degree of acetylation (DA), neutral sugar content, and composition. The *extrinsic parameters* which determine the gelation process may include pectin concentration, pH, ionic strength, water activity, and temperature.

The most important properties of pectin preparations depend on their molecular mass and DE and DA. The proportion and nature of neutral sugars in the side chains as well as in the "ballast" are also of significance. Pectin with a high degree of polymerization is more viscous in solution compared to a pectin with a lower degree of polymerization. Further, the gel strength will typically increase with the degree of polymerization, i.e., less pectin is needed with a high molecular mass pectin [44]–[46]. The DE strongly influences the functional properties of pectin with the two main groups, HM- and LM-pectin being influenced differently. At the typical conditions in HM-pectin applications, high DE means high gelling temperature whereas at typical use conditions for LM-pectins, low DE means high gelling temperature. No experimental evidence is available to demonstrate the influence of the hairy regions (see p. 3976) on the functionality of pectin. It could be speculated that by being bulky and providing kinks in the molecular chain, these parts will prevent molecules from aligning throughout their entire length. This may contribute to preventing precipitation and reducing potential syneresis (spontaneous exudation of solvent from the gel).

Chemical Reactions. By treating pectin with ammonia under alkaline conditions in alcohol suspensions, ca. 20% of the methyl ester groups are converted to acid amide groups, and amidated pectins are obtained [129]. Amidated pectins have a higher calcium reactivity than LM-pectins, and gels can be obtained with very few Ca^{2+} ions [134]. Carboxyl groups in pectin can be esterified easily with methanol [126], glycol, and glycerol but poorly with ethanol [135]. By using polyols, cross-linked, insoluble systems are obtained. Insoluble pectates can also be prepared by cross-linking with epichlorohydrin; these pectates have ion-exchange properties with a certain selectivity for calcium and heavy-metal ions. They are successfully used for the isolation of pectolytic enzymes [136], [137]. Pectins are readily degraded by oxidants [138] except for chlorite and chlorine dioxide, which can selectively oxidize aldehyde groups at the reducing chain end [139].

3.4.1. Physical Properties

Pectin is water-soluble, exhibiting an increased solubility with increasing DE and decreasing degree of polymerization. In order to ensure complete dissolution of pectin, it is necessary that it is properly dispersed without lumping. Once formed, lumps are extremely difficult to dissolve. Pectin, like any other gum or gelling agent, will not dissolve in media where gelling conditions exist. In order to add pectin to complex formulations such as food systems three alternative procedures are recommended: (1) dissolve the pectin in pure water and add the solution; (2) dry blend the pectin with five parts of sugar and add the mixture; (3) disperse the pectin in a liquid in which it is not soluble and add the dispersion.

Aqueous dilute pectin solutions, i.e., with a pectin content below 0.5% are almost Newtonian whereas more concentrated pectin solutions exhibit pseudoplasticity, i.e., shear-thinning behavior. From dilute viscosity data, the intrinsic viscosity, [η] (dL/g) may be determined. [η] indicates the hydrodynamic volume of a polymer molecule and depends primarily on the molecular mass, however viscosity is also influenced in a complex manner by many other factors such as the DE, pH (dissociation), and ionic strength. [η]-values for pectins typically lie in the range of 1.0 to 6.0 dL/g [47]–[51].

Figure 6. Network of a gel

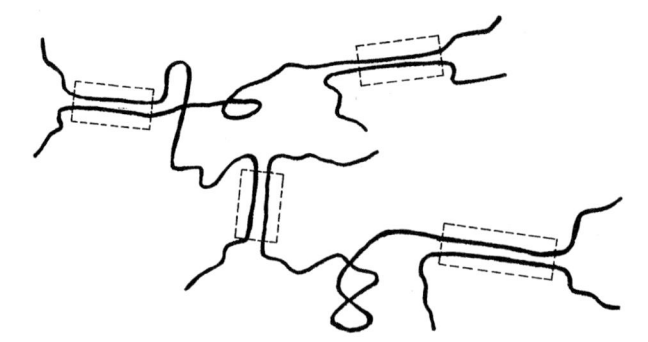

The molecular mass of pectin is often estimated by intrinsic viscosity methods using the Mark–Houwink relationship $[\eta] = KM^\alpha$ (see Section 3.5). Originally an α-value of 1.34 [52] was suggested corresponding to a rigid rod-like molecular structure, however more recent findings estimate α to be in the range 0.7–0.8 indicating a random coil structure [47], [49], [53], [54].

Pectin is insoluble in most organic solvents such as alcohols and acetone. Pectin can also be precipitated from aqueous solutions by quarternary detergents, water-soluble cationic polymers including proteins, and multivalent cations. LM-pectins can be precipitated by calcium ions; pectates by alkali cations and by acid.

Pectin is a polycarboxylic acid. Dissolved pectin is negatively charged at neutral pH and approaches zero charge at low pH. Since pectin is a polyprotic acid it is not possible to determine an exact value of the apparent dissociation constant, pK_a. Rather, pK_a is different for varying carbohydrate concentrations and for varying degree of dissociation, α. The negative charge density, in turn, is dependent on DE which implies that pK_a-values are increasing with increasing content of unesterified galacturonic acid units. Typically, pK_a-values at 50% dissociation, i.e., $\alpha = 0.5$, lie in the range 3.5–4.5 [55]–[59]. The usual dependence of pK_a with polymer concentration and ionic strength is observed, i.e., pK_a is lowered with increasing concentration and ionic strength.

3.4.2. Gel Properties

Pectin is used mainly as a gelling agent in industry; therefore its gelling properties are most important [4], [6], [106], [117]. The gel formation mechanism of pectin is similar to that of other gelling polysaccharides: Some regions of the polymer molecules associate in junction zones to form a three-dimensional network, which traps the solvent with cosolutes; free stretches of the molecules provide elasticity to the gel obtained [111], [140], [141] (Fig. 6). Irregularities in the pectin molecule, such as the distribution pattern of methyl ester or O-acetyl groups, rhamnosyl residues in the backbone, or the presence of side chains, limit the length of the junction zones and give shape to the free stretches of the macromolecules [142].

A comprehensive, coherent description of pectin gelation does not exist. Conventionally, a distinction is made between HM- and LM-pectin gelation, however, in reality this paradigm is too simple. It applies that different gelation mechanisms may act simultaneously. This is e.g., illustrated by the fact that gelation of LM-pectin, which is normally claimed to gel in the presence of certain metal ions, is further favored by a decrease in pH. If the gelation was solely determined by the formation of calcium-bridges between molecules, the opposite effect would be expected, an increase in pH leading to an increase in gel strength.

In order to adapt the conventional and still widely accepted theories for pectin gelation, the gelation phenomenon will, however, be treated as two distinct mechanisms, ie HM- and LM-pectin gelation. Gels used for jams and jellies are typically formed with HM-pectin at an acidic pH and require the presence of a high concentration of sugar. LM-pectin is typically used for yogurt fruit preparations; these gels can be formed without sugar over a wide pH range, however, the presence of a divalent cation is necessary. In most cases the cation, i.e., calcium, is inherently present in the fruit material. LM-pectin gels can be remelted whereas usually it is not possible to melt an HM-pectin gel, i.e., with HM-pectin gel preparations, the difference between the apparent temperatures of setting and melting is so large that the gels are said to be thermo-irreversible. Further, LM-pectin gels solidify almost immediately after gelling conditions have been introduced, while an HM-pectin gel will build up over time.

HM-pectins form so-called *low-water-activity* or *pectin – sugar – acid gels* and are used in jam, jelly, and marmalade production. The basic galacturonan chain (smooth regions) of the pectin molecule apparently contains blocks with conformational regularity to provide opportunities to build up junction zones. The homogalacturonan part of the molecule is configured as helices with three anhydrogalacturonic acid units per turn, with the methyl ester groups protruding from the helix. According to OAKENFULL and SCOTT [143], junction zones are stabilized by different forces between pectin chain molecules: hydrogen bonds between undissociated carboxyl and secondary alcohol groups, and hydrophobic interactions between methoxyl groups (Fig. 7). Both types of forces are fortified by sucrose; the low pH suppresses the dissociation of carboxyl groups. Sugar reduces the water activity of the system and thereby influences hydrogen bonding by decreasing polymer – water interactions and increasing polymer – polymer interactions. To a certain extent, sugar and acid are interchangeable: at lower sugar concentration, lower pH is required, but at higher sugar concentration, higher pH values are possible and necessary to avoid setting during the boil. The lower limit for the sugar concentration is 55%. At this concentration, the pH should not be higher than 2.8. At a sugar concentration of 80% (jellies), the mass will also gel at pH 3.5. This means that within the gelling range, at the same sugar concentration, more acid will give a stronger gel and, at the same pH value, this is achieved by adding more sugar. Addition of urea to a gel cancels out hydrogen bonds and results in a weaker gel with a lower setting temperature [144].

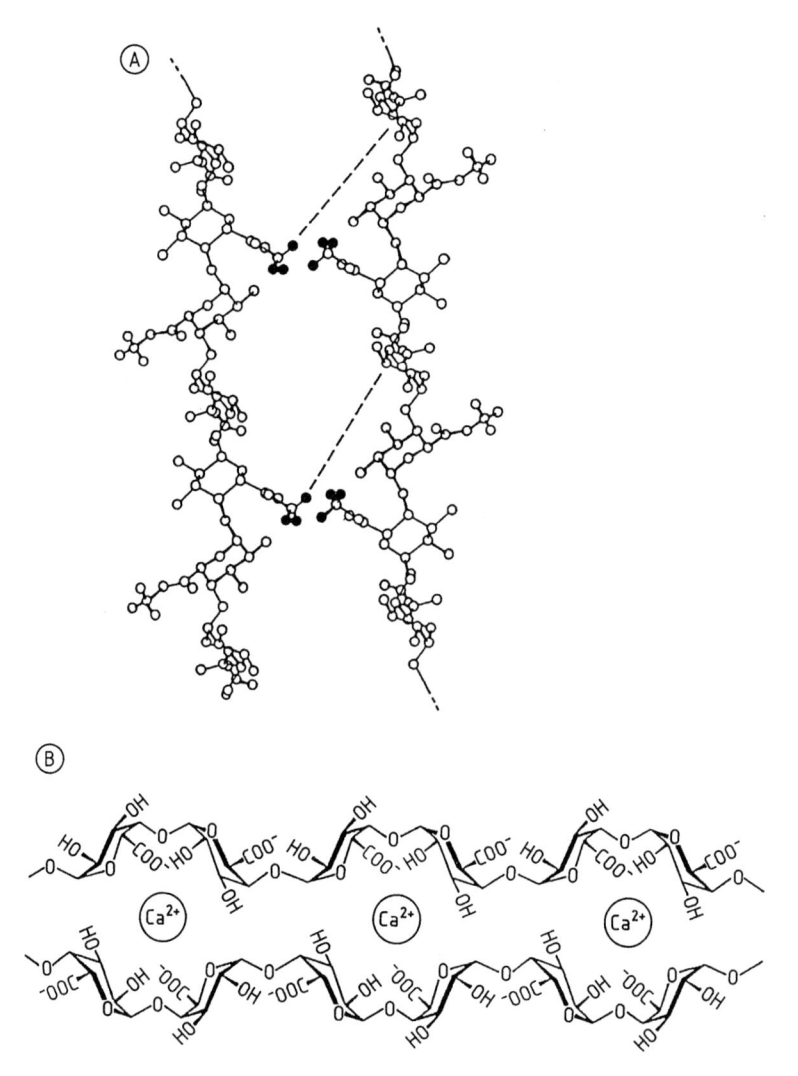

Figure 7. A) Junction zone in high ester pectin gel by hydrogen bonds (dotted lines) and hydrophobic interactions (filled circles); B) Junction zone in low ester pectin gel by dimerization of polygalacturonate (polyguluronate blocks) induced by their strong binding power for Ca^{2+} ions which fill the oxygen-lined cavities between the polysaccharide chains [110]

In general, pectin–sugar–acid gels are prepared by a boiling process followed by cooling. At a certain temperature the system sets to a gel. The food technologist is interested in the rate of setting. At the same rate of cooling the rate of setting determines the setting temperature or the setting time. The parameters that contribute to stronger gels also accelerate the setting rate. Based on the pH limits of gelling as well as on the setting rate, HM-pectins can be subdivided into rapid-set (DM > 70%) and slow-set (DM 60–65%) pectins. At the same sugar concentration, rapid-set pectins have a higher pH limit for gelling because these very highly methyl esterifiedpectins have few carboxyl groups that must be protonated. A fully methyl esterified pectin gels with sugar alone and does not need acid. The concentration of pectin in the gel influences the rheological properties of the gel, not the gelling rate.

LM-pectins form so-called *calcium pectate gels*. Theories about the chemical structure of calcium pectate gel junction zones were first developed in comparison with alginate gels (see Chapter 4). In calcium alginate gels, the junction zones are formed by α-1,4-L-polyguluronate blocks in which the diaxial configuration of the glycosidic linkages leads to a buckled ribbon with limited flexibility and a strong binding power for Ca^{2+} ions, which induce dimerization of alginate chains by filling the oxygen-lined cavities between them. This has evoked the picture of an eggbox, and the expression eggbox-type junction zones has become universally accepted [111], [141], [145], [146] (Fig. 7). In comparing the primary structures of poly-L-guluronate and poly-D-galacturonate (pectin), they are seen to be stereochemically analogous mirror images of each other, except at C-3. The two-fold helix, however, has not been observed with X-ray diffraction techniques, but molecular modeling calculations indicate that it can exist [60]. Circular dichroism data [61] suggest that a conformational change takes place when dissolved or Ca^{2+}-gelled pectate is dried and it is suggested that a transition takes place from a 2_1 ribbon-like to a 3_1 helical symmetry. Similar to the low-water-activity gels, junction zones in calcium pectate gels are terminated by rhamnosyl residues in the backbone, side chains attached to the backbone, or acetyl groups. The presence of some methoxyl groups does not inhibit formation of eggbox-type junction zones. With LM-pectins, acids or sugars are not so important. Gels of acidic fruit juices with LM-pectins can be made by addition of a calcium salt (low-sugar jams) or of milk with its neutral pH and calcium ions (desserts). The amount of calcium necessary for gelation depends on the degree of esterification, the way the LM-pectin has been prepared, and the types and amounts of other ingredients. Coagulation as a result of the addition of calcium salts must be absolutely avoided. Slow availability of Ca^{2+} for the pectin molecules is a prerequisite for obtaining a gel network. This can be accomplished in various ways: (1) an insoluble calcium salt (phosphate, citrate, tartrate) may be used resulting in a slow exchange of Ca^{2+} ions with the LM-pectin and formation of a gel; (2) use of calcium chelating agents such as diphosphates help to retard the availability of Ca^{2+} ions. The fact that calcium pectate gels and precipitates are often thermoreversible (i.e., they are soluble under conditions of gel formation at high temperature) can also be used to advantage. Soluble calcium salts such as calcium lactate or calcium chloride can therefore be added at boiling temperature, and gelling occurs on cooling. LM-pectins can be solubilized in milk by heating; the calcium caseinate of the milk provides the calcium necessary for the system to gel on cooling. Addition of sugars to such gels gives stronger gels; however, at higher concentrations the risk of coagulation increases. A solution to this problem is offered by *amidated pectins*. Gel formation with amidated pectins is also explained by chain associations via eggbox junction zones. Eggbox junction zones can accommodate amide groups, which, however, provide less drive for Ca^{2+} ion binding [146], [147]. On the one hand, they need fewer Ca^{2+} ions for

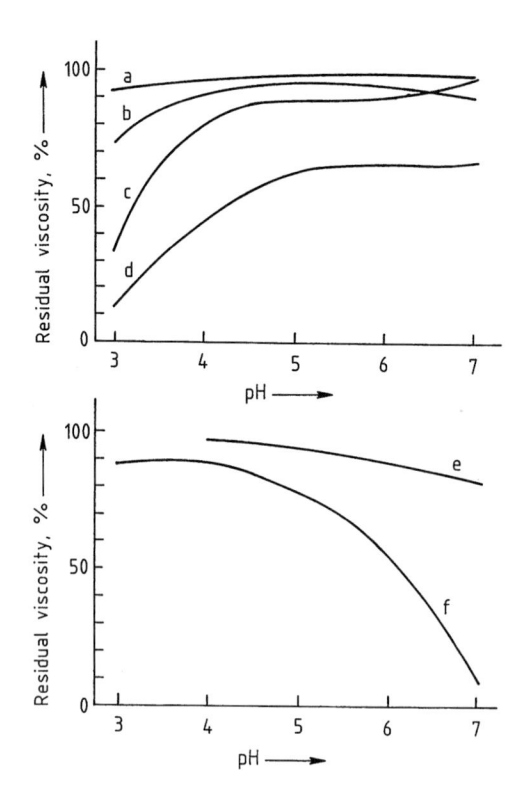

Figure 8. Stability of some polysaccharides at various pHs [131]: Residual viscosity after 10 min incubation at 90 °C
a) Carboxymethylcellulose – methyl cellulose;
b) Locust bean gum; c) Agar; d) Carrageenan;
e) Pectate; f) Pectin

gelation, and on the other hand in the presence of excess Ca^{2+} ions they are not as sensitive to coagulation [148]. In the United States, amidated LM-pectin is used for all applications of LM-pectins.

3.4.3. Stability and Chemical Reactions

Stability of pectin molecules in aqueous solution depends upon the temperature and the pH. Pectin has, in contrast to most other hydrocolloids, optimal stability at pH 3.5 to 4. Figure 8 shows the stability of pectin and some other thickening agents at various pH values, expressed as residual viscosity of buffered solutions after heating for 10 min at 90 °C [131]. As is evident from the figure, highly esterified pectin is vulnerable to high pH. Even at pH 5, depolymerization is considerable, in particular at elevated temperatures [133]. Consequently, it is difficult to raise the pH of pectin solutions without causing a decline in the average degree of polymerization, because when trying to mix an alkaline solution into a pectin solution, too high pH cannot be avoided locally. It is recommended to ensure good agitation and low temperature, and to avoid the use of hydroxides.

Low pH hydrolyzes ester bonds causing a decline in DE as well as in the content of O-acetyl groups, and it causes a decline in degree of polymerization by hydrolysis of

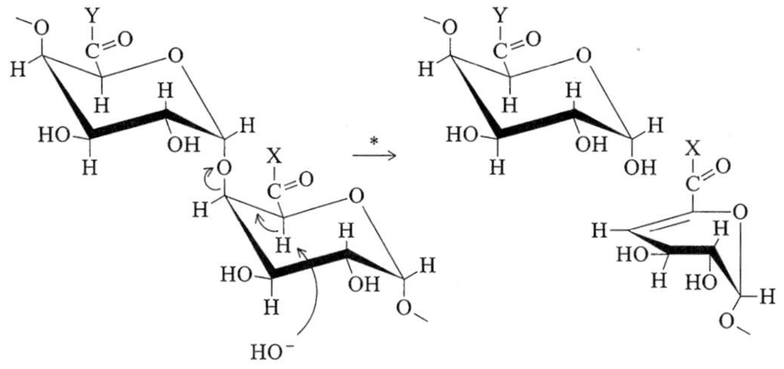

Figure 9. β-Eliminative depolymerization of a galacturonan chain by pectin lyase (PL) or pectate lyase (PAL) or chemically at pH ≥ 5

* Enzymatically
X = OCH$_3$, Y = OCH$_3$ Pectin lyase (PL)
X = OH, Y = OH or OCH$_3$ Pectat lyase (PAL)

Chemically
X = OCH$_3$, Y = OCH$_3$ or OH   pH > 5

glycosidic bonds [64], in particular at rhamnose insertions in the molecular backbone [65]. Very high acid concentration may degrade galacturonic acid to CO_2, furfural, reductic acid (2,3-dihydroxy-2-cyclopenten-1-one; $C_5H_6O_3$) and alginetin. Carbon dioxide production which is quantitative by boiling in 12 mol/L HCl has in the past been used for the quantitative determination of pectin [85], [132].

High pH depolymerization is due to β-elimination [66] (Fig. 9), it requires the presence of a methyl ester group at the anhydrogalacturonic acid residue which has its 4-C attached to the bond being split. Since the presence of methyl esters is required for β-elimination, vulnerability to this degradation mechanism is related to the DE. High pH further reduces DE (whereby the β-elimination becomes incomplete) as well as the content of O-acetyl groups.

Derivatization. The low-pH hydrolysis of natural ester linkages may be reversed under conditions of low water activity, e.g., in mixtures of methanol and concentrated sulfuric acid [67], [68], [126], or, for introduction of O-acetyl groups, mixtures of concentrated sulfuric acid and acetic anhydride [69]. Further, the carboxyl group may readily be esterified with glycol or glycerol but poorly with ethanol [135]. By using polyols, cross-linked, insoluble systems are obtained.

At high pH, ammonia may convert methyl esterified carboxylate groups to amides [42], [129]. This is used industrially since amidated pectins are of commercial importance.

Insoluble pectates can be prepared by cross-linking with epichlorohydrin; these pectates have ion-exchange properties with a certain selectivity for calcium and heavy-metal ions. They are successfully used for the isolation of pectolytic enzymes [136], [137].

Pectins are readily degraded by oxidants [138] except for chlorite and chlorine dioxide which can selectively oxidize aldehyde groups at the reducing chain end [139].

3.5. Analysis

3.5.1. Measurement and Standardization of Gel-Forming Capacity

HM-pectins are generally standardized to uniform strength at specified constant conditions. Expressing the sugar binding capacity of the pectin, the USA – SAG method suggested by the IFT Committee for Pectin Standardization, has been universally accepted for grading HM-pectins [149]. A standard gel is prepared in conical test glasses with the following conditions: soluble solids 65%, pH 2.20–2.40, gel strength 23.5%, SAG measured with a *ridgelimeter* [106], [152]. After 24 h at 25 ± 3 °C the gel is deposited on a glass plate, and the sagging of the gel under its own weight is measured after 2 min. From the SAG value and the pectin quantity, the grade can be calculated. Most commercial HM-pectins are standardized to 150 grade USA – SAG. (A gel strength of 150 grade SAG implies that 1 part of pectin is able to transform 150 parts of sucrose into a jelly with above standard properties.)

LM-pectin may be standardized by closely analogous procedures, however, no universally accepted method exists. With LM-pectins it is difficult to set up a single universal test because the conditions under which LM-pectins are used may differ widely with respect to soluble solids, calcium content, and pH (see also Section p. 3982.

3.5.2. Chemical Analysis [70]

In addition to gel-forming ability, the analysis of pectin preparations is concerned particularly with the *degree of of esterification, DE*. This is determined by converting the pectin to its acid form by passing it over an ion-exchange resin or washing it in an alcohol suspension, first with hydrochloric acid – alcohol and then with neutral alcohol. The acid and saponification equivalent is determined by titration, and from these values the anhydrogalacturonic acid content and the DE (in percent) can be calculated [63]. This principle for determination of DE has been adopted by the major legislative bodies [71]–[73]. Presence of *O*-acetyl groups in the pectin will result in an overestimation of DE as well as the anhydrogalacturonic acid content, therefore, it should be evaluated in a separate aliquot of the preparation. The pectin can also be precipitated with copper ions before and after saponification to determine the copper in the well-washed precipitate. From the amount of copper ions bound to the original pectin and the saponified pectin the anhydrogalacturonic acid and the DM can then be calculated without the interference of acetyl groups [150], [151]. Various methods exist for the separate determination of methyl ester and acetyl groups. Methanol is released on saponification of methyl ester and can be determined by GLC either directly [152] or in the headspace of a closed vial after conversion of the methanol to volatile methyl nitrite

[153], [154]. *Acetyl groups* can be conveniently determined by an enzymatic spectrophotometric method (supplier Boehringer), by GLC [155], or by distillation and titration after alkaline saponification [156]. A convenient new method is to saponify the pectin preparation in alkaline alcohol, which is then analyzed by HPLC for methanol and acetic acid [157]. Methods measuring the relative content of carboxylate groups to total material, either using size exclusion chromatography with combined detection by conductivity and refractive index [75], or using capillary electrophoresis [76], have also been published. The DE can be inferred from calibration curves if it can be assumed that the anhydrogalacturonic acid content is the same in the samples compared. Previously, the analytical methods used by legislative bodies did not comprise a correction for *O*-acetyl groups since the content of those is small in citrus pectin and modest in apple pectin, but it has now been included in the latest version of the FAO/WHO specification for pectin [71].

In commerical pectins, up to 25% of the carboxyl groups may be amidated. The *degree of amidation* is calculated from the amount of ammonia released on alkaline distillation [63], [158]. Amidated pectic acids undergo β-elimination reactions, whereas pectic acid does not, which permits quantitative analysis of mixtures of amidated and nonamidated pectins [159]. The anhydrogalacturonide content of pectins can also be determined from aqueous solutions by colorimetric methods with carbazole [160], the more specific *m*-hydroxydiphenyl [161], or sulfamate – *m*-hydroxydiphenyl [162]. Sometimes corrections must be made for interfering compounds (neutral sugars, amide groups, azide). The colorimetric methods can be automated easily and used for routine analyses of large series of samples [130].

With the above-mentioned methods, only the galacturonide residues in the backbone can be analyzed. The *neutral sugars* can be analyzed conveniently by gas chromatography after acid hydrolysis and conversion to volatile derivatives. By a preceding precipitation of pectins with copper ions the neutral sugars covalently attached to the galacturonan can be analyzed specifically. If starch is present it can be removed by enzymatic degradation [163].

To analyze the *pectin of plant material* the so-called alcohol-insoluble residue is usually prepared first by washing the plant material with refluxing alcohol. This inactivates endogenous enzymes and removes alcohol-soluble constituents. The pectin content is then determined in extracts of the alcohol-insoluble residue. The total pectin is determined in an alkaline extract or in the combined extracts of enzymatic and acid extraction. Another approach is the gradual extraction first with cold or hot water (HM-pectin), oxalate, ethylenediaminetetraacetic acid (EDTA), or cyclohexanediamine tetraacetate (CyDTA) (LM-pectin), and then acid or alkali (protopectin) [164], [165]. The total pectin content and the average degree of methoxylation can also be determined in the alcohol-insoluble residue when this is converted to the acid form before and after saponification. When treated with alcoholic calcium acetate solutions, the free carboxyl groups of pectin set free an equivalent amount of acetic acid that can be determined by titration. An alternative is determination of the bound copper ions from copper solutions [150].

Molecular mass of pectin may be determined with viscosimetry [44], [77], membrane osmometry [78], size exclusion chromatography [47], [48], [54], light scattering [54], [77], [79], [80], ultracentrifugation [34], and analysis of reducing end-groups [80]. Quoted results vary, partly because the intermolecular distribution of molecular mass is broad, partly because pectin molecules aggregate and may contain slight amounts of insoluble material, and, of course, partly because samples are different. As an example, a weight average molecular mass of 90 000 ±10 000 was reported by HARDING et al. [34]. Quoted values for the Mark–Houwink exponent, relating intrinsic viscosity to molecular mass, are generally in the vicinity of 0.8 [47], [49], [53], [54], suggesting a random coil molecular shape. Integrated systems combining high performance size exclusion chromatography, viscosity detection and light scattering detection are commercially available, including software with which data for molecular mass and Mark–Houwink parameters can be extracted.

3.6. Pharmaceutical and Nutritional Characteristics [166]

Pectin is not significantly degraded in the upper digestive tract of humans, and it can be recovered almost intact from the small intestine [81]. In the cecum and colon, it is fermented by microorganisms mainly to short-chain fatty acids, as can be concluded from *in vitro* studies [82], rat studies [83] and studies comparing the degradation patterns of humans and rats [84]. Since pectin is a dietary fiber, much attention has been paid to the possible health benefits of pectin which are: (1) reduced glycemic response [86]–[88] (2) prolonged gastric residence time [89], [90], [91], (3) reduced serum cholesterol level [86], and (4) effect against diarrhoea [92]. Pectic polysaccharides from certain botanical sources like ginseng root (*Panax ginseng*) [93], eel grass (*Zosteraceae*), and *Bupleurum falcatum* [94], [95] have shown healing effect on gastric and duodenal ulcers. Most studies, e.g., [97]–[99], conclude that pectin, in spite of its metal-binding ability, apparently does not inhibit the uptake of minerals from the diet. All of the above effects must be thought of as general tendencies in a vast amount of published findings which are not all mutually consistent. Some discrepancies may be explicable because it is attempted to generalize results achieved with different systems (*in vitro*, animal, or human) and with different pectins or pectin-rich plant materials. Publications often fail to specify important details about the pectin being used for the study, such as botanical origin, DE, etc.

3.7. Application in the Food Industry [167]

Indigenous manufacture of jams and marmalades, before commercial stabilizers were available, involved the use of pectin-rich fruit and in situ extraction of the pectin during prolonged cooking. Partly owing to this tradition, but mostly due to its superior stability and gelling ability at relatively low pH, pectin is the dominating gelling agent in modern production of jams as well as other products which are gelled, acidulous, and sweet. Examples are jelly fruits, and fruit preparations for industrial production of fruit-containing yogurt (Table 2 groups 1 and 2). Commercial pectins for these applications are tailor-made to yield specific gelation temperatures or gelling rates under specified conditions, and to exhibit specific functionalities such as heat reversibility, heat resistance, firmly gelled textures, pumpable semi-gelled textures, etc. *HM pectin* is used as a stabilizer in yogurt beverages and beverages in which milk proteins are heat-treated at relatively low pH during the production [100] – [102]. LM pectin finds use for thickening spoonable yogurt (Table 2 group 6). Particles of calcium pectinate are used as a substitute for fat in low-calorie foods [103] (Table 2 group 12). Other applications from Table 2 include groups 4, 8, 9, 10, and 11.

3.8. Market

Annual pectin production is estimated at 25 000 t (80% citrus pectin), sold mostly in standardized form. Pectin production takes place in Brazil, Denmark, France, Germany, Mexico, Switzerland, and United Kingdom. Smaller amounts are produced in Eastern Europe and the former Soviet Union. Average prices on the world market are $10 – 11 per kilogram HM-pectin and $ 12 – 13 per kilogram LM-pectin.

4. Alginates

Alginate is a collective term for a family of linear 1,4-linked α-L-gulurono-β-D-mannuronans of widely varying composition and sequential structures [168]. Commercial preparations are usually designated as alginates and include alginic acid, its salts, and derivatives [2], [3], [5], [169] – [174].

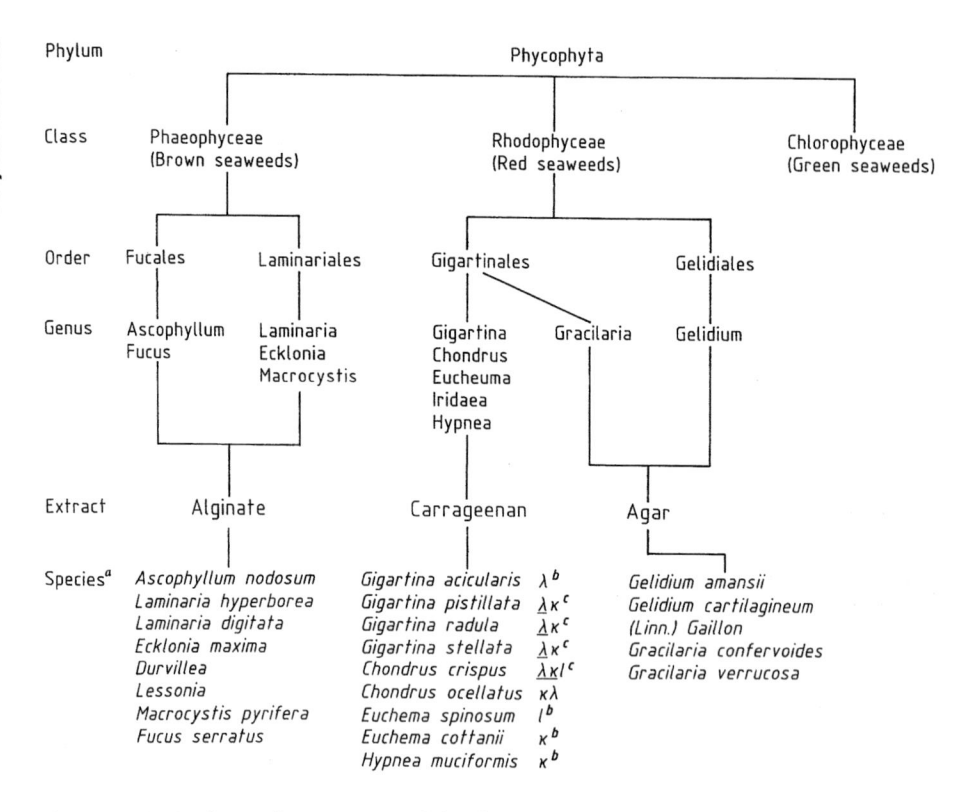

Phylum		Phycophyta		
Class	Phaeophyceae (Brown seaweeds)		Rhodophyceae (Red seaweeds)	Chlorophyceae (Green seaweeds)
Order	Fucales	Laminariales	Gigartinales	Gelidiales
Genus	Ascophyllum Fucus	Laminaria Ecklonia Macrocystis	Gigartina Gracilaria Chondrus Eucheuma Iridaea Hypnea	Gelidium
Extract	Alginate		Carrageenan	Agar
Species[a]	Ascophyllum nodosum Laminaria hyperborea Laminaria digitata Ecklonia maxima Durvillea Lessonia Macrocystis pyrifera Fucus serratus		Gigartina acicularis λ[b] Gigartina pistillata $\underline{\lambda}\kappa$[c] Gigartina radula $\underline{\lambda}\kappa$[c] Gigartina stellata $\underline{\lambda}\kappa$[c] Chondrus crispus $\underline{\lambda}\kappa\iota$[c] Chondrus ocellatus $\kappa\lambda$ Euchema spinosum ι[b] Euchema cottanii κ[b] Hypnea muciformis κ[b]	Gelidium amansii Gelidium cartilagineum (Linn.) Gaillon Gracilaria confervoides Gracilaria verrucosa

Figure 10. Origin of seaweed extracts — general classification
[a] Species of economic significance; [b] Contains only component mentioned; [c] Contains predominantly underlined component

4.1. Occurrence and Structure

Alginates form a group of polysaccharides that occur as structural components of the cell walls of brown seaweed (Phaeophyceae; Fig. 10) in which they make up to 40% of the total dry matter and play an essential role in maintaining the structure of the algal tissue. They were isolated for the first time in 1880 by STANFORD [175], who also introduced the names algin for the soluble substances in an aqueous sodium carbonate extract and alginic acid for the material that could be precipitated from this extract by addition of acid. Until 1954, alginate was considered a polymannuronic acid. In 1955, FISCHER and DÖRFEL [176] discovered L-guluronic acid in addition to D-mannuronic acid, and since ca. 1964, alginic acid has been known to be a copolymer of these two residues. In principle, alginates are composed of three structural elements: the homopolysaccharides α-1,4-L-guluronan and β-1,4-D-mannuronan, and a heteropolysaccharide consisting of alternating 1,4-linked α-L-guluronic (G) and β-D-mannuronic acid (M) residues [177] (Fig. 11). The structure of alginate can therefore be represented schematically as

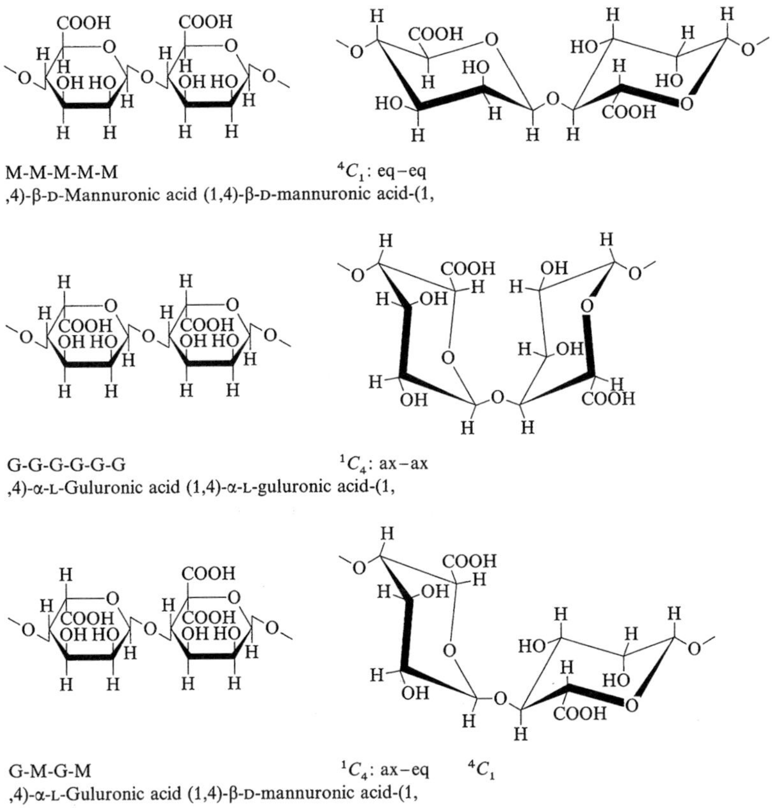

M-M-M-M-M
,4)-β-D-Mannuronic acid (1,4)-β-D-mannuronic acid-(1,
$^4C_1$: eq–eq

G-G-G-G-G-G
,4)-α-L-Guluronic acid (1,4)-α-L-guluronic acid-(1,
$^1C_4$: ax–ax

G-M-G-M
,4)-α-L-Guluronic acid (1,4)-β-D-mannuronic acid-(1,
$^1C_4$: ax–eq $^4C_1$

Figure 11. Structural units in alginates

$$-M-G-M-(M-M)_n-M-G-(M-G)_q-M-G-(G-G)_p-G-M-G-$$

Alginates, like pectins, are linear polymers. The D-mannuronic acid residues are in the $^4C_1$ configuration, and therefore the glycosidic linkage between them is equatorial–equatorial. Since the L-guluronic acid residues are in the $^1C_4$ configuration, they are connected by axial–axial glycosidic linkages similar to the linkages between galacturonosyl residues in pectin. However, unlike pectin, alginates do not contain neutral sugar residues. The properties of alginates are determined largely by their molecular mass and by the ratio in which the three structural elements occur in the polymer. This ratio is determined by the variety of seaweed from which the alginate is extracted [171], [174], [178]–[181].

Alginate production has also been reported in bacteria (i.e., *Pseudomonas aeruginosa* (pathogenic), *P. mendocina*, *P. putida*, *P. fluorescens*, and *Azetobacter vinelandii*). These alginates are also formed by the same three structural elements and are further characterized by the presence of *O*-acetyl groups. For *A. vinelandii* it has been shown that polymannuronic acid is synthesized first; mannuronic acid residues can then be

3995

Table 3. Species of brown seaweeds (Phaeophyceae); alginate content and ratio between mannuronic and guluronic acid residues (data from various sources

Species	% Alginate DM*	Man : Gul	
Macrocystis pyrifera	13 – 24	1.4 – 1.8	
Ascophyllum nodosum	20 – 30	1.4 – 1.9	
Laminaria digitata	15 – 27	1.3 – 1.6	
Laminaria (leaf)	15 – 26	1.3 – 6.0	
Hyperborea (stem)	27 – 33	0.4 – 1.0	
Ecklonia maxima	30 – 38	1.4 – 1.8	
	Mannuronan	Guluronan	Alternating
Macrocystis pyrifera	40.6%	17.7%	41.7%
Ascophyllum nodosum	38.4%	20.7%	41.0%
Laminaria hyperborea	12.7%	60.5%	26.8%
Azetobacter vinelandii (acetylated)	17.8%	0.5%	81.7%

* DM = degree of methoxylation.

transformed to guluronic acid residues by a mannuronan C-5 epimerase. With this enzyme, bacterial and algal alginates can be epimerized. The activity of the enzyme depends strongly on the Ca^{2+} concentration [168], [182], [183].

The *biosynthesis in plants* starts from D-mannose which, via guanosine diphosphate mannose, is oxidized to guanosine diphosphate mannuronic acid by a NAD dehydrogenase system and then polymerized to polymannuronic acid by a transferase enzyme. In homogenates of brown seaweed, guanosine diphosphate guluronic acid has also been identified, so at the end of the pathway copolymerization may also occur [172].

4.2. Production

The production of alginates is shown schematically in Figure 12. Current processes are still based on patents that were filed in the 1930s by U.S. companies [184], [185]. Alginate production is based on a series of ion-exchange processes: The water-insoluble calcium alginate in the raw material is, for the purpose of extraction and purification, converted first to the soluble sodium form and then, possibly via the calcium form, to insoluble alginic acid, which is neutralized to obtain the finished product. Various brown seaweed species can be used as raw materials (Table 3). On the west coast of North America the giant kelp *Macrocystis pyrifera* is harvested mechanically by special ships, which cut the weeds at a depth of 1 – 2 m. The weeds grow so fast that they can be harvested several times a year. Other species are harvested semimechanically by fishing boats, collected by reaping hooks, harvested from shore at low tide, or collected directly from the beach. Many regions have a tradition of harvesting or collecting brown seaweed, which in the past was used for the extraction of iodine and sodium carbonate.

The weeds are dried mechanically, ground, and then transported to factories. Salts, which might influence the solubility, and other impurities such as the high-polymeric

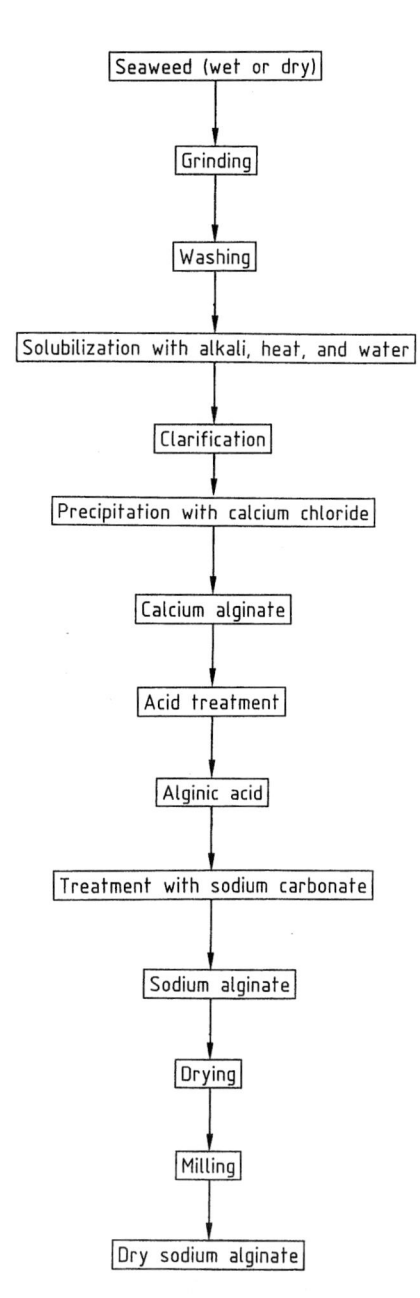

Figure 12. Flow sheet for the production of sodium alginate [170]

```
Seaweed (wet or dry)
        ↓
    Grinding
        ↓
    Washing
        ↓
Solubilization with alkali, heat, and water
        ↓
   Clarification
        ↓
Precipitation with calcium chloride
        ↓
  Calcium alginate
        ↓
   Acid treatment
        ↓
    Alginic acid
        ↓
Treatment with sodium carbonate
        ↓
   Sodium alginate
        ↓
      Drying
        ↓
      Milling
        ↓
  Dry sodium alginate
```

laminarin (a β-1,3; β-1,6-D-glucan) are removed by washing. The alginate remains in an insoluble form as the calcium salt or as alginic acid. Extraction of the material with cold or hot soda follows, combined with a mechanical disintegration. A homogeneous mass with a pH of ca. 10–11 is obtained. After dilution with water, the slurry separates into a liquid phase (sodium alginate) and a solid phase consisting predominantly of

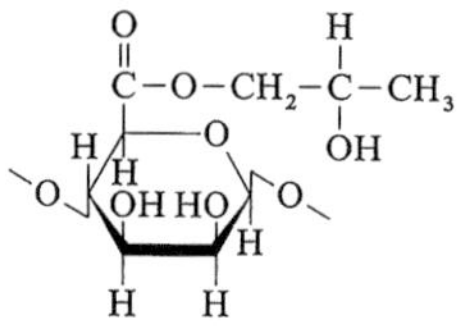

Figure 13. Propylene glycol mannuronate unit

cellulose. For removal of these solid impurities, a flotation step may precede the classical separation processes such as sieving and filtration. Alginic acid is obtained from purified sodium alginate by direct precipitation with acid or by precipitation with Ca^{2+} as calcium alginate and subsequent conversion to alginic acid by washing with acid. For the destruction of chloroplasts, either the sodium alginate solution or the calcium alginate precipitate is bleached with hypochlorite. Washed and mechanically dewatered alginic acid can be dried as such; in general, however, it is treated with sodium carbonate or other bases to produce alginate salts. At this stage, *propylene glycol alginate* can be prepared by treating partially neutralized alginic acid with propylene oxide [186] (Fig. 13). The finished products are highly refined, odorless white powders, which are permitted as food additives. For industrial purposes, less purified preparations are often sufficient.

4.3. Properties [187]

The properties of alginates depend largely on the degree of polymerization and the ratio of guluronan:mannuronan blocks in the molecules. The mannuronan regions in the molecules with their diequatorial glycosidic linkages are stretched and flat, while the diaxial glycosidic linkages in the guluronan regions give this part of the alginate molecule a buckled ribbon shape with limited flexibility. Alginic acid itself is insoluble in water and precipitates at pH <3.5. Ammonia, alkali and magnesium salts, and alginate salts with organic bases are water soluble; calcium ions and other multivalent ions precipitate with alginates. Depending on the degree of esterification, *propylene glycol alginate* can be water soluble at a pH as low as 2.5 and less sensitive to cations. *Alginates* are insoluble in organic solvents and can be precipitated with alcohols. They are, however, slightly miscible with simple alcohols, glycerol, and glycols, and these compounds can therefore be included to a certain content in alginate solutions. The dissociation constant pK of monomeric mannuronic acid is 3.38, and of guluronic acid 3.65. Alginic acid with a high proportion of guluronan has a pK of 3.74, when the proportion of mannuronan is high, this value is 3.42. Alginates are very stable in the pH range 5–10; only at higher pH values is viscosity loss by a β-eliminative degradation observed. *Propylene glycol alginate* behaves differently due to the presence of ester groups, which facilitate β-eliminative degradation starting at pH 5 and during heating, similar to pectins [188]. They are also more stable at acid pH values, and therefore can be used to thicken and stabilize acidic products. The acid stability of propylene glycol alginate increases with degree of esterification, but in strong acid the alginate chains depolymerize, with a resultant loss of viscosity [170].

Glycosidic linkages between mannuronic and guluronic acid residues are known to be less stable to acid hydrolysis than the linkages between two mannuronic or two guluronic acid residues [189].

Oxidation and Degradation. By using high acid concentrations and high temperatures, alginates and alginate esters can be quantitatively decarboxylated. Oxidants such as halogens or periodate, and redox systems such as polyphenols, ascorbic acid, or thiols, are able to degrade alginates. For production of low viscosity alginates, hydrogen peroxide is used. Bacterial alginate depolymerases have also been described [172], [190]. These enzymes have been used to study the distribution of L-guluronic and D-mannuronic residues along the backbone. They have no commercial significance; however, they condition the chemical conservation of alginate solutions. The degree of polymerization of alginates is very high and amounts to 800 in commercial preparations.

Viscosity. The high viscosity of aqueous alginate solutions is one of their most important properties (Table 4). By mixing alginates of different sources, and by oxidative degradation, manufacturers can offer products with viscosities varying from 100–5000 mPa · s. Viscosity numbers alone are meaningless without the shear stress or shear rate because alginate solutions are Newtonian fluids only at very low shear rates and low concentrations. Under other conditions, they are pseudoplastic. For *propylene glycol alginate* the pseudoplastic flow behavior starts only at concentrations >3 wt%. Table 4 also shows the strong concentration dependence of viscosity. Cations have a significant effect on viscosity. Low concentrations of Ca^{2+} ions considerably increase the viscosity as a result of complex formation. In measuring the viscosity of alginate solutions a distinction is made between the "direct" viscosity and the "sodium hexametaphosphate" viscosity. The latter is always lower because hexametaphosphate sequesters Ca^{2+} ions. In very pure alginate preparations, this effect is absent. The use of sequestrants as a protective measure is common in most applications of alginates.

The high ability to *stabilize suspensions* is often ascribed to the high viscosity of alginate solutions. However, such systems could also be stabilized by a yield stress value that exceeds the very low shear stresses originating from the action of gravity on the suspended particles. This is certainly the case for the thixotropic, almost imperceptibly gelled systems, that form in the presence of low concentrations of Ca^{2+} ions.

Gel Properties. In food applications, sodium alginates are used as gelling agents, which in many aspects resemble LM-pectins and also form gels with Ca^{2+} ions. As pure polyuronic acids, alginates are, however, more sensitive to acid than pectins and are insoluble at pH <3.5. Alginates also have a higher sensitivity to calcium than pectins, which is probably due to their higher degree of polymerization. For the preparation of acid gels, calcium salts are used that are soluble only under acidic conditions, together with compounds that release H^+ ions slowly, for instance, slowly soluble fumaric acid or

Table 4. Dynamic viscosities (in mPa · s) of polysaccharide solutions at different concentrations (data from various sources [4], [41], [62], [63])*

Concentration	Pectin	Alginate	Carrageenan	Agar	Gum arabic	Gum tragacanth	Gum karaya	Gum ghatti	Xanthan	Locust bean gum	Guar gum
1 %	50	214	57	4		54	3 000	6	2 000	59	3 025
2 %	200	3 760	397	25		906	8 500	10	7 000	1 114	25 060
3 %	550	29 400	4 411			10 605	20 000	40	11 500	8 260	111 150
4 %		39 660	25 356	400		440 265	30 000	60		39 660	302 500
5 %			51 425		7	111 000	45 000			121 000	510 000
10 %					17						
30 %					200						
50 %					4 163						

* Reported value give only general trends. Depending on preparation and measuring conditions differences may occur.

slowly hydrolyzing glucono-1,5-lactone. The use of sequestrants has also been mentioned. The gel-forming ability of alginates is related mainly to the proportion of L-guluronan blocks. These blocks bind with Ca^{2+} ions according to a cooperative mechanism, giving rise to the formation of junction zones, which are referred to as the eggbox model (see Section 3.4.1). The junction zones in alginate gels are stronger than those in LM-pectin gels, and they are not thermally reversible even in the presence of sugar. Alginates rich in guluronan blocks form strong, brittle gels with a tendency to syneresis, whereas alginates rich in mannuronic acid form weaker more elastic gels that are less prone to syneresis [192]. Single mannuronic acid residues or mannuronan blocks interrupt the guluronan junction zones. *Propylene glycol alginates* form weaker gels. They can be prepared more easily because propylene glycol alginate is less sensitive to acid and Ca^{2+} ions and, therefore, less liable to coagulation. Alginates also gel in the presence of equimolar quantities of an HM-pectin. The strongest gels are obtained with HM-pectin and alginates with a high proportion of guluronic acid residues at low pH. Ca^{2+} ions are not required, and sugar is not essential for gelation; it does, however, affect gel properties [144], [193], [194].

4.4. Analysis

The alginate content in alginate preparations can be determined by titration as described for pectins. From this analysis the degree of esterification can be deduced if propylene glycol alginate is present. In mixtures or foods, the uronic acid content can be determined by the colorimetric methods used for pectins [161]. If necessary this analysis can be preceded by a purification step. Titration as well as colorimetric methods do not, however, distinguish between pectins and alginates. Pectins can be degraded in solution with specific enzymes whereupon they become alcohol soluble. Alginates can then be determined in the alcohol-insoluble residue. Polyuronides can also be precipitated from preparations, mixtures, or foods with Ca^{2+} ions and thus separated from other polysaccharides. After hydrolysis to monomeric uronic acid residues [180], the latter can be analyzed by gas chromatography [29] or HPLC [195]. This method also allows analysis of the uronic acid composition of alginates and, after partial hydrolysis, the proportions of mannuronan and guluronan blocks. The proportions of M and G residues in an alginate preparation and their distribution in M, G, and MG blocks can also be determined by ^{1}H-NMR [181], [196] and circular dichroism [197]. As a result of their wide range of applications, no internationally recognized standard methods exist to assess the quality of alginates. Manufacturers therefore usually specify the viscosity or gel strength of their products by methods agreed upon with clients.

4.5. Applications

Food Industry. The applications of *alginates* in the food industry are based on

1) Their potential to form highly viscous solutions with outstanding suspension-stabilizing properties
2) Their stability at high temperature and high pH
3) Their reactivity with Ca^{2+} ions, which enables them to gel
4) The thermal stability of these gels [167], [170]

As a result alginates are used in almost all groups listed in Table 2. For acid products, *propylene glycol alginates* are used. In all of these applications, lump formation must be considered when adding the alginate to an aqueous system, similar to pectins. When sugar is part of the ingredients, some may be admixed with the alginate. When the application allows the addition of alginate as a solution, the alginate can best be solubilized by dosing it gradually in the vortex of a powerful mixer. For laboratory use the alginate can be moistened with alcohol. For all listed applications, only a few grams of alginate are added per kilogram of product; for the stabilization of foams a few milligrams per kilogram suffice.

Other Applications. Twice as much alginate is used in applications outside the food industry. Except for groups 17, 20, 23, and 26, all applications in Table 2 have been described. Of greatest impact is group 31, which claims to cover ca. one-third of alginate sales. Groups 27 and 29 also represent a substantial market. One of the first applications of alginates was in the treatment of boiler water (group 32). The calcium is flocculated as calcium alginate and not deposited as scale. For this application, alginate paste is used, which does not have to be refined to the extent necessary for food applications. The strong swelling power of alginic acid makes it a good binding agent for tablets (group 24), and its calcium reactivity enables denture replicas to be made (group 25). The triethanolamine salt is often used as a thickening agent in cosmetic and pharmaceutical ointments. Another interesting application mentioned in the commercial literature is as a gel for embedding the roots of plants to protect them from mechanical damage and from dehydration after replanting. In the last decade alginates have been used in biotechnology research to immobilize bioactive cells and to purify enzymes [137].

4.6. Market

The alginate industry is limited to a small number of companies. Production plants are found in locations where suitable weed species are present (i.e., North America, Scotland, Brittany, and Japan). Alginate is also produced in the former Soviet Union. About 20 000 t of alginate is produced annually; ca. one-third of this is used in the food industry. The price of sodium alginate for food applications is ca. $ 9.6 – $ 12 per kilogram.

5. Carrageenan

Carrageenan is a collective name for the galactan polysaccharides that can be extracted from red seaweed (Rhodophyceae; Fig. 10). They contain galactosyl residues along with 3,6-anhydrogalactosyl residues; both sugar units may be partially esterified with sulfuric acid. Carrageenan is a relatively new product; industrial carrageenan production started after 1945 as a substitute for agar, which was at that time in short supply [11], [146], [172], [173], [198]–[200].

5.1. Occurrence and Structure

The seaweed species *Chondrus crispus* has been known for many centuries in the United States and Canada since the immigration of the Irish during the first part of the 19th century. These weeds were harvested, washed, bleached in the sun, dried, and boiled with milk to a pudding (blancmange). The expression "Irish moss" points to this origin, and *Irish moss* or *Irish moss extract* is also often used for the extracted and purified weed polysaccharide. The word carragheen is also of Irish origin; it is the name of an Irish coastal town.

Structure [168]. In commercial carrageenans, three types can be distinguished: λ-carrageenan, ϰ-carrageenan, and ι-carrageenan. All three types occur in commercial preparations; their amounts vary depending on the weed source, geographical origin, season of harvest, and extraction procedure. *λ-Carrageenan* is a linear chain molecule composed of the dimeric repeating unit β-D-Gal-(1,4)-α-D-Gal (Fig. 14). These units are 1,3-glycosidically linked. The primary hydroxyl group of the α-galactosyl residue is esterified with sulfuric acid, and 70% of the hydroxyl groups at C-2 in both galactosyl residues are also esterified with sulfuric acid. λ-Carrageenan therefore has a sulfate content in the range of 32–39%. Because the galactose molecules exist in the $^4C_1$ configuration, the 1,3-glycosidic linkage is axial–equatorial and the 1,4-glycosidic linkage is equatorial–axial.

ϰ- And ι-*carrageenan* are composed of the dimer *carrabiose* in which a β-D-galactosyl residue is 1,4-linked to an α-D-3,6-anhydrogalactosyl residue [β-D-Gal-(1,4)-α-D-3,6-Anhydrogal] (Figure 14). The carrabiose units are 1,3-linked to form a linear polymer. Because the anhydrogalactose has the $^1C_4$ configuration, all glycosidic linkages are diequatorial. The difference between ϰ- and ι-carrageenan is in the esterification with sulfuric acid. ϰ-*Carrageenan* is esterified at the hydroxyl group at C-4 of the galactosyl residue and has a sulfate content of ca. 25–30%. In ι-carrageenan the hydroxyl group at C-2 of the anhydrogalactosyl residue is esterified as well, and it therefore has a sulfate content of 28–35%.

Furcellaran, once considered a different red seaweed polysaccharide, also has carrabiose repeating units in its chain and is now classified as ϰ-carrageenan because of its

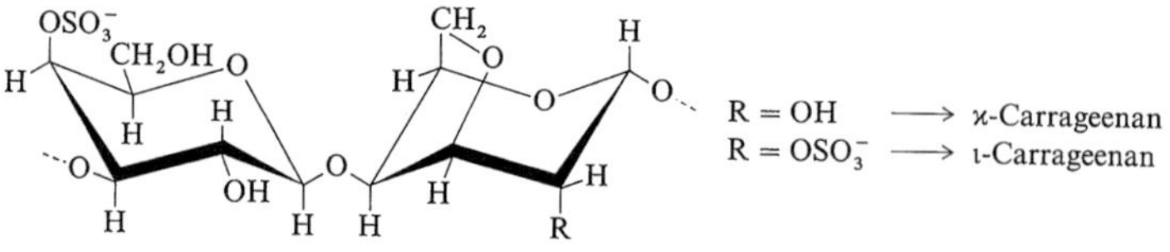

,3)-β-D-Galactose-4-sulfate (1,4) α-D-3,6-anhydrogalactose-(1,
$^4C_1$ Conformation $^1C_4$ Conformation
eq−eq

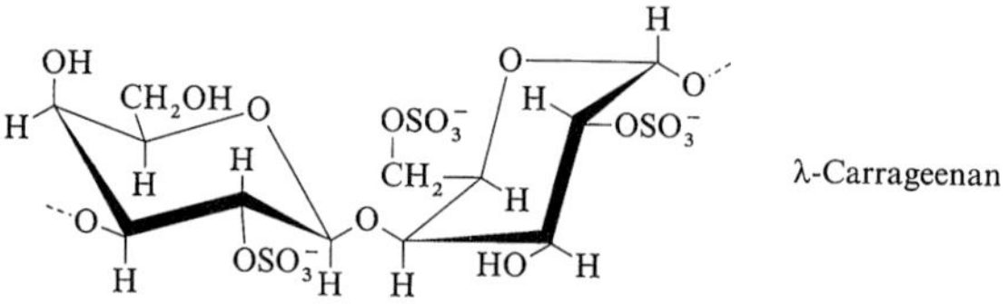

,3)-β-D-Galactose-2-sulfate (1,4) α-D-galactose-2,6-disulfate-(1,
$^4C_1$ Conformation $^4C_1$ Conformation
eq−ax

Figure 14. Dimeric repeating units in *λ*-, *ϰ*-, and *ι*-carrageenan

many similarities to this polysaccharide. It has, however, fewer sulfate ester groups. The various types of carrageenans can be separated on the basis of differences in solubility of their salts. This is also how they were discovered. Carrageenans have been shown to contain 0.1 – 1 wt% nitrogen, present as protein that is presumably covalently attached to the polysaccharide. Carrageenans must therefore be considered proteoglycans.

5.2. Production

The production of carrageenans is shown schematically in Figure 15. Carrageenan-yielding species are listed in Figure 10. The raw materials are harvested at low tide with special rakes or collected from the beach when washed ashore. Algae are also grown in tanks with circulating seawater or attached to nylon ropes. The latter method is practiced particularly along the coast of the Philippine islands to grow *Euchema cottonii*, an important raw material for *ϰ*-carrageenan [199]. After being harvested, the algae are washed and dried to a dry matter content of ca. 25 wt%; the carrageenan content is ca. 15% of the fresh weight. The dried algae are treated with alkali and ground to a paste. Alkaline conditions facilitate extraction of the macerated algae; they retard acid-catalyzed depolymerization of the galactan and catalyze the conversion of C-6 sulfated galactopyranosyl residues ($^4C_1$ conformation) to 3,6-anhydrogalactopyranosyl residues ($^1C_4$ conformation) and thus increases the yield of *ϰ*-carrageenan [201], [202]. The change in configuration has an important effect on the gelling properties of the carrageenan. Some authors believe that *ϰ*-carrageenan in algae is formed by a similar process, the only difference being that an enzyme acts as catalyst. Such an enzyme was

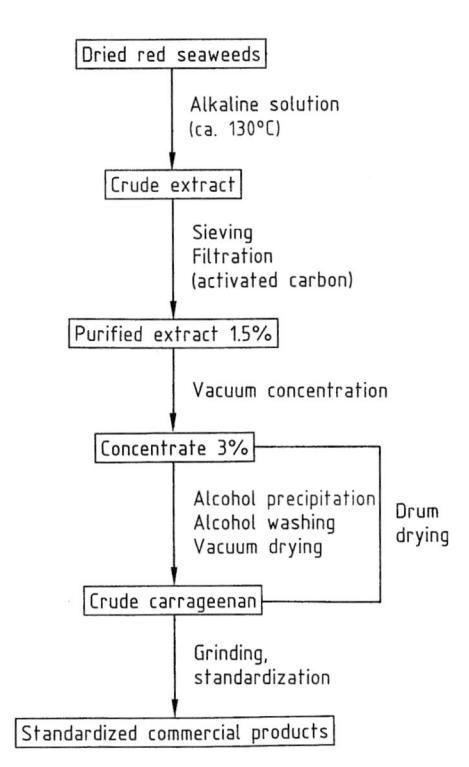

Figure 15. Flow sheet for the production of carrageenans

```
Dried red seaweeds
        │
        │ Alkaline solution
        │ (ca. 130°C)
        ▼
  Crude extract
        │
        │ Sieving
        │ Filtration
        │ (activated carbon)
        ▼
Purified extract 1.5%
        │
        │ Vacuum concentration
        ▼
  Concentrate 3% ─────┐
        │              │
        │ Alcohol      │ Drum
        │ precipitation│ drying
        │ Alcohol      │
        │ washing      │
        │ Vacuum drying│
        ▼              │
 Crude carrageenan ◄───┘
        │
        │ Grinding,
        │ standardization
        ▼
Standardized commercial products
```

indeed found in algae containing ϰ-carrageenan. This desulfatase enzyme has also been called "dekinkase" because it effects a change in the configuration of the transformed galactosyl unit and, as a result, the kink in the backbone by the equatorial–axial glycosidic linkage disappears when this changes to a diequatorial glycosidic linkage [203].

Indeed, ϰ-carrageenan is not an ideal copolymer; the theoretical anhydrogalactosyl content of 35% is never observed. ϰ-carrageenan of *Chondrus crispus* contains ca. 28% anhydrogalactosyl groups. The crude extract is further purified by sieving and filtration using a filter aid. If drum drying is anticipated, pigments are first removed with activated carbon. To prevent gelling, all of these operations must be carried out at higher temperatures. Because as with pectin and alginates, dilute extracts must be processed to a dry product, the energy cost for dewatering and recovery of alcohol is an important factor. Isopropanol is used as the alcohol for precipitation of the carrageenan.

To enable drum drying, mono- or diglycerides must be added to facilitate removal of the dried product from the drum. Because in this process, all soluble materials (including salts) remain in the product, a lower quality is obtained that can be used only for certain applications.

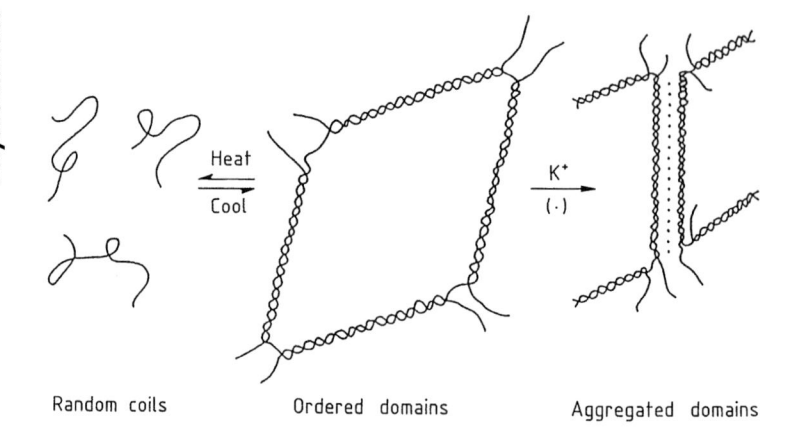

Figure 16. Gelation of carrageenans by formation of double helices and aggregation of helices [200]

The finished product is standardized by the addition of sugars. Other additives are sodium or calcium salts or locust bean gum for specific functional properties such as gelling and thickening.

5.3. Properties

All forms of λ-carrageenan as well as the sodium salts of $\varkappa$- and ι-carrageenan are soluble in cold water. The potassium and calcium salts of $\varkappa$- and ι-carrageenans, however, dissolve only at 70 °C and form gels or—depending on the ionic strength—viscous systems upon cooling (Table 4). $\varkappa$-Carrageenan is more sensitive to potassium ions, whereas ι-Carrageenan is more sensitive to calcium ions. From a solution of the three major types of carrageenans, $\varkappa$- and ι-carrageenan can be precipitated by potassium chloride (in the older literature $\varkappa$-carrageenan is distinguished from λ-carrageenan as the potassium-sensitive fraction). Evidence indicates that the diequatorial glycosidic linkages in ι- and $\varkappa$-carrageenan enable the association of the molecular chains into double helices that form upon cooling of hot solutions. In the presence of specific cations, these ordered domains aggregate into junction zones and form an extended network (i.e., a gel) (Fig. 16) [146], [204]. Another theory states that on cooling, intramolecular conformational changes occur in the carrageenan molecules and that gelation is based on selective cation interchain binding [205]. Upon heating, the helices unfold, the molecules go in solution again as random coils, and the gel melts. In the gel state the aggregation of double helices may continue, the network contracts, and the gel becomes brittle (short) and shows syneresis. A progressive retrogradation is halted by the presence of 1,4-linked D-galactosyl 6-sulfate or 2,6-disulfate residues in the $^4C_1$ configuration, which form equatorial–axial glycosidic linkages and produce a kink in the backbone. The *inability of λ-carrageenan to form*

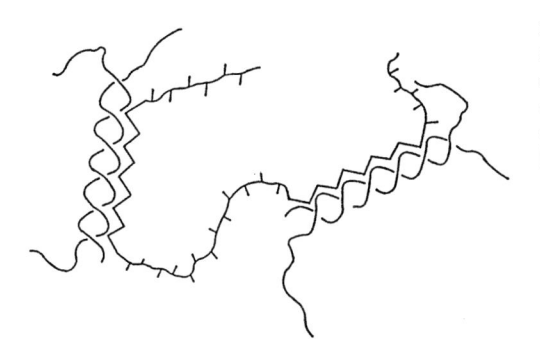

Figure 17. Synergistic gelation between helix-forming carrageenans and agar with galactomannans (locust bean gum) by direct association of smooth regions of the mannan chain with double helices in the seaweed polysaccharides [208]

gels can be explained by the occurrence of 1,4-linked D-galactose 2,6-disulfate units in its backbone. A solubilized preparation rich in *ϰ-carrageenan* gives a firm gel with a setting point of 40 °C and a melting point of 55 °C at a concentration of 0.5%, in the presence of 0.2% potassium salt. The gel strength increases strongly with increasing concentration of potassium ions; however, potassium ion concentrations >0.2% are unacceptable because of taste. Also, addition of sugar increases the gel strength. Both additons also increase the setting temperature, as well as the melting temperature, of the gels. The hysteresis (i.e., the difference between melting and setting temperature) remains small and is ca. 15 °C. The gels are brittle and have a tendency to become opaque and show syneresis. This can be prevented by adding ι-carrageenan. *Gels of ι-carrageenan* alone are transparent; they show no syneresis and little hysteresis. Due to the presence of 3,6-anhydro-D-galactose 2-sulfate residues, which may act as a wedging group and prevent tight aggregation of double helices, the gels are rather weak.

Of particular interest is the interaction of *ϰ-*carrageenan with locust bean gum galactomannan. By partially replacing the *ϰ-*carrageenan with locust bean gum, which does not gel on its own, a stronger gel with improved properties is obtained. It becomes more elastic, and has a lower tendency to syneresis and to become opaque. Similar observations were made with konjac glucomannans. The smooth regions of the mannan chain (i.e., regions with no galactose or glucose side groups) are thought to bind to the double helices of the *ϰ-*carrageenan, forming mixed junction zones [206], [207] (Fig. 17). These mixed junction zones have a lower tendency to form tightly packed aggregates. The industry has placed preparations on the market that exhibit constant gel strength and gel characteristics by blending different lots and adding potassium salts, sugar, and possibly locust bean gum. Because carrageenans may form lumps when added to water, premixing with sugar or adding the carrageenan in a powerful mixer is recommended.

Stability. The β-1,4-glycosidic linkages between galactosyl and anhydrogalactosyl residues are quite weak; as a result, carrageenans are very unstable in acid systems. At pH >4.5 they are, however, extremely stable, even under sterilization conditions. Their molar masses are reportedly $10^5 - 10^6$. Because of their gelation at intermediate temperature, their viscosity must be measured at 75 °C. At this temperature values up to

800 mPa · s have been measured for a 1.5 % solution; since carrageenan solutions have pseudoplastic behavior the shear rate must also be reported.

Reactivity with Proteins. The negatively charged carrageenans exhibit electrostatic interactions with positively charged proteins (pH below isoelectric point of proteins) or with positively charged sites on proteins (pH above the isoelectric point). Of commercial importance is their interaction with specific milk proteins, which provides the basis for the formation of milk gels or the stabilization of milk products. They have also been considered as precipitants of proteins from industrial waste streams, enzyme inhibitors, blood anticoagulants, lipemia-clearing agents, and immunologically active substances [200]. In milk systems a highly specific interaction between ϰ-casein and carrageenans has been established that causes ϰ-casein particles or casein micelles to attach to the carrageenan chain. When the molar mass of the carrageenans is sufficient, helical regions can form and aggregate, and a gel network is obtained. This can occur only with the gel-forming ϰ- and ι-types. Weak networks form at carrageenan concentrations as low as 0.02 %, which can fix casein particles. In chocolate milk, for instance, this network holds the cocoa suspension and, in cream, the lipid globules. Below its isoelectric point (pH 4.6), casein is positively charged and can interact directly with the negatively charged carrageenan. Above its isoelectric point, ϰ-casein can interact via its positively charged sites; in this system, calcium ions play a complex role involving aggregation of helices, neutralization of negatively charged sites on the peptide chain, and maintaining the integrity of the casein micelle in milk. Addition of 0.2 % ϰ-carrageenan to milk at the pasteurization temperature, on cooling to 43 °C, gives a firm gel that melts again at 60 °C.

The interaction of other proteins with carrageenan above their isoelectric point is mediated by Ca^{2+} ions, which cross-link negatively charged sites on the carrageenan with negatively charged sites on the protein.

5.4. Analysis

Carrageenan preparations can be characterized by graded fractionation with potassium chloride, giving fractions with typical sulfate content. These fractions also yield well-resolved *^{13}C-NMR spectra* that can be used for their characterization. Pyrolysis–capillary gas chromatography has been used for taxonomic characterization [172]. The constituent sugars galactose and anhydrogalactose can be analyzed *gas chromatographically* after methanolysis and trimethylsilyl derivatization. The presence of mannose in the chromatogram allows an estimation of the galactomannan content [29]. *IR spectroscopy* is also very useful in carrageenan analysis for establishing the location of the sulfate groups. Typical tests to demonstrate the presence of carrageenan are the formation of a blue coagulate with methylene blue, positive sulfate reaction after acid hydrolysis, and the formation of much stronger gels by the addition of potassium chloride. The 3,6-anhydrogalactose content can be determined colorimetrically by using resorcinol [172].

5.5. Applications [200]

Similar to pectins, carrageenans have their main applications in food manufacture; this includes their use as binders in pet food. Their heat stability at high pH is the decisive factor. The "milk reactivity" gives carrageenans a monopoly in group 7 of Table 2. Carrageenans are also used in groups 6–11, and 14. Carrageenans are used for application in water, as in dessert gels; galactomannans are often included. Because of the rapid hydrolysis at pH < 4.5, carrageenans are less suitable for acid products. Outside the food industry carrageenans have found uses in groups 17 and 22.

5.6. Market

The carrageenan industry is limited to a few companies located in regions where red seaweed occurs; namely, in Denmark, Brittany, and the United States. Smaller factories are located in Spain, South Korea, and Japan. Some carrageenan is also believed to be produced in the former Soviet Union. Worldwide annual production probably does not exceed 15 000 t, only 2000 – 3000 t is used outside the food industry (the substantial use in pet foods is included in the food sector). Prices differ greatly; selected and blended lots of standarized carrageenans with special gelling properties are priced at ca. $ 10 –$ 13 per kilogram. However, preparations half this price are also on the market.

In the Philippines, a semirefined carrageenan is produced from *Eucheuma cottinii* and *E. spinosum* by soaking the cleaned seaweed in 1 N alkali for ca. 2 h at 75 °C. The material is then thoroughly washed to remove residual salts, followed by sun or bin drying, and milling to a powder. The processed *Eucheuma* seaweed is a substance with hydrocolloid properties consisting of pulverized plant materials, including 10 – 15 % insoluble cellulosic material. It is used as a thickener, gelling agent, and stabilizer [62].

6. Agar

Agar, also called agar-agar or gelose, is a dried extract of red algae, long known in oriental kitchens for the preparation of fruit jellies, sweet mashed bean jellies, etc. In the 19th Century, agar was introduced in Europe and the United States, and it acquired its reputation as a solid microbiological culture medium in 1881 through the work of KOCH. It is now indispensable in any microbiological and biotechnological laboratory [168], [172], [173], [199].

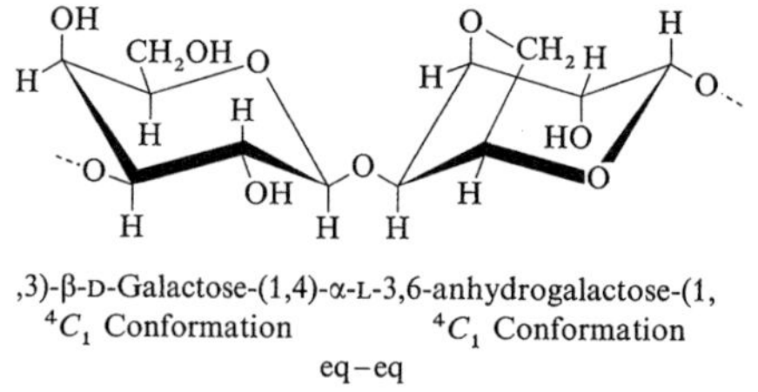

,3)-β-D-Galactose-(1,4)-α-L-3,6-anhydrogalactose-(1,
$^4C_1$ Conformation $^4C_1$ Conformation
eq–eq

Figure 18. Dimeric repeating units in agarose structure

6.1. Occurrence and Structure

Agar is a structural polysaccharide of the cell walls of a variety of red seaweed. It consists of two groups of polysaccharides: agarose and agaropectin. The amounts in which they occur depend on the weed source of the agar (agarose content 55 – 66%) and can be further changed by the conditions used for extraction. *Agarose* is a neutral, linear polysaccharide with no branching and has a backbone consisting of 1,3-linked β-D-galactose-(1-4)-α-L-3,6-anhydrogalactose repeating units. This dimeric repeating unit, called agarobiose, differs from carrabiose in that it contains 3,6-anhydrogalactose in the L-form and does not contain sulfate groups (Fig. 18). Since L-anhydrogalactose has the $^4C_1$ conformation, all glycosidic linkages are diequatorial as in $\varkappa$- and ι-carrageenan. The D-galactose residues may be 6-O-methylated up to 50%, depending on their origin.

The sugar skeleton of *agaropectin* is also composed of agarobiose modified with sulfate groups and in addition contains D-glucuronic acid and pyruvic acid. Also, small amounts of D-xylose, L-galactose, 4-O-methyl-L-galactose, and 6-O-methyl-D-galactose have been isolated [209].

6.2. Production [209]

Gelidium species, one of the principal traditional sources of agar, grow on rocky seabeds and are hand-picked by special divers in Japan. *Gracilaria* species grow in rather shallow sandy areas throughout the world and can be harvested at low tide or collected directly from the beach.

After harvesting, the agarophytes are washed and air-dried. On a dry-matter basis, refined seaweed may contain more than 30 wt% agar. In the traditional process, complex blends of seaweeds (so-called seaweed formulas for agar) are boiled in an excess of water with careful addition of 0.01 – 0.02% sulfuric acid or 0.05% acetic acid to promote good extraction. Extraction is also carried out under pressure, which results in a reduced processing time and increased yields, and is especially effective for rigid-type seaweed. The acid cooking and the pressurized cooking extraction are potentially destructive to the extracted agar and must be optimized for each seaweed type and other processing steps. *Gracilaria* species are successfully extracted by alkali treatment at 85 – 90 °C. Agar-like mucilaginous substances with no gelling ability (diagar sulfates) are converted to agar substances by treatment in aqueous sodium hydroxide containing minor amounts of Ca^{2+} ions. In this process, L-galactose 6-sulfate moieties apparently are converted to 3,6-anhydro-L-galactose moieties as in the case of carrageenans. Increased extractability of agar has been reported by the addition of polyphosphates or by acid pretreatment of seaweed (pH 1, < 15 °C) to liberate agar from nonagar substances with which they are combined.

The hot sol extract is then filtered, generally with a filter aid. On cooling, a gel is obtained which is then frozen and dehydrated. Historically, natural freezing combined with weather drying has been used in areas with suitable climate. Water is eliminated by sublimation, vaporization, and defrosted drip; the proportions of water eliminated in each of these steps depend on the climate. More industrialized factories have mechanical freezing systems or remove the water by vacuum evaporation. Water formed during thawing is removed by pressing or rotating filters. In this way, gel flakes are obtained that are further dried to ca. 20 % moisture content by air or drum drying. Agar is available on the market as bar-style agar, stringy agar, and powdered agar.

6.3. Properties and Analysis

At pH > 5, agar is extremely heat stable. Agar is insoluble in cold water, solubilizes on heating, and gels on cooling, as do carrageenans. The formation of an extended network is based on the association of agarose molecules from the disordered random coil into double helices, which aggregate into junction zones containing many chains [210]. The occasional L-galactose residues in the $^1C_4$ conformation function as kinking residues. Agar gels form from 0.5 % solutions, independent of the presence of salt. These gels are stronger than carrageenan gels and exhibit stronger hysteresis; depending on the type of agar and the concentration, setting points are in the range of 35 – 39 °C and melting points between 85 and 95 °C. The strength of agar gels increases with agarose content; at < 10 % agarose, no gels are obtained. Sulfation increases the critical concentration for gelation and decreases the degree of hysteresis [211].

Agar also interacts with locust bean gum, resulting in stronger gels and is compatible with most other plant gums. At 45 °C, 1 % agar solutions have viscosities of 2 – 10 mPa · s. Molecular masses of 5000 – 150 000 have been reported.

Analysis. The presence of agar can be determined from the presence of galactose and anhydrogalactose after methanolysis, trimethylsilylation, and gas chromatography. It can be distinguished from carrageenan by its much lower sulfate content, its ability to gel in the absence of cations, and the much greater hysteresis of agar gels compared to carrageenan gels.

6.4. Applications

The most typical application of agar is in microbiological and plant tissue culture media. Agar has numerous applications in the Japanese, Korean, and Chinese kitchen, varying from plain gels, noodle-like gels, and sweet desserts, to jellies. The stability of agar under neutral conditions, and high temperature, makes them attractive for use in foods of groups 13 and 14 (Table 2). Other applications are in groups 2, 8, 9, and 11. Outside the food industry, groups 16, 21 and 31 are of interest for agar use.

By precipitation with polyethylene glycol, purified agarose can be obtained from agar [212]. *Agarose* has found substantial application in laboratories as a support for chromatographic separation, enzmye immobilization, and electrophoresis.

6.5. Market

The production of agar is estimated at 4000 t/a, mainly as powder. Depending on quality, prices vary between $ 21 and $ 37 per kilogram; the latter refers to bacteriological agar. Production of stringy and bar-style agar is virtually a Japanese monopoly; Korea is the second largest producer. Japan and Spain produce almost half of the powdered agar; Korea and affiliates of Spanish producers in Chile, Morocco, and Portugal must also be mentioned.

7. Gum Arabic

Gum arabic, also called gum acacia or — according to its country of origin — Turkey, Indian, Senegal or Sudan gum, is the dried, gummy exudation produced in breaks or wounds in the bark of various species of acacia trees (*Acacia senegal* L. Wild). For commercial production the trees are tapped or drilled. After drying in air, the exudate tears are harvested by hand and transported to central collecting stations where they are sorted according to size, color, and contamination with impurities. Harvesting takes place during October to June. A young tree yields ca. 0.9 kg and an older tree up to 12 kg of gum annually. The trees grow in the tropical and subtropical areas of Africa, India, Australia, the southwestern United States, and Central America. World production has risen to more than 50 000 t/a.

Raw gum is still sold in substantial quantities; however, a precleaning by air classification to remove sand, fines, and bark is required. The product can be further processed to a kibbled, granular, or powdered form. Importers in Western Europe and the United States produce speciality products for particular applications. For this purpose the gum is dissolved, and the gum solution is exposed to sieving or centrifugation and then drum or spray dried [1]–[4], [213].

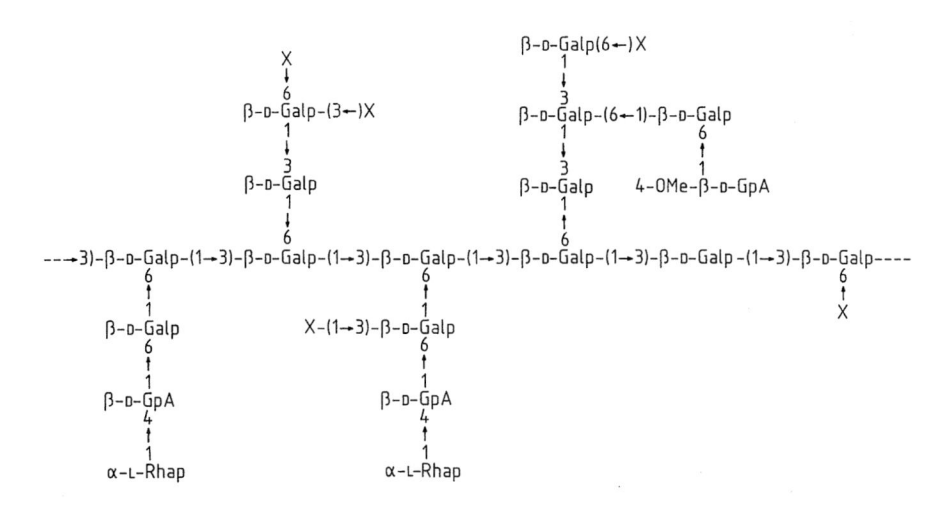

```
                                                    β-D-Galp(6←)X
                        X                                1
                        ↓                                ↓
                        6                                3
                 β-D-Galp-(3←)X              β-D-Galp-(6←1)-β-D-Galp
                        1                        1               6
                        ↓                        ↓               ↑
                        3                        3               1
                     β-D-Galp                  β-D-Galp      4-OMe-β-D-GpA
                        1                        1
                        ↓                        ↓
                        6                        6
---→3)-β-D-Galp-(1→3)-β-D-Galp-(1→3)-β-D-Galp-(1→3)-β-D-Galp-(1→3)-β-D-Galp-(1→3)-β-D-Galp----
           6                            6                                                6
           ↑                            ↑                                                ↑
           1                            1                                                X
        β-D-Galp              X-(1→3)-β-D-Galp
           6                            6
           ↑                            ↑
           1                            1
        β-D-GpA                      β-D-GpA
           4                            4
           ↑                            ↑
           1                            1
        α-L-Rhap                     α-L-Rhap
```

X = L-Araf-(1,
  or α-D-Galp-(1,3)-L-Araf-(1,
  or β-L-Arap-(1,3)-L-Araf-(1,
  or   L-Araf-(1,3)-L-Araf-(1,
  or   L-Araf-(1,3)-L-Araf-(1,3)-L-Araf-(1,
  or β-L-Arap-(1,3)-L-Araf-(1,3)-L-Araf-(1,

β-D-Galp = β-D-Galactopyranose
β-D-GpA = β-D-Glucopyranosyluronic acid
4-OMe-β-D-GpA = β-D-Glucopyranosyluronic acid-4-methylether
α-L-Rhap = α-L-Rhamnopyranose
L-Araf = L-Arabinofuranose

Figure 19. Structural features of gum arabic

Structure. Gum arabic consists of three principal fractions, an arabinogalactan fraction (AG) with very little protein, representing 88% of the gum und having a molecular mass of 279 000; a high molecular mass (1.45×10^6) arabinogalactan – protein complex (AGP) representing about 10% of the gum; and a glycoprotein (GI) with a molecular mass of 250 000, making up 1.2% of the gum. By treatment with proteolytic enzymes the AGP fraction is degraded to fragments with a molar mass similar to AG. The wattle blossom model has been proposed to describe the structure of AGP: Carbohydrate blocks with molecular mass of ca. 280 000 are linked together by a main polypeptide chain. The carbohydrate blocks consist of a β-1,3-linked galactopyranose backbone chain, with numerous side chains linked through β-1,6-galactopyranose residues and containing arabinofuranose, arabinopyranose (Ara), rhamnopyranose (Rha), glucuronic acid (GlcA), and 4-O-methylglucuronic acid. The sugar moieties occur in the ratio Gal 36, Ara 31, Rha 13, and GlcA 18. The structure of a representative segment is shown in Figure 19 [214] – [217]; depending on growth conditions and age, some structural variations exist.

Properties. The molecular mass of gum arabic averages around 500 000; the glucuronic acid content (including 4-O-methylglucuronic acid) of ca. 18% makes gum arabic an acidic polysaccharide. With a cation exchanger or by washing with acidified alcohol, gum arabic can be converted to arabinic acid, which exhibits the titration curve of a strong acid. A 5% solution has an apparent dissociation constant of 10^{-3}. Solutions in water have a pH of 2.2–2.7.

The most striking feature of gum arabic is its *extreme solubility in water*. Due to its highly branched, compact structure, solutions containing up to 40% gum arabic can be obtained, which still exhibit typical Newtonian flow behavior. Compared to other polysaccharides of similar molar mass, the viscosity of these solutions is very low (Table 4). Viscous solutions are obtained only at concentrations >30%, where effective molecular overlap begins to occur and solutions gradually assume pseudoplastic behavior. The *viscosity* of gum arabic solutions is highest in the pH range 4.5–5.5; in this range the carboxyl groups are ionized to a large degree. Addition of salts strongly decreases the viscosity due to suppression of the electrostatic charge. Gum arabic is compatible with most other hydrocolloids except for alginates and gelatin. It is also surface active and has been widely used to *stabilize oil-in-water emulsions*. The AGP fraction is responsible for its emulsifying ability. The polypeptide chain is believed to be at the periphery of the molecule and to adsorb onto hydrophobic surfaces. Gum arabic has been shown to be heat sensitive; proteinaceous material (AGP and GI) precipitates on prolonged heating, and as a consequence the solution viscosity drops considerably and the emulsifying ability is lost [213], [214], [217], [218].

Gum arabic can be *precipitated* with trivalent cations and by salts such as silicates, borates, and mercury nitrate. It is insoluble in oils and most organic solvents but soluble in aqueous ethanol up to ca. 60% ethanol. In gum arabic preparations enzyme activities have often been demonstrated, for example, oxidases, peroxidases, amylases, and pectinases. For applications in foods, these activities are undesirable, and the enzymes must be inactivated by heat [2].

Applications, Market. Gum arabic was a commodity 4000 years ago in Egypt, and was used for the manufacture of cosmetics, dyes, and tinctures. Half of the amount imported by the United States is used in the food industry; this makes gum arabic one of the most useful plant hydrocolloids. The annual production has risen to >50 000 t/a [213], [216]. Gum arabic is used in groups 2, 4, 8, and 11 (Table 2). Dried, powdered flavors (group 5) constitute a major application. High gum concentrations allow spray drying without loss of flavor. Its emulsifying and stabilizing ability, in combination with its use in high concentrations, makes gum arabic attractive for pharmaceutical and cosmetic applications (groups 16 and 21). In industrial applications such as an adhesive on gummed envelopes and stamps (group 20), and as a dispersive agent in dyes and tinctures, gum arabic is being replaced by starch and cellulose derivatives. Prices vary between $ 3 and $ 8.5 per kilogram.

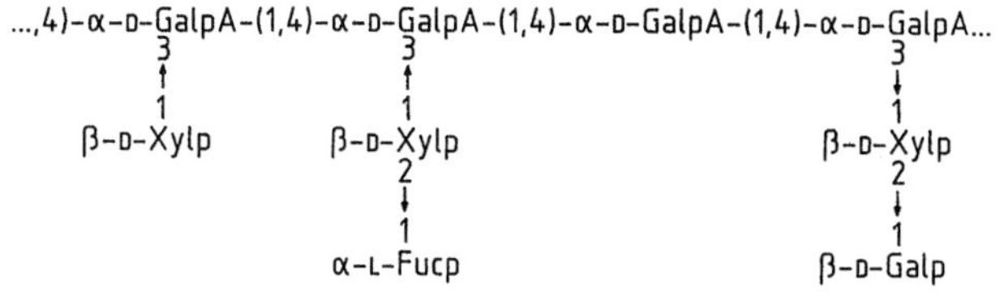

```
...,4)-α-D-GalpA-(1,4)-α-D-GalpA-(1,4)-α-D-GalpA-(1,4)-α-D-GalpA...
              3                    3                    3
              ↑                    ↑                    ↑
              1                    1                    1
         β-D-Xylp              β-D-Xylp             β-D-Xylp
                                   2                    2
                                   ↑                    ↑
                                   1                    1
                               α-L-Fucp             β-D-Galp
```

α-D-GalpA = α-D-Galactopyranosyluronic acid
β-D-Xylp = β-D-Xylopyranose
α-L-Fucp = α-L-Fucopyranose
β-D-Galp = β-D-Galactopyranose

Figure 20. Structural features of gum tragacanth

8. Gum Tragacanth

Gum tragacanth is the exudate from breaks or wounds inflicted in the bark of shrubs of the *Astragalus* species, particularly *Astragalus gummifer* Labillardière (Leguminosae family), which are found in the mountainous regions of Iran, Syria, and Turkey. The gum was described by THEOPHRASTUS in the third century B.C. The name tragacanth is of Greek origin and means goat horn. It probably refers to the curved shape of the 2–3-mm-thick ribbons that represent the better quality gum. Gum tragacanth also exists as flakes, which are of poorer quality. The bushes are carved to stimulate gum production. Ribbons are obtained from April to September; flakes are formed in the following months until November. The translucent, whitish, elastic ribbons, which originate predominantly in Iran, are considered the best quality. The harvest is collected at central stations and is exported mainly to the United States and the United Kingdom. Importers process the raw materials to standardized powders [1]–[4], [214], [215], [219]. Prices range from $ 16 – $ 21 per kilogram; the annual harvest is ca. 3000 t/a.

Structure. Gum tragacanth is composed of tragacanthic acid, which is the major component, and a nearly neutral *arabinogalactan* that is water insoluble but swells to a gel. The constituent sugar residues of this arabinogalactan are: 75 % L-arabinose, 12 % D-galactose, 3 % D-galacturonic acid methyl ester, and L-rhamnose. *Tragacanthic acid* is water soluble and consists of a backbone of 1,4-linked α-D-galactopyranosyluronic acid residues (polygalacturonic acid chain as in pectins) with short side chains including single-unit β-1,3-linked D-xylopyranose, dimeric β-1,3-linked D-xylopyropyranosyl-1,2-α-L-fucopyranose, and dimeric β-1,3-linked D-xylopyranose-1,2-D-galactopyranose. The structure is shown schematically in Figure 20. The sugar moieties are present in the relative proportions D-galacturonic acid 43 %, D-xylose 40 %, L-fucose 10 %, and D-galactose 4 %. Tragacanthic acid has a high molecular mass (840 000) and gives accordingly highly viscous solutions (Table 4) that exhibit pseudoplastic behavior at concentrations > 0.5 %. A 1 wt % solution can have a viscosity of 3600 mPa · s. Accompanying Ca^{2+} and Mg^{2+} ions may contribute to the high viscosity and gelling properties of gum tragacanth solutions. The viscosity changes little in the pH range 2–10 and is stable during heating in a 1% acetic acid solution [1]–[4], [214], [215], [219].

Applications, Market. The application of gum tragacanth in foods and acid products is based on its heat stability, and its stabilizing ability in emulsions and suspensions. Therefore, gum tragacanth is of particular interest for group 12 (Table 2). Because of the slimy consistency of gum tragacanth dispersions, this gum can also function as a fat replacer. Other applications are mentioned in groups 8 and 9. Outside the food industry gum tragacanth is also used in groups 16, 18, and 21. A special application is its use in spermicidal jelly.

9. Gum Karaya

Gum karaya is also named Sterculia gum, Kadaya gum, or Indian gum; it is the dried exudate from the bark of *Sterculia urens* or other *Sterculia* species (Sterculaceae family), trees occurring throughout northern and central India. The formation of exudate is artificially initiated by carving the bark. After a few days the exudate can be harvested as large irregularly shaped tears, which may weigh 1 kg or more. From April to June, 4.5 kg can be obtained per tree. The gum is sold to merchants at central trade stations and precleaned by removing pieces of bark; the tears are broken and graded according to purity and color. In consumer countries, the gum is then milled to a homogeneous powder. In the past, gum karaya was sold in a blend with gum tragacanth; in the meantime, however, it has found many applications because of its specific characteristics. The annual production is estimated at 5000 t, the larger part of which is used in the United States. Prices are ca. $ 4.8–$ 6.4 per kilogram. Expansion of the production requires afforestation and takes at least 10 years [1]–[4], [214]. Gum karaya is a highly acetylated, acidic polysaccharide with a backbone chain of alternating 1,4-linked α-D-galacturonopyranosyluronic acid residues and 1,2-linked α-L-rhamnopyranosyl residues. Singleunit β-D-galactopyranosyl- and β-D-glucuronopyranosyluronic acid side chains are attached to O-2 and O-3 of the galacturonic acid residues, whereas half of the rhamnose residues have 1,4-linked β-D-galactopyranosyl residues at O-4 (Fig. 21). The molecule also has 13 wt% acetyl groups. Gum karaya is poorly water soluble; however, it swells readily in cold water to many times its original volume, particularly when finely ground (<200 mesh), giving a dispersion that appears homogeneous. By using an autoclave, solutions of up to 20% can be obtained; their viscosity is, however, decreased as a result of degradation. Gum karaya solutions deviate from Newtonian behavior and exhibit viscoelasticity. Under alkaline conditions, deacetylation occurs and converts a gum karaya suspension into a ropy mucilage [214], [215]. The ability to swell and its resistance to enzymic and microbial degradation make powders of gum karaya good adhesives for dentures. Particles of ca. 0.6 mm swell to 60–100 times their volume and form a mucilage that allows drainage. Food applications are in groups 9, 11,

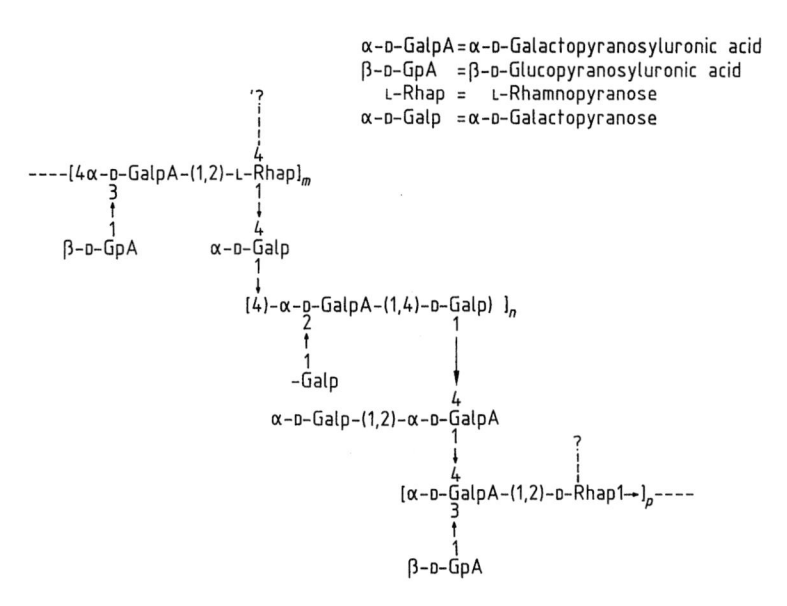

```
                        α-D-GalpA = α-D-Galactopyranosyluronic acid
                        β-D-GpA  = β-D-Glucopyranosyluronic acid
                 '?      L-Rhap  = L-Rhamnopyranose
                  ↓      α-D-Galp = α-D-Galactopyranose
                  4
----[4α-D-GalpA-(1,2)-L-Rhap]ₘ
      3             1
      ↑             ↓
      1             4
   β-D-GpA       α-D-Galp
                    1
                    ↓
             [4)-α-D-GalpA-(1,4)-D-Galp)]ₙ
                  2               1
                  ↑               
                  1               
                -Galp             
                                  ↓
                                  4
                     α-D-Galp-(1,2)-α-D-GalpA
                                  1
                                  ↓           ?
                                  4           ↑
                         [α-D-GalpA-(1,2)-D-Rhap1→]ₚ----
                                  3
                                  ↑
                                  1
                               β-D-GpA
```

Figure 21. Structural features of gum karaya

12, and 13 (Table 2). Gum karaya has also found use in the paper industry (group 29) and textile painting (group 30).

10. Gum Ghatti

Gum ghatti, also called India gum, is an exudate of *Anogeissus latifolia* (Combretaceae family), trees that grow in the deciduous forests of India and Sri Lanka. The Indian word gath means mountain pass and refers to the trade routes for this gum. The exudate usually has the shape of 1-cm-thick beads or tears, which may be colored by absorption of tannins from the bark. Also, dust and other particles may stick to the gum. Gum ghatti is harvested mainly in April and graded by traders according to color and impurities. The highest quality is pure and only lightly colored. Importers, particularly in the United States, purify and standardize the gum by grinding, sieving, and blending or by solubilization in water, filtration, and spray drying. It is of minor importance, with an annual production of ca. 1000 t [4].

Gum ghatti is composed of D-galactose, D-mannose, D-glucuronic acid, and L-rhamnose. A complex array of single-unit and oligomeric side is attached to a chain of alternating 1,4-β-D-glucuronic acid and 1,2-α-D-mannopyranose residues. The side chains comprise several 1,6-linked D-galactopyranose units, some terminated by glucuronic acid, joined to the mannopyranose residues at O-3 through 1,3-linked L-arabinopyranose residues. Single-unit L-arabinofuranose residues can also be found, some attached to the O-6 of mannopyranose residues, and other short sequences of L-

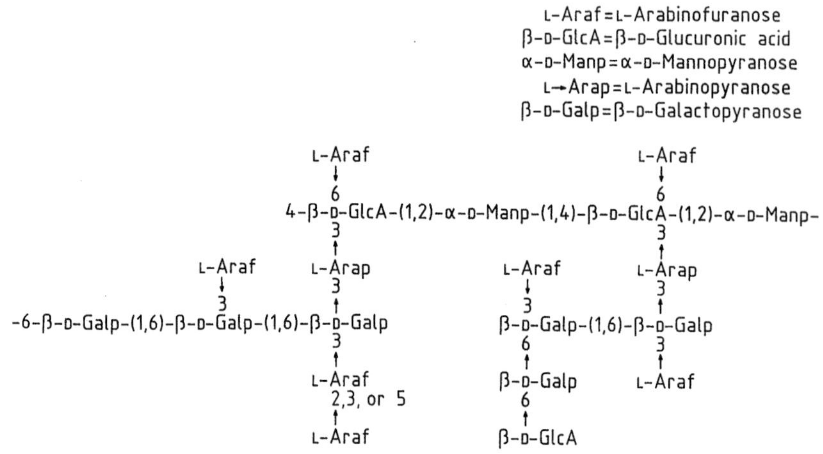

```
                                    L-Araf = L-Arabinofuranose
                                    β-D-GlcA = β-D-Glucuronic acid
                                    α-D-Manp = α-D-Mannopyranose
                                    L→Arap = L-Arabinopyranose
                                    β-D-Galp = β-D-Galactopyranose

                              L-Araf                            L-Araf
                                ↓                                 ↓
                                6                                 6
             4-β-D-GlcA-(1,2)-α-D-Manp-(1,4)-β-D-GlcA-(1,2)-α-D-Manp-
                                3                                 3
                                ↑                                 ↑
        L-Araf        L-Arap           L-Araf        L-Arap
          ↓             ↓                ↓             ↓
          3             3                3             3
-6-β-D-Galp-(1,6)-β-D-Galp-(1,6)-β-D-Galp    β-D-Galp-(1,6)-β-D-Galp
                    3                             6             3
                    ↑                             ↑             ↑
                  L-Araf                        β-D-Galp       L-Araf
                  2,3, or 5                       6
                    ↑                             ↑
                  L-Araf                        β-D-GlcA
```

Figure 22. Structural features of gum ghatti

arabinofuranose [215], [220]. *O*-acetyl groups have also been found. A schematic presentation of the structure is shown in Figure 22. In its acid form, gum ghatti has an equivalent mass of 1067; the gum occurs as the calcium or magnesium salt. The gum is not entirely soluble in cold water, but 90% can be solubilized by maceration and heating at 90 °C. Removal of the divalent cations enhances solubility. The remaining 10%, which is chemically identical to the soluble portion, is strongly swollen and also contributes to the viscosity [191]. Spray-dried preparations are completely soluble in water; they give, however, a lower viscosity of the aqueous solutions. The viscosity of gum ghatti solutions is higher than that of gum arabic solutions; viscosity increases quadratically with concentration. Like gum arabic, gum ghatti is an excellent stabilizer of emulsions and suspensions, and has found application in wax suspensions and pastes, oil emulsions, polishes, syrups, and as a dispersive agent in dried, vitamin-rich oil suspensions.

11. Xanthan Gum

Xanthan gum is the extracellular, high molecular mass heteropolysaccharide produced by the aerobic fermentation of *Xanthomonas campestris* NRRL-B1459 [1]–[5], [221]. The production of xanthan gum goes back to the investigations of JEANES and coworkers in the early 1960s at the Northern Regional Research Laboratory of the United States Department of Agriculture, Peoria, Illinois, in the framework of a large screening program for slime-producing microorganisms of industrial interest [222].

Figure 23. Flow sheet for the manufacture of xanthan

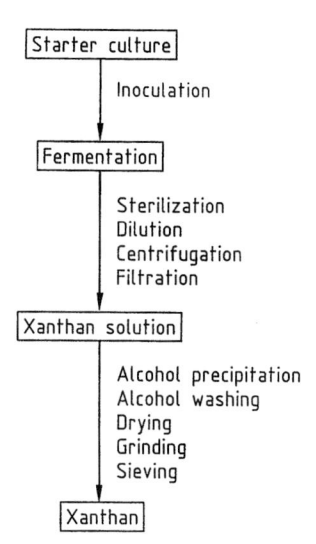

```
[Starter culture]
      │ Inoculation
      ▼
[Fermentation]
      │ Sterilization
      │ Dilution
      │ Centrifugation
      │ Filtration
      ▼
[Xanthan solution]
      │ Alcohol precipitation
      │ Alcohol washing
      │ Drying
      │ Grinding
      │ Sieving
      ▼
[Xanthan]
```

11.1. Production

Commercial production is carried out by a batchwise, submersed fermentation under strong aeration. During fermentation the bacterial cells are kept under a constant stress to direct their metabolism to metabolite production instead of growth. This is achieved by proper selection of medium composition. The medium cost forms a significant part (ca. 25%) of the total production cost; the sugar concentration is adjusted so that it falls as low as possible at the end of the fermentation cycle, but depletion must be avoided because otherwise the microorganisms will use the formed polysaccharide as a substrate [223]. For xanthan production, a typical medium contains a nitrogen source, phosphate and magnesium ions, trace elements, and glucose, which is kept at < 5%. The pH must not be below 7. After 96 h at 30 °C, less than 0.1% glucose is left in the medium, and a conversion to polysaccharide of more than 50% can be obtained.

In the patent literature [224], much attention is devoted to propagation of the starter culture. In this stage, nitrogen is supplied by organic sources (e.g., soy peptone or corn steep liquor), which also contain some growth factors. During the main fermentation, nitrogen is supplied by mineral salts (e.g., ammonium nitrate). A typical composition of the medium at this stage is 0.06% ammonium nitrate, 0.5% potassium dihydrogenphosphate, 0.01% magnesium sulfate heptahydrate, 2.25% glucose, and 97.18% water (Fig. 23). After inoculation of 5300 L of medium with 227 L of starter culture and vigorous aeration for 72 h at 28–31 °C, 77 kg of hydrocolloid can be recovered. The remaining medium contains < 0.1% glucose. This means a conversion of 64.7% and a final concentration of the polysaccharide of 1.45%, which gives the medium a viscosity of 3000 mPa · s at 25 °C.

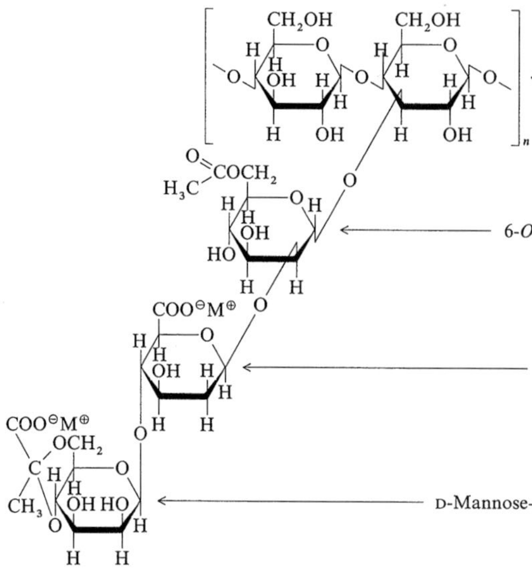

Figure 24. Structural features of xanthan

← β-1,4-Glucan

← 6-O-Acetyl-D-mannose

← D-Glucuronic acid

← D-Mannose-pyruvic acid-acetal

Xanthan

The usual precautions must be taken to preclude contamination of the batch fermentation through the culture medium, equipment and accessories, air, and neutralizer; thus it is doubtful if the batch operation can be replaced by a continuous process. Also the stability of the bacterial strain is a point of consideration. Since *Xanthomonas* bacteria do not form spores, the fermentation can be stopped by heat treatment in a heat exchanger. Further downstream processing involves centrifugation, filtration, and precipitation with ethanol or isopropanol as used for other hydrocolloids. The high viscosity of the fermentation liquor must be reduced by dilution with water or alcohol. The precipitate, dried by vacuum or with hot air, is processed to a marketable article by grinding and sieving.

11.2. Structure and Properties

Chemical Structure. Xanthan is a heteropolysaccharide composed of D-glucose, D-mannose, and D-glucuronic acid. It has a β-1,4-D-glucan (cellulosic) backbone substituted through C-3 on alternate glucose residues with a trisaccharide side chain consisting of β-D-mannose-(1,4)-β-D-glucuronic acid-(1,2)-α-D-mannose. The terminal mannose moiety may have pyruvate residues linked to the 4- and 6-positions, the internal mannose unit is acetylated at C-6 (Fig. 24) [225]. By modifying the biosynthetic pathway for xanthan production, the carbohydrate structure and the substitution pattern of the polymer can be genetically controlled. In this way, polysaccharides with quite different properties can be obtained [226].

Properties. Xanthan has a *molecular mass* of ca. 2×10^6. Since the structure and the molecular mass are genetically controlled, hydrocolloids with similar molecular mass and similar properties are obtained. *Aqueous solutions* of xanthan are extremely viscous (Table 4) and pseudoplastic (i.e., they exhibit a reversible, highly shear-thinning behavior) [227], a 0.25% solution in water containing 1000 µg/g NaCl and 147 µg/g $CaCl_2 \cdot H_2O$ has a viscosity of ca. 1000 mPa · s at a shear rate of $1\ s^{-1}$ which decreases to ca. 50 mPa · s at a shear rate of $100\ s^{-1}$. The viscosity of xanthan solutions is stable over a wide pH and temperature range. Also in the presence of salts the viscosity is maintained; polyvalent cations, however, can precipitate xanthan under alkaline conditions. Strong oxidants such as hypochlorite and persulfate *degrade* xanthan, particularly at high temperature. Xanthan is quite resistant to enzymic degradation, although it can be partly degraded by endo-β-1,4-glucannases after removal of calcium ions [228].

Gel Formation. Xanthan is very compatible with alginates and starch; it shows a synergistic viscosity increase with the galactomannan guar, and it forms strong transparent thermoreversible gels with certain glucomannans (konjac mannan) and certain galactomannans (carob and tara). If guar is modified through partial removal of galactose substituents with an α-galactosidase, it also gels with xanthan [229]. Two models have been proposed for weak gel formation in xanthan solutions; both models are based on the occurrence of xanthan molecules in helical structures (Fig. 25). In the model of Norton et al., single helical regions associate to form a network [230]. The model of Morris [231] is based on a double-helical model for xanthan. The primary mechanism is considered to be end-to-end association into fibrous aggregates via double-helix formation. The formation of thermoreversible gels by synergistic mixtures of xanthan with certain galacto- and glucomannans has recently been ascribed to intermolecular binding through cocrystallization of denatured xanthan chains within the mannan crystallite [229]. The cellulosic backbone of xanthan and the stereochemically similar mannan backbone both form ribbonlike structures. The mixed crystallites probably act as strong junction zones to consolidate the weak xanthan network.

11.3. Analysis

The presence of xanthan can be indicated by the formation of a gel during cooling of a 1:1 mixture with carob gum in a 1% solution. Its presence can also be indicated from the reaction of pyruvate with 2,4-dinitrophenylhydrazone in a hydrolysate.

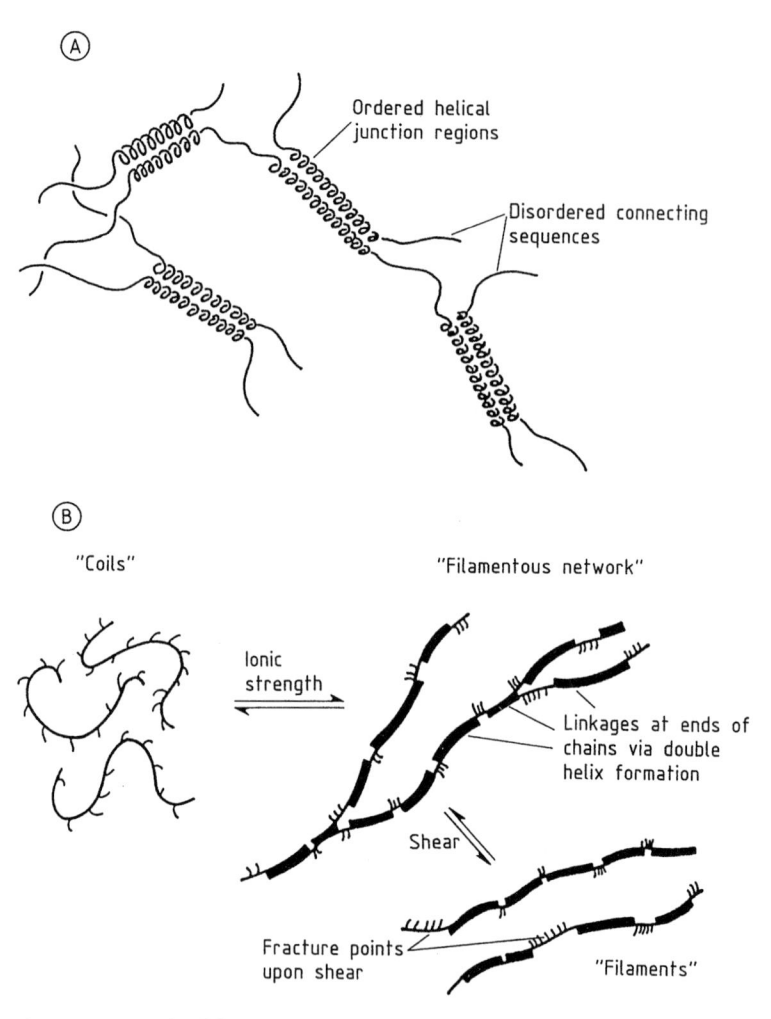

Figure 25. A) Weak gel formation in xanthan solutions by self-association of single helices and B) End-to-end association via double-helix formation into fibrous aggregates [229]

11.4. Applications, Market

The applications of xanthan are based especially on the viscous behavior and stabilizing properties of this hydrocolloid. In industry, xanthan is used in groups 18, 19, 26, 27, 28, 29, and 33 (Table 2). These applications enabled Kelco, the first producer of xanthan gum, to carry out a comprehensive toxicological evaluation that resulted, in 1969, in the acceptance of this hydrocolloid in foods, where it has found applications especially in groups 4, 5, 8, and 12.

Xanthan is produced by only two companies, one in the United States (Kelco) and one in France (Sanofi). Present production is estimated at ca. 10 000 t/a. For food applications prices are ca. $ 11 – $ 13 per kilogram. Because of its specific properties the potential for this polysaccharide is great.

12. Gellan Gum

Gellan gum is produced by the bacterium *Pseudomonas elodea* and has been approved recently for food uses [229], [232], [233]. It is a broad-spectrum, multipurpose gelling agent, which forms viscous solutions even at low concentrations. Gellan gum is both cold setting and thermosetting; its melting point (thermoreversible and thermoirreversible) depends on ionic strength. Gelled products are stable to retorting and microwave cooking.

Gellan gum is a linear anionic polysaccharide with a tetrasaccharide repeating unit:

-3)-β-D-Glc-(1,4)-β-D-GlcA-(1,4)-β-D-Glc-(1,4)-α-L-Rha-(1-

The native polysaccharide is esterified at the 1,3-linked glucosyl residue. Some of the hydrox-yl groups at C-2 are esterified with L-glycerate, and ca. 50% of the hydroxyl groups at C-6 with acetate. The commercially available product is saponified and known as Gelrite (Kelco). Gellan gum is priced at about $ 53 per kilogram but there are also blends available at $ 12 – $ 13 per kilogram.

13. Galactomannans [2]–[6], [11], [234]

Galactomannans are plant reserve carbohydrates, like starch, that occur in the endosperm of the seeds of many Leguminoseae. During sprouting, the galactomannans are enzymatically degraded and used as nutrition. Of the many known galactomannans, up to now only three have been processed and used on an industrial scale:

1) **Locust bean gum (carob gum)** has been known for a long time. It is derived from seeds of the carob tree (*Ceratonia siliqua* L.) which grows around the Mediterranean Sea, in Spain, Portugal, and Morocco, where more attention is paid to its systematic cultivation. The annual production is estimated at 10 000 – 12 000 t.
2) **Guar gum** is at present the most important galactomannan. It is derived from the seeds of the guar plant (*Cyamopsis tetragonoloba* L. Taub.), which grows in India and Pakistan. Since 1944, guar plants have also been cultivated on a large scale in the southern United States. Recently they have been introduced into Brazil, Africa, and Australia, where two crops a year can be obtained. The industrial production of guar gum began around 1940. The annual production is estimated at ca. 125 000 t.

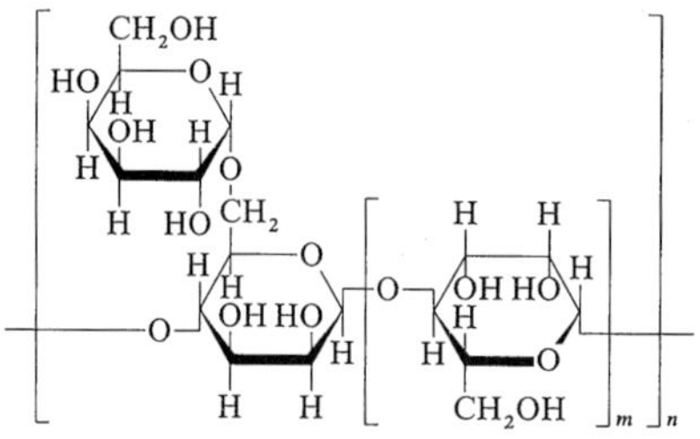

Figure 26. Structural features of galactomannans

$m = 3$: Locust beam gum
$m = 1$: Guar gum
$m = 2$: Tara gum

3) **Tara gum** has been produced only since the 1970s, to a much lesser degree (1000 t/a), from the seeds of the tara tree (*Cesalpina spinosa* L.), which occurs mainly in Peru.

The galactomannans on the market are flours prepared from the endosperm of the corresponding seeds. The carob and tara kernels are particularly difficult to process because of their tough, hard hulls. Remnants of the hulls and seeds are, therefore, always present in small amounts. The pure polysaccharide extracted from the flour of locust bean gum is often designated carobin gum and the polysaccharide from guar flour, guaran.

13.1. Structure

Galactomannans have a backbone of β-(1,4)-glycosidically linked mannopyranosyl residues. This backbone is substituted with single-unit α-D-galactopyranosyl residues linked to the *O*-6 of certain mannose moieties. This substitution renders the mannans soluble. Within the various mannans a wide spectrum of chemical structures occurs. This diversity of structure includes not only a wide variation in their mannose : galactose ratio, but also significant differences in the distributon of galactose units along the mannan backbone (Fig. 26) [235], [236]. For *locust bean gum* a mannose : galactose ratio of 3.5 has been established, compared to a ratio of 3 for *tara gum* and 1.5 for *guar gum*. A measure of the differences in the distribution of galactose side groups has been obtained from an examination of the degree of hydrolysis and the characteristic array of oligosaccharides obtained by digestion with an endomannanase. In this way, DEA et al. [237] established that a *tara gum* has a more statistically random distribution of side chains while *locust bean gum* has a nonregular, nonstatistically random distribution with a certain proportion of unsubstituted blocks of intermediate length. *Guar gum* was found to have few, if any, regions that were unsubstituted (smooth regions) with galactose. Galactomannans differ further in their molar mass distribution; the highest values were established for guar gum. These differences in fine structure between the galactomannans, involving the extent and pattern of galactosyl substitution and molar mass distribution, account for their very different functional properties.

13.2. Production [4], [11], [234]

Guar Gum. For the separation of the germ and the endosperm halves, guar seeds are first screened to remove foreign matter and then fed to an attrition mill to split the seeds in two endosperm halves; finally, the germ material is sifted off. Despite their relatively low galactomannan content the remaining endosperm halves, covered with hull (splits), are suitable for many technical applications. For higher-quality demands, and particularly for food applications, the endosperm must be further purified and liberated from the adhering hull. Various processes are used for this; in general, the splits are treated with moist or dry, hot air, which results in the loosening of the hull as a result of different swelling properties. After removal of the hull by sifting, pure endosperm is obtained with a galactomannan content of 85 – 95 % based on dry matter. By suitable milling and screening techniques the endosperm can be worked up to the desired commercial products.

Locust Bean and Tara Gum. Since the hull of carob and tara gum is very hard and tough, these gums are difficult to process. For removal of the hulls, two different processes exist. In the *acid process* the hulls are carbonized by treating the kernels with moderately dilute sulfuric acid at elevated temperature. Similar to the guar gum process, the remaining hull fragments are then removed by washing and brushing operations. After a drying step the dehulled kernels are cracked and the germs sifted off from the endosperm. Commercial products are obtained by milling and screening.

In the vanishing *roasting process* the kernels are roasted in a rotating furnace, where the hulls pop. The endosperm is then obtained as described above. This process yields products of somewhat darker color. However, the use of sulfuric acid is avoided.

13.3. Properties

The most prominent properties of galactomannans are their high water-binding capacity, their potential to form highly viscous solutions even at low concentrations, and the interaction of locust bean gum with carrageenans, agar, and xanthan gum. In double logarithmic plots of zero shear viscosity versus concentration multiplied by intrinsic viscosity, galactomannans exhibit a different type of viscosity behavior from most polysaccharides for which nonspecific physical entanglement between fluctuating random coil molecules is the underlying mechanism. Galactomannans show an earlier onset of concentrated solution behavior and a substantially greater concentration dependence thereafter [235], [236]. This behavior is attributed to the occurrence of interchain associations inherent to β-1,4-D-mannan chain segments. Such self-association has also been observed for β-1,4-D-glucan and β-1,4-D-xylan structures.

Locust bean gum and guar gum are both *water soluble*; they differ, however, in their dissolution behavior. Whereas guar gum is readily soluble in cold water, locust bean gum can be solubilized only by boiling. Solutions of both galactomannans are, depending on their purity, somewhat cloudy because of the presence of some proteins and crude fiber.

Even 1 wt% solutions are highly *viscous* (2000–6000 mPa · s, measured with an RTV Brookfield viscometer at 20 rpm); 2–5 wt% swollen dispersions show gellike behavior. The viscosity of solutions is hardly influenced by ionic strength. Locust bean gum solutions can gel at high sodium hydroxide concentrations. Also boric acid and alkaline copper salt solutions cause flocculation or gelling of the galactomannans. Mixing of solutions of locust bean gum and xanthan gum results in an increase of the viscosity or even in gel formation (after boiling [208]). Such synergistic interactions have also been observed for combinations of locust bean gum with carrageenan or agar (Section 5.3). Dilute galactomannan solutions form elastic, transparent films upon dehydration.

Galactomannans are rapidly *degraded* in acidic, aqueous solutions at elevated temperature, resulting in a rapid loss of viscosity. They can also be degraded by oxidants and by microbial enzymes (mannanases); under alkaline conditions, they are rather stable [238]. When stored for longer periods of time, solutions of galactomannans must therefore be protected against microbial degradation by addition of preservatives.

13.4. Analysis and Composition of Commercial Preparations

Commercial galactomannan preparations still may contain remnants of germs and hull that are rich in protein and fiber, and reduce the galactomannan content. The galactomannan content is therefore an important factor in quality. This *gum content* is determined by analyzing the impurities and subtracting their total from 100%. Depending on the application, products with different galactomannan content are used.

Another quality factor is the *viscosity* of the aqueous solution, usually measured in 1 wt% systems. Locust bean gum must first be solubilized by heating.

The *presence of galactomannans* can be demonstrated by various precipitation reactions (e.g., gelation with borax). For determination of the *mannose : galactose ratio*, the preparation is hydrolyzed and the sugar moieties are analzyed by using gas chromatography [15], [29].

13.5. Derivatives

As with cellulose, galactomannans contain primary and secondary hydroxyl groups, which, in principle, can be derivatized and substituted like cellulose. Only hydroxyalkyl, carboxymethyl, and cationic derivatives, as well as partially depolymerized products, are of industrial significance, however.

Hydroxyalkyl derivatives are obtained by treating galactomannans with ethylene or propylene oxide in an alkaline medium. These derivatives differ from the parent material by having a faster rate of hydration and being almost completely soluble.

Carboxymethylated products are obtained by "etherification" of a certain portion of the hydroxyl groups with monochloric acid or sodium monochloroacetate. Unlike carboxymethyl celluloses, which are soluble only at degrees of substitution in the range of 0.7–1.0, galactomannans—if not already soluble prior to derivatization—become completely soluble at a degree of substitution of <0.05.

Cationic derivatives are obtained by treatment of galactomannans with suitable reactive organic amines such as 2-hydroxy-3-chloropropyltrimethylammonium chloride or its reaction product 2,3-epoxypropyltrimethylammonium chloride, which forms in alkaline medium.

Partially depolymerized galactomannans with reduced viscosities are obtained by oxidative, hydrolytic (acid), or enzymatic degradation. Conversely, by reaction with bifunctional molecules such as epichlorohydrin, *cross-linked galactomannans* can be obtained, which—depending on their degree of cross-linking—are only partly water soluble or fully insoluble. Plant α-galactosidase that can remove galactose from galactomannans has been cloned in yeast and can be produced conveniently. This enzyme is active in systems containing up to 35% galactomannan [239].

13.6. Applications

Guar gum has found application in many industries; its use is still increasing. Most uses are based on its good solubility in water and the high viscosity of aqueous solutions. In many applications, its inertness against high concentrations of salts of monovalent cations, its ability to form gels with certain multivalent cations, and its flocculating action are also decisive. *Derivatives of guar gum* are used in the oil industry in secondary or tertiary oil recovery of nearly exhausted oil wells. For applications in the textile industry (Table 2, groups 30 and 31), a variety of native and derivatized types are produced. They are also used in the manufacture of paper (group 29), explosives (group 26), and cables. In the latter applications, guar gum is added in advance to immobilize the inevitably penetrating water. They are also used in the pharmaceutical industry and as flocculating agents in the processing of minerals and effluent purification (group 32). In the food industry, guar gums are used as efficient thickeners of aqueous systems, thereby controlling the mobilization of water. This influences the consistency, body, and shelf life of aqueous food systems, as well as the stability of oil–water and water–oil emulsions. They are used in groups 3, 5, 6, 8, 9, and 12, and have also found application in pet food (group 13) and cattle feed.

Due to its higher price, *locust bean gum* is used mainly in the food industry, in the same groups as the guar gums. Of special interest for locust bean gum is its synergistic action with other hydrocolloids such as carrageenans, agar, and xanthan gum.

13.7. Market

Of the 125 000 t of *guar gum* produced annually, ca. one-third is used in the food industry; the remainder has found technical application. *Locust bean gum* is used almost completely in the food and pet food industry. The prices on the world market average ca. $ 1.1 – $ 1.3 per kilogram for guar gum and $ 6.4 – $ 7.5 per kilogram for locust bean gum.

14. References

[1]   M. Glicksman: *Gum Technology in the Food Industry*, Academic Press, New York 1969.
[2]   R. L. Whistler, J. N. BeMiller (eds.): *Industrial Gums*, 2nd ed., Academic Press, New York 1973.
[3]   H. D. Graham (ed.): *Food Colloids*, AVI, Westport, Connecticut 1977.
[4]   H. Neukom, W. Pilnik: *Gelier- und Verdickungsmittel in Lebensmitteln*, Forster Verlag, Zürich 1980.
[5]   G. O. Aspinall: *The Polysaccharides*, vol. **2**, Academic Press, New York 1983.
[6]   P. Harris: *Food Gels*, Elsevier Applied Science Publishers, London 1990.
[7]   J. R. Mitchell, D. A. Ledwàrd: *Functional Properties of Food Macromolecules*, Elsevier Applied Science Publishers, London 1986.
[8]   M. L. Fishman, J. J. Jen: "Chemistry and Functions of Pectin," *ACS Symp. Ser.* **310** (1986).
[9]   D. Oakenfull: "Gelling Agents," *CRC Crit. Rev. Food Sci. Nutr.* **26** (1987), 1–25.
[10] M. Yalpani: "Industrial Polysaccharides; Genetic Enginnering, Structures/Property Relations and Applications," *Prog. Biotechnol.* **3** (1987).
[11] M. Glicksman: *Food Hydrocolloids*, CRC Press, Boca Raton, Fla. 1986.
[12] R. H. Walter: *The Chemistry and Technology of Pectin*, Academic Press, New York 1991.
[13] W. Schmolck, E. Mergenthaler: "Beiträge zur Analytik von Polysacchariden, die als Lebensmittelzusatzstoffe verwendet werden,"*Z. Lebensm. Unters. Forsch.* **152** (1973) 87–99.
[14] M. E. Endean: The Separation and Identification of Polysaccharide Stabilizers, Thickeners and Gums from Foods, Technical circular no. 575, The British Food Manufacturing Industrie Research Association, Leatherhead, Surrey 1974.
[15] *Schweizerisches Lebensmittelbuch*, vol. **3**, chap. 40, "Gelier- und Verdickungsmittel," Eidgen. Drucksachen- und Materialzentrale, Bern 1989.
[16] R. G. Morley, G. O. Phillips, D. M. Power, *Analyst (London)* **97** (1972) 315–319.
[17] J. E. Scott, D. Glick (eds.): *Methods of Biochemical Analysis*, Interscience, New York 1960, pp. 145–197.

[18] C. Rolin in R. L. Whistler, J. N. BeMiller (eds.): *Industrial Gums*, 3rd ed., Academic Press, San Diego 1993, p. 257–293.

[19] R. Ilker, A. S. Szcezesniak, "Structural and chemical bases for texture of plant foodstuffs," *J. Text. Studies*. **21** (1990) 1.

[20] J. P. Van Buren, "The Chemistry of Texture in Fruits and Vegetables," *J. Text. Studies*. **10** (1979) 1.

[21] M. McNeil, A. G. Darvill, S. C. Fry, P. Albersheim, "Structure and Function of the primary Cell Walls of Plants," *Ann. Rev. Biochem*. **53** (1984) 625.

[22] R. R. Selvendran: "The Chemistry of Plant Cell Walls," in G. G. Birch, J. Parker (eds.):*Dietary Fibre*, Applied Sci. Publ., London 1995.

[23] M. C. McCann, K.Roberts: *The Cytoskeletal Basis of Plant Growth and Form*, Oxford, 1991, pp. 109–129

[24] C. M. G. C. Renard, H. A. Schols, A. G. J. Voragen, J. F. Thibault, W. Pilnik, "Studies on Apple Protopectin. III. Characterization of the Material Extracted by Pure Polysaccharidases from Apple Cell Walls," *Carbohydr. Polym*. **15** (1991) 13.

[25] C. M. G. C. Renard, A. G. J. Voragen, J. F. Thibault, W. Pilnik, "Comparison between enzymatically and chemically extracted pectins from apple cell walls," *Animal Feed Sci. Technol*. **32** (1991) 69.

[26] J. A. De Vries, F. M. Rombouts, A. G. J. Voragen, W. Pilnik, "Enzymic Degradation of Apple Pectins," *Carbohydr. Polym.* **2** (1982) 25.

[27] R. R. Selvendran, "Development in the Chemistry and Biochemistry of Pectic and Hemicellulosic Polymers,"*J. Cell Sci. Suppl.* **2** (1985) 51.

[28] T. P. Kravtchenko, A. G. J. Voragen, W. Pilnik, "Analytical comparison ot three industrial pectin preparations," *Carbohydr. Polym*. **18** (1992) 17.

[29] H. Thier: "Identification and Quantification of Natural Polysaccharides in Food Stuffs," in G. O. Phillips, D. J. Wedlock, P. A. Williams (eds.): *Gums and Stabilisers for the Food Industry 2, Applications of Hydrocolloids*, Pergamon Press, Oxford 1984.

[30] T. P. Kravtchenko, M. Penci, A. G. J. Voragen, W. Pilnik, "Enzymic and chemical degradation of some industrial pectins," *Carbohydr. Polym*. **20** (1993) 195.

[31] R. H. Walter (ed.): *The Chemistry and Technology of Pectins*, Academic Press, New York 1991, p. 113.

[32] J.- F. Thibault, C. M. G. C. Renard, M. A. V. Axelos, P. Roger, M.- J. Crepeau, "Studies of the lenght of homogalacturonic regions in pectins by acid hydrolysis," *Carbohydr. Res*. **238** (1993) 271.

[33] J. A. de Vries: "Repeating units in the structure of pectin," in G. O. Phillips, P. A. Williams, D. J. Wedlock (eds.): *Gums and Stabilisers for the Food Industry* vol. **4** IRL Press, Oxford, 1988, p. 25.

[34] S. E. Harding, G. Berth, A. Ball, J. R. Mitchell, J. Garcia de la Torre, "The Molecular Weight Distribution and Conformation of Citrus Pectins in Solution Studies by Hydrodynamics," *Carbohydr. Polym*. **16** (1991) 1.

[35] Z. I. Kertesz, *The Pectic Substances*, Interscience Publishers, New York 1951.

[36] O. P. Beerh, B. Raghuramaiah, G. V. Krisnamurthy, "Utilization of mango waste: peel as a source of pectin," *J. Food Sci. Technol*. **13** (1976) 96.

[37] R. M. Weightman, C. M. G. C. Renard, J.- F. Thibault, "Structure and properties of the polysaccharides from pea hulls. Part 1: chemical extraction and fractionation of the polysaccharides," *Carbohydr. Polym*. **24** (1994) 139.

[38] M. J. Y. Lin, E. S. Humbert, "Extraction of pectins from sunflowers heads," *Can. Inst. Food Sci. Technol. J*. **11** (1978) 75.

[39] W. J. Kim, F. Sosulski, S. C. K. Lee, "Chemical and gelation characteristics of ammonia-demethylated sunflower pectins," *J. Food Sci.* **43** (1978) 1436.

[40] N. M. Ptitchkina, I. A. Danilova, G. Doxastakis, S. Kasapis, E. R. Morris, "Pumpkin pectin: gel formation at unusually low concentration," *Carbohydr. Polym.* **23** (1994) 265.

[41] EC Proposal for Counsil Directive on Food Additives Other than Colours and Sweeteners (III/ 3624/91/EM-REV3) 1992.

[42] G. H. Joseph, A. H. Keiser, E. F. Bryant, "High-polymer ammonia-demethylated pectinates and their gelation," *Food Technol.* **3** (1949) 85.

[43] J. C. E. Reitsma, J.-F. Thibault, W. Pilnik, "Properties of amidated pectins. I. Preparation and characterization of amidated pectins and amidated pectic acids," *Food Hydrocoll.* **1** (1986) 121.

[44] P. E. Christensen, "Methods of Grading Pectin in Relation to the Molecular Weight (Intrinsic Viscosity) of Pectin," *Food Res.* **19** (1954) 163.

[45] A. Kawabata, "Studies on chemical and physical properties of pectic substances from fruits," *Mem. Tokyo Univ. Agr.* **19** (1977) 115.

[46] W. J. Kim, V. N. M. Rao, C. J. B. Smit, "Effect of chemical composition on compressive mechanical properties of low ester pectin gels," *J. Food Sci.* **43** (1978) 572.

[47] H. A. Deckers, C. Olieman, F. M. Rombouts, W. Pilnik, "Calibration and Application of High-Performance Size Exclusion Columns for Molecular Weight Distribution of Pectins," *Carbohydr. Polym.* **6** (1986) 361.

[48] M. L. Fishman, D. T. Gillespie, S. M. Sondey, Y. S. El-Atawy, "Intrinsic Viscosity and Molecular Weight of Pectin Components," *Carbohydr. Res.* **215** (1991) 91.

[49] M. A. V. Axelos, J.- F. Thibault, J. Lefebvre "Structure of Citrus Pectins and Viscometric Study of their Solution Properties," *Int. J. Biol. Macromol.* **11** (1989) 186.

[50] J.-F. Thibault, M. Rinaudo "Interactions of mono- and divalent counterions with alkali- and enzyme-deesterified pectins in salt-free solutions," *Biopolymers* **24** (1985) 2131.

[51] C. Garnier, M. A. V. Axelos, J.- F. Thibault, "Phase Diagrams of Pectin-Calcium Systems: Influence of pH, Ionic Strenght, and Temperature on the Gelation of Pectins with Different Degrees of Methylation," *Carbohydr. Res.* **240** (1993) 219.

[52] H. S. Owens, H. Lotzkar, T. H. Schultz, W. D. Maclay, "Shape and Size of Pectinic Acid Molecules Deduced from Viscometric Measurements," *J. Am. Chem. Soc.* **68** (1946) 1628.

[53] G. Berth, H. Anger, F. Linow, "Streulichtphotometrische und Viskosimetrische Untersuchungen an Pektinen in wäßrigen Lösungen zur Molmassenbestimmung," *Nahrung* **21** (1977) 939.

[54] H. Anger, G. Berth, "Gel Permeation Chromatography and the Mark Houwink Relation for Pectins with Different Degrees of Esterification," *Carbohydr. Polym.* **6** (1986) 193.

[55] G. Ravanat, M. Rinaudo, "Investigation on Oligo- and Polygalacturonic Acids by Potentiometry and Circular Dichroism," *Biopolymers* **19** (1980) 2209.

[56] F. Michel, J.- F. Thibault, J.- L. Doublier, "Viscometric and Potentiometric Study of High-Methoxyl Pectins in the Presence of Sucrose," *Carbohydr. Polym.* **4** (1984) 283.

[57] E. Racapé, J.- F. Thibault, J. C. E. Reitsma, W. Pilnik, "Properties of Amidated Pectins. II. Polyelectrolyte Behavior and Calcium Binding of Amidated Pectins and Amidated Pectic Acids," *Biopolymers* **28** (1989) 1435.

[58] I. G. Plaschina, E. E. Braudo, V. B. Tolstoguzov, "Circular-Dichroism Studies of Pectin Solutions," *Carbohydr. Res.* **60** (1978) 1.

[59] A. Cesàro, F. Delben, S. Paoletti, "Thermodynamics of the Proton Dissociation of Natural Polyuronic Acids," *Int. J. Biol. Macromol.* **12** (1990) 170.

[60] M. D. Walkinshaw, S. Arnott, "Conformations and interactions of pectins, I. X-ray diffraction analysis of sodium pectate in neutral and acidified forms," *J. Mol. Biol.* **153** (1981) 1055.

[61] M. J. Gidley, E. R. Morris, E. J. Murray, D. A. Powell, D. A. Rees, "Spectroscopic and stoichiometric characterization of the calcium-mediated association of pectate chains in gels and in the solid state," *J. Chem. Soc. Chem. Commun.* **22** (1979) 990.

[62] FAO Food and Nutrition Papers 34 and 37, Specifications for Identity and Purity of Certain Food Additives. Distribution and Sales Section, FAO, Rome, Italy, 1986.

[63] Food Chemicals Codex, 3rd ed., National Academy of Sciences, Washington, D.C. 1992.

[64] J. N. BeMiller, "Acid-catalyzed hydrolysis of glycosides," *Adv. Carbohydr. Chem.* **22** (1967) 25.

[65] D. A. Powell, E. R. Morris, M. J. Gidley, D. A. Rees, "Conformations and Interactions of Pectins, II. Influence of Residue Sequence on Chain Association in Calcium Pectage Gels," *J. Mol. Biol.* **155** (1982) 517.

[66] P. Albersheim, H. Neukom, H. Deuel, "Splitting of pectin chain molecules in neutral solutions," *Arch. Biochem. Biophys.* **90** (1960) 46.

[67] W. Heri, H. Neukom, H. Deuel, *Helv. Chim. Acta* **44** (1961) 1939.

[68] R. Kohn, I. Furda, "Distribution of free carboxyl groups in the molecule of pectin after esterification of pectic and pectinic acid by methanol," *Collect. Czech. Chem. Commun.* **34** (1969) 641.

[69] E. L. Pippen, R. M. McCready, H. S. Owens, "Gelation properties of partially acetylated pectins," *J. Am. Chem. Soc.* **72** (1950) 813.

[70] L. W. Doner, "Analytical methods for determining pectin composition," *A.C.S. Symp. Ser.* **310** (1986) 13.

[71] FAO, Food and Nutrition Paper 52, Addendum 1, Rome, (1992), 87.

[72] *Food Chemicals Codex, FCC III Monographs,* 3. ed. (including supplements)National Academy Press, Washington D.C., 1981, 215.

[73] EEC, Council Directive 78/663, Off. J. EEC 14 08 78 (plus Updates), 1978.

[74] Z. I. Kertesz: *The Pectic Substances,* Interscience, New York 1951.

[75] A. Plöger, "Conductivity detection of pectin: a rapid HPLC method to analyze degree of esterification," *J. Food Sci.* **57** (1992) 1185.

[76] H.- J. Zhong, M. A. K. Williams, R. D. Keenan, D. M. Goodall, C. Rolin, *Carbohydr. Polym.* **32** (1997) 27.

[77] G. Berth, "Studies on the Heterogeneity of Citrus Pectin by Gel Permeation Chromatography on Sepharose 2B/Sepharose 4B," *Carbohydr. Polym.* **8** (1988) 105.

[78] R. C. Jordan, D. A. Brant, "An Investigation of Pectin and Pectic Acid in Dilute Aqueous Solution,"*Biopolymers* **17** (1978) 2885.

[79] S. Sawayama, A. Kawabata, H. Nakahara, T. Kamata, "A Light Scattering Study on the Effects of pH on Pectin Aggregation in Aqueous Solution," *Food Hydrocoll.* **2** (1988) 3.

[80] R. H. Walter, H. L. Matias, "Pectin Aggregation Number by Light Scattering and Reducing End-Group Analysis," *Carbohydr. Polym.* **15** (1991) 33.

[81] W. D. Holloway, C. Tasman-Jones, K. Maher, "Pectin digestion in humans," *Am. J. Clin. Nutr.* **37** (1983) 253.

[82] E. C. Titgemeyer, L. D. Burquin, G. C. Fahey, Jr. K. A. Garleb, "Fermentability of various fiber sources by human fecal bacteria in vitro," *Am. J. Clin. Nutr.* **53** (1991) 1418.

[83] M. Nyman, N.- G. Asp, "Fermentation of dietary fibre components in the rat intestinal tract," *Br. J. Nutr.* **47** (1982) 357.

[84] M. Nyman, N.- G. Asp, J. Cummings, H. Wiggins, "Fermentation of dietary fiber in the intestinal tract: comparison between man and rat," *Brit. J. Nutr.* **55** (1986) 487.

[85] J. J. Doesburg: "Pectic Substances in Fresh and Preserved Fruits and Vegetables," I.B.V.T. Communication no. 25, Sprenger Institute, Wageningen 1965.

[86] R. A. Baker, "Potential Dietary Benefits of Citrus Pectin and Fiber," *Food Technol.* (1994) Nov., 133.

[87] A. Siddhu, S. Sud, R. L. Bijlani, M. G. Karmarkar, U. Nayar, "Nutrient interaction in relation to glycaemic and insulaemic response,"*Indian. J. Physiol. Phamacol.* **36** (1992) 21–28.

[88] D. J. A. Jenkins, et al. "Pectin and complications after gastric surgery: normalisation of postprandial glucose and endocrine responses," *Gut* **21** (1980) 574.

[89] S. E. Schwartz, R. A. Levine, A. Singh, J. R. Scheidecker, N. S. Track, "Sustained Pectin Ingestion Delays Gastric Emptying," *Gastroenterology* **83** (1982) 812.

[90] S. Satchithanandam, M. Vargofcak-Apker, R. J. Calvert, A. R. Leeds, M. M. Cassidy, "Alteration of gastrointestinal mucin by fiber feeding in rats," *J. Nutr.* **120** (1990) 1179.

[91] C. D. Lorenzo, C. M. Williams, F. Hajnal, J. E. Valenzuela, "Pectin Delays Gastric Emptying and Increases Satiety in Obese Subjects," *Gastroenterology* **95** (1988) 1211.

[92] D. M. Zimmaro, et al. "Isotonic Tube Feeding Formula Induces Liquid Stool in Normal Subjects: Reversal by Pectin," *J. Parenteral and Enteral Nutr.* **13** (1989) 117.

[93] X.-B. Sun, T. Matsumoto, H. Yamada, "Anti-Ulcer Activity and Mode of Action of the Polysaccharide Fraction from the Leaves of Panax ginseng," *Planta Med.* **58** (1993) 432.

[94] H. Yamada, M. Hirano, H. Kiyohara, "Partial structure of an anti-ulcer pectic polysaccharide from the roots of Bupleurum falcatum L.," *Carbohydr. Res.* **219** (1991) 173.

[95] X.-B. Sun, T. Matsumoto, H. Yamada, "Effects of a Polysaccharide Fraction from the Roots of Bupleurum falcatum L. on Experimental Gastric Ulcer Models in Rats and Mice," *J. Pharm. Pharmacol.* **43** (1991c) 699.

[96] A. Kawabata: *Studies on Chemical and Physical Properties of Pectic Substances from Fruits*, Memoirs of the Tokyo University of Agriculture XIX, 1977, pp. 115–200.

[97] L. M. Drews, C. Kies, H. M. Fox, "Effect of dietary fiber on copper, zinc, and magnesium utilization by adolescent boys," *Am. J. Clin. Nutr.* **32** (1979) 1893.

[98] K. Y. Lei, M. W. Davis, M. M. Fang, L. C. Young, "Effect of pectin on zinc, copper and iron balances in humans," *Nutr. Rep. Int.* **22** (1980) 459.

[99] J. H. Cummings, et al., "The digestion of pectin in the human gut and its effect on calcium absorption and large bowel function,"*Br. J. Nutr.* **41** (1979) 477.

[100] P.- E. Glahn, "Hydrocolloid stabilization of protein suspensions at low pH," *Prog. Fd. Nutr. Sci.* **6** (1982) 171.

[101] A. Parker, P. Boulenguer, T. P Kravtchenko: "Effect of the addition of high methoxyl pectin on the rheology and colloidal stability of acid milk drinks," in K. Nishinari, E. Doi (eds.): *Food Hydrocolloids, Structures, Properties and Functions,* Plenum Press, New York 1994, p. 307.

[102] P.- E. Glahn, C. Rolin, "Casein-pectin interaction in sour milk beverages," *Food Ingred. Eur. Conf. Proc.* 1994, 252.

[103] B. U. Nielsen:"Fiber-Based Fat Mimetics: Pectin," in S. Roller, S. Jones (eds.): *Handbook of Fat Replacers,* CRC Press, Boca Raton, FL 1996, p. 161.

[104] W. Pilnik, P. Zwiker: "Pektine," *Gordian* **70** (1970) 202–204, 252–257, 302–305, 343–346.

[105] D. B. Nelson, C. J. B. Smit, R. R. Wiles in H. D. Graham (ed.): *Food Colloids,* Avi, Westport, Connecticut 1977, p. 418.

[106] C. Rolin, J. De Vries in P. Harris (ed.): *Food Gels,* Elsevier Applied Science, London 1990, pp. 401–434.

[107] G. O. Aspinall in J. Preiss (ed.): *The Biochemistry of Plants,* vol. **3**, Academic Press, New York 1980, p. 473.

[108] J. N. BeMiller, *ACS Symp. Ser.* **310** (1986) 12.

[109] J. K. Pederson in R. L. Davidson (ed.): *Handbook of Water Soluble Gums and Resins*, Chap. 15–1, McGraw-Hill, New York 1980.
[110] S. Hojgaard Christensen: "Pectins" in M. Glicksman (ed.): *Food Hydrocolloids*, vol. **3**, CRC Press, Boca Raton, Fla. 1986, p. 205.
[111] W. Pilnik: "Pectin – a Many Splendoured Thing" in G. O. Phillips, D. J. Wedlock, P. A. Williams (eds.): *Gums and Stabilisers for the Food Industry 5*, IRL Press, Oxford 1990, pp. 209–221.
[112] H. Braconnot, *Ann. Chim. Phys. Ser. 2* **28** (1825) 173–178.
[113] W. Pilnik, A. G. J. Voragen: "Pectic Substances and Other Uronides," in A. C. Hulme (ed.): *The Biochemistry of Fruits and their Products*, vol. 1, Academic Press, London 1970, pp. 53–87.
[114] F. M. Rombouts, W. Pilnik: "Pectic Enzymes" in A. H. Rose, (ed.): *Economic Microbiology*, vol. 5, "Enzymes and Enzymic Conversions," Academic Press, London 1980, pp. 225–280.
[115] A. G. J. Voragen, W. Pilnik: "Pectin-Degrading Enzymes in Fruit and Vegetable Processing," *ACS Symp. Ser.* **389** (1989) 93–115.
[116] W. Pilnik, A. G. J. Voragen: "The Significance of Endogenous and Exogenous Pectic Enzymes in Fruit and Vegetable Processing" in P. F. Fox (ed.): *Food Enzymology*, vol. **1**, Elsevier Applied Science, London 1991, pp. 303–337.
[117] C. D. May: *Industrial Pectins: Sources, Production and Application, Carbohydr. Polymers* **12** (1990) 79–90.
[118] E. M. McComb, R. M. McCready, *Anal. Chem.* **29** (1957) 819–821.
[119] S. C. Frey: "Cross-Linking of Matrix Polymers in the Growing Cell Wall of Angiosperms," *Ann. Rev. Plant. Physiol.* **37** (1986) 165–186.
[120] D. H. Northcote, *ACS Symp. Ser.* **310** (1986) 134.
[121] L. Rexova-Benkova, O. Markovic, *Adv. Carbohyd. Chem. Biochem.* **33** (1976) 101–130.
[122] H. A. Schols et al., *Carbohydr. Res.* **206** (1990) 105–115.
[123] M. J. F. Searle-van Leeuwen et al., *Eur. J. Appl. Microbiol. Biotechnol.* **38** (1992) 347–349.
[124] W. Pilnik: "Pektine" in R. Heiss (ed.): *Lebensmitteltechnologie*, Springer Verlag, Berlin 1988, pp. 228–234.
[125] M. A. Joslyn, G. de Luca, *J. Colloid Sci.* **12** (1957) 108–130.
[126] M. Manabe, *Nippon Nogei Kagaku Kaishi* **45** (1971) no. 4, 195–199.
[127] Sunkist Growers, US 1 332 985, 1971.
[128] California Fruit Growers Exchange, US 2 480 710, 1949.
[129] W. J. Kim, C. J. B. Smit, V. N. M. Rao, *J. Food Sci.* **43** (1978) 74–78.
[130] W. Pilnik, A. G. J. Voragen: "Gelling Agents (Pectins) from Plants for the Food Industry," *Adv. Plant Cell Biochem. Biotechnol.* **1** (1992) 219–270.
[131] W. Pilnik, R. A. MacDonald, *Gordian* **68** (1968) 531–535.
[132] M. Byland, A. Donutzhuber, *Sven. Papperstidn.* **15** (1968) 505–508.
[133] T. P. Kravtchenko, I. Arnould, A. G. J. Voragen, W. Pilnik, *Carbohyd. Polym.* (in press).
[134] G. H. Joseph, A. H. Kieser, E. F. Bryant, *Food. Technol (Chigago)* **3** (1949) 85–90.
[135] M. Manabe, *Nippon Nogie Kogaku Kaishi* **45** (1971) no. 9, 417–422.
[136] F. M. Rombouts, A. K. Wissenburg, W. Pilnik, *J. Chromatogr.* **168** (1979) 151–161.
[137] W. Somers, J. Visser, F. M. Rombouts, K. van t'Riet: "Developments in Downstream Processing of (Poly)saccharide Converting Enzymes," *J. Biotechnol.* **11** (1989) 199–222.
[138] H. Neukom, *Schweiz. Landwirtsch. Forsch.* **2** (1963) 112–122.
[139] H. F. Launer, Y. Tomimatsu, *Anal. Chem.* **31** (1959) no. 9, 1569–1574.
[140] D. A. Rees, *Adv. Carbohydr. Chem. Biochem.* **24** (1969) 267–332.

[141] D. A. Rees, E. R. Morris, D. Thom, J. K. Madden in G. O. Aspinall (ed.): *Polysaccharides*, vol. **1**, Academic Press, New York 1982, pp. 195–290.
[142] D. A. Rees, A. W. Wight, *J. Chem. Soc. B* 1971, 1366.
[143] D. Oakenfull, A. Scott: "New Approaches to the Study of Food Gels," in G. O. Philips, D. J. Wedlock, P. A. Williams (eds.): *Gums and Stabilisers for the Food Industry*, vol. 3, Elsevier Applied Science Publishers **1985**, London, pp. 465–475.
[144] E. R. Morris et al., *Int. J. Biol. Macromol.* **2** (1980) 237.
[145] D. Thom, I. C. M. Dea, E. R. Morris, D. A. Powell, *Progr. Fd. Nutr. Sci.* **6** (1982) 87–108.
[146] V. J. Morris in J. R. Mitchell, D. A. Ledward (eds.): *Functional Properties of Food Macromolecules*, Elsevier, Amsterdam, 1986, pp. 121–169.
[147] Racapé et al., *Biopolymers* **28** (1989) 1435–1448.
[148] B. Lockwood, *Food Process. Ind.* **41** (1972) 493, 47–51.
[149] I. F. T. Pectin Standardization, *Food Technol. (Chicago)* **13** (1959) 496–500.
[150] M. J. H. Keijbets, W. Pilnik, *Potato Res.* **17** (1974) 169–177.
[151] M. F. Katan, P. v.d. Bovenkamp, *Basic Clin. Nutr.* 3 (1981) 217–239.
[152] R. F. McFeeters, S. A. Armstrong, *Anal. Biochem.* **139** (1984) 212–217.
[153] L. G. Bartolome, J. E. Hoff, *J. Agric. Food Chem.* **20** (1972) 266–270.
[154] M. A. Litchman, R. P. Upton, *Anal. Chem.* **44** (1972) 1495–1497.
[155] G. C. Cochrane, *J. Chromatogr. Sci.* **13** (1975) 440–447.
[156] E. L. Pippen, R. M. McCready, H. S. Owen, *Anal. Chem.* **22** (1950) 1457–1458.
[157] A. G. J. Voragen, H. A. Schols, W. Pilnik, *Food Hydrocolloids* **1** (1986) 65–70.
[158] FAO Food and Nutrition Paper 31/2 (Rome) 1984, p. 75.
[159] J. C. E. Reitsma, W. Pilnik: "Analysis of Mixtures of Pectins and Amidated Pectins," *Carbohydr. Polym.* **10** (1989) 315–319.
[160] T. Bitter, H. M. Muir, *Anal. Biochem.* **4** (1962) 330.
[161] N. Blumenkranz, G. Asboe-Hansen, *Anal. Biochem.* **54** (1973) 484–489.
[162] T. M. C. C. Filisetti-Cozzi, N. C. Carpita, *Anal. Biochem.* **197** (1991) 157–162.
[163] T. P. Kravtchenko, A. G. J. Voragen, W. Pilnik, *Carbohydr. Polym.* **18** (1992) 17–25.
[164] R. R. Selvendran, P. Ryden: "Isolation and Analysis of Plant Cell Walls," in P. M. Dey, J. B. Harborne (eds.): *Methods in Plant Biochemistry* **2** (1990) 549–579.
[165] C. M. G. C. Renard, A. G. J. Voragen, J. F. Thibault, W. Pilnik, *Carbohyd. Poly.* **12** (1990) 9–25.
[166] H. U. Endress: in [12] pp. 251–267.
[167] P. A. Sanderson, J. Baird: "Industrial Utilization of Polysaccharides" in G. O. Aspinall (eds.): *The Polysaccharides*, vol. **2**, Academic Press, London 1983, pp. 411–490.
[168] T. J. Painter in G. O. Aspinall (ed.): *The Polysaccharides*, vol. **2**, Academic Press, London 1983, pp. 257–275.
[169] P. A. Sandford in R. R. Colwell, E. R. Pariser, R. J. Sinskey (eds.): *Biotechnology of Marine Polysaccharides* Hemisphere, Washington 1985, pp. 454–516.
[170] W. J. Sime. Alignates in P. Harris (ed.): *Food Gels*, Elsevier Applied Science Press, New York 1990, pp. 53–78.
[171] W. Pilnik, A. G. J. Voragen: "Pektine und Alginate" in H. Neukom and W. Pilnik (eds.): *Gelier- und Verdickungsmittel in Lebensmitteln*, Forster Verlag, Zürich 1980, pp. 67–94.
[172] E. Percival, R. H. McDowell: "Algal Polysaccharides," *Methods in Plant Biochemistry* **2** (1990) 523–548.
[173] V. J. Chapman: *Seaweeds and Their Uses*, 2nd ed., Methuen & Co. Ltd., London 1970.

[174] A. Haug: "Composition and Properties of Alignates," Report no. 30, Norwegian Institute for Seaweed Research, Trondheim 1964.
[175] E. C. C. Stanford, GB 142, 1881.
[176] F. G. Fischer, H. Dörfel, *Hoppe Seyler's Z. Physiol. Chem.* **302** (1955) 186–203.
[177] A. Haug, B. Larsen, O. Smidsrod, *Acta Chem. Scand.* **21** (1967) 691–704.
[178] W. A. P. Black, *J. Mar. Biol. Assoc. U.K.* **29** (1950) 45–72.
[179] H. A. Hoppe, O. J. Schmid, *Bot. Mar. Suppl.* **3** (1962) 16–66.
[180] A. Haug, B. Larsen, *Acta Chem. Scand.* **16** (1962) 1908–1918.
[181] A. Penman, G. R. Sanderson, *Carbohydr. Res.* **25** (1972) 273–285.
[182] D. F. Pindar, C. Bucke, *Biochem. J.* **152** (1975) 617–622.
[183] G. Skjäk-Break, B. Larsen, *Carbohydr. Res.* **139** (1985) 273–283.
[184] Marine Colloids, US 2 128 551, 1938.
[185] Kelco, US 2 036 934, 1936.
[186] A. B. Steiner, W. H. McNeely, US 2 494 911, 1950.
[187] A. P. Imeson in G. O. Phillips, D. J. Wedlock, P. A. Williams (eds.): *Gums and Stabilisers for the Food Industry 2*, Pergamon Press, Oxford 1984, p. 189.
[188] A. Haug, B. Larsen, O. Smidsrod, *Acta Chem. Scand.* **20** (1966) 183–190.
[189] A. Haug, B. Larsen, O. Smidsrod, *Acta Chem. Scand.* **21** (1967) 2859–2870.
[190] J. Preiss, G. Ashwell, *J. Biol. Chem.* **237** (1962) 309–316.
[191] M. Jefferies, G. Pass, G. O. Phillips, *J. Sci. Fd. Agric.* **28** (1977) 173–179.
[192] O. Smidsrod, *Faraday Discuss. Chem. Soc.* **57** (1974) 263.
[193] D. Thom. I. C. M. Dea, E. R. Morris, D. A. Powell, *Progr. Food Nutr. Sci.* **6** (1982) 97.
[194] K. Taft, *Progr. Food Nutr. Sci.* **6** (1982) 89.
[195] H. A. Schols, *Analyt. Biochem.* (in press).
[196] H. Grasdalen, *Carbohydr. Res.* **118** (1983) 225–260.
[197] J. S. Craigie, E. R. Morris, D. A. Rees, D. Thom, *Carbohydr. Polym.* **4** (1984) 237–252.
[198] G. A. Towle in, [2] pp. 237–252.
[199] J. K. Pedersen in, [4] pp. 113–133.
[200] N. F. Stanley in, [6] pp. 79–119.
[201] Marine Colloids, US 3 094 517, 1963.
[202] O. Smidsrod, B. Larsen, A. Pernas, A. Haug, *Acta Chem. Scand.* **21** (1967) 2585–2598.
[203] C. J. Lawson, D. A. Rees, *Nature (London)* **227** (1970) 390–395.
[204] E. R. Morris, D. A. Rees, G. R. Robinson, *J. Mol. Biol.* **138** (1980) 349.
[205] O. Smidsrod, H. Grasdalen, *Carbohydr. Polym.* **2** (1982) 270.
[206] I. C. M. Dea et al., *Carbohydr. Res.* **57** (1977) 249.
[207] B. V. McCleary, *Carbohydr. Res.* **71** (1977) 249.
[208] E. R. Morris: "Mixed Polymer Gels," in [6] pp. 291–359.
[209] T. Matsukashi in P. Harris (ed.): *Food Gels*, Elsevier Applied Science, London 1990, pp. 1–51.
[210] S. Arnolt et al., *J. Mol. Biol.* **90**, (1974) 269.
[211] I. C. M. Dea, A. Morrison, *Adv. Carbohydr. Chem. Biochem.* **31** (1975) 241.
[212] B. Russel, H. Mead, A. Polson, *Biochem. Biophys. Acta* **86** (1964) 169–174.
[213] C. R. Williams: "The Processing of Gum Arabic to give Improved Functional Properties," in G. O. Phillips, D. J. Wedlock, P. A. Williams (eds.): *Gums and Stabilisers for the Food Industry*, 5th ed., IRL Press, Oxford 1990, pp. 37–40.
[214] A. P. Imeson: "Exudate Gums," in A. Imseon (ed.): *Thickening and Gelling Agents for Food*, Blackie Academic and Professional, London 1992, pp. 66–97.

[215] A. M. Stephen: "Structure and Properties of Exudate Gums," in G. O. Phillips, D. J. Wedlock, P. A. Williams (eds.): *Gums and Stabilisers for the Food Industry*, 5th ed., IRL Press, Oxford 1990, pp. 3–16.

[216] S. Conolly, J. C. Fenyo, M. C. Vandevelde, *Food Hydrocolloids* **1** (1987) 477–480.

[217] J. C. Fenyo, M. C. Vandevelde: "Physico-Chemical Properties of Gum Arabic in Relation to Structure," in G. O. Phillips, D. J. Wedlock, P. A. Williams (eds.): *Gums and Stabiliserer for the Food Industry*, 5th ed., IRL Press, Oxford 1990, pp. 17–23.

[218] P. A. Williams, G. O. Phillips, R. C. Randall: "Structure-Relationships of Gum Arabic," in G. O. Phillips, D. J. Wedlock, P. A. Williams (eds.): *Gums and Stabilisers for the Food Industry*, 5th ed., IRL Press, Oxford 1990, pp. 25–36.

[219] D. M. W. Anderson: *Food Hydrocolloids* **2** (1988) 417–423.

[220] G. O. Aspinall, V. P. Bhavanadan, T. B. Christensen, *J. Chem. Soc.* (1965) 2677–2684.

[221] B. Urlacher, B. Dalbe, "Xanthan Gum," in A. Imeson (ed.): *Thickening and Gelling Agents for Food*, Blackie Academic and Professional, London 1992, pp. 202–226.

[222] A. R. Jeanes, J. E. Pittsley, F. R. Santi, *J. Appl. Polymer Sci.* **5** (1961) 519–526.

[223] P. Delest: "Fermentation Technology of Microbial Polysaccharides," in G. O. Phillips, D. J. Wedlock, P. A. Williams (eds.): *Gums and Stabilisers for the Food Industry*, 5th ed., IRL Press, Oxford 1990, pp. 301–313.

[224] A. A. Lawrence: *Edible Gums and Related Substances*, Noyes Data Corp., Park Ridge, N.J. 1973.

[225] P. E. Janson, L. Keene, B. Lindberg, *Carbohydr. Res.* **45** (1975) 275–282.

[226] M. R. Betlach et al.: "Genetic Engineering, Structure/Property Relations and Applications," in M. Yalpani (ed.): *Industrial Polysaccharides*, Elsevier, Amsterdam 1987, pp. 35–50.

[227] P. J. Whitcomb, C. W. Makosko, *J. Rheol.* **22** (1978) no. 5, 493–505.

[228] M. Rinaudo, M. Milos, *Int. J. Biol. Macromol.* **2** (1980) 45–48.

[229] V. J. Morris: "Science, Structure and Applications of Microbial Polysaccharides," in G. O. Phillips, D. J. Wedlock, P. A. Williams (eds.): *Gums and Stabilisers for the Food Industry*, 5th ed., IRL Press, Oxford 1990, pp. 315–328.

[230] I. T. Norton, D. M. Goodall, S. A. Frangon, E. R. Morris, D. A. Rees, *J. Mol. Biol.* **175** (1984) 371.

[231] V. J. Morris, *Food Biotechnol.* **4** (1990) 45–57.

[232] G. R Sanderson: "Gellan Gum," in, [6] pp. 201–232.

[233] W. Gibson: "Gellan Gum," in A. Imeson (ed.): *Thickening and Gelling Agents for Food*, Blackie Academic and Professional, London 1992, pp. 227–249.

[234] J. E. Fox: "Seed Gums," in A. Imeson (ed.) *Thickening and Gelling Agents for Food*, Blackie Academic and Professional, London 1992, pp. 153–170.

[235] I. C. M. Dea: "Structure/Function Relationships of Galactomannans and Food Grade Cellulosics," in G. O. Phillips, D. J. Wedlock, P. A. Williams (eds.): *Gums and Stabilisers for the Food Industry*, 5th ed., IRL Press, Oxford 1990, pp. 373–382.

[236] A. H. Clark, I. C. M. Dea, B. V. McCleary: "The Effect of the Galactomannan Fine Structure on Their Interaction Properties," in G. O. Phillips, D. J. Wedlock, P. A. Williams (eds.): *Gums and Stabilisers for the Food Industry*, 3rd ed., Elsevier Applied Science Publishers, London 1985, pp. 429–440.

[237] I. C. M. Dea, A. H. Clark, B. V. McCleary, *Carbohyd. Res.* **142** (1986) 275–294.

[238] P. M. Dey, *Adv. Carbohydr. Chem. Biochem.* **35** (1978) 341–376.

[239] B. V. McCleary, *Int. J. Biol. Macromol.* **8** (1986) 349–354.

Propanal

ANTHONY J. PAPA, Union Carbide Chemicals and Plastics Company, Inc., South Charleston, WV 25303, United States

1. Introduction 4037
2. Physical Properties 4037
3. Chemical Properties and Uses. 4038
4. Production 4041
5. Quality Specifications....... 4041
6. Economic Aspects 4042
7. Storage and Transportation .. 4042
8. Toxicology and Occupational Health.................. 4042
9. References............... 4043

1. Introduction

Propanal [*123-38-6*], propionaldehyde, CH_3CH_2CHO, M_r 58.08, is a low-boiling, colorless, flammable liquid with a sharp, suffocating odor. It is found in vegetables such as onions, in cheese and other dairy products, and as an active ingredient in coffee beans, contributing to the aroma of coffee [1].

Propanal is highly reactive, and is employed primarily as a chemical intermediate to prepare C-3 and C-6 compounds. It is generally produced by hydroformylation of ethylene. Annual world production was estimated at 154×10^3 t in 1988 [2].

2. Physical Properties

Propanal is completely miscible with organic solvents such as alcohols, ether, and benzene. Solubility in water decreases at elevated temperature. For example, propanal is miscible with water in all proportions below 15 °C, but becomes less soluble and separates from solution in aqueous mixtures containing ca. 20 – 70 wt% propanal as the temperature increases. Propanal forms azeotropic mixtures with water (98 wt% propanal, *bp* 47.5 °C) and such other compounds as propyl nitrite (82 wt% propanal, *bp* 47.3 °C), 1-chloropropane (*bp* 46.4 °C), and cyclopropyl methyl ether (*bp* 43 °C). Some physical properties of propanal follow [3] – [7]:

Critical temperature	235.95 °C
Critical pressure	4.660 MPa
mp	−80.0 °C
bp	
101.3 kPa	48.0 °C
40.0 kPa	23.6 °C
1.33 kPa	−38.0 °C
Vapor pressure (20 °C)	34.391 kPa
Refractive index (25 °C)	1.3593
Liquid density (20 °C)	0.797 g/cm^3
Liquid viscosity (20 °C)	0.346 mPa · s
Surface tension (20 °C)	22.95 mN/m
Flash point	
Closed cup	−30.0 °C
Open cup	−7.2 – 9.4 °C
Autoignition temperature	206.85 °C
Lower flammability limit	2.9 vol%
Upper flammability limit	17.0 vol%
Solubility in water (20 °C)	35.6 wt%
Solubility of water in propanal (20 °C)	21.1 wt%

3. Chemical Properties and Uses

Propanal exhibits the characteristic high reactivity of low molecular mass aldehydes such as ease of oxidation and polymerization (trimerization). The reactions are caused by the polarity of the carbonyl group and acidity of the α-hydrogen atoms.

Uses. Propanal has no uses as such; it serves as a source of a propyl group for chemical synthesis. Propanal is primarily converted to 1-propanol, propionic acid, and trimethylolethane (trihydroxymethylethane). Minor applications include the production of pharmaceuticals (e.g., meprobmat), agricultural chemicals, rubber additives, and corrosion inhibitors.

Reduction. 1-Propanol [71-23-8] is the major product of catalytic reduction of propanal (→Propanols). Reduction is carried out most economically by a continuous vapor-phase process over a heterogeneous catalyst of supported reduced nickel, copper, and/or zinc and manganese metals [8]–[10]. Reaction is normally conducted at 125–200 °C and a pressure of about 700 kPa. An advantage of the Cu–Zn catalyst is that it promotes hydrogenolysis of the propanal Tischenko ester (propyl propionate) byproduct to 1-propanol [11]. Liquid-phase reduction on a fixed bed of nickel and copper catalyst is also highly effective in reducing aldehyde to alcohol, but the technique suffers from low efficiency and high cost on a commercial scale.

Oxidation. Propionic acid [79-09-4] is produced commercially by liquid-phase noncatalytic air oxidation of propanal [12]–[14]. Oxidation proceeds readily at relatively low temperatures (50–100 °C), and high aldehyde conversions and acid selectivities (>90%) are achieved. Continuing efforts are being made to improve the process, particularly oxidation rates, using transition-metal catalysts [15]–[17]. Major byproducts result from reactions associated with the oxidation process itself, and only a few arise from propanal condensation reactions. Propionic acid is used as a preservative for animal feed and as a chemical intermediate for preparation of sodium and calcium salts added to baked goods, tobacco, and certain vegetables to inhibit the growth of mold and fungus. It is also used to prepare diethyl ketone by passing hot vapors over heterogeneous catalysts such as manganese(II) carbonate [18].

Peroxypropionic acid [4212-43-5] ($C_2H_5CO_3H$), useful for peroxidations of olefins, can be obtained by oxidation of propanal with oxygen over a complex-metal catalyst in acetone at 10–30 °C and atmospheric pressure in yields >85% without decomposition of the peroxy acid [19].

Condensation and Addition. Propanal is a key intermediate for the manufacture of trimethylolethane [77-85-0] (2-hydroxymethyl-2-methyl-1,3-propanediol) by aldol condensation with excess formaldehyde in the presence of an alkaline catalyst [20]–[22]. In 1988, 5.4×10^3 t of trimethylolethane was produced and used largely to impart hardness, gloss retention, and durability in alkyd-resin formulations [2].

The two α-hydrogens of propanal can be readily replaced by formaldehyde in a modified Perkin-type reaction catalyzed with a mixture of secondary amine and a carboxylic acid to produce methacrolein [78-85-3] [23]–[26]. Methacrolein yields >90% are obtained under relatively mild conditions (50–100 °C and <1 MPa). Atmospheric oxidation of methacrolein to methacrylic acid and subsequent esterification produces methyl methacrylate [80-62-6]. This route to methyl methacrylate appears to be more economical than similar processes starting with propionic acid. If successful, it could open a new major market for propanal [27], [28].

Propanal undergoes self-aldol condensation or cross-aldol condensation with aldehydes or ketones in the presence of aqueous strong alkali in a conventional two-phase liquid process [29]. Self-condensation gives initially the aldol 3-hydroxy-2-methylpentanal [30], which can be dehydrated by heating to 2-methyl-2-pentenal [623-36-9]. Typical impurities from the aldol condensation of propanal in alkali at ca. 80–100 °C have been identified, and include 2,4,6-triethyl-1,3,5-trioxane, 2,4-dimethyl-2,4-heptadienal, 5-hydroxy-2,4-dimethyl-2-heptenal, 7-hydroxy-2,4,6-trimethyl-2,4-nonadienal, and 3-ethyl-2,4-dimethylvalerolactone [30]. Most of these impurities result from chain lengthening by linear aldolization with 5-hydroxy-2,4-dimethyl-2-heptenal, which is present in significant amounts. The 2-methyl-2-pentenal obtained from dehydration of the propanal aldol intermediate can be readily reduced, oxidized, or made to undergo a variety of addition and condensation reactions. The most useful of these is reduction to 2-methyl-1-pentanol [105-30-6] (trade name UCAR Hexanol, Union Carbide Chemicals and Plastics Company, Inc.). Because of its branching, 2-methyl-1-pentanol exhibits lower evaporation rates and lower volatility than 1-pentanol or "primary amyl alcohol" (an industrial blend of 1-pentanol and 2-methylbutanol), making it useful as a solvent extender and solution viscosity reducer. About 900 t of 2-

methyl-1-pentanol was produced in 1990. The corresponding acetate ester was commercially available in 1985–1988, but has since been withdrawn from the market.

A continuous process for specific production of the aldol intermediate (97–99% yield) has also been developed [30]. Because of its difunctionality, this β-hydroxyaldehyde is useful for the preparation of pharmaceuticals, plasticizers, and polymers.

Patents describing vapor-phase heterogeneous processes using catalysts such as Pd–rare earth oxides [31], anatase TiO_2 [32], group 11 and/or 12B metals supported on SiO_2 [33], B_2O_3 and/or P_2O_5 supported on TiO_2 or La_2O_3 [34], and La_2O_3, Nd_2O_3, and Sm_2O_3 [35] have been reported. This technique usually gives lower yields (ca. 90%) and lower aldehyde conversions (<50%), and it does not challenge the separation simplicity of the two-phase alkali process. Improved conversions (95%) and yields (88%) to 2-methylpentanal [123-15-9] were reported with a heterogeneous vapor-phase reductive condensation of propanal in the presence of Pd–PrO_3 catalyst on Al_2O_3 [31]. Other vapor-phase reductive condensation catalysts have been described for direct preparation of saturated ketones from cross-aldol condensations of propanal, including Ni, Co, and Zn supported on Al_2O_3 [36] and Mg, Al, Zn, or Te oxides or phosphates and a group 8 metal [37].

The Tischenko reaction can be used for one-step preparation of propyl propionate [106-36-5] (a specialty solvent) from propanal, but problems are associated with the aluminum alkoxide catalysts employed in this process. Propyl, butyl, and pentyl propionates are available as specialty solvents from conventional esterification of propionic acid with the corresponding alcohol [38].

A propanal trimer, 2-methyl-1,3-pentanediol monopropionate, used as a solvent modifier for paints, can be obtained from Tischenko rearrangement of the adduct formed between the aldol of propanal and a third molecule of propanal [39]. The nature of the Tischenko reaction of aldols of propanal is discussed in [40].

Propanal is converted to 2,4,6-triethyl-1,3,5-trioxane in the presence of acidic catalysts [41]. The ease of thermal reversion of the cyclic trimer has precluded an utility for materials of this class.

Propanal undergoes reductive catalytic amination with ammonia or primary, secondary, or tertiary amines to the corresponding propyl amine derivatives, which are used as corrosion inhibitors, fuel additives, pharmaceuticals, agricultural chemicals, and rubber additives [42]–[44]. Substituted pyridines, propionitrile (ethyl cyanide), or acrylonitrile can all be prepared by reaction of ammonia and propanal in the vapor phase depending on the temperature and catalyst employed [45]–[48].

Propanal can be converted to diethyl ketone, a herbicide intermediate, at high temperature (450–500 °C) in the presence of steam and a rare-earth metal-oxide catalyst [49], but propionic acid is the preferred starting material for this ketone [2].

4. Production

Low-pressure rhodium-catalyzed hydroformylation of ethylene with carbon monoxide and hydrogen is the most important process for the production of propanal. Some is still produced in the United States (and all in Europe) by the cobalt-catalyzed hydroformylation process at higher pressure, and a small fraction is obtained in Japan as a byproduct from Wacker oxidation of propene during acetone production [2].

The low-pressure hydroformylation process for production of propanal uses a soluble rhodium catalyst complex with excess triphenylphosphine. Reaction occurs in the liquid phase at 90–130 °C and a total pressure of <2.8 MPa with CO pressure <0.38 MPa and H_2 pressure <1.38 MPa [50]. Selectivities are >90% because ethylene can lead to only one isomer and impurities can be minimized. Major impurities include 2-methyl-2-pentenal, which arises by aldol condensation; trimers from the Tischenko reaction afford 2-methyl-1,3-propanediol monopropionate, which can in turn disproportionate to a mixture of 2-methyl-1,3-propanediol dipropionate and 2-methyl-1,3-propanediol. Byproduct formation during propanal production can be inhibited by exclusion of oxygen, organic acids, and metal contaminants such as iron, which act as Lewis-acid catalysts for its condensation to higher products.

5. Quality Specifications

Propanal purity is determined by gas chromatography using a capillary column. Care should be taken to ensure that the injection port is clean and low in temperature (<200 °C) to avoid condensation of propanal in the port.

Common impurities include butanal, 2-methylpropanal, aldol products, and various trimers. Typical specifications for high-purity-grade propanal follow [51], [52]:

Propanal	99.5 wt%, min. (dry basis)
Acidity	0.1 wt%, max. (calculated as propionic acid)
Distillation (103 kPa)	initial *bp* 46.0 °C, min.; after 97 mL, 50.0 °C max.
Water content	1.0–2.5 wt%
Color	20 Pt–Co, max. (ASTM)
Iron content	20 ppm by weight, max.

Water is generally added to stabilize the pure concentrated aldehyde against condensation to cyclic trimers or acyclic oligomers, catalyzed by Lewis acids such as iron, and to decrease peroxide formation on standing [53].

6. Economic Aspects

Propanal is produced in the United States by Eastman Kodak, Hoechst Celanese, and Union Carbide Chemicals and Plastics Company, Inc. About 138 000 t was produced in 1988 in the United States [54]. In Europe, production is lower. The average annual growth rate for propanal during 1988–1993 is estimated to be ca. 0.8% [2]. There is currently an excess capacity for propanal manufacture of ca. 25%.

Traditionally, propanal production has depended on the demand for 1-propanol and propionic acid, which account for more than 90% of the aldehyde consumed (Table 1).

7. Storage and Transportation

Propanal should be stored in stainless steel, baked phenolic-lined steel, or aluminum containers to avoid contamination with metal salts present in the construction materials, which at >3 ppm can cause propanal condensation. Stainless steel is most commonly employed. Condensation can also be caused by light and oxygen, and it is particularly troublesome in high-purity samples. Storage at low temperatures is not protective.

Propanal is highly reactive and forms explosive peroxides on exposure to air. Storage under a nitrogen atmosphere is recommended; addition of water also reduces peroxide formation and provides stabilization against metal-catalyzed condensations [53]. Contact with other oxidizing materials should be avoided. Caution is advisable with emptied containers containing residual vapors, which may explode on ignition.

Propanal is a volatile and flammable liquid that can present a flammability hazard under normal storage conditions. It has a relatively high vapor pressure and low flash point. Important flammability data are given in Chapter 3.

8. Toxicology and Occupational Health

Acute exposure through inhalation of propanal vapor causes extreme irritation of the eyes and may irritate the respiratory tract. High concentrations may also cause drowsiness, headache, nausea, and difficulty in breathing. Contact of the liquid with the eyes or skin causes burns. Ingestion is irritating to the gastrointestinal tract, causing burning in the mouth and throat, nausea, abdominal discomfort, and diarrhoea [55].

The oral acute toxicity data reported for rats is LD_{50} 1410 mg/kg; specific toxic effects have not been described [56]. The oral LDLo for mice is 800 mg/kg [55]. The

Table 1. Production of propanal derivatives in 1988

Compound	Production, 10^3 t	
	United States	Western Europe
1-Propanol	88.5	11
Propionic acid	47.2	
Trimethylolethane	2.7	
Others	1.4	2

* Courtesy of SRI International.

dermal LD_{50} for rabbits is 5040 mg/kg, showing slight absorption; no toxic effects were noted [56]. Severe skin irritation resulted with guinea pigs.

Propanal is a severe irritant to rabbit eyes [56]. The inhalation LCLo for rats is 8000 ppm (4 h) and in humans the LC_{50} is 21.8 g/m^3 [55]. Propanal is an anesthetic when inhaled in relatively high amounts by rats [57]. Preliminary edema was the cause of death in mice and rabbits after inhalation of high levels of propanal [57]. Exposure of rats to vapor (0, 90, or 1300 ppm, 6 h/d, for 20 consecutive exposures) showed the lowest-observed-effect level (LOEL) at 1300 ppm (liver damage) and the no-observed-effect level (NOEL) at 90 ppm [53].

TLV and OSHA PEL values have not been established for propanal.

9. References

[1] Comline News Service: Comline Industrial Machinery & Mechanical Engineering, Tokyo, Feb. 18, 1988, p. 3.
[2] SRI International: *Chemical Economics Handbook*.
[3] Design Institute for Physical Property Data (DIPPR), ONLINE data base, Project 801, Source File Tape, revision dates Aug. 1988, 1989, 1990.
[4] Thermodynamics Research Center: *TRC Thermodynamic Tables – Non-Hydrocarbons (Loose-leaf data sheets)*, The Texas A & M University System, College Station, Texas 1980.
[5] J. M. Sorensen, W. Arlt: "Liquid-liquid Equilibrium Data Collection, Binary Systems," *DECHEMA Chemistry Data Series*, vol. **V**, part 1, Schon & Wetzel, Frankfurt/Main 1979.
[6] Thermodynamics Research Center: *TRC Thermodynamic Table*, 23-2-1-(1.1100)-k, p. 1, Dec. 31, 1961, p. k-5300.
[7] N. I. Sax, R. J. Lewis, Sr.: *Dangerous Properties of Industrial Materials*, 7th ed., Van Nostrand Reinhold, New York 1989, p. 2892.
[8] Union Carbide, US 4 762 817, 1988 and US 4 876 402, 1989 (J. E. Logsdon, R. A. Loke, J. S. Merriam, R. W. Voight).
[9] Davy McKee, EP 73 129, 1983 (M. W. Bradley, A. G. Hiles, J. W. Kippax).
[10] Union Carbide, CA 1 137 519, 1982 (C. C. Pai).
[11] Davy McKee, EP 74 193, 1983 (M. W. Bradley, N. Harris, K. Turner).
[12] Centrul de Chimie Timisoara, RO 84 438, 1984 (R. Valceanu, V. D. Gazdac).
[13] W. K. Langdon, E. J. Schwoegler, *Ind. Eng. Chem.* **43** (1951) 1011.
[14] Imperial Chemical Industries, DE 1 237 092, 1967 (C. B. Cotterill).

[15] Intreprinderea "Plafar", RO 87 470, 1985 (E. V. Ceausescu, O. Costisor, T. Cadariu, M. Brezeanu).
[16] Babcock-Hitachi, JP 60 181 032, 1984 (T. Kamiguchi, K. Arikawa, R. Yamada, H. Tanimoto).
[17] Institute of Chemical Physics, Academy of Sciences, SU 793 988, 1981 (S. A. Maslov, E. A. Blyumberg, G. N. Gvozdovskii, Yu. N. Koshelev).
[18] F. C. Whitmore: *Organic Chemistry*, Van Nostrand Company, New York 1951, pp. 123, 220.
[19] Intreprinderea "Plafar", RO 89 953, 1986 (E. V. Ceausescu, M. Brezeanu, O. Costisor, T. Cadariu).
[20] BASF, DE 2 507 461, 1975 (F. Merger, S. Winderl, H. Toussaint).
[21] VEB Leuna-Werke "Walter Ulbricht", DE 1 952 738, 1970 (H. D. Eilhauer, G. Reckling).
[22] Commercial Solvents, US 3 478 115, 1969 (J. B. Bronstein, Jr.).
[23] Bayer, DE 2 813 201, 1979 (O. Immel, H. H. Schwarz, H. Quast).
[24] Ruhrchemie, DE 2 855 504, 1980 (W. Bernhagen et al.).
[25] BASF, EP 58 927, 1982 (F. Merger, H. J. Foerster).
[26] Hoechst, DE 3 740 293, 1989 (G. Diekhaus, H. Kappesser).
[27] *Chem. Week* (1988) Jan. 20, 64.
[28] A. S. Bakshi, *Oil Gas J.* **83** (1985) no. 47, 99.
[29] BASF, DE 2 727 330, 1979 (W. Schoenlebe, H. Hoffmann, W. Lengsfeld).
[30] P.-Y. Blanc, A. Perret, F. Teppa, *Helv. Chim. Acta* **47** (1964) no. 2, 567–575.
[31] BASF, US 4 270 006, 1981 (G. Heilen, A. Nissen, O. Woertz).
[32] Eastman Kodak, US 4 316 990, 1982 (D. L. Morris).
[33] Amoco Corp., JP 61 15737, 1986.
[34] Rohm & Haas, US 4 490 476, 1984 (R. J. Piccolini, M. J. Smith).
[35] Kh. M. Minachev, O. K. Atal'yan, M. A. Markov, *Izv. Akad. Nauk SSSR, Ser. Kim.* **12** (1978) 2731–2737.
[36] BASF, DE 2 625 541, 1977 (A. Nissen et al.).
[37] BASF, DE 3 319 430, 1984 (W. Gramlich et al.).
[38] Union Carbide Chemicals and Plastics, "UCAR Esters for Coating Applications," F-48589, Danbury, Conn., Sept. 1984.
[39] Eastman Kodak, US 4 225 726, 1980 (D. L. Morris, A. W. McCollum).
[40] G. Fouquet, F. Merger, R. Platz, *Liebigs Ann. Chem.* (1979) 1591–1601.
[41] Marathon Oil, US 4 169 110, 1979 (J. M. Holovka, E. Hurley, Jr.).
[42] Mitsubishi Chemical Ind., JP 62 252 746, 1987 (S. Imaki, Y. Takuma, M. Oishi).
[43] Nippon Oils and Fats Co., EP 142 868, 1985 (M. Koyama et al.).
[44] Nippon Oils and Fats Co., JP 62 164 653, 1987 (M. Koyama, F. Takahashi).
[45] Nepera, EP 371 615, 1990 (D. Feitler, H. Wetstein).
[46] E. I. Du Pont de Nemours & Co., US 2 452 187, 1948 (W. F. Gresham).
[47] Phillips Petroleum Co., US 2 412 437, 1946 (C. R. Wagner).
[48] VEB Leuna-Werke "Walter Ulbricht", DE 1 272 915, 1968 (K. Smeykal, H. J. Naumann).
[49] Eastman Kodak, FR 1 524 596, 1968 (J. C. Fleischer, J. W. Reynolds, S. H. Young, C. R. Hargis).
[50] Union Carbide, US 4 277 627, 1981 (D. R. Bryant, E. Billig).
[51] Union Carbide Chemicals and Plastics Company, Product Specification, 1-4A3-1h, Danbury, Conn., Aug. 1, 1989.
[52] Eastman Kodak, Eastman Chemical Products, Sales Specification No. 3827-7, Kingsport, Tenn., Sept. 1989.
[53] Eastman Kodak, Eastman Chemical Products, Material Safety Data Sheet for Propanal, Kingsport, Tenn., April 24, 1990.

[54] U.S. Production and Sales, U.S. Tariff Commission, "Synthetic Organic Chemicals," U.S. Government Printing Office, Washington, D.C., 1988.
[55] Union Carbide Chemicals and Plastics Company, Solvents and Coatings Materials Division, Material Safety Data Sheet, Danbury, Conn., Aug. 29, 1990.
[56] D. V. Sweet (ed.): "Registry of Toxic Effects of Chemical Substances," *DHHS (NIOSH) Publ. U.S.* **4** (1987) no. 87–114, 3862.
[57] *Patty's* 3rd ed., vol. **II A,** pp. 2629–2669.

Propanediols

CARL J. SULLIVAN, ARCO Chemical Company, Newtown Square, Pennsylvania 19073, United States

1.	1,2-Propanediol and Higher Propylene Glycols 4047	1.7.	Uses	4053
1.1.	Physical Properties 4048	1.8.	Derivatives	4055
1.2.	Chemical Properties 4049	2.	1,3-Propanediol..........	4055
1.3.	Production 4051	2.1.	Properties	4055
1.4.	Quality Specifications....... 4052	2.2.	Production	4056
1.5.	Transportation and Storage .. 4053	2.3.	Uses	4057
1.6.	Economic Aspects 4053	3.	Toxicology...............	4057
		4.	References...............	4059

1. 1,2-Propanediol and Higher Propylene Glycols

1,2-Propanediol, [57-55-6], propylene glycol, $HOCH_2CH(CH_3)OH$, is very similar to ethylene glycol in its physical and chemical properties. The first reported description of propylene glycol was by WURTZ in 1859 [1].

Industrial-scale synthesis of propylene glycol from propylene oxide and water began in the 1930s. Current production uses this same process, which leads simultaneously to di- and tripropylene glycols. The current worldwide capacity for propylene glycol is estimated to be 9×10^5 t [2]. Production may be lower because some propylene glycol production units are designed to permit switching to ethylene glycol output depending upon relative demand for the glycols.

Propylene glycol is used in many diverse applications such as unsaturated polyesters for thermoset composites, food chemistry, food processing equipment, cosmetics, and pharmaceuticals. A growing application is in the field of deicers and automotive antifreeze components.

Table 1. Physical properties of propylene glycols

Property	Propylene glycol	Dipropylene glycol*	Tripropylene glycol*
M_r	76.10	134.18	196.26
bp, °C	187.9	232.8	271
mp, °C	<-60	<-40	<-30
d_{20}^{20}	1.036	1.023	1.019
Refractive index n_D^{20}	1.4328	1.4415	1.4449
Vapor pressure (20 °C), kPa	0.011	<0.001	<0.001
Heat of vaporization at bp, kJ/kg	711	400	330
Specific heat capacity (20 °C), kJ kg^{-1} K^{-1}	2.49	2.36	2.26
Electrical conductivity (20 °C), µSm	4.4	0.19	
Cubic coefficient of expansion (20 °C), K^{-1}	7.2×10^{-4}	7.4×10^{-4}	7.7×10^{-4}
Heat conductivity (20 °C), Wm^{-1} K^{-1}	0.20	0.16	0.11
Dynamic viscosity (20 °C), mPa · s	56	107	84
Surface tension (20 °C), mNm	38	35	35
Dielectric constant (20 °C)	28	21	14
Flash point (DIN 51 758), °C	103	120	149
Ignition temperature (DIN 51 794), °C	410	350	260
Lower explosion limit in air, vol%	2.6	1.26	1.3

*Physical properties may be affected by isomer ratios.

1.1. Physical Properties

Propylene glycol is a clear, colorless, strongly hygroscopic liquid. The higher propylene glycols (di- and tripropylene glycol) have similar properties. Dipropylene glycol [110-98-5] exists as three isomers differing in the positions of the methyl substituents, and tripropylene glycol [24800-44-0] is typically a mixture of several isomers. All three glycols are moderately viscous liquids that do not readily crystallize, but solidify to vitreous masses upon cooling. Fundamental physical properties of these diols are listed in Table 1 [3]–[9].

Propylene glycol is readily miscible with water and other polar solvents (e.g., alcohols and acetone). Propylene glycols are good solvents for many polar organic substances, such as phenols, alcohols, dyes, natural products, and some resins. Propylene glycol is insoluble to sparingly soluble in such nonpolar solvents as petroleum ethers, benzene, aromatic hydrocarbons, and carbon tetrachloride.

Propylene glycol greatly reduces the freezing point of water and is therefore being developed as an alternative to ethylene glycol in automotive antifreeze systems [10], [11] and aircraft deicers [12], [13]. The solidification temperatures of propylene glycol and its higher homologues at varying water ratios are plotted in Figure 1. The viscosities and densities of aqueous propylene glycol solutions are plotted in Figures 2 and 3 [4].

Vapor pressure versus temperature curves for the three propylene glycol homologues are plotted in Figure 4. Propylene glycol forms azeotropes with the solvents listed in Table 2 [14].

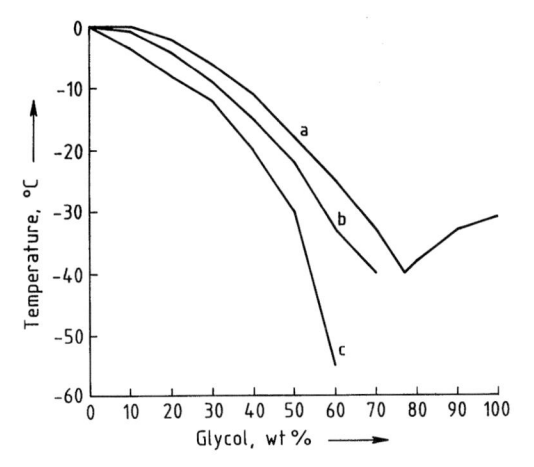

Figure 1. Solidification temperatures for aqueous propylene glycols
a) Tripropylene glycol; b) Dipropylene glycol; c) Propylene glycol

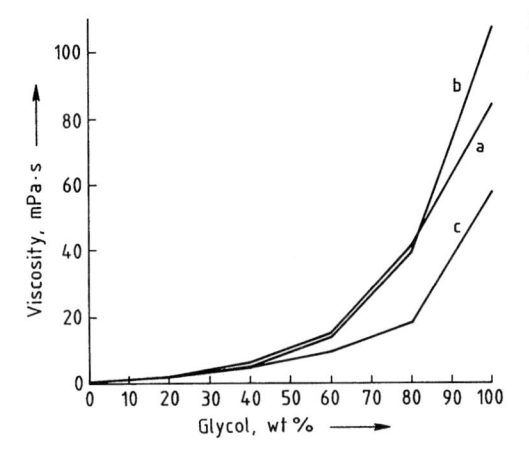

Figure 2. Viscosity of aqueous propylene glycols
a) Tripropylene glycol; b) Dipropylene glycol; c) Propylene glycol

1.2. Chemical Properties

The chemical properties of propylene glycol are determined predominantly by its hydroxyl groups, and its reactions are typical of alcohols (→ Alcohols, Aliphatic). Propylene glycol condenses with carboxylic acids at elevated temperature to yield esters and water. It also reacts readily with isocyanates and acid chlorides to yield carbamates and esters, respectively.

Propylene glycol undergoes polycondensation reactions with diacids to yield polyesters. Predominant among such applications is the reaction with maleic anhydride and other diacid components to yield unsaturated polyesters [15], [16]:

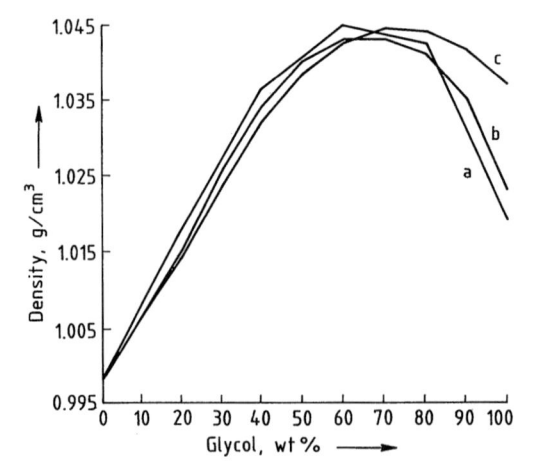

Figure 3. Density of aqueous propylene glycols at 20 °C
a) Tripropylene glycol; b) Dipropylene glycol; c) Propylene glycol

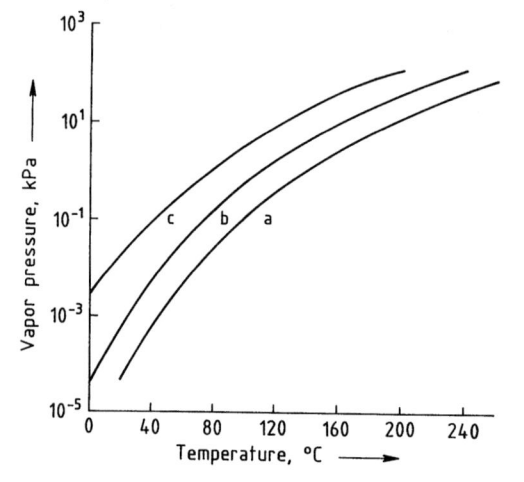

Figure 4. Vapor pressure versus temperature for propylene glycols
a) Tripropylene glycol; b) Dipropylene glycol; c) Propylene glycol

Table 2. Azeotropes with propylene glycol

Cosolvent	Cosolvent bp, °C	Azeotrope bp, °C	Propylene glycol, wt%
o-Nitrophenol	217	186	>62
Aniline	184	179.5	43
Toluene	110.6	110.5	1.5
o-Xylene	144.4	135.8	10
Dodecane	216	175	67
N,N-Dimethyl-p-toluidine	185.3	174	37
Dibutyl ether	142	136	
Tetradecane	252	179	70
Methyl hexyl ketone	173	<170	
N-Methylaniline	196	<181	>46
N,N-Dimethylaniline	194	<177	>45
Propylene glycol	187.9		

$$\text{HO-CH}_2\text{-CH(CH}_3\text{)-OH} + n\,\text{HO}_2\text{C-CH=CH-CO}_2\text{H} \longrightarrow$$

$$\text{HO-CH}_2\text{-CH(CH}_3\text{)-O}\left(\text{C(=O)-CH=CH-C(=O)-O-CH}_2\text{-CHO(CH}_3\text{)}\right)_{n-1}\text{C(=O)-CH}_2\text{=CH}_2\text{-}$$

Propylene glycol reacts with propylene oxide to form dipropylene glycol, tripropylene glycol, and polyether polyols. The resulting polypropylene glycol ethers are important industrial building blocks for polyurethane foams and elastomers [17], [18].

$$\text{HOCH}_2\text{-CH(CH}_3\text{)-OH} + n\,\text{CH}_3\text{-CH(O)CH}_3 \longrightarrow$$

$$\text{HO-CH(CH}_3\text{)-CH}_2\text{-(O-CH}_2\text{-CH(CH}_3\text{))}_n\text{OH}$$

Because of its 1,2-diol structure, propylene glycol also undergoes a variety of interesting cyclization reactions. Cyclic acetals and ketals are formed with aldehydes and ketones, respectively [19].

At elevated temperature in the presence of an acidic catalyst, propylene glycol can undergo dehydration to yield cyclic ethers. Very low levels of these dehydration products are common in aqueous byproduct streams of polycondensation reactions involving propylene glycol [20].

Propylene glycol is unusual in that the commercial product is a racemic mixture of two chiral isomers. Only small quantities of chiral material are available from specialty chemical distributors.

1.3. Production

The only industrial process for manufacturing propylene glycol is *direct hydrolysis of propylene oxide* with water. Dipropylene glycol and tripropylene glycol are formed by sequential addition of propylene oxide to propylene glycol. Consequently, all three products are produced simultaneously and separated by distillation.

Propylene oxide and water are combined in the initial stage of the process at a molar ratio of 1:15 at an initial temperature of 125 °C and a pressure of approximately 2 MPa (Fig. 5). Due to the exothermic reaction [21], the reactor effluent temperature will typically rise to 190 °C. At the above ratio of water to oxide, the resulting propylene glycol – dipropylene glycol – tripropylene glycol mixture is approximately in the ratio of 100:10:1. Higher propylene glycol ratios are attained by increasing the water-to-oxide

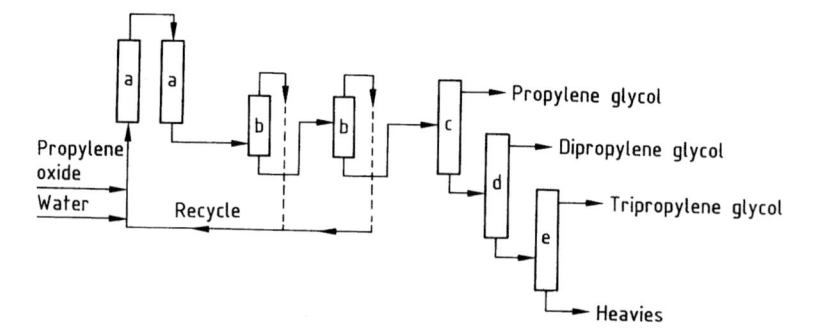

Figure 5. Schematic of propylene glycol production
a) Reactor; b) Evaporator (typically 2–4 columns; c) Propylene glycol distillation column; d) Dipropylene glycol distillation column; e) Tripropylene glycol distillation column

ratio; however, such an increase also increases recycle rates, lowers throughput, and raises energy costs.

The water–glycol mixture exits the reactor zone (a) at about 200 °C and is stripped of water in dehydration columns (b). Three successive vacuum distillations (c–e) separate high-purity propylene glycol, dipropylene glycol, and tripropylene glycol. The residue represents a mixture of higher glycols, which have some limited commercial use. Waste from this process is extraordinarily low. Because of the very low toxicity of propylene glycol, dipropylene glycol, and tripropylene glycol, these materials represent a minimal hazard to the environment. All are readily biodegradable.

Alternative routes to propylene glycol have been investigated. Acetoxidation of propene to yield glycol acetates followed by hydrolysis to the glycol has been reviewed [22]. This technique involves lower production costs than the propylene oxide route, but fixed capital investment costs are much higher. Direct hydroxylation of propene to propylene glycol using oxygen in the presence of osmium catalysts has been reported [23]–[25]. However, because of the ready availability of propylene oxide, the simplicity of the current process, and the extremely low level of waste generated by the propylene oxide route, no alternative process has yet been commercialized.

1.4. Quality Specifications

At least two commercial grades of monopropylene glycol exist—a high-purity industrial grade, and a higher-purity grade called "super pure" or "U.S.P." grade. Typical assays for the two grades of propylene glycol are the same; however, the U.S.P. grade is used in applications such as pharmaceuticals, cosmetics, and food additives, and entails higher quality standards (see Table 3) [4].

Table 3. Typical properties of commercial propylene glycol

Property	Industrial grade	U.S.P. grade
Purity, %	99.5	99.5
d_{20}^{20}	1.0377	1.035 – 1.037
Distillation initial bp, °C	185	185
dry point, °C	190	189
Acidity, wt% as acetic acid	0.005	0.002
Chlorides, ppm as Cl (max.)	10.0	1.0
Iron, ppm (max.)	1.0	0.5
Heavy metals, ppm [a]		5.0
Arsenic, ppm as As (max.)		3.0
Sulfate, wt% (max.)		0.006
Water, wt% (max.)		0.2

[a] All metal residues are calculated as if they were lead.

1.5. Transportation and Storage

Propylene glycol, dipropylene glycol, and tripropylene glycol are stable materials that require no special handling procedures. They are noncorrosive and can be transported or stored in stainless steel, aluminum, and lined steel containers. Carbon steel is also acceptable, although slight iron contamination may occur upon extended periods of storage. U.S.P.-grade propylene glycol is typically stored in stainless steel. The glycols are hygroscopic, and should be protected from unnecessary exposure to the atmosphere.

1.6. Economic Aspects

Worldwide capacity for propylene glycol production is summarized in Table 4. These capacity figures are flexible, because the same facilities can also be used to produce ethylene glycol depending upon market pressures. Actual production in the United States in 1989 is estimated to have been 3.65×10^5 t, approximately 95% of capacity [27].

1.7. Uses

Propylene Glycol. The single largest use of propylene glycol is in the manufacture of unsaturated polyester resins. It is reacted with saturated and unsaturated carboxylic diacids such as isophthalic acid and maleic anhydride. The resulting resins are dissolved in styrene or another polymerizable monomer, then combined with filler, chopped glass, a peroxide polymerization initiator, and other additives, and cured to yield a hard, cross-linked, thermoset composite [15], [16], [21]. The end-use applications for

Table 4. Worldwide capacity for propylene glycol production in 1990

Region	Number of manufacturers	Capacity, 10^3 t/a
North America	5	384
Western Europe	6	361
East Asia	7	118
Total	18	863

such products include automotive plastics, fiberglass boats, and construction. Roughly 45 % of the propylene glycol produced goes to this one broad application [5].

Propylene glycol has many diverse applications in the food and pharmaceutical industries because of the "Generally Regarded As Safe" (GRAS) status conferred on it by the FDA in the United States. Propylene glycol is used as a humectant, solvent, and preservative in food and pet food products. In the manufacture of food products, propylene glycol is used as a lubricant for machinery, as a solvent in food processing, in food wraps, and as an antifreeze agent in machinery cooling water. Propylene glycol is also used as an emollient, softening agent, and humectant in skin-care products and cosmetics, and as a carrier for livestock medicinal products [4], [7].

Propylene glycol is utilized in aircraft deicing and anti-icing fluids because of its proven performance, low toxicity, ready biodegradability, and environmental acceptance [13]. It offers freeze protection similar to that of ethylene glycol, the principle component of automotive antifreeze solutions. Automotive antifreeze compositions incorporating propylene glycol are a current research interest [10], [11], [28].

Propylene glycol is useful as a lubricant in combination with di- and tripropylene glycols. It is commonly employed as a humectant for tobacco. Propylene glycol is also introduced into many latex paints as a freeze–thaw protector, and to provide evaporation control in hot and dry environments. Propylene glycol is reacted with fatty acids or long-chain carboxylic acids to produce ester lubricants, emulsifiers, and plasticizers [29]–[31].

Propylene glycol is a precursor of many polyether polyols used in the urethane foam, elastomer, adhesives, and sealants industry, whereby the compound is reacted with propylene oxide and/or ethylene oxide in order to generate a low molecular mass polyether [17], [18].

Dipropylene and Tripropylene Glycols. Both glycols are used in hydraulic fluids and brake fluids because of their high stability and low corrosion properties. They are also used in inks and some paints because of their good solvency for pigments and binders, very low evaporation rates, and low toxicity. Stamp pad inks are a prime example [32]. Dipropylene glycol has some application in refining operations because it facilitates the separation of aromatic and aliphatic hydrocarbons. Di- and tripropylene glycols are precursors of low molecular mass esters and polyesters, which have utility as plasticizers and lubricants. Both glycols are also used to make polyester polyols for urethane applications.

1.8. Derivatives

The major end-use derivatives of propylene glycol, dipropylene glycol, and tripropylene glycol are unsaturated polyester resins as well as low molecular mass ester and polyester plasticizers. Propylene glycol-initiated polyethers are also modest-volume commodity products.

The class of materials known as glycol ethers (propylene glycol alkyl ethers) is commonly referred to as propylene glycol derivatives, but most of these are actually propylene oxide-derived products because of the ease with which they can be prepared from the oxide. Nevertheless, some glycol ether products actually are derived from the glycol — most notably propylene glycol mono-*tert*-butyl ether, which is produced by reacting propylene glycol with isobutene [33]. These derivatives find application predominantly in coatings and cleaners.

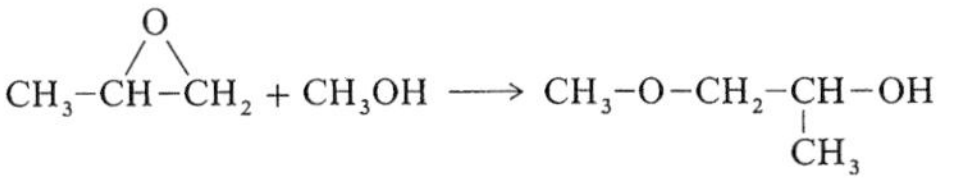

$$CH_3-CH-CH_2 + CH_3OH \longrightarrow CH_3-O-CH_2-CH-OH$$
$$\underset{\text{Propylene glycol methyl ether}}{CH_3}$$

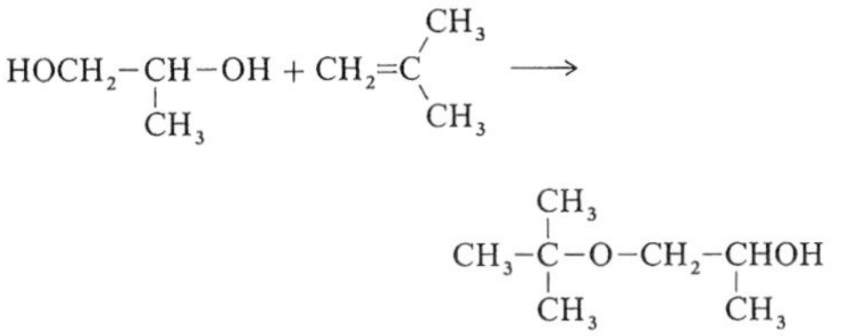

$$HOCH_2-CH-OH + CH_2=C(CH_3)_2 \longrightarrow$$
$$\underset{}{CH_3}$$

$$CH_3-\underset{CH_3}{\overset{CH_3}{C}}-O-CH_2-\underset{CH_3}{CHOH}$$
Propylene glycol *tert*-butyl ether

2. 1,3-Propanediol

1,3-Propanediol [*504-63-2*], trimethylene glycol, $HOCH_2CH_2CH_2OH$, is produced on a much smaller scale than its isomer, propylene glycol. In spite of some interesting areas of application, total production remains relatively small. The difficulties of manufacturing the product result in a poorly competitive price structure relative to other diols.

2.1. Properties

Key physical properties of 1,3-propanediol are as follows [3], [34]:

M_r 76.10
mp   −27 °C
bp (101.3 kPa)   214 °C

bp (0.7 kPa)	94 °C
Vapor pressure (20 °C)	0.008 kPa
Vapor pressure (100 °C)	0.98 kPa
d_{20}^{20}	1.0554
d_4^{20}	1.0529
Refractive index n_D^{20}	1.4389
Kinematic viscosity	46 m^2/s
Flash point	80 °C

1,3-Propanediol is a clear, colorless, odorless liquid that is miscible with water, alcohols, ethers, and formamide. It is sparingly soluble in benzene and chloroform.

The chemical properties of 1,3-propanediol are typical of alcohols. Like 1,2-propanediol, 1,3-propanediol condenses with carboxylic acids at elevated temperature to yield esters. It also reacts with isocyanates and acid chlorides to yield urethanes and esters, respectively. Unlike 1,2-propanediol, 1,3-propanediol has two primary hydroxyl groups with equivalent reactivity.

1,3-Propanediol readily forms ethers. 3,3′-Dihydroxydipropyl ether forms upon continued reflux of the diol. 1,3-Propanediol reacts with aldehydes and ketones, often in the presence of acidic catalysts, to form 1,3-dioxanes [35], [36]:

$$\underset{R'}{\overset{R}{>}}C=O + HOCH_2CH_2CH_2OH \xrightarrow{H^+}$$

$$\underset{R'}{\overset{R}{>}}C\underset{O-CH_2}{\overset{O-CH_2}{<}}CH_2 + 2\,H_2O$$

R = alkyl, H; R^1 = alkyl, H

Over aluminum oxide at temperatures above 250 °C, 1,3-propanediol decomposes to allyl alcohol, propanol, and other products.

1,3-Propanediol reacts with diacids to form polyesters. The polyester with terephthalic acid is reported to have a crystalline melting point of 220 °C [37]. Polyurethanes can also be synthesized from 1,3-propanediol.

2.2. Production

1,3-Propanediol is prepared by a two-step process involving the hydrolysis of acrolein to 3-hydroxypropanal followed by hydrogenation [3], [38]:

$$CH_2=CH-CH=O + H_2O \longrightarrow HO-CH_2CH_2CH=O$$

$$\xrightarrow[\text{Cat.}]{H_2} HO-CH_2CH_2CH_2-OH$$

Hydrolysis is carried out under weakly acidic conditions in water containing initially ca. 20% acrolein. Higher concentrations of acrolein tend to lead to greater amounts of undesired byproducts as a result of reaction between acrolein and hydroxypropanal.

3-Hydroxypropanal can be hydrogenated in the aqueous phase directly; however, the preferred technique is to extract the aldehyde into an organic solvent—particularly 2-methylpropanol—and then hydrogenate the aldehyde to yield the diol. Hydrogenation is conducted with Raney nickel under pressure in the aqueous phase and with nickel-supported catalysts at 2–4 MPa and 110–150 °C in the organic phase [39]. The diol is subsequently separated from solvent and water by distillation.

The yield of desired product by this route is relatively low—approximately 45%. Alternative techniques for synthesizing the diol have appeared in the patent literature. Hydroformylation of ethylene oxide followed by hydrogenation yields 1,3-propanediol in good yield (92%), but a high catalyst concentration and a very large excess of solvent render the process uneconomical [40]. More recently, hydroformylation of ethylene oxide directly to 1,3-propanediol with a rhodium–phosphine catalyst system has been disclosed [41]. The reaction of ethylene with formaldehyde and carboxylic acids has also not been commericialized because of low selectivity [42].

Handling of 1,3-propanediol is similar to that of propylene glycol. The material is noncorrosive and can be transported in stainless steel, aluminum, or lined steel vessels.

Quality analysis is readily accomplished by gas chromatography. A good technical grade generally displays a purity between 98.5 and 99.5%.

The environmental impact of this product is expected to be low because of its low toxicity and ready biodegradability. The handling of acrolein in the production process must be carefully controlled.

2.3. Uses

As a diol, 1,3-propanediol is subject to many of the same polymeric applications as other low molecular mass diols (e.g., ethylene glycol, propylene glycol, and 1,4-butanediol). However, its relatively high price limits its use to applications requiring very specific performance characteristics. It is a raw-material source for 1,3-dioxanes. 1,3-Propanediol-bis(4-aminobenzoate) [*57609-64-0*] can be used as a chain extender in polyurethane elastomers. This bisbenzoate, which can also be synthesized from 1,3-dichloro-propane, finds other applications as a cross-linking agent in epoxy formulations and as a rubber additive.

3. Toxicology

1,2-Propanediol. The toxicological properties of 1,2-propanediol have been reviewed [3], [4], [43]. The substance has such a low order of acute toxicity that it is considered practically nontoxic. Since 1942 it has been classified as an acceptable ingredient in pharmaceutical applications, and the United States FDA permits its use in food products and cosmetics. Official requirements for 1,2-propanediol used as an adjuvant in pharmaceuticals can be found in international and national pharmacopeias.

Acute oral LD_{50} values for rats range from 21.0 to 33.7 g/kg. For mice the values range from 23.9 to 31.8 g/kg. Tests with other animals (guinea pig, rabbit, mouse, and dog) yield LD_{50} values within the same range [43]. These values correspond to ingestion of 1.5–2.0 L by a typical adult human.

Administration of single, almost lethal doses to laboratory animals have caused balance disorders resulting from central nervous system depression. Significant impairment of motion was observed in rats after oral administration of 19.5 g/kg. A similar level of impairment was observed with an ethanol dosage of only 1.5 g/kg.

Propylene glycol is not injurious to the eye of rabbits, and it is not known to have caused any eye injuries in humans. Slight transitory stinging and tear formation may result from direct contact with the eye.

Propylene glycol does not produce any significant irritation effects in skin contact with rabbits or humans, although skin irritation has been reported in some individuals when high concentrations were held in contact with skin under closed conditions [44], [45].

Subchronic and chronic studies also indicate that propylene glycol has a low order of toxicity. A two-year study on rats demonstrated that incorporation of 5% propylene glycol into the diet did not increase or accelerate mortality. No tissue damage was noted at 7.5% of the diet for 20 weeks, but at 20% of the diet the rats exhibited nonspecific minor changes in liver and kidney cell structure [46].

Studies also indicate that propylene glycol does not present a health risk after inhalation. Monkeys and rats exposed to propylene-glycol-saturated atmospheres for 12–18 months showed no adverse effects [47].

In vivo and in vitro mutagenicity tests show no mutagenic behavior [48]. Reproductive studies with mice, rats, hamsters, and rabbits show no effects upon fetal development, and no teratogenic effects. In a reproductive study with rats, a level of 7.5% propylene glycol in food showed no effects over six generations [49].

In the body, propylene glycol is metabolized mainly to carbon dioxide via pyruvic acid, but it also enters into carbohydrate metabolism and promotes glycogen formation. At higher dose levels, some propylene glycol is excreted in the urine.

No permissible exposure levels have been established, because none appear to be necessary.

Dipropylene Glycol and Tripropylene Glycol. These two glycols have not been studied as extensively as propylene glycol, but both are considered to possess a low order of toxicity. The oral LD_{50} values in rats for dipropylene glycol and tripropylene glycol are 15 g/kg and 12 g/kg, respectively. In a subchronic study involving 5 wt% dipropylene glycol in drinking water, rats were not affected. However, at 10% some rats died, and there was evidence of liver and kidney involvement.

Both di- and tripropylene glycol demonstrate negligible eye and skin irritation. Continuous or repetitive application of the two glycols provides no indication that toxic quantities are absorbed through the skin. Because of low vapor pressures and low systemic toxicity, inhalation hazards appear negligible.

1,3-Propanediol. 1,3-Propanediol is mildly toxic. The oral LD_{50} for rats is 14–15 g/kg, whereas intraperitoneal LD_{50} values for rats and rabbits are 6–7 g/kg and 3 g/kg, respectively [50]. Prolonged administration to chickens inhibits growth [51]. Unlike its isomer, propylene glycol, 1,3-propanediol is not permitted for use in food or livestock feed products.

4. References

[1] A. Wurtz, *Justus Liebigs Ann. Chem.* **55** (1859) no. 3, 406.
[2] *Directory of Chemical Producers,* SRI International, 1990.
[3] *Ullmann,* 4th ed., **19**, 425–432.
[4] ARCO Chemical Company, Product Brochure, Propylene Glycols, USA, 1983.
[5] Dow Chemical, Product Brochure, Propylene Glycol U.S.P., USA, 1979.
[6] *Kirk-Othmer,* 3rd ed., **11**, 933–956.
[7] E. W. Flick (ed.): *Industrial Solvents Handbook,* Noyes Data Corp., Park Ridge, NJ, 1985.
[8] G. O. Curme, F. Johnston: *Glycols,* Reinhold Publ. Co., New York 1952.
[9] J. Mellon: *Polyhydric Alcohols, Spartan Books,* Washington 1962.
[10] F. E. Mark, W. Jetten, *ASTM Spec. Tech. Publ.* **88** (1986) 61–77; *Chem. Abstr.* **105** (1986) 45 989.
[11] C. Fiaud, P. Netter, M. Tadjamoli, M. Tainmann, *ASTM Spec. Tech. Publ.* **887** (1986) 162–175; *Chem. Abstr.* **105** (1986) 45 990 a.
[12] S. H. Bloom, M. A. Bloom, US 4 585 571, 1986.
[13] Kilfrost, Ltd., Product Brochure, Kilfrost ABC-3, 1989.
[14] L. H. Horsley: "Azeotropic Data-III," *Adv. Chem. Ser.* **116** (1973).
[15] J. Selley: "Polyesters, Unsaturated," *Encyclopedia of Polymer Science and Engineering,* 2nd ed., vol. **12**, Wiley Interscience, New York 1988, pp. 156–290.
[16] Amoco Chemicals Company, Product Brochure, Processing Unsaturated Polyesters on Amoco Isophthalic Acid, Bulletin IP-430, USA, 1989.
[17] S. D. Gagnon: "1,2-Epoxide Polymers," *Encyclopedia of Polymer Science and Engineering,* 2nd ed., vol. **6**, Wiley Interscience, New York 1986, pp. 275–307.
[18] G. Woods: *The ICI Polyurethane Book,* 1987.
[19] F. A. Meskens, *Synthesis* **7** (1981) 501.
[20] E. E. Parker, *Ind. Eng. Chem.* **58** (1966) no. 4, 54.
[21] T. McMilland (ed.): *Process Economics Program Yearbook International,* SRI International, 1990.
[22] H. W. Scheeline, H. Naka: "Propylene Glycol," *Process Economics Review,* Report No. PEP77-3, SRI International, 1978.
[23] Exxon, US 4 533 772, 1985 (R. C. Michaelson, R. G. Austin).
[24] Exxon, EP 77 201, 1981 (R. C. Michaelson, R. G. Austin).
[25] Gulf Research & Developmemt, US 4 308 409 (C. Y. Wu, T. P. Kobylinski, J. E. Bozik).
[26] The United States Pharmacopeia, 21st Revision, Official January 1, 1985, United States Pharmacopeial Convention, Inc., Rockville, MD, 1984.

[27] Synthetic Organic Chemicals, United States Production and Sales, 1989, United States International Trade Commission Publication no. 2338, 1990.
[28] R. D. Hercamp, R. D. Hudgens, G. E. Coughenour: "Aqueous Propylene Glycol Coolant for Heavy Duty Engines," SAE Technical Paper Ser. no. 900 434, SAE International, 1990.
[29] VEB Petrochemisches Kombinat Schwedt, DD 208 478, 1984 (R. Steinhauer, et al.); *Chem. Abstr.* **101** (1984) P 194 972 q.
[30] Meiji Seika Kaisha, Ltd., JP-Kokai 61 149 064, 1986 (Y. Takahashi, T. Yoshida); *Chem. Abstr.* **105** (1986) P 189 731 q.
[31] Institut Francais du Petrole, FR 2 560 884, 1985 (A. Bre, M. Mollard, M. Osgan); *Chem. Abstr.* **104** (1986) P 150 114 v.
[32] Dow Product Brochure, "The Glycol Ethers Handbook," 1981.
[33] L. R. Nudy, W. A. Johnston. *Household Pers. Prod. Ind.* **27** (1990) no. 4, 90.
[34] *Aldrich Catalog Handbook of Fine Chemicals,* Aldrich Chemical Company, Milwaukee, WI, 1988.
[35] J. L. Mateo, O. RuizMurillo, R. Sastre, *An. Quim. Ser C* **80** (1984) no. 2, 178.
[36] L. F. Lapuka et al., *Khim Geterotsikl. Soedin.* 1981,no. 9, 1182; *Chem. Abstr.* **95** (1981) 20 039 e.
[37] R. W. Lenz: *Organic Chemistry of Synthetic High Polymers,* Interscience, New York 1967, p. 93.
[38] R. Hall, E. S. Stern, *J. Chem. Soc.* 1950, 490.
[39] Ruhrchemie AG, US 4 094 914, 1978 (W. Rottig et al.).
[40] Shell Oil Company, US 3 463 910, 1970 (C. N. Smith, G. N. Schrauzer, K. F. Koetitz, R. J. Windgassen).
[41] Hoechst Celanese, US 4 873 378, 1989 (M. Murphy et al.).
[42] National Distillers and Chemical Corp., US 4 322 355, 1980 (D. Horvitz, W. D. Bargh).
[43] J. A. Ruddick, *Toxicol. Appl. Pharmacol.* **21** (1972) 102.
[44] K. Motoyoshi et al.: "The Safety of Propylene Glycol and Other Humectants," *Cosmet. Toiletries* **99** (1984) 83–91.
[45] R. J. Trancik, H. I. Maibach: "Propylene Glycol: Irritation or Sensitization," *Contact Dermatitis* **8** (1982) 185–189.
[46] F. Gaunt et al., *Food Cosmet. Toxicol.* **10** (1972) 151.
[47] O. H. Robertson et al., *J. Pharmacol. Exp. Ther.* **91** (1947) 52.
[48] FDA report PB-245 450 (1974).
[49] *Fed. Regist.* **42** (1977) June 17, 30 865.
[50] W. Van Winkle, *J. Pharmacol.* **72** (1941) 227.
[51] R. D. Creek, *Poult. Sci.* **49** (1970) no. 6, 1686; *Chem. Abstr.* **74** (1971) 74 453.

Propanols

ANTHONY J. PAPA, Union Carbide Chemicals and Plastics Company, Inc., South Charleston, WV 25303, United States

1. Introduction 4061
2. Physical Properties 4062
3. Chemical Properties 4063
4. Production 4065
4.1. Production of 1-Propanol.... 4065
4.2. Production of 2-Propanol.... 4067
5. Uses 4070
6. Specifications 4072
7. Economic Aspects 4073
8. Storage and Transportation .. 4074
9. Toxicology and Occupational Health................. 4075
10. References.............. 4076

1. Introduction

The propanols C_3H_7OH, M_r 60.10, comprise two isomers, 1-propanol [71-23-8] and 2-propanol [67-63-0], also called isopropyl alcohol, of which the latter is industrially the more important. Both are clear, colorless, flammable liquids with a slight odor resembling that of ethanol. They occur in nature in crude fusel oils and as fermentation and decomposition products of various vegetables.

The propanols are used mainly as solvents for coatings; in antifreeze compositions and household and personal products; and as chemical intermediates for the production of esters, amines, and other organic derivatives. 2-Propanol is produced by hydration of propene, while 1-propanol is manufactured by the hydrogenation of propanal, in turn derived from hydroformylation of ethylene. Annual U.S. production of 1-propanol and 2-propanol is estimated in 1988 to have been 98 t and 630 t, respectively [1].

Table 1. Physical properties of propanols

	1-Propanol	2-Propanol (anhydrous)	2-Propanol (91 vol% in water)
M_r	60.096	60.096	60.096
Critical temperature, °C	263.63	235.15	
Critical pressure, kPa	5175	4762	
fp, °C	−126.1	−88.5	−50
bp, °C			
101.325 kPa	97.15	82.26	80.40
39.997 kPa	74.479	60.66	
2.0 kPa	20.04		
1.3333 kPa		2.49	
Vapor pressure at 20 °C, kPa		4.4136	4.5
Refractive index at 25 °C	1.3837	1.3752	1.3769
Liquid density at 20 °C, kg/m^3	803.78	785.39	
Liquid viscosity at 20 °C, mPa·s (cP)	2.21	2.37	2.1 (25 °C)
Surface tension at 20 °C, mN/m (dyn/cm)	23.7	21.32	21.40
Flash point (closed cup), °C	15	11.85	18.3
Flash point (Tag open cup), °C	27	17.2	21.7
Autoignition temperature, °C	370.85	398.85	
Lower flammability limit in air, vol%	2.1	2.5	
Upper flammability limit in air, vol%	13.5	12	
Solubility in water at 20 °C, wt%	miscible	miscible	miscible

Table 2. Binary azeotropic mixtures with 1-propanol and 2-propanol

Component	1-Propanol		2-Propanol	
	bp at 101.3 kPa, °C	Alcohol, wt%	bp at 101.3 kPa, °C	Alcohol, wt%
Water	88.1	71.8	80.4	87.8
Ethyl acetate	nonazeotrope		75.9	25
Cyclohexane	74.3	20	69.4	32
2-Butanone	nonazeotrope		77.9	32
3-Pentanone	96.0	63	nonazeotrope	
Benzene	77.12	16.9	71.9	33.3
Propyl formate	80.6	9.8	75.5	∼36
Hexane	65.65	4	62.7	23
Dioxane	95.3	55	nonazeotrope	
Propyl ether	85.8	32.2	78.2	52

2. Physical Properties

The propanols are completely miscible with water and readily soluble in a variety of common organic solvents (e.g., ethers, esters, acids, ketones, and other alcohols). Physical properties of anhydrous 1- and 2-propanol as well as a 91 vol% azeotropic mixture of 2-propanol with water are provided in Table 1. Physical proper-ties of the propanols reflect the position of the hydroxyl group. Associative properties in solution cause the propanols to form azeotropes with a variety of compounds, including aromatics, esters, amines, and ketones. Examples of binary azeotropes of 1- and 2-propanol are given in Table 2.

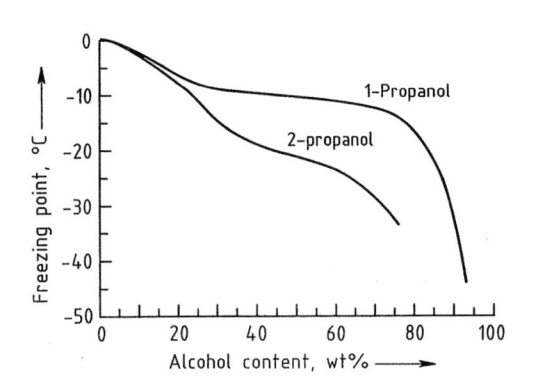

Figure 1. Freezing points of 1- and 2-propanol – water mixtures [2], [17]

Freezing points of 1-propanol – water and 2-propanol – water mixtures are plotted in Figure 1. These plots show that the advantages of a substantially lower freezing point for pure 1-propanol relative to 2-propanol are lost in aqueous solutions of the two alcohols.

3. Chemical Properties

The differences in reactivity between 1-propanol and 2-propanol reflect the influence of the nature of the hydrocarbon radicals in primary and secondary alcohols, respectively. Characteristic reactions of the hydroxyl group include dehydrogenation, oxidation, esterification, ammination, and dehydration. The chemical properties of greatest commercial importance are discussed below.

Dehydrogenation (Oxidation). 2-Propanol can be readily dehydrogenated to acetone, although the process is being replaced by the cumene – phenol process and by direct oxidation of propene (as developed by Wacker – Hoechst).

$$(CH_3)_2CHOH \xrightarrow{\text{Catalyst}} (CH_3)_2CO + H_2 \quad \Delta H = 67 \text{ kJ/mol}$$

Pure dehydrogenation of 2-propanol can be accomplished in the liquid or gas phase with a zinc – copper catalyst at 300 – 500 °C and 300 kPa to give ca. 90 % selectivity to acetone and 98 % 2-propanol conversion. Selectivity can be increased to 96.8 % by employing a Cu – Zn – Cr catalyst [18]. Dehydrogenation is endothermic and requires high temperatures.

Dehydrogenation can also be conducted oxidatively in the presence of silver or copper catalysts at 400 – 600 °C [19].

Recent patents describe a 2-propanol dehydrogenation process for the production of a mixture of acetone, methyl isobutyl ketone, and higher ketones [20], [21]. For example, vapor-phase dehydrogenation of an azeotropic mixture of 2-propanol and water over a copper-based catalyst at 220 °C yields a product mixture containing acetone (52.4%), 2-propanol (11.4%), methyl isobutyl ketone (21.6%), diisobutyl ketone (6.5%), and 4-methyl-2-pentanol (2.2%) [20].

Partial oxidation of 1-propanol can be achieved with air in the presence of catalysts to form propanal and propionic acid [79-09-4] [22]. Strong oxidizing agents such as nitric acid react exothermically with both alcohols, giving a complex mixture of products.

Esterification. The propanols can be converted into propyl esters by reaction with organic and inorganic acids using the conventional method for esterification in the presence of mineral-acid catalysts. The commercially important n-propyl acetate [109-60-4] and isopropyl acetate [108-21-4] are produced by direct esterification of the corresponding alcohols with acetic acid in the presence of sulfuric acid, p-toluenesulfonic acid, methanesulfonic acid [23], [24], or a strong cationic resin [25] as catalyst. A higher-pressure (170–310 kPa) continuous process at 110–160 °C with continuous removal of product ester and water has been described, which should offer higher production rates [26]. 1-Propanol can also undergo ester interchange with methyl or ethyl acetate in the presence of a strong cationic exchange resin to give n-propyl acetate [27].

Reaction of 2-propanol with CS_2 in alkaline solution produces xanthate esters. The method simply involves refluxing the alcohol with alkali-metal hydroxide in trichlorofluoromethane solvent while adding CS_2 [28], [29]. Sodium isopropyl xanthate, $(CH_3)_2CHOCSSNa$ [140-93-2], is a herbicide used for weed control.

Reaction of 2-propanol with phosphorus halides proceeds readily to form phosphite esters. Generally, reaction is conducted in a solvent containing an acid scavenger, but a new solventless spray method could have merit [30].

Titanium(IV) isopropoxide [546-68-9] and titanium(IV) propoxide [3087-37-4] are made by reaction of the corresponding alcohol with $TiCl_4$ in refluxing heptane with removal of HCl coproduct [31]. These esters are especially useful as polymerization and transesterification catalysts.

The propanols also form esters upon reaction with active metals. For example, aluminum isopropoxide [555-31-7] and aluminum propoxide can be prepared by contacting aluminum with alcohol vapor [32], [33]. These materials find use as catalysts, in the preparation of aluminum soaps, as dispersants in paint formulations, and as starting materials for ceramics.

Reaction with Ammonia and Amines. Both 1- and 2-propanol can be reductively aminated by reaction with ammonia or lower amines to give propylated amine derivatives. The reactions are normally conducted at 1.0–10.0 MPa and 190–250 °C using catalysts consisting of nickel, aluminum, cobalt, molybdenum, and/or chromium. Alcohol conversions are usually ca. 90% [34]. For example, H_2–NH_3–2-propanol (12:8:1 mole ratio) was passed over a nickel-based catalyst at 110 °C in the vapor phase at a liquid hourly space velocity (LHSV) of 1 h^{-1} to give a mixture of 93.5%

isopropyl amine [75-31-0] and 6.3% diisopropyl amine [108-18-9] with a 2-propanol conversion of 93% [35].

Etherification and Dehydration. 1-Propanol and 2-propanol can be reacted with one, two, or three moles of ethylene oxide, propylene oxide, or both to give a family of glycol ethers having broad utility as solvents. Reaction is generally catalyzed by alkali hydroxide.

Diisopropyl ether [108-20-3] can be prepared by liquid-phase dehydration of 2-propanol at 130–190 °C and 1.96–7.85 MPa over acidic catalysts containing aluminum [36], [37]. Dehydration of 1-propanol to di-*n*-propylether [111-43-3] can be accomplished with a cation-exchangeable layered clay catalyst containing titanium or zirconium at 180 °C [38].

The dehydration of 1- or 2-propanol to give propene has no practical value. Nevertheless, dehydration is most facile with 2-propanol in the presence of mineral acid catalysts at room temperature or higher, conditions that should otherwise be avoided in most instances.

Others. The propanols give hemiacetal adducts readily on addition to aldehydes. The products can in turn be converted to acetals with a second mole of alcohol under dehydration conditions in the presence of an acidic catalyst [39]. Acetals find use as intermediates in the preparation of pharmaceuticals.

2-Propanol can be condensed with such aromatic compounds as toluene and phenol to produce isopropoxylated derivatives.

4. Production

1-Propanol and 2-propanol are produced by very different processes starting from ethylene and propene, respectively.

4.1. Production of 1-Propanol

Propanal, obtained by hydroformylation of ethylene, is the primary commercial source of 1-propanol. In a second step the propanal is hydrogenated to 1-propanol.

$$CH_2 = CH_2 + CO + H_2 \underset{\text{pressure}}{\overset{\text{Catalyst, heat,}}{\rightleftharpoons}} CH_3CH_2CHO$$

$$CH_3CH_2CHO + H_2 \underset{\text{pressure}}{\overset{\text{Catalyst, heat,}}{\rightleftharpoons}} CH_3CH_2CH_2OH$$

Hydroformylation of Ethylene. Union Carbide Chemicals and Plastics Company, Inc. [40], and Hoechst Celanese practice the two-step oxo process based on rhodium-substituted phosphine-catalyzed low-pressure hydroformylation technology, while Eastman Kodak Company uses a higher-pressure cobalt-based process for production of the intermediate propanal [1]. Typical oxo conditions employed in the low-pressure rhodium-catalyzed oxo process are 90–130 °C, <2.8 MPa total pressure, with CO at <380 kPa, H_2 at <1.4 MPa, and <500 ppm rhodium [41]. Conditions of the cobalt oxo process are 110–180 °C, 20–30 MPa (200–300 atm), 1:1 to 1:2 H_2:CO ratios, and 0.1–1.0 wt% cobalt based on ethylene [42]. The crude propanal is removed by vapor stripping with excess synthesis gas, and the rhodium–triphenylphosphine catalyst is recycled [40]. Traces of CO are removed in a stripping column from the crude, condensed propanal before hydrogenation to prevent poisoning of the hydrogenation catalyst in the second step. The higher selectivity and milder conditions of the rhodium-catalyzed process provide crude 1-propanol for hydrogenation containing fewer impurities, such as higher aldol-condensation products and aldehyde trimers (Tischenko esters).

Hydrogenation of propanal is conventionally carried out as a heterogeneous process in either the vapor or liquid phase over a variety of metal catalysts. Heterogeneous vapor-phase processes are effective at ca. 110–150 °C, and 0.14–1.0 MPa at a 20:1 mole ratio of hydrogen to propanal [43]–[45]. Reductions are accomplished in excess hydrogen, and the heat of reaction is removed by circulating the vapor through external heat exchangers or by cooling the reactor internally [46].

Hydrogen efficiencies are >90%, with aldehyde conversions as high as 99.9% and alcohol yields >99%. The commonly used commercial hydrogenation catalysts include combinations of copper, zinc, nickel, and chromium compounds. Major impurities at vapor-phase process temperatures are dipropyl ether, ethane, and propyl propionate. In addition to reduced zinc oxide and copper oxide catalysts, selectivity enhancers such as alkali and transition metals can be added to significantly reduce the formation of esters (by Tischenko reaction of the starting aldehyde) and ethers [44]. Addition of 1–10% water to the reactor feed also suppresses ether formation [47].

The propyl propionate byproduct formed in the catalytic hydrogenation can be separated and hydrogenolyzed in the presence of reduced CuO–ZnO catalyst at 75–300 °C and 9.8 kPa–9.8 MPa to give 1-propanol as the major product [48].

Liquid-phase heterogeneous processes are conducted at higher pressures. For example, hydrogenation at 95–120 °C and 3.5 MPa of hydrogen gives 1-propanol with >99.9% purity. Nickel- and copper-based catalysts containing molybdenum, manganese, and sodium promoters are preferred [49]. The catalysts are usually supported on alumina.

Other processes have been described for the preparation of 1-propanol; for example, in two steps by isomerization and hydrogenation of propylene oxide [50]; from synthesis gas [51], [52]; from methanol [*67-56-1*] and synthesis gas [53]; and by the Guerbet reaction [54].

4.2. Production of 2-Propanol

There are two major commercial processes for the production of 2-propanol: indirect and direct hydration of propene. Smaller quantities are produced by hydrogenation of acetone.

Indirect hydration involves two steps. In the first, a mixture of mono- and diisopropyl sulfate esters is formed from reaction of propene with sulfuric acid. The reaction is exothermic, yielding about 50 kJ/mol.

$$3\ CH_3CH=CH_2 + 2\ H_2SO_4 \rightleftharpoons (CH_3)_2CHOSO_3H + [(CH_3)_2CHO]_2SO_2$$

The sulfate esters are then hydrolyzed to product 2-propanol in a second step.

$$(CH_3)_2CHOSO_3H + [(CH_3)_2CHO]_2SO_2 \xrightarrow{H_2O} 3\,(CH_3)_2CHOH + 2\ H_2SO_4$$

A major byproduct in the indirect process is diisopropyl ether, which is formed from reaction of the sulfate esters with product 2-propanol.

$$(CH_3)_2CHOSO_3H + [(CH_3)_2CHO]_2SO_2 \xrightarrow{(CH_3)_2CHOH} 2\,[(CH_3)_2CH]_2O + 2\ H_2SO_4$$

Other impurities include hydrocarbons, char, polymeric residues (oils), propanal, acetone, and odorous sulfur-containing compounds. Isopropyl sulfate esters are particularly unstable, decomposing at temperatures as low as 50–100 °C [55]. Mono- and disulfate esters can be thermally desulfonated to give SO_2, SO_3, and a mixture of propene and acetone.

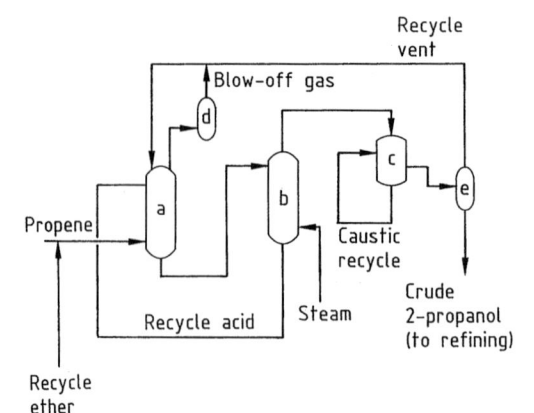

Figure 2. 2-Propanol production by indirect, weak-acid process
a) Absorbers; b) Strippers; c) Caustic scrubber; d) Scrubber; e) Liquifier

$$(CH_3)_2CHOSO_3H \xrightarrow{Heat} CH_3CH=CH_2 + SO_3 + H_2O$$

$$(CH_3)_2CHOSO_3H \xrightarrow{Heat} CH_3COCH_3 + SO_2 + H_2O$$

The indirect hydration process can be conducted with either strong or weak acid. A schematic for an indirect *weak-acid hydration* is shown in Figure 2.

Propene-containing feed gas (propene content $\leq 60\%$) is reacted with ca. 60% H_2SO_4 in a series of absorption columns (a) at 75–85 °C and 0.59–1.0 MPa. Production rates and efficiencies are linked to a balance of initial sulfuric acid concentration, temperature, total pressure, and propene partial pressure. High absorber temperatures favor ether formation. The propene is sparged into the lower portion of the absorbers while the H_2SO_4 is fed to the top to establish a countercurrent flow pattern. The "sulfates" product stream is taken from the bottom of the absorbers and fed to vacuum strippers (b) where it is hydrolyzed with steam and the resulting 2-propanol is flashed as overhead product. Product-free acid bottoms (recycle acid) from the strippers are concentrated to about 60% and recycled to the absorbers. The required reconcentration of acid in this process is less costly than in the strong-acid process, where reconcentration to >90% is required. Vaporized stripper product is neutralized by scrubbing with caustic (c), and uncondensed vapors (vent gas) are combined with the blow-off gas from the scrubbers and recycled to the reaction.

Crude 2-propanol is fed to a multicolumn refining system where a constant-boiling mixture (CBM) of 2-propanol is obtained. Byproduct diisopropyl ether is recovered as a lights stream from the first column in the refining train and recycled to the reaction absorbers, where it is hydrolyzed back to product, thus preventing ether accumulation in the process:

$$[(CH_3)_2CH]_2O + H_2O \xrightleftharpoons[]{H_2SO_4} 2\,(CH_3)_2CHOH$$

Heavy-end concentrated oils are also collected and discarded.

The main problem encountered in purification of 2-propanol is its separation from water. Enrichment of the CBM to >99% or absolute 2-propanol is accomplished by use of an azeotroping agent, which forms a ternary constant-boiling mixture. Diisopropyl ether and cyclohexane are common entraining agents for 2-propanol enrichment [56]. Undesirable intense odors inevitably arise from sulfur-containing compounds, and this can cause difficulties in such commercial uses as cosmetic products, spray products, and medicinal formulations. Odor is generally removed by contacting the 2-propanol with ion-exchange resins, activated carbon, activated alumina, or metals (e.g., copper, nickel) [57]–[59].

The weak-acid indirect hydration process suffers from high corrosion rates and disposal problems with water, acid, caustic, and off-gas wastes. A major advantage is the ability to use low-purity propene feed. Recent advances have been made in waste minimization, including replacement of the lead-lined absorbers with corrosion-resistant ceramics and synthetics.

Indirect hydration is still the primary process conducted in the United States by Exxon, Shell Oil, and Union Carbide Chemicals and Plastics [60], while Lyondell Petrochemical produces a smaller amount by hydrogenation of acetone [60]. There are indications that a few companies in Europe and Japan may also be employing this older technology [19].

In the *strong-acid process* reaction is conducted with a high sulfuric acid concentration (> 90 wt%), while the pressure and temperature are low compared to the weak-acid process (1.0–1.2 MPa and 20–30 °C, respectively). Hydrolysis is accomplished in a separate second stage. Both processes offer 2-propanol selectivities of > 90 wt%.

It is reported that for 1 t of 2-propanol produced, 0.8 t propene, 0.35 t H_2SO_4, 3.5 t steam, and 40–50 kW·h of electricity are required in an optimized process [61]. It is believed that the strong acid process is not employed as a major commercial source of 2-propanol in the world today because of the requirement for high-purity propene feedstock and waste-disposal problems.

Direct Hydration. The direct hydration of propene has been in commercial use since 1951.

$$CH_3CH=CH_2 + H_2O \underset{}{\overset{Catalyst}{\rightleftharpoons}} (CH_3)_2CHOH \quad \Delta H = -50 \text{ kJ/mol}$$

High pressures and low temperatures over an acidic fixed-bed catalyst characterize this process, causing the exothermic equilibrium reaction to be displaced to the right. Three versions of the direct hydration process are practiced commercially today.

1) Low-temperature (130–160 °C), high-pressure (8.0–10.0 MPa) vapor–liquid phase hydration over a sulfonated polystyrene ion-exchange resin catalyst was pioneered by Deutsche Texaco [62]–[69]. The feed, consisting of propene gas and liquid water, is fed in a supercritical state to the top of a fixed-bed reactor and allowed to

trickle downward. Feed containing about 92% propene can be used, resulting in propene conversions of about 75% per pass. About 5% diisopropyl ether and about 1.5% oxygenates (alcohols) of propene oligomers (hexenes) are formed as byproducts. This technique requires high-pressure equipment, and could suffer from problems of short catalyst life. This technology and plants based on it are licensed outside Europe [70].

2) High-temperature, high-pressure (270–300 °C, 20 MPa) vapor–liquid hydration of propene over a reduced tungsten oxide catalyst was developed by Tokuyama Soda [71]–[73]. This process utilizes a molar ratio of water to propene of about 2.5:1. Water is present in both the gas and liquid phases, which increases conversion (because equilibrium is shifted farther to the right due to the solubility of 2-propanol in water). This technology requires high-pressure equipment, but features high propene conversions of 60–70% per pass and 2-propanol selectivities of 98–99% based on converted propene. The catalyst must be durable (stable in the presence of water). The process has low gas-recycle requirements, so propene of only 95% purity can be used.

3) ICI developed technology for vapor-phase hydration of propene involving medium to high pressures. This process uses a WO_3–SiO_2 catalyst at 250 °C and 25 MPa [19], [74], and it gives yields of about 95%. In Germany, VEBA developed a similar hydration process based on a phosphoric acid catalyst supported on SiO_2 and operated at 180–260 °C and 2.5–6.5 MPa [75]–[77]. Typically, these processes require high propene recycle (less than 10% conversion per pass) and utility costs are probably high. High-purity propene(∼99%) is required. The phosphoric acid process is commercial in Germany, the Netherlands, the United Kingdom, and Japan.

Hydrogenation of Acetone. Hydrogenation can be conducted in the liquid phase over a fixed catalyst bed of a Raney-nickel catalyst to give 99.9% selectivity and 99.9% conversion of the acetone [78], [79]. Hydrogenation over copper oxide–chromium oxide at 120 °C and 196 kPa gives lower selectivities and conversions (98% and 94%, respectively) [80]. It is not essential that the acetone be pure. This process is particularly advantageous where excess acetone is available as a byproduct from another process (e.g., the cumene–phenol process).

Other Processes. Patents have been issued describing the manufacture of 2-propanol from cellulosic materials (e.g., cotton, corn, and wood) [81], [82].

5. Uses

Uses of the two propanols are dictated by their solvent properties, their high water miscibility, and by their potential for introducing the propyl group into chemical intermediates.

1-Propanol. In 1988 over 75% of the 1-propanol in the United States was employed in solvent applications, either directly or in the form of acetate ester or glycol ether derivatives [1]:

Solvent uses	24 900 t
1-Propyl acetate	19 500 t
n-Propylamines	11 300 t
Glycol ethers	5 400 t
Others	2 300 t
Total	63 400 t

As a solvent, 1-propanol is used principally in printing inks, paint, cosmetics, pesticides, and insecticides [83].

The EC used about 100 000 t of 1-propanol in 1988 [84]. In Germany, BASF converts most of its propanal into 1-propanol for printing inks, cosmetics, solvents, and intermediates for propylamines used in pharmaceuticals and pesticides [1]. In Japan, 1000–2000 t of 1-propanol was consumed in 1988 for printing inks and paints, all of which was imported [1].

1-Propanol is used commercially to produce glycol ethers. These are characterized by dual functionality, which imparts high solvency, chemical stability, and water compatibility. Glycol ethers such as ethylene glycol monopropyl ether (from 1-propanol and 1 mol of ethylene oxide), diethylene glycol monopropyl ether (from 1-propanol and 2 mol of ethylene oxide), propylene glycol monopropyl ether (from 1-propanol and 1 mol of propylene oxide) and dipropylene glycol monopropyl ether (from 1-propanol and 2 mol of propylene oxide) are marketed by Union Carbide Chemicals and Plastics [85] and Eastman Chemical Products [86].

n-Propyl propionate [*106-36-5*] is a new solvent available from Union Carbide Chemicals and Plastics and developed for use in automotive refinishing, appliance coatings, and high-solids coating systems [85]. *n*-Propyl propionate is touted as a replacement for the isomeric *n*-butyl acetate in coatings, where improved odor characteristics are desirable.

2-Propanol. 2-Propanol is used primarily as a solvent in inks and surfactants [70]. Other applications include its role as an antiseptic alcohol, as a reaction solvent for cellulose carboxymethyl ether [*9000-11-7*] (CMC), in the production of cosmetic base materials and pesticide carriers, as a source of material for organic synthesis, for washing of a flux used in soldering electrical circuits, and for removal of water from gasoline tanks in cars [70]. Table 3 shows approximate consumption data for 2-propanol in both 1986 and 1990 in the United States.

The coatings use of 2-propanol has reached a plateau due to a shift toward waterborne and high-solids coatings. Its use as an electronics solvent is growing, but the market is small [89]. 2-Propanol is a particularly versatile solvent for producing high-performance (e.g., gelation-resistant and polyamide-containing) printing inks [90]. In the pharmaceutical industry, 2-propanol enjoys favor as a processing solvent during manufacture of drugs. Exports have been very strong in recent years.

Table 3. 2-Propanol consumption in the United States (10^3 t)

	Year	
	1986	1990
Coating solvents	8.2	9.1
Processing solvents	6.4	6.8
Household/personal	6.4	5.4
Pharmaceuticals	6.4	5.4
Acetone	4.5	3.6
Miscellaneous solvent applications and chemical intermediates	4.5	3.2
Exports	9.1	11.8
Total	45.5	45.0

Use of 2-propanol for the production of acetone has decreased dramatically. There remains only one acetone plant in the United States (compared to four in 1985), and the market is still oversupplied [91]. However, since the cumene route to phenol provides equal amounts of phenol and acetone (→Acetone), a strong demand for acetone relative to phenol can increase the need for acetone production from 2-propanol [88].

2-Propanol is widely used as a chemical intermediate, for example in reductive amination to produce monoisopropyl amine (for herbicide and pesticide production), and as a source of isopropyl acetate. Some diisopropyl amine is utilized as an intermediate in the synthesis of diisopropylammonium nitrate, a corrosion inhibitor [91].

Nippon Petrochemical has begun marketing pure 2-propanol (99.99%) for use in high-purity large-scale integration (LSI) and silicon wafers for the electronics industry [92]. Another attractive new application is as an octane enhancer, carburetor anti-icing additive, and methanol cosolvent in motor gasoline blends [93].

2-Propanol is useful in extraction processes; thus, aqueous solutions are utilized in liquid–liquid extractions of fatty acids from vegetable oil at temperatures as low as −2 °C [94].

6. Specifications

ASTM standard specifications for propanol are given in Table 4.

Product purity is generally determined by capillary gas chromatography. Typical purities specified by manufactures for 1- and 2-propanol are 99.7 – 99.9 wt% and 99.8 wt%, respectively. Because of its use in packaging, 1-propanol may sometimes be required to pass a specific odor test conducted for consumers by an odor panel. Occasionally, alkalinity and propanal content are included in the requirements for 1-propanol.

Table 4. ASTM Specifications for propanols

	1-Propanol	2-Propanol	Specification method
ASTM standard	D 3622-90	D 770-90	
d_{20}^{20}	0.804 – 0.807	0.785 – 0.7870	
d_{25}^{25}	0.801 – 0.804	0.782 – 0.784	D 268 or D 4052
Color, Pt-Co (ASTM)	10	10	D 1209
Distillation range, 101.3 kPa	distill within a 2 °C range that includes 97.2 °C	distill within a 1.5 °C range that includes 82.3 °C	D 1078
Nonvolatile matter, mg/100 mL	5	5	D 1353
Water, max, wt%	0.1	0.2	D 1364 or D 1476
Acidity, acetic acid, max, wt%	0.003	0.002	D 1613

Table 5. 1-Propanol and 2-propanol manufacture in major producing countries (10^3 t)

Country	Production of 1-propanol in 1988	Capacities of 2-propanol in 1991
United States	98	791
France		115
Germany	5 [a]	215
Japan	0	209
Netherlands		250
United Kingdom		190

[a] Production for merchant market.

7. Economic Aspects

Manufacture of the propanols in major industrial countries is summarized in Table 5.

Since 2-propanol is no longer the primary feedstock for acetone it will see little, if any, growth in the United States unless new applications are found. Production units there are currently operating at about 60 – 80% of capacity [96]. United States output of 2-propanol was 649×10^3 t in 1989 [97], and forecasted to grow to only 726×10^3 t by the year 2000 [98]. Demand fell by 4.4%/a in 1977 – 1986 to 578×10^3 t in 1986 due to the decrease in use for acetone production [99]. The capacity for 2-propanol in Western Europe is roughly comparable to that in the United States. It is estimated based on U.S. import/export data that Western Europe is currently (1991) operating at or near full 2-propanol capacity.

2-Propanol production in Japan has been rising by about 5%/a. Demand in Japan reached 130×10^3 t in 1989 [100], with steady growth (3 – 4%) in paints and solvents and rapid growth (15 – 20%) in detergent applications [101]. Currently (1990) there is a shortage in Japan [100].

Nippon Petrochemicals (50%), Tokuyama Soda (25%), and Mitsui Toatsu Chemicals (25%) are the three major producers of 2-propanol in Japan [70]. Tokuyama Soda

Table 6. Safety data and transport regulations for propanols

	1-Propanol	2-Propanol
Flash point (open cup), °C	15.0	11.7
Flammability limits in air, vol%	2.0–12.0	2.0–12.0
Autoignition temperature, °C	371.11	398.89
Vapor pressure at 20 °C, kPa	1.9854	4.4053
Percent in saturated air, 25 °C	2.7	5.8
Density in saturated air, (air = 1), 25 °C	1.028	1.06
Heat of combustion, MJ/kg	−33.60	−33.35
U.N. No.	1274	1219
RID/ADR	Class 3, number 3 b	Class 3, number 3 b
IMDG Code	Class 3.2	Class 3.2
CFR 49	172.101 flammable liquid	172.101 flammable liquid

employs its own direct hydration technology based on a tungsten catalyst, while Nippon Petrochemicals uses technology developed by VEBA [70].

8. Storage and Transportation

The propanols are flammable liquids, and should be considered dangerous fire risks when exposed to heat or flame. Explosive limits for propanols in air are 2–12%. Some important data for the safe handling of propanols are given in Table 6.

Care should be taken during heating (e.g., distillation) of old samples of 2-propanol. Several violent explosions have been reported during attempts to concentrate old samples to small volume, suggesting that this compound should be classed as peroxidizable (via autoxidation of the tertiary hydrogen) [103].

Baked phenolic-lined steel or stainless steel tanks are recommended for storage and shipping of the propanols in order to maintain high quality. Mild steel tanks are also adequate providing a filtering system is installed to remove rust [104]. Storage and transport under dry nitrogen is also recommended to protect against moisture pickup and to minimize flammability hazards. Aluminum is not recommended. The flash points of the propanols are low enough so that the compounds can be stored underground under a nitrogen atmosphere. Pipes and pumps can be constructed from the same metals as the storage facility. Centrifugal pumps with explosion-proof electric motors are suitable [104].

Table 7. Toxicity of propanols in animals

	1-Propanol	2-Propanol
LD_{50} rat (oral), mg/kg	1870[a]	5045[b]
LD_{50} rabbit (intravenous), mg/kg	483[a]	1184[a]
LCLo rat (inhalation), ppm/4 h	4000[a]	12000[a], ppm/8 h
Eye injury, rabbit	20 mg/24 h, moderate 4 mg, open severe	10 mg, moderate
LD_{50} rabbit (cutaneous), mg/kg	5040[a]	12800[a]

[a] No description of toxic effects.
[b] Behavioral (altered sleep time).

Table 8. Odor detection limits for propanols

	Absolute odor threshold[a], ppm	50% Odor recognition[b], ppm	100% Odor recognition[b], ppm
1-Propanol	<0.03	0.08	0.13
2-Propanol (anhyd.)	3.20	7.50	28.2

[a] This is the concentration at which 50% of an odor panel observed an odor.
[b] The concentrations at which 50% and 100% of the panel defined the odor as due to propanol, respectively.

9. Toxicology and Occupational Health

The propanols are considered to have generally low toxicity in humans [102]. Ingestion of excessive dosages of 2-propanol by alcoholics and suicidals has been reported to cause poisoning due to depression of the CNS and respiratory irritation. Repeated or prolonged contact with propanols may cause drying of the skin. The propanols display an exposure warning because they can cause mild to severe irritation to the eyes, nose, and throat. However, adverse effects are rare in light of widespread use by the general public and considerable industrial exposure [102], [105].

Animal studies reveal irritation to the eyes, low acute oral toxicity, and low irritation to the skin, as shown by the data in Table 7.

The mean exposure limits for 1- and 2-propanol in air at the workplace by ACGIH (TLV) and OSHA (PEL) are 200 ppm and 400 ppm, respectively. The MAK value for 2-propanol is 400 ppm. Odor threshold data are provided in Table 8.

The propanols are registered in the Toxic Substance Control Act (TSCA) and the European Inventory of Existing Commercial Substances (EINECS).

2-Propanol is the subject of an EPA-TSCA Section 4 Test Rule requiring manufacturers and producers of 2-propanol to test the substance for health and environmental effects [108]. Under the provisions of this rule, the following toxicological testing assessments are being conducted:

1) Subchronic toxicity in rats and mice by inhalation

2) Developmental toxicity in rats and rabbits by gavage
3) Two-generation reproductive toxicity in rats by gavage
4) Acute neurotoxicity and neurobehavioral toxicity in rats by inhalation
5) Repeated neurotoxicity and neurobehavioral toxicity in rats by inhalation
6) Developmental neurotoxicity by gavage in rats
7) In vivo and in vitro genotoxicity assays
8) 2-Year oncogenicity bioassay in rats and mice

10. References

[1] SRI International: *Chemical Economics Handbook*, "Oxo Chemicals Report," Jan. 1991.
[2] Unpublished physical property data, Union Carbide Chemicals and Plastics Company Inc., Danbury, Conn.
[3] Data from DIPPR (Design Institute for Physical Property Data) (ONLINE data base) Project 801 Source File Tape, revision dates Aug. 1988, 1989, 1990.
[4] Thermodynamics Research Center: "TRC Thermodynamic Tables – Non-Hydrocarbons," Loose-leaf Data Sheets, The Texas A & M University System, College Station, Texas.
[5] N. I. Sax, R. J. Lewis, Sr.: *Dangerous Properties of Industrial Materials*, 7th ed.,Van Nostrand Reinhold, New York 1989.
[6] G. H. Tryon (ed.): *Fire Protection Handbook*, 12th ed., National Fire Protection Association, Boston, Mass. 1962.
[7] G. S. Ross et al.: "Polymorphism in Isopropyl Alcohol," *Science (Washington D.C.)* **141** (1963) 1043.
[8] Thermodynamics Research Center: *TRC Table* 23-2-1-(1.1000)-k, p. 1, Dec. 31, 1960; June 30, 1965; Dec. 31, 1976; p. k-5000.
[9] Thermodynamics Research Center: *TRC Table* 23-2-1-(1.1001)-k, p. 1, June 30, 1965, p. k-5010.
[10] Thermodynamics Research Center, "Selected Values of Properties of Chemical Compounds," Data Project, Loose-leaf Data Sheets, Texas A & M University, College Station, Texas 1980.
[11] R. C. Wilhoit, B. J. Zwolinski, "Physical and Thermodynamic Properties of Aliphatic Alcohols," *J. Phys. Chem. Ref. Data Suppl.* **2** (1973) no. 1.
[12] J. J. Jasper, "The Surface Tension of Pure Liquid Compounds," *J. Phys. Chem. Ref. Data* **1** (1972) no. 4, 841.
[13] J. S. Riddick, W. B. Bunger: *Organic Solvents: Physical Properties and Methods of Purification*, 3rd ed., Wiley-Interscience, New York 1970.
[14] Thermodynamics Research Center, "TRC Thermodynamic Tables – Nonhydrocarbons," The Texas A & M University System, College Station, Tex 1987.
[15] D. Ambrose, J. Walton: "Vapor Pressures Up to Their Critical Temperatures of Normal Alkanes and Alkanols," *Pure Appl. Chem.* **61** (1989) no. 8, 1395.
[16] L. H. Horsley, *Azeotropic Data–III*, American Chemical Society, Washington, D.C., 1973.
[17] J. Timmermans: *The Physico-chemical Constants of Binary Systems in Concentrated Solutions*, vol. 4, Interscience, New York 1960.
[18] Shell Oil Co., US 4 472 593, 1984 (L. H. Slaugh, G. W. Schoenthal, J. D. Richardson).
[19] K. Weissermel, H. J. Arpe: *Industrial Organic Chemistry*, Verlag Chemie, Weinheim 1978.
[20] Z. Hejda, R. Zidek, J. Kozuch, CS 241 425, 1988.

[21] J. Pasek, V. Pexidr, R. Zidek, J. Hajek, CS 234 604, 1987.
[22] Polska Akademia Nauk, PL 137 060, 1987 (A. Kowai, J. Haber).
[23] BP Chemicals Ltd., EP 158 499, 1985 (J. Russell, J. A. Stevenson).
[24] BP Chemicals Ltd., EP 9 886, 1980 (D. C. Buttle).
[25] Chemische Werke Huels AG, EP 66 059, 1982 (H. Alfs, W. Boexkes, E. Vangermain).
[26] BP Chemicals Ltd., JP 55 038 399, 1978.
[27] Kuraray Co., Ltd., JP 59 137 444, 1984.
[28] CIL Inc., ZA 8 004 227, 1981 (D. J. Gannon, D. T. F. Fung).
[29] Institutul de Cercetari, RO 89 624, 1986 (G. Duda et al.).
[30] Borsodi Vegyi Kombinat, HU 40 450, 1986 (J. Kalacska et al.).
[31] Huels AG, DE 3 739 577, 1988 (U. Hdo, H. G. Srebny, H. J. Vahlensieck).
[32] Mitsubishi Mining and Cement Co., Ltd., JP 60 218 344, 1985 (Y. Fukuda, M. Shimura, K. Miyahara).
[33] International Business Machines Corp., EP 248 195, 1987 (J. J. Cuomo, P. A. Leary, J. M. Woodall).
[34] Peti Nitrogenmuvek, HU 35 241, 1985 (G. Kincses et al.).
[35] Daicel Chemical Industries, Ltd., JP 58 174 348, 1983.
[36] Mitsui Toatsu Chemicals, Inc., JP 03 041 042, 1991 (A. Hiai, J. Ono, H. Kato, N. Kitano).
[37] Mitsui Toatsu Chemicals, Inc., JP 03 074 343, 1991 (A. Hiai, J. Ono, H. Kato, N. Kitano).
[38] Showa Denko K.K., JP 60 215 642, 1985 (T. Kametaka, T. Nozawa).
[39] Nippon Synthetic Chemical Industry Co., Ltd., JP 62 019 556, 1987 (T. Kakimoto).
[40] Union Carbide Corp., US 4 247 486, 1981 (E. A. V. Brewester, R. L. Pruett).
[41] Union Carbide Corp., US 4 277 627, 1981 (D. R. Bryant, E. Billig).
[42] J. Falbe in E. G. Hancock (ed.): *Propylene and its Derivatives*, J. Wiley & Sons, New York 1973, p. 333.
[43] Union Carbide Corp., EP 8 767, 1980 (C. C. Pai).
[44] Union Carbide Corp., US 4 762 817, 1988 (J. E. Logsdon, R. A. Loke, J. S. Merriam, R. W. Voight).
[45] Union Carbide Chemicals and Plastics Company Inc., US 4 876 402, 1989 (J. E. Logsdon, R. A. Loke, J. S. Merriam, R. W. Voight).
[46] J. B. Cropley, L. M. Burgess, R. A. Loke, *CHEMTECH* **14** (1984) 374.
[47] W. L. Faith, D. B. Keyes, R. L. Clark: *Industrial Chemicals*, 3rd ed., Wiley & Sons, New York 1965, p. 183.
[48] Davy McKee Ltd., EP 74 193, 1983 (M. W. Bradley, N. Harris, K. Turner).
[49] BASF, DE 3 228 881, 1984 (K. Baer et al.).
[50] VEB Leuna-Werke, DD 240 742, 1986 (P. Franke et al.).
[51] VEB Leuna-Werke, DD 273 163, 1989 (G. Kohl et al.).
[52] N. L. Holy, T. F. Carey, Jr., *Appl. Catal.* **19** (1985) no. 2, 219.
[53] Ruhrchemie AG, EP 84 833, 1983 (B. Cornils, C. D. Frohning, H. Bahrmann, W. Lipps).
[54] W. Ueda, T. Kuwabara, T. Ohshida, Y. Morikawa, *J. Chem. Soc. Chem. Commun.* 1990, no. 22, 1558.
[55] M. H. Carr, H. P. Brown, *J. Am. Chem. Soc.* **69** (1947) 1170.
[56] Tokuyama Soda, JP 77 012-166, 1977 (T. Sato, R. Ohuji, H. Yamanovchi).
[57] Exxon Research and Engineering Co., GB 2 004 538, 1979.
[58] Exxon Research and Engineering Co., US 4 219 685, 1980 (C. Savini).
[59] Takeda Chemical Ind. KK, JP 55 091 359, 1980.

[60] *Chemical Economics Handbook,* "Isopropyl Alcohol", United States Data Summary, SRI International, July 1990.
[61] V. R. Gurevick et al., *Azerb. Neft. Khoz.* 1990, no. 6, 54.
[62] *Hydrocarbon Process.* **68** (1989) no. 11, 111.
[63] *Hydrocarbon Process.* **70** (1991) no. 3, 185.
[64] W. Neier, J. Woellner, *CHEMTECH* 1973, Feb., 95.
[65] *Hydrocarbon Process.* **60** (1981) no. 11, 173.
[66] W. Neier, J. Woellner, *Erdoel Kohle Erdgas Petrochem.* **28** (1975) no. 1, 19.
[67] W. Neier, J. Woellner, *Hydrocarbon Process.* **5** (1972) no. 11, 113.
[68] Deutsche Texaco AG, DE 86-3628007, 1987 (R. R. Carls, G. Osterburg, M. Prezlj, W. Webers).
[69] Deutsche Texaco AG, DE 3 512 518, 1986 (F. Henn, W. Neier, G. Strehlke, W. Webers).
[70] *Jpn. Chem. Week* (1987) Dec., 10, 6, 8.
[71] Y. Onoue, Y. Mizutani, S. Akiyama, Y. Izumi, *CHEMTECH* 1978, July, 432.
[72] Y. Onoue, Y. Izumi, *CEER Chem. Econ. Eng. Rev.* **6** (1974) no. 7, 48.
[73] Tokuyama Soda, US 3 758 616, 1973 (Y. Izumu, Y. Kawasaki, M. Tani).
[74] J. C. Fielding in E. G. Hancock (ed.): *Propylene and its Industrial Derivatives,* J. Wiley & Sons, New York 1973.
[75] Hibernia-Chemie GmbH, BE 683 923, 1966.
[76] VEBA-Chemie AG, US 3 955 939, 1976 (A. Sommer, M. Urban).
[77] *Eur. Chem. News* (1970) July 24, 32.
[78] Mitsui Petrochemical Ind., Ltd., JP 03 133 941, 1991 (H. Fukuhara, Y. Shibuta, T. Tate, T. Isaka).
[79] Mitsubishi Petrochemical Co., Ltd., JP 03 141 235, 1991 (H. Fukuhara, K. Taniguchi).
[80] Mitsui Toatsu Chemicals, Inc., JP 03 041 038, 1991 (A. Hiai, J. Ono, H. Kato, N. Kitano).
[81] New Fuel Oil Development Technology Research Assoc., JP 61 067 493, 1986 (S. Onuma, M. Naganuma, H. Oiwa).
[82] Services de Consultation D et B Plus Ltee., CA 1 162 867, 1984 (H. C. Rothlisberger).
[83] *Manuf. Chem.* **60** (1989) July, 7.
[84] *Financ. Times (North Am. Ed.)* (1989) June, 7, 8.
[85] UCAR n-Propyl Propionate, Brochure F-60741 A, Union Carbide Chemicals and Plastics Corporation, Danbury, Conn., Oct. 1989.
[86] *The World of Eastman Chemicals,* Brochure ECP 3131, Eastman Chemical Products, Inc., Kingsport, Tenn. Jan. 1989.
[87] *Chem. Mark. Rep.* **232** (Aug. 31, 1987) 46.
[88] *Chem. Mark. Rep.* **238** (Sept. 3, 1990) 50.
[89] *Chem. Mark. Rep.* **232** (Aug. 31, 1987) 19.
[90] *Polym. Paint Colour J.* **172** (1989) June, 462.
[91] G. T. Austin, Isopropyl Acetate Production, *Chem. Eng.* **141** (1974) May 101.
[92] *Jpn. Chem. Week* (1987) Aug. 13, 3, 5.
[93] M. Prezelj, *Hydrocarbon Process.* **66** (1987) no. 9, 68.
[94] K. J. Shah, T. K. Venkatesan, *J. Am. Oil Chemists Soc.* (1989) June, 783.
[95] Directory of Chemical Producers – United States, Western Europe, and East Asia, SRI International, 1991.
[96] CPI Purchasing (Apr. 1989) 38–43.
[97] *Chem. & Eng. News* **68** (1990) April 9, 12.
[98] *Chem. Week* **143** (1989) Nov. 1, 15.
[99] *Chem. Mark. Rep.* **232** (1987) Aug. 31, 46.

[100] *Jpn. Chem. Week* **144** (1990) Aug. 16, 3.
[101] *Jpn. Chem. Week* (Apr. 18, 1991), 2.
[102] *Patty*, 3rd ed., **2 C.**
[103] L. Bretherick: *Bretherick's Handbook of Reactive Chemical Hazards,* Butterworth, London 1990, p. 391.
[104] "UCAR Alcohols for Coatings Applications," Brochure No. F-48588 A, Union Carbide Chemicals and Plastics Corp., Danbury, Conn., July 1989.
[105] N. I. Sax, R. J. Lewis, Sr.: *Dangerous Properties of Industrial Materials,* 7th ed., vol. **3,** Van Nostrand Reinhold, New York 1988.
[106] D. V. Sweet (ed.): *Registry of Toxic Effects of Chemical Substances,* Publication No. 87-114, vol. **4,** National Institute for Occupational Safety and Health, DHHS (NIOSH), Washington, D.C., Apr. 1987.
[107] T. M. Hellman, F. H. Small, *J. Air Pollut. Control Assoc.* **24** (1974) no. 10, 979.
[108] *Fed. Regist.* **54** (1989) no. 203, 43252–43264.

Propene

PETER EISELE, Linde AG, Werksgruppe Verfahrenstechnik und Anlagenbau, Höllriegelskreuth, Federal Republic of Germany (Chaps. 2, 3, 5; Sections 4.1, 4.2)

RICHARD KILLPACK, Shell International Chemical Company Ltd., Shell Centre, London, United Kingdom (Chaps. 6, 7; Section 4.3)

1.	Introduction 4081	4.3.2.	Catofin Process 4091
2.	Physical Properties 4082	4.3.3.	Phillips STAR Process 4092
3.	Chemical Properties. 4082	4.3.4.	Linde Process 4093
4.	Production 4082	5.	Storage, Transportation, Quality Requirements. 4093
4.1.	Production as Byproduct of Ethylene Production 4084	6.	Uses 4094
4.1.1.	Variants of Propene – Propane Separation 4086	6.1.	Thermal Uses. 4095
		6.2.	Motor Gasoline Uses 4095
4.1.2.	Apparatus and Construction Materials 4088	6.2.1.	Polygasoline Production – Dimerization 4095
		6.2.2.	Alkylation. 4096
4.2.	Production as Byproduct of Refinery Processes 4089	6.3.	Chemical Uses 4096
4.3.	Propane Dehydrogenation . . . 4089	7.	Economic Aspects 4098
4.3.1.	Oleflex Process 4090	8.	References. 4098

1. Introduction

Propene, $CH_3-CH=CH_2$, M_r 42.081, [115-07-1] was the first petrochemical raw material to be employed on an industrial scale and was used more than 60 years ago in the production of isopropanol. For a long time, propene was to some extent overshadowed by its olefin homologue ethylene. However, part of propene's turbulent development and expansion since 1965 is also due to the success of ethylene: Being a byproduct of ethylene production, many important areas of application were opened up to propene by the chemical industry. Secondary products of propene, such as polypropylene, acrylonitrile, propylene oxide, and in Europe cumene, have now clearly overtaken the classic secondary product isopropanol in importance.

2. Physical Properties

Important physical properties are as follows:

mp	−185.25 °C
bp	−47.70 °C
t_{crit}	91.76 °C
p_{crit}	4.621 MPa
ϱ_{crit}	0.22 g/cm^3
Molar volume	21.976 L (STP)
Gas density	1.9149 g/L (STP)
Density relative to air	1.49
Heat of fusion	71.37 kJ/kg
Lower heating value	45 813 kJ/kg
Enthalpy of formation ΔH_{298}	20.43 kJ/mol
Entropy S_{298}	0.227 kJ mol^{-1} K^{-1}
Free energy of formation ΔG_{298}	62.65 kJ/mol
Explosion limits in air (at 1 bar and 20 °C)	
Lower	2.0 vol% (35 g/m^3)
Upper	11.1 vol% (200 g/m^3)
Ignition temperature	455 °C

The temperature dependence of typical properties is given in Table 1.

3. Chemical Properties

The chemical properties of propene are, like those of ethylene, characterized by the reactivity of its double bond. Propene undergoes a number of industrially important polymerization, addition, and oxidation reactions (see Chap. 6).

4. Production

Although propene is one of the most important feedstocks for the organic chemicals industry, it is produced almost entirely as a byproduct because it is obtained in sufficient amounts in ethylene production by steam cracking and in some refinery processes (primarily cat cracking).

Whereas in Europe, *refineries* satisfy on average only 20% of the chemical industry's consumption of propene [5], in the United States they meet more than 40% of the consumption demand [6], [7]. Despite this high figure, refineries in the United States consume ca. 75% of their propene production in in-house, nonchemical applications [8]. This propene excess is utilized in gasoline production (alkylate and polymer gasoline; see Section 6.2), to produce liquefied petroleum gas (LPG), and as heating gas.

Table 1. Temperature dependence of some physical properties of propene, according to

Temperature, °C	Vapor pressure, kPa	Heat of vaporization, kJ/kg	Specific heat, kJ kg^{-1} K^{-1} c_{p_L}	c_{p_g}	Density, g/cm³	η_L, mPa·s	η_G, 10^{-4} mPa·s	Surface tension, mN/m	Thermal conductivity, mW m^{-1} K^{-1} λ_L	λ_G
−185.3		568	2.17		0.759	6.76		37.7	237	
−175		560	2.16		0.749	3.52		36.0	231	
−150		540	2.12		0.725	1.14	33	32.1	216	
−125	0.33	518	2.07		0.699	0.53	40	28.2	201	
−100	4.0	494	2.04		0.672	0.33	48	24.4	186	5.8
−75	24	469	2.04		0.644	0.235	55	20.7	171	8.1
−50	91	441	2.09		0.614	0.180	62	17.1	156	10.6
−47.7	101	438	2.09		0.611	0.176	63	16.8	154	10.8
−25	256	409	2.20		0.582	0.140	70	13.6	140	13.1
0	584	374	2.40	1.41	0.546	0.108	77	10.2	125	15.7
25	1150	331	2.70	1.52	0.506	0.081	84	7.0	110	18.4
50	2060	277	3.12	1.62	0.457	0.058	91	4.1	95	21
75	3400	196		1.71	0.389	0.039	98	1.4	80	24
91.9	4620			1.78	0.225		102			26
100				1.81			105			27
125				1.90			111			30
150				1.99			118			33
175				2.08			125			36
200				2.16			131			40
400				2.75			181			67
600				3.20			225			96
800				3.54			264			125
1000				3.80			298			152

A trend toward less severe cracking conditions and thus to increasedpropene production has been observed in *steam cracker plants* using liquid feedstock that have been designed since the mid-1980s. The increased consumption of propene has also boosted the demand for processes for propene production by *catalytic dehydrogenation* of propane. Various processes suitable for this purpose have been developed and in some cases tested.

4.1. Production as Byproduct of Ethylene Production

The influences of cracking feedstocks and cracking conditions on the composition of the cracked gas are described in detail in Ethylene.

A sour-gas-free, anhydrous C_3 fraction is obtained as overhead product from the depropanizer (a) in the processing of cracked gas. This fraction contains the C_3 hydrocarbons of the cracked gas (i.e., propane,propene, propadiene, and propyne) as well as traces of C_2 and C_4 hydrocarbons. Because of its propadiene and propyne content (depending on the cracking conditions, these can total up to 8 mol%), this C_3 fraction does not meet product specifications and therefore requires further treatment.

Figure 1 shows a typical flow diagram for the further processing of the C_3 fraction in a steam cracker plant. The C_3 fraction is fed to a selective hydrogenation unit (b) to remove propadiene and propyne. This hydrogenation can be performed in the gas and liquid phases (pressure ca. 18 bar), palladium catalysts being used in both. The amount of hydrogen added is calculated so that, on the one hand, complete conversion of C_3H_4 to C_3H_6 is achieved, and on the other, the smallest possible amount of propene is hydrogenated to propane. In practice a molar $H_2:C_3H_4$ ratio of ca. 1.5 has proved suitable.

The reaction conditions of C_3 hydrogenation in the gas phase differ from those in the liquid phase. In gas-phase hydrogenation the reaction is controlled by means of the operating temperature, which may be between 50 and 120 °C, depending on the preparation and aging state of the catalyst. With liquid-phase hydrogenation the reaction is controlled by the hydrogen partial pressure. The operating temperature of 15–25 °C is considerably lower in this case. Adiabatic fixed-bed reactors with intermediate coolers, as well as isothermal tubular reactors, have proved suitable for the exothermic hydrogenation reaction in both phases.

Characteristic of a hydrogenation with metered addition of hydrogen is the formation of smaller amounts of oligomers (mainly dimers and trimers). If required by the propene product specification these can be scrubbed out following hydrogenation with a small amount of C_3 in scrubber (c). The mixture of oligomers and scrubbing agent is recycled to the depropanizer for C_3 recovery, and the oligomers are subsequently led into the pyrolysis gasoline fraction.

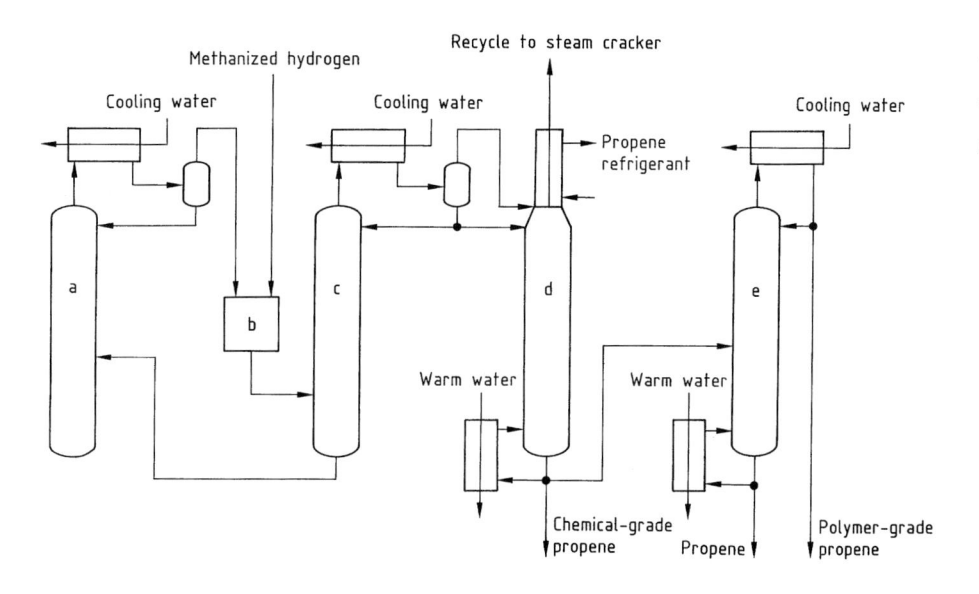

Figure 1. Typical flow diagram for work-up of the C_3 fraction in a steam cracker plant
a) Depropanizer; b) C_3 hydrogenation; c) Polymer scrubbing; d) C_3 stripper; e) C_3 splitter

The methane introduced with the hydrogen (in liquid-phase hydrogenations, the unconsumed hydrogen as well) is then stripped off (d) and recycled to the cracked gas processing unit of the steam cracker to recover entrained propene. The bottom product of the C_3 stripper meets the product specification of *chemical-grade propene.*

Chemical-grade propene is purified to polymer-grade propene in a downstream propene–propane separation column (e). A single water-cooled column designed for this purpose is illustrated in Figure 1. Such a process is only one of several variants that are discussed in more detail in Section 4.1.1. At this point, reference may be made only to one advantage, which provides for the (not essential) integration of propene–propane separation in the steam cracker plant. In modern plants the cracked gases are usually cooled to ambient temperature in a water scrubbing tower after the oil fractionation. The circulating water is heated to 80 to 85 °C in this operation and must be cooled again to be reused in the water scrubbing tower. The heated circulating water can therefore be used to heat the propene–propane separation stage and thus save energy. Also, other heat consumers (e.g., the reboiler of the C_3 stripper) may be supplied economically from this energy source. The propane separated in the C_3 splitter is usually (like ethane) recycled as cracking feedstock.

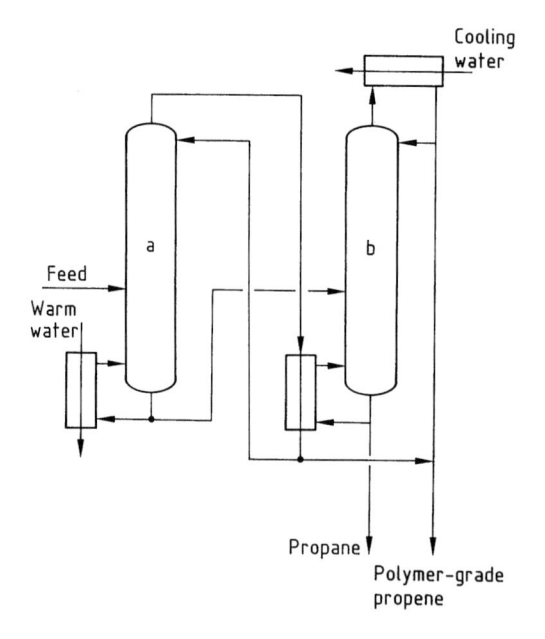

Figure 2. Propene–propane separation by the double-column process
a) High-pressure C$_3$ splitter (ca. 25 bar); b) Low-pressure C$_3$ splitter (ca. 18 bar)

4.1.1. Variants of Propene–Propane Separation

The closely similar boiling curves of propene and propane (bp of propane −42.1 °C) require highly complex separation units. The internal reflux ratio is generally between 0.90 and 0.97 (depending on the feedstock composition, product purities, column pressure, and number of trays). For this reason the propene–propane separation is a process unit in which the development of energy-saving solutions is worthwhile.

Single-Column Process. In principle, separation can be carried out in a single column (generally containing 150–200 trays), as illustrated in Figure 1 (e). The reflux can be condensed with cooling water (column pressure 16–19 bar) or in air coolers (column pressure 21–26 bar). However, for the throughputs that are common nowadays, huge column diameters are required, and transportation of these huge units (or parts of them) by rail or road is difficult. Thus the double-column process has been developed where the separation is carried out in two parallel columns that are smaller in diameter and can therefore be transported more easily. This double-column process requires only ca. 55 % of the original amount of cooling water.

Double-Column Process. The double-column process is illustrated in Figure 2. In this process, only the reflux from the second separation column (b) is condensed with cooling water. The pressure of the first column (a) is so high (ca. 25 bar) that its overhead vapors (ca. 59 °C) can be liquefied in the reboiler of the second column (ca. 18 bar, 51 °C) and serve as heat medium. Each column produces ca. 50 % of the propene

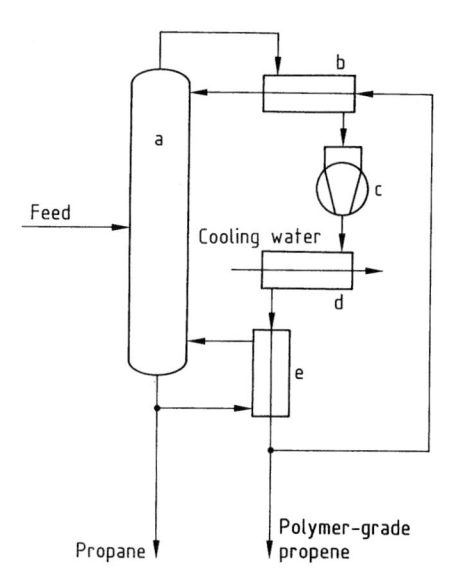

Figure 3. Propene–propane separation by the heat pump process
a) C_3 splitter; b) Reflux subcooler; c) Compressor; d) Post-cooler; e) Condenser and reboiler

product. The bottom product from the first column is the feedstock for the second column. Heating the first column with warm water is still possible with this process.

Heat Pump Process. If no cost-free (or even cost-saving) heating medium such as the warm circulating water of the steam cracker is available, then separation according to the heat pump process is possible (see Fig. 3). In this process the higher propene–propane ratio in the gas phase at lower pressure is utilized. The overhead vapors of the C_3 splitter (a), which operates at ca. 10 bar, are heated slightly in the reflux subcooler (b) and then compressed to a pressure that enables them to be liquefied in the reboiler (e). Apart from the energy needed to drive the compressor (c), only a small amount of cooling water is required in the aftercooler (d).

Lower Product Quality. Energy savings in the propene–propane separation can be achieved not only by appropriate process design, but also by restricting the purity of the polymer-grade propene to the absolute minimum necessary. Figure 4 shows the relative complexity of separation required for higher purities, relative to a minimum propene purity of 99.0 mol%. The complexity of separation and thus the energy consumption increases considerably with propene purity, and even more sharply as the number of column trays decreases.

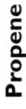

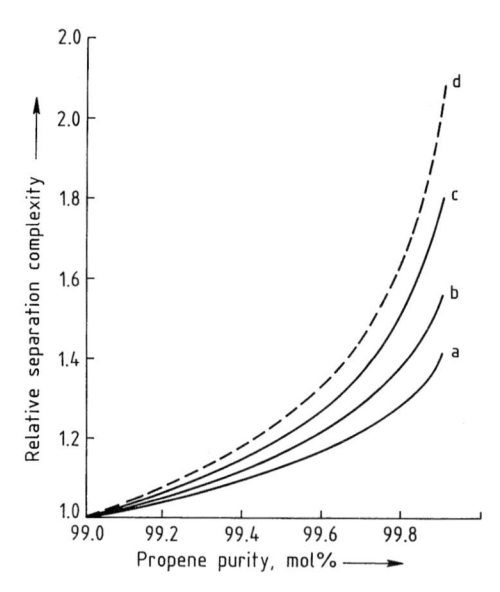

Figure 4. Relative complexity of the propene–propane separation as a function of enhanced propene purity (relative to a minimum propene purity of 99 mol%)
Base: Feedstock containing 93 mol% propene, 7 mol% propane; bottom product containing 30 mol% propene, 70 mol% propane
— single column with 200 trays
--- single column with 150 trays
a) Separation at 10 bar (heat pump process); b, d) Separation at 18 bar (reflux condensation with cooling water); c) Separation at 25 bar (reflux condensation with air cooler)

4.1.2. Apparatus and Construction Materials

In steam cracker plants, work-up of the C_3 fraction occurs above 0 °C and below 100 °C (with the exception of gas-phase hydrogenation, where the temperature can rise to ca. 200 °C). The C_3-treatment section of the plant thus does not place any special requirements on materials. Normal carbon steels and fine-grained steels (such as WSt 36) are used; aluminum is employed for special heat exchangers (such as propene-heated reboilers of propene–propane splitters).

In general, turbocompressors are used as compressors for the *heat pump variant* of the propene–propane separation. Their power requirements for a propene capacity of 100 000 t/a and a product purity of 99.5 mol% are 1400–1500 kW. The compressors are driven by electric motors or steam turbines, which can be integrated into the steam system of the steam cracker.

Highly chargeable sieve trays spaced relatively close together are incorporated in the generally 70–90-m-high C_3 splitters. The apparatus used in steam cracker plants is discussed in more detail in → Ethylene.

4.2. Production as Byproduct of Refinery Processes

The propene produced in refineries also originates from cracking processes. However, these processes can be compared to only a limited extent with the steam cracker for ethylene production because they use completely different feedstocks and have different production objectives.

Refinery cracking processes operate either purely thermally or thermally–catalytically. By far the most important process for propene production is the *fluid-catalytic cracking* (FCC) process in which the powdery catalyst flows as a fluidized bed through the reaction and regeneration phases. This process converts heavy gas oil preferentially into gasoline and light gas oil.

Purely thermal cracking processes, which in refineries contribute to propene production, are employed in coking and visbreaker units. In *coking units* (*delayed coking* and *fluid coking*), residues from the atmospheric and vacuum distillation of the crude oil undergo relatively severe cracking and are thereby converted into gas oil, coke, gasoline, and smaller amounts of cracked gas (6–12 wt% of C_4 and lighter). The cracked gas from the coking unit normally contains 10–15 mol% C_3, mostly propane. In *visbreaker units*, vacuum residues are subjected to mild cracking, with the object of reducing the viscosity of the residue oil. Smaller amounts of gas oil, gasoline, and cracked gas (2–3 wt% of C_4 and lighter) are formed here.

The working-up of refinery gases to isolate the C_3 fraction generally occurs as follows: first, the light components are separated at ca. 15 bar in a deethanizer, whose upper part is operated as an absorber with gasoline and oil as scrubbing agents. The C_5 and heavier hydrocarbons are removed from the bottom product in a connected debutanizer. The C_3–C_4 fraction obtained as overhead product is normally desulfurized on molecular sieves and dried, before being split into individual fractions in a final depropanizer.

4.3. Propane Dehydrogenation

The principal sources of propene are the steam cracking of hydrocarbon feedstocks (Section 4.1) and refinery conversion processes [e.g., fluid catalytic cracking, visbreaking, and coking (Section 4.2)]. Despite the magnitude of the sources, in these cases propene is a byproduct of processes for the manufacture of other products, such as ethylene in the case of steam cracking and motor gasoline in the case of catalytic cracking. The availability of propene is determined primarily by the demand for the main products, although factors such as feedstock and operating conditions have a significant influence on propene yield. The increasing demand for propene derivatives throughout the 1980s, especially for polypropylene, outstripped the availability from these established sources, and processes for the "on-purpose" production of propene by the dehydrogenation of propane from natural LPG fields have been developed commercially.

Propane dehydrogenation is an endothermic equilibrium reaction that is generally carried out in the presence of a noble- or heavy-metal catalyst such as platinum or chronium.

$$C_3H_8 \longrightarrow C_3H_6 + H_2$$

The process is highly selective; overall yields of propene from propane of ca. 90% are claimed for commercially available processes. Higher temperature and lower pressure increase propene yield. However, increased process temperature also causes pyrolysis (cracking) of propane to coke in addition to its dehydrogenation to propene (i.e., reduced selectivity), whereas lower operating pressure increases selectivity. Coke formation wastes feedstock and deactivates the dehydrogenation catalyst. Consequently, propane dehydrogenation processes are operated near atmospheric pressure at around 500–700 °C.

A number of technologies are available commercially for the dehydrogenation of propane to propene: these include *Oleflex* developed by UOP, Des Plaines, Illinois, United States; *Catofin* developed by Air Products and Chemicals, Allentown, Pennsylvania, United States; and *STAR* developed by Phillips Petroleum, Bartlesville, Oklahoma, United States. These processes differ in their modes of operation, the dehydrogenation catalyst, and the methods of catalyst regeneration.

4.3.1. Oleflex Process

The Oleflex process was developed from the *Pacol* process (UOP), which is used to dehydrogenate C_{10}–C_{14} paraffins to olefin feedstocks for the production of intermediates for synthetic detergents [9]. The Oleflex process is an adiabatic process in which the heat of reaction is supplied by reheating the process stream between the different reaction stages. The process is operated at a slight positive pressure, and a proprietary platinum catalyst is used. Fresh propane feed is mixed with recycled hydrogen and unconverted propane and admitted to a train of three radial-flow moving-bed catalytic reaction vessels (Fig. 5). The process is continuous, and overall selectivities for propene of 89–91% are claimed. The dehydrogenation catalyst circulates through the reactor section before passing to a separate regeneration vessel where coke is removed from the surface of the catalyst by combustion in air (UOP Continuous Catalyst Regeneration technology). The regenerated catalyst is returned to the first of the dehydrogenation reactors. Propene is recovered by conventional deethanizer–depropanizer splitting (see Section 4.1). Advantages claimed by the Oleflex process include continuous operation; a uniform, time-invariant catalyst activity profile; and isolation of the oxidative catalyst regeneration phase from the dehydrogenation reactor [9].

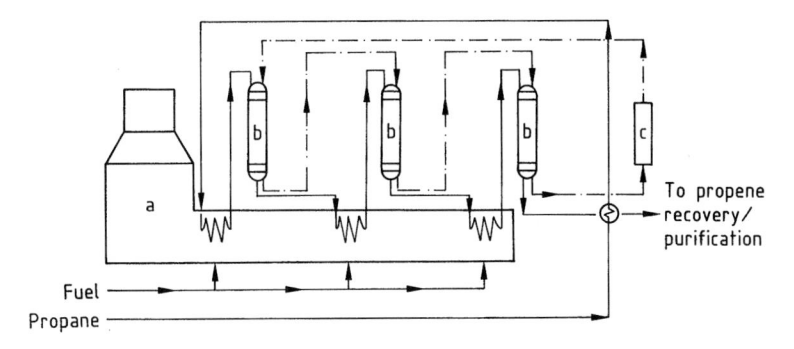

Figure 5. UOP Oleflex process
a) Reheat furnace; b) Moving-bed reactors; c) Continuous catalyst separation system

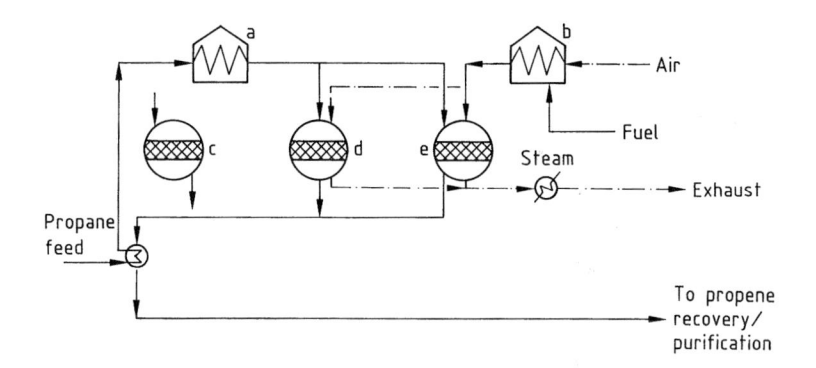

Figure 6. Houdry (Air Products) Catofin process
a) Charge heater; b) Air heater; c) Reactor on purge; d) Reactor on stream; e) Reactor on regeneration

4.3.2. Catofin Process

The Houdry (Air Products) Catofin process (see Fig. 6) operates under a slight vacuum at 550–750 °C [10]. The process utilizes adiabatic, fixed-bed multiple reactors. The dehydrogenation catalyst consists of activated alumina pellets impregnated with 18–20 wt% chromium. The process is cyclic and includes a reaction period, discharge of the reactor, and regeneration of the catalyst in situ. Multiple reactors are used in parallel to achieve continuous plant throughput, with the reactor containing the coked catalyst being taken off stream for the regeneration step. The mixed fresh and recycle propane stream is preheated to 600–700 °C and fed to the reactor at ca. 30 kPa (one-third atmospheric pressure): the influence of reactor pressure and temperature on propene yield is shown in Figure 7. Combustion of the deposited coke on the catalyst during regeneration heats the catalyst bed, and this energy is released to the endothermic dehydrogenation reaction when that reactor is returned to operation. The overall selectivity of propane to propene is reported to be about 87% [10]. Propene is recovered

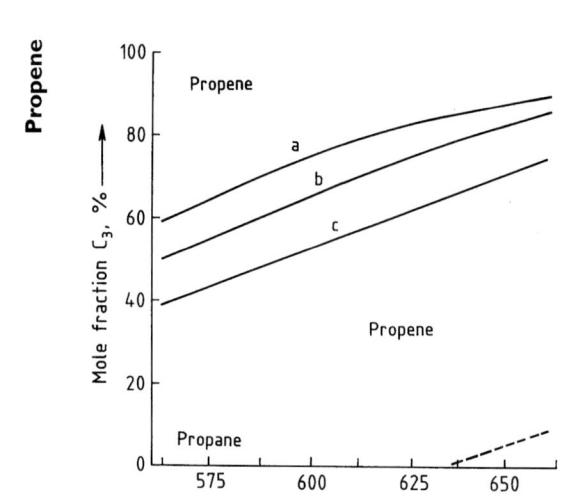

Figure 7. Influence of reactor pressure and temperature on propene yield by propane dehydrogenation
a) 30 kPa (one-third atmospheric pressure);
b) 50 kPa (half atmospheric pressure);
c) 101.3 kPa (atmospheric pressure)
—— mole fraction of methyl acetylene

by a conventional propane–propene splitter. Advantages claimed for the Catofin process include extended catalyst life and utilization of the heat generated in the exothermic regeneration step to assist dehydrogenation.

4.3.3. Phillips STAR Process

The Phillips Steam Active Reforming (STAR) process for the dehydrogenation of paraffins ($\leq C_5$) [11] differs significantly from other dehydrogenation processes. Here, steam is used as a diluent to maintain an overall positive pressure in the process reactor and simultaneously reduce the partial pressure of the hydrocarbons and hydrogen present. Thus, the equilibrium is shifted toward increased conversion. In addition, the process is performed isothermally. The preheated feedstream containing steam is admitted to a chain of multiple fixed-bed reactors. Each reactor consists of multiple catalyst-packed tubes in a furnace firebox, which supplies heat to the catalyst. Reactor operation is cyclic (i.e., one reactor is taken offline sequentially for catalyst regeneration), whereas dehydrogenation is maintained continuously. Catalyst deactivation occurs due to coke deposition, and after about 7 h on-line, an off-line catalyst regeneration by combustion for ca. 1 h is required. The catalyst is completely recovered, and overall catalyst lifetimes of one to two years have been reported [11]. Advantages claimed for the STAR process include 80% yield of propene on propane; isothermal operation that ensures sufficient heat input to promote dehydrogenation; and the steam diluent that reduces the hydrocarbon pressures while maintaining a more practical process pressure. Carbon dioxide (formed by side reactions) must be removed from the reactor product stream prior to propene separation.

4.3.4. Linde Process

The Linde process [12] operates at low reaction temperature and nearly isothermal conditions which are designed to minimize thermal cracking and coke formation. The process uses a chromium oxide catalyst in a fixed-bed tubular reactor. The catalyst has a relatively long cycle time (9 h) before regeneration is necessary. The propane feedstream is not diluted with hydrogen or steam as in other processes, resulting in high selectivity (e.g., 91%). The process is claimed to have low power requirements and low capital costs. Product separation yields polymer-grade propene.

5. Storage, Transportation, Quality Requirements

There is an extensive propene pipeline network in Texas and Louisiana in the United States and a very limited system in Belgium in Western Europe. In other locations propene transportation is largely by road, rail, and ship. This discontinuous delivery and supplying of consumers requires large storage capacities on the part of both propene producers and propene consumers.

Liquid propene is normally *stored* at ambient temperature in spherical pressure tanks with diameters up to 20 m. Propene is however also stored virtually pressureless at −47 °C, particularly for very large amounts. The latter form of storage requires reliquefaction devices.

Propene is *transported* by road and rail in cylindrical pressurized tanks at ambient temperature. The standard railway tanker holds 42 t of propene. In the case of road transport the maximum permitted overall truck weight of ca. 40 t restricts the propene load to ca. 20 t (corresponding to 40 m^3). In transportation by ship, batteries of smaller pressurized tanks as well as large atmospheric tanks with reliquefaction devices are used.

Quality Requirements. The quality requirements placed on propene vary widely depending on its subsequent use. Corresponding to production in steam cracker plants, two qualities are basically employed namely, the less pure chemical-grade propene and the highly pure polymer-grade propene. Examples of specifications for these two product qualities are given in Table 2. Depending on different posttreatment processes (in the case of polymer-grade propene) or different intended main uses (in the case of chemical-grade propene) the permissible concentrations of impurities may vary within relatively wide limits.

Table 2. Typical product specification of propene

	Chemical grade	Polymer grade	Detection method *	Detection limit, ppm
Propene, mol%	92–95	99.5–99.8		
Acetylene, mol ppm	< 10	< 2	GC	0.5
Ethylene, mol ppm	< 20	< 20	GC	0.5
Ethane, mol ppm	< 2000	< 100	GC	0.5
Propyne, mol ppm	< 20	< 5	GC	1
Propadiene, mol ppm	< 20	< 5	GC	1
C_{4+}, mol ppm	< 1000	< 10	GC	2
Hydrogen, mol ppm	< 10	< 10	GC	2
Nitrogen, mol ppm	< 50	< 50	GC	2
Oxygen, mol ppm	< 5	< 5	colorimetry	1
Carbon monoxide, mol ppm	< 5	< 5	GC (via methanation)	0.5
Carbon dioxide, mol ppm	< 5	< 5	GC (via methanation)	1
Sulfur, mass ppm	< 5	< 1	according to Wickbold (DIN EN 41)	1
Water, mol ppm	< 25	< 10	Karl Fischer	1
Propane	remainder	remainder	GC	1

* GC = gas chromatography.

Table 3. Propene production for chemicals in 10^3 t

Region	1980	1982	1984	1986	1988	1990
North America	6 797	6 378	7 714	8 350	9 577	10 003
South America	660	732	982	1 094	1 182	1 466
Western Europe	5 823	5 984	7 191	7 333	8 702	9 142
Eastern Europe	1 686	2 068	2 272	2 557	2 571	2 706
Japan	2 637	2 565	2 981	3 147	3 682	4 214
Asia–Pacific	920	1 014	1 369	1 654	2 035	2 399
Others	76	72	55	86	271	390
Total	18 599	18 813	22 564	24 165	28 020	30 320

6. Uses

Global propene production for chemical uses increased strongly through the 1980s [13], equivalent to a global average annual increase of 5% (see Table 3).

The extent of steam cracker and refinery production of propene varies from region to region [13]. In the United States, refinery production has a much greater share of total availability than in other regions due to a large motor gasoline market that requires an adequate FCC capacity and a preponderance of ethane feedstock for steam crackers, resulting in lower yields of steam cracker propene. The proportion of propene produced by steam cracking between 1980 and 1990 is given in Table 4.

The three commercial grades of propene are used for different applications. *Refinery-grade* propene (i.e., 50–70% pure propene in propane) is obtained from refinery processes. The main uses of refinery propene are in LPG for thermal use or as an

Table 4. Proportion of propene produced by steam cracking (in wt%)

Region	1980	1982	1984	1986	1988	1990
North America	59	46	54	58	58	56
South America	64	65	72	73	72	68
Western Europe	92	87	86	84	83	78
Eastern Europe	86	87	87	85	88	87
Japan	95	88	85	79	83	82
Asia–Pacific	88	86	89	88	89	88

octane-enhancing component in motor gasoline. Refinery-grade propene can also be used in some chemical syntheses (e.g., of cumene or isopropanol).

Chemical-grade propene is used extensively for most chemical derivatives (e.g., oxo alcohols, acrylonitrile, or polypropylene).

Polymer-grade propene contains minimal levels of impurities such as carbonyl sulfide that can poison catalysts used in polypropylene and propylene oxide manufacture.

6.1. Thermal Uses

Propene has a calorific value of 45 813 kJ/kg, and refinery grade propene can be used as fuel if more valuable uses are unavailable locally (i.e., propane–propene splitting to chemical-grade purity). Refinery-grade propene can also be blended into LPG for commercial sale, provided no low limits exist on the olefin content of LPG (unsaturated hydrocarbons produce more smoke than saturated hydrocarbons and have a lower calorific value).

6.2. Motor Gasoline Uses

Propene is used as a motor gasoline component for octane enhancement, via dimerization–formation of polygasoline or alkylation.

6.2.1. Polygasoline Production–Dimerization

Polymerization of refinery-grade propene at high temperature and pressure with a catalyst such as phosphoric acid produces *polygasoline* containing dimers, trimers, and tetramers. Polygasoline is used as a blendstock in motor gasoline. In 1977, *dimerization* of propene to isohexenes at lower temperature and pressure by using nickel complex–alkylaluminum halide catalysts was developed by Institut Français de Petrole (Dimersol process) [14]. Propene conversions of > 90 % are achieved. The product, *dimate*, has a low vapor pressure and a high blending octane rating (R + M)/2 of 89 (for a definition

Table 5. Consumption of propene (in wt%)

Product	1980	1982	1984	1986	1988	1990
Polypropylene	30	33	36	38	40	44
Acrylonitrile	19	18	18	17	15	14
Oxo alcohols	19	18	16	16	16	15
Propylene oxide	9	9	8	9	9	9
Cumene	10	9	9	8	8	8
Oligomers	8	7	6	5	5	4
Others	5	6	7	7	7	6

of blending octane rating, compared to 86 for cat cracker gasoline and 87 for polygasoline.

6.2.2. Alkylation

Alkylation is an acid-catalyzed reaction between propene (or other light olefins) and isobutane to yield a highly branched higher alkane (2,3-dimethylpentane in the case of propene) of low vapor pressure and high octane rating [15]. The reaction catalyst is either hydrofluoric acid, which gives octane ratings of 90–92, or sulfuric acid, which gives octane ratings of 88–91 [16]. If *hydrofluoric acid* is used as catalyst a second reaction is promoted in which hydrogen is transferred from isobutane to propene to yield propane and isobutene. The latter then alkylates isobutane to produce trimethylpentanes in parallel with dimethylpentane from the direct alkylation. The choice of catalyst depends on feedstock and process conditions as well as octane rating.

6.3. Chemical Uses

The principal chemical uses of propene are in the manufacture of polypropylene, acrylonitrile, oxo alcohols, propylene oxide, butanal, cumene, and propene oligomers [13]. Polypropylene is the leading global consumer of propene (see Table 5). Other uses include acrylic acid derivatives and ethylene–propene rubbers.

Polypropylene. Isotactic polypropylene is of great commercial importance. Polymerization of chemical-grade or polymer-grade propene is carried out by using Ziegler–Natta catalysts [17], in either a liquid-phase or a gas-phase process. Polypropylene has a high melting point, good rigidity, and good chemical and abrasion resistance. Weaknesses include poor thermal and UV stability, but these can be improved by appropriate stabilizers. Polypropylene is used widely, and global production exceeded 13×10^6 t in 1990.

Acrylonitrile. Acrylonitrile is obtained by catalytic oxidation of chemical-grade propene in the presence of ammonia (ammoxidation).

A number of commercial processes are available but the most significant is the Sohio process [18] in which a stoichiometric mixture of propene, ammonia, and oxygen is reacted over a catalyst at 400–500 °C. World consumption of acrylonitrile in 1990 was 3.7×10^6 t.

Propylene Oxide. Propene (polymer grade) can be converted into the ether propylene oxide either by hydrochlorination and epoxidation, or by reaction with an organic hydroperoxide [19], [20].

Propylene oxide is used as a raw material in the production of poly(propylene glycols), propanolamines, and polyether polyols, which themselves are intermediates for flexible and rigid polyurethane foams. Propylene oxide production in 1990 was 2.75×10^6 t, consuming some 2.2×10^6 t of propene.

2-Propanol. 2-Propanol is produced by the hydration of chemical-grade propene either in the presence of sulfuric acid or directly as a vapor-phase reaction in the presence of a catalyst at 180–270 °C [21]. Yields > 70 % based on propene are achieved at high selectivity (e.g., > 90 %). 2-Propanol is used extensively as a solvent in paint, cosmetics, and pharmaceuticals. Global production of 2-propanol in 1990 was 1.4×10^6 t, representing 1.2×10^6 t of propene.

Cumene. Propene can be used to alkylate benzene to produce 1-methylethylbenzene (cumene) [22]. The reaction is usually undertaken with refinery-grade propene in the presence of phosphoric acid catalyst. Cumene is then oxidized to cumene hydroperoxide, which decomposes to phenol and acetone ($\rightarrow$ Phenol). Global production of cumene was 2.3×10^6 t in 1990, representing 0.9×10^6 t of propene.

Oligomers. Propene oligomers are liquid polymers of propene with a low degree of polymerization. They are intermediates between polygasoline and the high polymer polypropylenes, and contain ten or more carbon atoms per molecule. Propene oligomers are prepared by acid-catalyzed polymerization. Depending on the degree of polymerization the products are used as lube oil additives or as detergent intermediates.

Hydroformylation. Propene reacts with carbon monoxide and hydrogen (synthesis gas) at elevated temperature and pressure in the presence of a transition-metal catalyst (e.g., a rhodium carbonyl catalyst) to give butanals (oxo process) [23]. *n*-Butanal is converted into the aldol and hydrogenated to produce 2-ethylhexanol.

Allylics. Propene can undergo many reactions to produce allylic compounds. In these reactions a methyl hydrogen atom is substituted and the carbon–carbon double bond is preserved (see ammoxidation p. 4097). The more important allylic derivatives include allyl chloride (3-chloropropene) [24] and acrolein (allyl aldehyde) [21]. Propene can be reacted with *acetic acid* in an acidic medium to produce allyl acetate (i.e., the double bond is retained in the molecule), or acetic acid is added to the carbon–carbon double bond to yield isopropyl acetate.

7. Economic Aspects

Propene is a byproduct of the production of ethylene by steam cracking (Section 4.1) or the production of motor gasoline by catalytic cracking (Section 4.2), so no direct manufacturing costs are attributed to these sources. Propene demand exceeds that from steam cracker sources, and marginal supplies are taken from refinery sources. *Propene economics* are based on its value in marginal applications: generally dimerization (Section 6.2.1) or alkylation (Section 6.2.2) for gasoline. The motor gasoline valuation of propene is set by the octane enhancement of motor gasoline by alkylate or Dimersol, both obtained from refinery-grade propene. Generally, motor gasoline absorbs incremental propene supplies, and all other uses must pay the equivalent octane value plus the costs of distilling refinery grade to the purity required. When propene demand in motor gasoline is weak, surplus refinery propene is returned to LPG uses or to refinery fuel (Section 6.1). When demand for propene is very high the value of the marginal supply rises above the octane value to reflect the new marginal use (e.g., in polypropylene).

Propene from the *dehydrogenation of propane* does have a direct manufacturing cost structure based on feedstock, utility, plant fixed costs, etc. In an environment in which propene demand exceeds the supply from steam crackers and catalytic crackers, propene values should cycle between a minimum of the cash cost of propane dehydrogenation and either a level necessary to justify the investment in further dehydrogenation plants or a higher level set by the demand for derivatives.

8. References

General References

[1] F. Andreas, K. Gröbe: *Propylenchemie*, Akademie-Verlag, Berlin 1969.
[2] F. Asinger: *Die Petrolchemische Industrie*, parts I and II, Akademie-Verlag, Berlin 1971.
[3] L. F. Albright, B. L. Crynes: *Industrial and Laboratory Pyrolyses*, American Chemical Society, Washington 1976.

Specific References

[4] C. L. Yaws, *Chem. Eng. (Int. Ed.)* **82** (1975) no. 7, 101–109; **83** (1976) no. 17, 79–87; **83** (1976) no. 23, 127–135; **83** (1976) no. 25, 153–162.

[5]   M. Stam: *Western European Petrochemicals into the 1980's* (unpublished, Euroeconomics, July 31st 1978).
[6]   E. J. Debreczeni, *Chem. Eng. (Int. Ed.)* **84** (1977) no. 12, 135–141.
[7]   D. S. Sanders, D. M. Allen, W. T. Sappenfield, *Chem. Eng. Prog.* **73** (1977) no. 7, 40–45.
[8]   P. H. Spitz, *Hydrocarbon Process.* **55** (1976) no. 7, 131–135.
[9]   P. R. Pujado, B. V. Vora, *Hydrocarbon Process.* **69** (1990) no. 3, 65.
[10] R. G. Craig, T. J. Delaney, J. M. Duffalo: *Catalytic Dehydrogenation Performance of the Catofin Process*, DeWitt Petrochemical Review, Houston 1990.
[11] R. O. Dunn et al.: *The Phillips Steam Active Reforming (STAR) Process of C_3, C_4, and C_5 Paraffins for the Dehydrogenation*, DeWitt Petrochemical Review, Houston 1992.
[12] *Eur. Chem. News* (1992) March 9, 19.
[13] World Petrochemical Industry Survey: *Olefins*, Parpinelli Tecnon srl, Milan 1991.
[14] P. M. Kohn, *Chem. Eng.* **114** (1977) May 23.
[15] L. F. Albright, *Oil Gas J.* (1990) Nov., 79.
[16] L. E. Chapin, G. C. Lilios, T. M. Robertson, *Hydrocarbon Process.* **64** (1985) no. 9, 67.
[17] G. Natta et al., *J. Am. Chem. Soc.* **77** (1955) 1708.
[18] J. L. Callahan, R. K. Grasselli, E. C. Milberger, H. A. Strecker, *Ind. Eng. Chem. Prod. Res. Dev.* **9** (1970) 134.
[19] R. B. Stobaugh et al., *Hydrocarbon Process.* **52** (1973) no. 1, 102.
[20] J. Poloczek, J. Bobinski, *Int. Chem. Eng.* **11** (1971) no. 1, 87.
[21] E. Wilhelan, R. Battino, R. T. Wilcock, *Chem. Rev.* **77** (1977) no. 2.
[22] P. R. Pujado, J. R. Salazar, C. V. Berger, *Hydrocarbon Process.* **55** (1976) no. 3, 91.
[23] P. Pino, F. Piacenti, M. Bianchi: *Organic Synthesis via Metal Carbonyls*, vol. **2**, J. Wiley and Sons, Inc., New York 1977, p. 43.
[24] A. W. Fairbairn, H. A. Cheney, A. J. Chemiavsky, *Chem. Eng. Prog.* **43** (1947) 280.

Propionic Acid and Derivatives

ULF-RAINER SAMEL, BASF Aktiengesellschaft, Ludwigshafen, Federal Republic of Germany (Chaps. 2–7, 9, Sections 8.2, 11.1–11.3)

WALTER KOHLER, BASF Aktiengesellschaft, Ludwigshafen, Federal Republic of Germany (Section 8.1)

ARMIN O. GAMER, BASF Aktiengesellschaft, Ludwigshafen, Federal Republic of Germany (Chap. 10)

ULLRICH KEUSER, BASF Aktiengesellschaft, Ludwigshafen, Federal Republic of Germany (Section 11.4)

1.	Introduction	4102	10. Toxicology	4115
2.	Physical Properties	4102	11. Derivatives	4116
3.	Chemical Properties	4104	11.1. Salts	4116
4.	Production	4105	11.2. Propionic Anhydride and Propionyl Chloride	4117
4.1.	Carbonylation of Ethylene (BASF Process)	4106	11.3. Esters	4118
4.2.	Oxidation of Propanal	4107	11.4. Chloropropionic Acid and Derivatives	4118
4.3.	Direct Oxidation of Hydrocarbons	4108	11.4.1. Properties and Production	4119
4.4.	Construction Materials	4109	11.4.1.1. 2-Chloropropionic Acid and Derivatives	4119
5.	Storage and Transportation	4110	11.4.1.2. 3-Chloropropionic Acid and Derivatives	4120
6.	Quality Specifications	4110	11.4.1.3. 2,2-Dichloropropionic Acid	4120
7.	Environmental Protection	4111	11.4.2. Uses	4121
8.	Uses	4111	11.4.2.1. 2-Chloropropionic Acid and Derivatives	4121
8.1.	Use as Food and Feed Preservative	4111	11.4.2.2. 3-Chloropropionic Acid	4123
8.2.	Other Uses	4114	11.4.2.3. 2,2-Dichloropropionic Acid	4123
9.	Economic Aspects	4114	11.4.3. Toxicology	4123
			12. References	4124

1. Introduction

Propionic acid [79-09-4], CH$_3$CH$_2$COOH, M_r 74.08, was first described by J. GOTTLIEB in 1844. In 1848 J. J. DUMAS recognized that it belonged to the group of compounds known as fatty acids. Since it is the first fatty acid which can be salted out of aqueous solution, it was named propionic acid ($\pi\rho\omega\tau o\sigma$ = first, $\pi i\omega\nu$ = fat).

Propionic acid occurs naturally, predominantly in the form of its esters in some essential oils. It is also formed as the free acid in various enzymatic and fermentation processes, and is produced as a result of anaerobic carbohydrate fermentation in the stomachs of ruminants. Propionic acid derivatives are also important intermediates in various physiological cycles.

Industrially, propionic acid is used in the production of cellulose esters, plastic dispersions, and herbicides. It is also used to a limited extent in pharmaceuticals and in flavors and fragrances. Propionic acid is gaining increasing importance for the preservation of forage cereals and animal feeds because many putrefying and mold-forming microorganisms cannot survive in its presence. This fungicidal and bactericidal action is exploited to protect bakery products and cheese from attack by bacteria and mold.

Propionic acid is produced specifically by carbonylation of ethylene and oxidation of propionaldehyde; in addition, it is obtained in large quantities as a byproduct in the oxidation of hydrocarbons.

2. Physical Properties

Propionic acid, mp −20.8 °C, bp 141.3 °C (at 101.3 kPa) is a clear, colorless liquid with a slightly pungent odor (odor threshold in air 0.03 vol%). Solid propionic acid forms monoclinic crystals and undergoes an increase in volume on melting of 12.2% [1]. In the liquid state and in the gas phase a considerable portion of the molecules of propionic acid are dimerized as is the case with its lighter homologues acetic [64-19-7] and formic [64-18-6] acids [2]–[5]. Propionic acid is miscible with water and most organic solvents in all proportions. Propionic acid–water mixtures show a positive deviation from Raoult's law [3]. The heats of mixing of various binary and ternary mixtures of propionic acid are given in [6], [7].

Propionic acid forms azeotropic mixtures with many liquids [8]–[11], for example with water at 82.3 wt% water [bp (101.3 kPa) 99.98 °C]. The liquid–vapor equilibria of the system propionic acid–water at various pressures have been investigated [12]. Further data on the liquid–vapor system propionic acid–water and also on other binary mixtures of propionic acid are given in [13]. The solubility of propionic acid in ternary mixtures is described in [14], [15].

Table 1. Vapor pressure of pure propionic acid

t, °C	p, kPa	t, °C	p, kPa
0.7	0.1	100.9	25
13.5	0.25	120.1	50
24.0	0.5	140.9	100
35.0	1.0	159.5	200
50.8	2.5	185.7	500
64.5	5.0	203.0	1000
79.3	10.0	219.7	2000

The temperature-dependence of the vapor pressure of pure propionic acid is shown in Table 1 and that of the density of the pure acid in Table 2. Table 3 shows the effect of temperature on the dynamic viscosity.

Other physical data are as follows:

Refractive index n_D^{20}	1.3865 [18]
Surface tension toward air	
(at 20 °C)	26.7×10^{-3} N/m [19]
(at 90 °C)	19.7×10^{-3} N/m
Thermal conductivity	
(at 0 °C)	153 W m^{-1} K^{-1} [20]
(at 20 °C)	150 W m^{-1} K^{-1}
(at 50 °C)	144 W m^{-1} K^{-1}
(at 100 °C)	136 W m^{-1} K^{-1}
Electrical conductivity (at 2 °C)	10^{-4} µS/cm
Dynamic dielectric constant	
(at 10 °C)	3.30 [9]
(at 40 °C)	3.44
Electric dipole moment, μ	1.75 D $\pm$ 5% [9]
Dissociation constant in H$_2$O	
(at 0 °C)	1.274×10^{-5} mol/L [21]
(at 20 °C)	1.338×10^{-5} mol/L [22]
(at 40 °C)	1.284×10^{-5} mol/L
(at 60 °C)	1.160×10^{-5} mol/L
Heat of neutralization with NaOH (at 25 °C)	56.82 $\pm$ 0.08 kJ/mol [23]
Ionization potential	10.24 $\pm$ 0.03 eV [9]
Heat of fusion	101.65 kJ/kg [24]
Heat of vaporization	418.7 kJ/kg [25]
Heat of formation	− 164.5 kJ/mol [25]
Heat of combustion	− 1528.3 kJ/mol [9]
Specific heat (liquid), c_p	
(at 0 °C)	2.077 kJ kg^{-1} K^{-1} [16]
(at 50 °C)	2.299 kJ kg^{-1} K^{-1}
(at 100 °C)	2.516 kJ kg^{-1} K^{-1}
Specific heat (vapor), c_p	
(at 100 °C)	2.823 kJ kg^{-1} K^{-1}
(at 200 °C)	3.245 kJ kg^{-1} K^{-1}
(at 300 °C)	3.647 kJ kg^{-1} K^{-1}
Isentropic compressibility (at 20 °C)	0.750 GPa [26]
Critical data	
t_{crit}	338.9 °C [16]
p_{crit}	5.37 MPa [16]
V_{crit}	0.322 kg/L [16]

Flash point   50 °C [27]
Ignition temperature   485 °C
Explosion limits in air (at 20 °C)   2.1 – 12.0 vol%

Table 2. Change in density of liquid propionic acid with temperature

t, °C	ϱ, g/cm^3	t, °C	ϱ, g/cm^3
−20	1.038	80	0.9286
−10	1.027	90	0.9176
0	1.016	100	0.9066
10	1.005	110	0.8954
20	0.9935	120	0.8841
30	0.9827	130	0.8726
40	0.9719	140	0.8609
50	0.9611	150	0.8490
60	0.9503	160	0.8367
70	0.9395	170	0.8241

Table 3. Dynamic viscosity of pure propionic acid as a function of temperature

t, °C	η, mPa · s
0	154.1
20	109.4
50	73.8
100	45.0
150	31.0

The specific heat of aqueous propionic acid solutions has been measured by ACKERMANN [28]. The values increase with increasing temperature, but decrease with increasing propionic acid concentration.

A graphical presentation of the temperature dependence of vapor pressure, heat of vaporization, specific heat, density, viscosity, surface tension, and thermal conductivity of gaseous and liquid propionic acid is given in [29].

3. Chemical Properties

As the third member of the homologous series of aliphatic monocarboxylic acids, propionic acid is still quite reactive and undergoes all reactions typical of these acids. It readily forms salts, amides, acid halides, and an anhydride (see Chap. 11). Esters are formed with alcohols and olefins, and with acetylenes the corresponding vinyl esters (→ Vinyl Esters).

The reactions of propionic acid differ from those of acetic acid because of the inductive effect of the methylene group. Thus, propionic acid is attacked by alkaline permanganate solution more readily than acetic acid, whereby it is degraded nonselectively into carbon dioxide, oxalic acid, and lactic acid [30].

Ionic substitution reactions also preferably occur at the methylene group. Chlorination in the presence of halogen-transfer reagents (e.g., PCl$_3$) gives α-chloropropionic acid. Radical chlorination (presence of peroxides, UV radiation) yields β-chloropropionic acid [31], [32]. Both acids are important intermediates in the production of herbicides and pharmaceuticals (see Section 11.4).

Thermal decomposition of calcium propionate or passage of propionic acid vapor over metal oxides at 300–400 °C gives diethyl ketone, which is used in fairly large quantities as an industrial solvent.

4. Production

Industrially, propionic acid is currently produced almost exclusively by three different processes:

1) Carbonylation of ethylene with carbon monoxide and water
2) Oxidation of propanal
3) Direct oxidation of hydrocarbons

Processes such as production of propionic acid as a byproduct in the synthesis of hydroxylammonium salts from 1-nitropropane [108-03-2] [33], by wood distillation, or by nitric acid oxidation of 1-propanol [62309-51-7] have become obsolete, although the latter gives propionic acid in 90% yield [34]. Synthesis of propionic acid via the alkali melt of an n- and isopropanol mixture could never compete successfully with other processes, although propionic acid was obtained with > 98% yield [35], [36]. The carbonylation of ethanol [64-17-5] [37] and acetic acid [64-19-7] [38] have also not yet been carried out industrially.

The *Koch synthesis* has been investigated intensively [39], [40]. As a carbonylation reaction of ethylene [74-85-1] in a strongly acidic medium, it is a variant of the Reppe synthesis. However, compared with the latter it never achieved much industrial significance and was carried out with little success.

Other possible sources of propionic acid, which are not used industrially for economic reasons, are its formation as a byproduct in the high- and low-pressure carbonylation of methanol [67-56-1] to give acetic acid (2% formation of propionic acid), the atmospheric oxidation of n-butene [25167-67-3] (4% formation of propionic acid), and the direct reaction of ethylene, carbon monoxide, and water over noble-metal catalysts [41]–[49].

Processes for making propionic acid accessible specifically by C_1-*chemistry* have yet to be assessed for their future importance. Two-step reactions have been described, in which propionic acid is obtained directly from synthesis gas (20–60 bar, 150–160 °C,

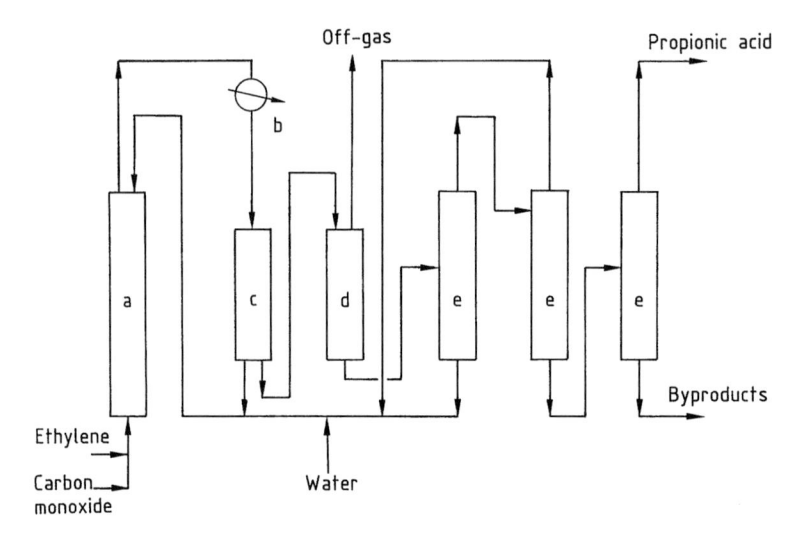

Figure 1. Production of propionic acid by the BASF process
a) High-pressure reactor; b) Heat exchanger; c) Separator; d) Expansion vessel; e) Distillation column

Rh catalysts) [50], [51]. With the current availability of ethylene and naphtha these processes are, however, not yet competitive.

It is sometimes desirable, to use propionic acid produced by *natural methods*, particularly for the use of propionic acid in flavors and fragrances. Appropriate microbiological and enzymatic processes have been developed, which are usually based on the anaerobic fermentation of starch or sugars [52]–[54]. However, the expensive production of this "natural" propionic acid limits its use to special areas of application.

4.1. Carbonylation of Ethylene (BASF Process)

In the Reppe synthesis ethylene is reacted with carbon monoxide [630-08-0] and water in the presence of $Ni(CO)_4$ [13463-39-3] according to the following equation:

$$CH_2=CH_2 + CO + H_2O \xrightarrow{Ni(CO)_4} CH_3CH_2COOH$$
$$\Delta H = -164.5 \text{ kJ/mol}$$

The reaction takes place at high pressure and is characterized by low raw material costs, high conversion, high yield, and a simple workup.

Process Description (Fig. 1) [25]. Ethylene and carbon monoxide are compressed and continuously pumped into the high-pressure reactor (a) together with feed solution. The crude propionic acid formed at 100–300 bar and 250–320 °C is drawn off at the head of the reactor and cooled in a heat exchanger (b) with production of steam. Part of the cooled reaction product is recycled to the reactor for temperature regulation (c), the main quantity is allowed to expand and is separated into an off-gas and a crude acid stream (d). Nickel is recovered from the off-gas and led back into the reactor. The off-gas is incinerated with recovery of heat. The crude acid stream is subsequently dehydrated and worked up by distillation in several columns (e). The nickel salts thus formed are recycled into the process. The pure propionic acid is finally obtained by distillation. The product residue is channeled out of the process.

Variants of the Reppe and Koch syntheses are described in [40], the effect of inorganic salts on the reaction conditions and yield is discussed. Boric acid is reported to accelerate the reaction and hinder the precipitation of catalyst salts [55].

4.2. Oxidation of Propanal

The oxidation of propanal [123-38-6] is an important route to propionic acid. This route is economically attractive although two steps (production of propanal and subsequent oxidation) are required: (1) Propanal is formed in large quantities as an intermediate in the production of n-propanol [62309-51-7] by hydroformylation of ethylene, and (2) the oxidation takes place in plants in which other aliphatic carboxylic acids (n-butyric and isobutyric acids, n-valeric and isovaleric acids) are also produced. A high utilization of available capacity is therefore guaranteed in both steps.

Propanal Production. The production of propanal is carried out by the hydroformylation of ethylene (→ Propanal). It is favored because, unlike the case of the higher aldehydes, n-/iso-mixtures cannot be formed.

Two syntheses currently compete: the classical cobalt-catalyzed high-pressure carbonylation at 200–280 bar, and 130–150 °C [56], and the rhodium- or iridium-catalyzed low-pressure carbonylation at ca. 20 bar, and ca. 100 °C [57]–[59].

In the *high-pressure synthesis* the yield is generally impaired by partial hydrogenation of propanal to give propanol. The isolation of the aldehyde takes place, after removal of cobalt, by distillation of an azeotrope with ca. 98% aldehyde content [56].

In the *low-pressure synthesis* (Union Carbide) [60], the aldehyde can be distilled directly from the reaction mixture in 99% purity.

Oxidation. Propanal is subsequently oxidized under very mild conditions at 40–50 °C to propionic acid with high selectivity.

In the United States propionic acid is produced by the oxidation of propanal by Union Carbide and Eastman Kodak (see Chap. 9).

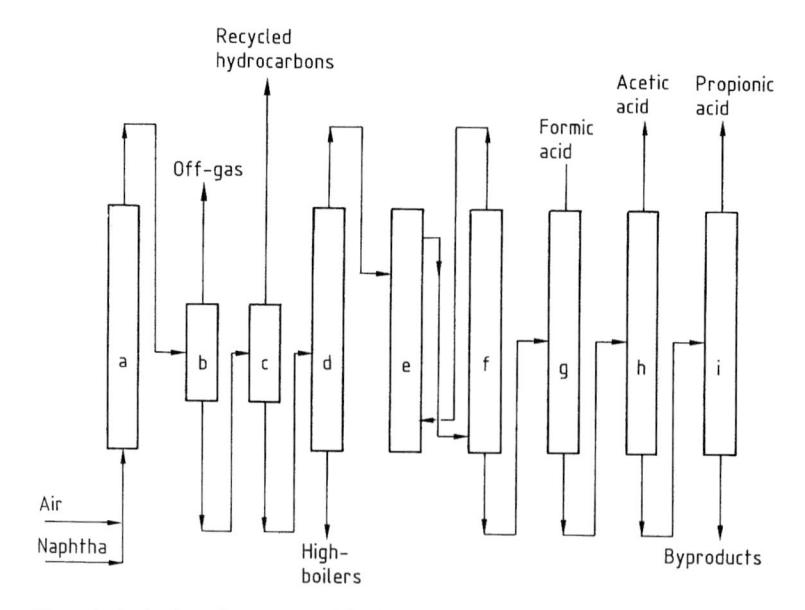

Figure 2. Production of propionic acid by the BP Chemicals process
a) Reactor; b) Gas–liquid separator; c) Liquid–liquid separator; d) Distillation column; e) Extraction; f) Separation of extracting agent; g) Formic acid distillation; h) Acetic acid distillation; i) Propionic acid distillation

4.3. Direct Oxidation of Hydrocarbons

A large quantity of propionic acid is obtained by the direct oxidation of hydrocarbons, predominantly naphtha. In this process, which is principally used for acetic acid production (→ Acetic Acid), formic acid [64-18-6], propionic acid, and an isomeric mixture of butyric acids are formed as byproducts. Whether this process is considered economic for propionic acid production is principally a question of the market evaluation of the different products. The process has favorable raw material costs, but requires a relatively complex workup of the product mixture [61], [62].

The composition of the acid mixture formed in the oxidation can be affected by the reaction conditions (pressure, temperature), the type of reactor (tubular reactor, column reactor), the raw material (naphtha, liquefied petroleum gas), and by the catalyst [61], [62]. If naphtha is used, the proportion of propionic acid in the mixture is ca. 10–15%. In other cases, e.g., with 2-methylpentane as raw material, the proportion of propionic acid can increase to 31% [63]. The total yield of acids does, however, vary considerably with the starting material [64]–[66].

BP Chemicals Process. (Fig. 2) [61], [67]. Naphtha is preheated to 170 °C and oxidized with air at 40–45 bar in several reactors (a) in series. The heat of reaction is used for steam generation. The cooled discharge from the reactors is separated from the reaction off-gas in a separator (b). Entrained liquid is recovered from the off-gas and recycled to the reactor; the off-gas is then incinerated. The liquid reactor discharge is separated into an organic phase, which contains unreacted hydrocarbons, and an aqueous phase which contains the product mixture (c). The unreacted hydrocarbons are recycled to the reactor. The low- and high-boilers are separated from the aqueous phase which then gives the crude acid (d). From this the C_1- to C_4-acids are obtained by extractive dehydration (e) followed by fractional distillation (g)–(i).

Propionic acid is produced by direct oxidation at BP Chemicals (UK), Hoechst – Celanese (United States), and Daicel (Japan) (see Chap. 9).

4.4. Construction Materials

The corrosiveness of propionic acid is not only determined by the water content, pressure, and temperature, but also by the degree of purity. The corrosion resistance of the materials used is therefore not always predictable and is frequently dependent on individual conditions.

Ordinary steel is totally unsuitable for the handling of propionic acid. Aluminum is only resistant at room temperature and to anhydrous, concentrated propionic acid; at higher temperature corrosiveness toward aluminum varies with the concentration of acid. Thus the corrosion maximum for 50 °C lies at 75 % acid, whereas there are two maxima at the boiling point, a weaker one at 1 % and a strong one at 99.8 % acid [68].

Copper and copper alloys are stable toward propionic acid up to its boiling point, but only if the solutions are free from air or oxidizing substances [68].

Of the stainless steels, the ferritic chromium steels (13 – 17 % Cr content, UNS no.: S 40 000) are unsuitable because propionic acid tends to cause pitting. Generally, austenitic CrNi steels and CrNiMo steels are used for handling propionic acid. At ambient temperature the CrNi steel X 10 CrNiTi 189 (DIN Materials no. 1.4541, UNS no.: S 30 400) is suitable. The CrNiMo steel X 10 CrNiMoTi 1810 (DIN no. 1.4571, UNS no.: S 31 600) is stable toward pure propionic acid up to its boiling point. However, impurities in propionic acid can make the use of titanium or Hastelloy (e.g., Hastelloy C NiMo 16 Cr 15 W, DIN no. 2.4819, UNS no.: N 10 276) equipment necessary, even below the boiling point [68].

Above the boiling point, up to ca. 230 °C Hastelloy C has shown excellent resistance even under oxidizing and reducing conditions. Above 230 °C only silver has proved to be a suitable material [25].

Table 4. Sales specification of propionic acid

Specification	Limits	Analytical method
Propionic acid content, wt%	99.5	gas chromatography
Water content, wt%	0.1	DIN 51777
Aldehydes, wt%	0.05	gas chromatography
Readily oxidizable substances, wt%	0.05	DAB 7
Evaporation residue, wt%	0.01	DIN 53172
Iron, ppm (mg/kg)	1	
Other heavy metals, ppm (mg/kg)	10	
Platinum–cobalt color, APHA	10	DIN 53409
Density (20 °C), g/cm^3	0.990–0.998	DIN 51757
bp, °C	140.7–141.6	DIN 51751

5. Storage and Transportation

Containers made of aluminum with a purity of 99.5% (DIN no. 3.0255) or alloyed steels (DIN no. 1.4541/UNS no.: S 30400 and 1.4571/S 31600) are suitable for storing pure propionic acid. Aluminum is unstable toward aqueous propionic acid. Polyethylene containers can be used for temporary storage of propionic acid and as small packing drums. Plastics are not recommended for long-term storage. Glass-reinforced plastics are unsuitable even for short-term storage.

According to GefStoffV, propionic acid is classified as "corrosive", R-phrase 34, and S-phrase 2-23-26. Transport classification is as follows:

RID/ADR	class 8, no. 32 c
GGVE/GGVS	class 8, no. 32 c
ADNR	class 8, no. 32 c
IMDG Code (D-GGVSee)	class 8
IATA–DGR	class 8 1848
UN no.	1848

6. Quality Specifications

Propionic acid is only sold in pure form with a content of 99.5 wt%, determined acidimetrically. The maximum water content is 0.1 wt%; the sulfate, chloride, and heavy metal content is generally < 1 ppm. The color index is normally below 5 APHA. Oxidizable substances (ketones, aldehydes, esters) can be determined from the reaction time in the KMnO$_4$ test (DAB 7). Their proportions should be < 0.05 wt%.

The standard method for quality control is gas chromatography, e.g., with Chromosorb AW/DM 6 S as the carrier and LAC 446 + H$_3$PO$_4$ as the stationary phase. Detection is carried out with a flame ionization detector (FID).

The acid corresponding to the requirements of the Food Chemical Codex (FCC) is given in Table 4 [69].

7. Environmental Protection

Like most lower carboxylic acids, propionic acid is rapidly degraded biologically. The biodegradability in the Zahn–Wellens test is 95% [71]. The biological oxygen demand for degradation after 5 days (BOD_5) is 1.3 g/g [72] which is 86% of the total oxygen demand (TOD) of 1.513 g/g. The chemical oxygen demand (COD) is 1.4 g/g [72].

The fish toxicity test with the golden orfe resulted in no mortality with 4000 mg/L propionic acid in 48 h [73]. For the blue gill sunfish (*Leponius macrochirus*) an LD_{50} of 188 mg/L was determined in 24 h [72], for *Daphnia magna* an LD_{50} of 130 mg/L in 24 h and an LD_{50} of 50 mg/L in 48 h [72].

The limiting concentration of bacterial toxicity (*Photobacterium phosphoricum*) is 1240 mg/L [74]. For rats an oral LD_{50} of 3500 mg/kg was determined [73].

On the basis of the toxicity data for the rat, bacteria, and golden orfe, propionic acid is classified in Germany as mildly hazardous to water (WGK 1 = water endangerment class 1) [74].

The growth-inhibiting effect of the nonneutralized propionic acid (DIN 38 412/9) on *Scenedesmus suspicatus* was determined in 96 h as follows (EC = effective concentration and 0, 50, and 90 denote no, 50%, and 90% effect, respectively) [75]:

EC 0 = 12 mg/L
EC 50 = 43 mg/L
EC 90 = 79 mg/L

The bioaccumulation in an *n*-octanol–water mixture is log P_{OW} = 0.025/0.33 [72].

In the TA Luft propionic acid is classified in class 2 [76], i.e., if a chemical plant emits a quantity stream of ≥ 2 kg/h, the propionic acid concentration may not exceed 0.1 g/m^3.

The maximum workplace concentrations for propionic acid in Germany (MAK) and in the United States (TLV) in 1992 were 10 mL/m^3 (30 mg/m^3).

8. Uses

8.1. Use as Food and Feed Preservative

Propionic acid and its salts are broad-spectrum preservatives because of their bactericidal, fungicidal, insecticidal, and antiviral effects, and their acaricidal effect which is found at higher concentrations. Effects at lower inclusion rates are more of a static nature. Minimum inhibition concentration (MIC) against fungi is 0.05–0.50 wt%, against bacteria 0.25–0.50 wt%, and against yeasts 0.10–1.0 wt% [77]. Efficacy of propionic acid against molds which cause spoilage of organic material such as foods and feeds is pronounced and of particular interest to industry. Gram-negative microorganisms such as *Escherichia coli* and *Salmonella ssp.* are more sensitive to propionic acid than gram-positive microorganisms. Minimum inhibition concentration against *Salmonella typhimurium*, *S. eimsbüttel*, *S. anatum*, *S. enteritidis*, and *S. montevideo* is 1.0 wt% after three days of action [78]. Propionates are ineffective against Salmonella in petri dishes but ammonium propionate and calcium propionate were successfully used to decontaminate Salmonella in meat meal at a 4.5% inclusion rate with monitoring after four weeks of application [79]. Vaccinia virus was inactivated within 1 h by 1.0 wt% and within 7.5 min by 2.0 wt% propionic acid [80].

Mites in cereals (*Acarus siro*) are killed by 1.0 wt%. A repellent effect is achieved by adding only 0.5 wt% propionic acid. Corn beetles (*Sitophilus granarius*) are killed by 1.0 wt% propionic acid. Development of their eggs is impaired by 0.7 wt%.

Although the mode of action of propionic acid is not yet fully understood, the preservative effect is certainly based on the undissociated acid. At pH 4.88 the undissociated form is present to 50% and at pH 6.0 to 7%. This pH range is relevant for practical food and feed preservation. The undissociated acid penetrates into the living cell like a nutrient. Propionic acid inhibits intracellular enzymes which are essential for carbohydrate metabolism [81]. Morphological changes in Salmonella bacteria after treatment with 1–3 wt% propionic acid are lysis of the bacteria and loss of flagellum.

The broad efficacy spectrum of propionic acid forms the basis for a wide variety of applications. Its salts must be applied at concentrations equivalent to that of the acid but their effect is slower and not as long lasting. This is because the salts must be hydrolyzed to release the acid. Hydrolysis is incomplete and slow under practical conditions in feeds and foods, e.g., at 15% moisture content and a pH value between 4 and 6.

Calcium, sodium, and ammonium salts of propionic acid have the advantages of an almost neutral taste and smell. Therefore they can also be applied in food preservation whereas propionic acid is not suitable in this field because its odor is pungent and not acceptable to humans. Most animals prefer the acid taste of feed preserved with propionic acid and increase their feed intake [82]. Recently a new feeding concept for pigs was introduced [83] in which organic acids such as propionic acid are of major importance to control digestion. The organic acids reduce the buffering capacity of feeds and influence the microflora content in the stomach and digestion parameters.

The various uses of propionic acid and its salts are listed in Table 5. Seeds and barley for malting process should not be treated with propionic acid because it inhibits germination.

All aspects regarding application technique, equipment, corrosive effects of propionic acid and propionates, storage and appropriate storage material, dosage, and efficacy are reviewed in [84]. The described effects can only be achieved by homogeneous distribution of the preservative in foods or feeds.

Regulatory status of propionic acid and its ammonium, sodium and calcium salts is as follows:

Table 5. Application of propionic acid (PA) and its salts as preservatives for foods and feeds *

Type of material	Application rate of PA or its salts	Purpose/effect
Bread, bakery products	0.3% Ca or Na salt; higher concentrations inhibit yeast fermentation	prevention of molds and formation of mycotoxins
Feed grain including corn	0.50% PA at 16% moisture 0.65% PA at 20% moisture 0.95% PA at 26% moisture 1.25% PA at 32% moisture 1.50% PA at 36% moisture	preservation for 6 months storage; reduction of microorganisms and nutrient losses; inhibition of mycotoxin formation and liberation of free fatty acids; protection from lumpiness and mold
Single feeds and complete feeds	0.2% PA or 0.4% salt at moisture level up to 14%; higher rates at unfavorable conditions	increase of storage time up to 2 months; improvement of weight gain and feed conversion with all types of animal
Feeds of animal origin and complete feeds	0.5% PA or 0.7–0.8% Ca- or NH_4-salt	reduction of pathogenic flora, e.g., Clostridia, Enterobacteriacea
Feeds of animal origin	1.0% PA or 2.0–3.0% Ca- or NH_4-salt 2.5–3.5% PA or 4.0–6.0% Ca- or NH_4-salt	protection against recontamination with Sal-monella after sterile manufacturing decontamination of heavily infected material
Brewer's yeast, liquid (ca. 10% dry matter)	1.0% PA	inactivation of yeast cells; direct feeding is possible after this treatment
Whey, liquid	0.3% PA 0.8–1.0% PA	increase of storage time by 2 days increase of storage time up to 1 week
Silage of grass, corn, and other forage	0.4% PA for treatment of all material 1.5–2.0 L PA/m² silage surface, diluted in water by 1 : 4	improvement of lactic acid fermentation; inhibition of butyric and acetic acid producing bacteria; reduction of losses and mold on surface; prevention of nutrient losses

* Regulatory status in different countries for foods and feeds must be observed.

FAO/WHO: ADI is not limited according to WHO Food Additives Series 1974, no. 5.

United States: GRAS status was reconfirmed in 1984 for propionic acid as a preservative for various foods, e.g., bread and other bakery products. The same applies to calcium and sodium propionates. Maximum inclusion rates are limited only by GMP (good manufacturing practice) according to CFR 21, §§ 184.1081, 184.1221, 184.1784 of April 1, 1989.

EC: Approved without upper limit and without specifying any fields of application according to EC Directive of Council 64/54 of Nov. 11, 1963.

Federal Republic of Germany: Approval for food preservation was withdrawn on March 31, 1988; a revision due to EC legislation is expected. According to the German feeds regulations there is no restriction or limitation for using propionic acid and its salts in all kind of feeds (feeds law 1989).

Other countries: There is no other country known in which propionic acid or its salts are not approved for food and feeds.

Table 6. Worldwide production capacities in 1989

Producer	Country	Capacity, t/a
BASF	FRG	60 000
Union Carbide	USA	68 000 [a]
Eastman–Kodak	USA	25 000 [a]
BP Chemicals	UK	30 000
Hoechst–Celanese	USA	7000
Daicel	Japan	2000
Total		192 000

[a] Used for production of various carboxylic acids.

8.2. Other Uses

Zinc, cadmium, lead, and mercury propionates can be used as vulcanization regulators in rubber production [32].

Propionic acid is used as an intermediate, frequently in the form of its anhydride, for the production of esters. The most important esters are cellulose acetate propionate [9004-39-1] (CAP), from which thermoplastics are produced, and vinyl propionate which is used as a basic monomer for dispersions.

Methyl, ethyl, propyl, and butyl propionates are used as solvents for resins and paints [32]. Because of their characteristic fruity aromas various propionate esters are also used in flavors and fragrances.

Small quantities of propionic acid are also further processed to propionyl chloride, which is used as a reactive intermediate for the introduction of the propionyl group in syntheses (e.g., in the production of pharmaceuticals).

Until the end of the 1980s a considerable proportion of propionic acid was chlorinated and further processed to form herbicides (see Chloropropionic acids, Section 11.4). With the arrival of more potent, optically active herbicides and of products with a broader spectrum of activity, the importance of racemic herbicides based on propionic acid has decreased considerably.

9. Economic Aspects

The production capacity in 1989 for propionic acid was 190 000 t/a worldwide. Of this, however, ca. 90 000 t/a were only partially usable for propionic acid, because these plants were also used for the production of other acids. Details of the distribution of the capacities in 1989 are given in Table 6.

Since 1984 the consumption of propionic acid has been stationary at ca. 130 000 t/a. No change is expected in the near future.

The areas of use of propionic acid and its distribution in 1989 are listed below.

Preserving animal feeds	45 %
Preserving bakery products	10 %
Herbicides	20 %
Plastics	10 %
Pharmaceuticals	10 %
Other	5 %

In the medium-term an increase in its use as a preservative for animal feeds is predicted. In contrast, the use in herbicides will decrease because of the increasing use of optically active substances (see Section 11.4.2.1).

10. Toxicology

Biochemistry, Biokinetics, and Metabolism. Propionic acid, or its metabolically activated form propionyl-CoA, occurs in mammals as a metabolic intermediate in the degradation of odd-numbered fatty acids, the side-chain oxidation of cholesterol, or in the oxidative degradation of certain amino acids (valine, isoleucine, threonine) [85], [86]. The proportion of propionic acid in the volatile fatty acids in the blood (total, contents 0.18 – 1.6 mmol/L) is up to 5 % [85]. Orally administered propionic acid or propionate is rapidly absorbed [87]. This can also be assumed for inhaled propionic acid.

Propionate is rapidly and completely metabolized after activation to propionyl-CoA by means of carboxylation to methylmalonyl-CoA and isomerization to succinate, which then enters the tricarboxylic acid cycle [85], [86]. Thus, conversion rates of 4.5 g propionic acid per hour for the liver of a 70 kg human can be calculated from in vitro investigations on slices of liver [85].

Acute Toxicity. The acute oral and inhalatory toxicity of propionic acid is low (LD_{50} oral rat: 2600 – 4290 mg per kilogram body weight, LC_{50} rat inhalatory >19.7 mg L^{-1} 4 h^{-1}). The only study on dermal application showed a higher toxicity (LD_{50} dermal rabbit: 500 mg per kilogram body weight), which may be caused by the strong irritant effect of the substance. Sodium and calcium propionates show a lower oral and a clearly lower dermal action in comparison with the acid [87] – [91].

Even after intravenous and subcutaneous administration there was no toxicity, but only a sedative effect shortly after the injection [85].

Propionic acid has an irritant/corrosive effect on the skin and an irritant effect on the rabbit eye [88], [89]. Its calcium salt, however, does not show these effects [88].

Subacute, Subchronic, and Chronic Toxicity. When $\leq 5\%$ propionic acid or sodium propionate was administered to rats with their feed (estimated substance uptake ≤ 4000 mg per kilogram body weight and day) from 28 days up to the total lifetime of the animals no toxic effects as a whole were found [88], [92] – [97]. However, with acid concentrations of ca. 0.4 % and higher, local changes in the epithelium of the forestomach occurred, caused by cell proliferation (increase in the basal cell and spindle cell layer). On longer administration, wart-like manifestations and ulcers developed [95], [96], [98]. On administration of 4% propionic acid in animal feed for 21 days the cell division rate of the forestomach epithelium increased, as measured by ^{3}H-thymidine incorporation [99]. Similar local effects are also known from other short-chain fatty acids such as acetic, butyric, valeric, or caproic acids [92], [95]. Since the forestomach of the rat is a special feature of this species, 90-day investigations on beagle dogs, which have no forestomach, were carried out. Here propionic acid produced a proliferation in the epithelium of the gullet (esophagus), with a no-effect concentration in the feeds of 1 wt%. On feeding with calcium propionate no such effect was found up to the highest concentration used of 4.35 wt% [88].

Genotoxicity (Mutagenicity). In many in vitro investigations with bacteria, yeasts, and cell cultures no indication of a mutagenic effect of propionic acid was found [100]–[103]. The same applies to sodium [104], [105] and potassium [104] propionates. In vivo investigations on the induction of chromosomal changes by propionic acid in the Chinese hamster [101] or by sodium propionate in the bone marrow of the rat [105] also gave negative results.

Teratogenicity. Investigations on developmental toxicity of calcium propionate in rats, mice, hamsters, and rabbits (300 and 400 mg per kilogram body weight and day) for 5–13 days (duration of test depending on animal species) [90] showed no effects.

Effects on Humans. Taking 6 g sodium propionate orally over several days gave slightly alka-line urine without any further effects [85]. Sodium propionate solutions (10 wt%, pH 7.2) have been used for the treatment of eye infections without irritant effects; 20 wt% solutions (pH 7–8.5) did not produce irritation or sensitization of the skin [106].

11. Derivatives

11.1. Salts

Ammonium propionate [17496-08-1], $C_3H_9NO_2$, M_r 105.11, is a strongly hygroscopic salt, which melts with absorption of water at 45 °C [107] (bp 70–75 °C at 1.3 kPa) [108]. Ammonium propionate is being increasingly used in the form of an aqueous solution as a preservative for animal feeds and cut grass.

Sodium propionate [137-40-6], $C_3H_5O_2Na$, M_r 96.06, melts at 281–284 °C to give an anisotropic liquid [109]. For its production and use see calcium propionate.

Calcium propionate [4075-81-4], $(C_3H_5O_2)_2Ca$, M_r 186.18, crystallizes as the monohydrate in monoclinic plates and as the trihydrate. The anhydrous salt dissolves up to 41.7 wt% in water and is insoluble in ethanol [110]. The aqueous solutions can cause inflammation [111].

Production. The industrial production of calcium and sodium propionates is carried out by the neutralization of propionic acid with the corresponding hydroxides and subsequent spray-drying of the concentrated aqueous solutions.

Uses. Calcium and sodium propionate have fungistatic action. They are added to bread and other bakery products to prevent rope of bread caused by bacteria (Na propionate) and mold formation (Ca propionate). They are used as preservatives for animal feeds and for treatment of ketosis in domestic pets. Mixtures of sodium and calcium salts in an aqueous solution containing 2.5% propionic acid are recommended for external use in treating mycosis [112].

11.2. Propionic Anhydride and Propionyl Chloride

Propionic anhydride [123-62-6], $(CH_3CH_2CO)_2O$, M_r 130.14, is a colorless liquid with an unpleasant pungent odor, which severely irritates the eyes and mucous membranes. Physical properties are as follows [113]:

bp	167 °C
mp	−45 °C
Flash point (open cup)	74 °C
Ignition temperature	315 °C
Density at 0 °C	1.0336 g/cm³
at 50 °C	0.9791 g/cm³
Kinematic viscosity (20 °C)	1.144 mPa·s
(50 °C)	0.7511 mPa·s

Toxicological data are as follows [114]:

LD$_{50}$ (oral in rats)	2.36 g/kg
Skin (rabbit), irritation dose	500 mg (open, moderate)
Eye (rabbit), irritation dose	750 µg (severe)

Production. Propionic anhydride is produced industrially either by heating propionic acid at 8 bar to 235 °C and distilling off the water formed [115], or by heating propionic acid with propionyl chloride in an inert solvent to 160 °C [116].

Other possible routes involve the reaction of ethylene, carbon monoxide, and propionic acid in the presence of nickel carbonyl (Reppe synthesis [117]–[120]) or cobalt or nickel iodide [121].

Uses. Propionic anhydride is used as an esterifying agent for fatty acids and cellulose, for the production of alkyd resins and perfume oils, and for the introduction of the propionyl group into dyes and pharmaceuticals.

Propionyl chloride [*79-03-8*], CH_3CH_2COCl, M_r 92.53, is a colorless, volatile, and readily flammable liquid, whose vapors hydrolyze in moist air. It is corrosive, lachrymatory, and has a strongly pungent odor. With water and lower alcohols it reacts vigorously under solvolysis to give the acid or ester. Important physical properties are:

bp	80 °C
mp	−94 °C
Flash point (open cup)	12 °C
d_4^{20}	1.0646

Production. Propionyl chloride is produced industrially by treatment of propionic acid with phosgene [*75-44-5*] or thionyl chloride [*7719-09-7*]. The reaction occurs at ca. 50 °C in the liquid phase in the presence of dialkylformamides. The product is then separated by distillation. Propionyl chloride can also be obtained by the reaction of propionic acid with PCl_5 [*10026-13-8*] [122].

Uses. Propionyl chloride is used for the introduction of the propionyl group and for the synthesis of propionate esters because of its high reactivity.

11.3. Esters

The esters of monohydric alcohols with propionic acid are formed in yields of > 90 % in the reaction of ethylene with carbon monoxide and the corresponding anhydrous alcohols in the presence of $Ni(CO)_4$ at 160–200 °C and 50 bar [123], [124].

The *ethyl ester* is also obtained in very good yields by treatment of ethanol with methylketene [*463-51-4*] [125].

Uses. The esters of the $C_1–C_4$ alcohols are used as solvents for resins and paints [32]. Because of their specific, fruity aromas propionate esters are used as flavorings for foods and in the perfume industry [126]. Some esters and their properties are listed in Table 7.

11.4. Chloropropionic Acid and Derivatives

The 2-chloro- and 3-chloropropionic acids and their derivatives are still important industrially, while 2,2-dichloropropionic acid is virtually no longer used.

Table 7. Properties of some propionate esters

Ester	CAS registry no.	Molecular formula	bp, °C	Flash point, °C	Aroma/use
Methyl	[554-12-1]	$C_4H_8O_2$	79.8	6	rum-like/flavoring, fragrance
Ethyl	[105-37-3]	$C_5H_{10}O_2$	99.2	12	butter- or rum-like/flavoring, fragrance
Propyl	[106-36-5]	$C_6H_{12}O_2$	121.1	19	pear-like/flavoring, fragrance, solvent
Butyl	[590-01-2]	$C_7H_{14}O_2$	145.5	32	Solvent
Isopentyl	[2438-20-2]	$C_8H_{16}O_2$	160	41	apricot- or pineapple-like/flavoring, fragrance
Benzyl	[122-63-4]	$C_{10}H_{12}O_2$	222	94	jasmine-like/fragrance

2-Chloropropionic acid is produced by direct chlorination of propionic acid, whereas 3-chloropropionic acid predominantly by addition of hydrogen chloride to acrylic acid.

11.4.1. Properties and Production

11.4.1.1. 2-Chloropropionic Acid and Derivatives

2-Chloropropionic Acid. Racemic 2-chloropropionic acid [598-78-7], $C_3H_5ClO_2$, is a colorless liquid that is miscible with water and most common organic solvents. Important *properties* are as follows: bp (101.3 kPa) 186 °C, bp (1.5 kPa) 84 °C; mp ca. −12 °C; vapor pressure at 60 °C 0.4 kPa, at 100 °C 3.0 kPa; density at 20 °C (DIN 51 757) 1.27 g/cm^3; flash point (DIN 51 758) ca. 107 °C; ignition temperature (DIN 51 794) >200 °C; explosion limits 3.7–14.3 vol%; water endangerment class (WGK) 1 [127]; degradability (DIN 38 412 part 25) >70% [127]; TA Luft class I [127].

Commercial grade acid is ca. 95% pure. The impurities are 2,2-dichloro- and 2,3-dichloropropionic acids, 2-chloropropionic anhydride, and unreacted propionic acid.

Production. 2-Chloropropionic acid is produced by reaction of propionic acid with chlorine at 110–120 °C in the presence of acid catalysts such as phosphorus trichloride, thionyl chloride, chlorosulfuric acid, or propionyl chloride [128], [129]. Since the reaction mixture is highly corrosive enamel, silver, glass (or leaded steel) are used as construction materials. One mole equivalent of hydrogen chloride is formed in the reaction and appears in the waste gas. From there it must be isolated and recycled on account of the economy of the process and disposal problems.

Methyl 2-chloropropionate [17639-93-9], $C_4H_7ClO_2$, is a colorless liquid that is miscible with almost all common organic solvents. Important *properties* are as follows: bp (101.3 kPa) 127–129 °C; mp <−10 °C; vapor pressure at 20 °C 0.9 kPa, at 50 °C 4.5 kPa; density at 20 °C (DIN 51 757) 1.14 g/cm^3; flash point (DIN 51 755) ca. 36 °C; ignition temperature (DIN 51 794) 450 °C; explosion limits 2.8–13.9 vol%; water solubility at 20 °C: 22.9 g/L [130]; water endangerment class (WGK) 1 [130]; degradability > 60 % [130]; TA Luft class III [130]; LD_{50} (oral, rat) 1080 mg/kg [130]; LC_{50} (by inhalation, rat) 15.3 mg L^{-1} 4 h^{-1} [130].

Methyl 2-chloropropionate is *produced* by esterification of 2-chloropropionic acid with methanol at 40–50 °C in the presence of acid catalysts such as sulfuric or hydrochloric acid.

2-Chloropropionyl chloride [*7623-09-8*], $C_3H_4Cl_2O$, is a colorless to pale yellow liquid that is miscible with almost all common aprotic solvents. *Physical properties*: bp (101.3 kPa) ca. 110 °C; *bp* (10 kPa) 45 °C; vapor pressure at 20 °C 3.0 kPa, at 60 °C 11.5 kPa; density at 20 °C (DIN 51 757) 1.27 g/cm^3; flash point (DIN 51 755) 34.5 °C; ignition temperature (DIN 51 794) 510 °C.

2-Chloropropionyl chloride is *produced* by reaction of propionyl chloride with chlorine at 70–90 °C in the presence of acid catalysts such as sulfuric or chlorosulfuric acid; or by reaction of 2-chloropropionic acid with, for example, thionyl chloride at 60–70 °C or with phosgene.

11.4.1.2. 3-Chloropropionic Acid and Derivatives

3-Chloropropionic acid [*107-94-8*], $C_3H_5ClO_2$, forms colorless plates that are readily soluble in water and alcohol; mp 41–42 °C; bp (101.3 kPa) 202–205 °C (decomp.); bp (2.1 kPa) 108 °C.

3-Chloropropionic acid is *produced* industrially by reaction of acrylic acid with hydrogen chloride at 40–45 °C or by reaction of propionic acid with chlorine under UV light at 20 °C [131].

3-Chloropropionyl chloride [*625-36-5*], $C_3H_4Cl_2O$, is a pale yellow liquid with bp 144 °C.

3-Chloropropionyl chloride is *produced* by reaction of acrylic acid with hydrogen chloride and phosgene in the presence of, for example, dimethylformamide as catalyst [132], or by reaction of propiolactone with thionyl chloride [133].

11.4.1.3. 2,2-Dichloropropionic Acid

2,2-Dichloropropionic acid [*75-99-0*], $C_3H_3Cl_2O_2$, is a colorless liquid that is miscible with water and alcohols; bp (101.3 kPa) 185–190 °C; bp (1.9 kPa) 90–94 °C; mp 13–15 °C; density at 20 °C (DIN 51 757) 1.40 g/cm^3.

2,2-Dichloropropionic acid is *produced* by reaction of propionic acid or 2-chloropropionic acid with chlorine at ≤ 180 °C in the presence of, e.g., phosphorus trichloride [134], thionyl chloride, organic acid chlorides [135], or sulfamic acid [136].

11.4.2. Uses

Chloropropionic acids and their derivatives mainly serve as intermediates. They are used to introduce the propionyl group in the production of biocides, dyes, and pharmaceuticals. These alkylation reactions take place with elimination of chlorine. In the production of cosmetics and plastics additives the chlorine group in the chloropropionic acid is replaced by other active groups, such as thiol groups.

11.4.2.1. 2-Chloropropionic Acid and Derivatives

Racemic 2-chloropropionic acid and 2,2-dichloropropionic acid have been used predominantly for the production of herbicides. The demand for both acids is decreasing sharply, partly as a result of the development of new herbicides with different chemical structures. 2,2-Dichloropropionic acid is seldom used. Racemic 2-chloropropionic acid is being replaced by the optically active L-enantiomer in the herbicide field.

By far the greatest quantities of 2-chloropropionic acid and its derivatives are still used for the production of *phenoxypropionic acids* (**1**, **2**). Like the phenoxyacetic acids they are highly active herbicides of the growth regulator type [137]. Other derivatives with herbicidal action are phenoxyphenoxypropionates. (**3**) [138], naphthoxypropionamides (**4**) [139] and phenylaminopropionates (**5**) [140].

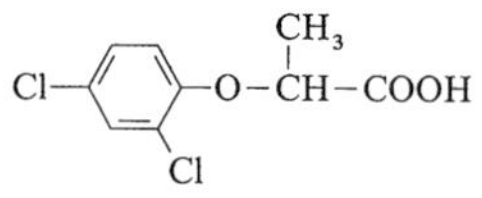

$$Cl-\underset{Cl}{\bigcirc}-O-\underset{CH_3}{CH}-COOH \quad Cl-\underset{CH_3}{\bigcirc}-O-\underset{CH_3}{CH}-COOH$$

Dichlorprop or 2.4 DP (**1**) 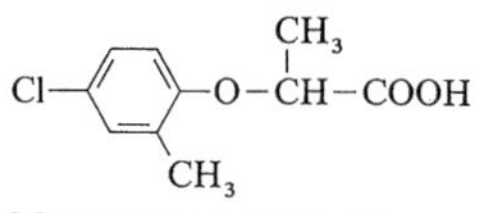   Mecoprop or MCPP (**2**)

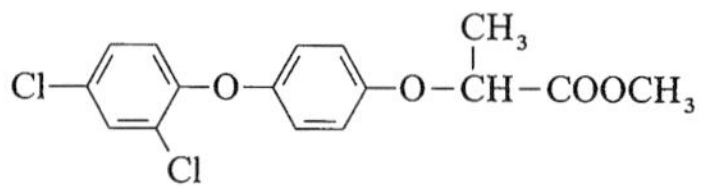

$$Cl-\underset{Cl}{\bigcirc}-O-\bigcirc-O-\underset{CH_3}{CH}-COOCH_3$$

Illoxan (**3**)

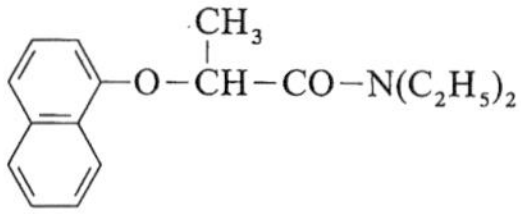

$$\underset{}{\bigcirc\bigcirc}-O-\underset{CH_3}{CH}-CO-N(C_2H_5)_2$$

Devrinol (**4**)

$$Cl-\underset{Cl}{\bigcirc}-\underset{}{\overset{O=C-\bigcirc}{N}}-\underset{CH_3}{CH}-COOC_2H_5$$

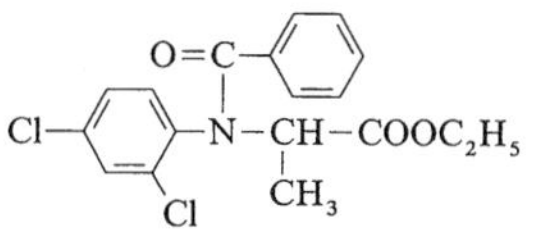

Suffix (**5**)

2-Chloropropionic acid, produced as described in Section 11.4.1, is racemic. Of the phenoxypropionic acids only the D-(+)-isomer shows herbicidal action. Its synthetic precursor is L-(−)-2-chloropropionic acid. Since the 1980s there has been a tendency to produce this class of herbicides from pure L-(–)-2-chloropropionic acid which is available, for example, in the form of its esters. These esters are produced, for example, by chlorination of the corresponding D-lactate esters with thionyl chloride. D-Lactic acid is obtained by a biotechnological process (→ Lactic Acid).

Activity. Phenoxypropionic acids are predominantly used to combat broad-leaved weeds in grasses, e.g., cereals. The spectrum of activity can, however, be shifted by appropriate substitution. The phenyl ethers of phenoxypropionate esters exhibit selective action towards damaging grasses and do not attack cultivated ones [138].

Trade Names of Herbicides based on Propionic Acid. U-46 range, racemates (**1**, **2**) (BASF), Duplosan range, optically active compounds (**1**, **2**) (BASF), Illoxan (**3**) (Hoechst), Devrinol (**4**) (Stauffer/ICI), Suffix (**5**) (Shell), Fusilade (**6**) (ICI), Verdict (**7**) (Dow).

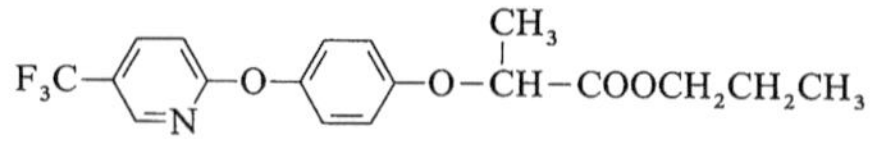

$$F_3C-\text{pyridyl}-O-\text{phenyl}-O-\underset{CH_3}{\underset{|}{CH}}-COOCH_2CH_2CH_3$$

Fusilade (**6**)

$$F_3C-\underset{Cl}{\text{pyridyl}}-O-\text{phenyl}-O-\underset{CH_3}{\underset{|}{CH}}-COOCH_3$$

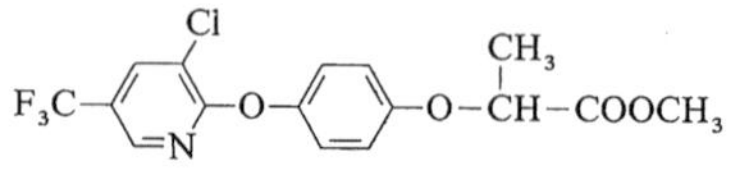

Verdict (**7**)

Esters of 2-Chloropropionic Acid. Methyl 2-chloropropionate is a building block of the fungicide Ridomil (**8**) (Ciba–Geigy) used in plant protection.

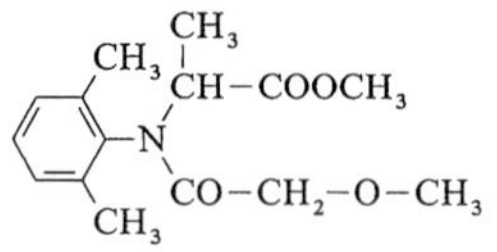

Ridomil (**8**)

2-Chloropropionyl Chloride. The Friedel–Crafts reaction of isobutylbenzene with chloropropionyl chloride provides a synthetic route for the antiphlogistic and antirheumatic ibuprofen (**9**).

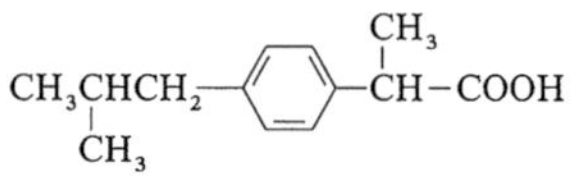

$$CH_3CHCH_2-\text{phenyl}-\underset{CH_3}{\underset{|}{CH}}-COOH$$
$$|$$
$$CH_3$$

Ibuprofen (**9**)

11.4.2.2. 3-Chloropropionic Acid

Both 3-chloropropionic acid and its chloride are building blocks for the production of pharmaceuticals such as the antiepileptic beclamide (**10**) and the antitussive oxolamine (**11**) [141]. The acid is also used for the production of 3-mercaptopropionate esters, which are employed as stabilizers, for example in plastics and coatings.

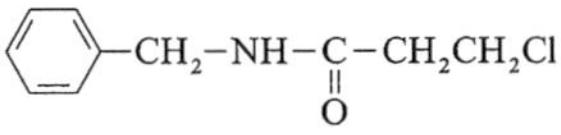

Beclamide (**11**)

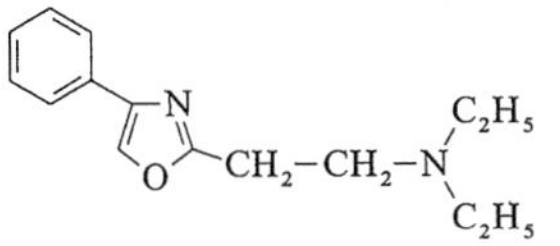

Oxolamine (**11**)

11.4.2.3. 2,2-Dichloropropionic Acid

The sodium salt of this acid was formerly an important herbicide (generic name delapon). It is now no longer used in Western Europe and the United States.

Trade Names. Basfapon (BASF), Dowpon (Dow).

11.4.3. Toxicology

Biochemistry, Biokinetics, and Metabolism. The sodium salt of 2-chloropropionic acid is an activator of the pyruvate dehydrogenase complex in both rats and dogs, which results in a lower blood glucose (in rats only), lactate, and pyruvate level. The sodium salt stimulates leucine metabolism in the perfused rat heart [142].

Acute Toxicity. The oral LD_{50} values of 2-chloropropionic acid in rats are in the range 500–800 mg/kg [142]–[144]. The LD_{50} in the mouse (i.p.) is 102 mg/kg [143], and the LC_{50} (rat, inhalation) > 3.38 mg/L at 4 h exposure [143]. Exposure of rats to an enriched or saturated atmosphere of chloropropionic acid (20 °C) leads to their death after an exposure time of more than 2 h [143]. 2-Chloropropionic acid has a corrosive effect on rabbit's skin and an irritant effect on the rabbit's eye [143].

The sodium salt shows a somewhat lower oral and inhalatory toxicity. Its LD_{50} (rat, dermal) is > 2000 mg/kg [143], [145], [146]. The sodium salt is not irritant to the skin but irritates the eye [143]. No sensitizing effects were observed in guinea pigs [143].

Toxicity on Long-Term Exposure. No clinical effects were observed if 0.1 wt% (2-chloropropionic acid) was administered to rats with their feed (estimated substance uptake up to 78 mg/kg per day) over 42 days. Higher concentrations (0.25, 0.5, and 1 wt%) lead to an increased sensitivity and ataxia after the sixth day of administration. Ataxia — reversible in some cases during a recovery period of 36 days — and tremor were observed in the groups that received the higher concentrations. If the *neutralized acid* (neutralizing agent NaOH) was administered at a concentration of 0.04 mol/kg (ca. 0.5 wt%) for 12 weeks, the same symptoms occurred, however, with temporary delay. Foci of necrotic cells in the inner granular cell layer of the cerebellum (concentration 0.25–1.0 wt%), atrophy of the testicular germinal epithelium (concentration 0.1–1.0 wt%), and thymic lymphoid necrosis (concentration 1.0 wt%) were observed [142]. In another experiment involving rats (concentration in feed 100, 800, and 4000 ppm over 3 months) with the *sodium salt* the same toxic effects occurred. Neurotoxicity and testicular changes were present in the group administered 4000 ppm. At a concentration of 800 ppm only marginal clinical signs of neurotoxicity remained and at 100 ppm no toxic symptoms were observed [143].

Mutagenicity. 2-Chloropropionic acid and its sodium salt showed no mutagenic activity in the Ames test [143], [147].

12. References

[1] H. Sackmann, F. Sauerwald, *Z. Phys. Chemie (Leipzig)* **195** (1950) 295–312.
[2] C. H. D. Calis-van Ginkel et al., *J. Chem. Thermodyn.* **10** (1978) 1083–1088.
[3] A. Apelblat, E. Manzurola, *Fluid-Phase Equilib.* **32** (1987) 163–193.
[4] M. Blinc, R. Blinc, *J. Polym. Sci.* **32** (1958) 506–508.
[5] K. Tyuzyo, *Bl. Chem. Soc. Jpn.* **30** (1957) 782–789, 851–856.
[6] J. J. Christensen, R. W. Hanks, R. M. Izatt: *Handbook of Heats and Mixing*, Wiley & Sons, New York 1982, pp. 1173–1175, 1449.
[7] [6] Supplementary vol. **1988**, pp. 858–860.
[8] J. Timmermans: *The Physico-Chemical Constants of Binary Systems in Concentrated Solution*, Interscience, vol. **2**, New York 1959, p. 941.
[9] D. R. Linde: *Handbook of Chemistry and Physics*, 70th ed., CRC Press, Boca Raton, 1989, D–15.
[10] G. Glaxton: *Physical and Azeotropic Data*, National Benzol and Allied Product Association, W. Heffer & Sons, Cambridge 1958.
[11] L. H. Horsley: "Azeotropic Data III," *Adv. Chem. Ser.* **116** (1973) p. 4 ff.
[12] T. Ito, F. Yoshida, *J. Chem. Eng. Data* **8** (1963) 315–320.
[13] J. Gmehling, U. Onken, J. R. Rarey-Nies: "Vapor–Liquid–Equilibrium Data Collection," *Chemistry Data Series*, vol. **I**, part 1 (1977), **1a** (1981), **1b** (1981), **2d** (1982), **3+4** (1979), **5** (1982).
[14] H. Stephen, T. Stephen: *Solubilities of Inorganic and Organic Compounds*, vol. **1**, part 2, Pergamon Press, London 1963.
[15] [14] vol. **2**, part 2 (1964).

[16] *VDI-Wärmeatlas*, 5th ed., Düsseldorf 1988.
[17] The Texas A & M University System, Table 23-2-1-(1.1210)-d, p. 1, College Station 1984.
[18] R. R. Dreisbach, R. A. Martin, *Ind. Eng. Chem.* **41** (1949) 2875.
[19] J. J. Jasper, *J. Phys. Chem. Ref. Data* **1** (1972) 851.
[20] W. Jobst, *Int. J. Heat Mass Transfer* **7** (1964) 725.
[21] G. Kortüm, W. Vogel, K. Andrussow, *Pure Appl. Chem.* **1** (1961) 243.
[22] D. D. Perrin, *J. Chem. Soc.* 1959, 1710, 1714.
[23] W. J. Canady, H. M. Papée, K. J. Laidler, *Trans. Faraday. Soc.* **54** (1958) 502.
[24] *Landolt*–Börnstein, 6th. ed., vol. IV/4a.
[25] *Ullmann*, 4th ed., **19**, 454 ff.
[26] M. A. Goodman, S. L. Whittenburg, *J. Chem. Eng. Data* **28** (1983) 350–351.
[27] BASF AG Ludwigshafen, Result of Techn. Lab., Ludwigshafen 1969.
[28] T. Ackermann, F. Schreiner, *Z. Elektrochem.* **62** (1958) 1143–1151.
[29] R. W. Gallant, *Hydrocarbon Process* **47** (1968) no. 6, 139–148.
[30] E. Przewalski, *J. Prakt. Chem.* **88** (1913) 500.
[31] *Kirk-Othmer*, 2nd ed., **16**, 554.
[32] *Ullmann*, 3rd ed., **14**, 383.
[33] *Ullmann*, 3rd ed., **8**, 744.
[34] Celanese, BP 771 583, 1957.
[35] SRI-Process Economics Program, no. 42, A 1, Menlo Park 1975, pp. 18.
[36] Dow Chemical, US 1 926 068, 1933 (C. J. Strosacker, C. C. Kennedy, E. L. Pelton).
[37] *Ullmann*, 4th ed., **9**, 161 ff.
[38] Texaco Dev., DE 3 124 720, 1981 (J. F. Knifton).
[39] M. Sittig, *Chem. Proc. Monogr.* no. 8 (1965) p.76.
[40] [39] no. 23 (1966) pp. 105.
[41] Monsanto, BE 793 203, 1971 (M. D. Forster, D. E. Morris).
[42] Monsanto, US 3 989 748, 1968 (F. E. Paulik, A. Hershman, J. F. Roth).
[43] Monsanto, US 3 989 747, 1968 (J. H. Craddock, J. F. Roth, A. Hershman, F. E. Paulik).
[44] BP Chemicals, DE-OS 2 101 909, 1970 (G. E. Foster, J. R. Bethell).
[45] Monsanto, BE 796 294, 1972 (D. E. Morris).
[46] BP Chemicals, GB 1 363 961, 1972 (M. J. Wriglesworth, D. J. Westlake).
[47] Monsanto, US 3 944 603, 1972 (D. E. Morris).
[48] Monsanto, DE-AS 2 263 442, 1972 (D. Forster, D. E. Morris).
[49] BASF, DE-AS 2 604 545, 1976 (K. Schwirten, W. Disteldorf, W. Eisfeld, R. Kummer).
[50] E. Drent, *ACS Symp. Ser.* **328** (1987) 154–175.
[51] Texaco, US 4 362 822, 1981 (J. F. Knifton).
[52] P. Blanc, G. Goma, *Biotechnol. Lett.* **11** (1989) 189–194.
[53] M. J. T. Carrondo, J. P. S. G. Crespo, M. J. Moura, *Appl. Biochem. Biotechnol.* **17** (1988) 295–312.
[54] H. Dellweg: *Biotechnology*, vol. **3**, Verlag Chemie, Weinheim 1983, p. 472.
[55] Dow Chemical, US 3 151 155, 1964 (J. W. McKoy, N. Swanson).
[56] *Ullmann*, 4th ed., **7**, 123 ff.
[57] D. Powergas, J. Matthey, *Chem. Eng. (N.Y.)* **84** (1977) Dec. 5, 110–115.
[58] D. Forster, A. Hershman, D. E. Morris, *Catal. Rev. Sci. Eng.* **23** (1981) 89–105.
[59] B. Cornils, R. Payer, K. C. Traenckner, *Hydrocarbon Process.* **54** (1975) no. 6, 83–91.
[60] *Hydrocarbon Process* **54** (1975) no. 11, 87.
[61] SRI-Process Economic Program no. 37, 1968, pp. 83 ff.

[62] SRI-Process Economic Program no. 37 a, 1973, pp. 33 ff.
[63] Distillers, GB 771 991, 1957 (A. Elce, I. K. M. Robson, D. P. Young).
[64] BP Oil, GB 743 989, 1956 (J. Habeshaw et al.).
[65] Distillers, GB 743 991, 1956 (A. Elce, I. K. Miles, D. P. Young).
[66] Distillers, FR 1 091 333, 1954 (A. Elce, I. K. Miles, D. P. Young).
[67] *Hydrocarbon Process* **54** (1975) no. 11, 102.
[68] G. B. Elder in B. J. Monitz, W. I. Pollock (eds.): *Process Ind. Corros.*, NACE Houston Tex., 1986, pp. 287 – 296.
[69] *Food and Chemical Codex*, 3rd ed., National Academy Press, Washington, DC, 1981.
[70] BASF, data sheet D 030 d, e, Ludwigshafen, Aug. 1988.
[71] OECD Guidelines for Testing of Chemicals, 302 B, Paris 1981.
[72] K. Verschueren: *Handbook of Environmental Data of Organic Chemicals*, 2nd ed., 1983, p. 1023.
[73] BASF, Entwurf zum Sicherheitsdatenblatt der BASF, Ludwigshafen 1988.
[74] Data sheet no. 140, Katalog wassergefährdender Stoffe, Gemeinsames Ministerialblatt, Bonn 1985.
[75] BASF AG, unpublished results.
[76] T. A. Luft, Gemeinsames Ministerialblatt, ed. A, no. 7, Bonn, Feb. 1986.
[77] BASF, Periodical Animal Nutrition in Research and Practice, vol. **2,** no. 4 (1973), no. 5 (1974), no. 7 (1975), no. 11 (1979).
[78] I. Brglez, *Prax. Vet. Br.* 1977 no. 1, 17 – 22.
[79] V. Haßling: *Rekontaminationsprobleme von Tiermehlen gegen Salmonellen durch Propionsäure, Propionate und Ameisensäure*, Dissertation, Universität Gießen 1985.
[80] I. Wekerle, *Tierärztl. Umsch.* **43** (1988) 646 – 654.
[81] K.-H. Bässler, *Z. Lebensm. Unters. Forsch.* **110** (1959) 28 – 42.
[82] H. Giessler, S. Dammert, *Landwirtsch. Forsch. Sonderh.* **38** (1981) 484 – 489.
[83] G. Bolduan et al., *J. Anim. Physiol.* **59** (1988) 72 – 78.
[84] G. Gaus: BASF Documentation MEA 3310 AD, Oct. 89, Technical Symposium Aug. 9, 1989, Whitestone, S.C., USA.
[85] K. H. Bässler, *Z. Lebensm. Unters. Forsch.* **110** (1959) 28 – 42.
[86] P. Karlson: *Kurzes Lehrbuch der Biochemie*, Thieme Verlag, Stuttgart 1987.
[87] WHO Food Additives Series no. 5, Genf 1974, pp. 110 – 118.
[88] BASF AG, Department of Toxicology, unpublished results.
[89] H. F. Smyth et al., *Am. Ind. Hyg. Assoc. J.* **23** (1962) 95 – 107.
[90] *Patty's* 2nd, pp. 1779 – 1780 and 3rd ed., pp. 4911 –4913.
[91] NIOSH, Registry of Toxic Effects of Chemicals, CD-Rom Version, update 9001.
[92] K. Mori, *GANN* **44** (1953) 421 – 427.
[93] S. Imai et al., *Nara Igaku Zasshi* **32** (1981) 715 – 722.
[94] W. D. Graham et al., *J. Pharm. Pharmacol.* **6** (1954) 534 – 545.
[95] H.-J. Altmann, W. Grunow, unpublished report BGA (1988).
[96] W. Griem, *Bundesgesundheitsbl.* **28** (1985) 322 – 327.
[97] *Bundesgesundheitsbl.* no. 30, Bonn 1987, 370 – 371.
[98] BP International, unpublished results.
[99] C. Rodriguez et al., *Toxicology* **38** (1986) 103 – 117.
[100] Report of Litton Bionetics for FDA, *Nat. Tech. Inf. Serv.* USA PB-266 897 (1976).
[101] A. Basler et al., *Food Chem. Toxic.* **25** (1987) 287 –290.
[102] W. v.d. Hude et al., *Mutat. Res.* **203** (1988) 81 – 94.
[103] W. Szybalski, *Ann. N.Y. Acad. Sci.* **76** (1958) 475 –489.

[104] M. Ishidate et al., *Food Chem. Toxic.* **22** (1984) 623–636.
[105] T. Kawachi et al. in R. Montesano et al. (eds.): *IARC Sci. Publ.* **27** (1980) 323–330.
[106] W. W. Heseltine, *J. Pharm. Pharmacol.* **4** (1952) 120–122.
[107] S. Zuffanti, *J. Am. Chem. Soc.* **63** (1941) 3124.
[108] R. Escales, H. Koepke, *J. Prakt. Chem.* **87** (1913) 264.
[109] G. E. Symons, A. M. Buswell, *Ind. Eng. Chem.* **24** (1932) 460.
[110] A. Renard, *C. R.* **104** (1887) 914.
[111] B. Behrens, J. Wajzer, *Biochem. Z.* **264** (1933) 124.
[112] *Index Merck*, 9th ed., Verlag Chemie, Weinheim 1961, p. 633.
[113] *The Merck Index*, 10th ed., Merck & Co Inc., Rahway 1983, p. 1127.
[114] NIOSH, Registry of Toxic Effect of Chemical Substances, 1980.
[115] Knapsack AG, DE 951 809, 1954 (W. Vogt).
[116] G. Leiderer, DE-AS 1 932 303, 1969.
[117] W. Reppe, H. Kröper, *Ann.* **582** (1953) 47, 61.
[118] R. E. Brooks, W. F. Gresham, J. v. Hardy, J. M. Lupton, *Ind. Eng. Chem.* **49** (1957) 2004.
[119] BASF, DE 848 355, 1943 (W. Reppe, H. Kröper).
[120] BASF, DE 863 194, 1943 (W. Reppe, H. Kröper).
[121] Monsanto, US 3 989 751, 1974 (D. Forster, A. Hershman).
[122] J. D. Roberts, *J. Am. Chem. Soc.* **72** (1950) 4237.
[123] IG Farben, DE 765 969, 1940 (W. Reppe, H. Kröper).
[124] BASF, DE 915 567, 1953 (W. Reppe, H. Kröper).
[125] Rohm & Haas, US 1 898 687, 1931 (F. O. Rice).
[126] *Kirk-Othmer*, 3rd ed., **10**, 456 ff.
[127] BASF AG, Company data from Aug. 15, 1991.
[128] Dow Chem., US 2 010 685, 1932.
[129] Dow Chem., US 1 993 713, 1935.
[130] BASF AG, Company data from Aug. 15, 1991.
[131] 3 M, US 2 682 504, 1954.
[132] BASF, DE 1 140 185, 1963.
[133] Goodrich, US 2 548 161, 1951.
[134] Dow Chem., US 2 809 992, 1957.
[135] Brit. Celanese, GB 892 564, 1962.
[136] BASF, DE 1 133 710, 1963.
[137] R. Wegler: *Chemie der Pflanzenschutz- und Schädlingsbekämpfungsmittel*, vol. **V**, Springer Verlag, Berlin 1977, p. 188 ff.
[138] Hoechst, DE-OS 2 223 894, 1973.
[139] Stauffer Chem., GB 1 066 606, 1967.
[140] Shell Int., GB 1 164 160, 1969.
[141] A. Kleemann, I. Engel: *Pharmazeutische Wirkstoffe*, Thieme Verlag, Stuttgart 1978.
[142] J. L. O'Donoghue, *Neurotoxicity of Industrial and Commercial Chemicals*, vol. **II**, CRC Press, Inc., Boca Raton, Florida 1985, pp. 115–116.
[143] BASF AG, Department of Toxicology, unpublished results, Ludwigshafen 1960, 1988, 1990, 1992.
[144] RTECS, update 8910.
[145] J. L. Morrison, *J. Pharmacol. Exp. Pharmaceut.* **86** (1946) 336–338.
[146] E. Yount et al., *J. Pharmacol. Exp. Therap.* **222** (1982) 501–508.

[147] Japan Chemical Industry Ecology—Toxicology & Information Center (JETOC), *Newsletter* **4** (1985) 15 (14–20).

Propylene Oxide

DIETMAR KAHLICH, Dow Deutschland Inc., Stade, Federal Republic of Germany (Chaps. 1–3, 4.4, 8–10)

UWE WIECHERN, Dow Deutschland Inc., Stade, Federal Republic of Germany (Chaps. 4.1, 4.4, 5–7)

JÖRG LINDNER, Dow Deutschland Inc., Stade, Federal Republic of Germany (Chaps. 4.2–4.4)

1.	Introduction 4129	5.	Environmental Protection and Ecotoxicology 4149
2.	Physical Properties 4130		
3.	Chemical Properties. 4132	6.	Quality Specifications and Analysis 4151
4.	Production and Raw Materials 4134		
4.1.	Production Route via Propylene Chlorohydrin. 4135	7.	Handling, Storage, and Transportation. 4152
4.1.1.	Chlorohydrin Process 4135	8.	Uses 4154
4.1.2.	Other Chlorohydrin Technologies 4142		
4.2.	Indirect Oxidation Routes . . . 4143	9.	Economic Aspects 4156
4.3.	Direct Oxidation Routes. 4148	10.	Toxicology and Occupational Health. 4156
4.4.	Comparison of the Various Propylene Oxide Technologies 4149	11.	References. 4159

1. Introduction

Propylene oxide (PO) [75-56-9], 1,2-epoxypropane, methyloxirane, is a very reactive substance and one of the most important chemical intermediates.

$$H_2C\underset{O}{-}\overset{CH_3}{\underset{|}{CH}}$$

Since the early 1950s it has become increasingly important to the chemical industry. The demand for PO continues to grow throughout the world; annual worldwide production capacity in 1991 was ca. 3.9×10^6 t.

Today PO is a commodity that can be produced economically only in world-scale plants with capacities of 100 000 t/a or above. It is the starting material for a broad spectrum of products, including polymers (polyurethanes, polyesters), oxygenated solvents (propylene glycol ethers), and industrial fluids (monopropylene glycol and polyglycols).

2. Physical Properties

Propylene oxide, C_3H_6O, M_r 58.081, is a colorless, low-boiling liquid with a sweet, ethereal odor. It is also reactive and highly flammable, with a wide explosive range. Flammable concentrations of vapor form readily at room temperature. Important physical constants are as follows [7]–[20]:

mp (101.3 kPa)	−111.93 °C
bp (101.3 kPa)	34.23 °C
Flash point	− 37 °C
Critical temperature	209.1 °C
Critical pressure	4920 kPa
Critical density	312 kg/m³
Critical compressibility factor	0.2284
Autoignition temperature in air at 101.3 kPa	430 °C
Explosive limits in air (STP)	
lower	2.3 vol%
upper	36.0 vol%
Heat of combustion (25 °C, 101.3 kPa)	−33 035 kJ/kg
Heat of polymerization	−1500 kJ/kg
Heat of fusion	112.6 kJ/kg
Heat of solution in water at 25 °C	−45 kJ/kg
Heat of formation of the ideal gas (25 °C)	−1600 kJ/kg
Heat of formation of liquid PO (25 °C)	−2080 kJ/kg
Standard enthalpy (298.15 K)	248 kJ/kg
Standard entropy (298.15 K, 1 atm)	4.94 kJ kg⁻¹ K⁻¹
Free energy of formation (25 °C, 101.3 kPa)	459 kJ/kg
Cubic expansion coefficient at 20 °C	0.00151 1/K
Solubility of PO in water at 20 °C	40.5 wt%
Solubility of water in PO at 20 °C	12.8 wt%
Dipole moment	2.010 Debyes
Evaporation rate (butyl acetate = 1)	33.7
Molecular diffusion volume	4.937×10^{-5} m³/mol
n_D^{20}	1.36610
n_D^{25}	1.36335
n_D^{30}	1.36060
Henry constant at 25 °C	8.65×10^{-5} Pa m³/mol

Table 1 provides temperature-dependent physical properties for both gaseous and liquid PO.

PO has unlimited miscibility with most common organic solvents, but it is subject to a miscibility gap with water (Fig. 1). The binary system PO – water is characterized in Figure 2. This and other binary and ternary vapor-liquid equilibria systems are described in [22].

PO forms azeotropes with dichloromethane, diethyl ether, cyclopentene, isoprene, cyclopentane, 2-methyl-1-butene, pentene, pentane, cyclohexane, hexene, and hexane [22], [23].

Table 1. Physical properties of propylene oxide at various temperatures

Temperature, °C	Density		Heat of vaporization, kJ/kg	Vapor pressure kPa	Specific heat, kJ kg^{-1} K^{-1}		Thermal conductivity, 10^{-4} W cm^{-1} K^{-1}		Surface tension, mN/m	Dynamic viscosity, mPa·s	
	Vapor, g/L	Liquid, kg/m^3			Vapor	Liquid	Vapor	Liquid		Vapor	Liquid
−20	0.24	877	522	8.6	1.10	1.95	0.78	16.6	(29.8)	0.0076	0.487
0	0.63	854	504	24.4	1.18	2.00	0.91	15.8	(26.8)	0.0082	0.388
20	1.44	830	485	58.8	1.27	2.06	1.05	15.0	23.9	0.0087	0.319
40	2.90	804	465	124.5	1.37	2.13	1.21	14.2	21.1	0.0093	0.269
60	5.33	778	444	238.0	1.48	2.20	1.39	13.4	18.3	0.0099	0.232
80	9.14	749	421	418.0	1.61	2.28	1.60	12.5	15.5	0.0106	0.203
100	14.81	719	396	686.0	1.75	(2.37)	1.84	11.7	12.8	0.0112	0.180
120	23.06	686	368	1068.0	1.93	(2.47)	2.11	10.8	10.2	0.0120	0.162
140	34.93	650	336	1590.0	2.15	(2.61)	2.42	(9.8)	(7.6)	0.0128	0.147
160	52.24	609	(298)	2280.0	2.47	(2.84)	(2.79)	(8.9)	(5.2)	0.0137	0.135
180	78.81	599	(250)	3175.0	3.07	(3.27)	(3.25)	(7.9)	(2.9)	0.0149	0.125
200	126.5	489	(180)	4325.0	5.08	(4.56)	(3.90)	(6.9)	(1.0)	0.0170	0.116

Extrapolated values are given in parentheses.

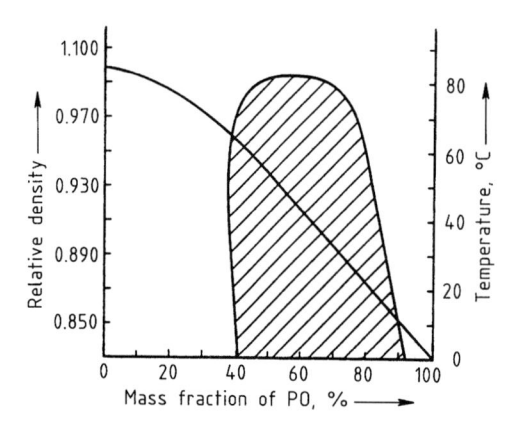

Figure 1. Miscibility gap and density of the propylene oxide – water system [19]

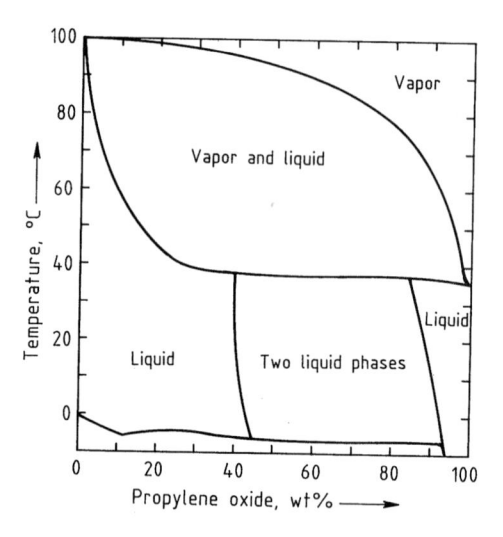

Figure 2. Phase diagram for the propylene oxide – water system [20], [21]

PO exists in two optical isomers [24], [25], but commercial PO is a racemic mixture, and it is to this mixture that the listed properties refer.

3. Chemical Properties

PO is a highly reactive, versatile chemical intermediate. Its polarity and the strained three-membered epoxide ring cause it to be opened easily by reaction with a wide variety of substances. In this respect it is analogous to ethylene oxide (→Ethylene Oxide), and the number of possible derivatives is very large.

Most industrial reactions of PO are catalyzed, although PO alone reacts readily with all compounds containing active hydrogen atoms (including the hydrogen halides), as well as chlorine and ammonia. Alkaline catalysts are preferred (e.g., sodium hydroxide,

Table 2. Summary of the most important PO reactions

Reactant	Product(s)	Structural formula
Polymerization (ROH as initiator)	polyether polyols	R[−O−CH(CH$_3$)−CH$_2$]$_n$−OH
Water	monopropylene glycol ($n = 0$) dipropylene glycol ($n = 1$) tripropylene glycol ($n = 2$)	CH$_3$−CH(OH)[−CH$_2$−O−CH(CH$_3$)]$_n$−CH$_2$OH (and isomers)
Alcohols and phenols (ROH)	propylene glycol ethers	CH$_3$−CH(OH)−CH$_2$(OR) (and isomers/highers)
Ammonia	isopropanolamine	CH$_3$−CH(OH)−CH$_2$(NH$_2$) (and isomers/highers)
Carbon dioxide	propylene carbonate	CH$_3$−CH−CH$_2$ with O−C(=O)−O ring
Isomerization	allyl alcohol acetone propanal	CH$_2$=CHCH$_2$OH CH$_3$COCH$_3$ CH$_3$CH$_2$CHO

sodium acetate, sodium alkoxide, or potassium alkoxide). Acidic materials, such as sulfuric acid, boron trifluoride, stannic chloride, or acid clays (which are sometimes used at lower reaction temperatures) often form byproducts by converting oxides to cyclic ethers, or by isomerizing them to aldehydes.

The most important type of reaction is the addition of PO to compounds containing labile hydrogen, but it also condenses with a number of substances that do not contain reactive hydrogen.

The most important reaction of PO from a commercial standpoint is its violent polymerization to form poly(ether polyols) in the presence of catalysts such as bases, acids, or salts. Chain initiators are substances containing active hydrogen atoms (e.g., water, glycols, phenols, amines, or carboxylic acids). PO forms mixed polyols with ethylene oxide, tetrahydrofuran, 3,4-epoxy-1-butene (butadiene monoxide), and carbon dioxide.

PO is hydrolyzed by water in an uncatalyzed, liquid-phase reaction to give monopropylene glycol, dipropylene glycol, tripropylene glycol, and polyglycols. PO reacts with ketones or aldehydes to form cyclic ketals or acetals, and with carbon dioxide to form propylene carbonate. Hydrogen halides convert PO to propylene halohydrins, and ammonia gives mono-, di-, and triisopropanolamine.

Over alumina PO isomerizes to propanal and acetone, and over Li$_3$PO$_4$ catalysts to allyl alcohol. Hydrogenation over nickel leads to 1-propanol. Many other reactions are also possible, including reactions with natural products like starch or cellulose. Table 2 summarizes the most important PO reactions.

All the reactions involving PO may become violent if not properly controlled, and possible applications should be investigated very thoroughly.

Under pyrolytic conditions PO isomerizes to propanal, acetone, methyl vinyl ether, and allyl alcohol [26]. Decomposition of PO does not result in hazardous products.

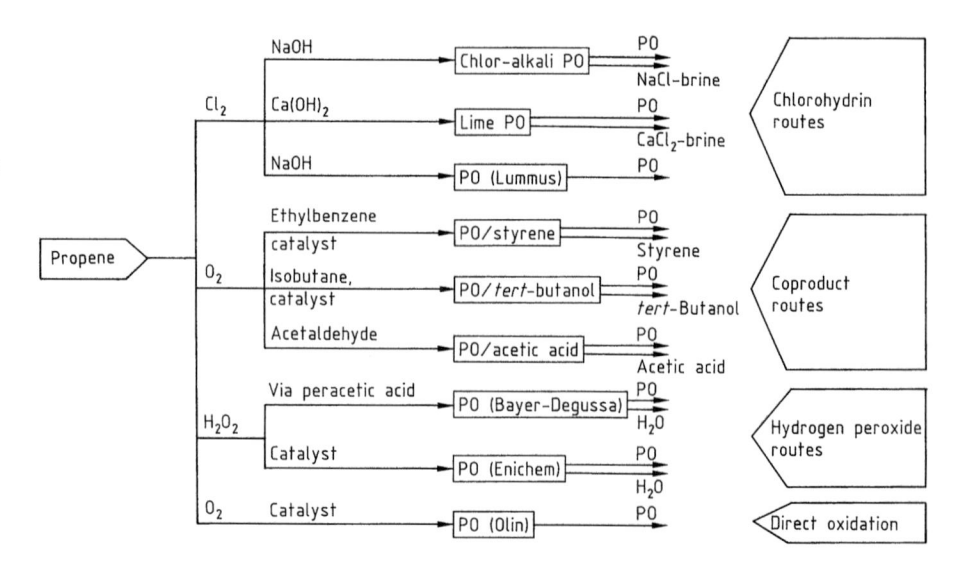

Figure 3. Summary of propylene oxide production routes

PO is a stable compound under moderate conditions, but polymerization may occur on contact with highly active catalytic surfaces such as iron, tin, or aluminum chlorides, peroxides of iron and aluminum, or alkali-metal hydroxides and clay-based absorbent materials. Anything that destroys the neutrality of PO may also cause polymerization. The chemical stability of PO and its polymerization under storage conditions are discussed in [27].

4. Production and Raw Materials

The selection of production routes is decisively influenced by the application and market potential of coproducts, as well as by availability of raw materials and possibilities for byproduct management [28], [29].

Technologies developed up to this point can be divided into:

1) Chlorohydrin processes (Section 4.1)
2) Indirect oxidation processes (Section 4.2)
3) Direct oxidation processes (Section 4.3)

Only the chlorohydrin and the indirect oxidation processes are practiced currently on an industrial scale. The following sections highlight the various processes and describe in more detail those currently in use (see Fig. 3).

It has so far not been possible to produce PO in a technically and economically satisfactory manner by direct gas-phase oxidation (with silver catalysts) in analogy with the process used for ethylene oxide. Common to all current PO process technologies is the fact that significant amounts of coproducts are always generated. For the conventional chlorohydrin process this means up to two moles of salt (sodium or calcium chloride) in the form of dilute brine solution per mole of PO. In the case of the direct oxidation methods, complex organic mixtures arise, which must then be separated via distillation and either sold, further converted, or incinerated. Indirect oxidation technologies also lead to PO and coproducts.

4.1. Production Route via Propylene Chlorohydrin

4.1.1. Chlorohydrin Process

Conventional production of PO is based on the dehydrochlorination of propylene chlorohydrin [78-89-7] with a base [30]. Technically, such units are similar to the chlorohydrin units used to produce ethylene oxide (EO) prior to development of the direct oxidation process. Because the demand for PO has increased rapidly, old plants have been steadily improved and new state-of-the-art, grass-roots PO plants have been built throughout the world.

The present commercial chlorohydrin process is carried out in two main steps: synthesis of propylene chlorohydrin (PCH), and subsequent dehydrochlorination of PCH to PO. These steps are followed by PO purification and wastewater treatment.

Chlorohydrination. In the PCH synthesis reaction (chlorohydrination), propene and chlorine are mixed in an aqueous solution. A propene–chloronium complex is formed as an intermediate (Eq. 1) [31].

$$CH_3-CH=CH_2 + Cl_2 \xrightarrow{H_2O} \underset{Cl^+ \quad Cl^-}{CH_3-CH\text{---}CH_2} \quad (1)$$

Propene–chloronium complex

The chloronium ion then reacts with water to form hydrochloric acid and the propylene chlorohydrin isomers 1-chloro-2-propanol (90%) and 2-chloro-1-propanol (10%) (Eq. 2).

$$CH_3-CH\text{---}CH_2 + H_2O \longrightarrow$$
$$\underset{Cl^+ \quad Cl^-}{}$$

$$\underset{(90\%)}{\underset{\text{1-Chloro-2-propanol}}{\underset{\text{(PCH 1)}}{CH_3-\overset{OH}{\underset{|}{CH}}-\overset{Cl}{\underset{|}{CH_2}}}}} + \underset{(10\%)}{\underset{\text{2-Chloro-1-propanol}}{\underset{\text{(PCH 2)}}{CH_3-\overset{Cl}{\underset{|}{CH}}-\overset{OH}{\underset{|}{CH_2}}}}} + HCl \quad (2)$$

Formation of the chloronium complex is accompanied by two side reactions of chlorine and propene. In the gas phase they react to form the main byproduct, 1,2-dichloropropane (DCP) (Eq. 1.1).

$$CH_3-CH=CH_2 + Cl_2 \longrightarrow \underset{\underset{\text{1,2-Dichloropropane}}{\text{(DCP)}}}{CH_3-\underset{Cl}{\underset{|}{CH}}-\underset{Cl}{\underset{|}{CH_2}}} \quad (1.1)$$

Propene and chlorine also react via 1-chloropropene (allyl chloride) (Eq. 1.2) to give two dichlorohydrin isomers 1,3-dichloro-2-propanol and 2,3-dichloro-1-propanol (DCH 1 and 2, Eq. 1.3).

$$CH_3-CH=CH_2 + Cl_2 \longrightarrow \underset{\text{Allyl chloride}}{CH_2=CH-\underset{Cl}{\underset{|}{CH_2}}} + HCl \quad (1.2)$$

$$CH_2=CH-\underset{Cl}{\underset{|}{CH_2}} + H_2O + Cl_2 \longrightarrow$$

$$\underset{\text{1,3-Dichloro-2-propanol}}{\underset{\text{(DCH 1)}}{\underset{Cl}{\underset{|}{CH_2}}-\underset{OH}{\underset{|}{CH}}-\underset{Cl}{\underset{|}{CH_2}}}} + \underset{\text{2,3-Dichloro-1-propanol}}{\underset{\text{(DCH 2)}}{\underset{OH}{\underset{|}{CH_2}}-\underset{Cl}{\underset{|}{CH}}-\underset{Cl}{\underset{|}{CH_2}}}} + HCl \quad (1.3)$$

A simplified overview of all propene-chlorine reactions is thus:

$$\text{Propene + Chlorine} \begin{cases} \rightarrow \text{Chlorium ion} & \rightarrow \text{PCH 1 + PCH 2} \\ \rightarrow \text{DCP} & \\ \rightarrow \text{Allyl chloride} & \rightarrow \text{DCH 1 + DCH 2} \end{cases}$$

In aqueous solution the chloronium ion can also undergo other reactions, leading to formation of DCP with chloride ions alone (Eq. 2.1) and formation of 2,2'-dichlorodiisopropyl ether (DCIPE) in the presence of PCH (Eq. 2.2).

$$\underset{Cl^- \quad Cl^+}{CH_3-CH\cdots CH_2} + Cl^- \longrightarrow \underset{\underset{DCP}{}}{CH_3-\underset{Cl}{\underset{|}{CH}}-\underset{Cl}{\underset{|}{CH_2}}} \quad (2.1)$$

$$\underset{Cl^+ \quad Cl^-}{CH_3-CH\cdots CH_2} + \underset{\text{PCH 1}}{CH_3-\underset{OH}{\underset{|}{CH}}-\underset{Cl}{\underset{|}{CH_2}}} \longrightarrow$$

$$\underset{\text{2,2'-Dichlorodiisopropyl ether}}{\underset{(DCIPE)}{\begin{matrix} CH_3-CH-CH_2Cl \\ | \\ O \\ | \\ CH_3-CH-CH_2Cl \end{matrix}}} + HCl \quad (2.2)$$

In a simplified overview, the chloronium ion is therefore responsible for the following reactions:

Chlorium ion
$\rightarrow$ + H$_2$O $\rightarrow$ PCH 1 + PCH 2 + HCl
$\rightarrow$ + Cl$^-$ $\rightarrow$ DCP
$\rightarrow$ + PCH $\rightarrow$ DCIPE

Other chlorinated compounds (e.g., monochloroacetone) are produced in very small amounts. Economic considerations require that coproduct formation be reduced as far as possible by using the best available technology. Based on propene, the most expensive raw material for the chlorohydrin PO process, modern industrial-scale plants achieve the following yields:

PCH   88 – 96 %
DCP   3   – 10 %
DCH   0.3 – 1.2 %
DCIPE   0.2 – 0.8 %
Others 0   – 1.0 %

Besides the design of the reactors, the amount of water entering into the chlorohydrination has a significant impact on coproduct formation, especially DCP and DCIPE. Excess water has two major effects on side reactions. Firstly, it reduces the propylene chlorohydrin and chloride ion concentrations in the reactor, resulting in less DCP and DCIPE. Secondly, excess water prevents the formation of an organic liquid phase in which chlorine and propene react to form more DCP [32].

Dehydrochlorination. In the second reaction step, dehydrochlorination (also called epoxidation/saponification), the aqueous PCH stream is treated with a base—either lime, Ca(OH)$_2$, or caustic soda (as chlorine cell liquor)—to form crude PO and a dilute CaCl$_2$–NaCl brine stream (Eq. 3).

$$2\ \underset{\text{PCH}}{CH_3-\overset{OH}{\underset{|}{CH}}-\overset{Cl}{\underset{|}{CH_2}}} + \underset{\text{(or 2 NaOH)}}{Ca(OH)_2} \longrightarrow$$

$$2\ CH_3-\overset{O}{\overset{\triangle}{CH-CH_2}} + \underset{\text{(or 2 NaCl)}}{CaCl_2} + 2\ H_2O \quad (3)$$

In a subsequent reaction a minor part of the resulting PO is hydrolyzed to 1,2-propanediol (monopropylene glycol, MPG) (Eq. 3.1).

$$CH_3-\overset{O}{\overset{\triangle}{CH-CH_2}} + H_2O \longrightarrow \underset{\text{MPG}}{CH_3-\overset{OH}{\underset{|}{CH}}-\overset{OH}{\underset{|}{CH_2}}} \quad (3.1)$$

Other byproducts formed in the epoxidation step are 1-chloro-2,3-epoxypropane (epichlorohydrin, Eq. 3.2) and its reaction products 3-chloropropane-1,2-diol (glycerol monochlorohydrin, GMCH), 1,2,3-propanetriol (glycerol), propanal, acetone, and 1-hydroxypropanone (acetol).

$$\underset{\underset{\text{(or DCH 2)}}{\text{DCH 1}}}{\underset{|||}{\text{CH}_2\text{-CH-CH}_2}}\xrightarrow[-\text{NaCl}-\text{H}_2\text{O}]{+\text{NaOH}}\underset{\text{Epichlorohydrin}}{\text{CH}_2\text{-CH-CH}_2} \quad (3.2)$$

$$\xrightarrow{+\text{H}_2\text{O, energy}}\underset{\text{GMCH}}{\underset{|||}{\text{CH}_2\text{-CH-CH}_2}}\xrightarrow[-\text{NaCl}]{+\text{NaOH}}\underset{\text{Glycerol}}{\underset{|||}{\text{CH}_2\text{-CH-CH}_2}}$$

Chlorohydrination. In a typical commercial chlorohydrin PO process, gaseous propene (either chemical grade, ca. 8% propane, or polymer grade, ca. 0.3% propane) and gaseous chlorine are mixed in roughly equimolar amounts with an excess of water to form a dilute solution of PCH (ca. 4 wt%), aqueous HCl, and minor amounts of chlorinated hydrocarbon byproducts, mainly DCP.

Since the reactor design (Fig. 4) is based upon propene and chlorine vapor dispersion into water, and because all reaction products remain in aqueous solution, the role of water is very important. It is used not only as reactant (one mole per mole of propene) and direct cooling medium, but also as a diluent to minimize the formation of byproducts. Numerous patents claim technology for boosting chlorohydrin yields to 90–95% [32]–[38].

Depending on the capacity of the plant, one or several parallel reactors (a) are used. The reaction temperature varies from 45 to 90 °C, depending on the reactor concept. The exothermic reaction itself raises the temperature by 20–30 °C. The operating pressure is usually slightly above atmospheric pressure (1.1–1.9 bar). The chlorine: propene ratio must be closely controlled to maintain the proper amount and composition of vent gas. In the separator (b) the reactor vent gas (a mixture of propane, excess propene or chlorine, oxygen, nitrogen, hydrogen, carbon dioxide, PCH, DCP, and other reaction products) is separated from the PCH solution and then fed to a scrubber (c). Vent gas from the scrubber can be sent to a thermal oxidizer to protect the environment and recover energy.

Dehydrochlorination (Saponification). The bottoms from the separator (b) are fed to a saponifier (d) where the PCH is mixed with NaOH or lime. The dehydrochlorination of PCH to PO is rapid. Optimum PCH conversion requires a slight excess of alkalinity. Half the base is consumed by neutralization of HCl produced in the chlorohydrination step.

The saponifier has two main functions: (1) conversion of PCH to PO and (2) stripping of organic impurities from the aqueous brine before final treatment in a biological oxidation plant. Crude PO leaves the saponifier as an overhead stream that also contains organic byproducts and water. If lime is used, solids enter the saponifier, so the column should be equipped with trays to minimize fouling. If caustic soda is used, either trays or packing are acceptable. As far as reaction kinetics are concerned [39], an ideal saponifier would be based on the following design criteria:

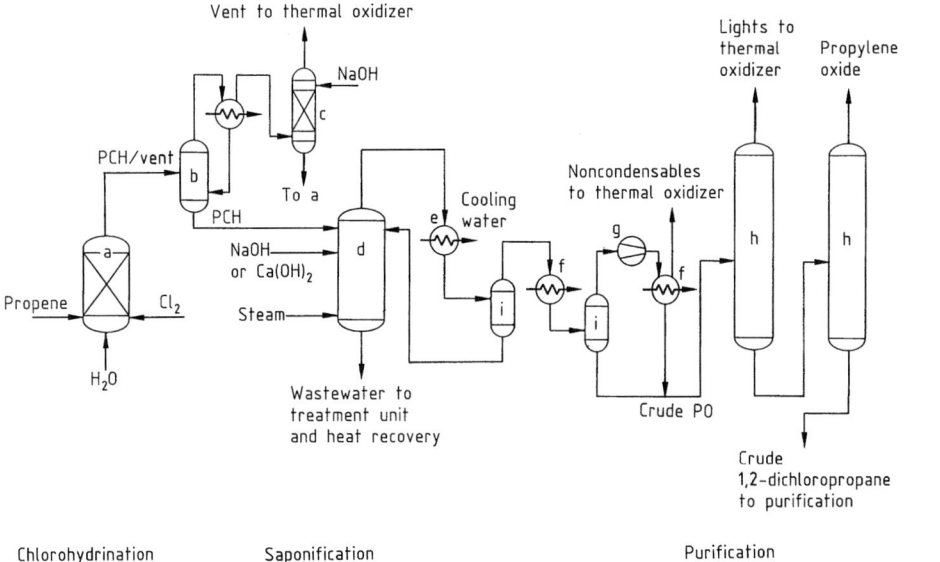

Figure 4. Typical arrangement for a propylene oxide chlorohydrin process
a) Propylene chlorohydrin reactor; b) Separator; c) Vent gas scrubber; d) Saponifier; e) Partial condenser; f) Cross exchanger; g) Compressor; h) Propylene oxide purification train; i) Drums

1) Short residence time in the top to flash PO as quickly as possible, thereby minimizing the formation of MPG
2) A stripping section to minimize PO losses
3) Long residence time to complete the conversion of DCH 2 to epichlorohydrin (DCH 1 conversion is as fast as the PCH conversion)
4) A stripping section to recover epichlorohydrin

These measures ensure that the formation of volatile products is optimized, and that such compounds are removed before subsequent products (e.g., GMCH, glycerol) are formed and lost in the diluted brine. An overall unit-ratio balance is shown in Figure 5.

Purification. After condensation (e-g) the crude PO is sent to distillation towers (h), in the first of which the lights are removed. Bottoms are sent to the finishing column, where PO product is taken off overhead. DCP, DCIPE, epichlorohydrin, and water remain in the bottom for further treatment which depends on the final utilization of coproducts.

Construction Materials. Since chlorine in aqueous solution as well as generated HCl and PCH are extremely corrosive, the construction materials must be selected very carefully [40]–[42]. The reactor (a), including the last segments of feed lines, separator (b), vent condensers, scrubber (c), saponifier (d), and all related piping are made of corrosion-resistant materials. Adequate for the purpose are poly(vinylidene fluoride)-

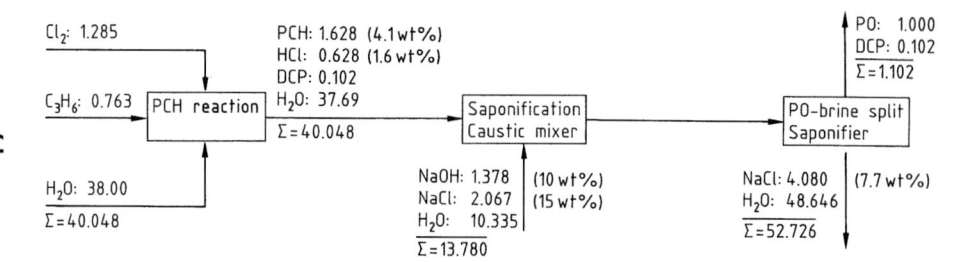

```
Cl₂: 1.285
                          PCH: 1.628 (4.1 wt%)                                    PO: 1.000
                          HCl: 0.628 (1.6 wt%)                                    DCP: 0.102
                          DCP: 0.102                                              Σ=1.102
C₃H₆: 0.763  ┌─────────┐  H₂O: 37.69        ┌──────────────┐        ┌──────────────┐
─────────────│ PCH     │──────────────────  │ Saponification│──────  │ PO-brine split│
             │ reaction│  Σ=40.048          │ Caustic mixer │        │ Saponifier    │
             └─────────┘                    └──────────────┘        └──────────────┘
                 │                 NaOH: 1.378  (10 wt%)              NaCl: 4.080   (7.7 wt%)
H₂O: 38.00       │                 NaCl: 2.067  (15 wt%)              H₂O: 48.646
Σ=40.048                           H₂O: 10.335                        Σ=52.726
                                   Σ=13.780
```

Figure 5. Unit ratios for a propylene oxide chlorohydrin process assuming a water–propene ratio of 50:1, caustic soda as cell liquor, DCP as the only byproduct, a DCP yield of 5%, and all steps stoichiometric.

fiberreinforced plastics, graphite, rubber- or brick-lined carbon steel, titanium and titanium–palladium alloys; Inconel, Monel, and Hastelloy.

Wastewater. The brine effluents of modern PO plants contain:

1) Ca. 3–5 wt% calcium chloride (lime) or ca. 8 wt% sodium chloride (NaOH)
2) Slight excess alkalinity
3) 100–400 ppm total organic carbon (TOC)
4) 30–60 ppm adsorbable organic halides (AOX)

The main components are MPG and GMCH–glycerol, which are biodegradable. The amount of brine effluent is 30–60 kg/kg PO. Downstream final biological treatment systems achieve an AOX reduction of >85% (see Chap. 5 for wastewater treatment).

Lime versus Caustic Soda as Alkalinity Source. The chlorohydrin PO process can be operated with caustic soda or slaked lime. Slaked lime is primarily used when chlorine is produced by the mercury or membrane processes to avoid "downgrading" high-purity caustic (Germany, Japan, Czechoslovakia, and Poland). Caustic is usually used in the form of cell effluent in combination with the diaphragm chlor-alkali process since the low-concentration catholyte from such a chlorine cell can be employed directly (United States, Germany). Although this process option in principle allows the spent PO brine to be recycled to the chlorine plant after resaturation, such a recycle system has never been applied at full technical scale. Major hurdles not yet overcome include hydraulic imbalances, as well as sensitivity of the chlor-alkali process to organic impurities. If an integrated diaphragm chlor-alkali site becomes short of NaOH, it is in principle possible to convert the PO process from cell effluent to lime.

A final detailed evaluation of the use of lime relative to caustic soda cannot be made in this article because the parameters to be considered are too diverse, and they are subject to frequent change (e.g., the global alkalinity supply/demand situation). However, the key advantages and disadvantages are summarized in Table 3 to provide a qualitative overview with respect to economic, environmental, technological, site, and site-integration issues.

Figure 6 provides a block diagram of a PO–lime process.

Table 3. Comparison of use of lime and sodium hydroxide in the chlorohydrin process [a]

Lime	NaOH (chlor-alkali/PO integrated complex)
+ Frees NaOH for sales (1.4 t/t PO)	+ no caustic evaporator capacity needed (NaOH/Cl_2 = 1.08)
− Higher capital investment; additional caustic evaporation capacity, handling, and infrastructure are needed; unloading of CaO, storage, slaker, thickeners, clarifiers, neutralizer, and centrifuges/filter are required	+ lower capital investment and operating cost (no solids handling required)
+ Up to 50 % less discharge of chloride (NaOH as cell liquid containing up to 15 wt% NaCl, wastewater with 5 wt% calcium chloride)	− higher discharge of chloride (wastewater with 8 wt% sodium chloride) +/− brine recycle theoretically possible, ca. 15 % more wastewater
− Disposal of separated solids, uncontaminated and contaminated with organics depending on CaO quality	
− More excess alkalinity in wastewater	
− Lime biox technology, dedicated or diluted (problems with $CaCO_3$/$CaSO_4$ precipitation)	

[a] − denotes disadvantage, + denotes advantage.

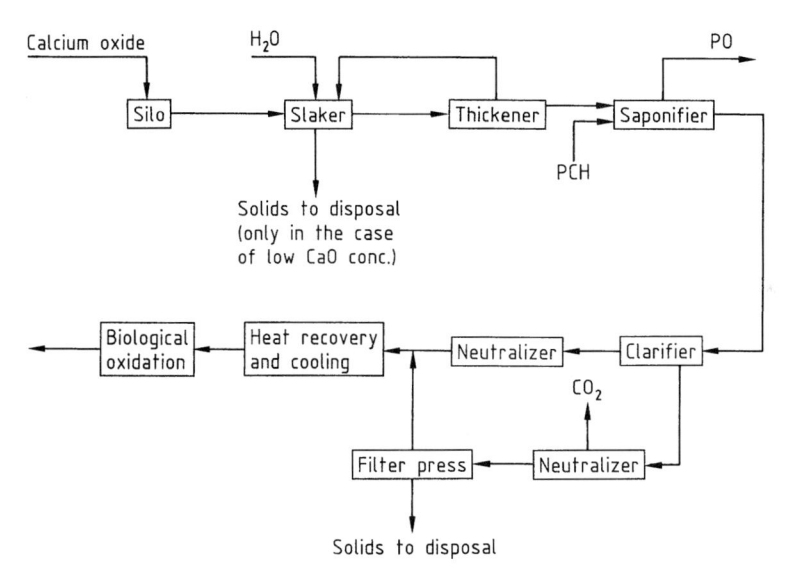

Figure 6. Block diagram for the PO/lime process

One very important issue is the quality of the calcium oxide used for the lime process. Normally, it has a purity of 85–95 wt%. The MgO and SiO_2 contents should be kept as low as possible to minimize solids handling.

4.1.2. Other Chlorohydrin Technologies

Modified chlorohydrin technologies have been proposed [29], but they are not currently applied on an industrial scale:

1) The formation of propylene chlorohydrin in two process steps using *tert*-butyl hypochlorite as a chloride-free chlorine-transfer agent (Lummus process)
2) Electrochemical production of PO—the formation of propylene chlorohydrin and subsequent dehydrochlorination to the oxide take place in chlor-alkali electrolysis cells
3) Electrochemical production of PO, in which the formation of propylene chlorohydrin and dehydrochlorination to the oxide occur outside chlor-alkali membrane electrolysis cells

All three process modifications are designed to permit complete recycle of PO brine to the chlor-alkali cells at constant or even reduced DCP byproduct levels. Achieving a balance of the requisite countercurrent conditions leads to (1) increased complexity of the unit operations (e.g., large, complex cascade reactors) and/or (2) high operating costs (e.g., energy-intensive distillation due to a low PO concentration in the catholyte). Furthermore, the selectivity with respect to DCP is at best comparable to that obtained with existing chlorohydrin technology. In fact, DCP formation is likely to be increased, and other chlorinated byproducts (e.g., chloroacetone) may be formed at the expense of DCP. Finally, the sensitivity of chlorine cells to organic contaminants in the concentrated recycle brine must be taken into account. The disadvantages of these process modifications are high in terms of both investment and energy costs.

Propylene Chlorohydrin with Recycle (Lummus Process). One chlorohydrin PO process not currently operated on an industrial scale uses *tert*-butyl hypochlorite instead of chlorine. This allows recycling of the major portion of the aqueous process streams. Process water from the hypochlorination can be recycled to the propylene chlorohydrin reactor. Brine, produced during saponification, can be appropriately treated and recycled to the chlorine cells. Resaturation is not required.

This integrated chlor-alkali/PO process is balanced on water; i.e., kg water in = kg water out. The level of organic byproducts is comparable to that of standard chlorohydrin technology. Lummus has several patents that disclose the technology of such a closed-loop process [43]–[45].

tert-Butyl hypochlorite is prepared from cell effluent, chlorine, and *tert*-butanol (Eq. 4). The key difference relative to established chlorohydrin processes results from the fact that *tert*-butyl hypochlorite is insoluble in water. It therefore separates as an organic phase from the aqueous brine solution and can be isolated by means of a phase separator. Consequently, propene can be hypochlorinated with *tert*-butyl hypochlorite in an aqueous system, thereby avoiding the presence of chloride ions (Eq. 5).

$$Cl_2 + NaOH + \textit{tert}\text{-}C_4H_9OH \rightarrow \textit{tert}\text{-}C_4H_9OCl + NaCl + H_2O \quad (4)$$
$$\textit{tert}\text{-}C_4H_9OCl + H_2O + C_3H_6 \rightarrow C_3H_6OHCl + \textit{tert}\text{-}C_4H_9OH \quad (5)$$
$$C_3H_6OHCl + NaOH \rightarrow C_3H_6O + NaCl + H_2O \quad (6)$$

No contact occurs between the chloronium complex and chloride ions (see p. 4135), so the ratio of water to propene in the process can be reduced without a concurrent increase in the level of byproducts. Furthermore it is possible to recycle a major portion of the nearly chloride-free process water to the front end of the chlorohydrin reactor. Because of the lower water consumption, the two brine solutions—one resulting from *tert*-butyl hypochlorite formation step (Eq. 4) and the other from the hydrolysis of propylene chlorohydrin to PO (Eq. 6)—are close to saturation. They can be recycled to a chlorine cell after removal of organic impurities by chlorinolysis.

By contrast, in the conventional chlorohydrin PO process it is essential to minimize reaction between the chloronium complex and chloride ions, because this leads to the formation of DCP. The high ratio of water to propene required to suppress this undesirable side reaction means that the aqueous brine streams have salt concentrations too low to permit complete recycle to the chlorine cells (see Section 4.1.1). A major disadvantage of the Lummus process is the slowness of the reaction between *tert*-butyl hypochlorite and propene. Huge, titanium-clad cascade reactors are required. The presence of an additional component (*tert*-butyl alcohol) also leads to a more complex byproduct spectrum. The treatment of recycle streams is very energy intensive, and investment costs for the process are high.

4.2. Indirect Oxidation Routes

Several reaction routes and process alternatives for the production of PO by indirect oxidation have been reported in the patent literature, but only a few are practiced on an industrial scale. Indirect oxidation is a two-step process:

1) Formation of hydrogen peroxide or an organic peroxide from a suitable alkane, aldehyde, or acid
2) Conversion of the peroxide to water or the corresponding alcohol or acid by epoxidation of propene to PO

Epoxidation of Propene with Hydrogen Peroxide. The epoxidation of propene can be accomplished with hydrogen peroxide in the liquid phase and in the presence of, for example, a molybdenum-based catalyst. Subsequent reactions can lead to formation of considerable amounts of glycols and glycol esters, and thus to separation problems [46]–[53]. One process modification, published by Enichem [54], makes use of a titanium silicate catalyst in a methanol–water mixture at 40–50 °C and a propene pressure of 4 bar. PO selectivities of 75–97% are reported. Disadvantages of this

Table 4. Reaction systems for PO – coproduct synthesis

Raw material	Peroxide	Intermediate	Coproduct
Acetaldehyde	peracetic acid		acetic acid
2-Propanol	peracetic acid	acetone	2-propanol
Isobutane	*tert*-butyl hydroperoxide	*tert*-butyl alcohol	isobutene
Isopentane	*tert*-pentyl hydroperoxide	*tert*-pentyl alcohol	isoprene
Ethylbenzene	ethylbenzene hydroperoxide	α-phenylethanol	styrene
Cumene	cumene hydroperoxide	dimethylphenylmethanol	α-methylstyrene
Cyclohexene	cyclohexene peroxide	cyclohexanol	cyclohexanone

process are the considerable hazard potential associated with concentrated H_2O_2 solutions and high H_2O_2 production costs.

Epoxidation of Propene with Peroxycarboxylic Acids. The required peroxy acids can be prepared by different routes. For example, peracetic acid can be formed by oxidation of acet-aldehyde with air or oxygen, or by reaction of acetic acid with H_2O_2.

Organic peracids such as performic acid, perphthalic acid, permaleic acid, perpropionic acid, and peracetic acid react with propene to give PO and the corresponding carboxylic acid. The rate of epoxidation depends on the type of solvent used. Peracetic and perpropionic acids are the preferred epoxidizing agents, and no catalyst is required.

If the starting material for production of peroxyacetic acid is acetaldehyde, the coproduct acetic acid must be removed from the production loop. Process economics depend strongly on the demand for acetic acid and its value to industry. Process experience with a 12 500 t/a plant is available at Daicel; the yield of PO is 90%. Typically 1.3 kg of acetic acid is produced per kg of PO [55]–[69]. Scaleup of this technology to industrial capacities ($\geq$ 100 000 t/a) is expensive and difficult, and the value of acetic acid is not sufficiently high to justify the process on a large scale.

The Bayer – Degussa process starts from acetic or propionic acid and H_2O_2, and proceeds via peracetic or perpropionic acid, respectively. The acid coproduct can be recycled to the front end of the process. PO selectivities $\geq$ 95% can be reached.

A common disadvantage of all processes based on H_2O_2 is that safety considerations require H_2O_2 production facilities to be restricted in size. Economic large-scale production of H_2O_2 for the production of PO is not possible with today's processes.

Coproduct Processes. Oxidation of propene with organic hydroperoxides in the presence of catalysts has developed into a successful indirect oxidation route to PO. Half the world PO production is now based on this route [70]–[101]. Various organic hydroperoxides can be used (Table 4).

Organic hydroperoxides are generated by catalytic oxidation of appropriate substrates with air or oxygen. Because of coproduct value and market demand, the *tert*-butyl hydroperoxide and ethylbenzene hydroperoxide routes are those practiced in industry today.

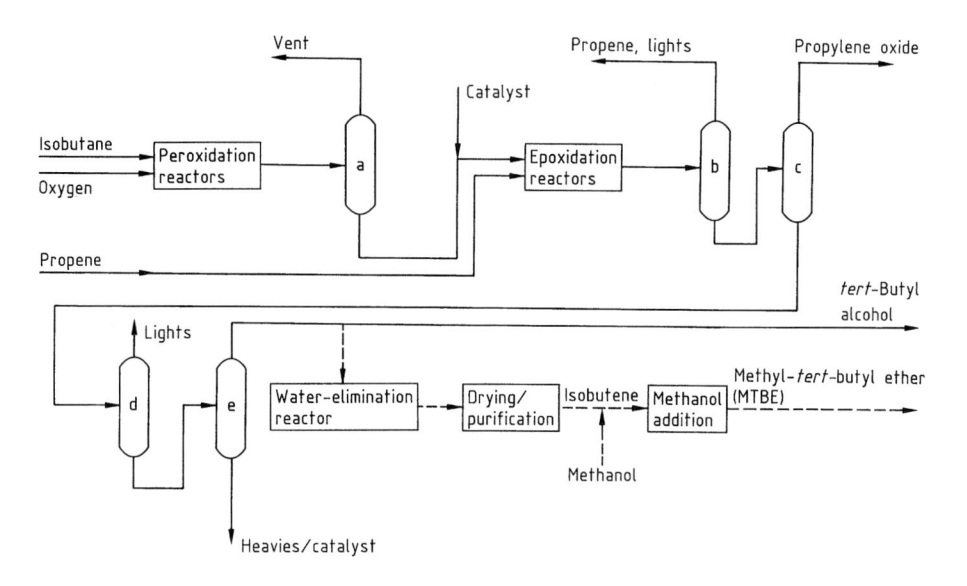

Figure 7. Flow scheme for PO–*tert* butyl alcohol
a) Vent column; b) Lights scrubber; c) PO column; d) *tert*-butyl alcohol lights column; e) *tert*-butyl alcohol column

The production of PO by indirect oxidation with organic hydroperoxides requires two steps. The *tert*-butyl alcohol or α-phenylethanol coproducts formed in parallel with PO are of considerable economic value, because they can be converted to methyl *tert*-butyl ether [1634-04-4] (MTBE) or styrene [100-42-5], respectively, in subsequent reaction steps. When styrene or MTBE is in great demand, coproduct process economics are competitive with those of the alternative chlorohydrin production routes. The coproducts are always formed in larger amounts than PO itself. In the case of the process with *tert*-butyl hydroperoxide, 2.5 – 3.5 kg of *tert*-butyl alcohol are formed per kilogram of PO. The ethylbenzene process produces 2.2 – 2.5 kg of styrene per kilogram of PO. Obviously, the presence of these coproducts can also become a disadvantage if the demands for PO and the respective coproduct are not properly balanced, requiring that PO production capacity be restricted as demand for the coproduct weakens. No such considerations apply to the chlorohydrin processes. A regional limitation of the coproduct processes results from raw material logistics. Especially for the *tert*-butyl hydroperoxide process, economic operation requires integration into a refinery complex, where mixed butanes and ethylbenzene are readily available.

Process with tert-*Butyl Hydroperoxide.* The main reactions in this process are:

$$4\ CH_3-\underset{CH_3}{\underset{|}{\overset{CH_3}{\overset{|}{C}}}}-H + 3\ O_2 \longrightarrow$$

$$2\ CH_3-\underset{CH_3}{\underset{|}{\overset{CH_3}{\overset{|}{C}}}}-OOH + 2\ CH_3-\underset{CH_3}{\underset{|}{\overset{CH_3}{\overset{|}{C}}}}-OH \quad (7)$$

$$2\ CH_3-\underset{CH_3}{\underset{|}{\overset{CH_3}{\overset{|}{C}}}}-OOH + 2\ CH_3-CH=CH_2 \longrightarrow$$

$$2\ CH_3-\underset{\underset{O}{\diagdown\!\diagup}}{CH-CH_2} + 2\ CH_3-\underset{CH_3}{\underset{|}{\overset{CH_3}{\overset{|}{C}}}}-OH \quad (8)$$

Figure 7 shows a simplified flow scheme for the PO–*tert*-butyl alcohol process. Isobutane is generated by isomerization of a mixed butane fraction. This is then converted with oxygen at 120–140 °C and 25–35 bar in an epoxidation reactor to *tert*-butyl hydroperoxide, which partially decomposes to *tert*-butyl alcohol [28]. This reaction takes place in the liquid phase and does not require a catalyst; however, additives can be used to optimize reaction conditions. Isobutane conversion is ca. 48 % at a selectivity of 50 % relative to hydroperoxide and 46 % relative to *tert*-butyl alcohol.

The crude peroxide solution is used for the epoxidation of propene. It consists of a mixture of *tert*-butyl hydroperoxide, *tert*-butyl alcohol, and smaller amounts of aldehydes and ketones. This is mixed with a homogeneous catalyst, dissolved in toluene; propene is added, and the mixture is fed to the epoxidation reactor system. The preferred catalyst is a molybdenum salt. Epoxidation is accomplished in a staged reaction system consisting of up to five consecutive reactors. The liquid-phase reaction is carried out at 110 °C at ca. 40 bar in the first part of the reactor system. In the second part the temperature is elevated to 120 °C to complete conversion of the hydroperoxide. The selectivity of the converted peroxide to PO is ca. 80 % with a 10-fold excess of propene; i.e., not all the peroxide reacts to form PO. It may decompose without transferring oxygen to propene. Propene conversion reaches ca. 9 % per pass, of which 98 % is converted to PO. Crude PO is then distilled from the reaction mixture and finished. The bottoms of this distillation consist of a mixture of *tert*-butyl alcohol, catalyst, and medium- and high-boiling components.

The crude PO contains hydrocarbons, carbonyl compounds, and ethylene oxide as impurities, which must be removed by rectification and extractive distillation. For further purification of the carbonyl-free PO rectification with *n*-octane is suggested; the overhead product is pure PO. The hydrocarbons present in the crude PO are successively removed from the *n*-octane, which is then reused.

The *tert*-butyl alcohol can be dehydrated to isobutene and then converted to methyl *tert*-butyl ether (MTBE), which is of significant use as a fuel additive for lead-free gasoline. Educt–product ratios are shown below:

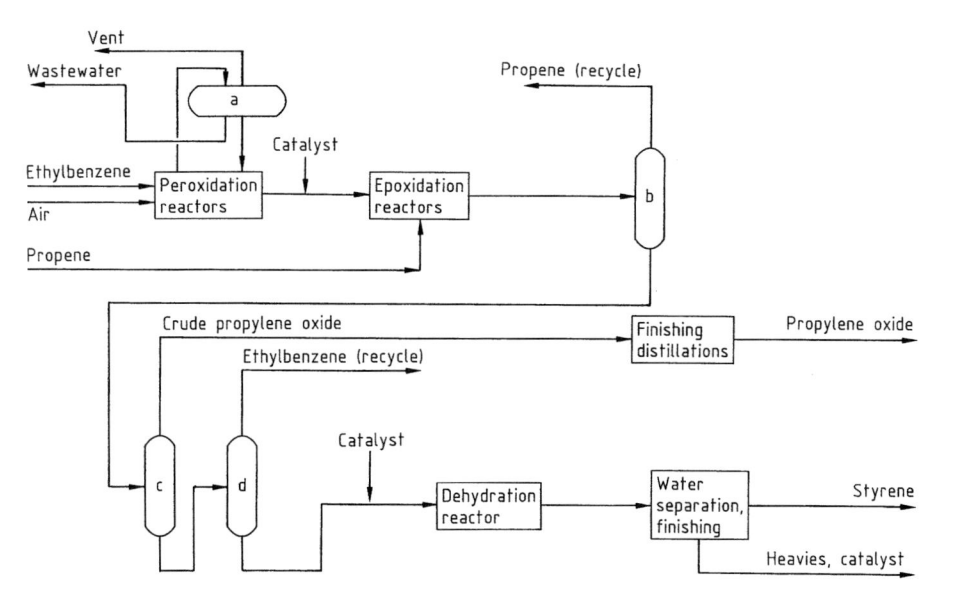

Figure 8. Flow scheme for PO–styrene process
a) Separator; b) Recycle column; c) Crude PO column; d) Ethylbenzene recycle column

Educts: 2.90 kg isobutane
0.88 kg propene
Products: 2.98 kg *tert*-butyl alcohol
0.15 kg acetone
1.00 kg PO

Process with Ethylbenzene Hydroperoxide. The main reactions of this process are:

$$C_6H_5CH_2CH_3 + O_2 \longrightarrow \underset{\text{Ethylbenzene hydroperoxide}}{C_6H_5\overset{\overset{\displaystyle CH_3}{|}}{C}H(OOH)} \quad (9)$$

$$C_6H_5\overset{\overset{\displaystyle CH_3}{|}}{C}H(OOH) + CH_3-CH=CH_2 \longrightarrow$$

$$\underset{\text{PO}}{CH_3-\underset{\underset{\displaystyle O}{\diagdown\diagup}}{CH-CH_2}} + \underset{\alpha\text{-Phenylethanol}}{C_6H_5\overset{\overset{\displaystyle CH_3}{|}}{C}HOH} \quad (10)$$

$$\downarrow \text{Dehydration}$$

$$\underset{\text{Styrene}}{C_6H_5CH=CH_2}$$

Figure 8 shows a simplified flow sheet of the process, based on patent data. A more detailed flow scheme has been published by Chem Systems Inc. [5].

In the first process step ethylbenzene is reacted with air at 146 °C and 2 bar to give a 12–14 wt% solution of ethylbenzene hydroperoxide in ethylbenzene. The reaction takes place in the liquid phase. For safety reasons conversion is restricted to 10%. The reactor configuration suggested in [102] consists of up to nine separate reaction zones, over which the temperature is reduced from 146 °C to 132 °C by controlled recycle gas addition to avoid decomposition of the peroxide. The resulting peroxide solution in ethylbenzene is mixed with a homogeneous molybdenum catalyst (Halcon) and propene [103]. Alternatively, a heterogeneous titanium-based catalyst can be used (Shell). The latter offers a similar performance, but may be easier to separate from the product stream [104].

For the Halcon route the reaction mixture consists of 100 parts peroxide solution, 2 parts molybdenum naphthenate, and 30 parts propene. Epoxidation of propene is carried out at 100 °C and 35 bar. The crude product stream contains PO, α-phenylethanol, acetophenone, propene, and other organic impurities. Catalyst removal requires a wash with aqueous alkali and phase separation. Propene can be recycled after distillation. Crude PO is purified by distillation. The finishing is similar to that in the *tert*-butyl hydroperoxide route. Ethylbenzene is recovered for recycle from the bottoms of the PO distillation. The remaining α-phenylethanol-rich mix-ture, which still contains up to 15% acetophenone, is fed to a dehydration reactor and mixed with an alumina catalyst and triphenylmethane as process solvent. Thermal treatment at 270 °C and 0.35 bar leads to dehydration. Styrene is obtained with a selectivity of 98%.

α-Phenylethanol and acetophenone are the main byproducts. Educt – product ratios for this process have been reported by Chem Systems Inc. [5]:

Educts:
    2.11 kg benzene
    0.76 kg ethylene
    0.88 kg propene
Products:
    2.54 kg styrene
    1.00 kg PO

4.3. Direct Oxidation Routes

There is no direct oxidation process for producing PO that is close to commercialization. Difficulties in controlling temperature variations over fixed catalyst beds are among the several hurdles preventing satisfactory selectivity for PO. Nevertheless, construction of a pilot plant has been announced for a new direct oxidation process involving a molten nitrate salt as catalyst [105]. An advantage of this particular direct oxidation variant may be the potential for maintaining a constant reactor temperature (454 °C) in the molten salt bed. However, in addition to PO, several byproducts arise in substantial amounts, including methanol, acetaldehyde, carbon dioxide, ethylene, and formaldehyde.

The breakthrough in direct oxidation of propene to PO has not yet been achieved. For commercial PO production during the next ten years, the only alternatives are the chlorohydrin process or coproduct technologies.

4.4. Comparison of the Various Propylene Oxide Technologies

Direct comparison of the chlorohydrin PO process with the PO/*tert*-butyl alcohol or PO/styrene coproduct process is very difficult, as the technologies are too diverse and have different objectives. For a PO producer with no market for the coproduct it is obvious that there is currently no alternative to the chlorohydrin PO process. A producer able to include MTBE or styrene in the product portfolio must decide whether a coproduct process or two independent plants represents the preferred solution with respect to economics, demand balances, and geographic flexibility. The fact that stand-alone units for chlorohydrin PO, styrene, and MTBE are still being announced and built indicates that the PO coproduct process is not always economically superior. A comparative economic model of chlorohydrin versus coproduct PO processes is presented in [5]. A decision in favor of one PO production route or the other also depends on the infrastructure, location, and integration of a particular site. Thus, a chlorohydrin PO process can be operated most economically if it is integrated within a chlor-alkali plant. The MTBE process requires integration into a refinery complex. The effluent load of a chlorohydrin PO plant is larger than that of a coproduct plant, so extensive water treatment facilities are required, and the salt load means that an environmentally acceptable geographical location (close to the sea) is desirable. On the other hand, the fire and explosion risk with a PO coproduct plant is significantly greater than in the case of a chlorohydrin facility, and product flexibility is considerably more limited with a coproduct plant than with a stand-alone process.

5. Environmental Protection and Ecotoxicology

Propylene oxide is recognized as a toxic substance (Chap. 10), so specific measures must be taken to avoid environmental pollution during PO production. Critical issues associated with water discharge, air emissions, and handling of PO are discussed briefly below.

Wastewater. State-of-the-art technology for wastewater treatment requires that chlorinated organic compounds first be separated via rectification, stripping, or extraction. The stream is then sent to a biological oxidation unit for eliminating biodegradable compounds like propylene glycol.

In the past, large concrete ponds were often used as aeration basins, but today more and more companies are installing biotowers. Significant advantages of such towers include:

1) Higher efficiency
2) Greatly reduced space requirements
3) Less danger of ground water contamination
4) Easier collection and treatment of the vent

The most critical parameters related to water discharge are:

1) Salt discharge at "upstream sweet-water" sites (chlorohydrin process only)
2) Total organic carbon (TOC)
3) Chemical oxygen demand (COD)
4) Adsorbable organic halides (AOX)
5) Temperature
6) Solids disposal (separated as filter cake)

Off-Gas. Environmental legislation varies considerably from country to country, one of the strictest being that in Germany. Discharge to the atmosphere of vent streams with a mass flow ≥ 25 g PO/h is permitted in Germany for PO concentrations up to 5 mg/m^3 [106], [107]. Normally, however, such streams are incinerated with state-of-the-art thermal oxidizers.

For a chlorohydrin PO process, thermal oxidizer treatment is required for the reactor vent, noncondensables in the crude PO, lights from the purification, and tank vent streams. Fugitive emissions represent the only remaining emission source.

In the coproduct processes it is advisable to use oxygen instead of air, since this reduces the volume of vent gas subject to treatment.

Fugitive Emissions. Fugitive emissions are caused mainly by: leaking mechanical seals of pumps, gaskets, and stuffing boxes of valves.

PO is not known to occur naturally. It has been suggested that, apart from PO production facilities, PO may be introduced into the atmosphere from combustion exhausts of sources that burn hydrocarbons [108], [109]. In the atmosphere PO reacts with photochemically produced hydroxyl radicals and has an estimated half-life of 19.3 d. PO is not expected to contribute to ozone depletion [110], [111]. In water, PO is hydrolyzed to propylene glycol with an estimated half-life of 11.6 d at 25 °C and pH 7. The chloride ions in salt water accelerate the chemical degradation to a half-life of 4.1 d [109]. PO is practically nontoxic to fish on a static acute basis (LC$_{50}$ >100 mg/L) [112]–[116]. Biodegradation of PO under aerobic static laboratory conditions is high (BOD$_{20}$/ThOD = 74.7%, where ThOD means theoretical oxygen demand). The biochemical oxygen demand (BOD) of PO is 0.35 g O$_2$/g PO and 1.65 g O$_2$/g PO for 5 d and 20 d, respectively [107].

PO evaporates relatively rapidly from dry soil surfaces, and it is moderately volatile from water and wet soils, but the evaporation rate is diminished by leaching. PO is expected to be very mobile in soil. In moist soils hydrolysis is the most significant degradation process. No appreciable concentrations of PO are expected to be found in the environment or aquatic organisms [118].

Table 5. Typical specifications for PO in urethane applications

Parameter	Chlorohydrin process	PO/*tert*-butyl alcohol process	PO/styrene
Purity, %	99.9	99.9	99.9
Color (APHA)	10	10	10
Water, mg/kg	100	100	100
Aldehydes, mg/kg	30	30	30
Total chlorides, mg/kg	30	30	30
Methanol, mg/kg	not present	50	not present
Furan, mg/kg	not present	not present	50

6. Quality Specifications and Analysis

PO obtained from different processes (Chap. 4) displays consistently high purity (>99.9%; see Table 5). The quality of the propene feed contributes significantly to the impurity content in the final product. Polymer-grade propene is preferable to reduce formation of acetaldehyde and ethylene oxide. Acetaldehyde causes problems in polyol applications, whereas ethylene oxide leads to undesirable contaminants in propylene glycol.

Other impurities may also be present depending on the nature of the production process (e.g., chlorinated organic compounds, hydrocarbons, ethers, aldehydes, acetone, methanol, butanol, *tert*-butyl alcohol, or furan).

Quality specifications are becoming increasingly strict, and customer satisfaction can best be assured by instituting statistical quality tools.

PO and its process-specific coproduct spectrum (e.g., aldehydes, chlorinated organic compounds, methanol, acetone, hydrocarbons, and ethers) can be determined by gas chromatography. Water can be analyzed by the Karl-Fischer method (DIN 51777, ASTM D 203). The platinum–cobalt standard is used as a color reference (DIN 53 409, ASTM D 1209).

Processes are monitored with on-line gas chromatographs and, more recently, with IR systems.

7. Handling, Storage, and Transportation

Detailed attention is required on issues of safety, health, and the environment. Product stewardship policies provide the basic tools for managing these vital business activities. For example, in 1988, the U.S. Chemical Manufacturers Association (CMA) adopted a "Responsible Care" program; details are provided in [119], [120]. Based on these principles, many companies have developed new product stewardship policies; e.g., [121].

Handling. Safety precautions related to the handling of PO (e.g., protective clothing, sampling, vessel entry, loading, unloading, and general issues) are described in detail in data sheets provided by the manufacturers [122], [123] and elsewhere [124]. Therefore, only the most important issues are discussed in this article.

Personnel handling PO must receive appropriate training. PO reacts readily with all compounds containing active hydrogen. Contact with materials such as acids, bases, oxidizing materials, and clay-based absorbents must be strictly avoided.

Spills, Leaks, and Fire. In case of a spill, leak, or fire the following measures should be taken:

1) Shut off all potential ignition sources.
2) Dilute liquid spills with copious amounts of water (H_2O:PO ratios $\geq 200:1$). Use water spray to reduce the extent of vapor. The effectiveness of such a spray can be checked by testing the area with an explosimeter.
3) Maintain the water spray, and shut off the flow of PO to the leak.
4) Do not shut off the water until it is certain that the leak has stopped and that flammable concentrations have dissipated.
5) Prevent runoff from entering sewers and/or natural waters.
6) Adequate fire extinguishing media include water, alcohol foam, CO_2, and dry powder.
7) Small fires (e.g., at a pump-seal leak) can be extinguished with a fire hose, deluge gun, CO_2, or dry powder.
8) Large fires (e.g., from a blown gasket) present a different and more serious problem. Do not attempt to extinguish the external fire, or to shut off the feed to the fire. Instead, change the nature of the feed to the fire. Start filling the container with water or nitrogen and continue to fill at a rate adequate to maintain pressure on the leak. With sufficient water or nitrogen dilution, fuel to the fire will be cut off from the inside, or the fire will be extinguished as a result of creating a nonflammable mixture. Water should be sprayed on all adjacent oxide- or flammable-containing equipment to keep it cool. Once the fire is out, the entire area should be sprayed with water to reduce the chance of reflash.

Table 6. Hazard classifications and transport information for propylene oxide [a]

Transport mode	Codes according to legal requirements				Danger class	Label
	International	Europe	United States	Germany		
Sea	IMO/IMDG			GGVSee	3.1 UN no. 1280 page no. 3047	3
Barge				ADNR		
Rail		RID [b]	AAR	GGVE	3-2(A)	3
Road		ADR [b]	DOT	GGVS	3-2(A)	3
Air	ICAO/IATA				3 UN no. 1280	LF

[a] Sample shipment not allowed by post.
[b] ADR/RID hazard identification: (Kemler Code) 33-1280 (UN no.).

Storage. Storage tanks for PO are usually made of ASTM A-516 grade steel or stainless steel. Because of the tendency of PO to polymerize, all equipment should be absolutely free of impurities. Therefore, PO tanks are constructed in such a manner that they can be completely drained and cleaned.

Tanks and piping systems must be purged with nitrogen to reduce their oxygen content to < 0.5 vol % [automatic N_2 pad and depad system (N_2 inertization)]. Pure N_2 should be used (max. 100 ppm O_2).

Fragile devices (e.g., glass or plastic sight and gauge glasses as well as expansion joints and flexible or screwed connections) should be avoided in PO service. Teflon spiral-wound stainless steel gaskets are preferred. Flat graphite and/or EPR rubber gaskets are also acceptable.

All equipment containing liquid PO should be protected with automatic deluge or sprinkler systems. The maximum storage temperature should be 35 °C.

Tanks and facilities for loading and unloading PO should be equipped with spill-retention walls or dikes (remotely drained) to direct spills into containment areas. Such catch basins should suffice for 110 % of the tank capacity, or be large enough to accomodate an adequate volume of deluge water.

The use of insulated tanks should be avoided because experience shows that corrosion underneath the insulation creates the potential for a spill. Other relevant safety considerations (e.g., minimum instrumentation) are described in [122], [123].

Transportation. PO is transported in gravity or gas ships, railcars, tank trucks, and containers. Hazard classifications and transport information are listed in Table 6.

Emergency response organizations, e.g., the "TUIS" (Transport Unfall Informationssystem) in Germany, have been established to provide local authorities with adequate information and help when requested. Special emergency response and transportation equipment data sheets (Tremcards) are available [125].

The European Chemical Industry Council (CEFIC) has approved a project called "International Chemical Environment" (ICE) in which emergency response will be further improved through close cooperation between national chemical federations, fire brigades, local authorities, and national emergency response centers. The goal is to provide professional assistance all over Europe in case of distribution incidents involving chemicals.

In order to achieve maximum conveyance safety the following minimum requirements for transportation equipment are recommended (Dow Chemical).

Sea Barges and Gravity Gas Vessels. For detailed information, see [126], [127]:

1) If a vessel under consideration is in nondedicated service, tanks and associated piping should be effectively cleaned to remove all traces of previous cargos and rust before PO is loaded. All tanks must be entered for final inspection.
2) Documentation must be provided for the three most recent cargos. No materials known to catalyze reactions of PO (e.g., acids, alkalis, amines, or ammonia) are permitted among the previous three cargos.
3) Independent vapor return should be used during loading and unloading.
4) Cargo tanks (on gravity ships, for example) with a design pressure of less than 0.6 bar (gauge) should be equipped with a cooling system to maintain PO far below its boiling point. Use of such a cooling unit is not required in colder regions of the world and on journeys of less than seven days.
5) Piping systems for tanks that are to be loaded with PO should be completely segregated from piping systems for all other tanks (removal of spool pieces and installation of blind flanges).
6) For other important points see the Section above on Storage.

Railcars. Design according to RID (Europe), AAR (USA), and GGVE (Germany). Top-loading/unloading nozzles are preferred, protected by a thick (2.5 cm) heavy-duty dome in the event of a roll over.

Tank Trucks. Design according to ADR (Europe), DOT MC 330/331 (USA), GGVS (Germany). Valves should be located at one side and in front of the rear wheels. Heavy-duty protection is recommended for the bottom line and valves.

Containers. Type 5 vessels (liquefied gas under pressure) are recommended for PO service. Valves should be located at the bottom (both sides), protected by an adequate safety system.

Other important criteria for railcars, tank trucks, and containers are that the maximum allowable working pressure should be at least 5 bar, design pressure at least 10 bar. Recommended minimum wall thickness is 6 mm.

8. Uses

PO is an important basic chemical intermediate. Virtually all the PO produced is converted into derivatives, often for applications similar to those of ethylene oxide (EO) derivatives (→Ethylene Oxide).

PO is used primarily to produce polyether polyols, propylene glycols, and propylene glycol ethers. The distribution of PO applications (derivatives) in the U.S. market is shown below, and is typical for the world market [128]:

Polyether polyols comprising	
Flexible foams	51%
Rigid foams	10%
Nonfoam uses	7%
Propylene glycols	22%
Glycol ethers	4%
Miscellaneous	6%

Most of the PO produced (65–70%) is used as an intermediate for *polyether polyols* that are mainly consumed in the manufacture of polyurethanes. Polyurethanes can be prepared with a wide range of hardness, rigidity, and den-sity characteristics. Flexible polyurethane foams are important in furniture and automobile seating, bedding, and carpet underlay. They are made from polyols with molecular masses >3000. Polyols with lower molecular masses lead to rigid foams for such applications as thermal insulation. Polyether polyols used in polyurethanes are based on di-, tri-, or polyhydric alcohols. They are usually PO/EO-copolymerized, with a PO content of ca. 70 wt% for rigid and 90 wt% for flexible foams.

Homo- or copolymerized polyethers are also used as surface-active agents in detergents, textiles, defoamers, hair-care preparations, brake fluids, and lubricants. Increasing the PO/EO ratio causes a surfactant to become more hydrophobic. Polypropylene glycols with molecular masses in the range 400–4000 are produced by polyaddition of PO and a difunctional alcohol. Polypropylene glycols become less hydrophilic with increasing molecular mass. Non-polyurethane applications include lubricants for rubber; in metal rolling, drawing, and machining; antifoaming agents; in deicing formulations for gasoline; and in hydraulic fluids.

Propylene glycols constitute the second largest application for PO. Monopropylene glycol (MPG) is the direct reaction product of PO with water. Di-, tri-, and higher propylene glycols are coproduced by the reaction of MPG with PO. Propylene glycol is consumed mainly as a raw material for unsaturated polyester resins, especially for the textile and construction industries. It is also used as a humectant, and as a solvent and emollient in food, drugs, and cosmetics. Further uses relate to coatings, plasticizers, heat transfer and hydraulic fluids, antifreezes, and aircraft deicing fluid. Applications of propylene glycol have benefitted from its lower toxicity relative to ethylene glycol. Di- and tripropylene glycols are used in hydraulic and drilling fluids, in cutting oils, and as lubricants. The main advantage of the polyglycols is their biodegradability.

Propylene glycol ethers are formed by the reaction of PO with alcohols, usually methanol, ethanol, propanol, or butanol. Certain traditional ethylene glycol ethers and their acetates are being replaced by their PO-based analogues. The demand for PO-based glycol ethers and acetates is therefore growing rapidly, and now constitutes the third largest market for PO (ca. 4%). Typical applications include solvent use in

Table 7. Comparison of global PO production capacities, consumption, and process shares

PO production	1986	1991	1996[a]
Chlorohydrin process, %	58	55	49
PO/styrene process, %	18	19	24
PO/*tert*-butyl alcohol process, %	24	26	24
Direct oxidation, %	0	0	2
World PO capacity, 10^3 t/a	3300	3900	4900
World PO consumption, 10^3 t/a	2600	3250	3900

[a] Forecast.

coatings, paints, inks, resins, cleaners, waxes, and electronic circuit board lamination. The glycol ethers are also found in heat-transfer fluids and anti-icing agents for jet fuel.

Various *specialty organic compounds* can be derived from PO, including allyl alcohol (glycerol synthesis), propylene carbonate (a specialty solvent for organic and some inorganic compounds), mono-, di-, and triisopropanolamines (detergent raw material), and hydroxypropylated cellulose.

9. Economic Aspects

PO is a raw material for major consumer goods in virtually all industrialized countries, and it occupies third place in the consumption of total propene (ca. 10 %) after polypropylene and acrylonitrile. Because of continuing expansion of its field of application, PO production is currently increasing (see Table 7).

Until 1969 essentially all the PO was produced by the chlorohydrin process. A plant using a version of the peroxidation process (indirect oxidation) was started by Arco in the United States in 1969. Currently ca. 50 % of world PO-production capacity is based on the chlorohydrin process, with the remainder coming from peroxidation (Table 7). Dow Chemical and Arco Chemical are dominant in PO production, representing more than 80 % of world capacity. World production of PO is currently (1992) ca. 3.2×10^6 t/a, with an expected annual growth rate of ca. 4 %.

Table 8 provides an overview of PO production capacities. The technical, economic, and environmental aspects of large-scale PO production are discussed in [5], [29].

10. Toxicology and Occupational Health

Extensive experience has shown that PO can be handled safely provided appropriate precautions are taken. However, because of its toxicity, PO should be handled only by persons familiar with the hazards involved and willing to observe appropriate precautions.

Table 8. World PO capacities in 1991

Country	Producer	Location	Process [a]	Capacity, 10^3 t/a
America				
Brazil	Dow Quimica	Aratu	CH	150
Canada	Dow	Sarnia, Ont	CH-NaOH	60
United States	Arco Chemical	Channelview, TX	styrene	270
	Arco Chemical	Bayport, TX	TBA	550
	Dow Chemical	Freeport, TX	CH-NaOH	500
	Dow Chemical	Plaquemine, LA	CH-NaOH	200
	Olin	Lake Charles, LA	DO	5
Total				1735
Asia				
China	Zhangdian Petrochem.	Zibo/Shandong	CH	10
	Jinling Petrochem. Corp.	Zhongshan	CH	10
	Tianjing Dagu	Tianjing	CH	16
	Gao Qiao	Shanghai	CH	16
	NJ Zhongshan	Guang Dong	CH	16
India	Manali Petrochem.	Baroda, Madras	CH	12
	UB Petroproducts	Manali, Madras	CH	12
Japan	Sumi-Arco	Chiba	styrene	140
	Tokuyama Soda	Tokuyama City	CH	50
	Showa Denko	Kawasaki	CH	30
	Mitsui Toatsu	Nagoya	CH	36
	Asahi Glass Co.	Kashima	CH	100
Korea (South)	Yukong Arco Chemical	Ulsan, Yukong	styrene	100
Taiwan	Chiung Long Petrochem.	Linyuan/Kaohsiung	CH	15
Total				563
Europe				
Bulgaria	Neftochim	Burgas	CH	12
Czechoslovakia	Novacke Chemicke Zavody	Novaky	CH	5
France	Arco Chimie France	Fos-sur-Mer	TBA	200
Germany	Dow Deutschland Inc.	Stade	CH-NaOH	420
	Erdoelchemie	Koeln-Worringen	CH	150
	Bunawerke	Schkopau	CH	50
	BASF	Ludwigshafen	CH	90
Italy	Enichem	Priolo, Sicily	CH	60
Netherlands	Arco Chemie Nederland	Botleck	TBA	250
	Shell Netherland Chemie	Moerdijk	styrene	140
Poland	State	Rokita	CH	20
Romania	Oltchim	Rimnicu Vilcea	CH	12
	State	Midia	CH	60
Spain	Repsol Quimica/Arco	Puertollano	styrene	50
Former Soviet Union	State	Sumgait	CH	25
	State	Nizhnekamsk	styrene	50
Yugoslavia	Sodaso	Tuzia	CH	20
Total				1614
Total worldwide				3912

[a] CH = chlorohydrin PO process (using lime unless stated otherwise); DO = direct oxidation of propene; styrene = styrene/PO process; TBA = *tert*-butyl alcohol/PO process.

In general, the health hazards of handling PO are associated with vapor inhalation and eye and skin contact.

Oral Toxicity. The acute oral toxicity of PO is moderate. LD_{50} values have been reported for rats, mice, and guinea pigs in the range 300–1000 mg/kg [129], [130].

Skin Contact. PO has only a low to moderate potential for absorption through the skin in toxic amounts. The dermal LD_{50} for skin absorption of PO in rabbits and guinea pigs is 1240 mg/kg [130] and 7200 mg/kg [131], respectively. PO is not irritating to the skin if allowed to evaporate freely, but contact for several minutes may result in mild local redness [132]. Aqueous solutions are more irritating than undiluted PO. More sustained contact results in the development of chemical burns, blistering, and swelling [133], [134].

Eye Contact. High vapor concentrations of PO are irritating to the eye. Liquid PO has caused serious local injury to the eyes of laboratory rabbits [135]. Cases of corneal burns have been reported in humans [129]. Permanent impairment of vision, even blindness, may result from such contact. Affected eyes should be immediately flushed with flowing water for at least 15 min.

Inhalation. Exposure to PO at several hundred ppm may result in headache, dizziness, drowsiness, nausea, chest discomfort, and cough [134]. Higher concentrations (>1000 ppm) may lead to edema, irritation to the respiratory tract, and loss of consciousness [136], [137]. One case of human poisoning resulted from vapor exposure to 1500 ppm for 10 min. Recovery from this poisoning incident occurred the following day [138].

Inhalation studies with laboratory animals suggest a "no-adverse-effect level" for all species of 150 ppm [129]. No evidence of neurotoxicity was detected in male rats exposed to PO by inhalation for 24 weeks at a level of 300 ppm [17].

At exposure levels $\geq$4000 ppm PO is considered to have mild central nervous system effects in animals, characterized by ataxia, uncoordination, and general depression [133], [139].

The odor threshold in air is 200 ppm (95% confidence limit: 114–353 ppm) [137], a value too high to prevent accidental overexposure. Nevertheless, the sweet, penetrating odor may be sufficient to signal the danger of acute exposure.

Teratogenicity. PO does not produce teratogenic effects in laboratory animals before or during gestation. In rats, fetotoxicity has been seen only at doses high enough to cause toxic effects to the mother. In rabbits, exposure has not proven substantially toxic either maternally or with respect to the fetus [140].

Mutagenicity. PO directly alkylates proteins and DNA, induces gene mutations in all microorganisms tested, and induces chromosomal aberrations in mammalian cells in vitro [141]–[147]. In vivo, an increased frequency of micronuclei was induced in mice after intraperitoneal PO exposure, but not after oral administration, even at near-lethal doses. No clear dominant lethal effects were found in rodents or monkeys [142], [148]. No chromosomal abnormalities or effects on the sister chromatid exchanges were observed in monkeys exposed for two years [139], [149], [150].

Carcinogenicity. Lifetime inhalation studies on laboratory rats and mice suggest a weak carcinogenic effect of PO in animals at high dose levels [151]–[156], but the evidence for carcinogenity in humans is inconclusive [157]–[159]. Evaluations of toxicity to warm-blooded species are described in [160]. PO has been classified by the U.S. Environmental Protection Agency as a class B-2 substance (probable human carcinogen) [159], and it is listed as a group III carcinogen in the German TA Luft [107].

Permissible Exposure Limits. PO is recognized as a toxic substance by the International Agency for Research on Cancer (IARC), and is classified as a class 2A carcinogen (TLV-TWA = 48 mg/m^3 = 20 ppm) [139]. The German MAK commission has listed PO as a group III A2 carcinogen (TRK = 6 mg/m^3 = 2.5 ppm) (at 20 °C/ 1013 hPa: 1 ppm = 2.42 mg/m^3).

11. References

General References

[1] *Beilstein*, 4th ed., Propylene Oxide.
[2] *Kirk Othmer*, 3rd ed., **19**, p. 246 ff.
[3] *Winnacker-Küchler*, 4th ed., **2**, p. 565 ff.
[4] K. Weissermel, J. J. Arpe: *Industrial Organic Chemistry*, 2nd ed., VCH Verlagsgesellschaft, Weinheim 1992.
[5] Chem SystemsPERP "Process Evaluation Research Planning," Propylene Oxide, Tarrytown, New York, Nov. 1990.
[6] SRI 1985PEP "Process Economics Program" EO/PO Edition 2D. Specific References

Specific References

[7] F. L. ötting: "Low Temperature Heat Capacity and Related Thermodynamic Functions of Propylene Oxide," *Journal Chem. Phys.* **41** (1964) 149–153.
[8] R. A. McDonald, S. A. Shrader, D. R. Stull: "Vapor Pressures and Freezing Points of 30 Organics," *J. Chem. Eng. Data* **4** (1959) 311–323.
[9] K. H. Simmrock et al.: "Critical Data of Pure Substances," *DECHEMA Chemistry Data Series*, vol. 2, Frankfurt 1986.
[10] C. L. Yaws, M. P. Rackley, *Chem. Eng. (N.Y.)* **83** (1976) 129.
[11] Dow Chemical Co., *Alkylene Oxides*, Product Bulletin, Form no. 110-551-77R, Midland, Mich. 1977.

[12] K. A. Kobe, A. E. Ravicz, S. P. Vohra, *J. Chem. Eng. Data*, **1** (1956) 50–56.

[13] R. W. Gallant: "Physical Properties of Hydrocarbons," *Hydrocarbon Process.* **46** (1967) 143–150.

[14] Dow Chemical Europe, Material Safety Data Sheet, Propylene Oxide, Horgen, Switzerland, 1987.

[15] B. G. Witt, M. W. Chase, *Experimental and Calculated Flammability Limits of 30 Chemicals*, NCT-577, 1975.

[16] U.S. Coast Guard, Department of Transportation, *CHRIS—Hazardous Chemical Data*, vol. **II**, Washington D.C., U.S. Government Printing Office, 1984–1985.

[17] The Dow Chemical Company, unpublished data.

[18] P. H. Howard et. al.: *Handbook of Environmental Fate & Exposure Data for Organic Chem.*, vol. **I**, Lewis Publishers 1989, pp. 483–489.

[19] *Ullmann*, 4th ed., **19**, 471–481.

[20] J. N. Wickert, W. S. Tamplin, R. L. Shank, *Chem. Eng. Prog. Symp. Ser.* **48** (1952) no. 2, 92.

[21] G. O. Curme, F. Johnston: *Glycols*, ACS Monograph No. 114, Chap. 11, Reinhold Publishing Corp., New York 1952.

[22] J. Gmehling et al., "Vapor-Liquid Equilibrium Data Collection," *DECHEMA Chemistry Data Series*, vol. **1**, Frankfurt 1977–1984.

[23] L. H. Horsley et al.: "Azeotropic Data,"*Adv. Chem. Ser.* **6** (1952); **35** (1962).

[24] Abderhalden, Eichwald, *Ber. Dtsch. Chem. Ges.* **51** (1918) 1312.

[25] Schurig et al., *Angew. Chem.* **90** (1978) 993–995.

[26] M. C. Flowers: "Kinetics of the Thermal Gas-Phase Decomposition of 1,2-Epoxypropane, "*J. Chem. Soc. Faraday Trans. 1* **73** (1977) 1927–1935.

[27] M. Kwasny, M. Syczewski, *Przem. Chem.* **67** (1988) no. 5, 228–231.

[28] K. H. Simmrock, *Chem. Ing. Tech.* **48** (1976) no. 12, 1085–1096.

[29] K. H. Simmrock, *Hydrocarbon Process.* **57** (1978) no. 11, 105–113.

[30] B. Oser, *Bull. Soc. Chim. Fr.* 1860, 235.

[31] R. Elm: "Die Kinetik der Reaktion zwischen gasförmigen Propylen und wässerigen Chlorlösungen," Dissertation, Universität Dortmund 1977.

[32] Shell, BE 630 446, 1963.

[33] Bayer, FR 1 357 443, 1964.

[34] Naphthachimie, DE-OS 2 101 119, 1971 (R. Bouchet).

[35] BASF, DE-OS 2 022 819, 1971 (E. Bartholomé).

[36] G. Mikula, CS 152 702, 1974; *Chem. Abstr.* **81** (1974) 63 135 r.

[37] Petrocarbon Developments, FR 2 194 673, 1974.

[38] Mitsui Toatsu Chemicals, JP-Kokai 77 48 606, 1977 (Y. Watanabe et al.).

[39] S. Carra, M. Morbidelli, *Chem. Eng. Sci.* **34** (1979) 1123–1232.

[40] G. N. Kirby, *Chem. Eng.* (1985) Febr. 4, 81–83.

[41] D. H. DeClerk, A. J. Patarcity, *Chem. Eng.* (1986) Nov. 24, 46–63.

[42] H. Busse, H. Schindler, *Chem. Ing. Tech.* (1990) no. 4, 271–277.

[43] Lummus, US 4 496 752, 1985 (A. B. Gelbein et al.).

[44] Lummus, DE 3 016 668, 1980 (A. B. Gelbein et al.).

[45] Lummus, US 4 008 133, 1977 (A. B. Gelbein et al.).

[46] Institut Français du Petrole, DT 1 815 998, 1968 (H. Mimoun et al.).

[47] Institut Français du Petrole, FR 2 114 752, 1970.

[48] Sumitomo GB 1 302 441, 1969 (J. Watanabe et al.); DT 2 012 049, 1969 (J. Watanabe et al.).

[49] Naphtachimie US 3 778 451, 1966 (M. Poite); DE 1 668 500, 1966 (M. Poite).

[50] Shell, DE-OS 1 940 205, 1969 (H. Fernholz).
[51] Ugine Kuhlmann, DE-OS 2 446 830, 1973 (J. P. Schirmann et al.).
[52] UCC, DE-OS 2 607 768, 1975 (Ch. McMullen).
[53] Union Oil, US 3 293 269, 1964 (L. G. Wolgemuth).
[54] *J. Catal.* **129** (1991) 159–167.
[55] Institut Français du Petrole, FR 1 549 184, 1964 (H. Mimoun et al.); DT 1 668 312, 1964 (H. Mimoun et al.).
[56] Laporte Chem., FR 1 519 147, 1966 (E. C. Kurby et al.); DT 1 618 625, 1966 (E. C. Kurby et al.).
[57] Knapsack, GB 1 076 288, 1946 (K. Sennewald et al.); DT 1 240 515, 1964 (K. Sennewald et al.).
[58] Celanese, US 3 341 556, 1959 (A. L. Stauzenberger et al.).
[59] Petrocarbon Development, DT 1 917 031, 1968 (J. V. Fletscher et al.).
[60] Petrocarbon Development, DT 1 925 378, 1968 (J. V. Fletscher et al.).
[61] Petrocarbon Development, GB 1 318 524, 1969 (J. V. Fletscher et al.); DT 1 066 920, 1969 (J. V. Fletscher et al.).
[62] Daicel, DT 1 923 392, 1968 (K. Jamagishi et al.).
[63] Daicel, US 3 663 574, 1968 (K. Jamagishi et al.).
[64] Daicel, US 3 838 020, 1971 (O. Kageyama et al.); DT 2 258 521, 1971 (O. Kageyama et al.).
[65] Asahi Chem. JP 49 43 926, 1969 (S. Imanura, Y. Wakasa, K. Kotaoka).
[66] Metallgesellschaft, DE-OS 2 334 982, 1973 (Th.Simo).
[67] Interox, DE-OS 2 502 776, 1977 (A. Hildonk, P. Greenhalgh).
[68] Bayer, DE-OS 2 519 297, 1975 (G. Prescher et al.).
[69] Olin, NL 7 601 976, 1975.
[70] E. J. Mistrik, *Chem. Tech. (Leipzig)* **27**, no. 1, 23 (1975).
[71] Sun Oil, GB 1 261 617, 1968.
[72] Sun Oil, BE 729 193, 1968 (K. M. Brownstein et al.); DE 1 910 258, 1968 (K. M. Brownstein et al.).
[73] Atlantic Richfield, FR 1 500 728, 1966 (H. R. Grane et al.); DT 2 159 605, 1971 (H. R. Grane et al.).
[74] Atlantic Richfield, GB 1 345 900, 1971; DT 2 159 605, 1971.
[75] Atlantic Richfield, US 3 907 902, 1966; DT 1 643 586, 1966.
[76] Halcon Int., NL 6 515 037, 1964.
[77] Halcon Int., BE 702 236, 1966 (J. P. Schmidt et al.); DT 1 668 204, 1966 (J. P. Schmidt et al.).
[78] Halcon Int., GB 1 308 340, 1969 (St. Herzog); DT 2 017 068, 1969 (St. Herzog).
[79] Akzo, DE-OS 2 159 764, 1971 (A. Wegerhoff et al.).
[80] Mitsui Petrochem., US 3 592 857, 1967 (Y. Shinohara); DE 1 668 480, 1967 (Y. Shinohara).
[81] Atlantic Richfield, GB 1 146 202, 1965 (M. N. Sheng et al.); DT 1 568 002, 1965 (M. N. Sheng et al.).
[82] Atlantic Richfield, GB 1 298 253, 1969.
[83] Atlantic Richfield, GB 1 339 296, 1969.
[84] Halcon Int., BE 665 240, 1964; DT 1 518 996, 1 543 002, 1964.
[85] Halcon Int., GB 1 074 330, 1962 (J. Kollar); DT 1 468 012, 1962 (J. Kollar).
[86] Halcon Int., GB 1 122 702, 1965.
[87] Halcon Int., US 3 360 584, 1965 (J. Kollar); DT 1 568 764, 1965 (J. Kollar).
[88] Halcon Int., US 3 360 585, 1965 (Ch. N. Winnick); DT 1 568 763, 1965 (Ch. N. Winnick).
[89] Halcon Int., US 3 449 219, 1966 (J. P. Schmidt).
[90] Halcon Int., US 3 458 534, 1964 (Ch. N. Winnick et al.); DT 1 543 029, 1964 (Ch. N. Winnick).
[91] Halcon Int., US 3 459 810, 1965 (Ch. Y. Choo et al.); DT 1 568 808, 1965 (Ch. Y. Choo et al.).

[92] Halcon Int., GB 1 218 560, 1966 (Th. W. Stein et al.); DT 1 668 234, 1966 (Th. W. Stein et al.).
[93] Institut Français du Petrole, FR 1 506 286, 1966 (I. Seree de Roch et al.); DT 1 618 532, 1966 (I. Seree de Roch et al.).
[94] Shell Int., GB 1 364 367, 1972 (F. Wattimena et al.); DE 2 334 315, 1972 (F. Wattimena et al.).
[95] Naphthachimie, FR 1 550 166, 1967 (M. Poite).
[96] Rhone-Poulenc, GB 1 156 531, 1965 (J. C. Brunie et al.); DT 1 593 306, 1965 (J. C. Brunie et al.).
[97] Mitsui Petrochem, JA 45-35 041, 1965.
[98] Sumitomo Chem., GB 1 141 725, 1966 (I. T. Yamachara et al.); DT 1 618 907, 1966 (I. T. Yamachara et al.).
[99] R. W. Serebryakov et al., SU 362 828, 1970.
[100] Kh. E. Khcheyan et al., SU 381 667, 1971.
[101] Kh. E. Khcheyan et al., SU 387 985, 1971.
[102] Halcon Int., DE 2 617 432, 1979 (J. P. Schmidt).
[103] Halcon Int., DE 1 939 791, 1979 (M. Becker et al.).
[104] Shell Oil, US 3 829 392, 1974 (H. P. Wulff).
[105] Olin Corp., WO 9 015 053, 1990 (J. L. Meyer et al.).
[106] Bundes-Immissionsschutzgesetz (BImSchG) 1991.
[107] Technische Anleitung zur Reinhaltung der Luft (TA Luft) 1986.
[108] IARC, Allyl Compounds, Aldehydes, Epoxides, and Peroxides 36:227, 1985.
[109] D. A. Bogyo et al., Investigation of Selected Potential Environmental Contaminants: Epoxides USEPA-560/11-80-005, 1980, p. 201.
[110] GEMS, Graphical Exposure Modeling System. Fate of Atmospheric Pollutants (FAP) Data Base. Office of Toxic Substances, USEPA, 1986.
[111] L. Cupitt, Fate of Toxic and Hazardous Materials in the Air Evironment, USEPA-600/S3-80-084, 1980, p. 7.
[112] Summary of Environmental Response Evaluation of Propylene Oxide. NTIS/OTS, 0509917 Doc, Dow, 1978.
[113] R. C. Crews, Effects of PO on Selected Species of Fishes, AD/A-003-637, 1974, in SPLAT (00 033 21A).
[114] J. W. Deneer et al., *Quant. SAR for Acute Tox. of some Epoxy Compounds to the Guppy*, aq. tox. **13** (3), 195–204, 1988.
[115] K. Verschueren, *Handbook of Environmental Data on Organic Chemicals*, 2nd ed., van Nostrand, Reinhold 1983.
[116] A. L. Bridie et al.: "Acute Toxicity of some Petrochemicals to Goldfish," *Water Res.* **13** (1974) 623–626.
[117] Beratergremium für umweltrelevante Altstoffe: "1,2-Epoxypropan," *BUA-Stoffbericht*, Umweltbundesamt UBA, Berlin, in print.
[118] P. H. Howard (ed.): *Handbook of Environmental Fate & Exposure Data for Organic Chem.*, vol. **I** Lewis Publishers, 1989, pp. 483–489.
[119] *A Guide to the Distribution Code of Management Practices*, Chemical Manufacturers' Association, Washington, D.C., 1991.
[120] *Distribution Code Risk Management Implementation Aid*, Chemical Manufacturers' Association, Washington D.C., 1991.
[121] Guideline No. 30, *The Distribution Safety Review Process*, 1st ed., Dow Europe, 1990 (only for internal use).

[122] *Safe Handling and Storage of Propylene Oxide,* Dow Chemical Company, 1988.
[123] *PO Product Stewardship, a Guide for Safe Handling,* Dow Europe, 1987.
[124] Hommel: *Handbuch der gefährlichen Güter* Springer Verlag, Heidelberg 1987, p. 172.
[125] *Emergency Response and Transportation Equipment Data Sheet for PO,* Dow Chemical 1989.
[126] IMO, *International Code for the Construction and Equipment of Ships Carrying Dangerous Chemicals in Bulk,* 1983.
[127] Guidelines for the Marine Chartering and Handling of Alkylene Oxides, Dow Chemical Company, April 1988.
[128] A. R. Kavaler, *Chem. Mark. Rep.* **237** (1990) no. 2, 42.
[129] Smyth et al., *Am. Ind. Hyg. Assoc. J.* **30** (1969) no. 5, 470–476.
[130] V. K. Rowe et al., *AMA Arch. Ind. Health* **13** (1956) 228–236.
[131] Smyth et al., *J. Ind. Hyg. Toxicol.* **30** (1948) 63–68.
[132] C. H. Hine et al.: *Industrial Hygiene and Toxicology,* 2nd revised ed. vol. **II**, Interscience, New York 1958, p. 1642.
[133] G. D. Clayton, F. E. Clayton (eds.): *Patty's Industrial Hygiene and Toxicology,* 3rd ed.,vol. **2 A, 2 B, 2 C,** Toxicology, J. Wiley & Sons, New York 1981, p. 2187.
[134] *Encyclopedia of Occupational Health and Safety,* vol. **I** & **II**, Geneva, Switzerland, International Labor Office, 1983.
[135] W. M. Grant: *Toxicology of the Eye,* 3rd ed., Charles C. Thomas Publisher, Springfield, IL, 1986.
[136] N. H. Proctor, J. P. Hughes: *Chemical Hazard of the Workplace,* J. B. Lippincott Company, Philadelphia, PA, 1978, p. 430.
[137] K. H. Jacobsen et al., "The Toxicity of Inhaled Ethylene Oxide and Propylene Oxide Vapors," *Arch. Ind. Health* **13** (1956) 237–244.
[138] R. E. Gosselin et al.: *Clinical Toxicology of Commercial Products,* 5th ed., Williams and Wilkins, Baltimore 1984, p. II–97.
[139] *Documentation of the Threshold Limit Values and Biological Exposure Indices,* American Conference of Governmental Industrial Hygienists, Cincinnati, Ohio, 1989.
[140] Hackett, et al., Report to NIOSH, Battelle Pacific Northwest, 1982.
[141] R. W. Pero et al., *Cell. Boil. & Tox.* **1** (1985) no. 4, 309–314.
[142] D. R. Wade et al., *Mutat. Res.* **58** (1978) 217–223.
[143] J. Bootmann et al., *Mutat. Res.* **67** (1979) no. 2, 101–112.
[144] R. E. McMahon, *Cancer Res.* **39** (1979) 682–693.
[145] E. H. Pfeiffer, *Food Cosmet. Toxicol.* **18** (1980) no. 2, 115–118.
[146] J. C. Cline et al., *9th Annual Meeting of Environmental Mutagen Society,* San Francisco, March 9–13, 1978.
[147] K. Hemminki et al., *Chem. Biol. Interact.* **30** (1980) 259.
[148] FAAT 10, *2-Generation Repro. Study in Rats,* 1988, pp. 82–88.
[149] Environmental Health Research and Testing Inc., 1981, published in *Toxicol. Appl. Pharmacol.* **76** (1984) 85–95.
[150] Environmental Health Research and Testing Inc.: *Toxic and Mutagenic Effects of PO and EO on the Spermatogenic Functions in Cynomolgus Monkeys,* 1982.
[151] A. R. Sellakuman, *J. Natl. Cancer Inst.* **79** (1987) no. 2, 285–289.
[152] C. F. Kupper et al., *Food Chemical Toxicol.* **26** (1988) 159–167.
[153] NTP Bioassay, Technical Report 267, 1983.
[154] D. Lynch et al., *Toxicol. Appl. Pharmacol.* **76** (1984) 69–84.
[155] A. L. Walpole, *Ann. N.Y. Acad. Sci.* **68** (1958) 750.

[156] H. Dunkelberg, *Br. J. Cancer* **39** (1979) 588.

[157] International Agency for Research on Cancer: "Propylene Oxide (Group 2a)," in *IARC Monographs on the Evaluation of Carcinogenic Risks to Humans, Overall Evaluations of Carcinogenicity, An Updating of IARC Monographs*, vol. **1** to **42**, Supplement 7, 1987, pp. 328–329.

[158] Propylene Oxide, Environmental Health Crit. 56, 1–53, World Health Organization, Geneva 1985.

[159] *IARC Monogr. Eval. Carcinog. Risk Chem. Man*, (1985) p. V26 238.

[160] Deutsche Forschungsgemeinschaft: *1,2-Epoxypropan, Toxikologisch-arbeitsmedizinische Begründung von MAK-Werten*, VCH Verlagsgesellschaft, Weinheim 1989.

Purine Derivatives

Hans H. Lenz, Knoll AG Ludwigshafen, Federal Republic of Germany

1. Introduction 4165
2. Properties 4166
3. Occurrence and Production .. 4167
4. Quality Specifications and Analysis 4170
5. Uses 4170
6. Economic Aspects 4172
7. References 4172

1. Introduction

Purine [120-73-0], 7H-imidazo[4,5-d]pyrimidine, M_r 120.12, is a bicyclic molecule composed of the fused heterocycles imidazole and pyrimidine.

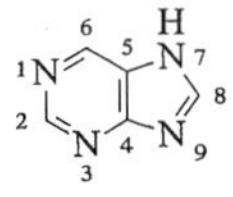

The name purine (from "purum uricum") as well as the ring-numbering technique still practiced today can be attributed to Emil Fischer. Oxygen-containing purine derivatives are often represented in the isomeric lactam form, as with xanthine (**1**):

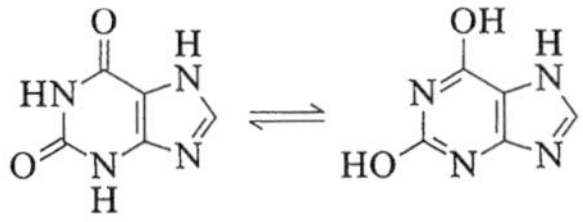

Infrared spectra reveal that oxygen-containing purine derivatives (including hypoxanthine and uric acid) exist mainly as lactams whereas free amine groups can be detected with aminopurines (e.g., adenine) [1]. Purine derivatives with an N-methylated pyrimidine ring (e.g., from compounds like xanthine) are also based on the lactam form.

Purines were among the first natural products identified. Thus, in 1776 Scheele extracted a substance from urinary calculi which Fourcroy named uric acid (*acide ourique*) in 1793. In 1834, Liebig and Mitscherlich established its correct empirical formula as $C_5H_4N_4O_3$. Runge isolated

Table 1. Melting points of representative purines

Common name	Systematic name	CAS registry no.	M_r	mp, °C
Purine	7H-imidazo[4,5d]pyrimidine	[120-73-0]	120.1	217
Adenine	6-aminopurine	[73-24-5]	135.1	360–365 (decomp.)
Guanine	2-amino-6-hydroxypurine	[73-40-5]	151.1	365 (decomp.)
Hypoxanthine	6-hydroxypurine	[68-94-0]	136.1	
Mercaptopurine	6-mercaptopurine	[50-44-2]	152.2	313–314 (decomp.)
Thioguanine	2-amino-6-mercaptopurine	[154-42-7]	167.2	>360
Xanthine	2,6-dihydroxypurine	[69-89-6]	152.1	
3-Methylxanthine		[1076-22-8]	165.1	380 (decomp.)
Theophylline	1,3-dimethylxanthine	[58-55-9]	179.2	269–272
monohydrate		[5967-84-0]	197.2	
Theobromine	3,7-dimethylxanthine	[83-67-0]	179.2	351
Caffeine	1,3,7-trimethylxanthine	[58-08-2]	193.2	234–235
monohydrate		[5743-12-4]	211.2	
Uric acid	2,6,8-trihydroxypurine	[69-93-2]	168.1	

caffeine from coffee beans in 1820, and in 1844 UNGER isolated guanine from guano. These and other studies led MEDICUS in 1875 to suggest a bicyclic structure for uric acid and its derivatives. Thorough investigations by EMIL FISCHER (1882–1897) clarified the relationships linking these compounds, and in 1884 the current name was assigned to the series of substances and the Medicus structure was confirmed. At about the same time A. KOSSEL isolated guanine (1883) and adenine (1885) from nucleic acids. Finally, around the turn of the century, this initial research era was brought to a close with a series of total syntheses by W. TRAUBE [2]. Extensive work was later resumed by BREDERECK [3].

2. Properties

Physical Properties. Purine and its C-alkyl derivatives are very soluble in water. Purines containing hydroxyl groups show only poor solubility in water and organic solvents, whereas aminopurines are fairly soluble. With the exception of uric acid, the compounds are amphoteric, yielding water-soluble salts with both acids and bases. The acidic character of purine derivatives increases with the number of hydroxyl groups in the molecule. The dissociation constant of uric acid (pK_a 5.7) is almost comparable to that of aliphatic carboxylic acids; aminopurines, on the other hand, are more basic. The melting points of purines are often very high, and they are of limited use for characterization. Melting points of several purines are listed in Table 1.

Chemical Properties. Purine derivatives undergo nucleophilic substitution at the 2-, 6-, and 8-positions due to delocalization of the pyrimidine and imidazole π-electrons. Appropriately activated derivatives are also subject to electrophilic substitution at position 8. Derivatives free of substituents in the 7-position readily yield salts under alkaline conditions (e.g., sodium theophylline). These salts remain very stable to hydrolytic ring cleavage. On the other hand, the pyrimidine ring in 1,3,7-peralkylated xanthines (e.g., caffeine) is highly susceptible to basic cleavage. For more information see [4] – [6].

3. Occurrence and Production

Purine derivatives are widespread in nature. Adenine and guanine, as components of nucleic acids and nucleosides, are fundamental building blocks of life. Purine itself is a component of the *N*-nucleoside nebularine [*550-33-4*] [7], which displays antibiotic properties. *N*-Alkylpurinediones (*N*-alkylxanthines) are also abundant in nature; the most common is caffeine, which occurs in concentrations of 1 – 1.5 % in coffee beans and up to 5 % in black tea leaves, as well as in a number of other plants.Caffeine is isolated by solvent extraction from green coffee beans and tea leaves. The Zosel high-pressure extraction process is a more sophisticated procedure involving supercritical carbon dioxide [8], resulting in mild, selective, and toxicologically safe decaffeination. Another purine, theobromine, is found in cacoa hulls at concentrations of 1.5 – 3 %.

Purine biosynthesis is a multistep enzyme-catalyzed process based on simple building blocks. 5′-Phosphoribosylglycinamide (**2**) is first produced from α-D-ribose and glycine. Subsequent steps lead to the formation of 5′-phosphoribosyl-5-aminoimidazole (**3**), onto which the pyrimidine ring is added, again in a multistep sequence.

The product inosinic acid (**4**) [*131-99-7*] is a starting material for the synthesis of other purine compounds [9].

The final product of human purine metabolism is uric acid (**5**). Thus, adenine (**6**) is converted to xanthine (**1**) via hypoxanthine (**7**).

Guanine (**8**) is converted to **1** directly. Xanthine oxidase then catalyzes oxidation of xanthine to uric acid.

Physical disorders related to purine metabolism include xanthinuria, an inherited metabolic disease caused by the absence of xanthine oxidase, and gout, which leads to deposits in the joints as a result of excessive uric acid concentration in the blood.

Other organisms are capable of degrading uric acid further to allantoin (most mammals and reptiles), allantoic acid, and ultimately urea (e.g., fish).

A number of synthetic pathways for purines have been described with such starting materials as hydrocyanic acid or formamide [10], but only moderate yields have been achieved. A prebiotic synthesis based on hydrocyanic acid has also been considered (e.g., for adenine, which is formally a pentamer of HCN) [11]. Industrial processes begin with simple starting materials. A stepwise approach is used to construct either the pyrimidine or imidazole ring, onto which is subsequently fused the second heterocycle. A synthetic approach originally developed by Traube [2] and later subjected to several improvements is well-adapted to the production of the commercially important xanthine derivatives caffeine, theophylline, and theobromine. A series of steps starting with N,N'-dimethylurea [12], [13] and cyanoacetic acid leads to theophylline (**9**), which can be methylated to give caffeine (**10**):

The same reaction path starting with monomethylurea [12] yields 3-methylxanthine (**11**), which undergoes preferential methylation to theobromine (**12**).

Other known routes to xanthine derivatives are not utilized commercially, either because the starting materials are expensive or because overall yields are less satisfactory (e.g., syntheses starting from imidazole derivatives, or the direct cyclization of 1,3-dimethyl-6-methylamino-5-nitrosouracil [14]).

Guanine and its derivatives can also be prepared from suitably substituted pyrimidine derivatives [15]. Guanine, adenine, xanthine, and hypoxanthine can be made from appropriate imidazole derivatives [16]:

Purine derivatives are also accessible from arylazomalonic acid derivatives [17]:

4. Quality Specifications and Analysis

Since purine derivatives find their most important use as pharmacological agents, quality standards and analytical practices are subject to the regulations of current pharmacopoeia (USP, BP, Ph. Eur., DAB, JP, etc.). The food industry also generally employs materials that comply with the specifications of the pharmacopoeia (e.g., caffeine for soft drinks). Purine derivatives employed as pharmacological agents must be prepared according to current good manufacturing practice (GMP) specifications. Product control must always comply with relevant national laws and international agreements (e.g., the Pharmaceutical Inspection Convention). Purity control by thin layer chromatography, customary in the past, is increasingly being replaced by high-pressure liquid chromatography.

5. Uses

Due to their broad spectrum of biological activity, many purine derivatives find application in medicine. Theophylline was previously utilized as a diuretic, but it no longer plays a role in this type of therapy. The most important current use for theophylline and certain of its salts (e.g., theophylline ethylenediamine [*317-34-0*], theophylline cholinate [*4499-40-5*]), and such derivatives as diprophylline (**13**) [*479-18-5*] [18], proxyphylline (**14**) [*603-00-9*] [19], etofylline (**15**) [*519-37-9*] [18], or bamifylline (**16**) [*2016-63-9*] [20] involves their bronchodilating antiasthmatic properties. Recent pharmaceutical developments have resulted in a trend toward increasing use of theophylline itself rather than its derivatives.

13: $R^1 = CH_2CHCH_2OH$, $R^2 = H$
       $\quad\quad\quad\;\;|$
       $\quad\quad\quad\;OH$

14: $R^1 = CH_2CHCH_3$, $R^2 = H$
       $\quad\quad\quad\;\;|$
       $\quad\quad\quad\;OH$

15: $R^1 = CH_2CH_2OH$, $R^2 = H$

16: $R^1 = CH_2CH_2N(C_2H_5)(CH_2CH_2OH)$

17: $R^1 = CH_2CH_2CH_2NHCH_2CH(OH)-C_6H_3(OH)_2$; $R^2 = CH_2-C_6H_5$; (alternatively $R^2 = H$)

19: $R^1 = CH_2CH_2NHCH(CH_3)CH_2-C_6H_5$, $R^2 = H$

20: $R^1 = CH_2CH_2NHCH(CH_3)CH(OH)-C_6H_5$, $R^2 = H$

Another theophylline derivative, reproterol (**17**) [*54063-54-6*] [21], in which the xanthine skeleton is combined with a sympathomimetically active structure, shows only sympathomimetic action. Pentoxifylline (**18**) [*6493-05-6*], a theobromine derivative, is employed as a vasodilator [22].

18: $CH_3C(O)(CH_2)_4$- at N, with caffeine-like xanthine core (1,3,7-trimethyl positions modified)

Fenetylline (**19**) [*3736-08-1*] is used as a CNS stimulant [23], [24]; cafedrine (**20**) [*58166-83-9*] as an analeptic agent [24], [25]. Dimenhydrinate [*523-87-5*], the salt of 8-chlorotheophylline [*85-18-7*] [26] and diphenhydramine [*58-73-1*] [26], is a commonly employed antiemetic.

A number of guanine and adenine derivatives are employed as chemotherapeutic agents in treatment of the herpes virus, including acyclovir [*59277-89-3*] [27] and vidarabin [*24356-66-9*] [28]. 6-Mercaptopurine [29] and 6-thioguanine [30] are applied in the treatment of leukemia.

Caffeine is the purine derivative produced industrially in the greatest quantity. Caffeine stimulates the central nervous system, causing increases in cardiac activity, metabolism, respiration, and blood pressure. It is frequently utilized in combination with analgesics (e.g., acetylsalicylic acid), thereby enhancing the effect of the analgesic [31].

The cosmetic industry exploits caffeine's ability to increase the flow of blood (e.g., in skin cosmetics). Another industrial use relates to the manufacture of copying papers (diazo paper). More important economically is caffeine's use in the soft drink industry. Cola beverages typically contain the compound at a level of about 100 mg/L. An extensive discussion and evaluation of caffeine metabolism and of potential health hazards associated with its use in beverages and pharmaceuticals is given in [32].

6. Economic Aspects

The worldwide demand for caffeine is ca. 10 000 t/a. Approximately one-third of the total demand is covered by naturally occuring caffeine, the rest by synthetic material. The major share of synthetic caffeine is produced in Germany, with smaller amounts in the United States, Mexico, and the People's Republic of China.

The second most widely distributed purine derivative, theophylline, is entirely synthetic. By far the major portion of the annual demand (ca. 2500 t) is supplied by Germany. The major suppliers of caffeine and theophylline are Knoll and Boehringer Ingelheim.

Theobromine is used mainly in the production of pentoxifylline. The major demand is met by synthetic material and a small part by extraction of cacoa hulls.

7. References

[1] C. L. Angell, *J. Chem. Soc.* 1961, 504.
[2] W. Traube, *Ber. Dtsch. Chem. Ges.* **33** (1900) 3035.
[3] H. Bredereck, H. G. v. Schuh, A. Martini, *Chem. Ber.* **83** (1950) 201. H. Bredereck, F. Effenberger, H. G. österlin, *Chem. Ber.* **100** (1967) 2280.
[4] J. H. Lister in D. J. Brown (ed.): *Heterocyclic Chemistry, Fused Pyrimidines*, Part II, Wiley-Interscience, New York 1971.
[5] G. Shaw in S. Coffey (ed.): *Rodd's Chemistry of Carbon Compounds*, 2nd ed., vol. **4**, part L, Elsevier, Amsterdam 1980, pp. 1 – 100.
[6] A. R. Katritzky, C. W. Rees in K. T. Potts (ed.): *Comprehensive Heterocyclic Chemistry*, vol. **5**, part 4 A, Pergamon Press, London 1984, pp. 499 – 605.
[7] N. M. Löfgren, B. Lüning, *Acta Chem. Scand* **7** (1953) 15, 225; **8** (1954) 670. K. Isono, S. Suzuki, *J. Antibiot. Ser. A* **13** (1960) 270.
[8] Studiengesellschaft Kohle, DE 2 005 293, 1970 (K. Zosel). HAG, DE 2 119 678, 1971 (W. Roselius). HAG, DE 2 127 642, 1971 (O. Vitzhum).Studiengesellschaft Kohle, AT 4 4003-71, 1971 (K. Zosel).
[9] J. H. Lister, *Rev. Pure Appl. Chem.* **13** (1963) 30.
[10] Ethyl Corporation, US 4 511 716, 1985 (K. H. Shin).
[11] J. Oro, A. P. Kimball, *Arch. Biochem. Biophys.* **94** (1961) 217. A. W. Schwartz, C. G. Bakker, *Science (Washington D.C.)* **245** (1989) 1102.

[12] Knoll AG, DE 896 640, 1942 (K. Kraft, L. Suranyi).
[13] BASF, GB 750 549, 1956.
[14] H. Goldner, G. Dietz, E. Carstens, *Justus Liebigs Ann. Chem.* **691** (1965) 142.
[15] Merck, DE 162 336, 1904. Hüls Troisdorf, DE 3 729 471, 1989 (M. Feld, H. a. d. Fünten, W. Voigt). C. Parkanyi, H. L. Yuan, *J. Heterocycl. Chem.* **27** (1990) 1409.
[16] Sagami Chemical Research Center, DE 2 166 506, 1971 (N. Asai).
[17] Kohin Co, EP 45 503, 1981 (J. Suzuki). Merck & Co., US 4 092 314, 1978 (M. L. Vander Zwan, D. F. Reinhold).
[18] H. J. Roth, *Arch. Pharm. (Weinheim Ger.)* **292** (1959) 234.
[19] Ganes Chem. Works, US 2 715 125, 1955 (R. V. Rice).
[20] A. Christiaens, BE 602 888, 1961 (R. de Ridder).
[21] K. H. Klingler, *Arzneim. Forsch.* **27** (1977) 4.
[22] Chemische Werke Albert, DE 1 235 320, 1964 (W. Mohler, M. Reiser, K. Popendiker); DE 2 234 202, 1972 (A. Söder); 2 302 772, 1973 (G. Nesemann, A. Söder, H. Thurm); 2 330 741, 1973 (A. Söder).
[23] Degussa, DE 1 123 329, 1958 (E. Kohlstaedt, K. H. Klingler).
[24] Degussa, US 3 029 239, 1962 (E. Kohlstaedt, K. H. Klingler).
[25] Degussa, DE 1 095 285, 1956 (E. Kohlstaedt, K. H. Klingler).
[26] Searle, US 2 499 058, 1950; 2 534 813, 1950 (J. W. Cusic).
[27] H. J. Schaeffer et al., *Nature (London)* **272** (1978) 583. Wellcome Foundation, BE 833 006, 1975.
[28] W. W. Lee et al., *J. Am. Chem. Soc.* **82** (1960) 2648. E. J. Reist et al., *J. Org. Chem.* **27** (1962) 3274; **29** (1964) 3725.
[29] G. B. Elion, E. Burgi, G. H. Hitchings, *J. Am. Chem. Soc.* **74** (1952) 411.
[30] G. B. Elion, G. H. Hitchings, *J. Am. Chem. Soc.* **77** (1955) 1676.
[31] W. T. Beaver: "Combination Analgesics," *Am. J. Med.* **77** (1984) no. 3 A, 38–53. E. M. Laska et al.: "Caffeine as an Analgesic Adjuvant," *JAMA J. Am. Med. Assoc.* **251** (1984) 13.
[32] P. B. Dews: *Caffeine*: Perspectives from Recent Research, Springer Verlag, Berlin 1984.

Pyridine and Pyridine Derivatives

SHINKICHI SHIMIZU, Koei Chemical Co. Ltd., Osaka 541, Japan
NANAO WATANABE, Koei Chemical Co. Ltd., Osaka 541, Japan
TOSHIAKI KATAOKA, Koei Chemical Co. Ltd., Osaka 541, Japan
TAKAYUKI SHOJI, Koei Chemical Co. Ltd., Osaka 541, Japan
NOBUYUKI ABE, Koei Chemical Co. Ltd., Osaka 541, Japan
SINJI MORISHITA, Koei Chemical Co. Ltd., Osaka 541, Japan
HISAO ICHIMURA, Koei Chemical Co. Ltd., Osaka 541, Japan

1.	Introduction	4176	3.2.	Bipyridines ... 4192
2.	Pyridine and Alkylpyridines	4176	3.3.	Quaternary Pyridinium Salts ... 4192
2.1.	Properties	4177	3.4.	Pyridine N-Oxides ... 4195
2.2.	Production	4179	3.5.	Piperidines ... 4196
2.2.1.	Separation from Tar	4179	3.6.	Halopyridines ... 4198
2.2.2.	Synthesis from Aldehydes or Ketones with Ammonia	4180	3.7.	Pyridinecarbonitriles, Carboxylic Acids, and Carboxamides ... 4202
2.2.3.	Synthesis from Acrylonitrile and Ketones	4184	3.8.	Aminopyridines ... 4206
2.2.4.	Synthesis from Dinitriles	4185	3.8.1.	2-Aminopyridine ... 4206
2.2.5.	Dealkylation of Alkylpyridines	4185	3.8.2.	Other Aminopyridines ... 4207
2.2.6.	Synthesis of 5-Ethyl-2-Methylpyridine from Paraldehyde and Ammonia	4185	3.9.	Pyridinols ... 4209
			3.10.	Pyridyl Alcohols ... 4211
2.2.7.	Synthesis from Nitriles and Acetylene	4187	3.11.	Pyridinecarbaldehydes ... 4212
2.2.8.	Other Synthetic Methods	4187	3.12.	Pharmaceuticals and Agrochemicals ... 4213
2.3.	Quality Specifications, Storage, and Transportation	4187	4.	Toxicology ... 4213
2.4.	Uses	4188	4.1.	Acute Toxicity ... 4213
2.5.	Economic Aspects	4189	4.2.	Subacute and Chronic Toxicity 4220
3.	Pyridine Derivatives	4189	4.3.	Mutagenicity and Ecotoxicity . 4220
3.1.	Vinylpyridines	4189	5.	References ... 4221

1. Introduction

Pyridine, C_5H_5N, is a six-membered heterocyclic compound containing one nitrogen atom. Pyridine and its homologues are commonly called pyridine bases. The first pyridine derivative, 2-methylpyridine (α-picoline; *pix*, Latin = pitch) was isolated from coal tar in 1846 by ANDERSON [1]. In 1851, ANDERSON obtained pyridine (*pyros*, Greek = fire) and dimethylpyridine (lutidine; *lutum*, Latin = dirt) from bone oil [2].

RAMSAY synthesized pyridine in 1876 by passing a mixture of acetylene and hydrogen cyanide through a red-hot tube [3]. Typical syntheses of pyridine derivatives are based on the work of HANTZSCH (1882) [4] and CHICHIBABIN (1906) [5]. The method of the latter is especially suitable for mass production and it is still an important industrial process.

Compounds containing a pyridine ring, such as vitamin B_6(pyridoxine) [6], nicotinamide [7], nicotinic acid, the coenzymes nicotinamide adeninedinucleotide (NAD) and reduced NAD (NADH), and many alkaloids, play important roles in metabolism. Pyridine bases are widely used in pharmaceuticals including nicotinamide and nicotinic acid. Similarly, pyridine derivatives are important insecticides and herbicides due to their high bioactivity. Further, they are used as adhesives for textiles and as chemicals, solvents, and catalysts.

2. Pyridine and Alkylpyridines

Pyridine and alkylpyridines are produced commercially by synthesis as well as by isolation from natural sources such as coal tar. Commercially important compounds are pyridine, 2-methylpyridine, 3-methylpyridine, 4-methylpyridine, 2,6-dimethylpyridine, 3,5-dimethylpyridine, and 5-ethyl-2-methylpyridine.

2.1. Properties

Pyridine is miscible in all proportions with water and most common organic solvents and has a boiling point 35 °C higher than benzene.

Physical data of pyridine and alkylpyridines are summarized in Table 1.

Since the pyridine ring has three double bonds, six π-electrons exist, which are sufficient for aromatic ring formation without involving the lone pair electrons of the nitrogen atom. Since the lone pair electrons remains free, quaternary salts retain the aromaticity. However, the nitrogen atom has a higher electronegativity than the carbon atoms and shows an electron-withdrawing effect. This is represented by the resonance hybrids:

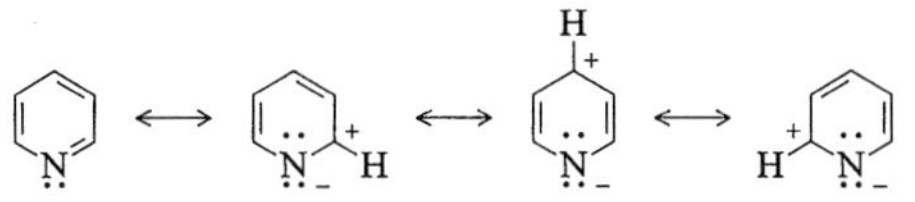

Therefore the electron densities at the 2- and 4-positions are low, and the ring is regarded as a π-electron-deficient aromatic ring. The weak basicity of pyridine and alkylpyridines is due to the lone pair of electrons on the ring nitrogen atom. The basicity of pyridine derivatives is increased by electron-donating substituents and decreased by electron-withdrawing substituents.

In oxidation and reduction reactions, the pyridine ring exhibits properties characteristic of π-electron-deficient aromatic rings: resistance to oxidation and facile reduction. In the oxidation of alkylpyridines with alkaline $KMnO_4$, the pyridine ring is not oxidized; instead the corresponding carboxylic acids are formed [8]. Furthermore in the oxidation of quinoline with alkaline $KMnO_4$, the main product is pyridine-2,3-dicarboxylic acid. This shows that the pyridine ring is more stable to oxidation than the benzene ring [9].

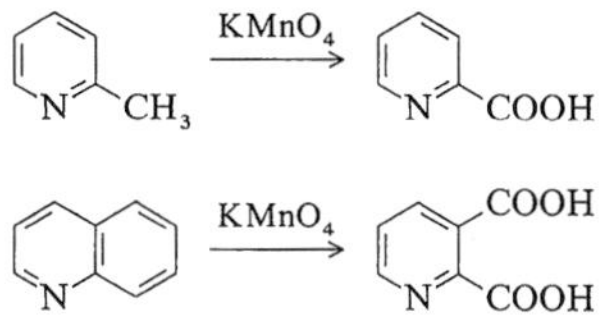

On reduction with hydrogen in the presence of catalyst, quinoline and 2-phenylpyridine are reduced preferentially at the pyridine ring [10].

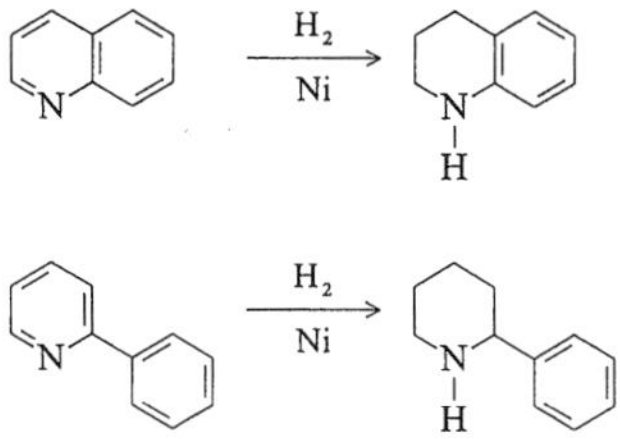

Table 1. Physical properties of pyridine and alkylpyridines

Compound	CAS registry no.	M_r	mp, °C	bp, °C	d_4^{20}	n_D^{20}	pK_a (H$_2$O, 25 °C)	Solubility in H$_2$O at 20 °C, g/100 g	Azeotrope with H$_2$O wt% H$_2$O	Azeotrope with H$_2$O bp, °C	Ignition temperature, °C	Explosion composition, %
Pyridine	[110-86-1]	79.10	−41.7	115.3	0.9819	1.5102	5.22	miscible	41.3	93.6	550	1.7 – 10.6
2-Methylpyridine	[109-06-8]	93.13	−66.7	129.4	0.9455	1.5010	5.96	miscible	48	93.5	535	1.4 – 8.6
3-Methylpyridine	[108-99-6]	93.13	−18.2	144.1	0.9564	1.5043	5.63	miscible	63	97		1.3 – 8.7
4-Methylpyridine	[108-89-4]	93.13	3.6	145.4	0.9546	1.5058	5.98	miscible	63.5	97.35	500	1.3 – 8.7
2,6-Dimethylpyridine	[108-48-5]	107.16	−6.1	144.5	0.9237	1.4977	6.72	27.2 (45.3 °C)	51.5	96.02		
3,5-Dimethylpyridine	[591-22-0]	107.16	−6.5	171.9	0.944	1.5049	6.15	3.3				
5-Ethyl-2-methylpyridine	[104-90-5]	121.18	−70.3	178.3	0.9208	1.4974		1.2	72	98.4		

Because of the low π-electron density at the ring carbon atoms, electrophilic reactions rarely occur on the pyridine ring, and they occur at the 3-position only under drastic conditions. Although nitration of pyridine bases is difficult [11], sulfonation occurs more readily [12].

Friedel–Crafts reactions and reactions catalyzed by $PdCl_2$ such as oxychlorination of benzene, which are useful for benzene derivatives, do not take place because of the coordination of the nitrogen atom to $AlCl_3$ or $PdCl_2$. Electron-donating substituents such as hydroxyl, amino, or alkyl groups attached to the pyridine ring facilitate electrophilic reactions because of the increased electron density in the ring.

In contrast, nucleophilic substitution reactions occur readily at the 2-, 4-, and 6-positions. Treatment of pyridine with sodium amide gives 2-aminopyridine (Chichibabin reaction) [13]. Reaction of pyridine bases with Grignard reagent, alkyllithium, or aryllithium gives 2-substituted pyridines [14]. Treatment of pyridine with methanol and hydrogen in the presence of a nickel catalyst gives 2-methylpyridine [15]. Substitution reactions of 2- or 4-halopyridines with nucleophiles, such as alkoxides, thiolates, amines, or carbanions, occur easily and provide important synthetic methods.

Radical reactions of pyridine—for example, the phenylation of pyridine with benzene diazonium chloride—give substituted products in the decreasing order 2- > 3- > 4- [16], and chlorination with chlorine above 270 °C or under UV irradiation gives a mixture of 2-, 4-, and 6- chloropyridines.

In the reaction of substituted groups in the pyridine ring, dealkylation and decarboxylation occur mainly at the 2-position [17], whereas hydroxymethylation of methyl groups with formaldehyde occurs at the 2- and 4-positions [18].

Like aliphatic tertiary amines, pyridine bases give N-oxides with hydrogen peroxide and peroxy acids, and form quaternary ammonium salts with alkyl halides.

2.2. Production

2.2.1. Separation from Tar

Pyridine bases are a constituent of tars. They were isolated from coal tar or coal gas before synthetic manufacturing processes became established. The amounts contained in coal tar and coal gas are small, and the pyridine bases isolated from them are a mixture of many components. Thus, with a few exceptions, isolation of pure pyridine bases was expensive. Today, almost all pyridine bases are produced by synthesis.

Table 2. Synthesis of 2- and 4-methylpyridine from acetaldehyde and ammonia

Company	Catalyst	Yield, %		Ref.
		2-Methylpyridine	4-Methylpyridine	
Koei Chemical	$Co_3Al_3(PO_4)_5^-$	45	9	[20]
Nippon Kayaku	$Al_2O_3-SiO_2-CdCl_2$	35	44	[21]

2.2.2. Synthesis from Aldehydes or Ketones with Ammonia

The reaction of aldehydes or ketones with ammonia is the most general synthetic reaction for the manufacture of pyridine bases and allows the preparation of various pyridines. This reaction was first studied in detail by CHICHIBABIN in 1924 [19] and since then been studied extensively for industrial manufacturing because of cheap access to raw materials. The reaction is usually carried out at 350–550 °C and a space velocity of 500–1000 h^{-1} in the presence of a solid acid catalyst (e.g., silica–alumina).

Aldehydes react with ammonia as follows:

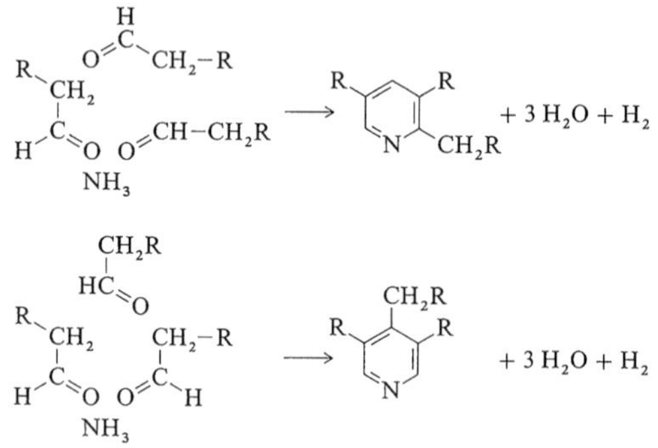

For example, acetaldehyde and ammonia give 2-methylpyridine (Eq. 1) and 4-methylpyridine (Eq. 2). Some examples of the synthesis are given in Table 2.

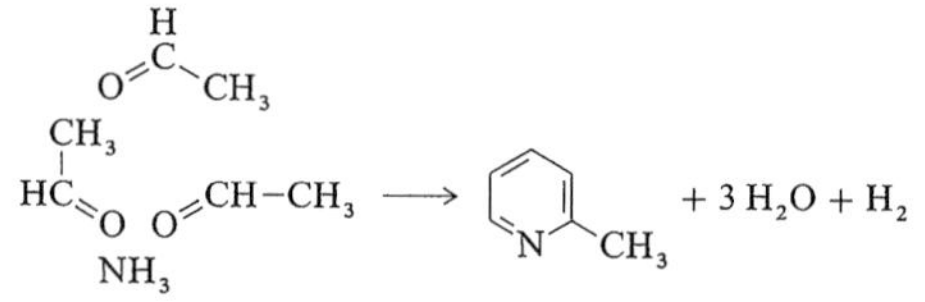

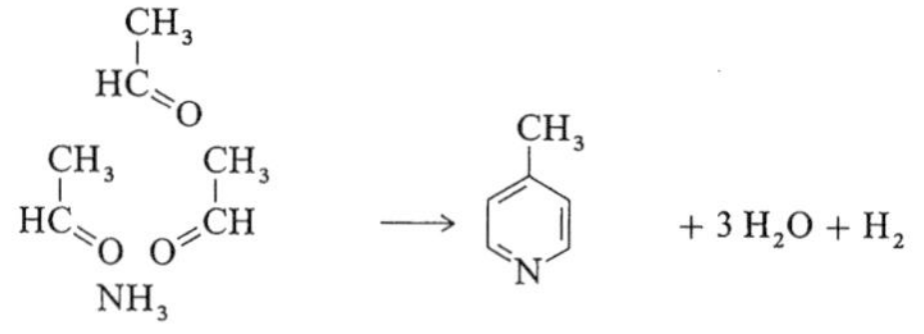

Table 3. Synthesis of 3-methylpyridine and pyridine from acrolein and ammonia

Company	Catalyst	Yield, %		Ref.
		Pyridine	3-Methylpyridine	
Degussa	$Al_2O_3 - MgF_2$	25	49	[22]
ICI	$SiO_2 - Al_2O_3 - H_2SiF_2$	62	15	[23]
Nippon Kayaku	$SiO_2 - Al_2O_3 - CdF_2$	26	56	[24]
Koei Chemical	$SiO_2 - Al_2O_3 - MnF_2$	20	45	[25]
Daicel Chemical	$SiO_2 - Al_2O_3$	22	49	[26]

With α, β-unsaturated aldehydes the reaction occurs according to the following schemes:

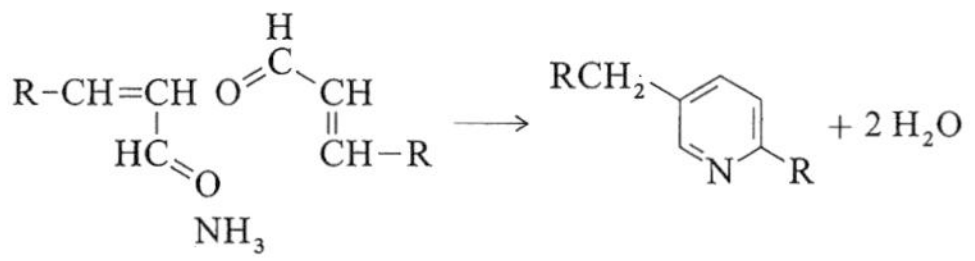

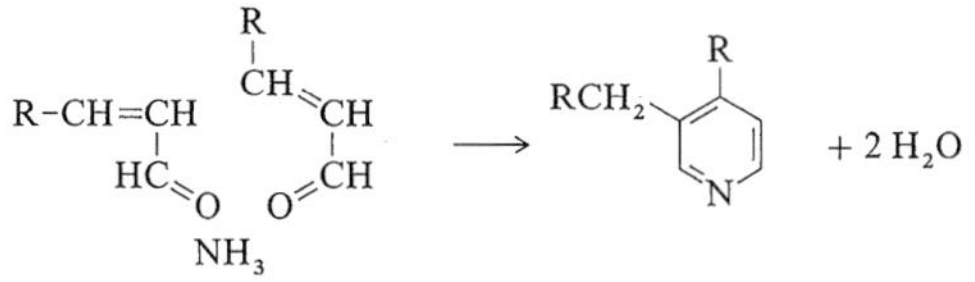

For example, acrolein and ammonia give 3-methylpyridine (Eq. 3); pyridine is simultaneously formed by demethylation. Examples of this synthesis are given in Table 3.

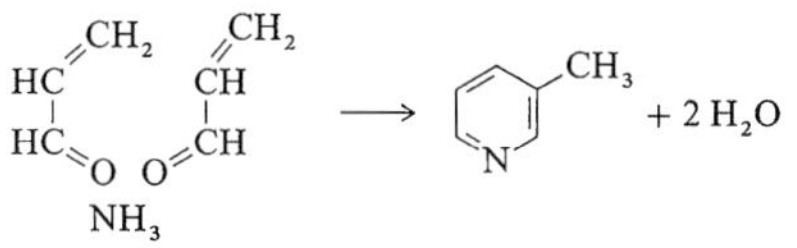

Acrolein and acetaldehyde react with ammonia mainly to form pyridine:

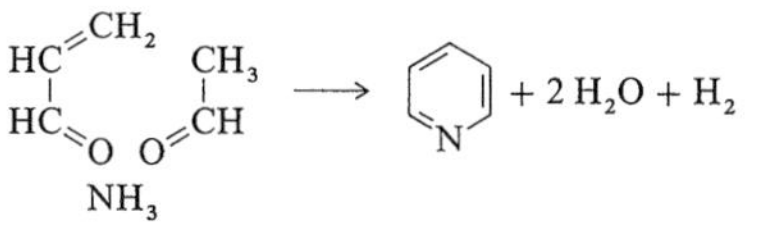

Acrolein and propionaldehyde react with ammonia to give primarily 3-methylpyridine (Eq. 5).

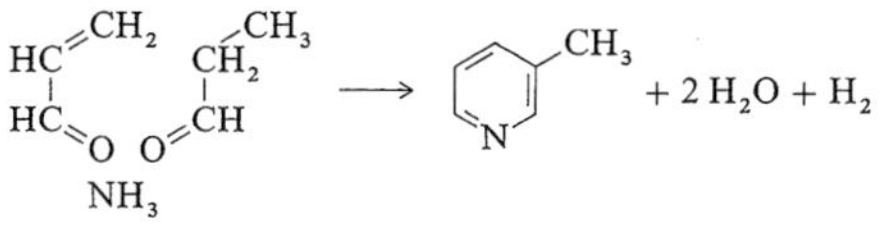

Acetaldehyde and formaldehyde react with ammonia to give mainly pyridine (Eq. 6): they appear to first form acrolein (Eq. 7), and then acrolein and formaldehyde react with ammonia to give pyridine. Simultaneously, 2-, 3-, and 4-methylpyridines are

4181

Table 4. Synthesis of pyridine and 3-methylpyridine from acetaldehyde and formaldehyde with ammonia

Company	Catalyst	Yield, %		Ref.
		Pyridine	3-Methylpyridine	
ICI	$SiO_2 - Al_2O_3 -$ coke	38	25	[27]
Rütgerswerk	$SiO_2 - Al_2O_3 - CdF_2$	57	29	[28]
Nepera	ZSM-5 *	54	28	[29]
Koei Chemical	Tl–ZSM-5	63	9	[30]

* Zeolite.

formed, as shown in Equations (1)–(4). This method is one of the most widely used for pyridine production. Table 4 lists some examples of this process, and Figure 1 illustrates the flow sheet of the plant.

A preheated gaseous mixture of acetaldehyde, 36% formaldehyde, and ammonia is passed through the reactor (a) packed with the catalyst at 400–450 °C. The reaction mixture is separated from ammonia and hydrogen by a collector (b) and extracted with solvent, e.g., benzene (c). The solvent is removed from the extract (d) and pyridine and 3-methylpyridine are isolated in continuous distillation columns (e). The catalyst is periodically regenerated by air.

$$\text{CH}_3\text{CHO} + \text{HCHO} + \text{NH}_3 \longrightarrow \text{pyridine} + 3\,H_2O + H_2 \qquad (6)$$

$$\text{CH}_3\text{CHO} + \text{HCHO} \longrightarrow \text{CH}_2{=}\text{CHCHO} + H_2O \qquad (7)$$

Propionaldehyde and formaldehyde react with ammonia to give 3,5-dimethylpyridine (Eq. 8) [31]; benzaldehyde and acetaldehyde give 2-phenylpyridine [1008-89-5] (Eq. 9) and 4-phenylpyridine [939-23-1] (Eq. 10) [32].

$$(8)$$

$$(9)$$

$$(10)$$

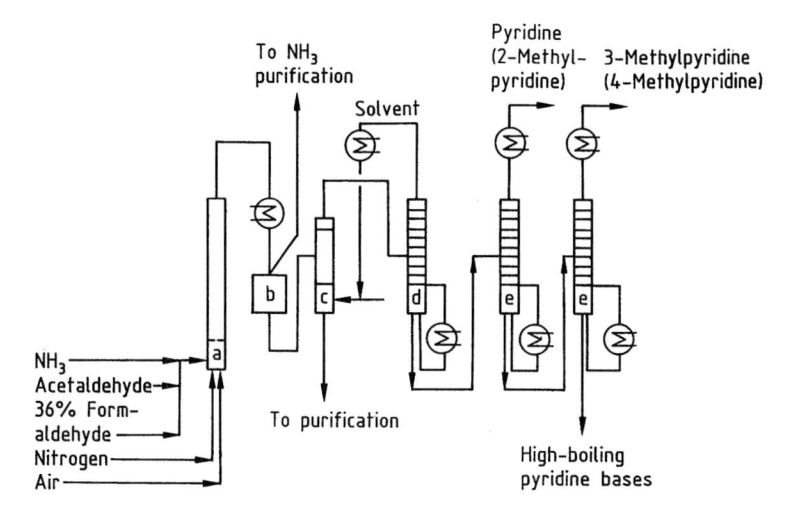

Figure 1. Flow sheet of pyridine and methylpyridine production from acetaldehyde and formaldehyde with ammonia
a) Reactor; b) Collector; c) Extraction; d) Solvent distillation; e) Distillation

Ketones and aldehydes react with ammonia according to the following general scheme:

$$R^1-CH_2-C(=O)R^2 + O=CH-R^3 + O=C(R^2)-CH_2-R^1 + NH_3 \longrightarrow \text{pyridine}(R^1, R^2, R^3) + 3 H_2O + H_2$$

Typically, acetone [67-64-1] and formaldehyde with ammonia give 2,6-dimethylpyridine (Eq. 11) [33]:

$$CH_3-C(=O)-CH_3 + O=CH_2 + CH_3-C(=O)-CH_3 + NH_3 \longrightarrow \text{2,6-dimethylpyridine} + 3 H_2O + H_2 \quad (11)$$

α, β-Unsaturated ketones or aldehydes react with ammonia according to the following scheme:

$$R^4-CH_2-C(=O)R^5 + HC(R^3)=C-R^2 + O=C(R^1) + NH_3 \longrightarrow \text{pyridine}(R^1, R^2, R^3, R^4, R^5) + 2 H_2O + H_2$$

For example, acrolein and acetone react with ammonia to give 2-methylpyridine (Eq. 12) [34]:

4183

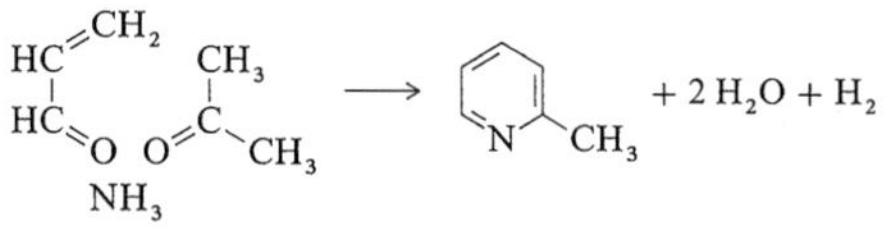

$$\text{HC}{=}\text{CH}_2\text{, HC}{=}\text{O, O}{=}\text{C(CH}_3\text{)CH}_3\text{, NH}_3 \longrightarrow \text{2-methylpyridine} + 2\,\text{H}_2\text{O} + \text{H}_2 \quad (12)$$

As a variant, acetone with ammonia gives 2,4,6-trimethylpyridine [108-75-8] with simultaneous demethylation (Eq. 13) [35]. Cyclopentanone [120-92-3] and acrolein with ammonia give 2,3-cyclopentenopyridine [533-37-9] (Eq. 14) [36]. Using aniline instead of ammonia results in formation of quinoline (Eq. 15):

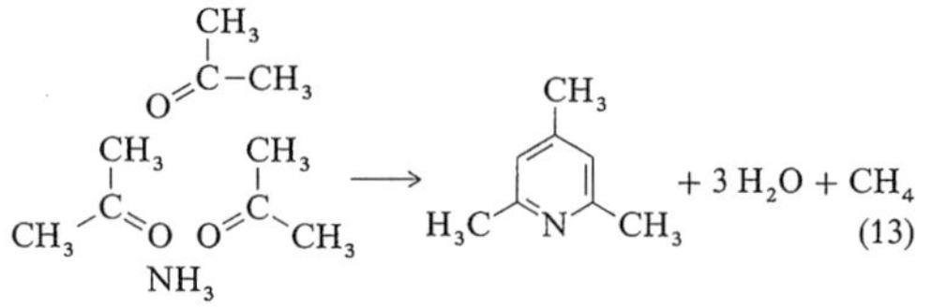

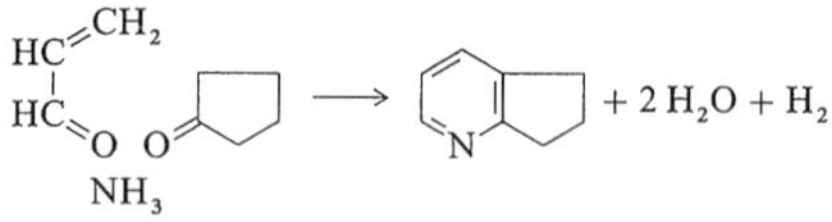

$$\text{3 acetone} + \text{NH}_3 \longrightarrow \text{2,4,6-trimethylpyridine} + 3\,\text{H}_2\text{O} + \text{CH}_4 \quad (13)$$

$$\text{acrolein} + \text{cyclopentanone} + \text{NH}_3 \longrightarrow \text{2,3-cyclopentenopyridine} + 2\,\text{H}_2\text{O} + \text{H}_2 \quad (14)$$

$$\text{aniline} + \text{acrolein} \longrightarrow \text{quinoline} + \text{H}_2\text{O} + \text{H}_2 \quad (15)$$

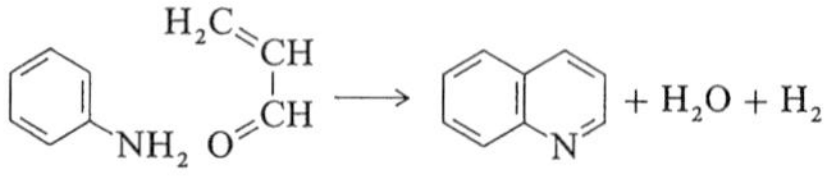

As shown above, various pyridines can be obtained by using different combinations of aldehydes, ketones, ammonia, and amines.

2.2.3. Synthesis from Acrylonitrile and Ketones

Synthesis from acrylonitrile and ketones is one of the current processes for manufacturing 2-methylpyridine. This process gives 2-methylpyridine selectively, in contrast to the process using acetaldehyde and ammonia, which gives 4-methylpyridine as a byproduct. First, the reaction of acrylonitrile and acetone, catalyzed by a primary aliphatic amine such as isopropylamine and a weak acid such as benzoic acid [65-85-0], occurs in the liquid phase at 180 °C and 2.2 MPa to give 5-oxohexanenitrile [10412-98-3], with 91 % selectivity (Eq. 16). The acrylonitrile conversion is 86 % [37]. Then cyclization and dehydration of the initial product are carried out in the gas phase in the presence of hydrogen over a palladium, nickel, or cobalt-containing catalyst at ca. 240 °C to give 2-methylpyridine in 84 % yield (Eq. 17) [38]. 4-Methyl-5-oxohexanenitrile [10413-01-1], formed from acrylonitrile and 2-butanone, gives 2,3-dimethylpyridine [583-61-9] in 89 % yield [38].

$$CH_3-\underset{\underset{O}{\|}}{C}-CH_3 + CH_2=CH-CN \longrightarrow$$

$$CH_3\underset{\underset{O}{\|}}{C}CH_2CH_2CH_2CN \quad (16)$$

$$CH_3\underset{\underset{O}{\|}}{C}CH_2CH_2CH_2CN \xrightarrow{H_2} \underset{N\ CH_3}{\bigcirc} + H_2O \quad (17)$$

2.2.4. Synthesis from Dinitriles

In a vapor-phase reaction over a nickel-containing catalyst in the presence of hydrogen, 2-methylglutaronitrile [4553-62-2] gives 3-methylpiperidine [626-56-2], which then undergoes dehydrogenation over palladium–alumina to give 3-methylpyridine [39]:

$$NC-CH_2CH_2\underset{\underset{CH_3}{|}}{CH}-CN + 4\,H_2 \longrightarrow \underset{\underset{H}{N}}{\bigcirc}^{CH_3} + NH_3 \quad (18)$$

$$\underset{\underset{H}{N}}{\bigcirc}^{CH_3} \longrightarrow \underset{N}{\bigcirc}^{CH_3} + 3\,H_2 \quad (19)$$

A one-step gas-phase reaction over a palladium-containing catalyst is reported to give 3-methylpyridine in 50% yield [40].

2.2.5. Dealkylation of Alkylpyridines

Alkylpyridines of low commercial value, obtained as byproducts of pyridine base synthesis, are occasionally converted into useful pyridine bases by dealkylation. The methods for dealkylation involve oxidative dealkylation by air over a vanadium oxide catalyst [23], steam dealkylation over a nickel catalyst [24], [25], and hydrodealkylation over a silver or platinum catalyst [44]. Examples are listed in Table 5.

2.2.6. Synthesis of 5-Ethyl-2-Methylpyridine from Paraldehyde and Ammonia

Reaction of paraldehyde with aqueous ammonia in the liquid phase is carried out at 200–300 °C and 12–13 MPa in the presence of an ammonium salt (e.g., ammonium phosphate) to give 5-ethyl-2-methylpyridine (MEP) in about 70% yield (Eq. 20) [45]. Figure 2 shows the reaction route, and Figure 3 illustrates the manufacture of MEP by the Montecatini–Edison process.

Table 5. Dealkylation of alkylpyridines

Starting material	Catalyst	Additives	Yield of pyridine, %	Ref.
3-Methylpyridine	V/Cr/Ag–Al$_2$O$_3$	air, H$_2$O	82	[41]
2-Methylpyridine	Ni–SiO$_2$	H$_2$, H$_2$O	93	[42]
2-Methylpyridine	Ni–ZrO$_2$	H$_2$O	50	[43]
Alkylpyridine	Ag	H$_2$	58	[44]

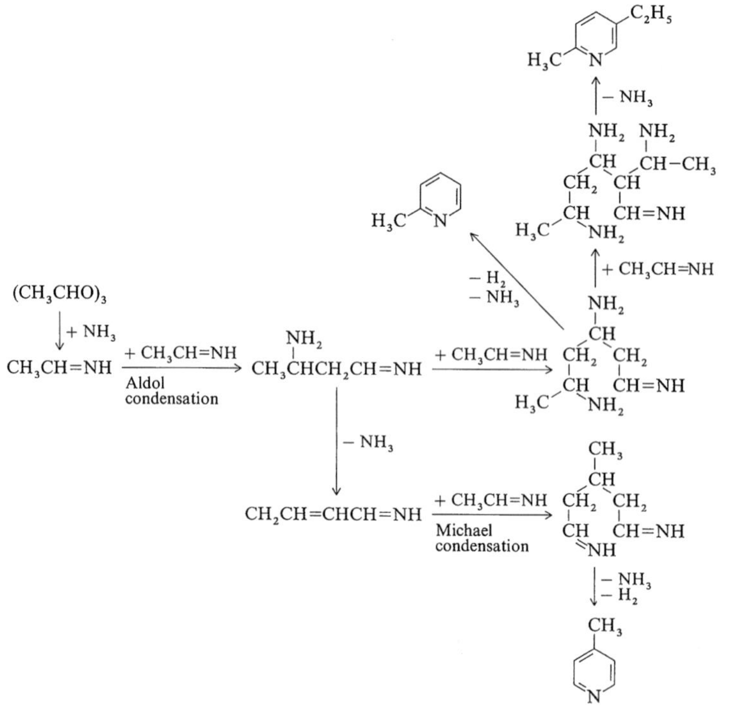

Figure 2. Mechanism of 5-ethyl-2-methylpyridine formation [45]

Paraldehyde, produced from acetaldehyde and sulfuric acid, is reacted with 30–40 % aqueous ammonia and acetic acid at 220–230 °C and 10–20 MPa. The reaction mixture is separated into two phases in a separator (c). Ammonia is recovered from the aqueous layer by a stripper (d). MEP, 2-methylpyridine, and 4-methylpyridine are isolated from the organic layer by distillation.

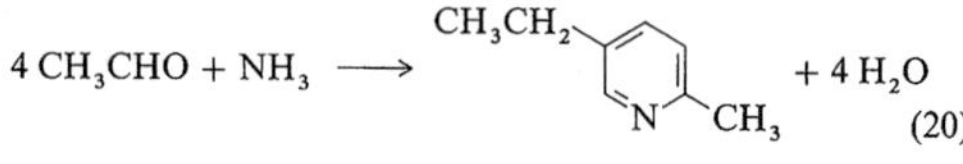

$$4\,CH_3CHO + NH_3 \longrightarrow \underset{\text{(5-ethyl-2-methylpyridine)}}{\text{MEP}} + 4\,H_2O \quad (20)$$

2.2.7. Synthesis from Nitriles and Acetylene

Liquid-phase reaction of nitriles with acetylene is carried out at 120–180 °C and 0.8–2.5 MPa in the presence of an organocobalt catalyst and gives 2-substituted pyridines [46]:

$$\begin{array}{c} \text{HC} \equiv \text{CH} \\ \text{HC} \equiv \text{CH} \\ \text{N} \equiv \text{C-R} \end{array} \longrightarrow \begin{array}{c} \\ \text{N} \quad \text{R} \end{array}$$

For example, acetonitrile [75-05-8] and acetylene react in the presence of cobaltocene as catalyst to give 2-methylpyridine in 76% yield [47]. Acrylonitrile and acetylene react in the presence of cyclopentadienylcobalt–cycloocta-1,5-diene catalyst to give 2-vinyl-pyridine with 93% selectivity [48].

2.2.8. Other Synthetic Methods

Ethylene [74-85-1] and ammonia react in the presence of a palladium complex catalyst to give 2-methylpyridine and MEP [49]. Pyridine can be prepared from cyclopentadiene by ammoxidation [50], or from 2-pentenenitrile [13284-42-9] by cyclization and dehydrogenation [51]. Furfuryl alcohol or furfural reacts with ammonia in the gas phase to give pyridine [52]. 2-Methylpyridine is also prepared from aniline [53].

2.3. Quality Specifications, Storage, and Transportation

Specifications of pyridine vary according to country but are usually > 99.8% purity by gas chromatographic analysis. Table 6 lists the standard specification for refined pyridine (ASTM) in the United States [54].

Pyridine bases should generally be stored under dark, cool conditions. They are transported in drums, tank cars, and bulk containers in accordance with the following regulations:

Road:	GGVS/ADR	Class 3, No. 15 b
Rail:	GGVE/RID	Class 3, No. 15
Sea:	GGVSee/IMDG	Class 3.2
Air:	IATA-DGR	Class 3
UN no.:	1282	

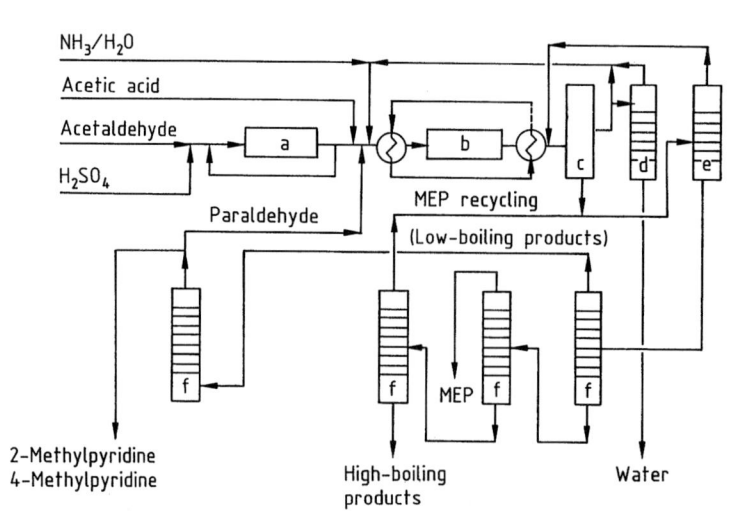

Figure 3. Flow sheet of 5-ethyl-2-methylpyridine (MEP) production by Montecatini–Edison process
a) Paraldehyde production; b) Pyridine reactor; c) Separator; d) Stripper; e) Dewatering column; f) Fractionating columns

2.4. Uses

Pyridine is an excellent solvent, especially for dehydrochlorination reactions and extraction of antibiotics. Large amounts of pyridine are used as starting material for pharmaceuticals and agrochemicals: for example, herbicides such as diquat and paraquat, insecticides such as chlorpyrifos, and fungicides such as pyrithione (see Section 3.12).

2-Methylpyridine. The major use of 2-methylpyridine is as a precursor of 2-vinylpyridine. The terpolymer of 2-vinylpyridine with butadiene and styrene is used as an adhesive for textile tire cord. 2-Methylpyridine is also used as a material for a variety of pharmaceuticals and agrochemicals: for example, chemicals such as nitrapyrin to prevent loss of ammonia from fertilizers, herbicides such as picloram, and coccidiostats such as amprolium (see Section 3.12).

3-Methylpyridine. A considerable amount of 3-methylpyridine is used as a starting material for pharmaceuticals and agrochemicals: for example, insecticides such as chlorpyrifos, feed additives such as nicotinic acid and nicotine carboxamide, and herbicides such as fluazifop-butyl (see Section 3.12).

4-Methylpyridine. The primary use of 4-methylpyridine is in the production of the antituberculosis agent isonicotinic hydrazide. Polymers containing 4-vinylpyridine, obtained from 4-methylpyridine, are used as anion exchangers (see Section 3.12).

Table 6. Standard specifications of refined pyridine (ASTM)

Appearance	clear liquid, free of extraneous matter and sediment
Odor	pyridine, characteristic
$d^{15.56}$	0.985–0.990
Color	not darker than no. 20 on platinum–cobalt scale
Distillation range at atmospheric pressure	
Total distillation range	$\leq 2\,°C$
Initial distillation temperature (first drop)	$\geq 114.0\,°C$
End point (dry point)	$\leq 117.0\,°C$
Water	≤ 0.20 wt %
Water solubility	clear solution, no turbidity or oil film

Polyalkylpyridines. Large amounts of MEP are used as a starting material for nicotinic acid. 2,6-Dimethylpyridine is used for the antiarteriosclerotic pyridyl carbamate, while 3,5-dimethylpyridine is used for producing the antiulcer medication omeprazole (see Section 3.12).

2.5. Economic Aspects

The amounts of pyridine bases produced worldwide are estimated roughly as follows: pyridine, ca. 26 000 t/a (1989); 2-methylpyridine, ca. 8000 t/a (1989); 3-methylpyridine, ca. 9000 t/a (1989); 4-methylpyridine, ca. 1500 t/a (1989); and MEP, ca. 8000 t/a (1989). Table 7 lists the primary manufacturers of pyridine and alkylpyridines.

3. Pyridine Derivatives

3.1. Vinylpyridines

Vinylpyridines have the following general formula:

R—[pyridine ring]—CH=CH$_2$

2-Vinylpyridine was first synthesized in 1887, and 4-vinylpyridine in 1920. However, not until 1950 did vinylpyridines begin to be used for various commercial purposes. Today only 2- and 4-vinylpyridines are of industrial importance; Table 8 lists some physical properties.

Due to the electron-withdrawing effect of the ring nitrogen atom, 2- and 4-vinylpyridines act as electrophiles. Nucleophiles such as methoxide, cyanide, hydrogen sulfide, and others add to 2- and 4-vinylpyridine at the vinylic site to give addition products.

Table 7. Manufacturers of pyridine and alkylpyridines

Country	Company	Products
United States	Reilly Industry	pyridine; 2-, 3- and 4-methylpyridine; dimethylpyridine; trimethylpyridine
	Nepera Chemical Co.	pyridine; 2-, 3- and 4-methylpyridine
	Kopper Co.	coal-tar-derived pyridine
Japan	Koei Chemical Co.	pyridine; 2-, 3- and 4-methylpyridine; dimethylpyridine; trimethylpyridine
	Daicel Chemical	pyridine; 3-methylpyridine
	Nippon Steel	coal-tar-derived pyridine
Belgium	Reilly Industry	pyridine; 2-, 3- and 4-methylpyridine
Netherlands	DSM	2-methylpyridine
Switzerland	Lonza	2-methyl-5-ethylpyridine

Table 8. Physical properties of vinylpyridines

	2-Vinylpyridine	4-Vinylpyridine
CAS registry no.	[100-69-6]	[100-43-6]
Density (20 °C), g/cm^3	0.977	0.988
bp, °C (20 kPa)	110	120
(4 kPa)	70	
(2 kPa)		65
n_D^{20}	1.5509	1.5525
Solubility in water (20 °C), g/L	27.5	29
Viscosity (20 °C), mPa · s	1.17	
pK_a	4.98	5.5

The reactions shown below have been carried out on a commercial scale [55]–[57]:

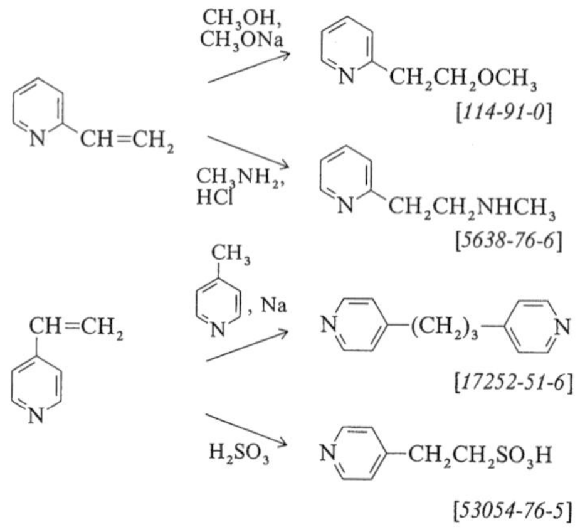

In addition to these Michael-type reactions, vinylpyridines are reduced at the side chain to give ethylpyridines [58]. Addition of chlorine to the vinyl group leads to (1,2-dichloroethyl)-pyridines [59].

Vinylpyridines are readily polymerized or copolymerized with styrene, butadiene, isobutylene, methyl methacrylate, and other compounds in the presence of radical, cationic, and anionic initiators. The homopolymer is soluble in organic solvents such as methanol and acetone, whereas cross-linked copolymers are insoluble in organic solvents.

Preparation. Industrially, 2- and 4-vinylpyridines are manufactured by treatment of 2- or 4-methylpyridine with aqueous formaldehyde, followed by dehydration of the resulting intermediate alcohol [60]:

$$\text{2-methylpyridine} + CH_2O \longrightarrow \text{2-(2-pyridyl)ethanol} \xrightarrow{-H_2O} \text{2-vinylpyridine}$$

For the manufacture of 2-vinylpyridine, the reaction is carried out at 150–200 °C in an autoclave. The conversion must be kept relatively low with short reaction time to suppress the formation of byproducts. After removal of unreacted 2-methylpyridine by distillation, concentrated aqueous sodium hydroxide is added to the residue and the resultant mixture is distilled under reduced pressure. During distillation, the dehydration of 2-(2-pyridyl)ethanol occurs to give 2-vinylpyridine as a distillate, which can be purified further by fractional distillation under reduced pressure in the presence of an inhibitor such as 4-*tert*-butylcatecohol. 4-Vinylpyridine is manufactured by a similar method.

Uses. Among vinylpyridines, only 2-vinylpyridine is in large demand, as much as 3000 t annually. The polymer of 2-methyl-5-vinylpyridine [*140-76-1*] was formerly used as a coating material for medicine tablets; today, it is not used at all. Although 4-vinylpyridine has been used for various purposes, its total annual demand is presumed to be only several hundred tons.

Most of the 2-vinylpyridine is used in the production of a latex terpolymer of 2-vinylpyridine, styrene, and butadiene, used as a tire-cord binder.

The cross-linked resin made from 4-vinylpyridine and divinylbenzene has been used to remove poisonous hexavalent chromium ions from wastewater [61] by formation of complexes with the nitrogen atoms of the pendant pyridine rings. Similarly, this type of resin can be used to remove or recover phenol from wastewater [62]. Furthermore, the poly(4-vinylpyridines) are used as ion-exchange membranes and electrodes for reversible cells.

The addition product of methanol to 2-vinylpyridine, 2-(2-methoxyethyl)pyridine, is a veterinary anthelmintic (methypidine) [63]. 4,4'-(1,3-Propanediyl)bispiperidine, made from 4-vinylpyridine, is used as a raw material for polyamide resins [64], and 2-(4-pyridyl)ethylsulfonic acid from 4-vinylpyridine is a coagulation accelerator for the gelatin layer of photographic plates [65].

3.2. Bipyridines

Bipyridines have become increasingly important as intermediates for herbicides since the 1960s. A significant part of pyridine is used for the manufacture of bipyridines (i.e., 2,2'- and 4,4'-bipyridine). The herbicides diquat [85-00-7] and paraquat (dichloride [1910-42-5]; bis(methylsulfate) [2074-50-2]) are produced by quaternization of 2,2'- and 4,4'-bipyridines, respectively. Their physical properties are listed in Table 9.

2,2'-Bipyridine is synthesized by the dimerization of pyridine in the presence of an oxidizing agent or catalyst, such as iron(III) chloride, iodine, or nickel–aluminum, or by the reaction of 2-bromopyridine with copper. For industrial production, Raney nickel is used, and the amount of catalyst employed is an indicator of the economy of the process. In Table 10, the historical development of the process is demonstrated in terms of the decrease in amount of catalyst employed.

Production of 4,4'-Bipyridine. As early as 1870, pyridine was known to react with sodium to give 4,4'-bipyridine after oxidation of the initial product [72]. The reduction of pyridine with zinc powder and acetic acid followed by oxidation was also known. Many patents have been applied for concerning the improvement of this procedure with respect to reaction temperature, choice of solvent, and isolation of the product. The ICI process, which seems at present to be the best, is described below [73].

Sodium is dissolved in liquid ammonia at $-45\,°C$, and the solution is successively diluted with pyridine and N,N-dimethylformamide. The resulting mixture is poured into N,N-dimethylformamide at $-25\,°C$, while air is passed through the solution. The mixture is then warmed to room temperature, and ammonia is allowed to evaporate. After work-up, 4,4'-bipyridine is obtained in 84 % yield.

Other variants are described in [74]–[77].

3.3. Quaternary Pyridinium Salts

Properties. Pyridine and some of its derivatives are readily converted to quaternary salts by alkylating agents such as alkyl halides. Quaternary salts can be regarded as being formed by the neutralization of strong bases (i.e., pyridinium hydroxides) with strong acids (e.g., hydrogen halides). Therefore, these salts are practically neutral when dissolved in water. Although stable under normal conditions, they are degraded into pyridine hydrohalides and alkenes on intense heating [78], or they undergo the Ladenburg rearrangement in the presence of copper as a catalyst [79]:

Table 9. Physical properties of bipyridines (M_r 156.18)

Compound	CAS registry no.	mp, °C	bp, °C	Solubility in water
2,2'-Bipyridine	[366-18-7]	70.1	272–273	sparingly soluble
2,3'-Bipyridine	[581-50-0]		295–296	practically insoluble
2,4'-Bipyridine	[581-47-5]	61.1–61.5	280–282	practically insoluble
3,3'-Bipyridine	[581-46-4]	68	291–292	readily soluble
3,4'-Bipyridine	[4394-11-0]	61	297	soluble in hot water
4,4'-Bipyridine	[553-26-4]	114	304.8	soluble in hot water

Table 10. Amount of Raney nickel employed in synthesis of 2,2'-bipyridine

Method	Raney nickel, g/100 g 2,2'-bipyridine
[66]	595
[67]	143
[68]	25
[69]	17
[70]	14
[71]	8

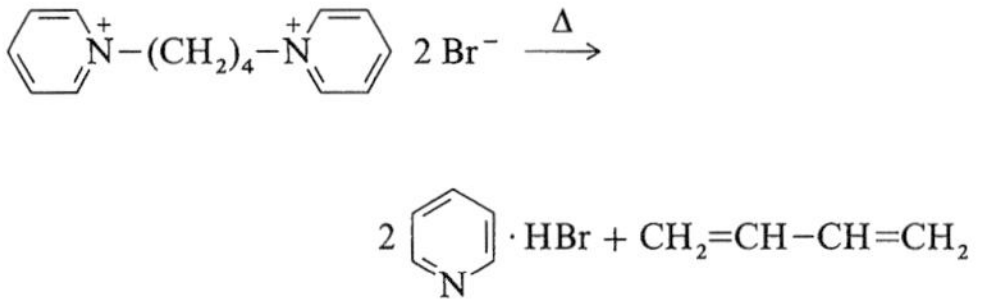

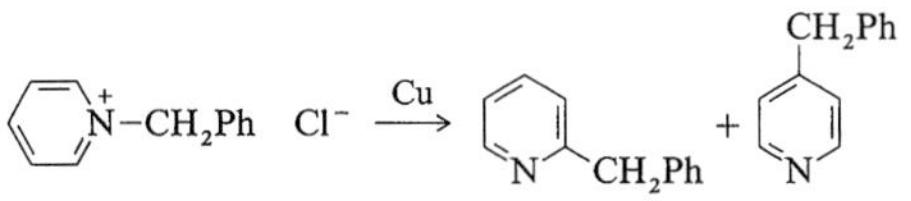

The pyridine ring of these salts is susceptible to nucleophilic attack as a result of the quaterniza-tion of the nitrogen atom. The reaction of 1-methylpyridinium salts with alkaline ferricyanide gives *N*-methylpyridone via oxidation of a pseudobase, which is considered to exist at low concentration in equilibrium with the pyridinium hydroxide [80], [81]:

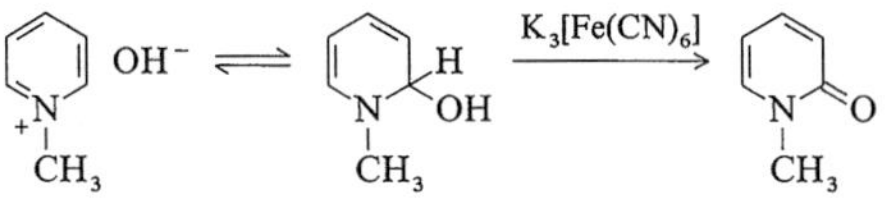

1-Pyridylpyridinium dichloride [5421-92-1], obtained from pyridine and thionyl chloride or chlorine, is particularly useful as a synthetic intermediate [82], [83]. It affords 4-amino- and 4-hydroxypyridine on ammonolysis and hydrolysis, respectively. Treatment with phenol and sodium phenoxide gives 4-phenoxypyridine, and reaction with hydrogen sulfide in pyridine gives 4-mercaptopyridine [84]–[87].

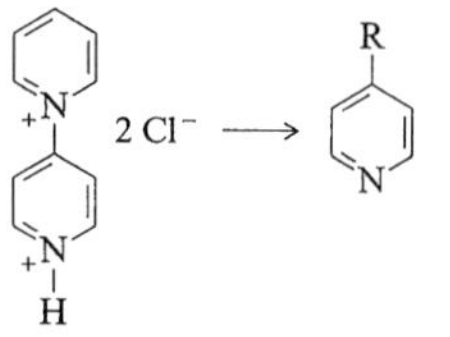

R = NH$_2$ [504-24-5]
OH [626-64-2]
OPh [4783-86-2]
SH [4556-23-4]

Uses. The major use of quaternary pyridinium salts is in the manufacture of the herbicides paraquat and diquat. These compounds are produced by the quaternization of 4,4′-bipyridyl and 2,2′-bipyridyl with methyl chloride and dibromoethane, respectively. Higher alkylpyridinium salts are used in the textile industry as dye auxiliaries and spin bath additives (antistatic agents and softeners). The higher alkylpyridinium salts also exhibit antimicrobial activity. Hexadecylpyridinium chloride [123-03-5] is a topical antiseptic, and amprolium [121-25-5], a quaternary salt of 2-methylpyridine, is a veterinary coccidiostat.

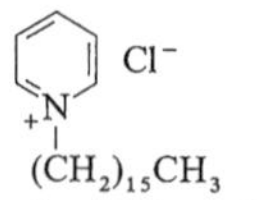

Hexadecylpyridinium chloride

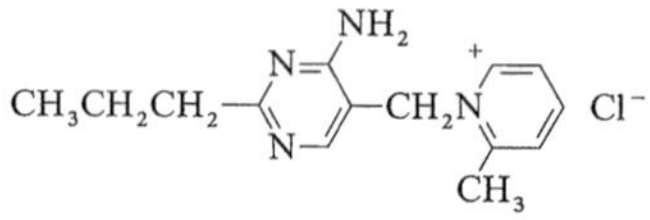

Amprolium

Of the many quaternary salts, 1-butylpyridinium bromide [1124-64-7] and other lower 1-alkyl homologues are of current interest. Although each component is solid at ambient temperature, the mixture of these salts with aluminum chloride leads to melts that can exist as liquids below room temperature in fairly wide proportions.

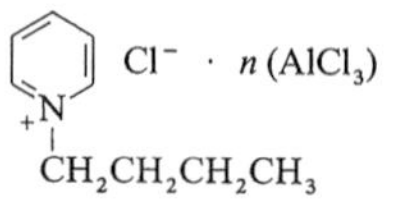

The molten salt with $n = 2$ exhibits a specific conductivity of ca. 7 mS/cm at 25 °C [88]. The utilization of these molten salts for battery electrolytes [89] and electroplating baths for aluminum has been proposed [90]. These binary salts are reportedly excellent solvents and catalysts for Friedel–Crafts reactions [91] and for the formylation of toluene by carbon monoxide [92].

3.4. Pyridine N-Oxides

Properties. Some physical properties of pyridine N-oxides are listed in Table 11. Resonance structures of pyridine N-oxide are as follows:

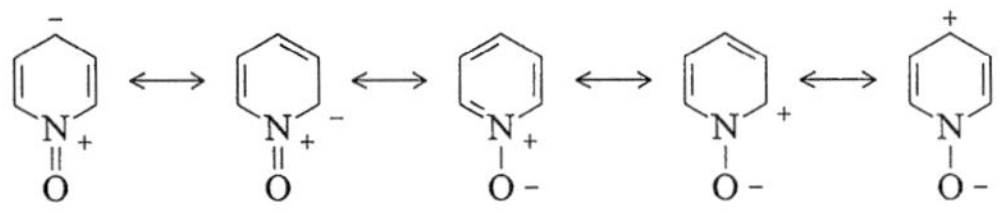

The N-oxide group in pyridine N-oxide has both electron-withdrawing and electron-donating effects. Consequently, pyridine N-oxide reacts with both electrophiles and nucleophiles, resulting in a more versatile reactivity of the pyridine ring compared to pyridine itself.

Nitration of pyridine N-oxide is a typical electrophilic reaction of the N-oxide. Pyridine N-oxide is nitrated under relatively mild conditions due to the contribution of structures **1** and **2**, giving 4-nitropyridine N-oxide in good yield [93], [94], in contrast to the nitration of pyridine, which requires drastic conditions and gives 3-nitropyridine only in poor yield. The nitro group of 4-nitropyridine N-oxide can be reduced to an amino group or displaced by various nucleophiles such as halides and alkoxides to give many useful 4-substituted pyridines. The oxygen of pyridine N-oxides can usually be removed by phosphorus trichloride.

Owing to the contribution of resonance structures **3** and **4**, the oxygen atom of the N-oxide group readily undergoes protonation, acylation, allylation, etc. Reaction with nucleophiles proceeds via quaternized intermediates, an example being the formation of 2-acetoxypyridine [3847-19-6] from the reaction of pyridine N-oxide with acetic anhydride [95]:

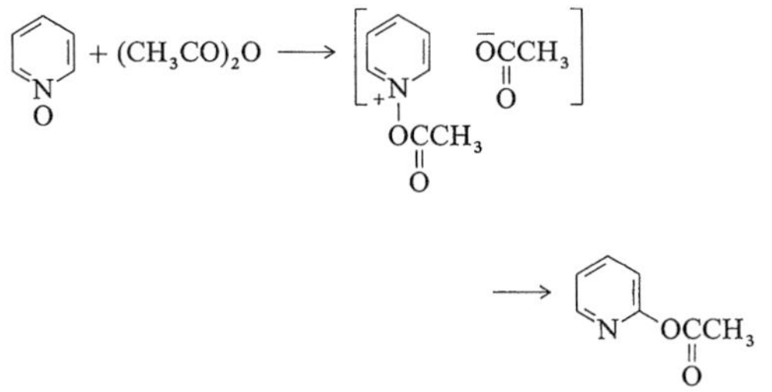

Trimethylsilyl cyanide, formed in situ from trimethylsilyl chloride and sodium cyanide, reacts with pyridine N-oxide to give almost exclusively 2-pyridinecarbonitrile [100-70-9] in 80% yield [96]. Although the chlorination of pyridine N-oxide with sulfuryl chloride gives a mixture of 2- [109-09-1] and 4- chloropyridines [626-61-9] [97], treatment of 3-methylpyridine N-oxide with phosphoryl chloride in the presence of diisopropylamine yields predominantly 2-chloro-5-methylpyridine [18368-64-4] [98]. In the chlorination of 3-pyridinecarbonitrile N-oxide and nicotinic acid N-oxide, 2-chloro-3-pyridinecarbo-

Table 11. Physical properties of pyridine N-oxides

Compound	CAS registry no.	mp, °C	bp, °C (kPa)	pK$_a$
Pyridine N-oxide	[694-59-7]	67	122–124 (0.67)	0.79 (24 °C)
2-Methylpyridine N-oxide	[931-19-1]		123–124 (2.0)	
3-Methylpyridine N-oxide	[1003-73-2]	37–38	146–149 (2.0)	1.08 (25°C)
4-Methylpyridine N-oxide	[1003-67-4]	186–188		1.29 (25°C)
Nicotinic acid N-oxide	[2398-81-4]	258		
3-Pyridinecarbonitrile N-oxide	[14906-64-0]	178		

nitrile [6602-54-6] [99] and 2-chloronicotinic acid [2942-59-8] [100] are obtained as the main products.

The N-oxide of 2-methylpyridine reacts with acetic anhydride at the methyl group to give 2-acetoxymethylpyridine [1007-49-4] [101], which on hydrolysis leads to 2-pyridylmethanol [586-98-1].

Preparation. Pyridine N-oxide can be prepared by treatment of pyridine with 30% hydrogen peroxide in acetic acid [102]. A similar and easier method is also known [103], in which molybdenum trioxide is used as catalyst and water as solvent. Pyridine N-oxide and its alkyl derivatives are isolated by distillation after complete decomposition of unreacted peroxides.

Uses. Pyridine N-oxides are important as synthetic intermediates in the manufacture of pharmaceuticals and agrochemicals. The antiulcer agent omeprazole [73590-58-6] is produced from 2,3,5-trimethylpyridine N-oxide [74409-42-0]. Niflumic acid [4394-00-7] and pranoprofen [52549-17-4] are analgesics and anti-inflammatories, which are manufactured from nicotinic acid N-oxide, obtained either by N-oxidation of nicotinic acid or by hydrolysis of 3-pyridinecarbonitrile N-oxide. Zinc pyrithione [13463-41-7], the zinc salt of 2-pyridinethiol N-oxide [1121-31-9], is a fungicide derived from 2-chloropyridine N-oxide [2402-95-1].

3.5. Piperidines

Physical properties of common piperidines are listed in Table 12.

Piperidines react with alkyl halides and acid anhydrides to give N-alkylpiperidines and amides, respectively.

Table 12. Physical properties of piperidines

Compound	CAS registry no.	M_r	*bp*, °C (kPa)	*mp*, °C	n_D^{20}	d_4^{20}	Flash point, °C
Piperidine	[*110-89-4*]	85.15	106		1.4525	0.861	4
2-Methylpiperidine	[*109-05-7*]	99.18	119		1.4459	0.844	8
3-Methylpiperidine	[*626-56-2*]	99.18	125		1.4470	0.845	17
4-Methylpiperidine	[*626-58-4*]	99.18	124		1.4458	0.838	7
cis-2,6-Dimethylpiperidine	[*766-17-6*]	113.20	126		1.4394	0.840	11
2-Piperidinecarboxylic acid	[*4043-87-2*]	129.16		282			
3-Piperidinecarboxylic acid	[*498-95-3*]	129.16		261			
4-Piperidinecarboxylic acid	[*498-94-2*]	129.16		> 300			
4-Piperidinol	[*5382-16-1*]	101.15	108 – 114 (1.3)				107
4-Benzylpiperidine	[*31252-42-3*]	175.28	279	6 – 7	1.5370	0.997	
2-Piperidylmethanol	[*3433-37-2*]	115.18		68 – 70			
2-(2-Piperidyl)ethanol	[*1484-84-0*]	129.20	234	38 – 40		1.010	102
1-Methylpiperidine	[*626-67-5*]	99.18	106 – 107		1.4378	0.816	3
1-Methyl-2-piperidylmethanol	[*20845-34-5*]	129.20	79 – 80 (0.9)		1.4823	0.984	81

Piperidine and most of its derivatives are easily produced by hydrogenation of the corresponding pyridine derivatives at elevated temperature and pressure over nickel, palladium, or ruthenium catalysts [104]–[108].

A major use of piperidine is in the production of dithiuram tetrasulfide [120-54-7], which is used as a vulcanization accelerator in the rubber industry. Other uses of piperidine are in the production of vasodilators such as dipyridamole and minoxidil, diuretics such as etozolin, and fungicides such as piperalin. Piperidine is also used as a solvent. 2-Methylpiperidine is used for the herbicide piperophos, and cis-2,6-dimethylpiperidine for the antiarrhythmic pirmenol. 4-Benzylpyridine is quaternized and hydrogenated to give a 4-benzylpiperidine derivative that is used as a cerebral vasodilator (ifenprodil tartrate).

3.6. Halopyridines

Physical properties of common halopyridines are listed in Table 13.

The 2- and 4-chloropyridines react with various nucleophiles to give alkyl ether [109], alkyl thioether [110], and alkylamine derivatives [111]. Furthermore, pyridylation of phenylacetonitrile can be achieved by using strong base [112].

Because of its relatively low reactivity, 3-chloropyridine does not undergo nucleophilic attack easily.

Haloalkylpyridines are useful intermediates for pyridylacetonitriles [113] or benzimidazolylthiomethylpyridines [114], [115].

Table 13. Physical properties of halopyridines

Compound	CAS registry no.	M_r	bp, °C (kPa)	mp, °C	n_D^{20}	d_4^{20}	Flash point, °C
2-Chloropyridine	[109-09-1]	113.55	166 (95)	−46.5	1.5320	1.200	65
3-Chloropyridine	[626-60-8]	113.55	148		1.5330	1.194	65
4-Chloropyridine hydrochloride	[7379-35-3]	150.01		210			
2-Bromopyridine	[109-04-6]	158.00	192 – 194		1.5720	1.657	54
3-Bromopyridine	[626-55-1]	158.00	173		1.5700	1.640	51
4-Bromopyridine hydrochloride	[19524-06-2]	194-46		270			
2-Fluoropyridine	[372-48-5]	97.09	126 (100)		1.4660	1.128	28
2-Chloro-3-methylpyridine	[18368-76-8]	127.57	192 – 193 (100)				
2-Chloro-6-methylpyridine	[23468-31-7]	127.57	184/100		1.5270	1.167	73
2,3-Dichloropyridine	[2402-77-9]	147.99	203 – 204	65 – 67			
2,5-Dichloropyridine	[16110-09-1]	147.99	193 – 194	59 – 62			
2,5-Dichloropyridine	[2402-78-0]	147.99	211 – 212	86 – 89			
3,5-Dichloropyridine	[2457-47-8]	147.99	178 – 179	65 – 67			
2-Chloromethylpyridine hydrochloride	[6959-47-3]	164.04		125 – 129			
3-Chloromethylpyridine hydrochloride	[6959-48-4]	164.04		137 – 143			
4-Chloromethylpyridine hydrochloride	[1822-51-1]	164.04		166 – 173			
2,3,5,6-Tetrachloropyridine	[2402-79-1]	216.88	251 – 252	91 – 92			
Pentachloropyridine	[2176-62-7]	251.33	279 – 280	124 – 126			
2-Chloro-5-trifluoromethylpyridine	[52334-81-3]	181.54	152 – 153	30 – 32			
2-Chloro-3-pyridinecarboxylic acid	[2942-59-8]	157.56		175			
6-Chloro-3-pyridinecarboxylic acid	[5326-23-8]	157.56		190			

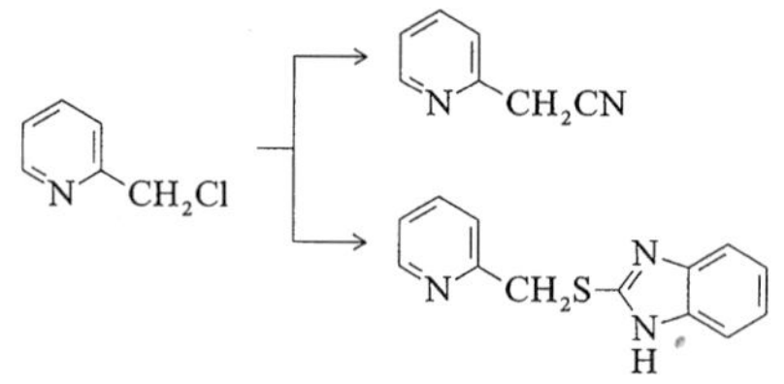

Halogenation of pyridines generally gives a mixture of chlorinated pyridines. For instance, direct chlorination of pyridine with molecular chlorine can be achieved above 270 °C to give 2-chloropyridine and 2,6-dichloropyridine [116]–[118].

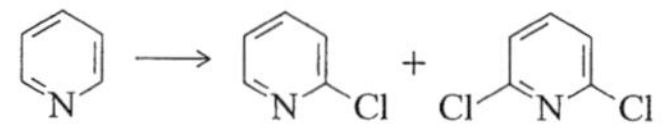

In the chlorination of methylpyridines, reaction occurs first at the side chain and then at a ring position. Liquid-phase chlorination generally leads to chloromethylpyridines [119].

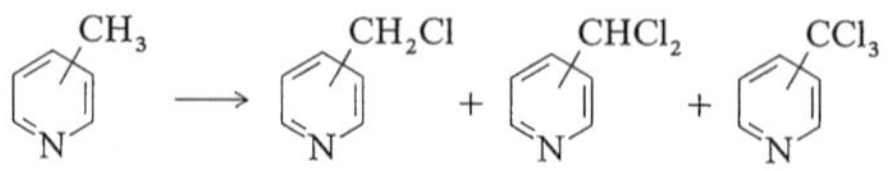

In contrast, gas-phase chlorination tends to give a mixture of trichloromethylchloropyridines [120]–[122]:

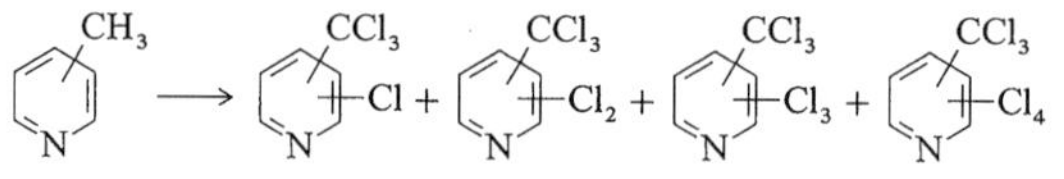

Exhaustive chlorination of pyridines gives pentachloropyridine [123], [124]:

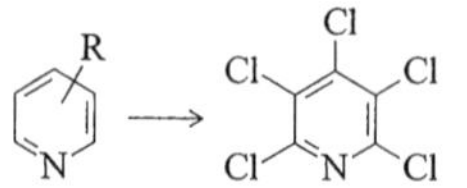

Pentachloropyridine can be reduced by zinc metal to 2,3,5,6-tetrachloropyridine because of enhanced reactivity at the 4-position [125], [126].

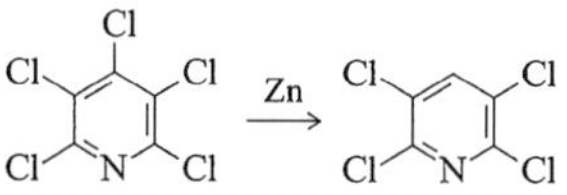

5-Trifluoromethyl-2-chloropyridine is produced directly from 3-methylpyridine by the combined action of chlorine and hydrogen fluoride [127]:

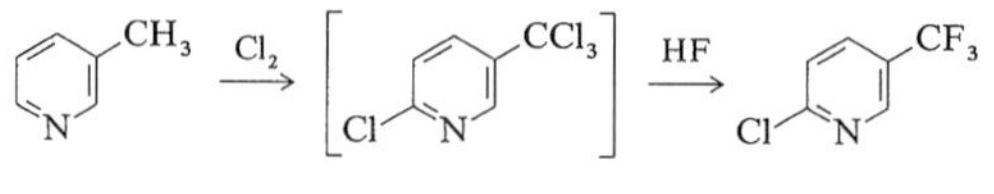

The preparation of 2- and 4-chloropyridines from pyridinols is achieved by the use of halogenating reagents such as phosphoryl chloride [128]–[130].

$$\text{Pyridine-OH} \longrightarrow \text{Pyridine-X} \quad X = Cl, Br, I$$

Pyridyl alcohols react with thionyl chloride or phosphorus halide reagents, such as phosphoryl chloride, phosphorus trihalide, or phosphorus pentahalide, to give the corresponding haloalkylpyridines [114], [115], [131].

$$\text{Pyridine-(CH}_2)_n\text{OH} \longrightarrow \text{Pyridine-(CH}_2)_n\text{X} \quad X = Cl, Br, I$$

Diazotization of aminopyridines in the presence of halide ions is used for the preparation of halopyridines [132]–[135].

$$\text{Pyridine-NH}_2 \longrightarrow \text{Pyridine-X} \quad X = F, Cl, Br, I$$

Chlorination of pyridine N-oxides leads to halopyridines; usually, 2- and 4-isomers are formed by the action of phosphorus halide reagents [98]–[100] (see Section 3.4):

$$\text{Pyridine}(R_n)\text{-N-oxide} \longrightarrow \text{Pyridine}(R_n)\text{-Cl}$$

In industry, vapor-phase chlorination products are important intermediates for pharmaceuticals and agrochemicals.

A major use of 2-chloropyridine is the production of pyrithione, which is used widely as a fungicide. Another use is in the production of insecticides such as pyripropoxyfen. Pyridylation products of phenylacetonitrile lead to antihistamines such as chlorpheniramine and antiarrhythmics such as disopyramide. Reaction of 4-chloropyridine with mercaptoacetic acid gives pyridylmercaptoacetic acid, which is a precursor for cephalosporin antibiotics (e.g., cephapirin sodium salt). 2,6-Dichloropyridine is used for quinoline antibiotics such as enoxacin.

2-Chloro-6-trichloromethylpyridine is used to prepare nitrapyrin, which prevents the loss of ammonia from fertilizers. 2,5-Dichloro-6-trichloromethylpyridine is converted to the herbicide clopyralid. 2,3,4,5-Tetrachloro-6-trichloromethylpyridine is a precursor of the herbicide picloram. 2,3,5,6-Tetrachloropyridine is used for producing the insecticide chlorpyrifos, and 2-chloro-5-trifluoromethylpyridine for the herbicide fluazifop-butyl.

Uses of liquid-phase chlorination products are limited at present. Major uses of 2-chloropyridine-3-carboxylic acid are for the production of herbicides such as diflufenican and anti-inflammatory agents such as niflumic acid and pranoprofen.

Table 14. Physical data of pyridinecarbonitriles, pyridinecarboxamides, and pyridinecarboxylic acids

Compound	CAS registry no.	M_r	mp, °C	bp, °C (kPa)
Nitriles				
2-Pyridinecarbonitrile	[100-70-9]	104.11	26	216–226
3-Pyridinecarbonitrile	[100-54-9]	104.11	51	198–202
4-Pyridinecarbonitrile	[100-48-1]	104.11	76–78	190–226
6-Methyl-2-pyridinecarbonitrile	[1620-75-3]	118.15	71–71.5	112–114 (2.0)
2,6-Pyridinedicarbonitrile	[1452-77-3]	129.12	104–105	143 (2.67)
Amides				
2-Pyridinecarboxamide	[2893-33-6]	122.13	126–127	
3-Pyridinecarboxamide	[98-92-0]	122.13	128–131	
4-Pyridinecarboxamide	[1453-82-3]	122.13	155–156	
Acids				
2-Pyridinecarboxylic acid	[98-98-6]	123.11	137–138	
3-Pyridinecarboxylic acid	[59-67-6]	123.11	236	
4-Pyridinecarboxylic acid	[55-22-1]	123.11	317	
2,3-Pyridinedicarboxylic acid	[89-00-9]	167.12	229–230	
2,4-Pyridinedicarboxylic acid	[499-80-9]	167.12	242–243	
2,6-Pyridinedicarboxylic acid	[499-83-2]	167.12	228	

3.7. Pyridinecarbonitriles, Carboxylic Acids, and Carboxamides

The 3-carbonitrile, 3-carboxylic acid, and 3-carboxamide are important from both a physiological and an industrial point of view.

3-Pyridinecarboxylic acid and 3-pyridinecarboxamide are known as niacin. Niacin is widely distributed in plants and animals. It forms coenzymes, nicotinamide–adenine dinucleotide (NAD) and nicotinamide–adenine dinucleotide phosphate (NADP), which participate in oxidation–reduction cycles in living cells. Since 3-pyridinecarboxylic acid is obtained by oxidation of nicotine with nitric acid, it is commonly called nicotinic acid. 3-Pyridinecarbonitrile and 3-pyridinecarboxamide are referred to as nicotinonitrile and nicotinamide. The prefix isonicotin- is often used to denote the 4-position of pyridinecarbonitrile, -carboxylic acid, and -carboxamide.

Physical data for pyridinecarbonitriles, pyridinecarboxylic acids, and pyridinecarboxamides are listed in Table 14.

Pyridinecarbonitriles. Pyridinecarbonitriles are important intermediates for a variety of pyridine derivatives[136]–[141]:

Table 15. Synthesis of pyridinecarbonitriles

Company	Catalyst	Starting material	Conversion, %	Yield, %	Ref.
Lonza	V_2O_5	2-methylpyridine	62.7	44.2	[142]
	V_2O_5	3-methylpyridine	89.3	83.5	
	V_2O_5	4-methylpyridine	98.4	81.3	
Degussa	Sb_2O_5–V_2O_5–TiO_2– montmorillonite – SiO_2	3-methylpyridine	94	90	[143]
Yuki Gousei	MoO_3–V_2O_5	3-methylpyridine	96.4	83.0	[144]
	MoO_3–Cr_2O_3–Al_2O_3	4-methylpyridine	97.3	82.3	
Takeda Chemical	V_2O_5–Sb_2O_5–Cr_2O_3–TiO_2	3-methylpyridine	100	98.6	[145]
Koei Chemical	V_2O_5–P_2O_5–SiO_2	2-methylpyridine	97.6	79.3	[146]
	V_2O_5–P_2O_5–SiO_2	3-methylpyridine	96.1	82.3	
	V_2O_5–P_2O_5	4-methylpyridine	99.5	94.0	
Nippon Shokubai	V_2O_5–Sb_2O_5–TiO_2–SiO_2–SiC	2-methylpyridine		73	[147]
	V_2O_5–Sb_2O_5–TiO_2–SiO_2–SiC	3-methylpyridine		85	
	V_2O_5–Sb_2O_5–TiO_2–SiO_2–SiC	4-methylpyridine		97	

Pyridinecarbonitriles are usually manufactured by catalytic vapor-phase ammoxidation of alkylpyridines :

$$\text{Py–CH}_3 + 3/2\, O_2 + NH_3 \xrightarrow{\text{cat.}} \text{Py–CN} + 3\, H_2O$$

Generally, 1 to 20 mol of ammonia and 2 to 20 mol of oxygen are used per mole of alkylpyridine. Reaction temperatures range between 280 and 500 °C. Catalysts containing vanadium oxide are commonly used. Examples of the process are given in Table 15, and a flow sheet is illustrated in Figure 4.

3-Methylpyridine is ammoxidized to 3-cyanopyridine in the reactor (a). The reaction gas is quenched with water in the absorber (b), and the condensed mixture is extracted with a solvent in the extraction column (c). 3-Cyanopyridine is separated from the solvent and 3-methylpyridine by three-stage distillation.

3-Pyridinecarbonitrile is used for the production of 3-pyridinecarboxamide (vitamin complex), and 4-pyridinecarbonitrile for antituberculosis agents such as isoniazid [148].

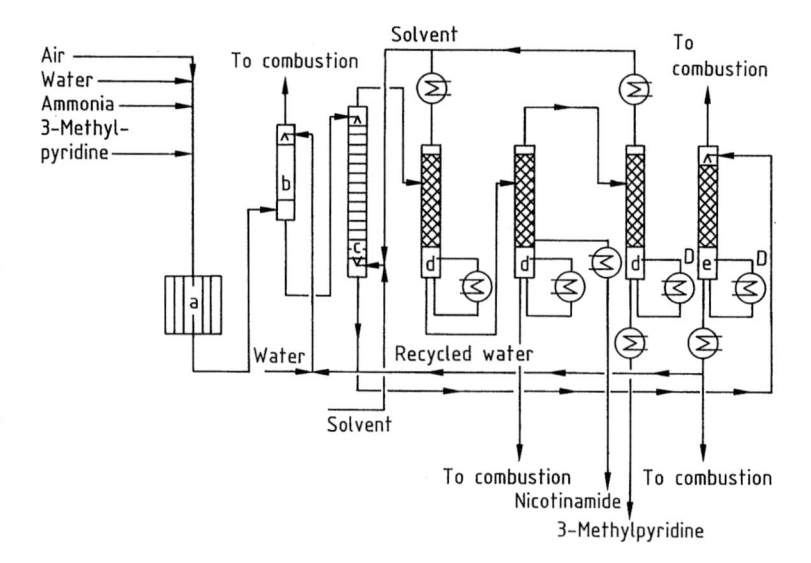

Figure 4. Ammoxidation of 3-methylpyridine
a) Multitubular reactor; b) Absorber; c) Extraction column; d) Fractionating columns; e) Treatment of process water

Pyridinecarboxylic Acids. Pyridinecarboxylic acids behave like ordinary carboxylic acids. Some important reactions are shown below [149], [150]:

Two basic methods are used for the production of pyridinecarboxylic acids: hydrolysis of pyridinecarbonitriles and nitric acid oxidation of alkylpyridines.

Pyridinecarbonitriles are hydrolyzed to pyridinecarboxamides under basic conditions at 0–100 °C; subsequent hydrolysis to pyridinecarboxylic acids is carried out under more severe conditions [151], [152].

Nitric acid oxidation of alkylpyridines (e.g., 5-ethyl-2-methylpyridine to 3-pyridinecarboxylic acid) has been developed commercially by Lonza.

The flow diagram of the process is shown in Figure 5 [153]. 5-Ethyl-2-methylpyridine (MEP) is mixed with nitric acid and fed to a tubular reactor (b). The reaction to form 3-pyridine carboxylic acid nitrate is carried out at 180–370 °C and 2–50 MPa. Nitrogen oxides gases are recovered by the absorber (d) and recycled. After removal of water (e), crystallization of the nitrate (f), and neutralization with MEP (i), the free acid is

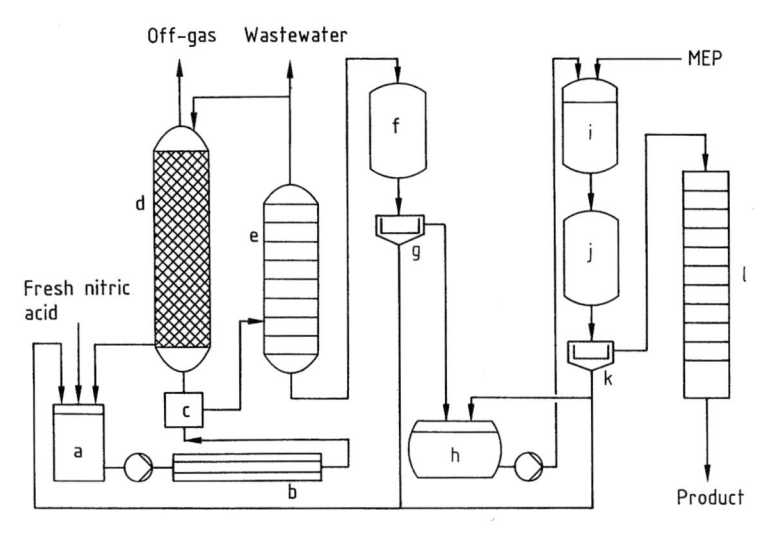

Figure 5. Lonza process for 3-pyridinecarboxylic acid
a) Mixing; b) Reactor; c) Pressure relief; d) Absorption column; e) Distillation column; f) Crystallization; g) Separator; h) Dissolution; i) Neutralization; j) Crystallization; k) Separator; l) Dryer

crystallized (j) and dried. Unreacted nitric acid from the separators (g, k) is recycled to the reactor.

$$\text{5-ethyl-2-methylpyridine} + 4\tfrac{1}{2}\, HNO_3$$

$$\longrightarrow \text{2,3-pyridinedicarboxylic acid} + 4\tfrac{1}{2}\, NO + CO_2 + 3\, H_2O$$

$$\longrightarrow \text{3-pyridinecarboxylic acid} + CO_2$$

3-Pyridinecarboxylic acid has been used mainly as vitamin B_3 or vitamin PP for treating pellagra. Furthermore, it is used as an intermediate for the vasodilator nicorandil. Other uses are for plant growth regulators (inabenfide); antihistamines (terfenadine); and antidepressants (nialamide). 2-Piperidinecarboxylic acid, derived from 2-pyridinecarboxylic acid, is used as intermediate for local anesthetics such as mepivacaine hydrochloride and bupivacaine hydrochloride. 2,3-Pyridinedicarboxylic acid is used as an intermediate for the herbicide imazapyr.

Pyridinecarboxamides. Pyridinecarboxamides are used for the production of pyridinecarboxylic acids [152] and aminopyridines [154].

Table 16. Physical properties of aminopyridines

Compound	CAS registry no.	M_r	mp, °C	bp, °C (kPa)
2-Aminopyridine	[504-29-0]	94.12	59–60	210
3-Aminopyridine	[462-08-8]	94.12	64–65	260
4-Aminopyridine	[504-24-5]	94.12	159	273
2-Amino-6-methylpyridine	[1824-81-3]	108.14	40	208–209
2,6-Diaminopyridine	[141-86-6]	109.13	121–122	170 (4.0)
N,N-Dimethyl-4-pyridinamine	[1122-58-3]	122.17	112–113	145–150 (4.0)

Pyridinecarboxamides are prepared by hydrolysis of pyridinecarbonitriles (see p. 4203) [155].

3-Pyridinecarboxamide and 3-pyridinecarboxylic acid are important as vitamin B_3 [156]. 4-Pyridinecarboxamide is used for producing antibiotics (e.g., cefsulodin sodium) [157].

3.8. Aminopyridines

Table 16 lists commercially available aminopyridines and their physical properties.

3.8.1. 2-Aminopyridine

2-Aminopyridine [504-29-0] exists as tautomeric amino and imino forms. The amino form predominates over the imino form, and their ratio is generally 1000:1.

2-Aminopyridine generally reacts with alkylating agents at the ring nitrogen to give derivatives of type **5**. In the presence of sodium amide or sodium methoxide, however, it reacts with alkylating agents at the exocyclic nitrogen. Such reactions are used to produce antihistamines of type **6**.

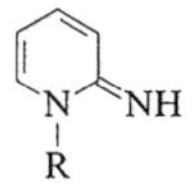

Electrophilic reactions such as halogenation and nitration occur at the positions *para* and *ortho* to the amino group [158] – [160]:

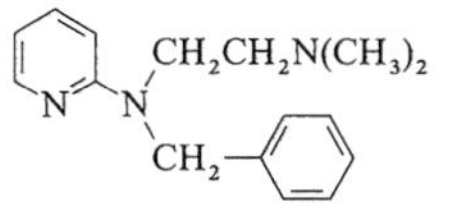

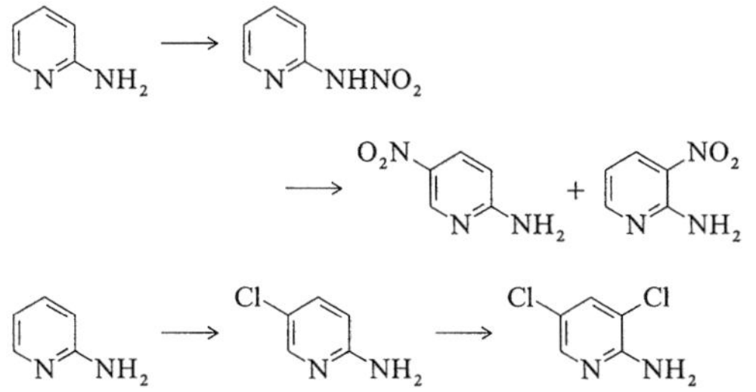

The most important method for the manufacture of 2-aminopyridine is the reaction of pyridine with sodium amide (Chichibabin amination). Upon hydrolysis of the intermediate sodium salt, 2-aminopyridine is obtained in high yield [161] – [164]:

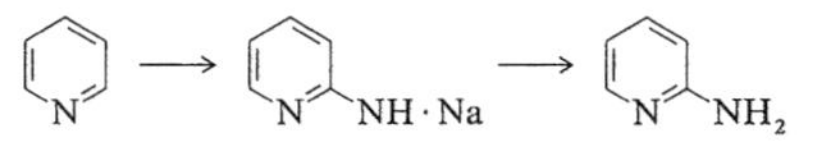

The major uses of 2-aminopyridine are as an intermediate for pharmaceuticals such as sulfapyridine [*144-83-2*], tripelennamine [*91-81-6*], piroxicam [*36322-90-4*], and tenoxicam [*59804-37-4*], as well as a variety of agrochemicals.

3.8.2. Other Aminopyridines

2-Amino-6-methylpyridine. 2-Amino-6-methylpyridine [*1824-81-3*] is produced from 2-methylpyridine and sodium amide. It is converted to nalidixic acid [*389-08-2*], an antibacterial agent [165].

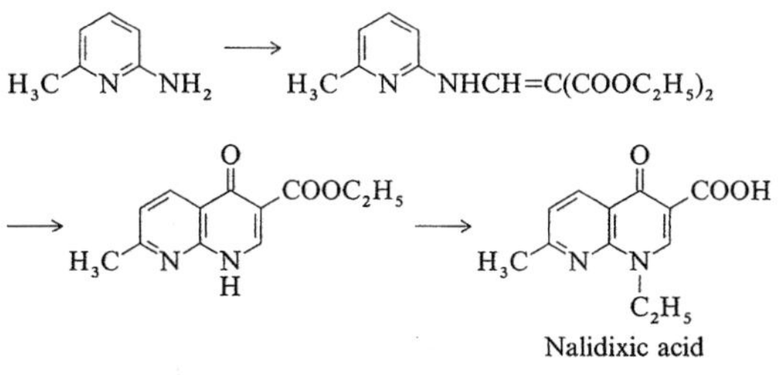

Nalidixic acid

2,6-Diaminopyridine. Amination of pyridine or 2-aminopyridine under severe conditions gives 2,6-diaminopyridine [*141-86-6*], the coupling of which with benzenediazonium salts gives the antiseptic phenazopyridine [136-40-3] [166]. 2,6-Diaminopyridine is also utilized for the production of polyamides.

3-Aminopyridine and 4-Aminopyridine. 3-Aminopyridine [462-08-8] and 4-aminopyridine [504-24-5] are produced from the corresponding pyridinecarboxamides and sodium hypochlorite in alkaline solution (Hofmann reaction).

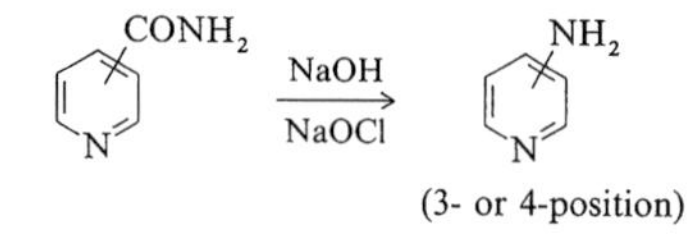

$$\text{Pyridine-CONH}_2 \xrightarrow[\text{NaOCl}]{\text{NaOH}} \text{Pyridine-NH}_2$$

(3- or 4-position)

These aminopyridines are used as intermediates for pharmaceuticals such as pinacidil [85371-64-8] and for agrochemicals.

N,N-Dimethyl-4-pyridinamine [1122-58-3] is widely employed as a supernucleophilic catalyst for many organic reactions. Two processes are used for its production:

1) N-(4-Pyridyl)pyridinium halide, prepared from pyridine and halogenating agents, is reacted with dimethylamine or N,N-dimethylformamide to give N,N-dimethyl-4-pyridinamine [167]–[170].

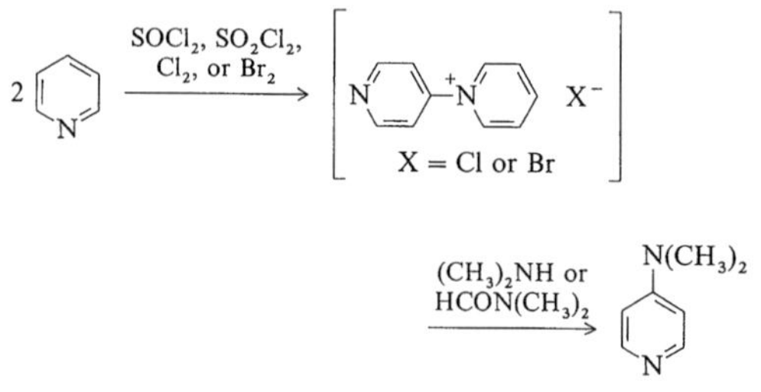

$$2\ \text{Pyridine} \xrightarrow[\text{Cl}_2,\text{ or Br}_2]{\text{SOCl}_2,\text{ SO}_2\text{Cl}_2,} [\text{N-pyridyl-pyridinium}]^+ X^-$$

X = Cl or Br

$$\xrightarrow[\text{HCON(CH}_3)_2]{\text{(CH}_3)_2\text{NH or}} \text{4-N(CH}_3)_2\text{-pyridine}$$

2) 1-[2-(2-Pyridyl)ethyl]-4-cyanopyridinium chloride, prepared from 4-pyridinecarbonitrile, 2-vinylpyridine, and hydrogen halide, gives N,N-dimethyl-4-pyridinamine by a reaction similar to that described above [171].

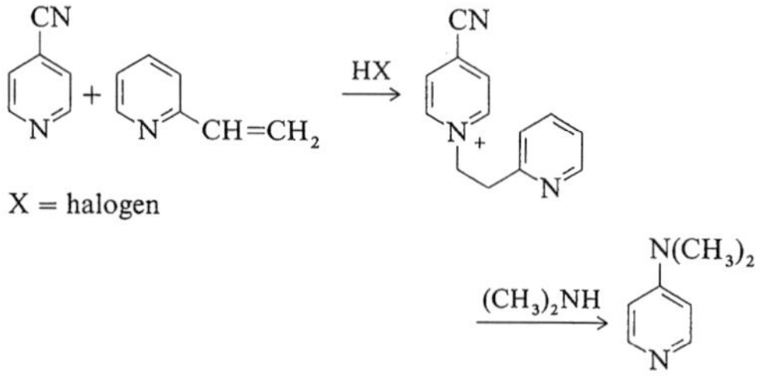

$$\text{4-CN-pyridine} + \text{2-vinylpyridine} \xrightarrow{\text{HX}} \text{pyridinium salt}$$

X = halogen

$$\xrightarrow{\text{(CH}_3)_2\text{NH}} \text{4-N(CH}_3)_2\text{-pyridine}$$

The major uses are as a catalyst for acylation, alkylation, halogenation, cyanation, and silylation, and as an accelerator in the manufacture of polyurethanes.

Table 17. Physical properties of pyridinols

Compound	CAS registry no.	M_r	bp, °C (kPa)	mp, °C
2-Pyridinol	[142-08-5]	95.10	280–281	105–107
3-Pyridinol	[109-00-2]	95.10	151–153 (0.4)	126–129
4-Pyridinol	[626-64-2]	95.10	230–235 (1.6)	146–149
6-Methyl-2-pyridinol	[3279-76-3]	109.13		158–160
6-Methyl-3-pyridinol	[1121-78-4]	109.13		168–170
5-Chloro-2-pyridinol	[4214-79-3]	129.55		163–165
2-Nitro-3-pyridinol	[15128-82-2]	140.10		69–71

3.9. Pyridinols

Physical properties of typical pyridinols are listed in Table 17.

Tautomerization of pyridinols is discussed in [172]; both 2- and 4-pyridiols prefer the ketonic pyridone tautomer to the hydroxylic tautomer.

In contrast, 3-pyridinol is a phenolic compound because the ketonic tautomer does not exist.

Chlorination of 2- and 4-pyridinols with phosphorus halides gives halopyridines [126]–[128].

Electrophilic substitution of pyridinols is facile because of the electron-donating nature of the hydroxyl group. Reaction occurs at positions *ortho* or *para* to the hydroxyl group, similar to phenols [173]–[176].

When 2- and 4-pyridinols are alkylated with alkyl halides, they give a mixture of the O- and N-alkylated products [177], but 3-pyridinol gives only the O-alkylated product [178].

Diazotization of aminopyridines is used for the production of pyridinols [179]–[182].

2-Pyridinol is also obtained from pyridine N-oxide by reaction with acid anhydrides and subsequent hydrolysis [183], [184].

2-Alkyl-3-pyridinols are reported to be formed from alkanoylfurans and ammonia under pressure [185]–[187].

2-Pyridinol is a synthetic intermediate for the tranquilizer amphenidone. 3-Pyridinol is used for synthesizing the cholinesterase inhibitors pyridostigmine bromide and distigmine bromide. 4-Pyridinol is a precursor of the X-ray contrast agent propylio-

Table 18. Physical properties of pyridyl alcohols

Compound	CAS registry no.	M_r	bp, °C (kPa)	mp, °C	n_D^{20}	d_4^{20}	Flash point, °C
2-Pyridylmethanol	[586-98-1]	109.13	112–113 (0.5)		1.5440	1.131	110
3-Pyridylmethanol	[100-55-0]	109.13	154 (2.9)		1.5460	1.124	130
4-Pyridylmethanol	[586-95-8]	109.13	107–110 (0.1)	57–59			112
2,6-Pyridyldimethanol	[1195-59-1]	139.15	185 (2.0)	114–116			
2-(2-Pyridyl)ethanol	[103-74-2]	123.16	114–116 (1.2)		1.5370	1.093	92
2-(4-Pyridyl)ethanol	[5344-27-4]	123.16	262				160

done. It is hydrogenated to give 4-piperidinol, which is used to produce the psychotropic agent pericyazine.

3.10. Pyridyl Alcohols

Physical properties of common pyridyl alcohols are listed in Table 18.

Reactivities of pyridyl alcohols are similar to ordinary alcohols; for example, chloroalkylpyridines are formed by the action of thionyl chloride or phosphorus halide reagents [130], [131], and dehydration gives alkenylpyridines.

Pyridylmethanols. The simple pyridylmethanols are produced by catalytic hydrogenation of pyridinecarbonitriles in acidic media [188].

Substituted 2-pyridylmethanols can be produced from 2-methylpyridine N-oxides by the action of acetic anhydride, followed by hydrolysis [130], [131], [189].

2-Pyridylmethanol is used in the preparation of the anti-inflammatory ibuprofen piconol. 3-Pyridylmethanol is used as an intermediate for the vasodilator nicotinic alcohol. 2,6-Pyridyldimethanol is carbamated to give the antiarteriosclerotic agent pyridinol carbamate.

Pyridylethanols. Condensation of 2- or 4-methylpyridines with formaldehyde or other aliphatic aldehydes gives the corresponding pyridylethanols due to the high acidity of the methyl group in the 2- or 4-position [18], [190]–[193].

$$\text{Py-CH}_3 \longrightarrow \text{Py-CH}_2\text{CH}_2\text{OH}$$

Pyridylethanols are used mainly for the manufacture of vinylpyridines (see Section 3.1). Other uses of 2-(2-pyridyl)ethanol are in the production of a psychotropic agent thioridazine and the antispasmodic tiquizium bromide.

3.11. Pyridinecarbaldehydes

Physical properties of typical pyridinecarbaldehydes are listed in Table 19.

Reactions of pyridinecarbaldehydes are similar to those of aromatic aldehydes: benzoin condensation [194], oxidation to pyridinecarboxylic acids [195], reduction to pyridylmethanols [196], oximation [197], and hydrazone formation [198].

Pyridinecarbaldehydes are prepared from pyridylmethanols by oxidation with manganese dioxide [199].

Table 19. Physical properties of pyridylcarbaldehydes

Compound	CAS registry no.	M_r	bp, °C (kPa)	mp, °C	n_D^{20}	d_4^{20}	Flash point, °C
2-Pyridinecarbaldehyde	[1121-60-4]	107.11	181		1.5370	1.126	54
3-Pyridinecarbaldehyde	[500-22-1]	107.11	95–97 (0.9)		1.5490	1.135	60
4-Pyridinecarbaldehyde	[872-85-5]	107.11	77–78 (1.6)		1.5440	1.122	54
2-Pyridine aldoxime	[873-69-8]	122.13		110–112			
3-Pyridine aldoxime	[1193-92-6]	122.13		150–153			
4-Pyridine aldoxime	[696-54-8]	122.13		130–133			

The oxidation of methylpyridines with selenium dioxide gives pyridinecarbaldehydes [200], which are also obtained by vapor-phase oxidation of methylpyridines over a vanadium–molybdenum–titanium oxide catalyst [201].

The catalytic partial reduction of 3-pyridinecarbonitrile with hydrogen over a palladium catalyst in acidic media gives 3-pyridinecarbaldehyde [202].

Quaternary salts of pyridinecarbaldehyde oximes, particularly pralidoxime methiodide, are used as antidotes for poisoning by organophosphate acetylcholinesterase inhibitors. Another use of 2-pyridinecarbaldehyde is in the production of bisacodyl, a laxative.

3.12. Pharmaceuticals and Agrochemicals

Typical pharmaceuticals and agrochemicals derived from pyridine are listed in Table 20.

4. Toxicology

4.1. Acute Toxicity

The acute toxicity of pyridine and its common derivatives is listed in Table 21.

Pyridine has a narcotic action. Toxic doses cause weakness, ataxia, unconsciousness, and salivation. In humans, pyridine is readily absorbed through the lungs, gastrointestinal tract, and skin. In 1893 a human fatality resulted from the accidental ingestion

Table 20. Pharmaceuticals and agrochemicals derived from pyridine

Compound	Common name	Developer	Use
Pharmaceuticals			
	omeprazole [73590-58-6]	Astra	antiulcer
	niflumic acid [4394-00-7]	Heydon	anti-inflammatory
	pranoprofen [52549-17-4]	Yoshitomi	anti-inflammatory
	dipyridamole [58-32-2]	Thomae	vasodilator
	minoxidil [38304-91-5]	Upjohn	vasodilator
	etozolin [73-09-6]	Warner-Lambert	diuretic
	pirmenol [61477-94-9]	Parke, Davis	antiarrhythmic
	ifenprodil tartrate [23210-56-2]	Funai	cardiovascular
	chlorpheniramine maleate [2438-32-6]	Schering	antihistamine
	disopyramide [3737-09-5]	Roussel Uclaf	antiarrhythmic
	cephapirin sodium [24356-60-3]	Bristol-Myers	antibiotic

Table 20. continued

Compound	Common name	Developer	Use
	enoxacin [74011-58-8]	Dainippon	antibiotic
	isoniazid [54-85-3]	Bayer	antituberculous
	nicorandil [65141-46-0]	Tyugai	vasodilator
	terfenadine [50679-08-8]	Merrell Dow	antihistamine
	nialamide [51-12-7]	Pfizer	monoamine oxidase
	mepivacaine hydrochloride [1722-62-9]	Astra	local anesthetic
	bupivacaine hydrochloride [14252-80-3]	Astra	local anesthetic
	cefsulodin sodium [52152-93-9]	Takeda	antibiotic
	sulfapyridine [144-83-2]	Visuvia	chemotherapeutic
	tripelennamine [91-81-6]	Ciba-Geigy	antihistamine
	tenoxicam [59804-37-4]	Roche	anti-inflammatory
	piroxicam [36322-90-4]	Pfizer	anti-inflammatory
	nalidixic acid [389-08-2]	Winthrop	chemotherapeutic
	pinacidil [85371-64-8]	Leo Denmark	antihypertensive
	amphenidone [134-37-2]	Wallace & Tiernan	tranquilizer

Table 20. continued

Compound	Common name	Developer	Use
	pyridostigmine bromide [101-26-8]	Roche	autonomic anticholinesterase
	distigmine bromide [15876-67-2]	Chemie Linz	autonomic anticholinesterase
	propyliodone [587-61-1]	Glaxo	X-ray contrast
	pericyazine [2622-26-6]	Rhône-Poulenc	psychotropic
	ibuprofen piconol [112017-99-9]	Hisamitu/Torii	anti-inflammatory
	nicotinyl tartrate [100-55-0]	Roche	vasodilator
	pyridinol carbamate [1882-26-4]	Banyu	antiarteriosclerotic
	thioridazine [50-52-2]	Sandoz	psychotropic
	tiquizium bromide [71731-58-3]	Hokuriku	antispasmodic
	pralidoxime iodide [51-15-0]	Wyeth-Ayerst	antidote
	bisacodyl [603-50-9]	Thomae	laxative
Agrochemicals			
	diquat dibromide [85-00-7]	ICI	herbicide
	paraquat dichloride [1910-42-5]	ICI	herbicide

Table 20. continued

Compound	Common name	Developer	Use
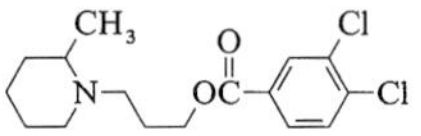	piperalin [3478-94-2]	Eli Lilly	fungicide
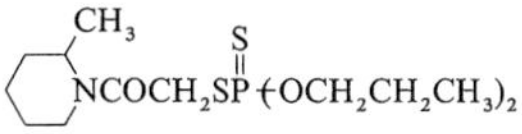	piperophos [24151-93-7]	Ciba-Geigy	herbicide
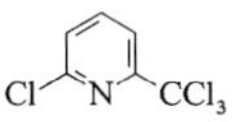	nitrapyrin [1929-82-4]	Dow	prevents loss of fertilizer
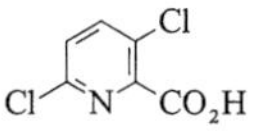	clopyralid [1702-17-6]	Dow	herbicide
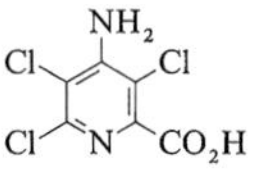	picloram [1918-02-1]	Dow	herbicide
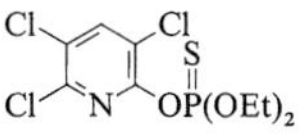	chlorpyrifos [2921-88-2]	Dow	insecticide
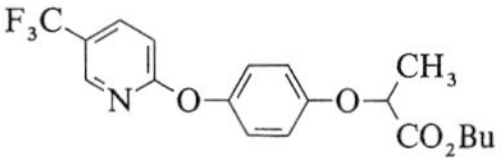	fluazifop-butyl [69806-50-4]	Ishihara	herbicide
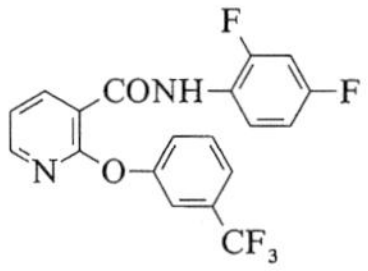	diflufenican [83164-33-4]	May & Baker	herbicide
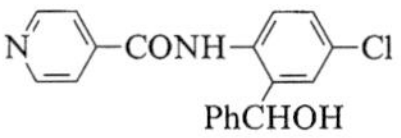	inabenfide [82211-24-3]	Tyugai	growth regulator
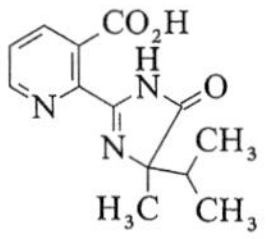	imazapyr [81334-34-1]	ACC	herbicide

Others

Compound	Common name	Developer	Use
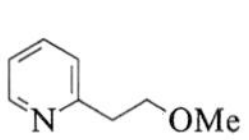	metyridine [114-91-0]	ICI	anthelmintic
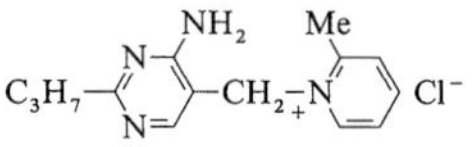	amprolium [121-25-5]	Merck	coccidiostat
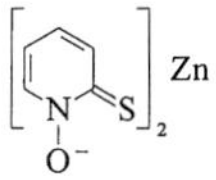	zinc pyrithione [13463-41-7]	Olin	fungicide
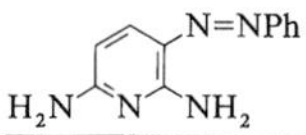	phenazopyridine [136-40-3]	Gödecke	antiseptic

Table 21. Acute toxicity of pyridine and derivatives*

Compound	CAS registry no.	LD$_{50}$ (oral), mg/kg	LCLo (inhalation), ppm/h	LD$_{50}$ (i.p.), mg/kg	LD$_{50}$(i.v.), mg/kg
Pyridine	[110-86-1]	R: 891	R (LC$_{50}$): 9010/1	R: 866	
Pyridine HCl	[628-13-7]	R (LDLo): 1600		R: 800	
2-Methylpyridine	[109-06-8]	R: 790	R: 4000/4	M: 529	
3-Methylpyridine	[108-99-6]	R: 400	R: 8700/2	M: 596	M: 298
4-Methylpyridine	[108-89-4]	R: 1290	R: 100/4	M: 335	
2,4-Dimethylpyridine	[108-47-4]	R: 200			
2,6-Dimethylpyridine	[108-48-5]	R: 400	R: 7500/1		
2-Methyl-5-ethylpyridine	[104-90-5]	R: 368	R: 1000/4		
2,4,6-Trimethylpyridine	[108-75-8]	R: 400	R: 2500/2		
2-Chloropyridine	[109-09-1]	M: 110	R: 100/4	M: 130	
2,3-Dichloropyridine	[2402-77-9]			M: 135	
2,5-Dichloropyridine	[16110-09-1]			M: 1690	
2,6-Dichloropyridine	[2402-78-0]	M: 176		M: 115	
Pentachloropyridine	[2176-62-7]			M: 235	
2-Chloro-6-(trichloromethyl)pyridine	[1929-82-4]	R: 940, M: 710			
2-Chloromethylpyridine HCl	[6959-47-3]	R: 316, M: 316			
3-Chloromethylpyridine	[6959-48-4]	R: 316, M: 316			
2-Aminopyridine	[504-29-0]	R: 200	Hum (TCLo): 5/5	M: 28	R: 29
3-Aminopyridine	[462-08-8]	Q: 178		M: 28	M: 24
4-Aminopyridine	[504-24-5]	R: 21, Hum (LDLo): 0.59		R: 6.5	
2,3-Diaminopyridine	[452-58-4]			M: 25	
2,6-Diaminopyridine	[141-86-6]			M: 100	M: 56
4-Dimethylaminopyridine	[1122-58-3]	M (LDLo): 470			M: 56
2-Aminomethylpyridine	[3731-51-9]	Q: 750			M: 340
2-Vinylpyridine	[100-69-6]	R: 100	R: 5500/1.5		
4-Vinylpyridine	[100-43-6]	R: 100	R: 2000/2		
3-Cyanopyridine	[100-54-9]	R: 1185		M (LD$_{25}$): 800	
Picolinic acid	[98-98-6]	Q: 562		M: 360	M: 487
Nicotinic acid	[59-67-6]	R: 7000		M: 358	M: 5000
Isonicotinic acid	[55-22-1]	R: 5000, M: 3123			M: 5000
Methyl nicotinate	[93-60-7]			M (LDLo): 2000	
Ethyl isonicotinate	[1570-45-2]				M: 56
Nicotinamide	[98-92-0]	R: 3500		M: 2050	R: 2200
Isonicotinic acid hydrazide	[54-85-3]	R: 1250		M: 100	M: 149
Pyridine 1-oxide	[694-59-7]	Q: 1000		M: 1425	M: 180
2-Pyridinemethanol	[586-98-1]	Q: 1000			M: 1000
2-Acetylpyridine	[1122-62-9]	R: 2280			
2-Benzoylpyridine	[91-02-1]			M: 475	
Nicotinaldehyde	[500-22-1]			M: 720	
2-Pyridinaldoxime	[873-69-8]			M: 200	
2-Hydroxypyridine	[142-08-5]			M: 410	M: 750
2-Pyridinethione	[2637-34-5]	M: 533		M: 250	M: 250
2,2'-Bipyridine	[366-18-7]	R: 100		M: 200	
Diquat dibromide	[2764-72-9]	R: 215			
4,4'-Bipyridine	[553-26-4]	R: 172			
Paraquat dichloride	[1910-42-5]	R: 57, Hum (LDLo): 43		R: 26	R: 21
4-Phenylpyridine	[939-23-1]				M: 89

Table 21. (continued)

Compound	CAS registry no.	LD$_{50}$ (oral), mg/kg	LCLo (inhalation), ppm/h	LD$_{50}$ (i.p.), mg/kg	LD$_{50}$ (i.v.), mg/kg
Hexadecylpyridinium chloride	[6004-24-6]	R: 200		M (LDLo): 3	
4-Nitropyridine N-oxide	[1124-33-0]	R: 107			
Nicotine	[54-11-5]	R: 50, Hum (LDLo): 0.88			
Niflumic acid	[4394-00-7]	R: 250, M: 375		M: 196	M: 152
Piperidine	[110-89-4]	R: 400, M: 30	R: 4000/4	M: 50	

* R = rat; M = mouse; Q = quail; Hum = human.

of half a cupful of commercial pyridine [203]. OSHA states that a vapor concentration of 3600 ppm in air is immediately dangerous to life and health [204]. The LDLo (oral) in humans is 500 mg/kg [205].

The acute toxicity of monoalkylpyridines is similar to that of pyridine, but some dimethyl- and trimethylpyridines are several times more toxic than pyridine. 2-Methylpyridine is one of the most important commercially produced alkylpyridines; its toxicity and safety data are summarized in [206].

Of aminopyridines, the toxicity of 4-aminopyridine has been thoroughly investigated because of its high toxicity and its use as a bird repellent. The acute toxicity of 4-aminopyridine has been determined for 41 species of birds and mammals. The oral LD$_{50}$ values are generally < 10 mg/kg [207], and the LDLo (oral) in humans is 0.59 mg/kg. Most of the pharmacological effects of 4-aminopyridine can be attributed to its neurotoxicity [208].

The calculated fatal dose of 2-aminopyridine for a man of 70-kg weight is ca. 5 g [209]. A fatal accident involving 2-aminopyridine was described as follows: an employee died 1.5 h after spilling 2-aminopyridine on his clothes during its distillation and continuing to work. Aminopyridines are toxic if inhaled, swallowed, or absorbed through skin [209].

The acute toxicity of 13 chlorinated pyridines has been investigated [210], including monochloro to pentachloro derivatives. In the comparison of relative toxicity to mice by intraperitoneal injection, 2,6-dichloropyridine shows the highest toxicity (LD$_{50}$ = 115 mg/kg) and 2,5-dichloropyridine the lowest (LD$_{50}$ = 1690 mg/kg).

The toxicity of 2-chloropyridine to rabbits after dermal application (LD$_{50}$ = 64 mg/kg) and by intraperitoneal injection (LD$_{50}$ = 48 mg/kg) indicates that 2-chloropyridine is readily absorbed through the skin.

3-Acetylpyridine causes central neurotoxic effects upon intraperitoneal injection of 70–80 mg/kg in rats [211]. The zinc salt of 1-hydroxypyridine-2-thione (HPT), known as zinc pyrithione (ZnPT), is used as an antidandruff agent. The oral LD$_{50}$ of HPT in mice is 533 mg/kg. The metal complexes (Zn, Cd) are more toxic than HPT itself. Percutaneous and eye toxicity is also discussed in [212].

Paraquat (1,1'-dimethyl-4,4'-bipyridinium dichloride), which consumes the largest amount of pyridine as starting material, exhibits very high toxicity. The oral LDLo in humans is 43 mg/kg. Nicotine is one of the well-known pyridine derivatives from natural sources, and its toxicity has been investigated in detail. The LDLo in humans is 0.882 mg/kg. Reports on the toxicity of paraquat and nicotine can be found in [213].

Vinylpyridines irritate the skin intensely, and contact with the vapor or liquid generally causes inflammation or blisters. They also show high toxicity by inhalation.

4.2. Subacute and Chronic Toxicity

Few long-term studies can be found on the toxicity of pyridine to animals. Six rats were fed a basal diet containing 0.1 % pyridine. Five rats died during the 14th to 32nd days. The livers and kidneys revealed acute lesions and some of the livers exhibited pronounced cirrhosis [214].

When humans are exposed to pyridine vapor, the toxicity affects mainly the central nervous system and the gastrointestinal tract. Symptoms are headache, dizziness, nausea, insomnia, anorexia, and weakness of the limbs. As examples, the cases of a chemist who had worked with pyridine for six months and of a man who had worked with it for 13 years are reported in [215]. Formerly, pyridine was used as an anticonvulsant, and an epileptic who received 1.8 – 2.5 mL/d of pyridine for one month died as a result of liver and kidney damage [216].

In 1983 LANGSTON first reported that 1-methyl-4-phenyl-1,2,3,6-tetrahydropyridine (MPTP) causes acute symptoms of parkinsonism [217]; more than 1000 papers have been published, mainly in the United States, in the following years [218].

4.3. Mutagenicity and Ecotoxicity

Mutagenic pyridine derivatives include 2,2'-bipyridine, 4-nitropyridine N-oxide derivatives, chloropyridine derivatives, chloromethylpyridine derivatives, 2,6-diaminopyridine, 4-phenylpyridine, 2-hydroxy-3-nitropyridine, isonicotinic acid hydrazide, and 2-pyridine aldoxime.

Of these compounds, chloromethylpyridine derivatives [219] and 4-nitropyridine N-oxide derivatives [220] have been investigated with respect to their carcinogenicity.

Pyridine has a characteristic disagreeable, nauseating odor, detectable at < 1 ppm. The odor threshold is 0.021 ppm [221]. TLV and MAK values (in ppm) in several countries are as follows:

Australia	5
Germany	5
Italy	2
Japan	5
CIS	1.7
United States	5

The acute toxicity of pyridine for *Daphnia magna* is 240 mg/L, and the 24-h LC$_{50}$ is 170 mg/L for newborns and 944 mg/L for adults. Reported acute toxic concentrations in fish vary widely and range from 26 to 1300 mg/L [204].

5. References

[1] T. Anderson, *Justus Liebigs Ann. Chem.* **60** (1846) 86.
[2] T. Anderson, *Justus Liebigs Ann. Chem.* **80** (1851) 44.
[3] W. Ramsay, *Ber. Dtsch. Chem. Ges.* **10** (1877) 736.
[4] A. Hanzsch, *Justus Liebigs Ann. Chem.* **215** (1882) 72.
[5] A. E. Chichibabin, *Zh. Russ. Fiz. Khim. O-va.* **37** (1905) 1229.
[6] R. Kuhn, G. Wendt, *Ber. Dtsch. Chem. Ges. B* **71** (1938) 1118.
[7] P. Karrer, H. Heller, *Helv. Chim. Acta* **22** (1939) 1292.
[8] A. W. Singer, S. M. McElvain, *Org. Synth.*, coll. vol. **3** (1955) 740.
[9] C. F. Koelsch, A. F. Steinhauer, *J. Org. Chem.* **18** (1953) 1516.
[10] J. Braun, A. Pelgold, J. Zeemann, *Ber. Dtsch. Chem. Ges.* **55** (1922) 3783.
[11] H. J. den Hertog, J. Overhoff, *Recl. Trav. Chim. Pays-Bas* **49** (1930) 552.
[12] H. Meyer, W. Ritter, *Monatsh. Chem.* **35** (1914) 765.
[13] A. E. Chichibabin, A. Zeide, *Zh. Russ. Fiz. Khim. O-va.* **46** (1914) 1216.
[14] R. A. Abramovich, F. Helmer, J. G. Saha, *Can. J. Chem.* **43** (1965) 725. R. A. Abramovich, J. G. Saha, *Adv. Heterocycl. Chem.* **6** (1966) 229.
[15] M. G. Reinecke, L. R. Kray, *J. Am. Chem. Soc.* **86** (1964) 5355.
[16] P. Riehm, *Justus Liebigs Ann. Chem.* **238** (1887) 1.
[17] J. Herzenberg, R. Covini, M. Pieroni, A. Nenz, *Ind. Eng. Chem. Prod. Res. Dev.* **6** (1967) no. 3, 195–197.
[18] Allied Chem. & Dye Co., US 2 556 845, 1947.
[19] A. E. Chichibabin, *J. Prakt. Chem.* **107** (1924) 122.
[20] Koei Chemical Co., DE 1 770 870, 1967.
[21] Nippon Kayaku, JP 71 39 873, 1969; *Chem. Abstr.* **76** (1972) 34112x.
[22] H. Beschke, H. Friedrich, *Chem. Ztg.* **101** (1977) 377–384.
[23] ICI, DE-AS 1 917 037, 1968.
[24] Nippon Kayaku, JP 70 39 545, 1967; *Chem. Abstr.* **74** (1971) 141558c.
[25] Koei Chemical Co., BE 758 201, 1969.
[26] Daicel Chemical, JP 86 14 859, 1979.
[27] ICI, BE 845 405, 1975.
[28] Rütgerswerke, DE 2 203 384, 1972.
[29] Nepera Chemical Co. Inc., US 512 834, 1983.
[30] Koei Chemical Co., EP-A 232 182, 1987.
[31] Koei Chemical Co., JP-Kokai 61 53 265, 1984.
[32] I. Liepina et al., *Latv. PSR Zinat. Akad. Vestis, Kim. Ser.* 1989, no. 3, 334–338; *Chem. Abstr.* **111** (1989) 232533q.
[33] Koei Chemical Co., DE-OS 2 064 397, 1984.
[34] Degussa, DE-OS 2 703 069, 1978.
[35] Distillers, GB 817 038, 1957.

[36] Degussa, DE-OS 2 639 702, 1976.
[37] Stamicarbon N.V., NL 7 013 453, 1970.
[38] Stamicarbon N.V., NL 7 809 552, 1978.
[39] Dynamit Nobel, DE-OS 2 514 004, 1975. Dynamit Nobel, DE-OS 2 519 529, 1975. The Rummus Co.; Dynamit Nobel, DE-OS 2 729 072, 1977.
[40] BASF, DE 3 104 765, 1981.
[41] Daicel Chemical, JP-Kokai 81 127 357, 1980; *Chem. Abstr.* **96** (1982) 35109b.
[42] Stamicarbon N.V., NL 7 017 718, 1970.
[43] Agency of Industrial Sciences and Technology, Japan Steel Works Ltd., JP-Kokai 89 70 464, 1987; *Chem. Abstr.* **111** (1989) 77861z.
[44] F. Mensch, *Erdoel Kohle Erdgas Petrochem.* **22** (1969) no. 2, 67 – 71.
[45] A. Nenz, M. Pieroni, *Hydrocarbon Process.* **47** (1968) no. 11, 139 – 144. R. L. Frank, F. J. Pilgrim, E. F. Riener, *Org. Synth.*, coll. vol. **4** (1963) 451.T. Shoji, N. Abe, *Petrotech (Tokyo)* **7** (1984) 195;*Chem. Abstr.* **101** (1984) 56791e.
[46] H. Boenemann, *Angew. Chem.* **90** (1978) 517 – 526; *Angew. Chem. Int. Ed. Engl.* **17** (1978) 505.
[47] Lonza, US 4 212 978, 1978.
[48] Studienges. Kohle mbH, DE 2 840 460, 1978.
[49] Y. Kusunoki, H. Okazaki, *Hydrocarbon Process.* **53** (1974) no. 11, 129 – 131.
[50] Hoechst, DE-OS 3 244 032, 1984.
[51] Du Pont, US 655 087, 1984.
[52] U. D. Nazirova et al., *Dokl. Akad. Nauk UzSSR* 1988, no. 2, 40 – 42; *Chem. Abstr.* **109** (1988) 230746w.
[53] Mobil Oil Corp., EP-A 82 613, 1983.
[54] *Reagent Chemicals,* American Chemical Society, Washington, D.C., 1977, p. 724.
[55] H. E. Reich et al., *J. Am. Chem. Soc.* **77** (1955) 5434.
[56] L. M. Jampolsky et al., *J. Am. Chem. Soc.* **74** (1952) 5222.
[57] Koei Chemical, JP-Kokai 60 178 863, 1984; *Chem. Abstr.* **104** (1986) 68760u.
[58] R. L. Frank et al., *J. Am. Chem. Soc.* **71** (1949) 2804.
[59] Eli Lilly, US 3 180 872, 1963.
[60] Allied Chemical & Dye, US 2 556 845, 1947.
[61] Sumitomo Chemical, JP-Kokai 75 73 898, 1973; *Chem. Abstr.* **83** (1975) 209277d.
[62] Koei Chemical, JP-Kokai 59 95 228, 1982; *Chem. Abstr.* **101** (1984) 193022u.
[63] A. M. J. Broome et al., *Nature (London)* **189** (1961) 59.
[64] Schering, DE-OS 2 942 680, 1979.
[65] Agfa-Gevaert, DE-OS 2 439 551, 1974.
[66] W. H. F. Sasse, *Org. Synth.* **46** (1966) 5 – 10.
[67] ICI, GB 960 176, 1962 (G. H. Lang et al.).
[68] ICI, DE 1 445 079, 1960.
[69] M. Cieslak, M. Kopepska, *Pr. Inst. Przem. Org.* **2** (1970) 21.
[70] ICI, DE-OS 2 230 560, 1971.
[71] ICI, GB 1 202 711, 1967.
[72] T. Anderson, *Justus Liebigs Ann. Chem.* **154** (1870) 270.
[73] ICI, NL-A 6 603 415, 1965.
[74] ICI, NL-A 6 512 461, 1964.
[75] ICI, FR 1 380 806, 1962.
[76] ICI, BE 639 705, 1962.
[77] ICI, GB 957 098, 1961.

[78] R. E. Lyle et al., *J. Am. Chem. Soc.* **77** (1955) 1291.
[79] C. H. G. Hands et al., *J. Soc. Chem. Ind.* **66** (1947) 407.
[80] A. H. Blatt, *Org. Synth.*, coll. vol. **2** (1943) 419.
[81] R. A. Abramovitch et al., *J. Chem. Soc. B* 1971,131.
[82] B. C. McKusick, *Org. Synth.* **43** (1963) 97.
[83] Nippon Soda Co., JP-Kokai 80 59 167, 1978; *Chem. Abstr.* **93** (1980) 168139w.
[84] J. P. Wibaut et al., *Recl. Trav. Chim. Pays-Bas* **73** (1954) 140.
[85] K. Bowden, P. N. Green, *J. Chem. Soc.* 1954, 1795.
[86] R. R. Renshaw et al., *J. Am. Chem. Soc.* **59** (1937) 297.
[87] D. Jerchel, H. Fisher, K. Thomas, *Chem. Ber.* **89** (1956) 2921.
[88] R. A. Carpio et al., *J. Electrochem. Soc.* **126** (1979) 1644.
[89] US Dept. of the Air Force, US 4 122 245, 1977.
[90] Nisshin Steel, JP-Kokai 62 70 593, 1985; *Chem. Abstr.* **107** (1987) 224963g.
[91] J. A. Boon et al., *J. Org. Chem.* **51** (1986) 480.
[92] Texaco, US 4 554 383, 1984.
[93] E. Ochiai et al., *J. Pharm. Soc. Japan* **67** (1947) 79; *Chem. Abstr.* **45** (1951) 9538a.
[94] A. R. Katritzky et al., *J. Chem. Soc.* 1957, 1769.
[95] M. Katada, *Yakugaku Zasshi* **67** (1947) 51; *Chem. Abstr.* **45** (1945) 9536d. J. H. Markgraf, *J. Am. Chem. Soc.* **85** (1963) 958.
[96] H. Vorbrugger et al., *Synthesis* 1983, 316.
[97] B. Bobranski et al., *Chem. Ber.* **71** (1938) 2385.
[98] Bayer, DE-OS 3 800 179, 1988.
[99] Koei Chemical, JP-Kokai 81 169 672, 1980; *Chem. Abstr.* **96** (1982) 162549v.
[100] Lonza, DE-OS 2 713 316, 1976.
[101] V. Boekelheide et al., *J. Am. Chem. Soc.* **76** (1954) 1286.
[102] E. Ochiai, *J. Org. Chem.* **18** (1953) 534.
[103] E. G. Novikov et al., *Khim. Prod. Koksovaniya Uglei Vostoka SSSR* (1970) 170; *Chem. Abstr.* **76** (1972) 153512n.
[104] H. Adkins et al., *J. Am. Chem. Soc.* **56** (1934) 2425.
[105] M. Freifelder et al., *J. Org. Chem.* **26** (1961) 3805.
[106] M. Freifelder et al., *J. Org. Chem.* **27** (1962) 287.
[107] O. Hibino et al., *Kogyo Kagaku Zasshi* **68** (1965) 1703; *Chem. Abstr.* **64** (1966) 3472h.
[108] H. K. Hall, Jr., *J. Am. Chem. Soc.* **80** (1958) 6413.
[109] M. Freifelder et al., *J. Org. Chem.* **30** (1965) 1319.
[110] A. Michaelis et al., *Justus Liebigs Ann. Chem.* **331** (1904) 245.
[111] Dow, BE 628 487, 1963.
[112] G. D. Searle & Co., US 3 225 054, 1965.
[113] G. M. Badger et al., *J. Chem. Soc.* 1956, 616.
[114] Haessel A.G., EP 103 553, 1982.
[115] Takeda Chemical Ind., EP 174 726, 1984.
[116] E. Klingsberg: *Pyridine and Its Derivatives*, vol. **2,** Interscience Publishers, New York 1961, p. 299.
[117] J. J. Eisch, *Adv. Heterocycl. Chem.* **7** (1966) 1.
[118] Degussa, US 3 920 657, 1975.
[119] W. Mathe et al., *Angew. Chem. Int. Ed. Engl.* **2** (1963) 144.
[120] Dow, US 3 420 833, 1969.
[121] Dow, US 3 424 754, 1966.

[122] Dow, US 3 732 230, 1973.
[123] ICI, EP 65 358, 1981.
[124] Dow, US 4 256 894, 1978.
[125] Dow, US 3 694 332, 1972.
[126] Dow, US 3 993 654, 1976.
[127] Ishihara Sangyo, JP-Kokai 55 85 564, 1980; *Chem. Abstr.* **94** (1981) 47146h.
[128] E. Koenig et al., *Chem. Ber.* **54** (1921) 1360.
[129] M. Conrad et al., *Chem. Ber.* **20** (1896) 1679.
[130] J. A. Leben, *Chem. Ber.* **29** (1896) 1679.
[131] T. Itai et al., *Yukagaku Zasshi* **75** (1955) 296; *Chem. Abstr.* **50** (1956) 1810g.
[132] C. F. H. Allen et al., *Org. Synth.*, coll. vol. **3** (1955) 136.
[133] C. Rath, *Justus Liebigs Ann. Chem.* **486** (1931) 95.
[134] O. Seide, *Chem. Zentralbl.* **3** (1923) 1022.
[135] H. J. Hertog, *Recl. Trav. Chim. Pays-Bas* **69** (1950) 673.
[136] Koei Chemical Co., JP-Kokai 86 251 663, 1985; *Chem. Abstr.* **106** (1987) 176178k.
[137] Lummus Co., US 40 082 441, 1975.
[138] Roche Product Ltd., GB 717 172, 1954.
[139] R. L. Frank et al., *J. Am. Chem. Soc.* **70** (1948) 3482.
[140] B. Lipka et al., PL 50 077, 1965; *Chem. Abstr.* **65** (1966) 5446d.
[141] Koei Chemical Co., JP-Kokai 89 175 968, 1987; *Chem. Abstr.* **112** (1990) 118662b.
[142] Lonza, DE-OS 2 435 344, 1974.
[143] Degussa, DE-OS 3 107 755, 1982.
[144] Yuki Gosei Kogyo, JP 70 13 572, 1965; *Chem. Abstr.* **73** (1970) 77059r.
[145] Takeda Chemical Ind., EP 290 996, 1988.
[146] Koei Chemical Co., US 4 778 890, 1987.
[147] Nippon Shokubai Kagaku Kogyo Co., EP 290 996, 1988.
[148] T. P. Sycheva et al., *Khim. Farm. Zh.* **6** (1972) 6 – 8.
[149] H. Spitzer et al., *Chem. Ber.* **59B** (1926) 1477 – 1486.
[150] S. M. McElvan, R. Adams, *J. Am. Chem. Soc.* **458** (1922) 2744.
[151] Degussa, DE 2 517 054, 1975; *Chem. Abstr.* **86** (1977) 43576k. J. E. Paustian et al., *CHEMTECH* 1981, 174 – 179.
[152] C. B. Rosas, G. B. Smith, *Chem. Eng. Sci.* **35** (1980) 330 – 337; *Chem. Abstr.* **93** (1980) 79696.
[153] R. A. Bergamin, *Chimia* **44** (1990) 255 – 257.
[154] C. F. H. Allen, C. N. Wolf, *Org. Synth.*, coll. vol. **4** (1963) 45.
[155] Degussa, DE-OS 2 517 054, 1942.
[156] H. Friedrich: *Kirk-Othmer*, 3rd ed., vol. **24**, Wiley-Interscience, New York 1983, pp. 59 – 93.
[157] Takeda Chemical Ind., DE-OS 2 619 243, 1976.
[158] W. Deady, M. R. Grimett, C. H. Potts, *Tetrahedron* **35** (1979) 2895.
[159] E. Klingsberg, R. H. Mizzoni: *Pyridine and Its Derivatives*, vol. **2**, Interscience Publishers, New York 1961, p. 469.
[160] R. A. Abramovitch, R. H. Mizzoni: *Pyridine and Its Derivatives*, vol. **3**, John Wiley & Sons, New York 1974, p. 1.
[161] M. T. Leffler, *Org. React. (N.Y.)* **1** (1942) 91.
[162] A. F. Pozharskii, A. M. Simonov, V. H. Doron'kin, *Russ. Chem. Rev. (Engl. Transl.)* **47** (1978) 1042.
[163] E. Klingsberg, A. S. Tomcufcik, L. N. Starker: *Pyridine and Its Derivatives*, vol. **3**, Interscience Publishers, New York 1962, p. 1.

[164] R. A. Abramovich, C. S. Giam: *Pyridine and Its Derivatives,* vol. **3,** John Wiley & Sons, New York 1974, p. 41.
[165] Sterling Drug Inc., US 3 149 104, 1961.
[166] Pyridium Corporation, US 1 680 108, 1928. Pyridium Corporation, US 1 680 109, 1928. Pyridium Corporation, US 1 608 111, 1928.
[167] R. F. Evans, H. C. Brown, H. C. van der Plas, *Org. Synth.* **43** (1963) 97.
[168] Nippon Soda Co., JP-Kokai 82 72 961, 1980. Nippon Soda Co., JP-Kokai 82 108 068, 1980. Nippon Soda Co., JP-Kokai 82 108 069, 1980. Nippon Soda Co., JP-Kokai 82 102 864, 1980.
[169] V. G. Hofle, W. Steglich, H. Vorbruggen, *Angew. Chem.* **90** (1978) 602.
[170] Schering AG, DE 2 517 774, 1975.
[171] Reilly Tar, US 4 158 093, 1977.
[172] R. C. Elderfield: *Heterocyclic Compounds,* vol. **1,** Wiley-Interscience, New York 1950, p. 435.
[173] C. A. Salemink et al., *Recl. Trav. Chim. Pays-Bas* **68** (1949) 1013.
[174] O. v. Schickh et al., *Chem. Ber.* **69** (1936) 2593.
[175] E. Koenigs et al., *Chem. Ber.* **57** (1924) 1187.
[176] P. Nantka-Namirski et al., *Acta Pol. Pharm.* **24** (1967) 228; *Chem. Abstr.* **69** (1968) 2827.
[177] L. Ruzicka et al., *Helv. Chim. Acta* **3** (1920) 806.
[178] E. Koenigs et al., *Chem. Ber.* **61** (1928) 1022.
[179] F. Freidel, *Chem. Ber.* **45** (1912) 428.
[180] M. Brash et al., *J. Chem. Soc.* 1959, 3530.
[181] W. Herz et al., *J. Org. Chem.* **26** (1961) 122.
[182] D. E. Parker, W. Shive, *J. Am. Chem. Soc.* **69** (1947) 62.
[183] S. Oae et al., *Tetrahedron* **21** (1965) 1971.
[184] B. M. Bain et al., *Chem. Ind. (London)* 1960, 402.
[185] B. R. Baker et al., *J. Org. Chem.* **20** (1955) 116.
[186] H. Ledetschke, *Chem. Ber.* **85** (1952) 202.
[187] H. Ledetschke, *Chem. Ber.* **86** (1953) 123.
[188] Koei Chemical Co., JP 60 132 959, 1983; *Chem. Abstr.* **103** (1985) 215195y.
[189] P. W. Ford et al., *Aust. J. Chem.* **18** (1968) 867.
[190] W. Koenigs et al., *Chem. Ber.* **35** (1902) 1343.
[191] K. Loffer et al., *Chem. Ber.* **40** (1907) 1325.
[192] R. L. Frank et al., *J. Am. Chem. Soc.* **68** (1946) 1368.
[193] K. Winterfeld et al., *Justus Liebigs Ann. Chem.* **573** (1951) 85.
[194] W. Mathes et al., *Chem. Ztg.* **80** (1956) 475.
[195] W. Mathes et al., *Chem. Ber.* **84** (1951) 452.
[196] C. Harries et al., *Justus Liebigs Ann. Chem.* **410** (1915) 95.
[197] S. Gisburg et al., *J. Am. Chem. Soc.* **79** (1957) 481.
[198] Ciba Geigy, DE-OS 2 262 780, 1972.
[199] E. P. Papodopoulous et al., *J. Org. Chem.* **31** (1966) 615.
[200] W. Bosche et al., *Chem. Ber.* **73** (1940) 839.
[201] Raschig AG, BE 710 192, 1951.
[202] Nepera, US 3 274 206, 1964.
[203] E. Browning: *Toxicity and Metabolism of Industrial Solvents,* Elsevier, Amsterdam 1965, pp. 304–309.
[204] Occupational Safety & Health Administration, Material Safety Data Sheets from the Occupational Health Services Database, OSHA, Washington, D.C., 1985.
[205] A. Jori et al., *Ecotoxicol. Environ. Saf.* **7** (1983) no. 3, 251–275.

[206] *Dangerous Prop. Ind. Mater. Rep.* **7** (1987) no. 4, 101–104.
[207] E. W. Schafer Jr., P. B. Brunton, D. J. Cunningham, *Toxicol. Appl. Pharmacol.* **26** (1973) 532.
[208] D. A. Spyker, C. Lynch, J. Shabanowitz, J. A. Sinn, *Clin. Toxicol.* **16** (1980) no. 4, 487–497.
[209] *Dangerous Prop. Ind. Mater. Rep.* **5** (1985) no. 5, 39–41.
[210] P. J. Gehring, T. R. Torkelson, F. Oyen, *Toxicol. Appl. Pharmacol.* **11** (1967) 361.
[211] C. D. Balaban, *Brain Res. Rev.* **9** (1985) no. 1, 21–42.
[212] J. G. Black, D. Howes, *Toxicol. Annu.* **3**(1979) 1–26.
[213] NIOSH, Registry of Toxic Effect of Chemical Substances (RTECS), Dialog Information Service Inc., File 336, July 1990.
[214] J. H. Baxter, M. F. Mason, *J. Pharmacol. Exp. Ther.* **91** (1947) 350.
[215] Comm. Eur. Communities, (Rep.) no. Eur 11553, Solvents in Common Use 1988, 215–234.
[216] L. J. Pollock, I. Finkelman, A. J. Arieff, *Arch. Intern. Med.* **71** (1943) 95–106.
[217] J. W. Langston et al., *Science (Washington, D.C.)* **219** (1983) 979.
[218] T. P. Singer, R. R. Ramsay, K. Mckeown, A. Trevor, *Toxicology* **49** (1988) no. 1, 17–23. H. Imai, *Brain Nerve* **40** (1988) no. 11, 1011–1024.
[219] *Carcinog. Tech. Rep. Ser. U.S. Nat. Cancer Inst.*, NCI-TR-95, 78.
[220] K. Takahashi, G. F. Huang, M. Araki, Y. Kawazoe, *Gann* **70** (1979) 799–806.
[221] O. Tada *Kogai Bunseki Shishin* **3** (1972) 49–62.

Pyrimidine and Pyrimidine Derivatives

ALAN P. CHORLTON, ICI Specialties, Blackley, Manchester, United Kingdom

1.	Introduction 4227	5.	Industrially Important Pyrimidines 4232
2.	Chemical Properties 4228	5.1.	Pharmaceuticals 4232
3.	Production 4229	5.2.	Agrochemicals 4233
4.	Important Naturally Occurring Derivatives of Pyrimidine 4230	5.3.	Dyes 4233
		6.	References 4234

1. Introduction

Pyrimidine was first isolated by GABRIEL and COLMAN in 1899 [1]–[4].

Pyrimidine [289-95-2], $C_4H_4N_2$, M_r 80.09, mp 20–22 °C, bp 123–124 °C, UV max (water): 240 nm (ϵ 2400) is a hygroscopic, water-soluble solid with a penetrating odor. It is also a weak base ($pK_a = 1.23$) that forms a number of solid derivatives (hydrochloride: mp 162 °C; picrate: yellow needles, mp 156 °C; methiodide: yellow plates, mp 136–137 °C). There are many synthetic approaches to the preparation of pyrimidine [4]. The most effective way to prepare pyrimidine in large quantity is to react 1,1,3,3-tetraethoxypropane with formamide over alumina at 200 °C [5].

2. Chemical Properties

The reactivity of the pyrimidine ring is approximately equivalent to that of 1,3-dinitrobenzene or 3-nitropyridine. This is because a double-bonded ring nitrogen is comparable in its electron-withdrawing ability to a nitro group.

Electrophilic substitution at the ring positions occurs only in the presence of one or more activating substituents such as amino or hydroxyl groups. Attack invariably proceeds at a vacant 5-position. Pyrimidines with a single activating substituent are subject only to halogenation; 2-and 4-aminopyrimidines all give 5-bromo derivatives.

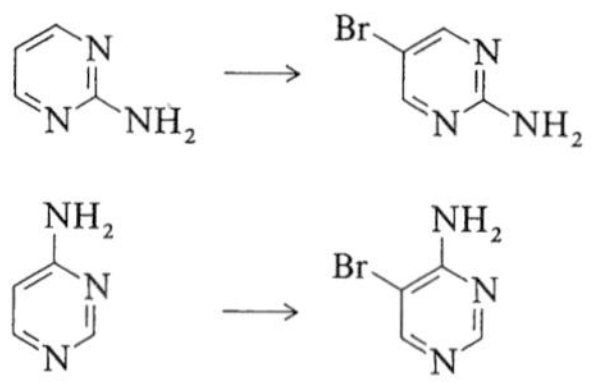

Pyrimidines bearing two activating substituents (e.g., pyrimidine-4,6-dione) readily undergo electrophilic substitution at the 5-position giving a series of compounds with the general formula (**1**).

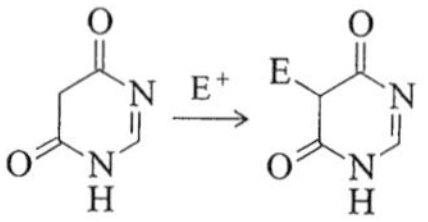

1

E = NO, NO$_2$, Ph–N=N, halogen

Nucleophilic attack occurs readily at positions 2, 4, and 6. In general, the 4-position is the most reactive. For example, the reaction of phenyl magnesium bromide or phenyl lithium with pyrimidine leads to 6-phenyl-1,6-dihydropyrimidine, which on oxidation gives 4-phenylpyrimidine.

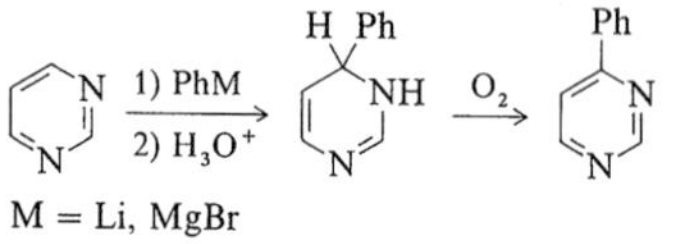

M = Li, MgBr

The reactivity of pyrimidine and its derivatives has been comprehensively reviewed [4].

3. Production

The most general types of ring closure appliable to the preparation of pyrimidines are illustrated below. Other approaches are possible but are rarely utilized.

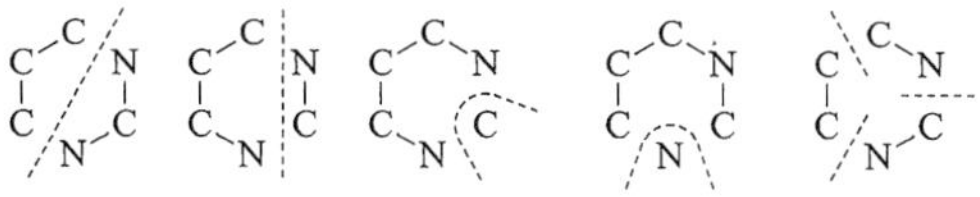

A   B   C   D   E

Type A syntheses are usually the most successful, and over 80% of all known pyrimidines have been prepared in this way. A typical example is the condensation of urea with diethyl malonate to give barbituric acid.

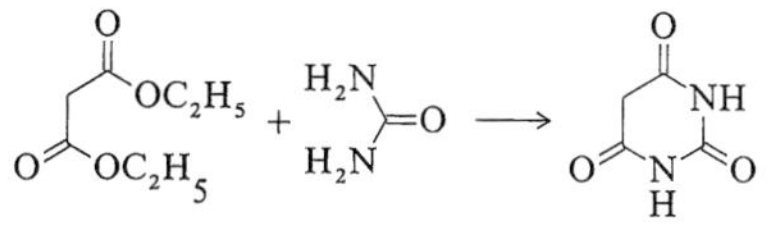

The synthetic utility of this reaction is greatly enhanced by the fact that any compound containing the group $H_2N-C-NH_2$ can take the place of urea. Thus, guanidines are precursors to 2-aminopyrimidines, amidines give 2-alkyl or aryl-substituted pyrimidines, and thioureas lead to pyrimidine-2-thiones. The ester group in diethyl malonate can also be replaced by other groups (e.g., keto, aldehyde, nitrile, acid chloride and imino moieties) opening the way to a wide variety of pyrimidine derivatives.

The most common example of a type B synthesis is the reaction of an isocyanate with a 3-aminoacrylate derivative to yield a uracil.

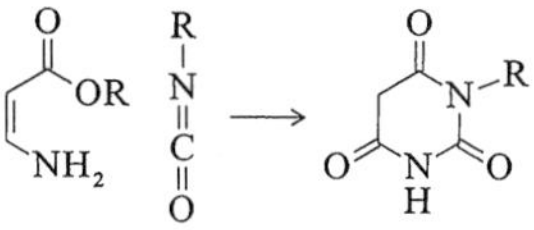

2-Alkyl pyrimidines are readily accessible by type C syntheses. Condensation of 1,3-diaminopropane with the appropriate carboxylic acid, followed by cyclization, results in the dihydropyrimidine **2**, which on dehydrogenation with palladium/alumina affords the 2-substituted pyrimidine **3** [6].

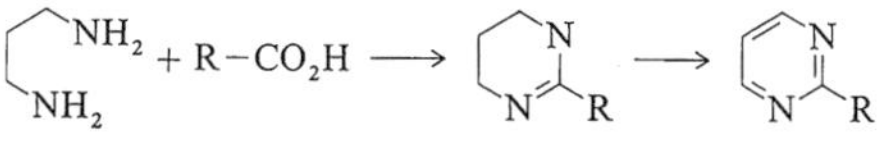

In a similar sequence diamines with a higher oxidation level lead to pyrimidines directly. Thus, malonodiamide condenses with ethyl formate to give 4,6-dihydroxypyridimines [7], and with diethyl carbonate to give 4,6-diaminopyrimidine-2-ones [8].

The Shaw synthesis is an example of a type D synthesis, and is used to prepare many uracil and thiouracil derivatives [9].

Base-catalyzed trimerization of acetonitrile is an example of a type E synthesis, and constitutes a very efficient way of preparing certain pyrimidines (e.g., 2,6-dimethyl-4-aminopyrimidine from acetonitrile) [10].

Other approaches to pyrimidines have been reviewed elsewhere [4].

4. Important Naturally Occurring Derivatives of Pyrimidine

Orotic acid [65-86-1], $C_5H_4N_2O_4$, M_r 156.10, mp 345–346 °C, is thought to be the key precursor in the biosynthesis of all naturally occurring pyrimidines. It can be synthesized directly from urea and oxalacetic acid ester [11]. It has been used in combination with 4-amino-5-imidazole carboxamide for the treatment of liver disease [12].

Uracil [66-22-8], $C_4H_4N_2O_2$, M_r 112.09, mp 335 °C, is widely distributed in nature, since it is a constituent of nucleic acids. Synthesis: reaction of malic acid and urea in concentrated sulfuric acid [13]. Uracil is an excellent starting material for the preparation of pyrimidines, pteridines, and purines.

Thymine [65-71-4], $C_5H_6N_2O_2$, M_r 126.11, mp 335–340 °C, unlike uracil, is not found in ribonucleic acid but in deoxyribonucleic acid. Synthesis: treatment of 3-methylmalic acid with fuming sulfuric acid to yield 2-formylpropionic acid, which in turn condenses with urea to give thymine [14].

Alloxan [50-71-5], $C_4H_2N_2O_4$, M_r 142.07, mp 256 °C (decomp.), turns pink at 230 °C and forms mono- and tetrahydrates which lose water at 150 °C. Alloxan induces permanent diabetes in many animals, but not in humans. The best synthetic route to alloxan is direct oxidation of barbituric acid with chromium trioxide [15].

Cytosine [71-30-7], $C_4H_5N_3O$, M_r 111.10, mp 320–325 °C (decomp.), is widely distributed in nature. Since it is a constituent of nucleic acids it can be isolated by hydrolyzing thymus nucleic acids. The principal synthesis is from 3,3-dihydroxypropionitrile and urea [16].

Thiamine [59-43-8], vitamin B_1, is a very important pyrimidine, manufactured currently at a level of about 2000 t/a. Synthetic thiamine is prepared in a multistage process. The starting material is acetamidine, which is condensed with the formyl derivative of ethyl 2-ethoxypropionate to give 5-ethoxymethyl-4-hydroxy-2-methylpyrimidine. The heterocyclic thiazole component is introduced in the form of 5-(2'-hydroxy)-4-methylthiazole [17].

Numerous bicyclic heterocycles contain the pyrimidine nucleus, including the purines adenine and guanine, which are important basic constituents of nucleic acids.

5. Industrially Important Pyrimidines [18], [19]

5.1. Pharmaceuticals

Thonzylamine hydrochloride [63-56-9], $C_{16}H_{23}ClN_4O$, M_r 322.83, mp 173 – 176 °C, is used as an antihistaminic drug. It is prepared by treating the sodium salt of 2-(4-methoxybenzyl)aminopyrimidine with N,N-dimethyl-2-chloroethylamine [20].

Piribedil [3605-01-4], $C_{16}H_{18}N_4O_2$, M_r 298.35, mp 98 °C, is a central dopaminergic agonist used as a peripheral vasodilator. It is prepared by treating 1-(3′,4′-methylenedioxybenzyl) piperazine with 2-chloropyrimidine in the presence of a base [21].

Citicoline [987-78-0], $C_{14}H_{26}N_4O_{11}P_2$, M_r 488.33, is a coenzyme in one of the biosynthetic pathways leading to lecithin. It can be isolated from liver and yeast. Citicoline is of clinical use as a central circulation stimulant. It is produced industrially by biochemical action of *Brevibacterium ammoniagenes* on cytidine 5′-monophosphate and choline [22].

Cytarabin [69-74-9], $C_9H_{13}N_3O_5$, M_r 243.22, mp 212 – 215 °C, is used for the treatment of leukemia and is also active against herpes. Cytarabin is synthesized by treatment of 2,4-dimethoxypyrimidine with 2,3,5-tri-2-benzyl-D-arabinofuranosyl chloride to give the 4-methoxypyrimidone (**4**: R = CH$_2$Ph, X = OCH$_3$Me), which undergoes aminolysis with ammonia to the corresponding amine (**4**: R = CH$_2$Ph, X = NH$_2$). Subsequent hydrogenolysis leads to cytarabin (**4**: R = H, X = NH$_2$) [23].

5.2. Agrochemicals

Bensulfuron-methyl [*83055-99-6*], $C_{16}H_{18}N_4O_7S$, M_r 410.4, *mp* 185–188 °C, is an example of a sulfonyl urea, a new class of selective systemic herbicides that inhibit biosynthesis of the essential amino acids valine and isoleucine, thereby interrupting cell division and plant growth. Bensulfuron-methyl is used in the control of annual and perennial weeds in rice [24].

Nuarimol [*63284-71-9*], $C_{17}H_{12}ClFN_2O$, M_r 314.7, *mp* 126–127 °C, is a systemic fungicide that acts by inhibiting ergosterol biosynthesis. It is used to control a wide range of pathogenic fungi, such as powdery mildews on fruit [25].

5.3. Dyes

Pyrimidines are also used in the dye industry [26]. The most important examples are the isoindoline pigments, *Pigment yellow 139* [*36888-99-0*], $C_{16}H_9N_5O_6$, M_r 367.27, is used for the coloration of paints and plastics. It is synthesized from 1,3-diiminoisoindoline and barbituric acid [27].

Pyrimidines of lesser industrial importance are discussed in [18], [19], [26].

6. References

[1] G. W. Kenner, A. Todd: "Pyrimidine and its Derivative," in R. C. Elderfield (ed.): *Heterocyclic Compounds,* vol. 6, Wiley-Interscience, New York 1957, pp. 234–324.
[2] D. J. Brown: "The Pyrimidines," in A. Weissberger (ed.): *The Chemistry of Heterocyclic Compounds,* vol. 16, Wiley-Interscience, New York 1962.
[3] D. J. Brown: "The Pyrimidines," Wiley-Interscience, New York 1970, [2] Suppl. I.
[4] D. J. Brown: "Pyrimidines and Their Benzo Derivatives," in A. J. Boulton, A. McKillop (eds.): *Comprehensive Heterocyclic Chemistry,* vol. **3,** Pergamon Press, New York 1984, pp. 57–155.
[5] H. Bredereck, R. Gompper, H. Herlinger, *Chem. Ber.* **91** (1958) 2832.
[6] G. E. Hibert, *J. Am. Chem. Soc.* **54** (1932) 2076–2083.
[7] Z. Budensinsky, F. Roubinek, E. Svatek, *Collect. Czech. Chem. Commun.* 1965, 3730.
[8] G. A. Howard, B. Lythgoe, A. R. Todd, *J. Chem. Soc.* 1944, 476–477.
[9] M. R. Atkinson, G. Shaw, R. N. Warrener, *J. Chem. Soc.* 1956, 4118–4123.
[10] E. C. Horning (ed.): *Org. Syn. Coll.,* vol. **III,** Wiley-Interscience, New York 1955, p. 71.
[11] R. Muller, *J. Prakt. Chem.* **56** (1897) 488–495.
[12] R. Haroka, T. Kamiya, US 3 271 398, 1966.
[13] O. Baudisch, D. Davidson, *J. Am. Chem. Soc.* **28** (1926) 2379–2383.
[14] H. W. Scherp, *J. Am. Chem. Soc.* **68** (1946) 912–913.
[15] E. C Horning (ed.): *Org. Syn. Coll.,* vol. **III,** Wiley-Interscience, New York 1955, p. 39.
[16] J. A. Hill, W. J. Le Quesne, *J. Chem. Soc.*1965, 1515–1516.
[17] H. G. Frank: *Industrial Aromatic Chemistry,* Springer Verlag, Berlin 1988, p. 411.
[18] M. Sittig: *Pharmaceutical Manufacturing Encyclopedia,* vols. **1** and **2,** Noyes Data Corp., Park Ridge, NJ, 1988.
[19] C. R. Worthing (ed.): *The Pesticide Manual,* 9th ed., British Crop. Protection Council, Farnham 1991.
[20] H. L. Friedman, A. V. Tolstouhov, US 2 465 865, 1949.
[21] G. Negnier, R. Canevari, M. Laubic, US 3 299 067, 1967.
[22] Takeda Chemical, US 3 687 932, 1972 (J. Nakamachi).
[23] Upjohn, US 3 116 282, 1963 (J. Hunter).
[24] Du Pont, US 4 420 325, 1984 (R. Saurs).
[25] Eli Lilley, GB 1 218 623, 1967 (J. Davenport).
[26] *The Color Index,*3rd ed., The Society of Dyers and Colorists, Bradford 1982.
[27] Bayer, DE, 3 022 839, 1980 (M. Lorenz).

Pyrrole

Albrecht Ludwig Harreus, BASF Aktiengesellschaft, Ludwigshafen, Federal Republic of Germany

1. Pyrrole 4235
2. Pyrrole Derivatives 4237
3. Toxicology and Occupational Health 4238
4. References 4239

1. Pyrrole

Pyrrole [109-97-7], 1H-pyrrole, azole, C_4H_5N, M_r 67.09, with its five-membered heterocyclic ring system has a fundamental role in the living world.

Pyrrole derivatives (cyclic tetrapyrroles) assume their most important role in respiration (hemoglobin, cytochromes) and photosynthesis (chlorophyll). Bile pigments are examples of natural linear tetrapyrroles. Low molecular mass pyrroles with antibiotic action have been isolated from microorganisms (pyrrolnitrin).

F. F. Runge detected pyrrole in 1834 as a constituent of coal tar. The very nonspecific reaction used for detection (red coloration of a spruce splint dampened with hydrochloric acid) gave pyrrole its name (Greek pyrros = fiery).

Physical Properties. Pyrrole, mp −23.4 °C, bp 130 °C (101.3 kPa), when freshly distilled, is a colorless liquid with a chloroform-like odor, which rapidly becomes brown in air and gradually resinifies. Pyrrole is hygroscopic, not very soluble in water (ca. 5 %), and miscible with most organic solvents. Further physical data are as follows:

Density d_4^{20}	0.9698
Refractive index n_D^{20}	1.5085
Viscosity (20 °C)	1.31 mPa · s
Flash point (DIN 51 755)	33 °C
Ignition temperature (DIN 51 794)	550 °C
Explosion limits	
lower	3.1 vol %
upper	14.8 vol %

The most important physical properties and spectroscopic data for pyrroles are summarized in [5].

The pronounced aromatic character of pyrrole is not consistent with the classical constitutional formula (**1**). The direction of the dipole moment, the bond lengths, and a very low basicity show that the free electron pair on nitrogen participates in the formation of an aromatic 6π-electron system, which is conveyed schematically by (**2**), whereas (**1**) represents only one of several possible mesomeric structures.

Chemical Properties. Pyrrole is unstable toward mineral acids, and is protonated in the 2-position. The resulting pyrrylium cation polymerizes very readily to give high molecular mass pyrrole resins (pyrrole red).

Because of its high π-electron density, the characteristic reactions of pyrrole are electrophilic substitutions. Substitution occurs preferentially in the 2-position. The reactions proceed under mild conditions, but must be adapted to the compound's sensitivity toward acid. Examples of electrophilic substitution include acetylation with acetic anhydride (which does not require a Friedel–Crafts catalyst), sulfonation with the SO_3–pyridine complex, nitration with acetyl nitrate (nitric acid/acetic anhydride), and halogenation with bromine, which leads directly to a tetrasubstituted product.

As a result of delocalization of the electron pair on the nitrogen, pyrrole is weakly acidic (pK_a ca. 17.5). Corresponding salts can be prepared with alkali metals in liquid ammonia or metal hydrides in inert solvents (e.g., tetrahydrofuran). These can be converted into N-alkylpyrroles by reaction with alkyl halides. Pyrryl potassium reacts with carbon dioxide to give the potassium salt of pyrrole-2-carboxylic acid.

Pyrrole can be reduced catalytically (Pd, Pt, Rh, Ni) to pyrrolidine. With less reactive reducing agents (e.g., zinc in glacial acetic acid) the intermediate 3-pyrroline [*109-96-6*] (2,5-dihydropyrrole) can be obtained. This is the only one of the three dihydropyrroles that is somewhat stable.

Production and Use. Suitable processes for the industrial production of pyrrole include the reaction of furan (→ Furan and Derivatives) with ammonia over SiO_2/

Al$_2$O$_3$ catalysts (zeolites) [6] and the catalytic dehydrogenation of pyrrolidine [*123-75-1*] [7].

The reaction of 2,5-dimethoxytetrahydrofuran with ammonia over zeolites [8] also leads to pyrrole, as does gas-phase cyclocondensation of butynediol with ammonia over an aluminum oxide/thorium oxide catalyst [9].

Mild steel drums lined with polyethylene are suitable as transport and storage vessels.

Today BASF and Shell produce pyrrole commercially, although its use remains limited. An electrically conducting polypyrrole is prepared from pyrrole by electrochemical polymerization.

2. Pyrrole Derivatives

There are many possible synthetic routes to substituted pyrroles [11]–[13]. Of the classical syntheses of *C-substituted pyrroles*, the Knorr pyrrole synthesis (**3**) together with the Paal–Knorr (**4**) and Hantzsch syntheses (**5**) are particularly important. Appropriate α-aminoketones for the Knorr synthesis are usually generated in situ by reduction of the corresponding oximino compound, most frequently with zinc in acetic acid. The methylene group of the carbonyl compound has to be activated, therefore β-dicarbonyl compounds such as β-ketoesters and β-diketones are preferred.

N-Substituted pyrroles can easily be prepared industrially by dehydrogenation of the corresponding pyrrolidines. This reaction is carried out in the gas phase at 250–300 °C over such catalysts as Pd/Al$_2$O$_3$ or Pd/MgO/Al$_2$O$_3$ [7], [14]. Yields of *N*-alkylpyrroles surpass 90%. Silicates are less suitable as carriers. *N*-Alkylpyrroles can also be produced by reaction of 2-butene-1,4-diol with amines in the gas phase (e.g., copper catalysts) [15].

Many *N*-alkyl and *N*-aryl pyrroles can be produced quite elegantly by reacting appropriate amines with 2,5-dialkoxytetrahydrofurans, in particular 2,5-dimethoxytetrahydrofuran [*696-59-3*] [16], which is readily accessible industrially through electrooxidation of furan with subsequent hydrogenation [8], [11], [12], [17].

$$CH_3O-\overset{}{\underset{O}{\frown}}-OCH_3 \xrightarrow{R^1-NH_2} \underset{R^1}{\overset{}{\boxed{N}}}$$

Special methods have been developed for the synthesis of *3-substituted pyrroles* [18].

3-Substituted 4-phenylpyrroles, such as fenpiclonil [*74738-17-3*] (Ciba-Geigy), display fungicidal properties, and are derived from the natural product pyrrolnitrin [19].

Fenpiclonil

Derivatives of pyrrole-3,4-dicarboxylate esters show herbicidal activity [20], and those of 1-aryl-3-cyanopyrroles have insecticidal effects [21]. *N*-Methyl-2-pyrrylmethyl sulfides can be used as flavorings [22].

N-Methylpyrrole [*96-54-8*], 1-methylpyrrole, 1-methyl-1*H*-pyrrole, C_5H_7N, M_r 81.12, mp −57 °C, bp 112–113 °C (101.3 kPa), d_4^{20} 0.9088, n_D^{20} 1.4889, flash point 16 °C (DIN 51755), ignition temperature 400 °C (DIN 51794), is a bright yellow liquid that must be protected from light and air during storage. *N*-Methylpyrrole is miscible with almost all common organic solvents and ca. 1 % soluble in water.

N-Methylpyrrole is added as a stabilizer to chlorinated hydrocarbons [23]. It serves as starting material for the synthesis of 1-methylpyrrole-2-acetic acid and its derivatives, which are building blocks for many anti-inflammatory drugs, including tolmetin (Cilag) [24].

N-Methylpyrrole is produced commercially by BASF.

3. Toxicology and Occupational Health

Pyrrole is toxic if swallowed. The acute oral LD_{50} in rats is 137 mg/kg [25]. The acute intraperitoneal LD_{50} in mice is 98 mg/kg [26]. The subcutaneous LD_{50} in mice is 61 mg/kg [27]. Pyrrole does not cause skin irritation when applied to intact skin of white rabbits for up to 4 h, but it does cause significant ocular lesions, with a risk of serious permanent damage to the eyes. The LC_{50} is 2.27 mg/L (rat, inhalation, 4 h). Rats exposed to an atmosphere highly enriched or saturated with pyrrole at room temperature died after 30 min. Irritation of eye and nasal mucosa were observed after overexposure [25].

Pyrrole showed no mutagenic activity in the Ames test (with or without metabolic activation) [28]–[32]. Pyrrole did not induce DNA damage in a test with *Escherichia coli* [32] and also gave a negative result in the DNA-repair test in rat hepatocytes [33], [34]. Pyrrole had positive genotoxic activity in the rec-assay with *Bacillus subtilis*, and exhibited the ability to cleave DNA in lambda phages [35]. After intraperitoneal administration of pyrrole to mice, an epoxide was identified as a toxic metabolite in the lung. The enzyme involved was P-450. Emphysema and bronchiolar necrosis were found in the animals [36], [37].

N-Methylpyrrole is harmful if swallowed. The acute oral LD_{50} in rats is about 1400 mg/kg. The dermal LD_{50} in rabbits is > 400 mg/kg. The substance causes irritation of healthy intact or scarified skin in white rabbits when it is applied for up to 2 h, but it does not lead to significant ocular irritation. The LC_{50} in rats after inhalation for 1 and 4 h is greater than 22.4 and 5.6 mg/l, respectively; there was no mortality. Rats exposed to an atmosphere highly enriched or saturated with N-methylpyrrole at room temperature died after 1 h. Both severe irritation of the mucosa and narcotic effects were observed [38].

N-Methylpyrrole showed no mutagenic activity in the Ames test with or without addition of S-9 mix [28], [38]. It was positive in the rec-assay with *Bacillus subtilis* ($\pm$ S-9 mix), and exhibited an ability to cleave DNA in lambda phages [35].

4. References

General References

[1] A. Gossauer: *Die Chemie der Pyrrole*, Springer Verlag, Berlin 1974.
[2] R. A. Jones, G. P. Bean: "The Chemistry of Pyrroles," *Organic Chemistry*, vol. **34**, Academic Press, New York 1977.
[3] A. H. Jackson in P. G. Sammes (ed.): *Comprehensive Organic Chemistry*, vol. **4**, Pergamon Press, Oxford–New York 1979, pp. 275.
[4] R. A. Jones (ed.): *Pyrroles. Chemistry of Heterocyclic Compounds*, vol. **48** part 1, Wiley Interscience, New York 1990.

Specific References

[5] R. A. Jones in A. R. Katritzky, A. J. Boulton (eds.): *Advances in Heterocyclic Chemistry*, vol. **11**, Academic Press, New York–London 1970, pp. 383.
[6] K. Hatada, M. Shimada, K. Fujita, Y. Ono, T. Keii, *Chem. Lett.* **1974**, 439–442. Daicel, JP-Kokai 83/090 548, 1982.Asahi, JP-Kokai 89/301 658, 1988 (T. Yamamoto, M. Iwasaki).
[7] BASF, EP 67 360, 1981 (N. Goetz, L. Hupfer, W. Franzischka).
[8] BASF, EP-A 303 206, 1987 (W. Hoelderich, H. Hesse, H. Siegel).
[9] W. Reppe, *Justus Liebigs Ann. Chem.* **596** (1955) 155.
[10] H. Naarmann, *Angew. Makromol. Chem.* **162** (1988) 1–17.
[11] J. M. Patterson, *Synthesis* 1976, 281–304.

[12] L. N. Sobenina, A. I. Mikhaleva, A. I. Trofimov, *Khim. Geterotsikl. Soedin.* 1989, 291–308.
[13] L. N. Sobenina, A. I. Mikhaleva, A. I. Trofimov, *Russ. Chem. Rev. (Engl. Transl.)* **58** (1989) 163–180.
[14] Ansul Chem. Company, US 3 008 965, 1959 (R. J. Zellner).
[15] BASF, DE-OS 3 309 355, 1983 (H. Menig, M. Fischer, K. Baer).
[16] BASF, DE-AS 2 710 420, 1977 (D. Degner, H. Nohe, H. Hannebaum).
[17] N. Elmig in R. A. Raphael, E. C. Taylor, H. Wynberg (eds.): *Advances in Organic Chemistry*, vol. **II**, Interscience Publishers, New York 1960, pp. 100–102.
[18] H. J. Anderson, C. E. Loader, *Synthesis* 1985, 353–364.
[19] H. P. Fischer, *Nachr. Chem. Tech. Lab.* **38** (1990) 732–740. Nippon Soda, DE 2 927 480, 1979 (K. Ahkuma, H. Takagi, A. Nakata). Bayer, EP-A 392 286, 1989 (D. Wollweber, W. Brandes, S. Dutzmann, G. Haenssler).
[20] Monsanto, EP-A 88 743, US 4 461 642, 1982 (R. K. Howe, L. F. Lee). Shell Int. Res., GB 2 214 180, 1988 (K. M. Patel, J. E. Powell).
[21] Rhône-Poulenc, EP-A 372 982, 1988 (P. Timmons et al.).
[22] Firmenich, US 3 985 906, 1975 (M. Winter et al.).
[23] Du Pont, US 2 795 623, 1954 (F. W. Starks). Hooder Chem. Corp., US 2 998 461, 1959 (D. H. Campbell).
[24] A. Kleemann, J. Engel: *Pharmazeutische Wirkstoffe*, 2nd ed., Georg Thieme Verlag, Stuttgart 1982.
[25] BASF, unpublished results (1987–1988).
[26] R. Rinaldi, Y. Bernard, *Prog. Biochem. Pharmacol.* **1** (1965) 542–549.
[27] C. F. Reinhardt, M. R. Britelli (eds.): *Patty's Industrial Hygiene and Toxicology*, 3rd ed., vol. **2 A**, Wiley-Interscience, New York 1981, pp. 2700–2701.
[28] H. U. Aeschbacher et al., *Food Chem. Toxicol.* **27** (1989) 227–232.
[29] L. R. Ferguson, W. A. Denny, D. G. MacPhee, G. Donald, *Mutat. Res.* **157** (1985) 29–37.
[30] I. Florin, L. Rutberg, M. Curvall, C. R. Enzell, *Toxicology* **15** (1979) 219–232.
[31] C. H. Ho et al., *Mutat. Res.* **85** (1981) 335–345.
[32] M. Riebe, K. Westphal, P. Fortnagel, *Mutat. Res.* **101** (1982) 39–43.
[33] G. M. Williams, H. Mori, J. Hirono, M. Nagao, *Mutat. Res.* **79** (1980) 1–5.
[34] G. M. Williams, *Regul. Toxicol. Pharmacol.* **5** (1985) 132–144. *Food. Addit. Contam.* **1** (1984) 173–178.
[35] E. H. Kim, K. Shinohara, H. Murakami, H. Omura, *J. Fac. Agric., Kyushu Univ.* **31** (1987) 279–285.
[36] M. R. Boyd, *CRC Crit. Rev. Toxicol.* **7** (1980) 103–176.
[37] J. W. Bridges, L. F. Chasseaud (eds.): *Progress in Drug Metabolism*, vol. **9**, Taylor & Francis, London–Philadelphia 1986, p. 83.
[38] BASF, unpublished results (1978–1990).

2-Pyrrolidone

Albrecht Ludwig Harreus, BASF Aktiengesellschaft, Ludwigshafen, Federal Republic of Germany

1.	Introduction 4241	5.	Other Pyrrolidone Derivatives	4247
2.	2-Pyrrolidone 4241	6.	Toxicology and Occupational	
3.	N-Methyl-2-pyrrolidone 4243		Health.	4248
4.	N-Vinyl-2-pyrrolidone 4246	7.	References.	4250

1. Introduction

Pyrrolidones became available industrially through the fundamental work of W. Reppe on acetylene chemistry. He synthesized butynediol and from it γ-butyrolactone, the starting material for the commercial production of pyrrolidones ($\rightarrow$ Butyrolactone).

2. 2-Pyrrolidone

Physical Properties. 2-Pyrrolidone [616-45-5], pyrrolidone, 2-pyrrolidinone, γ-butyrolactam, C_4H_7NO, M_r 85.12, *mp* 25.57 °C, *bp* 245 °C (101.3 kPa), is a colorless, hygroscopic liquid above its melting point. It is miscible with water and common organic solvents, but not with aliphatic or cycloaliphatic hydrocarbons. Further physical data follow:

Density (30 °C)	1.103 g/cm^3
Refractive index n_D^{30}	1.484
Viscosity (30 °C)	10.2 mPa·s
Enthalpy of fusion	135 kJ/kg
Enthalpy of evaporation (40 °C)	666 kJ/kg
Dielectric constant (35 °C)	27.1
Dipole moment (30 °C, dioxane)	3.8 D
Flash point (DIN 51758)	138 °C
Ignition temperature (DIN 51794)	390 °C

Chemical Properties. 2-Pyrrolidone has amphoteric properties: it forms salts with hydrogen chloride, hydrogen bromide, and alkalis. It exhibits typical properties of primary lactams [1]. On heating with strong aqueous alkalis or mineral acids it is hydrolyzed to 4-aminobutyric acid [2]. The reactions at the lactam nitrogen are industrially the most important, in particular viny-lation (see Chap. 4). Alkali-metal salts of 2-pyrrolidone react with alkyl halides or sulfates to give the corresponding *N*-alkyl-2-pyrrolidones. *N*-Alkylation can also be achieved by treatment of 2-pyrrolidone with alcohols in the gas phase on alumina [3] or copper chromite catalysts [4]. *N*-Acylation can be readily achieved with carboxylic acid anhydrides or chlorides [2]. Noteworthy among the condensations with aldehydes are the reactions with formaldehyde to give *N*-hydroxymethyl-2-pyrrolidone [5] and with acetaldehyde to give *N*-(1-hydroxyethyl)-2-pyrrolidone [6], [7], which can be used to produce *N*-methyl- and *N*-vinyl-2-pyrrolidone, respectively. 2-Pyrrolidone polymerizes to a polyamide (nylon 4) in the presence of alkali and a carboxylic acid chloride catalyst [8].

Production. 2-Pyrrolidone can be produced by the following methods:

1) Catalytic [9] or electrochemical reduction [2] of succinimide
2) Carbonylation of allylamine [10]
3) Hydrogenation of succinic acid dinitrile under hydrolytic conditions [11]
4) Reaction of maleic or succinic anhydride in aqueous ammonia with Pd–Ru catalysts [12]

Industrially, 2-pyrrolidone is almost exclusively produced by reacting aqueous γ-butyrolactone [*96-48-0*] with ammonia [13]:

$$\text{butyrolactone} + NH_3 \rightleftharpoons HO\text{-}(CH_2)_3\text{-}C(=O)NH_2 \xrightarrow[-H_2O]{\Delta} \text{2-pyrrolidone}$$

Formation of the undesired 4-(*N*-2-pyrrolidonyl)butyramide can be suppressed if the twostage reaction is carried out in the gas phase on a magnesium silicate catalyst (250–290 °C, 0.4–1.4 MPa) [14].

Quality Specifications and Analysis. Technical-grade 2-pyrrolidone has a minimum content of 99.5 wt% (gas chromatography), and a maximum water content of 0.1 wt% (Karl Fischer titration).

Handling, Storage, and Transportation. 2-Pyrrolidone is transported in steel drums. Other materials suitable for storage and transportation are aluminum and stainless steel. Moisture must be excluded. 2-Pyrrolidone can be kept for ca. 1 year in closed containers. Yellowing does not affect the quality. 2-Pyrrolidone is not self-igniting or explosive, but is flammable.

Uses. Over 95 % of the 2-pyrrolidone produced is processed into N-vinyl-2-pyrrolidone. It is also used in the pharmaceutical industry for the production of pyrrolidone nootropics, especially piracetam (see 5). 2-Pyrrolidone is also used as a solvent and reaction medium since it is high boiling, inert, and noncorrosive. It can be used to decolorize kerosene and other hydrocarbons. Aromatic hydrocarbons can be extracted from petroleum with 2-pyrrolidone. In the polish and cleaning materials industry it is added to styrene–acrylate copolymer dispersions as a film-forming agent.

Economic Aspects. Worldwide demand for 2-pyrrolidone was 14 000 t in 1983 [15] and has since risen considerably. Important producers are BASF and GAF.

3.   N-Methyl-2-pyrrolidone

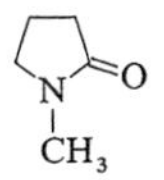

N-Methyl-2-pyrrolidone (NMP) [*872-50-4*], N-methylpyrrolidone, 1-methyl-2-pyrrolidone, C_5H_9NO, M_r 99.13, is an important, versatile solvent and reaction medium for the chemical industry because of its favorable properties [16], [17].

Physical Properties. NMP, *mp* –24.4 °C, *bp* 204.3 °C (101.3 kPa), is a colorless liquid with a weakly aminelike odor; it is completely miscible with water and most common organic solvents. Further physical data follow:

bp (1.3 kPa)	81 – 82 °C
Density (25 °C)	1.028 g/cm^3
Refractive index n_D^{25}	1.469
Viscosity (20 °C)	1.796 mPa·s
Surface tension (25 °C)	0.041 N/m
$t_{crit.}$	451 °C
$p_{crit.}$	4.78 MPa

Vapor pressure

t, °C	40	60	80	100	150
p, kPa	0.134	0.459	1.33	3.37	21.9

Specific heat capacity

t, °C	0	25	50	100
c_p (liquid), kJ kg^{-1} K^{-1}	1.70	1.78	1.86	2.03

Enthalpy of evaporation (20 °C)	550 kJ/kg
Thermal conductivity (20 °C)	1.8 W m^{-1} K^{-1}
Dielectric constant (23 °C)	32
Dipole moment	4.09 D
Flash point (DIN 51758)	91 °C
Ignition temperature (DIN 51794)	245 °C
Lower explosion limit	1.3 vol%
Upper explosion limit	9.5 vol%

Chemical Properties. NMP is a very weak base and forms a solid hydrochloride with anhydrous hydrogen chloride. A 10% aqueous solution has a pH of 7.7 – 8. NMP is chemically very stable. The lactam ring is only opened by strong aqueous acids and bases to give 4-methylaminobutyric acid. NMP has limited resistance toward oxygen. It is oxidized at the 5-position, after several intermediate stages N-methylsuccinimide is formed. This compound can also be produced directly from NMP, for example by oxidation with ruthenium tetroxide [18]. The carbonyl group of NMP reacts with phosgene, phosphorus pentachloride, and other chlorinating agents. 2-Chloro-1-methyl-1-pyrrolidinium chloride [15862-82-5] can be readily produced with phosgene [19] and reacts with various nucleophiles (amines, alkoxides) [19], [20]:

It can be used as a catalyst for the synthesis of carboxylic acid chlorides from carboxylic acids [19].

With strong bases (e.g., lithium diisopropylamide) NMP can be deprotonated in the 3-position to form amide enolates which can react with alkyl halides [21] and aryl bromides [22]. Condensations with esters at the CH$_2$ group in the 3-position are achieved in the presence of sodium ethoxide [23]. This route is also used in the Späth nicotine synthesis [24]. Numerous catalytic effects can be obtained with NMP as the reaction medium. For example, certain acetylides can only be alkylated in NMP [25]; ethylene glycol can only be produced from synthesis gas on rhodium carbonyls when NMP is used as the reaction medium [26].

Production. Large-scale production of NMP is predominantly carried out by reacting γ-butyrolactone with methylamine in a shaft reactor (high-pressure tube with special baffles) at 200 – 350 °C and ca. 10 MPa.

Other processes also analogous to those used for pyrrolidone synthesis can also be used, in particular hydrogenation of N-methylsuccinimide or mixtures of maleic or succinic anhydride and methylamine [27]. NMP can also be produced by hydrogenation of N-hydroxymethyl-2-pyrrolidone (see p. 4242) [28] or by reaction of acrylonitrile with methylamine in the presence of a peroxide radical initiator [29].

Quality Specifications and Analysis. Technical-grade NMP has the following specifications (gas chromatography): NMP content, min. 99.5 wt%; methylamine, max.

0.02 wt %; water, max. 0.1 wt % (Karl Fischer titration); color index, APHA 50 max. For special areas of use (e.g., electronics) higher purities are supplied with the following specifications: methylamine, max. 0.01 wt %; water, max. 0.05 wt %.

Handling, Storage, and Transportation. NMP is transported in tank cars or trailers and in drums. Transport and storage containers are generally made of mild steel. Stainless steel, nickel, and aluminum are also suitable as container and drum materials. Small quantities can be stored in polyethylene, polypropylene, or clear glass bottles. NMP shows unlimited shelf life in tightly closed containers, slight discoloration does not impair its quality. It is flammable.

Uses. NMP is an important solvent because of its low volatility, thermal stability, high polarity, and aprotic, noncorrosive properties. Its favorable toxicological and ecological properties account for the fact that NMP is replacing other solvents such as chlorinated hydrocarbons.

The most important areas of use are:

1) *Petrochemical processing*: acetylene recovery from cracked gas, extraction of aromatics and butadiene, gas purification (removal of CO_2 and H_2S), lube oil extraction
2) *Engineering plastics*: reaction medium for the production of high-temperature polymers such as polyethersulfones, polyamideimides, and polyaramids
3) *Coatings*: solvent for acrylic and epoxy resins, polyurethane paints, waterborne paints or finishes, printing inks, synthesis/diluent of wire enamels, coalescing agent
4) *Agricultural chemicals*: solvent and/or cosolvent for liquid formulations
5) *Electronics*: cleaning agent for silicon wafers, photoresist stripper, auxiliary in printed circuit board technology
6) *Industrial and domestic cleaning*: component in paint strippers and degreasers e.g., removal of oil, fat, and soot from metal surfaces, carbon deposits and other tarry polymeric residues in combustion engines

Economic Aspects. World production of NMP is currently estimated at 20 000 – 30 000 t/a. The most important producers are BASF and GAF.

Environmental Aspects. NMP is nontoxic to most aquatic life and shows good biodegradability [16], [30]. The most important data are: BOD_5 1100 mg/g NMP (DIN 38 409 part 51), COD 1600 mg/g NMP (DIN 38 409 part 41), TOC (total organic carbon) 600 mg/g NMP (DIN 38 409 part 3), biodegradability: > 90 % DOC (dissolved organic carbon) degradation (Zahn-Wellens static test, DIN 38 412 part 25).

4. N-Vinyl-2-pyrrolidone

Physical Properties. N-Vinyl-2-pyrrolidone (NVP) [88-12-0], N-vinylpyrrolidone, 1-vinyl-2-pyrrolidone, 1-ethenyl-2-pyrrolidinone, C_6H_9NO, M_r 114.14, is a colorless to yellowish liquid when freshly distilled, with a characteristic odor, mp 13.5 °C, bp 90 – 92 °C (1.3 kPa). It is completely miscible with water and most organic solvents, but only partially miscible with aliphatic hydrocarbons. Further physical constants follow:

Density (20 °C)	1.043 g/cm^3
Refractive index n_D^{20}	1.514
Viscosity (20 °C)	2.4 mPa·s
Vapor pressure (20 °C)	0.012 kPa
Flash point (DIN 51 758)	95 °C
Ignition temperature (DIN 51 794)	240 °C
Lower explosion limit	1.4 vol%
Upper explosion limit	10 vol%

Chemical Properties. NVP is stable toward alkalis at room temperature. Above 0 °C it is cleaved by aqueous mineral acids into acetaldehyde and 2-pyrrolidone; the latter reacts with an excess of NVP to give 1,1′-ethylidene-bis-2-pyrrolidone [31]. Protic compounds such as amides, thiols, alcohols, and phenols add to the double bond according to the Markownikoff rule, for example, N-(1-phenoxyethyl)-2-pyrrolidone is formed with phenol [32], [33]. Catalytic hydrogenation produces N-ethyl-2-pyrrolidone. Hydroformylation on rhodium catalysts forms 2-N-(2-pyrrolidonyl)propanal as the major product [34]. On prolonged standing, particularly in warm conditions, NVP tends to polymerize. Industrially, NVP is converted into polyvinylpyrrolidone using radical initiators.

Production. Industrially, NVP is produced by reacting 2-pyrrolidone with acetylene in high-pressure autoclaves at 130 – 160 °C and pressures of up to 2.6 MPa [32], [33], [35]. Vinylation proceeds in the liquid phase and is catalyzed by 2-pyrrolidone – potassium, which is obtained by dissolving 1 – 5 % potassium hydroxide (the water formed is removed by distillation in vacuo) or potassium in 2-pyrrolidone. To avoid the danger of explosions acetylene – nitrogen mixtures are used instead of acetylene. The crude product obtained by continuous or batch production is purified by vacuum distillation. Leading producers are BASF and GAF.

Other methods such as the elimination of N-(1-hydroxyethyl)- (see p. 4242) and N-(1-alkoxyethyl)-2-pyrrolidones are not industrially important [6], [7], [33].

Quality Specifications and Analysis. Composition is determined by gas chromatography and water is titrated with the Karl Fischer reagent. A typical commercial product has the following composition: NVP content, min. 99 wt%; 2-pyrrolidone, max. 0.2 wt%; water, max. 0.1 wt%; color index (freshly filled, APHA), 40 max.; stabilizers: 10 ppm *N,N'*-bis(1-methylpropyl)-1,4-benzenediamine or 0.1 wt% sodium hydroxide.

Handling, Storage, and Transportation. NVP must be protected from heat and direct sunlight. Even in closed containers NVP can only be stored for a limited period because it tends to polymerize. Stabilizers are therefore added. *N,N'*-Bis(1-methylpropyl)-1,4-benzenediamine [*101-96-2*] is an important, NVP-soluble stabilizer that does not interfere with further processing. Use of this stabilizer allows the product to be stored for ca. 300 d at 20 °C, and ca. 150 d at 25 °C. Stabilization with solid sodium hydroxide that can be easily filtered off, is not as efficient. Storage at temperatures close to the melting point results in separation of NVP from the stabilizer due to repeated solidification and melting. This shortens the shelf life considerably and can lead to uncontrolled polymerization. Solidified NVP must therefore be melted carefully in a water bath (max. 40 °C) or at room temperature (max. 30 °C) and homogenized by continuous agitation. NVP is flammable.

Uses. NVP is an important precursor and intermediate for process auxiliaries and additives. The main areas of use are the production of polyvinylpyrrolidone and copolymers, the preferred comonomers being vinyl acetate and methyl acrylate. Ca. 10 – 15 % of the monomer is used in the pharmaceutical industry for the production of a polyvinylpyrrolidone – iodine complex used as a disinfectant. NVP is also used as a reactive solvent for UV-curable resins for the production of printing inks and paints as paper and textile auxiliaries, and as an additive in the cosmetics industry.

5. Other Pyrrolidone Derivatives

For certain areas of application special *N*- or *C*-substituted 2-pyrrolidones have been developed, whose physiochemical properties are especially adapted for a specific use. For example, *N*-ethyl-2-pyrrolidone [*2687-91-4*] is used as a reaction medium for low- and high-temperature reactions; *N*-(2-hydroxyethyl)-2-pyrrolidone [*3445-11-2*] is used in the electronics industry, *N*-octyl- [*2687-94-7*] and *N*-dodecyl-2-pyrrolidone [*2687-96-9*] serve as formulation auxiliaries for triazole fungicides [36]. *N*-(2-Aminoethyl)-2-pyrrolidone [*24935-08-8*] derivatives are used as solubilizing and complexing aids [37]. A range of nootropic pharmaceuticals (e.g., piracetam [*7491-74-9*] **1**) are *N*-substituted 2-pyrrolidones [38]. *N*-1-(3-Trifluoromethylphenyl)-2-pyrrolidone derivatives (e.g., fluorochloridone [*61213-25-0*] **2**) are used as herbicides [39].

6. Toxicology and Occupational Health

2-Pyrrolidone does not have a particularly high acute toxicity: LD_{50} values are > 3.2 g/kg – ca. 8.5 mL/kg (rat, oral); 3.2 g/kg – 5.8 mL/kg (mouse, oral); 6.5 mL/kg (guinea pig, oral); 3 g/kg (rat, i.p.); 3.7 g/kg (mouse, i.p.); 3 g/kg (mouse, i.v.) [40] – [46]. In rabbits it is only mildly irritating to intact skin, if at all, and not irritating to the eyes [41], [42]; in another study clouding of the cornea has, however, been reported [43]. In guinea pigs the compound causes severe skin irritation but no sensitization [40]. It can be absorbed through the skin and causes death in rats (ca. 2 mL/kg) [41]. Inhalation (8 h) of air saturated with 2-pyrrolidone at 30 °C produced no symptoms in rats [41]. In a 13-d study a slight increase in the mean relative kidney weight was observed in rats fed with 10 g/kg [40]. 2-Pyrrolidone did not produce mutagenicity in the Ames test ($\pm$ S-9 mix) [41] but induced aneuploidy in *Saccharomyces cerevisiae* [46]. It is nonteratogenic in mice (oral, i.p.) and rats (oral) [40], [41].

2-Pyrrolidone has been identified as an endogenous component in the blood plasma of rats, mice, and humans; it is metabolized to 4-aminobutanoic acid and other compounds (e.g., succinimide) [47].

2-Pyrrolidone possesses psychotropic properties resembling those of tranquilizing drugs in mice, rabbits, rats, cats, and pigeons. It is said to have irritating properties to human skin [40].

N-Methylpyrrolidone is only slightly toxic if ingested or inhaled. LD_{50} values are 3.5 mL/kg – 4.3 g/kg (rat, oral); 7.5 mL/kg (mouse, oral); 3.5 g/kg (rabbit, oral); 2.2 mL/kg (rat, i.v.); 2.4 mL/kg (rat, i.p.); 1.9 mL/kg (mouse, i.p.), > 2.5 g/kg (rat, percutaneous); 4 – 8 g/kg (rabbit, percutaneous). Inhalation (6 – 8 h) of a highly enriched or saturated atmosphere at 20 or 50 °C produced no symptoms in cats, mice, rats, rabbits, or guinea pigs. The LC_{50} value is > 5.1 mg/L (rat, inhalation, aerosol, 4 h) [16], [48].

In rabbits, NMP produced slight skin irritation (method according to Draize); undiluted it caused moderate reddening and scaling after short exposure (5 – 15 min), 20-h exposure led to additional edema. In the rabbit eye it caused conjunctivities, redness, and reversible clouding of the cornea. NMP is not a skin sensitizer in guinea pigs [16], [48].

Rabbits administered NMP by gavage in doses of 0.4 mL kg^{-1} d^{-1} (5 times/week for 5 weeks) showed no adverse effects, yet death occurred at 1 and 2 mL kg^{-1} d^{-1}. Cats were less sensitive. In rats, 0.25 mL kg^{-1} d^{-1} (4 weeks) caused no effects; 0.5 and 1 mL kg^{-1} d^{-1} led to slightly retarded body weight gain in males. A slight decrease in leucocytes was observed at 1 mL kg^{-1} d^{-1} ; 2 mL kg^{-1} d^{-1} retarded body weight gain and led to a decrease in leucocytes and testicular damage. In a 90-d study on rats, 2.5 g/kg feed caused no changes; 5 g/kg increased thyroid gland weight. Dermal application (20 d, 0.4 and 0.8 mL kg^{-1} d^{-1}) to intact or abraded skin of rabbits produced slight local irritation.

Inhalation (17 exposures, 6 h/d, 5 d/week) of a saturated or highly enriched atmosphere produced no symptoms in cats, rabbits, guinea pigs, or rats, but caused death in mice. In a 6-week study (6 h/d, 5 d/week) air saturated at 25 °C caused no symptoms in rats. In a 4-week inhalation study (whole body exposure, 6 h/d, 5 d/week), 0.1 and 0.5 mg/L were tolerated whereas 1 mg/L caused lethality through additional dermal and oral absorption. In another study (head–nose exposure), no symptoms could be detected (0.9 mg/L, 6 h/d, 10 d) [16], [48].

NMP was found to be nonteratogenic in rats (oral, dermal, and inhalative) and mice (i.p., oral) in doses that were not maternally toxic[16], [48]. In several mutagenicity tests NMP did not demonstrate a genotoxic potential. It induced aneuploidy in *Saccharomyces cerevisiae*, but neither spindle disfunction nor chromosome damage was found in mice [16]. NMP was not carcinogenic in a 2-year inhalation study on rats (0.04 and 0.4 mg/L, 6 h/d, 5 d/week) [16].

NMP is rapidly absorbed and is mainly excreted in the urine within 24 h (rats, i.v.), excretory half-life 2.3 h. It is metabolized into three compounds, mainly 4-methylaminobutanoic acid [49].

The MAK value is 100 ppm (8 h) [48]. Additional information on toxicity is given in [16], [48].

N-Vinyl-2-pyrrolidone exhibits moderate acute toxicity if ingested, inhaled, or administered intraperitoneally or intravenously. The LD$_{50}$ values are > 0.8 – 2.5 g/kg (rat, oral); 0.9 g/kg (mouse, oral); ca. 0.3 – 0.47 g/kg (mouse, i.v.); > 0.25 – 0.605 g/kg (mouse, i.p.); 1 – 4 mL/kg (rat, dermal). The LC$_{50}$ is > 3.7 mg/L (rat, cat, rabbit, mouse, guinea pig, inhalation, 6 h) [41], [50]. NVP induces narcotic symptoms, irritation of the skin and eye, and damage to the liver, spleen, and kidneys. In rabbits, it is a mild irritant to intact skin; the undiluted compound causes redness, edema, and clouding of the cornea that lasts for more than 8 days. Inhalation (8 h) of air saturated with NVP at room temperature produced slight irritation of the mucosa. Repeated oral administration of NVP (drinking water or by gavage) caused liver damage in rats. Signs of liver damage were recorded for rabbits and cats. Intraperitoneal administration (5 d, 0.1, 0.25 mL/kg) caused death in mice but not in rats [41], [50].

Inhalation of NVP vapor (up to 15 days, 6 h/d, 180 ppm) caused death of mice, cats, and rabbits but not of rats and guinea pigs. Damage to the liver was seen in mice and cats. Inhalative exposure of rats to the aerosol (0.3 – 0.75 mL/kg, 4 h/d, 5 d/week for 4 weeks) or vapor (1–120 ppm, 6 h/d, 5 d/week for 3 months) caused damage to the nasal mucous membranes and adenoids above 0.75 mL/L (5 ppm) and foci and areas of cellular alterations in the liver above 15 ppm. 120 ppm caused premature death. In the Syrian gold hamster (45 ppm, 6 h/d, 5 d/week for 7 weeks to 3 months) symptoms of irritation were seen in the upper respiratory tract [41], [50].

Slight signs of liver toxicity but no carcinogenic impairment of the liver was noted. In several tests NVP did not show mutagenic or clastogenic activity. In a 2-year inhalation study in rats (5, 10, and 20 ppm, 6 h/d, 5 d/week) carcinomas of the liver, adenomas, and adenocarcinomas of the nasal cavity were observed. Squamatous carcinomas of the larynx were reported at 20 ppm [41], [50].

A single intravenous dose of ^{14}C-NVP is quickly metabolized and eliminated (mainly in the urine) in the rat. The plasma half-life is 1.9 h [51].

Epidemiological studies in humans revealed no adverse effects [52]. An MAK value cannot be defined on account of the carcinogenic potential. Additional information on toxicity is given in [50].

7. References

[1] BASF, *Pyrrolidon*, technical bulletin, Ludwigshafen 1987.
[2] J. Tafel, M. Stern, *Ber. Dtsch. Chem. Ges.* **33** (1900) 2224–2231.
[3] BASF, DE 830 194, 1950 (C. Schuster).
[4] Mitsubishi Petrochem., JP-Kokai 74 117 459, 1973 (T. Ayusawa, S. Fukami).
[5] W. Reppe, *Justus Liebigs Ann. Chem.* **596** (1955) 210.
[6] J. Falbe, H.-J. Schulze-Steinen, *Brennst. Chem.* **48** (1967) 136–139.
[7] Mitsubishi Chemical, DE-OS 3 544 134, 1985; US 4 837 337, 1988 (Y. Murao, M. Miyake).
[8] K. Dachs, E. Schwartz, *Angew. Chem.* **74** (1962) 540–545.
[9] Du Pont, US 2 187 745, 1934 (W. A. Lazier).
[10] J. Falbe, F. Korte, *Chem. Ber.* **98** (1965) 1928–1937. J. F. Knifton, *J. Organomet. Chem.* **188** (1980) 223–236.
[11] Chevron, US 4 181 662, 1978 (W. A. Sweeney). Standard Oil, US 4 036 836, 1975 (J. L. Greene).
[12] Phillips Petroleum, US 4 800 227, 1984 (M. S. Matson).
[13] E. Späth, J. Lintner, *Ber. Dtsch. Chem. Ges.* **69** (1936) 2727–2731. BASF, DE-AS 1 765 007, 1968 (W. Himmele, E. Hofmann, H. Hoffmann, R. Plaß).
[14] GAF, US 4 824 967, 1988 (K.-C. Liu, P. D. Taylor).
[15] *Inf. Chim.* **239** (1983) 65.
[16] BASF: *N-Methylpyrrolidone*, product brochure, Ludwigshafen 1990.
[17] GAF, *N-Methyl-2-pyrrolidone, Properties and Chemical Reactions*, 1972.
[18] S. Yoshifuji, Y. Arakawa, Y. Nitta, *Chem. Pharm. Bull.* **35** (1987) 357–363.
[19] H. Eilingsfeld, M. Seefelder, H. Weidinger, *Angew. Chem.* **72** (1960) 836–845; *Chem. Ber.* **96** (1963) 2671–2690.
[20] General Electric, US 4 353 830, 1979 (V. Mark). T. Jen et al., *J. Med. Chem.* **18** (1975) 90–99. McNeil, DE-OS 1 770 752, 1968 (G. I. Poos).
[21] P. Hullot, T. Cuvigny, M. Larchevêque, H. Normant, *Can. J. Chem.* **54** (1976) 1098–1104.

[22] J. D. Stewart, S. C. Fields, K. S. Kochhar, H. W. Pinnick, *J. Org. Chem.* **52** (1987) 2110–2113.
[23] G. M. Ksander, J. E. McMurry, M. Johnson, *J. Org. Chem.* **42** (1977) 1180–1185.
[24] E. Späth, H. Bretschneider, *Ber. Dtsch. Chem. Ges.* **61** (1928) 327–334.
[25] BASF, DE 944 311, 1955 (M. Kracht, H. Pasedach).
[26] E. Watanabe et al., *J. Chem. Soc. Chem. Commun.* 1986, 227–228.
[27] Mitsubishi Chemical, US 4 731 454, 1986 (M. Otake, J. Fukushima). Amoco, US 4 814 464, 1987 (R. J. Olson). Hüls AG, EP-A 252 242, 1986 (H. M. Zur, W. Otte). BASF, EP-A 460 474, 1990 (H.-J. Weyer, R. Fischer, W. Harder).
[28] Mitsubishi Chemical, JP-Kokai 87 120 360, 1985 (M. Otake, I. Fukushima, K. Fujita). Stamicarbon, EP-A 37 603, 1980 (P. J. N. Meyer, J. M. Penders).
[29] BASF, DE-OS 3 701 297, 1987 (F. Merger, J. Liebe, W. Harder).
[30] BASF, *N-Methylpyrrolidone, Biodegradability,* company brochure, Ludwigshafen 1991.
[31] E. Senogles, R. A. Thomas, *J. Chem. Soc. Perkin Trans. II* 1980, 825–828.
[32] W. Reppe, *Justus Liebigs Ann. Chem.* **601** (1956) 135–138. I. G. Farbenind., Chem. Zentralbl. 1941, II, 2735.
[33] S. A. Miller: *Acetylene. Its Properties, Manufacture and Uses,* vol. **2**, Ernest Benn, London 1966, pp. 338–341.
[34] Texaco, US 4 831 159, 1986 (J.-L. Lin).
[35] GAF, WO 89 09210, 1988 (K. C. Liu, P. D. Taylor); DE-OS 3 215 093, 1982 (R. Parthasarathy, E. V. Hort, P. M. Chakrabarti).
[36] Bayer, DE 3 910 921, 1989 (K. Reizlein et al.); EP-A 453 915, 1990 (H. O. Horstmann, K. Wangermann).
[37] GAF, US 4 918 198, US 4 924 006 , 1989 (L. R. Anderson, M. M. Hashem, R. B. Login).
[38] E. R. Gamzu, T. M. Hoover, S. I. Gracon, M. V. Nintemann, *Drug Dev. Res.* **18** (1989) 177–189.
[39] Stauffer, US 4 110 105, 1976 (E. G. Tech).
[40] CRCS Inc., working draft, EPA contract no. 68-01-650, TSCA Interagency Testing Committee, July 15, 1985.
[41] BASF AG, unpublished results (1953–1990).
[42] W. B. Deichmann, H. W. Gerade (eds.): *Toxicology of Drugs and Chemicals,* 4th ed., Academic Press, New York 1969, pp. 508.
[43] V. V. Aleshin, *Med. Probl. Okhr. Vneshn. Stredy* 1974, 4; *Chem. Abstr.* **86** 26605b.
[44] National Institute for Occupational Safety and Health, RTECS no. 20 387, Jan. 1987.
[45] G. H. Cocolas et al., *J. Pharm. Sci.* **72** (1983) 812–814.
[46] V. W. Mayer, C. J. Goin, R. E. Taylor-Mayer, *Environ. Mol. Mutagen.* **11** (1988) 31–40.
[47] E. F. Bandle et al., *Life Sci.* **35** (1984) 2205–2212.
[48] Deutsche Forschungsgemeinschaft: "N-Methylpyrrolidon,"*Gesundheitsschädliche Arbeitsstoffe,* VCH Verlagsgesellschaft, Weinheim 1978, 1988.
[49] D. A. Wells, G. A. Digenis, *Drug. Metab. Dispos.* **16** (1988) 243–249.
[50] Deutsche Forschungsgemeinschaft: "N-Vinylpyrrolidon,"*Gesundheitsschädliche Arbeitsstoffe,* VCH Verlagsgesellschaft, Weinheim 1991.
[51] G. A. Digenis, J. S. McClanahan, *The Toxicologist* **2** (1982) 165–166. J. S. McClanahan et al., *Drug. Chem. Toxicol.* **7** (1984) 129–148.
[52] A. Zober et al., to be published in *Proceedings of XIXth International Conference on Occupational Health in Chemical Industry,* Basel (Switzerland), Sept. 17–20, 1991, WHO, Copenhagen 1992.

Quinoline and Isoquinoline

GERD COLLIN, DECHEMA e.V., Frankfurt/Main, Federal Republic of Germany (Chaps. 1, 2)
HARTMUT HÖKE, Rütgerswerke AG, Frankfurt/Main, Federal Republic of Germany (Chap. 3)

1.	Quinoline	4253	2.1.	Properties	4256
1.1.	Properties	4253	2.2.	Production	4257
1.2.	Production	4254	2.3.	Uses	4257
1.3.	Uses and Economic Aspects	4254	3.	Toxicology	4257
1.4.	Derivatives	4255			
2.	Isoquinoline	4256	4.	References	4258

1. Quinoline

Quinoline was discovered in coal tar by F. F. RUNGE in 1834.

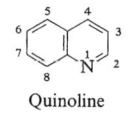

1.1. Properties

Quinoline [91-22-5], C_9H_7N, M_r 129.16, mp −15.6 °C, bp 238 °C (101.3 kPa), d^{20} 1.0929, n_D^{20} 1.6268, specific heat (17 °C) 1.516 kJ/kg, heat of vaporization 367.6 kJ/kg, heat of combustion (25 °C) 36.4 MJ/kg, is a colorless liquid with characteristic odor. It is sparingly soluble in cold water, readily soluble in hot water, miscible with many organic solvents, and volatile in steam.

Quinoline is a weak tertiary base. With strong acids, it forms readily crystallizable salts; with alkyl halides and dialkyl sulfates, it forms quaternary addition compounds. Halogenation gives preferably the 3-substituted product; nitration, the 5- and 8-substituted products; and sulfonation, the 6- or 8-substituted products, depending on reaction conditions. Quinoline is catalytically hydrogenated to 1,2,3,4- or 5,6,7,8-tetrahydroquinoline, and further to decahydroquinoline. Oxidation leads to quinoline N-oxide, 2-quinolinol or 2-quinolinone, quinolinic acid (pyridine-2,3-dicarboxylic acid), or nicotinic acid, depending on the oxidizing agent used and the reaction conditions. Biochemical oxidation yields 2-oxoquinoline as the primary metabolite, which is then hydroxylated in the 6- or 8-position [5]. Reaction with sodium amide or ammonia in the presence of a Raney nickel catalyst gives 2-aminoquinoline.

1.2. Production

High-temperature coal tar contains an average of 0.3% quinoline. Quinoline, together with isoquinoline and 2-methylquinoline, is recovered from the methylnaphthalene fraction of coal tar by extraction with sulfuric acid, followed by precipitation with ammonia. Then quinoline is separated from isoquinoline (bp 6 °C higher) by rectification of the crude base mixture (quinoline bases). It can be purified further by several processes, for example, selective oxidation, treatment with alkali, resinification of the byproducts with formaldehyde, or formation of the hydrate (in contrast to isoquinoline) [6]. Quinoline can also be obtained from the methylnaphthalene fraction by azeotropic distillation with ethylene glycol or diethylene glycol, and subsequent distillation [7].

Quinoline can be synthesized via the Skraup method by heating aniline and glycerol with sulfuric acid in the presence of a dehydrogenating agent (→ Amines, Aromatic), catalytic gas-phase reaction of aniline and acetaldehyde with formaldehyde – methanol [8], dehydrogenative cyclization of N-propylaniline in the gas phase [9], or pyrolysis of N,N'-diphenyl-1,1-propanediamine [10]. However, these syntheses are not yet industrially important because sufficient quantities of quinoline can be recovered from coal tar.

1.3. Uses and Economic Aspects

World production of quinoline is more than 2000 t/a. The main application of quinoline is the production of 8-quinolinol, which is obtained by alkali fusion of quinoline-8-sulfonic acid [85-48-3]. In addition, quinoline is used in the production of pyridine-2,3-dicarboxylic acid [89-00-9] (quinolinic acid) by oxidation [11]. Quinolinic acid is used to manufacture the herbicide Assert [81405-85-8] (imazapyr) [12]. Quinoline can be used to produce methine dyes (cyanines) and nicotinic acid [59-67-6] (via quinolinic acid).

Hydroquinolines can be employed as intermediates for pharmaceuticals such as antibiotics [13]. From quinoline, 4,5-dihydroimidazo[1,2-*a*]- quinoline-1-acetamides can be prepared as anticonvulsants, sedatives, and anxiolytics [14]. Quinoline can act as catalyst in the preparation of pharmaceuticals such as pyridazinone antihypertensives [15]. Fungicides can be obtained from 2-anilinoquinolines (from 2-aminoquinoline) [16]. Derivatives of 2-quinolinone [59-31-4] (carbostyril, prepared by oxidation of quinoline with hypochlorous acid, or selective, enzymatic hydroxylation of quinoline [17]) can be used as pharmaceuticals (e.g., cardiotonic drugs or antihistaminics) [18]. Quinoline alone, or as a mixture with isoquinoline and quinoline homologues, is an excellent solvent and extractant, especially for polycyclic aromatic compounds. Quinoline bases serve as acid-binding agents, corrosion inhibitors, or pickling inhibitors.

1.4. Derivatives

2-Methylquinoline [91-63-4], quinaldine, $C_{10}H_9N$, M_r 143.19, bp 247.6 °C (101.3 kPa), mp −2 °C, d^{20} 1.0582, n_D^{20} 1.6116, is a colorless liquid with a weak quinoline-like odor. It is soluble in organic solvents and very slightly soluble in water.

2-Methylquinoline

High-temperature coal tar contains an average of 0.2% quinaldine. It is recovered from the quinoline base mixture by rectification and hydration. Quinaldine can be synthesized by the Skraup method from aniline and crotonaldehyde with sulfuric acid in the presence of a dehydrogenating agent.

Quinaldine is used for the synthesis of several dyes such as the trimethinecyanine dye pinacyanol [605-91-4] and the corresponding vinylogous dye, the quinophthalone dye quinoline yellow [8004-92-0], and the indicator dye quinaldine red [117-92-0].

4-Methylquinoline [491-35-0], lepidine, $C_{10}H_9N$, M_r 143.19, bp 265.6 °C (101.3 kPa), mp 8 °C, d^{20} 1.0868, n_D^{20} 1.6200, is a colorless liquid with a weak quinoline-like odor. It is slightly soluble in water, miscible with many organic solvents, and volatile in steam.

4-Methylquinoline

4-Methylquinoline can be recovered from the highest-boiling fraction of the coal-tar quinoline bases via its o-cresol addition compound. It can be synthesized by condensation of aniline and methyl vinyl ketone [19].

4-Methylquinoline is used for the synthesis of several cyanine dyes.

8-Quinolinol [148-24-3], 8-hydroxyquinoline, oxine, C_9H_7ON, M_r 145.15, bp 266.6 °C (100.2 kPa), mp 75.8 °C, forms colorless crystals with a phenolic odor. It is readily soluble in ethanol, benzene, and chloroform; soluble in acetone, alkali, and acid; slightly soluble in diethyl ether; and very slightly soluble in water.

8-Quinolinol

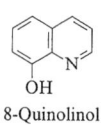

8-Quinolinol is produced by alkali fusion of quinoline-8-sulfonic acid or synthesized via the Skraup method by heating *o*-aminophenol (via benzoxazolone [20]) with *o*-nitrophenol, glycerol, and sulfuric acid.

8-Quinolinol and its salts with bivalent metals or mineral acids are used as fungicidal, bactericidal, insecticidal, and algicidal agents in crop protection products and disinfectants, as well as in preservatives for leather, textiles, plastics, coatings, paper pulp, and seeds. Several halogenated derivatives of 8-quinolinol show, in addition to a general antiseptic activity, a specific activity against amoebas. In azo dyes 8-quinolinol can serve as a heterocyclic coupling component. 8-Quinolinol, its dibrominated derivative Bromoxine [521-74-4], and the thiol analogue Thiooxine [491-33-8] are analytical reagents for the detection of numerous metals.

2. Isoquinoline

Isoquinoline was discovered in coal tar by S. Hoogewerff and W. A. van Dorp in 1885.

Isoquinoline

2.1. Properties

Isoquinoline [119-65-3], C_9H_7N, M_r 129.16, mp 26.5 °C, bp 243.25 °C (101.3 kPa), d^{20} 1.0986, n_D^{20} 1.6148, specific heat 1.507 kJ/kg, heat of fusion 57.8 kJ/kg, heat of vaporization 379.3 kJ/kg, heat of combustion (25 °C) 36.45 MJ/kg, is a colorless, hygroscopic liquid that forms crystals on solidifying, with an odor resembling that of benzaldehyde. It is very slightly soluble in water, but miscible with many organic solvents.

Isoquinoline is a stronger base than quinoline. It forms readily crystallizable salts and quaternary addition compounds. Catalytic hydrogenation leads to 1,2,3,4- or 5,6,7,8-tetrahydroisoquinoline and further to decahydroisoquinoline. Oxidation in the liquid or vapor phase gives a mixture of nicotinic and isonico-tinic acids (via pyridine-3,4-dicarboxylic acid by decarboxylation), whereas oxidation with peroxyacids gives isoquinoline *N*-oxide.

2.2. Production

High-temperature coal tar contains an average of 0.2% isoquinoline. It is separated by distillation from the lower-boiling quinoline and the higher-boiling 2-methylquinoline of the quinoline base mixture. Further refining is based on the fact that isoquinoline, in contrast to quinoline and 2-methylquinoline, cannot be hydrated but can be crystallized at low temperature.

Isoquinoline can by synthesized, for example, via the Bischler–Napieralski reaction by cyclodehydration of *N*-acyl derivatives of β-phenylethylamine with Lewis acids and subsequent dehydrogenation.

2.3. Uses

Isoquinoline is used predominantly in the production of pharmaceuticals such as the anthelmintic praziquantel [*55268-74-1*]. Isoquinoline and hydroisoquinolines can also be used as intermediates for fungicides [21] and dyes (e.g., isoquinoline red [*6359-40-6*]).

3. Toxicology

The oral LD_{50} of quinoline in rats is 270–460 mg/kg [22], [23]; the dermal LD_{50} is 1400 mg/kg in rats [24], and ca. 600 mg/kg in rabbits [23]. Inhalation of air saturated with quinoline for 7 h caused death of almost all rats [25]. Acute symptoms were unspecific: the rats exposed showed piloerection, ataxia, and lack of motility. Dissection revealed hemorrhaging and cellular edema in the gastrointestinal tract, liver, and lung. The 24-h exposure of the skin of rabbits caused slight to moderate, but reversible, edema and erythema; after exposure of the eyes of rabbits, slight to moderate, but reversible, irritation was found [26]. After resorption in mammals, quinoline is excreted rapidly after hydroxylation and conjugation [27], [28]. Long-term feeding (0.05–0.25% in the diet for 20–40 weeks) led to hepatocellular carcinomas and hemangiosarcomas in rats and mice [29], [30], but not in hamsters and guinea pigs [30]. When low doses (1.75 µmol per animal) were injected intraperitoneally into newborn mice on the first, eighth, and fifteenth day after birth, liver tumors were found 52 weeks after treatment [31]. Quinoline showed tumor-initiating activity on mouse skin at an effective total dose of > 2 mg per animal [32], which is ca. 250–300 times higher than the dose of benzo[*a*]pyrene required to obtain the same tumor incidence. Quinoline has toxic effects on aquatic life, with fish apparently being more sensitive than the aquatic invertebrates and algae tested; e.g., 5 mg/L was lethal to trout after 14 h and bluegill sunfish after 4 h [33], whereas toxic effects on daphnia were noticeable only above 50 mg/L [34]. The compound is readily degraded under natural conditions.

The oral LD_{50} of 2-methylquinoline (quinaldine) is 1230 mg/kg in rats [23], and the LD_{50} (i.p.) of 4-methylquinoline (lepidine) is 270 mg/kg in rats [35]. 4-Methylquinoline was found to be a tumor initiator on mouse skin, whereas 2-methylquinoline was not [32]. 4-Methylquinoline is tumorigenic when injected into newborn mice [31]. 2-Methylquinoline showed very weak mutagenicity with *Salmonella typhimurium* TA 100 [36], whereas the number of his-revertants was greater for 4-methylquinoline than for quinoline [37].

For 8-quinolinol an oral LD_{50} of 1200 mg/kg in rats and an i.p. LD_{50} of 48 mg/kg in mice are reported [38]. After i.p. injection of 100 mg/kg into mice, many died after several hours, whereas the same dose of quinoline caused no deaths [39]. In plant chromosomal studies, 8-quinolinol was found to act as a spindle inhibitor [40]. In two rodent studies by the U.S. National Toxicological Program, 8-quinolinol was classified as noncarcinogenic [40], [41] but was found to be mutagenic in the Ames test [42], [43], genotoxic in several other in vitro assays, and positive in the micronucleus assay [39].

The oral LD_{50} of isoquinoline is 360 mg/kg [23]. Isoquinoline shows no tumor-initiating activity [44] and proved to be nonmutagenic in the Ames test [37].

4. References

General References

[1] *Beilstein,* Quinoline **20** 339; **20** (1) 134; **20** (2) 222; **20** (3/4) 3334; 2-Methylquinoline **20** 387; **20** (1) 148; **20** (2) 238; **20** (3/4) 3454; 4-Methylquinoline **20** 395; **20** (1) 150; **20** (2) 244; **20** (3/4) 3477; 8-Quinolinol **21** 91; **21** (1) 221; **21** (2) 55; **21** (3/4) 1135; Isoquinoline **20** 380; **20** (1) 143; **20** (2) 236; **20** (3/4) 3410.

[2] H.-G. Franck, G. Collin: *Steinkohlenteer,* Springer Verlag, Berlin 1968, pp. 13, 23, 26, 62–64, 83, 97, 151, 158, 160–162.

[3] H.-G. Franck, J. W. Stadelhofer: *Industrial Aromatic Chemistry,* Springer Verlag, Berlin 1988, pp. 274, 419–423.

[4] *Kirk-Othmer,* 3rd ed., **19,** 532–572.

Specific References

[5] O. P. Shukla, *Biol. Mem.* **13** (1987) 115–131. G. Schwarz et al., *Hoppe Seyler Biol. Chem.* **370** (1989) 1183–1189.

[6] R. Oberkobusch, *Brennst. Chem.* **40** (1959) 145–151.

[7] Rütgerswerke, EP 100 109, 1983 (H. Hörmeyer). Sumikin Coke & Chemicals, JP 86 161 265, 1986 (K. Kageyama, S. Takeya, T. Nakamura). Allied, DE 3 227 492, 1982 (S. E. Belsky, C. T. Mathew).

[8] Reilly Tar & Chemical, US 3 020 280-2, 1957 (F. E. Cislak, W. R. Wheeler).

[9] ICI, GB 1 184 242, 1967 (W. H. Bell, R. A. C. Rennie).

[10] Snam Progetti, DE 2 223 018, 1971.
[11] Rütgerswerke, EP 149 857, 1984 (W. Orth, E. Pastorek, W. Fickert). Ruetgers-Nease Chemical, EP 232 118, 1987 (W. Michalowicz). Hilton-Davis Chemical, US 4 537 971, 1985 (R. W. J. Rebhahn, J. E. Kassner, R. E. Werner).
[12] American Cyanamid, EP 41 623, 41 624, 1981 (P. L. Orwick, A. R. Templeton).
[13] Sumitomo Chemical, JP 87 212 363, 1986 (S. Murahashi). Hoechst, DE 3 706 020, 1987 (R. Lattrell, W. Dürckheimer, R. Kirstetter).
[14] Synthelabo, FR 2 593 179, 1986 ;EP 231 138, 1987 (P. George, D. de Peretti).
[15] Diamond Shamrock, US 4 281 125, 1980 (M. F. Depompei, A. Hlynsky).Société Nationale des Poudres et Explosifs, EP 249 556, 1987 (J. P. Senet, G. Sennvey, G. Wooden).
[16] BASF, DE 3 716 512, 1987 (J. Schubert et al.).
[17] Rütgerswerke AG, DE 3 903 759, 1989 (F. Lingens, R. Bauder, H. Höke).
[18] Otsuka Pharmaceutical, JP 82 35 588, 1980; JP 83 88 314, 1981.
[19] BASF, DE 3 719 014, 1987 (H. Merkle, G. Reissenweber).
[20] Rütgerswerke, DE 3 601 024, 1986 (J. Haase).
[21] Nippon Steel, JP 87 294 679, 1986 (S. Watanabe et al.).
[22] Beratergremium für umweltrelevante Altstoffe der Gesellschaft Deutscher Chemiker (BUA): *Chinolin*, BUA-Stoffbericht, VCH Verlagsgesellschaft, Weinheim (to be published).
[23] H. F. Smyth jr., C. P. Carpenter, C. S. Weil, *AMA Arch. Ind. Hyg. Occup. Med.* **4** (1951) 119–122.
[24] Rütgerswerke AG: unpublished test report (1981).
[25] Rütgerswerke AG: unpublished test report (1981).
[26] Rütgerswerke AG: unpublished test reports (1979).
[27] L. Novack, B. B. Brodie, *J. Biol. Chem.* **187** (1950) 787–792.
[28] J. N. Smith, R. T. Williams, *Biochem. J.* **60** (1955) 284–290.
[29] K. Hirao et al., *Cancer Res.* **36** (1976) 329–335.
[30] Y. Shinohara et al., *Gann* **68** (1977) 785–796.
[31] E. J. La Voie et al. in M. Cooke, A. J. Dennis (eds.): "Polynuclear Aromatic Hydrocarbons," *Proc. 10th Int. Symp. 1985*, Battelle Press, Columbus/Richland 1988, pp. 503–518.
[32] E. J. La Voie et al., *Cancer Letters (Shannon Irel.)* **22** (1984) 269–273.
[33] V. C. Applegate et al., *Fish Wildl. Serv. (U.S.) Res. Rep.* **207** (1957) 1–157.
[34] K. Verschueren (ed.): Handbook of Environmental Data on Organic Chemicals, 2nd ed., Van Nostrand, New York 1983.
[35] National Institute for Occupational Safety and Health (NIOSH): Registry of Toxic Effects of Chemical Substances, US Government Printing Office, Washington, D.C. 1987, p. 13841.
[36] M. Dong, I. Schmeltz, E. J. La Voie, D. Hoffmann in P. W. Jones, R. I. Freudenthal (eds.): *Carcinogenesis*, vol. **3**, Raven Press, New York 1978, 97–108.
[37] M. Nagao et al., *Mutat. Res.* **42** (1977) 335–342.
[38] in [35], p. 21549.
[39] M. A. Hamoud, T. Ong, M. Petersen, J. Nath, *Teratog. Carcinog. Mutagen.* **9** (1989) 111–118.
[40] L. S. Gold et al., *EHP Environ. Health Perspect.* **74** (1987) 237–239.
[41] J. K. Hasemann, J. E. Huff, E. Zeiger, E. E. McConnel, *EHP Environ. Health Perspect.* **74** (1987) 229–235.
[42] J. L. Eppler et al., *Mutat. Res.* **39** (1977) 285–296.
[43] R. Talcott, M. Hollstein, E. Wei, *Biochem. Pharmacol.* **25** (1976) 1323–1328.
[44] M. Dong, I. Schmeltz, E. Jacobs, D. Hoffmann, *Anal. Toxicol.* **2** (1978) 21–55.

Resorcinol

KLAUS W. SCHMIEDEL, Hoechst Aktiengesellschaft, Frankfurt am Main, Federal Republic of Germany
DANIEL DECKER, Hoechst Aktiengesellschaft, Frankfurt am Main, Federal Republic of Germany

1.	Introduction	4261	4.3.	Other Processes ... 4266
2.	Physical Properties	4261	5.	Environmental Protection ... 4267
3.	Chemical Properties	4263	6.	Quality Specifications and Analysis ... 4267
4.	Production	4263	7.	Uses ... 4268
4.1.	Via Sulfonation of Benzene	4265	8.	Transport Classifications and Occupational Health ... 4268
4.2.	Via Hydroperoxidation of m-Diisopropylbenzene	4265	9.	References ... 4269

1. Introduction

Resorcinol [108-46-3], $C_6H_6O_2$, M_r 110.112, is a white crystalline compound with a weak odor and a bittersweet taste. Other names are 1,3- (or *meta*-) benzenediol, 1,3-dihydroxybenzene or dioxybenzene, and 3-hydroxyphenol. Resorcinol does not occur in nature as such. The first syllable of the name resorcinol is derived from the word resin because HLASIWETZ and BARTH obtained it by the destructive distillation of a natural resin in 1864 [1]. The structure of resorcinol

is similar to that of orcinol, 5-methyl-1,3-benzenediol [6153-39-5], which accounts for the second part of the name. Resorcinol has been produced industrially for more than 100 years.

2. Physical Properties [2]

The physical properties of resorcinol can be summarized as follows:

mp	(101.3 kPa)	109.8 °C
bp	(101.3 kPa)	276.5 °C
	(1.33 kPa)	151.5 °C
Vapor pressure, liquid		
	(110 °C)	0.10003 kPa
	(150 °C)	0.99438 kPa
	(190 °C)	6.1064 kPa
Density, solid		
	(α-phase, 20 °C)	1.278 g/cm^3
	(β-phase, 20 °C)	1.327 g/cm^3
Density, liquid		
	(110 °C)	1.1758 g/cm^3
	(150 °C)	1.1439 g/cm^3
	(190 °C)	1.1103 g/cm^3
Dynamic viscosity, liquid		
	(130 °C)	4.994 mPa · s
	(150 °C)	3.013 mPa · s
	(190 °C)	1.388 mPa · s
Dynamic viscosity, vapor		
	(110 °C)	0.00873 mPa · s
	(150 °C)	0.00967 mPa · s
	(190 °C)	0.01060 mPa · s
Specific heat capacity, solid		
	(20 °C)	1191 J kg^{-1} K^{-1}
	(50 °C)	1325 J kg^{-1} K^{-1}
	(70 °C)	1417 J kg^{-1} K^{-1}
Specific heat capacity, liquid		
	(at mp)	2185 J kg^{-1} K^{-1}
	(140 °C)	2151 J kg^{-1} K^{-1}
	(180 °C)	2287 J kg^{-1} K^{-1}
Specific heat capacity, vapor		
	(110 °C)	1402 J kg^{-1} K^{-1}
	(150 °C)	1512 J kg^{-1} K^{-1}
	(190 °C)	1609 J kg^{-1} K^{-1}
Entropy (ideal gas, 25 °C)		2888 J kg^{-1} K^{-1}
Heat of formation (ideal gas, 25 °C)		−274.7 kJ/mol
Gibbs free energy of formation (ideal gas, 25 °C)		−181.3 kJ/mol
Heat of fusion (at mp)		21.0 kJ/mol
Heat of vaporization		
	(110 °C)	78.2 kJ/mol
	(150 °C)	75.2 kJ/mol
	(190 °C)	72.0 kJ/mol
Heat of combustion (25 °C)		−2719 kJ/mol
Estimated critical data		
	t_{crit}	810 °C
	p_{crit}	7490 kPa
	v_{crit}	0.3 L/mol
Flash point (closed cup)		126.85 °C
Lower explosion limit in air (199 °C)		1.4 vol%
Density of aqueous solution		1.1101 g/cm^3
Solubility		
	in water (0 °C)	66.7 g/100 g solvent
	(20 °C)	141.0 g/100 g solvent
	(40 °C)	266.3 g/100 g solvent
	(60 °C)	506.0 g/100 g solvent

in CCl$_4$ (40 °C)	0.12 g/100 g solvent
in CHCl$_3$ (40 °C)	0.78 g/100 g solvent
in acetone (40 °C)	243.3 g/100 g solvent
in benzene (40 °C)	0.85 g/100 g solvent
in nitrobenzene (40 °C)	17 g/100 g solvent
pK_1 (20 °C)	9.4
pK_2 (20 °C)	11.4

Resorcinol exists in at least two crystalline modifications (phases) [3]. At normal pressure the α-phase is stable below ca. 71 °C, whereas the β-phase is stable above that temperature up to the melting point. Transition rates from the α- to the β-phase between 72 and 100 °C are given in [4]. Both phases exist as rhombic–pyramidal crystals (Pna2$_1$), but the unit cells differ in their dimensions: α-resorcinol: $a = 1.053$ nm, $b = 0.95$ nm, $c = 0.566$ nm; β-resorcinol: $a = 0.791$ nm, $b = 1.257$ nm, $c = 0.55$ nm. The unit cells of both phases contain four resorcinol molecules.

Interatomic distances in the resorcinol molecule are as follows: C–C 0.138 nm, C–O 0.134 nm, C–H 0.108 nm, O–H 0.102 nm.

Literature data on solubilities vary widely because of the presence of the α- and β-phases: according to [5], 198 g of resorcinol is soluble in 100 g of water at 50 °C; according to [6], 371.5 g is soluble in 100 g water at 50.4 °C.

The density of aqueous resorcinol solutions increases approximately proportionally to the resorcinol concentration.

3. Chemical Properties [7]

Resorcinol is a dihydric phenol and exhibits the typical reactivity of a phenol. Its most important reaction is with formaldehyde [50-00-0] to form phenolic resins.

The hydrogen atom surrounded by the two *meta*-hydroxyl groups can be substituted much more easily than the other ring hydrogens. Compared to catechol [120-80-9] and hydroquinone [123-31-9], resorcinol has the highest reactivity toward formaldehyde.

4. Production

Resorcinol is produced commercially worldwide in only a few specialized plants. All of these plants use benzene [71-43-2] as the main feedstock (see Fig. 1). In Japan, resorcinol is produced in two plants (Sumitomo Chemical and Mitsui Petrochemical) via 1,3-diisopropylbenzene [99-62-7]. The United States (INDSPEC Chemical Corp.) and Germany (Hoechst) each produce it in one plant, using the "classical" route via 1,3-benzenedisulfonic acid [98-48-6] (see Fig. 2).

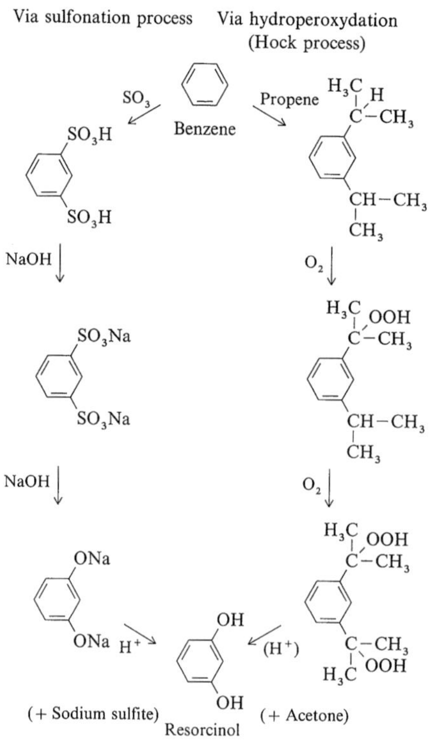

Figure 1. Resorcinol production reactions

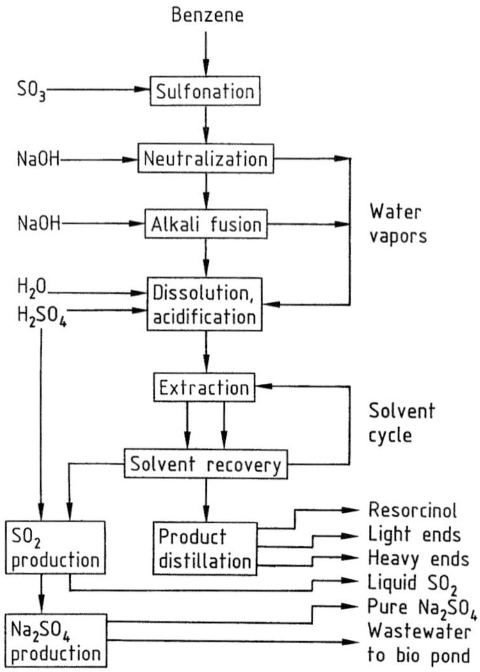

Figure 2. Resorcinol production via sulfonation

4.1. Via Sulfonation of Benzene [8]

The sulfonation of benzene formerly led to the production of considerable amounts of waste because it was carried out with mixtures of sulfuric acid and sulfur trioxide. Excess sulfuric acid was precipitated with lime to form gypsum, which had to be disposed of in landfills.

Process Description. In the process variation used in Germany today, the sulfonation of benzene is carried out continuously with sulfur trioxide [7446-11-9] alone so that only small amounts of sulfuric acid are contained in the sulfonation mixture and the addition of lime is not required. After neutralization with sodium sulfite, soda ash, or sodium hydroxide solution, the sulfonation product disodium benzene-1,3-disulfonate [831-59-4] is mixed with excess sodium hydroxide and fed to an alkali fusion reactor at 320–350 °C. The endothermic solid-state reaction yields a white powder, consisting chiefly of disodium resorcinate [6025-45-2], sodium sulfite, and some excess sodium hydroxide.

Depending on the method used, the product of the fusion reaction is treated with either a small quantity of water (in this case, solid sodium sulfite which contains organic impurities, is obtained as byproduct) or a large quantity of water forming an almost saturated solution of the product. In both processes the dissolved disodium resorcinate is then reacted with sulfur dioxide, sulfuric acid, or hydrochloric acid to give resorcinol. The dissolved resorcinol is extracted with an organic solvent. The preferred solvent is diisopropyl ether [108-20-3], but benzene, 4-methyl-2-pentanone (methyl isobutyl ketone) [108-10-1], or others can also be used. The solvent is then distilled off, and crude resorcinol is purified further by distillation in vacuum. The byproducts of this process are small amounts of light ends, essentially phenol, as well as cresols and 3-mercaptophenol [40248-84-8], and small amounts of heavy ends, mostly oligohydroxybiphenylenes.

The salt solution remaining after extraction is processed further. When sulfuric acid is used to neutralize disodium resorcinate, sulfur dioxide and sodium sulfate [7757-82-6] can be recovered as byproducts from this stream.

4.2. Via Hydroperoxidation of *m*-Diisopropylbenzene

Benzene, together with a benzene – cumene mixture recycled from the alkylation process, is alkylated with propene [115-07-1] in the liquid phase by using an AlCl$_3$ – HCl complex as catalyst (→ Phenol). After addition of recycled *p*-diisopropylbenzene (*p*-DiPB) and triisopropylbenzene (TriPB), the alkylate is subjected to isomerization/transalkylation. In this step most of *p*-DiPB and TriPB are converted into *m*-diisopropylbenzene (*m*-DiPB). The reaction mixture is then separated by distillation into three fractions (benzene – cumene, *m*-DiPB, and *p*-DiPB – TriPB). The subsequent autoxidation of pure *m*-DiPB proceeds according to a radical chain mechanism. It is accomplished in a cascade of aeration reactors by using compressed air in an aqueous alkaline medium at 90 – 110 °C and 0.5 – 0.7 MPa to yield [1,3-phenylenebis-(1-methylethylidene)]bishydroperoxide (*m*-diisopropylbenzene dihydroperoxide, DHP) [*721-26-6*] [9] (see Fig. 1).

After an overall residence time of 6 – 8 h the oxidate contains ca. 20% DHP and ca. 35% *m*-diisopropylbenzene monohydroperoxide (MHP). The heterogeneous oxidate is subjected to phase separation. *m*-Diisopropylbenzene dihydroperoxide is then crystallized, centrifuged, dissolved in acetone, and fed to a cleavage reactor (→ Phenol). Higher concentrations of hydroperoxides lead to the above-average formation of byproduct and increased safety hazards. The main byproducts of oxidation are *meta*-substituted acetophenones and dimethylphenyl carbinols. The latter are converted to recyclable *m*-DiPB by acid-catalyzed dehydration and hydrogenation in downstream unit operations.

The cleavage of DHP to resorcinol and acetone [*67-64-1*] is carried out under acid catalysis (preferably with ca. 1% H_2SO_4) in boiling acetone with a reaction time of 30 min [10], [11]. After neutralization, acetone is distilled off at normal pressure and resorcinol under vacuum. Further purification of resorcinol can be achieved by recrystallization or extraction. The overall process yield is ca. 75%, based on benzene. Alternatively, cleavage of the oxidate can be carried out in the presence of hydrogen peroxide [*7722-84-1*]. In this case the carbinols formed as byproducts are subsequently oxidized to DHP and thus eventually also converted to resorcinol [12].

4.3. Other Processes

Koppers in the United States and Mitsui Toatsu in Japan have developed a process for manufacturing resorcinol from *1,3-diaminobenzene* [*108-45-2*], which is produced by the hydrogenation of dinitrobenzene [13]. The *dehydrogenation of 1,3-cyclohexanedione* [*504-02-9*] to resorcinol, described by British Oxygen, was further developed by Hoechst in Germany to a four-stage process [14]. The starting materials are acetone and methyl acrylate [*96-33-3*] or acrylonitrile [*107-13-1*]. Neither process has been commercialized. The direct hydroxylation of phenol with hydrogen peroxide, for example, has been investigated extensively, but it yields only catechol and hydroquinone (→ Hydroquinone) [15].

5. Environmental Protection

Resorcinol must not enter rivers, lakes, or groundwater. Since it is readily biodegradable, small amounts in a biological wastewater treatment plant do not pose any problems. If possible, resorcinol should be collected and returned to the producer. When resorcinol is present in the air in the form of dust, smoke, or vapor, it can be removed by air scrubbing with a small amount of water in a wet filter equipped with a demister (a once-through operation).

6. Quality Specifications and Analysis

Resorcinol can be identified by IR [16] or Raman spectroscopy [17]. The UV and NMR spectra are given in [18].

A violet color is produced when an aqueous solution of resorcinol is treated with an iron(III) chloride ($FeCl_3$) solution.

Freezing-point measurement is the most useful test for determining product quality [19]. Freezing-point depression indicates impurities, most frequently the presence of water. Because of the existence of two solid resorcinol phases, determination of the melting point as a measure of product quality is not recommended. Thus, for instance, the melting range of 110–113 °C given in [20], does not indicate the presence of impurities. Impurities are best determined by gas chromatography of a solution of resorcinol in toluene or ethyl acetate, using a fused quartz capillary column with a nonpolar stationary phase, and a flame-ionization detector (FID). Freshly distilled technical-grade resorcinol exhibits approximately the following characteristics:

Appearance	white flakes, crystals, or powder
Freezing point	min. 109.5 °C
Purity	min. 99.5 % (wt %)
Water content	max. 0.1 % (wt %) (Karl Fischer)
Phenol	max. 0.1 % (area/area) (capillary GC)
o-Cresol	max. 0.1 %
m-, p-Cresols	max. 0.1 %
Catechol	max. 0.1 %
3-Mercaptophenol	max. 0.2 %
Heavy ends	max. 0.3 %

Resorcinol becomes darker during storage especially when exposed to light, but this does not impair its suitability for most applications.

7. Uses

Resorcinol is used in industry as an intermediate. Worldwide consumption is about 40 000 t/a (1990), the primary consumer (more than 50%) is the *rubber industry*. In the production of tires and other reinforced rubber products (conveyor belts, driving belts), resorcinol–phenol–formaldehyde condensates are used to enhance adhesion between cord and rubber (dip formulations, dry bonding agents). Furthermore, some rubber mixtures contain resorcinol to improve some properties after curing.

The second largest market for resorcinol (ca. 25%) is in high-quality wood adhesives, which are made from resorcinol, phenol, and formaldehyde. These resorcinol–phenol–formaldehyde resins are especially suitable for the manufacture of laminated wooden beams—which must be to some extent waterproof—at ambient temperature. The first such use was in the manufacture of wooden aircraft propellers ("propeller glue").

Further uses include

1) A new process for producing the intermediate *m*-aminophenol [*591-27-5*]
2) Production of light stabilizers for plastics
3) Production of sunscreen preparations for the skin
4) Production of dyes (fluorescein, eosin)
5) Manufacture of special pharmaceuticals (e.g., acne preparations)

8. Transport Classifications and Occupational Health [21]

Transport classifications and labeling are as follows:

UN no.	2876
EC no.	604-010-00-1
IMDG Code	6.1/2876/III
RID/ADR	6.1/14 C
DOT/IMO	poison B

Resorcinol must be labeled "harmful" in the EC and *mindergiftig* (less toxic) in Germany, with hazard label Xn.

Occupational Health. No MAK value has been set. Exposure to resorcinol must be kept below 10 ppm or 45 mg/m^3 TWA (OSHA-PEL and ACGIH-TLV).

Toxicology. For oral toxicity, an LD_{50} = 301 mg/kg (rat) has been found. Tests with rats show fast excretion (> 90% within 24 h), mainly via the urine as the glucuronide [22]. Resorcinol is less toxic than phenol or catechol if absorbed by ingestion or skin penetration. Ingestion of 8 g of resorcinol by a child resulted in hypothermia, hypotension, decrease of respiration, tremors, icterus, and hemoglobinuria.

In a two-year study, no evidence of carcinogenic activity was found when up to 225 mg/kg resorcinol in water was administered orally to male or female rats or mice. However, clinical signs suggestive of a chemical-related effect on the central nervous system, including ataxia, recumbency, and tremors, were observed in the rats and mice during the two-year study [23].

9. References

[1] H. Hlasiwetz, L. Barth, *Ann. Chem. Pharm.* **130** (1864) 354–359.
[2] Database DIPPR, American Institute of Chemical Engineers, host STN, Chem. Abstr. Service, Columbus, Ohio; FIZ, Karlsruhe.
[3] A. Kofler, *Arch. Pharm. Ber. Dtsch. Pharm. Ges.* **281** (1943) 8–22.
[4] B. V. Erofeev, L. T. Mendeleev, *Vestsi Akad. Navuk BSSR, Ser. Fiz. Tekh. Navuk* 1956, no. 4, 99, 101; *Chem. Abstr.* **51** (1957) 13 536.
[5] C. L. Speyers, *Am. J. Sci.* **14** (1902) no. 4, 294.
[6] W. H. Walker et al., *J. Phys. Chem.* **35** (1931) 3262.
[7] *Beilstein*, **6**, no. 554, 796; **E I** 398, **E II** 802; **E III** 4292; **E IV** 5658.
[8] K. Weissermel, H.-J. Arpe: *Industrielle Organische Chemie*, Verlag Chemie, Weinheim 1976.
[9] Mitsui Petrochemical Ind., EP 284 424, 1988 (H. Miki et al.); *Chem. Abstr.* **110** (1989) 78 057 h.
[10] Sumitomo Chemical Co., JP 6100 327, 1986 (H. Tomita et al.); *Chem. Abstr.* **105** (1986) 24 062 j.
[11] Mitsui Petrochemical Ind., EP 28 931, 1981 (I. Imai et al.); *Chem. Abstr.* **95** (1981) 186 840 p.
[12] Mitsui Petrochemical Ind., EP 327 361, 1989 (K. Yorozu, H. Ohno); *Chem. Abstr.* **112** (1990) 76 597 b.
[13] Process Evaluation/Research Planning, Third Quarterly Report (1977), Resorcinol from m-Phenylenediamine: The Koppers Process, Chem. Systems Inc., Tarrytown, N.Y. .
[14] Hoechst AG, DE 2 437 983 A 1, 1976 (W. H. Mueller, K. Riedel); *Chem. Abstr.* **84** (1976) 165 541 z.
[15] Anic S.p.A., DE-OS 3 309 669, 1983 (A. Esposito et al.); *Chem. Abstr.* **100** (1984) 22 409 a.
[16] *FT-IR-Atlas*, VCH Verlagsgesellschaft Weinheim 1988, no. 1007.
[17] D. Penot, J.-P. Mathieu, *J. Chim. Phys., Phys. Chim. Biol.* **52** (1955) 829–833.
[18] Sadtler Research Lab.: *Standard Spectra Collection*, H-NMR no. 6672, C-NMR no. 4428, C-13 no. 4428, UV no. 2572.
[19] H. Grimminger, H. Knorsch, A. Vollmer, *Pharm. Ind.* **52** (1990) no. 10, 1280.
[20] Aldrich Chemical Co. Inc.: *Catalogue Fine Chemicals*, Milwaukee, WI.
[21] Database HSDB of NLM; host DIMDI, Köln.
[22] Y. C. Kim, H. B. Matthews, *Fundam. Appl. Toxicol.* **9** (1987) no. 3, 409–414.
[23] *Chemical Regulation Reporter* **14**, March 15, 1991, no. 49, 1783.

Salicylic Acid

OLIVIER BOULLARD, Rhône-Poulenc Chimie S.A., Paris, France
HENRI LEBLANC, Rhône-Poulenc Chimie S.A., Paris, France
BERNARD BESSON, Rhône-Poulenc Chimie S.A., Paris, France

1.	Introduction 4271	5.	Uses and Economic Aspects . . 4276
2.	Physical and Chemical Properties 4272	6.	Salicylic Acid Derivatives 4277
3.	Production 4273	7.	Toxicology 4279
4.	Quality Specifications, Storage and Transportation, and Environmental Protection . . . 4275	8.	References 4280

1. Introduction

Salicylic acid [69-72-7], also called 2-hydroxybenzoic acid or *o*-hydroxybenzoic acid, is widely distributed in the plant kingdom in the form of esters.

Some 2400 years ago HIPPOCRATES prescribed decoctions of willow leaves as a treatment for fever and pain. The active principle in willow leaves is salicylic acid, whose biosynthesis is based on the deamination of phenylalanine to *trans*-cinnamic acid which, when hydrolyzed and oxidized at the β- carbon atom, gives salicylic acid.

Salicylate esters are found in several plant genera, the principal ones being *Salix*, *Spiraea*, and *Gaultheria*. For example, methyl salicylate is present in large quantity in the form of a glucoside in birch bark, various spiraeas (meadowsweet or *Spiraea ulmaria*), and wintergreen leaves.

Salicylic acid was first obtained in 1838 by R. PIRIA, an Italian chemist at the Institute of Pisa, by melting salicylaldehyde, obtained from meadowsweet, with caustic potash. In 1844 the French chemist CAHOURS obtained salicylic acid by hydrolysis of methyl salicylate. The final step, namely preparation from a natural source, was accomplished in 1874 by the German chemist H. KOLBE, who synthesized the acid by carboxylation of sodium phenoxide, a process still in use.

Salicylic acid and its derivatives are used mainly to synthesize pharmaceutical products and as intermediates in the production of dyes and agrochemical and perfumery products.

2. Physical and Chemical Properties

Physical Properties. Salicylic acid, $C_7H_6O_3$, M_r 138.12, crystallizes in the form of colorless needles (water) or monoclinic prisms (ethanol), *mp* 159 °C; salicylic acid begins to sublime at 76 °C; flash point (closed cup) 157 °C; heat of sublimation 81.8 kJ/mol; density d_4^{20} 1.443; dissociation constants K_1 1.05×10^{-3}, K_2 4.0×10^{-14} (19 °C); vapor pressure 1.66 mbar (110 °C) and 19.3 mbar (150 °C); solubility (per 100 g of solution) in methanol 38.46 g (21 °C), ethanol 34.87 g (21 °C), chloroform 1.55 g (30 °C), benzene 1.00 g (30 °C); solubility (per 100 mL of solution) in diethyl ether 23.4 g (17 °C), acetone 31.3 g (23 °C); extremely soluble in ammonia, insoluble in liquid sulfur dioxide. Solubility in water:

°C	0	10	20	40	60	80
g/L	0.8	1.2	1.8	3.7	8.2	20.5

Salicylic acid is an aromatic *o*-hydroxy carboxylic acid, and in contrast to its *meta* and *para* isomers (→ Hydroxycarboxylic Acids, Aromatic) it is subject to intramolecular hydrogen bonding and steam volatility. It also sublimes more readily than its isomers, and is substantially more acidic: dissociation constants for the *meta* and *para* isomers are 8.7×10^{-5} and 3.3×10^{-5}, respectively.

Chemical Properties. The difunctional salicylic acid molecule combines the properties of phenols with those of aromatic carboxylic acids. One equivalent of alkali hydroxide neutralizes only the carboxyl group. An excess of hydroxide is required to form the dialkali salt, from which the free OH group reforms in the presence of carbon dioxide. Chelation occurs with some metal ions such as Fe(III), leading to a violet coloration. Salicylic acid can be used as an indicator in EDTA determinations of Fe(III) [6].

Salicylic acid is esterified by alcohols in the presence of strong acids without significant etherification. Combined ether–esters can be prepared from dialkali salicylate in the presence of alkyl halide; these are converted by alkaline hydrolysis into the corresponding alkoxybenzoic acids. The phenolic OH group is etherified by an alkaline aqueous solution of dialkyl sulfate, and esterified by the action of acyl halides or acid anhydrides. Reduction with sodium and amyl alcohol affords tetrahydrosalicylic acid, whose oxidation product is pimelic acid [7]. Catalytic hydrogenation of salicylic acid esters over Raney nickel produces esters of *cis-trans*-2-hydroxycyclohexane carboxylic acid [8].

Since electrophilic reactants attack the less sterically hindered 5-position in preference to the 3-position, it is possible to obtain either 5-substituted or 3,5-disubstituted derivatives directly. For example, the monosubstituted products obtained on nitration, sulfonation, halogenation, alkylation, acylation, or coupling with diazonium salts are generally salicylic acids substituted at the 5-position. Derivatives substituted exclusively at the 3-position are obtained by indirect means, such as substitution of sulfosalicylic acid at the 5-position, followed by elimination of the sulfonic group. More severe conditions may lead to decarboxylation as well. Thus, 2,4,6-trinitrophenol (picric acid) is obtained upon treatment of salicylic acid with fuming nitric acid, and tribromophenol results from treatment with bromine in the presence of water.

When heated at or above its melting point salicylic acid decomposes into phenol and carbon dioxide. Under a carbon dioxide atmosphere at 230 °C the main product is phenyl salicylate. At 250 °C, xanthone is formed in parallel with phenol. For the behavior of salicylic acid as a function of temperature see [9].

Xanthone

Analysis. Quantitative determination of salicylic acid is generally accomplished by titration in an alkaline medium. Kolthoff's method permits salicylic acid also to be determined with a bromide solution, which converts it to tribromophenol [10]. For trace analysis it is convenient to use colorimetric methods with the reagent iron(III) chloride [11]. Thin-layer chromatography (TLC) provides a convenient test of purity, since each of the isomeric acids can be detected by alkaline coupling (caustic soda) with diazotized *p*-nitroaniline, spraying with Fe(III) chloride, or fluorescence under UV light. The TLC approach is often replaced today by reversed-phase high-performance liquid chromatography (HPLC).

3. Production

Kolbe–Schmitt Synthesis. Salicylic acid is prepared on an industrial scale by the Kolbe–Schmitt synthesis from dry sodium phenoxide in a stream of carbon dioxide at 150–160 °C and a pressure of 5 bar.

$$\text{C}_6\text{H}_5\text{-ONa} + \text{CO}_2 \rightleftharpoons \text{C}_6\text{H}_4(\text{COONa})(\text{OH})$$

The use of pressure (SCHMITT, 1884) results in a yield of about 90%, whereas without pressure (KOLBE, 1874) the yield does not exceed 50% because disodium salicylate and phenol are formed in equivalent amounts.

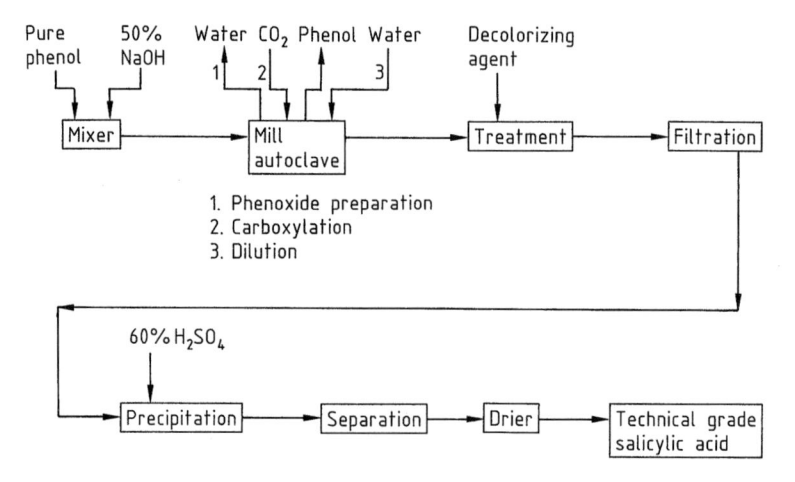

Figure 1. Simplified representation of salicylic acid production by the Kolbe–Schmitt method

Alternative reaction mechanisms for the carboxylation of phenoxide salts have been described in the literature [12]. 4-Hydroxyisophthalic acid, which yields salicylate by direct release of CO_2 (or by carboxylation of phenolate), has been discussed as a possible intermediate in the reaction process [13]. An electrophilic substitution mechanism via a complex formed between phenol, one molecule of CO_2, and an alkali metal, has also been suggested [14]. Temperature, the nature of the alkali metal, and the CO_2 pressure are all of decisive importance with respect to the reactivity and selectivity of the phenoxide [12], [15].

The various stages in the industrial synthesis of salicylic acid [16] are outlined in Figure 1. Carboxylation may be carried out if in autoclaves heated with steam or heat-exchange fluids and equipped with counterblades, or in powerful mill autoclaves. The process is still conducted mainly in a batchwise manner. To ensure that reaction proceeds satisfactorily, the reaction mass must not only be in a finely ground state during the carboxylation stage, but water must also be rigorously excluded. The presence of water reduces the yield by protonating the phenoxide and releasing alkali-metal hydroxide, which then converts CO_2 into carbonate with the regeneration of water. Water may also be formed in situ via a secondary etherification reaction [15]. Sodium phenoxide is prepared with a 1–2% molar excess of caustic soda; larger amounts of alkali lead to the formation of water, as described above. Anhydrous sodium phenoxide may be prepared either in the autoclave mixer itself by evaporation of an aqueous solution of phenoxide, starting at normal pressure and then gradually introducing vacuum, or in special drying equipment. In order to prevent discoloration and tar formation it is important that the carbon dioxide contains as little oxygen as possible ($< 0.1\%$).

The carboxylation step is exothermic: $\Delta H = -90.1$ kJ/mol. Using a 6 m^3 reactor, about 3 t of sodium phenoxide is converted into salicylic acid in 25 h. Phenol derived from the formation of disodium salicylate is recovered by distillation. The crude sodium salicylate is dissolved in a mixture of water and a decolorizing agent (e.g., activated charcoal, aluminum, or zinc powder [17]). Salicylic acid is then precipitated with sulfuric acid. This synthesis can be accomplished in a continuous manner by working with a solution of the anhydrous phenoxide in a suitable medium. Recommended solvents are phenol itself [18], higher alcohols [19], dialkyl ketones [20], nitrobenzene [21] and, as a dispersant [22], gasoline.

Other Processes. Consideration has been given to producing salicylic acid by air oxidation of o-cresolate at 230 °C in the presence of a copper-based catalyst [23] or copper benzoate (175 – 215 °C) [24]. Alkaline copper benzoate can also be converted by heat treatment into salicylate directly [25]. Benzoylsalicylic acid is an intermediate in the synthesis of phenol from toluene by the Dow process ($\rightarrow$ Phenol).

Salicylic acid can also be obtained by fermentation of such polycyclic aromatic compounds as naphthalene with the aid of microorganisms [26].

4. Quality Specifications, Storage and Transportation, and Environmental Protection

Quality Control. Technical-grade salicylic acid obtained from the Kolbe – Schmitt process is already extremely pure: salicylic acid content 99.5%; phenol, p-hydroxybenzoic acid, or 4-hydroxyisophthalic acid 0.05 – 0.1% (as impurity); ash $< 0.1\%$; water 0.2%.

An even higher quality acid (pharmaceutical grade) can be obtained by crystallizing the sodium salicylate from water at a temperature not exceeding 20 °C [27], or by sublimation of the acid at 20 mbar and a temperature of 154 °C [28] or with the aid of a carrier gas. A more modern process achieves sublimation directly by utilizing the heat of neutralization from the reaction of sodium salicylate with hydrogen chloride [29].

The quality specifications of the European Pharmacopoeia are limted to the following items: identity check; color (ethanolic solution); melting point (158 – 161 °C); assay (99.0 – 100.5%); if sulfated ash ($< 0.1\%$); heavy metals (< 20 ppm); chloride (< 100 ppm); if sulfates (< 200 ppm); and loss on drying $< 0.5\%$.

Storage and Handling. Salicylic acid dust is combustible in air. The low combustion energy for the process indicates a high level of combustion sensitivity, so appropriate measures must be adopted to avoid sources of ignition and to protect against potentially severe explosive effects. Explosive conditions can be avoided by maintaining the oxygen level at $< 8\%$.

The explosive characteristics of salicylic acid dust (particle size < 100 μm) are as follows: minimum combustion temperature, 490 °C; minimum combustion concentration, 30 g/m^3; minimum combustion energy, < 5 mJ; maximum pressure produced, 7.2 bar; maximum rate of pressure increase, 216 bar/s [unpublished work carried out by Rhône-Poulenc in accordance with ISO 6184/1].

Transportation. Salicylic acid is not subject to any transport restrictions. The acid is transported in bags (25 kg), drums (50 kg), big bags (500 – 1000 kg), or bulk containers (15 – 20 t).

Environmental Protection. Effluent and gaseous emission problems are comparable to those posed by phenol.

Freshwater ecotoxicity of salicylic acid on *Daphnia magna:* (ED$_{50}$ 24 h; immobilization): 180 mg/L (AFNOR T 90 301 Standard, French). Salicylic acid is readily biodegradable and very slightly bioaccumulable.

5. Uses and Economic Aspects

Salicylic acid is used mainly in the synthesis of acetylsalicylic acid, the most commonly dispensed pharmaceutical product. In the form of esters, amides, and salicylic acid salts it serves as a starting material for other pharmaceutical products. Technical-grade salicylic acid is used primarily as an intermediate in the production of agrochemical products, dyes, and colorants, as well as in the rubber industry and in the manufacture of phenolic resins.

Salicylic acid itself offers therapeutic benefits in the treatment of rheumatic disorders, for which purpose it is usually administered in the form of the readily soluble sodium salt. On account of its keratolytic action the acid is also widely used for cleaning the skin and removing scales.

By virtue of its bacteriostatic properties it is used as a disinfectant or preservative; however, its presence is not permitted in foods.

Economic Importance. The distribution of salicylic acid in terms of application can be estimated as follows: acetylsalicylic acid 55%, esters and salts 18%, resins 10%, dyes and colorants 10%. The recent development and introduction of new analgesics that compete aggressively with acetylsalicylic acid has had a direct effect on market evolution, and a steady decline in the consumption of acetylsalicylic acid has been observed. It is expected that the discovery of new uses will stabilize the situation.

6. Salicylic Acid Derivatives

Salts. The carboxyl group in salicylic acid is easily converted into salts by the action of metal carbonates. In order to prevent discoloration, aqueous salt solutions should be kept slightly acidic. Salts are obtained in solid form by concentrating their aqueous solutions.

Sodium salicylate [54-21-7], $C_7H_5NaO_3$, M_r 160.11, forms white, odorless, shiny crystalline flakes; solubility (in a solution of 100 mL): water 125 g (25 °C), ethanol 17 g (15 °C). The technical-grade product (99.5 %) is obtained by evaporating a solution of sodium salicylate; pharmaceutical-quality material [30] is prepared by two successive crystallizations of sodium salicylate hexahydrate from a 45 % aqueous solution at 10 °C. Sodium salicylate is used as an analgesic, antipyretic, and antineuralgic.

Other Salts. Besides sodium salicylate, a number of other common salts are known (e.g., ammonium, magnesium, calcium, aluminum), as is morpholine salicylate. Several of the salts are marketed under various trade names.

Esters. Several important salicylate esters are described below. These are formed by reaction with the appropriate alcohols. All are soluble in both ether and alcohol, but only sparingly soluble in water.

Methyl salicylate [119-36-8], oil of wintergreen, $C_8H_8O_3$, M_r 152.15, is a colorless, oily liquid with a characteristic odor, *mp* −9 °C, *bp* 222 °C, d_4^{20} 1.184. Methyl salicylate is synthesized by heating a mixture of salicylic acid and methyl alcohol in the presence of sulfuric acid. It is used to treat neuralgia and rheumatism, as well as to stimulate capillary blood circulation. It is also used as an insecticide, sunscreen, fragrance, and synthetic intermediate.

Benzyl salicylate [118-58-1], $C_{14}H_{12}O_3$, M_r = 228.25, is a clear liquid or colorless to opaque crystalline mass with a characteristic odor, *mp* 24 °C, *bp* 318 °C, d_4^{15} 1.180. It is obtained from a mixture of sodium salicylate and benzyl chloride, or by transesterification of methyl salicylate in the presence of benzyl alcohol. It is present in ylang-ylang and carnation oils, and is widely employed as an additive in soaps, in detergents, and in perfumery products (→ Benzyl Alcohol).

Isoamyl salicylate [87-20-7], $C_{12}H_{16}O_3$, M_r 208.26, is a colorless liquid with the fragrance of orchids, *bp* 270 °C, d_4^{20} 1.050. Isoamyl salicylate is used as a fragrance and stabilizer in perfumery and as an antirheumatic agent (topical application).

Phenyl salicylate [118-55-8], salol, $C_{13}H_{10}O_3$, M_r 241.22, is a colorless, crystalline powder, *mp* 43 °C, *bp* 172 °C at 16 mbar. It is obtained from a mixture of salicylic acid and phenol in the presence of sulfuric acid, or by transesterification of methyl salicylate in the presence of sodium phenoxide. Phenyl salicylate is used as an antiseptic, preservative, as a sunscreen and a general photoprotective agent for synthetic products, and as an emollient.

Acetylsalicylic Acid [50-78-2], $C_9H_8O_4$, M_r 180.15, is isolated as monoclinic colorless needles or crystalline powder (from water), or as flat platelets (from isoamyl alcohol), mp 143–144 °C depending on the heating rate and crystalline form; dissociation constant $K = 2.8 \times 10^{-4}$ (25 °C); solubility (in 100 mL of solvent): water 0.25 g (15 °C), ethanol 20 g (25 °C), ether 5 g (18 °C). The compound hydrolyzes to some extent during recrystallization from water.

Acetylsalicylic acid is prepared by reacting acetic anhydride with salicylic acid at a temperature of < 90 °C either in a solvent (e.g., acetic acid or aromatic, acyclic, or chlorinated hydrocarbons) or by the addition of catalysts such as acids or tertiary amines [31]; for quality specifications see [30]. Acetylsalicylic acid is used as an antipyretic, analgesic, and anti-inflammatory agent, and it has an antirheumatic effect. Acetylsalicylic acid also has antithrombosis and anticoagulant properties. It is marketed by Bayer as a consumer product under the trade name Aspirin and as a bulk product by Rhône-Poulenc (Rhodine).

Other Derivatives. *Salicylamide* [65-45-2], $C_7H_7NO_2$, M_r 137.13, is a white crystalline powder, mp 140 °C, sparingly soluble in water and soluble in alcohol. It is obtained by the ammonolysis of methyl salicylate [32]. Salicylamide is used as an analgesic, antipyretic, antirheumatic, sedative, and fungicide.

Salicylanilide [87-17-2], $C_{13}H_{11}NO_2$, M_r 213.33, forms colorless and odorless crystals, mp 136–137 °C. It is sparingly soluble in water and soluble in alcohol and chloroform. The substance is prepared from a mixture of salicylic acid and aniline in the presence of PCl_3 [33]. Salicylanilide and particularly its derivatives are powerful fungicides, and are used in the synthesis of dyes, colorants, lacquers, and textiles, and also as a disinfectant. It is distributed commercially under the trade name Shirlan (ICI).

3-Chlorosalicylic acid [1829-32-9], 3-chloro-2-hydroxybenzoic acid, $C_7H_5ClO_3$, M_r 172.57, mp 180 °C, is obtained either by the carboxylation of sodium 2-chlorophenoxide (Kolbe–Schmitt synthesis) or by chlorination of 5-sulfosalicylic acid, followed by displacement of the sulfonic group by means of superheated steam [34].

4-Chlorosalicylic acid [5106-98-9], 4-chloro-2-hydroxybenzoic acid, $C_7H_5ClO_3$, M_r 172.57, mp 211 °C, is prepared from 4-aminosalicylic acid by the Sandmeyer process [35] and used as a synthetic intermediate. The N-butylamide derivative of 4-chlorosalicylic acid (Jadit, Hoechst) is an antimycotic.

5-Chlorosalicylic acid [321-14-2], 5-chloro-2-hydroxybenzoic acid, $C_7H_5ClO_3$, M_r 172.57, mp 176 °C, is synthesized either by the chlorination of salicylic acid or by the Kolbe–Schmitt method starting from sodium 4-chlorophenoxide and CO_2 at 150 °C (higher temperature should be avoided since the decomposition reaction may be explosive) [34], [36]. It is used as an intermediate in the synthesis of agrochemical products and disinfectants. The ethanolamine salt of 5-chlorosalicylic acid 4-nitro-

2-chloroanilide (Bayluscid, Bayer) is used as an antiparasitic agent to combat schistosomiasis.

Bayluscid

5-(2,4-Difluorophenyl)salicylic acid [22494-42-4] (Diflunisal, Merck), $C_{13}H_8F_2O_3$, M_r 250.20, mp 212–213 °C.

Diflunisal

This compound can be prepared by acetylation of 2,4-difluorobiphenyl, followed by oxidation to 4-acetyloxy-2′,4′-difluorobiphenyl which, when heated under pressure in the presence of potassium carbonate and carbon dioxide, undergoes hydrolysis and carboxylation [37]. Diflunisal is a newly marketed analgesic [38].

7. Toxicology

Salicylic acid and its derivatives are resorbed by the skin and mucous membranes. Salicylic acid esters dissociate hydrolytically under the influence of esterases. Depending on the pH of the urine, salicylic acid is either oxidized to gentisic acid, or it is eliminated by the kidneys in the form of salicyluretic acid or glucuronide salicylate. Since the elimination rate of salicylic acid is lower than the rate of resorption, in certain cases there is a danger that it may accumulate in the body.

Salicylic acid has keratolytic activity and is a tissue irritant. In the stomach this irritant action affects mainly the mucous-producing cells and striated cells. Long-term treatment with salicylates slows the rate of blood coagulation by reducing platelet aggregation. Prostaglandin biosynthesis is also inhibited by salicylic acid and its derivatives, which partly explains their anti-inflammatory action. The acid and its derivatives display antipyretic and analgesic effects, and are fungicides and bacteriostatic agents [39].

Symptoms of cutaneous and pulmonary hypersensitivity to salicylates are well known, but it appears that these reactions are not always the result of true allergy. A decrease in the synthesis of prostaglandins and, consequently, other biologically active mediators is evidently involved here as well. Cross-sensitization has been established, for example, between methyl salicylate and acetylsalicylic acid [40]–[42].

One-time doses in excess of 10 g may prove fatal. The main cause of such *acute* intoxication is the disturbance of the acid–base equilibrium. Severe cases of intoxication may produce delirium, tremor, respiratory insufficiency, sweating, exsiccosis, hyperthermia, or coma. Symptoms of less severe intoxication include hyperventilation, tinnitus, nausea, vomiting, impaired vision and hearing, vertigo, and nervous disorders.

In the case of *chronic* intoxication, symptoms include digestive disorders and gastric and intestinal pain, sometimes with serious hemorrhaging, which nevertheless often remains hidden. Salicylate-induced anaemia, which may be observed in cases of chronic administriation, is an iron deficiency due to hidden bleeding. In the elderly the symptoms of chronic intoxication are often of a neuropsychiatric nature, including confusion and agitation. Despite the very harmful effects of salicylates on tissues, no hepatic or renal problems have been reported to date following chronic administration of pure forms of the compound [34], [43] – [45].

Experiments with various species of animals have shown that administration of relatively large doses of salicylic acid and its derivatives may have a teratogenic effect. The causative agent is probably salicylic acid. These findings have never been confirmed in humans, however, and there are no grounds for concern assuming the maintenance of appropriate hygiene and proper working conditions [46] – [49]. In addition to possible local symptoms of irritation due to salicylic acid and its derivatives, the major safety consideration is a risk of allergy. All renewed contact with these products should be avoided by those with a previous history of hyperallergic reactions.

8. References

General References

[1]   *Beilstein,* **10 H,** 43 ff.; **10 I,** 20 ff.; **10 II,** 25 ff.; **10 III,** 87 ff.; **10 IV.**
[2]   *Rodd's Chemistry of Carbon Compounds,* vol. **III B,** Elsevier, Amsterdam 1956, pp. 756 – 768.
[3]   A. S. Lindsey, H. Jeskey: "The Kolbe – Schmitt Reaction," *Chem. Rev.* **57** (1957) 583 – 620.
[4]   M. J. H. Smith, P. K. Smith: *The Salicylates,* 2nd ed., Wiley, New York 1967.
[5]   S. Negwer: *Organisch-chemische Arzneimittel und ihre Synonyma,* 5th ed., Akademie-Verlag, Berlin 1978.

Specific References

[6]   F. T. Welcher: *The Analytical Uses of EDTA,* D. van Nostrand, New York, p. 58.
[7]   A. Müller, *Org. Synth.* (collect. vol. **2**) (1943) 535.
[8]   J. Pascual et al., *J. Chem. Soc.* 1943, 1949.
[9]   M. Wesolowski, *Thermochim. Acta* **31** (1979) 133 – 146.
[10] I. M. Kolthoff, *Pharm. Weekbl.* **58** (1921) 699.
[11] J. E. Heestermann, *Chem. Weekbl.* **32** (1935) 463.
[12] A. S. Lindsey, H. Jeskey, *Chem. Rev.* **57** (1957) 592 – 600.
[13] K. Ota, *Mem. Kyushu Inst. Technol. Eng.* **6** (1976) 165 – 175.
[14] A. J. Rostron, A. M. Spivey, *J. Chem. Soc.* 1964, 39.
[15] Ueno, EP 254 596, 1986.
[16] BIOS, no. 116, 664, 1246; CIOS, XXIII-25; FIAT, no. 744. W. L. Hardy, *Ind. Eng. Chem.* **49** (1957) no. 6, 55 A.
[17] Bayer, DE-OS 2 644 318, 1976.

[18] Wacker, DE 624 318, 1931. Bayer, DE 955 598, 1954. Dow, DE-AS 1 215 173, 1961.
[19] Monsanto, US 2 824 892, 1956.
[20] Ilford, GB 638 196, 1948.
[21] J. Hirao, Y. Hara, *Yuki Gosei Kagaku Kyokaishi* **25** (1967) 577–581.
[22] Ueno, JP 76 23 494, 1968.
[23] Asahi, US 3 360 553, 1967.
[24] Dow, BE 665 631, 1965.
[25] W. W. Kaeding, A. T. Shulgin, *J. Org. Chem.* **27** (1962) 3551.
[26] Sun Oil, US 3 183, 1963.
[27] Monsanto, GB 353 921, 1930.
[28] BIOS, no. 1141.
[29] Dow, US 4 137 258, 1978.
[30] Pharmacopée Européenne, Monography 366, 1985.
[31] McKetta: *Encyclopedia of Chemical Processing and Design,* vol. **4,** Dekker, New York 1977, p. 28.
[32] E. R. Kline, *J. Chem. Educ.* **19** (1942) 332.
[33] C. F. H. Allen, J. van Allan, *Org. Synth.* (collect. vol. **3)** (1955) 765.
[34] N. W. Hirwe et al., *Proc. Indian Acad. Sci. Sect. A* **8** (1938) 208.
[35] J. T. Sheehan, *J. Am. Chem. Soc.* **70** (1948) 1665.
[36] E. Plazek, *Rocz. Chem.* **10** (1930) 761.
[37] Merck, US 4 131 618, 1977; DE-OS 2 532 559, 1975.
[38] *Chem. Eng. News* **56** (1978) April 10, 7; **58** (1980) April 14, 24.
[39] W. Forth, D. Henschler, W. Rummel: *Pharmakologie und Toxikologie,* Bibliographisches Institut, Mannheim 1975, p. 425.
[40] A. Szczeklik et al. *J. Allergy Clin. Immunol.* **60** (1977) 276.
[41] C. Hindson, *Contact Dermatitis* **3** (1977) 348.
[42] P. Kallos, H. D. Schlumberger, *J. Pharm. Pharmacol.* **30** (1978) 67.
[43] L. Lagplante, C. Beaudry, *Union Med. Can.* **108** (1979) no. 3, 280.
[44] W. Siegenthaler et al., *Med. Trib. (German ed.)* **13** (1978) no. 18, 66.
[45] R. Ludewig, K. Lohs: *Akute Vergiftungen,* G. Fischer Verlag, Stuttgart 1979, p. 381.
[46] K. T. Szabo et al., *Toxicol. Appl. Pharmacol.* **19** (1971) 371–372.
[47] C. A. Kimmel et al., *Teratology* **4** (1971) 15–24.
[48] A. B. G. Lansdown, P. Grasso, *Experienta* **25** (1969) 885–887.
[49] I. D. G. Richards, *Br. J. Prev. Soc. Med.* **23** (1969) no. 4, 218–225.

Saponins

MICHAEL W. SCHWARZ, E. Merck OHG, Darmstadt, Federal Republic of Germany

1.	Introduction 4284	3.1.	Steroid Saponins 4288
1.1.	General Properties. 4284	3.2.	Glycoalkaloids 4292
1.2.	Distribution 4285	3.3.	Triterpene Saponins. 4295
1.3.	Isolation 4286	4.	Animal Saponins 4299
1.4.	Structure Elucidation. 4286	4.1.	Asterosaponins 4299
1.5.	Analysis 4287	4.2.	Holothurins 4300
2.	Pharmacology 4287	5.	References. 4301
3.	Plant Saponins 4288		

Abbreviations
api   β-D-apiose
ara   α-L-arabinose
drib   β-D-2′-deoxyribose
f   furanoside
fuc   β-D-fucose
fru   β-D-fructose
gal   β-D-galactose
galA   β-D-galacturonic acid
glc   β-D-glucose
glcA   β-D-glucuronic acid
qui   β-D-quinovose
rha   α-L-rhamnose
xyl   β-D-xylose

1. Introduction

Saponins are glycosides occurring primarily in plants but also in starfish (*Asteroidea*) and sea cucumbers (*Holothuridea*). Properties generally considered to be shared by this group of natural products are surfactant activity, hemolytic action, steroid-complexing ability, and biocidal capability.

The characteristic soapy lather formed when saponin-containing plant extracts are agitated in water provides this group of secondary metabolites with its common name (Latin *sapo* = soap), although this property is shared with structurally related compounds. Other properties are characteristic of particular types of saponins rather than all members of the class.

Saponins are categorized according to the structure of the aglycone moiety (*sapogenin*) and the number of linked sugar chains. Aglycones can be divided into triterpenoid and steroid sapogenins. Steroid saponins with basic properties are termed glycoalkaloids.

Hexoses common in all saponins are β-D-glucose (glc), β-D-galactose (gal), and α-L-rhamnose (rha); pentoses are α-L-arabinose (ara) and β-D-xylose (xyl). Pentoses occur primarily as pyranosides, less frequently as furanosides (f).

Saponins linked to one or two sugar chains are called *monodesmosides* and *bisdesmosides*, respectively (Greek *desmos* = chain). Trisdesmosidic saponins are rare [1], [2]. Mild hydrolysis of saponins yields *prosapogenins*, neutral glycosides with less than three sugar units, or acidic or basic glycosides with one sugar unit. Prosapogenins are less soluble than saponins and do not have typical saponin properties.

1.1. General Properties

Surfactant Activity. The well-known ability of saponins to cause frothing has long been used to detect these materials in plants or plant extracts. The combination of a hydrophobic aglycone and hydrophilic substituents accounts for the amphiphilic nature of saponins. Increased length and branching of the sugar chain correlate with enhanced surfactant power. Hence, triterpenoidal bisdesmosides are more active in this sense than monodesmosides [3].

Hemolytic Activity. Saponins vary considerably in their ability to lyse erythrocytes. Different glycosides of the same sapogenin differ in activity as a function of the type and structure of substituents [4], [5]. Bisdesmosides are generally less potent than monodesmosides. Introducing polar hydroxyl groups into the aglycone decreases activity, whereas esterification restores it. The linkage position of the sugar chain in triterpenoids also influences reactivity; thus, C-3 glycosides are more reactive than saponins with a C-28 glycosidic substituent.

Steroid-Complexing Ability. Saponins form insoluble complexes with cholesterol [6] and other sterols. This affinity is more pronounced with steroid saponins and glycoalkaloids than with triterpenoids.

Biocidal Activity. Monodesmosidic saponins exhibit fungitoxic or fungistatic and weak antimicrobial activity [7]. Examples include the specific action of α-tomatine [8] and the oat (*Avena sativa*) root saponin avenacin [9]. Steroid saponins and glycoalkaloids generally demonstrate more pronounced action, whereas triterpenes exhibit a broader spectrum of effectiveness.

The permeabilization of crucial cell membranes probably explains the toxicity of saponins to mollusks [10] and other invertebrates [11]. Ecdysterone-like activity [12] and the repellent or antifeedant action of saponins against termites [13], leaf-cutting ants [14], and the potato beetle [15] have been reported. Aquatic vertebrates such as fish and tadpoles are affected through permeabilization of their gills [16], resulting in a rapid loss of physiological function.

1.2. Distribution

Saponins occur mainly in plants of the subdivision *Angiospermae* of the division *Spermatophytae*. Steroid saponins are restricted mostly to the class *Monocotyledoneae*, a known exception being the saponins of foxglove (*Digitalis* sp.). Triterpene and glycoalkaloid saponins are found primarily in plants of the class *Dicotyledoneae*, a recently discovered exception being saponins isolated from montbretia (*Crocosmia crocosmiiflora*) [17]. Certain families of the class *Dicotyledoneae* (e.g., *Solanaceae, Hippocastanaceae, Primulaceae, Rosaceae,* and *Caryophyllaceae*) are especially rich in saponin-containing genera.

Monodesmosides are concentrated in the outer tissues of seeds, roots, and bark, consistent with their function as a barrier against microorganisms, whereas bisdesmosides are more abundant in leaves and stems. Secretion of proteolytic enzymes by fungal hyphae triggers conversion of the more soluble (and, therefore, transportable) bisdesmosides by specific glycosidases [18] into the more active monodesmosidic saponins.

Animal saponins are restricted to representatives of the phylum *Echinodermata*. Species of the class Holothuridea (sea cucumbers) contain triterpene saponins, whereas saponins from species of the class *Asteroidea* (starfish) contain steroids. These saponins induce avoidance responses in susceptible predator species [19].

1.3. Isolation

Methods for saponin isolation and purification are covered in several reviews [20]–[22]. Ground dried or fresh specimens are extracted after optional defatting with solvents such as methanol, ethanol, their aqueous mixtures, or, in the case of glycoalkaloids, slightly acidified water. Formation of artifacts due to enzymatic or solvolytic cleavage of glycosidic bonds or acid-catalyzed elimination and rearrangement of the aglycone must be prevented by suitable choice of conditions.

Distribution of a crude extract between water and butanol separates the saponins from such water-soluble compounds as oligosaccharides. Precipitation with organic solvents (e.g., ether or acetone) or complexing agents (e.g., cholesterol) permits further concentration.

Saponins with free carboxylic acid groups can be purified by ion-exchange chromatography. For cases in which further enrichment by crystallization is not possible, chromatographic techniques are used for the separation of crude saponin mixtures. Methods such as high-performance thin-layer chromatography (HPTLC), high-performance liquid chromatography (HPLC), and droplet countercurrent chromatography (DCCC) [23] have gained wide acceptance. Methods have been reported for isolating saponins from roots of ginseng (*Panax schinseng*) [24] and *Bupleurum falcatum* [25] by preparative HPLC of crude extracts.

1.4. Structure Elucidation

Historically, saponin structures were deduced after methylation and acid hydrolysis to aglycones and partially methylated sugars. However, this method leads to artifacts, so milder hydrolysis methods have been developed. Relatively large amounts of purified material are still needed for classical structural analysis.

The advent of modern spectroscopic methods has simplified the task. Fast-atom bombardment mass spectrometry (FAB–MS) facilitates the sequencing of sugar chains, while ^{1}H and ^{13}C NMR [27] techniques are used to establish specific linkages and resolve ambiguities. Structures of saponins from the starfish *Asterias forbesii* [28], *Astragalus ernestii* [29], and *Allium giganteum* [30] were recently elucidated by modern two-dimensional NMR methods.

1.5. Analysis

Analytical procedures must be adapted to the particular application in question. Crude estimates of saponin content are obtained via gravimetric methods [31]. Most technical-grade saponins are mixtures rather than single substances. Quantitative estimation of their makeup is accomplished by measuring the surface tension of solutions, their foaming power, and their hemolytic activity [32].

Saponins or saponin mixtures for pharmaceutical applications require analytical methods more specific to particular active compounds, such as spectrophotometric quantitation of a saponin color reaction [33], [34]. Combinations of methods including thin-layer chromatography–densitometry (TLC–DM) [35], gas chromatography–mass spectrometry (GC–MS) [36], LC–MS [37], HPLC [38], and radioimmunoassay (RIA) [39], have been reported for ginseng saponins and other compounds.

2. Pharmacology

Toxicology. In contrast to poikilothermic animals, saponin toxicity to homeothermic species by oral intake is quite low (50–100 mg/kg). The significance of saponins as components of the human diet or of animal feeds has been reviewed extensively [20]. Saponins exhibit different spectra of toxicity on parenteral or intravenous application (LD_{50} 0.7–50 mg/kg). The hemolytic action of many saponins prevents their intravenous administration, although detoxification by complex formation with serum cholesterol, albumin, or other plasma constituents reduces this effect.

Membranolytic saponins combine irreversibly with cell membrane systems and produce lesions [40] with a pore diameter of about 8 nm. Various structural models for the complex have been proposed [40], [41]. The role of cholesterol as the primary binding site for saponins in cell membranes has been challenged [42], and various saponins have been shown to interact in different ways with erythrocytes and liposomal membranes with respect to cholesterol, phosphatidylcholine, and distearoyllecithin [43]. The theory that β-glucosidase-induced hydrolysis of the glycosidic bond is an essential step in membranolysis [4] cannot explain changes induced by saponins in the osmotic behavior of liposomes devoid of β-glucosidase [44]. A unifying theory of saponin action on membranes has still to be developed.

Pharmacology. Claims of therapeutic activity for saponins are numerous, with the result that saponins have been denounced by some as panaceas [45].

Resorption of saponins by the small intestine is generally low [46], but enzymatic or bacterial decomposition in the large intestine [47] has been demonstrated. Permeabilization of cells of the small intestinal mucosa [48] reduces their capacity to transport nutrients and increases secretion. A therapeutic benefit is increased secretion in the nasal pharyngeal cavity after administration of cough syrup containing extracts of roots from *Primula officinalis*, *Glycyrrhiza glabra*, and *Gala senega*.

Anti-inflammatory activity has been reported for saponins from the seed of the horse chestnut (*Aesculus hippocastanum*) [49] and *Panax inseng* (Greek *pan* = all, *axos* = cure) [50], [51]. In the former case, pharmacological action was blocked by adrenalectomy, hypophysectomy, or sympatholytic drugs. Antiexudative and antigranulomatous action was explained in terms of activation of corticosterone secretion.

The saponin from *Glycyrrhiza glabra* and the hemisuccinate of the corresponding aglycone (carbenoxolone) are used to treat gastric ulcers [52]. Chemical modifications of the aglycone have been reported to reduce side effects [53]. The ammonium salt of the aglycone is used as an antiallergic agent, and the same effect is reported for saponins from *Ilex crenata* [54].

Hypocholesterolemic effects of saponins might be due to a reduction in the uptake of dietary cholesterol by complex formation, increasing bile salt excretion, or direct interaction with sterols in the membranes of mucosal cells [20]. Active saponins have been found in soybean (*Glycine max*) [55] and *Quillaja saponaria* [56].

Psychotropic effects are reported for ginseng saponins [57], and patents have been filed for analgesic, sedative, and cardiovascular-activating effects [58] as well as antitumor activity [59].

Antiviral activity of saponins has been reported against the herpes- and poliovirus [60], Epstein–Barr virus [61], and human immunodeficiency virus (HIV-1) [62].

Bisdesmosidic saponins solubilize monodesmosides [63], an adjuvant effect that seems to increase immune response to vaccines [64] and the uptake of β-lactam antibiotics [65] after oral administration.

3. Plant Saponins

3.1. Steroid Saponins

Aglycones of monodesmosidic steroid saponins from plants consist of a hexacyclic spirostanol system, whereas bisdesmosides are derived from the pentacyclic furostanol system or a hexacyclic structure such as that in nuatigenin. Hydrolysis of C-26 glycosidic linkages of furostanol sapogenins results in cyclization to spirostanols. Therefore, furostanol saponins carry the designation "proto-" prefixed to the name of the corresponding spirostanol saponin. Especially in the older literature, the suffix "-oside" provides another indication of a furostanol saponin. Nuatigenin saponins rearrange to the spirostanol derivative isonuatigenin on acid hydrolysis [66].

Aglycones are named after the plant sources from which they were first isolated. C-25 (*S*)-epimers of known C-25 (*R*)-sapogenins are assigned the prefix "neo-", whereas C-25 (*R*)-epimers of known C-25 (*S*)-sapogenins are given the prefix "iso-". Epimers with a C-3 α-OH instead of a C-3 β-OH are designated by the prefix "epi-".

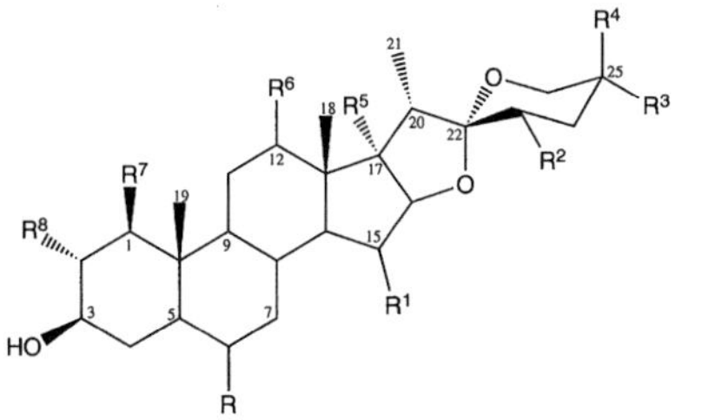

Sapogenin	CAS registry no.	Δ	C-5 H	R	R¹	R²	R³	R⁴	R⁵	R⁶	R⁷	R⁸
Diosgenin (1)	[512-04-9]	5		H	H	H	CH₃	H	H	H	H	H
Yamogenin (2)	[512-06-1]	5		H	H	H	H	CH₃	H	H	H	H
Pennogenin (3)	[507-89-1]	5		H	H	H	CH₃	H	OH	H	H	H
Ruscogenin (4)	[472-11-7]	5		H	H	H	CH₃	H	H	H	OH	H
Yuccagenin (5)	[511-97-7]	5		H	H	H	CH₃	H	H	H	H	OH
Kammagenin (6)	[564-44-3]	5		H	H	H	CH₃	H	H	O=	H	OH
Isonuatigenin (7)	[7050-41-1]	5		H	H	H	OH	CH₃	H	H	H	H
Tigogenin (8)	[77-60-1]		α	H	H	H	CH₃	H	H	H	H	H
Neotigogenin (9)	[470-01-9]		α	H	H	H	H	CH₃	H	H	H	H
Chlorogenin (10)	[562-34-5]		α	α-OH	H	H	CH₃	H	H	H	H	H
Neochlorogenin (11)	[511-91-1]		α	α-OH	H	H	H	CH₃	H	H	H	H
Hecogenin (12)	[467-55-0]		α	H	H	H	CH₃	H	H	O=	H	H
Gitogenin (13)	[511-96-6]		α	H	H	H	CH₃	H	H	H	H	OH
Paniculogenin (14)	[16750-37-1]		α	α-OH	H	OH	H	CH₃	H	H	H	H
Digitogenin (15)	[511-34-2]		α	H	β-OH	H	CH₃	H	H	H	H	OH
Smilagenin (16)	[126-18-1]		β	H	H	H	CH₃	H	H	H	H	H
Sarsasapogenin (17)	[126-19-2]		β	H	H	H	H	CH₃	H	H	H	H

Figure 1. Spirostanol-type sapogenins

Monodesmosides have linear or branched sugar chains linked in most cases to the C-3 β-OH group, whereas furostanol bisdesmosides display an additional glycosidic linkage at C-26. Besides the common sugars listed previously, less prevalent sugars include D-fucose (fuc), β-D-2′-deoxyribose (drib), and β-D-apiose [api(f)]. For a review of steroid saponins, see [67].

Spirostanol Saponins. Figure 1 provides the structures for a number of spirostanol aglycones. Most spirostanol saponins are monodesmosides with a single glycosidic bond at the C-3 β-OH. This is also true of hydroxylated sapogenins such as digitogenin (**15**) and paniculogenin (**14**). Monodesmosidic steroid saponins exhibit typical saponin properties, such as strong hemolytic activity.

Important saponins derived from diosgenin (**1**; and its C-25 (S)-epimer yamogenin (**2**), are dioscin [rha1 → 2(rha1 → 4glc)-**1**] (see Fig. 2 for an example of the signifance of the abbreviated notation used here and elsewhere) and gracillin [rha1 → 2(glc1 →3glc)-**1**], two glycosides first isolated from wild yam (*Dioscorea* sp.). Dioscin has been reported in a number of plant sources, including palms [68]. The aglycones are important sources for steroid synthesis, as are hecogenin (**12**), smilagenin (**16**), and its epimer sarsasapogenin (**17**). The latter were first isolated from *Agave* sp. and *Smilax crenata*, respectively. Sarsasapogenin (**17**) yields the monodesmoside parillin, glc1 → 2[rha1 → 4 (glc1 → 6glc)]-**17**, and the bisdesmoside sarsaparilloside (**18**).

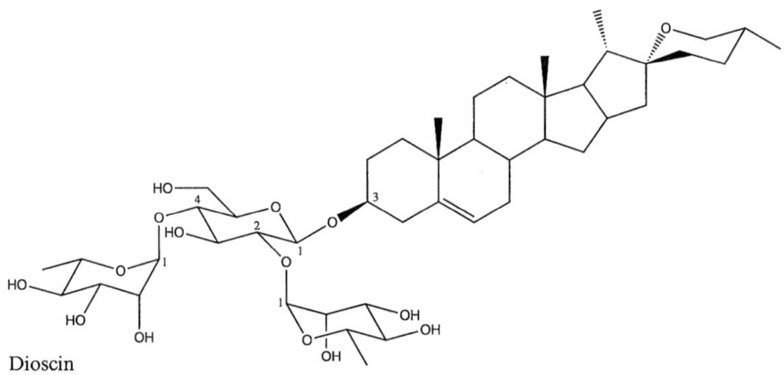

Dioscin

rha1 → 2(rha1 → 4g1c)−3−O−diosgenin

rha1 → 2(rha1 → 4g1c)−**1**

rha1
 ↘
   ₂⁴glc → diosgenin
 ↗
rha1

Figure 2. Alternative notational schemes for saponins, using diocin as an example

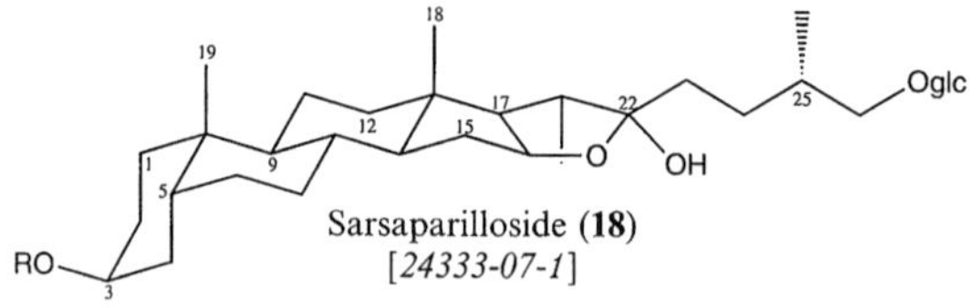

Sarsaparilloside (**18**)
[24333-07-1]

A number of steroid saponins have been isolated from foxglove (*Digitalis purpurea, D. lanata*), of which digitonin, glc1 → 3gal1 → 2 (xyl1 → 3glc)1 → 4gal-**15**, and its aglycone digitogenin (**15**) are the most important. Digitonin forms very insoluble complexes with cholesterol, and has been used as a reagent for the determination of this sterol. Minor sapogenins in *Digitalis* sp. are gitogenin (**13**), tigogenin (**8**), and its epimer neotigogenin (**9**). Recently, glc1 → 3glc-**13** has been isolated from Agave cantala [69].

Together with diosgenin (**1**) derivatives, the pennogenin (**3**) saponin dioscinin, rha1 → 4rha1 → 4(rha1 → 2glc)-**3** was isolated from the palm *Trachycarpus wagnerianus* [70]. Another saponin was isolated from *Dracaena mannii* [71] and characterized as rha1 → 2rha1 → 3glc-**3**, showing antimicrobial activity.

Saponins of ruscogenin (**4**) were isolated from *Ophiopogon* sp. [72] and identified as ophiopogonin B (rha1 → 2fuc-1-O-**4**) and D [rha1 → 2(xyl1 → 3fuc)-1-O-**4**]. In both cases the sugar chain is attached to the C-1-hydroxyl group.

Chlorogenin (**10**) was first isolated from *Chlorogalum pomeridianum*. Its saponins were used historically as a poison to stun fish, while at the same time leaving them edible by humans.

Furostanol Saponins. As already mentioned, furostanol saponins are readily converted to spirostanol saponins by dilute acid or β-glucosidases.

Leaves and stems often yield the protosaponins of spirostanols isolated from other parts of the plant.

The hydroxyl group at C-22 can be displaced by alkyl alcohols to yield alkylprotosaponins [73], or eliminated with acetic acid. The resulting $\Delta^{20,22}$-furostene saponins are termed pseudoprotosaponins. The isolation of furostene bisdesmosides [69] represents the first report of a genuine pseudoprotosaponin in nature.

Furostanols are easily distinguished from spirostanol saponins with Ehrlich's reagent [74], and also by IR spectroscopy, since furostanols lack the characteristic absorption bands of a C-22 spiroketal moiety.

The best known example of a bisdesmosidic steroid saponin is sarsaparilloside (**18**) (R = glc1 → 2[rha1 → 4(glc1 → 6glc)]−) from sarsaparilla, the dried root of *Smilax aristolochiaefolia*, used as a flavoring in beverages.

Furostanol glycosides isolated from garlic (*Allium sativum*) [75] were shown to be derived from chlorogenin (**10**). Mild hydrolysis yields antifungal spirostanol saponins.

Furostanols of tigogenin (**8**) were isolated from *Nicotiana tabacum* [76], of sarsasapogenin (**17**) from *Asparagus sp.* [77], of diosgenin (**1**) from *Balanites aegyptiaca* [78], and of gitogenin (**13**) from *Trigonella foenum-graecum* [79]. Trisdesmosidic furostanol glycosides have been reported for ruscogenin (**4**) saponins [72].

Avenacosides A and B were isolated from oat seeds [80] and identified as glc1 → 2(rha1 → 4glc)- and glc1 → 3glc1 → 2(rha1 → 4glc)-glycosides of nuatigenin-26-glucopyranoside (**19**), respectively. A highly specific β-glucosidase (avenacosidase) [81] yields the antifungal 26-desglucoavenacosides. Nuatigenin saponins from *Solanum aculeatissimum* have recently been proposed as a starting material for steroid synthesis [82].

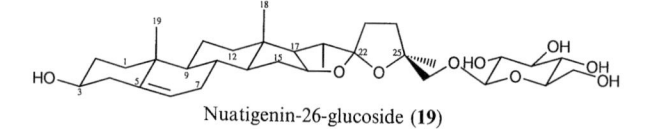

Nuatigenin-26-glucoside (**19**)

Other Steroid Systems. Saponins from *Polypodium vulgare* are derived from a rare type of cholestanone system with a hydroxytetrahydropyran ring at C-17 [83].

Uses. Spirostanol and furostanol saponins of diosgenin (**1**) and its epimer yamogenin (**2**) were the first natural starting materials for the industrial synthesis of pregnenolone and progesterone. Hecogenin (**12**), smilagenin (**16**), and sarsasapogenin (**17**) are also used as educts for steroid hormones. However, the growth in demand could not be satisfied by saponins alone, so the microbiological conversion of sterols such as stigmasterol to steroid precursors and total synthesis have both become important alternatives.

3.2. Glycoalkaloids

Glycoalkaloids are limited principally to the genera *Solanum, Lycopersicum, Cestrum*, and *Veratrum* of the family Solanaceae. They share a common C_{27} cholestane structure but differ in the structures of their side chains. Because of the presence of these materials in cultivated plants [84], [85], the corresponding structures have been known for some time.

Nitrogen analogues of the spirostanol steroid saponins, with nitrogen replacing an oxygen in the spiroketal ring, are called spirosolanes. Depending on the stereochemistry at C-22 they belong to either the solasodane (22R) or the tomatidane (22S) group. Analogues in which the 3-β-hydroxyl group has been replaced by an amino group are called aminospirostanes.

Solanidanes and *solanocapsines* have contiguous six-ring systems, whereas epinitrilocholestanes have a tetrahydropyridine ring bound to C-20 of a cholestane structure. Glycosides of the spirosolane and solanidane types are bound exclusively through the C-3 β-OH group.

The spirosolane-type aglycone solasodine (**20**) and its saponin α-solamargine (rha1 → 2 (rha1 → 4glc)-**20**) are analogues of diosgenin (**1**) and dioscin, respectively.

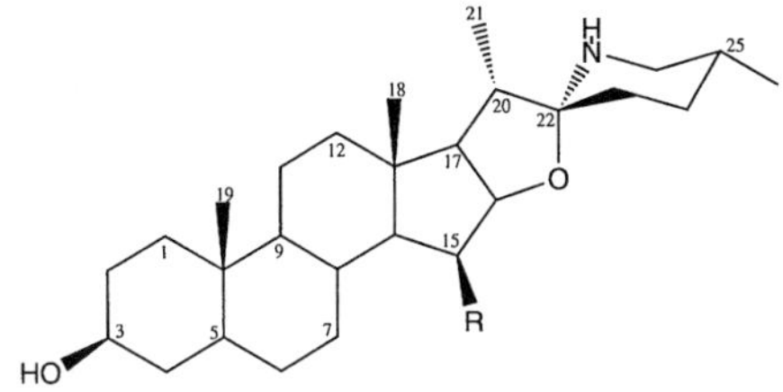

Figure 3. Solasodane-type glycoalkaloids

Sapogenin	CAS registry no.	Δ	C-5 H	R
Solasodine (**20**)	[126-17-0]	5		H
Soladulcidine (**21**)	[511-98-8]		α	H
Hydroxysoladulcidine (**22**)	[4912-43-0]		α	OH

Soladulcidine (**21**) and 15-hydroxysoladulcidine (**22**) correspond to tigogenin (**8**) and chlorogenin (**10**), respectively (cf. Fig. 3).

Solasodine (**20**) can be converted by N-acetylation and subsequent rearrangement to a $\Delta^{20,22}$-furostene, useful as a precursor in steroid synthesis [86]. This reaction is also feasible with the C-22 epimers tomatidine (**23**) and tomatidenol (**24**). Saponins such as α-tomatine (gal1 → 2(xyl1 → 3glc)1 → 4gal-**23**) can be used as natural sources for steroid precursors. The distribution of α-tomatine and its significance have been reviewed [87].

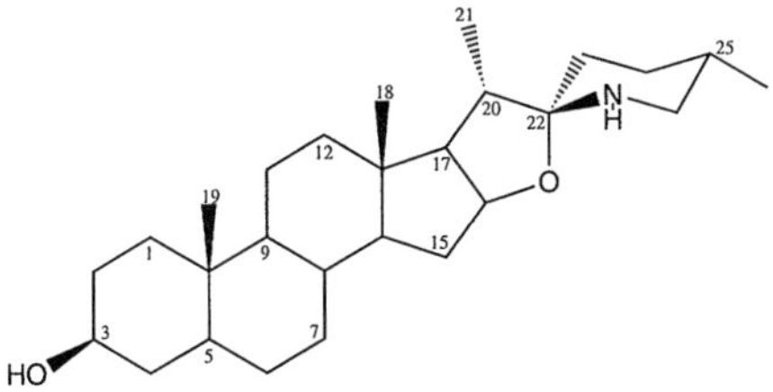

Tomatidine (**23**) [77-59-8]
Δ^5, tomatidenol (**24**) [546-40-7]

The glycoalkaloid solanine was found in the cultivated potato (*Solanum tuberosum*) and identified as consisting of saponins derived from solanidine (**25**), namely, α-chaconine (rha1 → 2(rha1 → 4glc)-**25**) and α-solanine (rha1 → 2(glc1 → 3gal)-**25**) [88]. Other sapogenins found in *Solanum* sp. (see Fig. 4) are leptinidine (**26**) [89] and demissidine (**27**), with the saponins demissine (glc1 → 2(xyl1 → 3glc)1 → 4gal-**27**) and commersonine [glc1 → 2(glc1 → 3glc)1 → 4gal-**27**]. Chemical degradation of solanidanes to steroids has not been achieved [90].

Solanocapsine (**28**) has been isolated as a sapogenin from *S. pseudicapsicum* [91]; no glycoside has yet been reported.

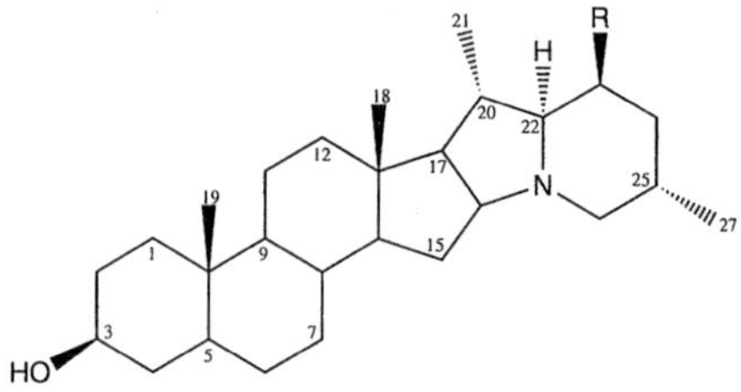

Figure 4. Solanidane-type glycoalkaloids

Sapogenin	CAS registry no.	Δ	C-5 H	R
Solanidine (**25**)	[*80-78-4*]	5		H
Leptinidine (**26**)	[*2448-17-1*]	5		OH
Demissidine (**27**)	[*474-08-8*]		α	OH

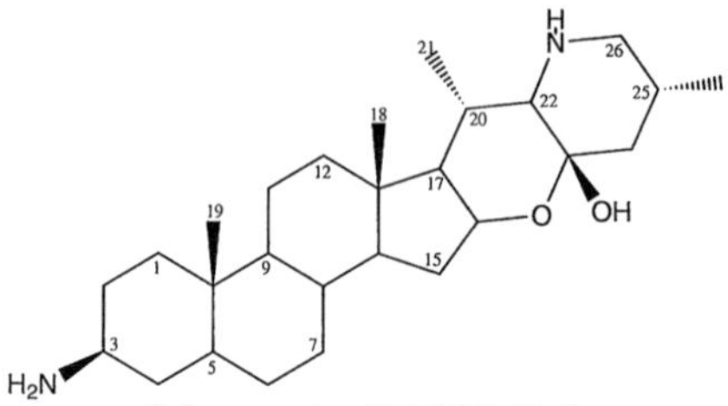

Solanocapsine (**28**) [*639-86-1*]

Natural epinitrilocholestanes (Fig. 5) include verazine (**29**) and tomatillidine (**30**). Intermediates in the conversion of spirosolanes to steroids [92] and in the transformation of spirosolanes to solanidanes [93] have similar structures.

The aminospirostanes jurubidine (**31**) and paniculidine (**32**) from *S. paniculatum* [94] are sapogenins of furostane glycosides such as jurubine (**33**).

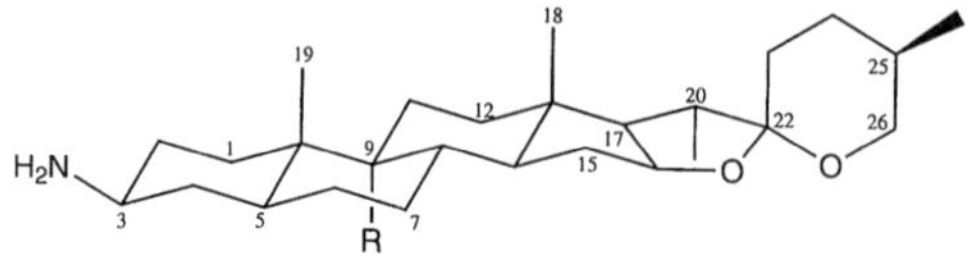

R = H, jurubidine (**31**) [*6084-44-2*]
R = OH, paniculidine (**32**) [*138464-37-6*]

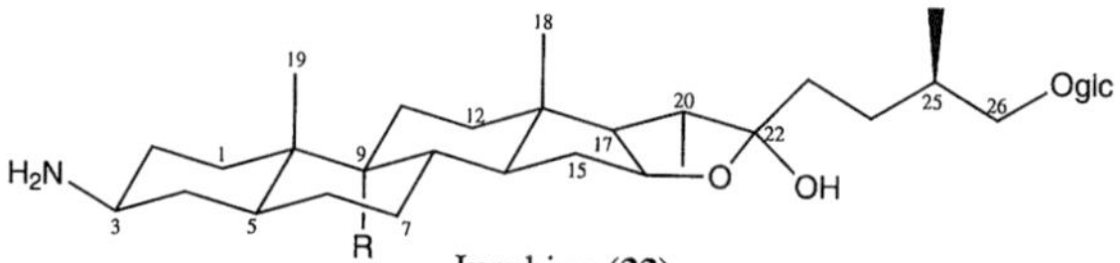

Jurubine (**33**)

Uses. The spirosolanes solasodine (**20**), tomatidine (**23**), and tomatidenol (**24**) serve as natural sources for steroid synthesis. The use of enzymes [95] for glycolysis of solasodine saponins produced by plant cell cultures has been proposed [96].

Other glycoalkaloids or sapogenins have gained little commercial interest.

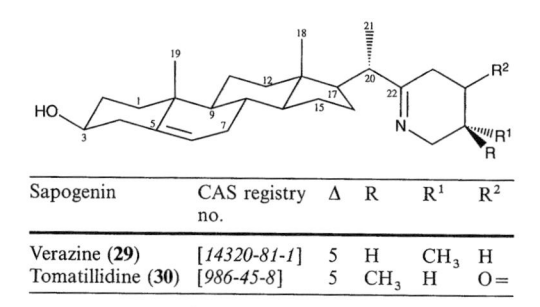

Figure 5. Epinitrilocholestane-type glycoalkaloids

Sapogenin	CAS registry no.	Δ	R	R¹	R²
Verazine (**29**)	[*14320-81-1*]	5	H	CH$_3$	H
Tomatillidine (**30**)	[*986-45-8*]	5	CH$_3$	H	O=

3.3. Triterpene Saponins

Triterpene glycosides are the most common saponins in nature, and they have been reviewed extensively [20]–[22]. A derivative of triterpenoids has recently been found in tertiary sediments [97].

Monodesmosides have linear or branched sugar chains, usually linked to the C-3 β-OH group. The common sugars listed previously are supplemented by such less frequently encountered sugars as β-D-fucose (fuc), β-D-fructose [fru(f)], β-D-quinovose (qui), β-D-glucuronic acid (glcA), and β-D-galacturonic acid (galA).

Oleananes. Saponins derived from β-amyrin (**34**) are the most common triterpene glycosides (cf. Fig. 6).

The variability of substituents is immense. The most frequently hydroxylated carbon centers of the ring system are C-2, C-16, C-21, and C-22. Saponins of hydroxylated aglycones are esterified with organic acids (e.g., acetic, butyric, 2-methylbutyric, tiglic, angelic, and benzoic acids). One example is the principal saponin glc1 → 2(glc1 → 4glcA)-**47** of horse chestnut, which is esterified at C-22 with acetic acid and at C-21 with tiglic or angelic acid. The complete mixture of saponins is called escin.

Of the methyl substituents, C-23, C-24, C-28, C-29, and C-30 are most frequently oxidized to hydroxymethyl, aldehyde, and carboxylic acid groups. Sapogenins with a C-28 carboxylic acid function are present in plant leaves and stems in the form of bisdesmosides with an acyl–glycosidic linkage to a second sugar chain. Enzymatic cleavage of the acyl–glycosidic bond yields the biologically more active monodesmosides. One example is found in a report on the isolation of two oleanolic acid (**54**) saponins from *Momordica cochinchinensis* [98], momordins Id [xyl1 → 2 (xyl1 → 3glcA)-3-*O*-**54**] and IId [xyl1 → 2(xyl1 → 3glcA)-3-*O*-**54**-28-*O*-glc]. Quinoside A from *Chenopodium quinoa* has been reported [2] to be a trisdesmoside of hederagenin (**59**), 3,23-bis (glc)-**59**-28-*O*-ara3 ← 1glc.

Saponins from licorice, the extract of *Glycyrrhiza glabra*, have a keto group at C-11. The resulting enone structure facilitates UV detection during HPLC separation. The composition, uses, and analysis of licorice have been reviewed [99]. The main sapogenin from this source is glycyrrhetic acid (**38**), with the principal saponin being glycyrrhizin (glcA1 → 2glcA-**38**).

Figure 6. Oleanane-type sapogenins

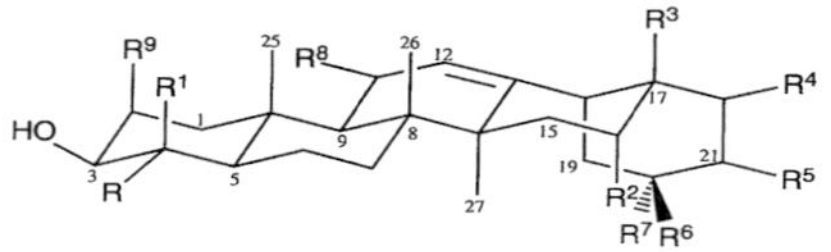

Sapogenin	CAS registry no.	R	R^1	R^2	R^3	R^4	R^5	R^6	R^7	R^8	R^9
β-Amyrin (34)	[559-70-6]	CH_3	CH_3	H	CH_3	H	H	CH_3	CH_3	H	H
Sophoradiol (35)	[6822-47-5]	CH_3	CH_3	H	CH_3	α-OH	H	CH_3	CH_3	H	H
Canoniensistriol (36)	[83718-68-7]	CH_3	CH_3	H	CH_3	α-OH	α-OH	CH_3	CH_2OH	H	H
Glycyrrhetol (37)	[14226-18-7]	CH_3	CH_3	H	CH_3	H	H	CH_3	COOH	O=	H
Glycyrrhetic acid (38)	[471-53-4]	CH_3	CH_3	H	CH_3	H	H	CH_3	CH_3	O=	H
Liquiritic acid (39)	[10379-72-3]	CH_3	CH_3	H	CH_3	H	H	COOH	CH_3	H	H
Epikatonic acid (40)	[76035-62-6]	CH_3	CH_3	H	CH_3	H	H	COOH	CH_3	H	H
Liquiridiolic acid (41)	[20528-70-5]	CH_3	CH_2OH	H	CH_3	H	β-OH	COOH	CH_3	H	H
Melilotigenin (42)	[114702-59-9]	CH_3	CH_2OH	H	CH_3	O=	H	COOH	CH_3	H	H
Soyasapogenol B (43)	[595-15-3]	CH_3	CH_2OH	H	CH_3	α-OH	H	CH_3	CH_3	H	H
Soyasapogenol E (44)	[6750-59-0]	CH_3	CH_2OH	H	CH_3	O=	H	CH_3	CH_3	H	H
Soyasapogenol A (45)	[508-01-0]	CH_3	CH_2OH	H	CH_3	α-OH	α-OH	CH_3	CH_3	H	H
Azukisapogenol (46)	[86425-21-0]	CH_3	CH_2OH	H	CH_3	H	α-OH	COOH	CH_3	H	H
Protoescigenin (47)	[20853-07-0]	CH_3	CH_3	OH	CH_2OH	β-OH	H	CH_3	CH_3	H	H
Primulagenin A (48)	[465-95-2]	CH_3	CH_3	OH	CH_2OH	β-OH	H	CH_3	CH_3	H	H
Theasapogenol D (49)	[53227-91-1]	CH_3	CH_3	OH	CH_2OH	β-OH	α-OH	CH_3	CH_3	H	H
Theasapogenol B (50)	[13844-01-4]	CH_3	CH_3	OH	CH_2OH	β-OH	H	CH_3	CH_3	H	H
Theasapogenol C (51)	[14440-27-8]	CH_2OH	CH_3	OH	CH_2OH	β-OH	α-OH	CH_3	CH_3	H	H
Theasapogenol A (52)	[13844-22-9]	CH_2OH	CH_3	OH	CH_2OH	β-OH	α-OH	CH_3	CH_3	H	H
Theasapogenol E (53)	[15399-41-4]	CHO	CH_3	OH	CH_2OH	β-OH	α-OH	CH_3	CH_3	H	H
Oleanolic acid (54)	[508-02-1]	CH_3	CH_3	H	COOH	H	H	CH_3	CH_3	H	H
Echinocystic acid (55)	[510-30-5]	CH_3	CH_3	H	COOH	H	α-H	CH_3	CH_3	H	H
Entagenic acid (56)	[5951-41-7]	CH_3	CH_3	H	COOH	β-OH	β-OH	CH_3	CH_3	H	H
Serratagenic acid (57)	[6488-64-8]	CH_3	CH_3	H	COOH	H	H	COOH	CH_3	H	H
Sperculagenic acid (58)	[18671-48-2]	CH_3	CH_3	H	COOH	H	H	CH_3	COOH	H	H
Hederagenin (59)	[465-99-6]	CH_2OH	CH_3	H	COOH	H	H	CH_3	CH_3	H	H
Bayogenin (60)	[6989-24-8]	CH_2OH	CH_3	OH	COOH	H	H	CH_3	CH_3	H	OH
Polygalacic acid (61)	[22338-71-2]	CH_2OH	CH_3	OH	COOH	β-OH	H	CH_3	CH_3	H	OH
Platycodigenin (62)	[22327-82-8]	CH_2OH	CH_3	OH	COOH	β-OH	H	CH_3	CH_3	H	OH
Esculentic acid (63)	[56283-68-2]	CH_3	CH_3	H	COOH	H	H	COOH	CH_3	H	H
Jaligonic acid (64)	[51776-39-7]	CH_2OH	CH_3	OH	COOH	H	H	COOH	CH_3	H	OH
Gypsogenin (65)	[639-14-5]	CHO	CH_3	H	COOH	H	H	CH_3	CH_3	H	H
Quillaic acid (66)	[631-01-6]	CHO	CH_3	H	COOH	H	α-OH	CH_3	CH_3	H	H
Dianic acid (67)	[91652-29-8]	CHO	CH_3	H	COOH	H	H	CH_3	CH_3	H	H
Gypsogenic acid (68)	[5143-05-5]	COOH	CH_3	H	COOH	H	H	CH_2OH	CH_3	H	H
Medicagenic acid (69)	[599-07-5]	COOH	CH_3	H	COOH	H	H	CH_3	CH_3	H	OH

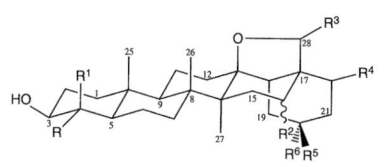

Sapogenin	CAS registry no.	Δ	R	R¹	R²	R³	R⁴	R⁵	R⁶
Protoprimulagenin A (**70**)	[*2611-08-7*]		CH₃	CH₃	a-OH	H	H	CH₃	CH₃
Priverogenin B (**71**)	[*20054-97-1*]		CH₃	CH₃	a-OH	H	β-OH	CH₃	CH₃
Cyclamiretin A (**72**	[*5172-34-9*]		CH₃	CH₃	a-OH	H	H	CH₃	CHO
Anagalligenin A (**73**)			CH₃	CH₃	a-OH	OH	β-OH	CH₃	CH₃
Anagalligenin B (**74**)	[*33722-92-8*]		CH₂OH	CH₃	a-OH	H	H	CH₃	CH₃
Saikogenin E (**75**)	[*13715-23-6*]	11	CH₃	CH₃	e-OH	H	H	CH₃	CH₃
Saikogenin F (**76**)	[*14356-59-3*]	11	CH₂OH	CH₃	e-OH	H	H	CH₃	CH₃
Saikogenin G (**77**)	[*18175-79-6*]	11	CH₂OH	CH₃	a-OH	H	H	CH₃	CH₃

Figure 7. Epoxyoleanane-type sapogenins

A subgroup of the oleanane-type saponins consists of glycosides with an ether bridge between C-13 and C-18 referred to as epoxyoleanane saponins (Fig. 7).

Examples are protoprimulagenin A (**70**), found in *Primula* sp., cyclamiretin A (**72**) and its saponin cyclamin [xyl1 → 2(glc1 → 3glc)1 → 4(glc1 → 2-ara)-**72**] from *Cyclamen europeaum*, and the saikogenins (**75**)–(**77**) from *Bupleurum falcatum*, used in traditional oriental medicine [100].

Epoxyoleanane-type saponins are very sensitive to extraction conditions, and they yield artifacts upon acidic cleavage of the ether linkage. Protoprimulagenin A (**70**) yields primulagenin A (**48**), whereas saikogenins (**75**)–(**77**) rearrange to $\Delta^{11,13(18)}$-dienes not found in natural sources.

Ursanes. Aglycones of the ursane type (cf. Fig. 8) are derived from α-amyrin (**78**). Recent reports describe the isolation of saponins of ursolic acid (**79**) from both maté (*Ilex paraguayensis*) [101] and *Cynara cardunculus* [102]. Ilexgenin A (**82**) was first isolated from *I. pubenscens* [103]. An example of acyl–glycosidic linkages in ursane-type saponins is the asiaticoside rha1 → 4-glc1 → 6glc-28-O-**84**.

Dammaranes. In addition to saponins derived from oleanolic acid (**54**), the roots of *Panax schinseng* contain saponins originating from the tetracyclic dammarane-type aglycones (see Fig. 9). Elucidation of their true structures was complicated by rapid epimerization at C-20 during isolation. Progress in the identification of saponins from *Panax* sp. has been reviewed [104].

Ginsenosides of protopanaxadiol (**87**) are bisdesmosides with glycosidic linkages at C-3 and C-20, whereas protopanaxatriol (**88**) has glycosides linked to C-6 and C-20. Major saponins are ginsenosides Rb1 (glc1 → 2glc-3-O-**87**-20-O-glc6 ← 1glc) and Rg1 (glc-3-O-**88**-20-O-glc).

Lupanes and Hopanes. Sapogenins of the pentacyclic lupane and hopane systems are not very common compared to other triterpenes. Although unsubstituted ring

Sapogenin	CAS registry no.	R	R^1	R^2	R^3	R^4	R^5	R^6	R^7
Amyrin (78)	[638-95-9]	CH_3	CH_3	H	CH_3	CH_3	H	CH_3	H
Ursolic acid (79)	[77-52-1]	CH_3	CH_3	H	COOH	CH_3	H	CH_3	H
Pomolic acid (80)	[13849-91-7]	CH_3	CH_3	H	COOH	CH_3	OH	CH_3	H
Cincholic acid (81)	[5948-32-3]	CH_3	CH_3	H	COOH	CH_3	OH	CH_3	OH
Ilexgenin A (82)	[108524-94-3]	CH_3	COOH	H	COOH	CH_3	OH	CH_3	H
Rotundic acid (83)	[20137-37-5]	CH_2OH	CH_3	H	COOH	CH_3	OH	CH_3	H
Asiatic acid (84)	[464-92-6]	CH_2OH	CH_3	H	COOH	CH_3	H	CH_3	OH
Madecassic acid (85)	[18449-41-7]	CH_2OH	CH_3	OH	COOH	CH_3	H	CH_3	OH

Figure 8. Ursane-type sapogenins

Sapogenin	CAS registry no.	R	R^1	R^2	R^3	R^4	R^5
Dammarenediol (86)	[14351-29-2]	CH_3	CH_3	H	H	OH	CH_3
Protopanaxadiol (87)	[7755-01-3]	CH_3	CH_3	H	OH	OH	CH_3
Protopanaxatriol (88)	[1453-93-6]	CH_3	CH_3	OH	OH	OH	CH_3

Figure 9. Dammarane-type sapogenins

systems are present in fruit waxes and other nonpolar plant materials, saponins are scarce. A saponin from *Asparagus gonocladus* has been isolated and identified as the glc1 → 2rha-glycoside of betulinic acid (90) [105]. Known hopane-type sapogenins include **91** and **92**.

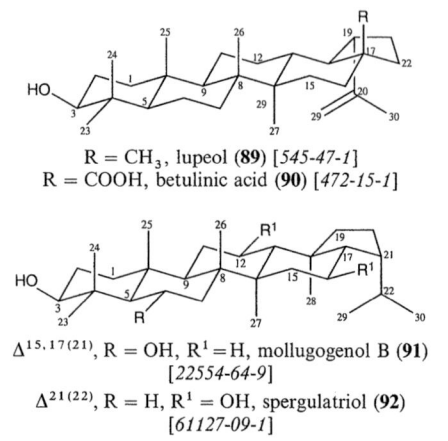

R = CH_3, lupeol (89) [545-47-1]
R = COOH, betulinic acid (90) [472-15-1]

$\Delta^{15,17(21)}$, R = OH, R^1 = H, mollugogenol B (91)
[22554-64-9]

$\Delta^{21(22)}$, R = H, R^1 = OH, spergulatriol (92)
[61127-09-1]

Uses. Commercial applications of triterpene saponins are based largely on their surfactant properties and their ability to form oil-in-water emulsions. Formerly used as emulsifiers in X-ray film manufacture and as components of detergents or cosmetics, saponins have been largely replaced in these applications by synthetic compounds.

The extract of *Glycyrrhiza glabra* containing glycyrrhetic acid (**38**) and glycyrrhetin is used by the tobacco and food industries for flavoring purposes [99]. As already mentioned, this sapogenins and saponins also have pharmaceutical value for the treatment of ulcers and gastritis [52]. Saponins used as expectorants are isolated from *G. glabra, Gala senega, Primula officinalis,* and *Hedera helix* [106].

A vast number of folk medicines contain saponins, and they are used as tonics and stimulants. The best-known examples are extracts of *Panax schinseng,* containing ginsenosid saponins, and *Bupleurum falcatum,* containing saikosaponins, which are important in traditional oriental medicine.

Molluscicidal saponins have been proposed as a means of controlling the spread of schistosomiasis in tropical countries through elimination of the snail vector *Biomphalaria glabrata* [10].

4. Animal Saponins

Marine invertebrates of the phylum Echinodermata are categorized into five classes [107]: *Holothuridea* (sea cucumbers), *Asteroidea* (starfish), *Echinoidea* (sea urchins, sand dollars), *Ophiuroidea* (brittle stars, basket stars), and *Crinoidea* (sea lilies, feather stars). Only members of the classes *Holothuridea* and *Asteroidea* produce saponins. Asterosaponins are based on a steroid cholestane system, whereas holothurins have a triterpenoid lanostane framework. Besides the common sugars listed previously, less common sugars encountered are the 6-deoxy compounds β-D-quinovose (qui) and β-D-fucose (fuc), both in their pyranose forms.

4.1. Asterosaponins

Hydrolysis of crude saponin isolated from starfish yields the steroid asterone (**93**) [107] together with asterogenol, the corresponding 20S-hydroxyl reduction product [108]. A noteworthy feature is the $\Delta^{9(11)}$ double bond. Asterone (**93**) is derived from the genuine sapogenins thornasterol A (**94**) and B (**95**) [109] via retro-aldol cleavage.

Asterone (**93**) [*37717-02-5*]

R = H, thornasterol A (**94**) [*55897-77-3*]
R = CH₃, thornasterol B (**95**) [*55897-78-4*]

Most saponins from starfish are sulfated at the 3-β-OH group and glycosylated at the 6-α-OH moiety. Their sugar chains consist of five to six monomeric units, with branching at the second sugar. Examples are forbesides A, B, and C, isolated from *Asterias forbesi* and characterized as gal1 → 3fuc1 → 2gal1 → 4(qui1 → 2xyl)1 → 3qui-6-O-**94**-3-O-SO₃Na⁺, qui1 → 2gal1 → 4 (qui1 → 2xyl)1 → 3qui-6-O-**94**-3-O-SO₃Na⁺, and fuc1 → 2gal1 → 4(qui1 → 2qui)1 → 3(6-deoxy-β-D-xylo-4-hexosulopyranosyl)-6-O-**94**-3-O-SO₃Na⁺, respectively [28].

Sapogenins with an epoxy function in place of the keto group in the thornasterol side chain [110] or penta- and hexahydroxylated cholestane ring systems [111] have been reported.

Uses. Biological activities reported for asterosaponins [112] include cytotoxicity to tumor cells and antiviral activity. However, the high toxicity of these compounds prevents their pharmaceutical use. Use as precursors for steroid synthesis [113] is not technically feasible.

4.2. Holothurins

Holothurins are similar to the asterosaponin glycosides isolated from sea cucumbers. They have a $\Delta^{9(11)}$ double bond, and are prone to yield artifacts. The genuine aglycone holothurigenol (**96**) yields a $\Delta^{7,9(11)}$-diene elimination product on acidic hydrolysis.

Holothurigenol (**96**) [*72244-90-7*]

Aglycones are glycosylated at the 3-β-OH group with sugar chains containing two to six

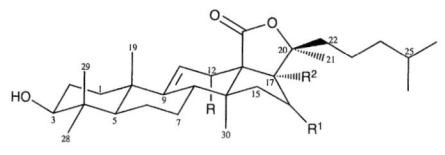

Sapogenin	CAS registry no.	Δ	R	R¹	R²
Deoxybivittogenin (**97**)	[77394-02-6]		H	H	H
Bivittogenin (**98**)	[67797-17-5]		OH	H	H
Echinogenol (**99**)			OH	H	OH
Holotoxigenol (**100**)		25	H	O=	H

Figure 10. Holothurinogen-type sapogenins

monomer units. In contrast to asterosaponins, hydroxyl groups on the aglycone system are not sulfated. Instead, hydroxyl moieties in the glycoside chain are methylated and sulfated.

Holothurin B, isolated from *Holothuria leucospilata*, was identified as qui1 → 2($Na^{+-}O_3S$-O-4'-xyl)-3-O-**96** [114], whereas holothurin A was shown to be CH_3-O-3'-glc1 → 3glc1 → 4-holothurin B [115].

Other examples are saponins derived from aglycones with linear side chains at C-20 (cf. Fig. 10). Bivittosides A through D are derived from **97** and **98** [116], echinosides from **99**, and holotoxins A and B from holotoxigenol (**100**) [117].

Sapogenins with a Δ^7—instead of $\Delta^{9(11)}$—double bond have been reported in *Cucumaria frondosa* [118] and *C. echinata* [119].

Uses. Although interesting biological activities have been reported (e.g., cytotoxicity to tumor cells [119] and positive inotropic and chronotropic action on atrial muscle [120]), the pharmaceutical use of holothurins is prevented by their severe toxic effects.

5. References

[1] I. Kitagawa, H. K. Wang, M. Yoshikawa, *Chem. Pharm. Bull.* **31** (1983) 716–722.
[2] B. N. Meyer et al., *J. Agric. Food Chem.* **38** (1990) 205–208.
[3] C. Adler, K. Hiller, *Pharmazie* **40** (1985) 676–693.
[4] R. Segal, P. Shatkowsky, I. Milo-Goldzweig, *Biochem. Pharmacol.* **23** (1974) 973–981.
[5] H. Abe et al., *Planta Med.* **34** (1978) 160–166.
[6] R. Schönheimer, H. Dam, *Z. Physiol. Chem.* **215** (1933) 59.
[7] B. von Wolters, *Planta Med.* **14** (1966) 392.
[8] P. A. Arneson, R. D. Durbin, *Plant. Physiol.* **43** (1968) 683.
[9] W. M. L. Crombie, L. Crombie, *Phytochemistry* **25** (1986) 2069–2073.
[10] A. Marston, K. Hostettmann, *Phytochemistry* **24** (1985) 639–652.
[11] Y. Birk in I. E. Liener (ed.): *Toxic Constituents of Plant Foodstuffs,* Academic Press, New York 1969, p. 169.
[12] C. Arnault, K. Sláma, *J. Chem. Ecol.* **12** (1986) 1979–1986.

[13] W. Sanderman, H. Funke, *Naturwissenschaften* **57** (1970) 407.
[14] G. Febvay, P. Bourgeois, A. Kermarrec, *Agronomie (Paris)* **5** (1985) 439–444.
[15] R. Kuhn, I. Löw, *Chem. Ber.* **94** (1961) 1088.
[16] I. Ishaya, Y. Birk, A. Bondi, Y. Tencer, *J. Sci. Food Agric.* **20** (1969) 433.
[17] Y. Asada, T. Ueoko, T. Furuya, *Chem. Pharm. Bull.* **37** (1989) 2139–2146.
[18] R. Segal, E. Schlösser, *Arch. Microbiol.* **104** (1975) 147.
[19] H. M. Feder, A. M. Christensen in R. A. Boolootian (ed.): *Physiology of Echinodermata*, J. Wiley & Sons, New York 1966, p. 87.
[20] K. R. Price, I. T. Johnson, G. R. Fenwick, *CRC Crit. Rev. Food Sci. Nutr.* **26** (1987) 27–135.
[21] S. B. Mahato, S. P. Sarkar, G. Poddar, *Phytochemistry* **27** (1988) 3037–3067.
[22] T. Schöpke, K. Hiller, *Pharmazie* **45** (1990) 313–342.
[23] K. Hostettmann, *Planta Med.* **39** (1980) 1–18.
[24] H. Kanazawa et al., *Chem. Pharm. Bull.* **38** (1990) 1630–1632.
[25] S. Sakuma, H. Motomura, *J. Chromatogr.* **400** (1987) 293–295.
[26] D. Fraisse, J. C. Tabet, M. Mecchi, J. Raynaud, *Biomed. Environ. Mass Spectr.* **14** (1986) 1–14.
[27] P. K. Agrawal, D. C. Jain, R. K. Gupta, R. S. Thakur, *Phytochemistry* **24** (1985) 2479–2496.
[28] J. A. Findlay, M. Jaseja, D. J. Burnell, J.-R. Brisson, *Can. J. Chem.* **65** (1987) 1384–1391, 2605–2611.
[29] H. K. Wang, *Chem. Pharm. Bull.* **37** (1989) 2041–2046.
[30] Y. Sashida, K. Kawashima, Y. Mimaki, *Chem. Pharm. Bull.* **39** (1991) 698–703.
[31] M. E. Wall, R. C. Eddy, M. C. McClennan, M. E. Klump, *Anal. Chem.* **24** (1952) 1337.
[32] R. Wasicky, M. Wasicky, *Qual. Plant. Plant Foods Hum. Nutr.* **8** (1961) 65.
[33] W. Winkler, *Arzneim. Forsch.* **18** (1968) 1031.
[34] V. H. Honerlagen, H.-R. Trelter, *Dtsch. Apoth. Ztg.* **119** (1979) 1483.
[35] L. E. Liberti, A. D. Marderosian, *J. Pharm. Sci.* **67** (1978) 1487–1489.
[36] E. Bombardelli, A. Bonati, B. Gabetta, E. M. Martinelli, *J. Chromatogr.* **196** (1980) 121–132.
[37] H. Hattori et al., *Chem. Pharm. Bull.* **36** (1988) 4467–4473.
[38] H. Yamaguchi et al., *Chem. Pharm. Bull.* **36** (1988) 4177–4188.
[39] U. Sankawa et al., *Chem. Pharm. Bull.* **30** (1982) 1907–1910.
[40] P. Seeman, *Fed. Proc. Fed. Am. Soc. Exp. Biol.* **33** (1974) 2116–2124.
[41] S. Takagi, H. Ohtsuka, T. Akiyama, U. Sankawa, *Chem. Pharm. Bull.* **30** (1982) 3485–3492.
[42] R. Segal, I. Milo-Goldzweig, *Biochim. Biophys. Acta* **512** (1978) 223.
[43] T. Nakamura et al., *J. Pharmacobio. Dyn.* **2** (1979) 374–382.
[44] B. S. Yu et al., *Chem.-Biol. Interact.* **56** (1985) 303–319.
[45] G. Reznicek, J. Jurenitsch, *Pharm. uns. Zeit* **20** (1991) 278–281.
[46] T. Odani, H. Tanizawa, Y. Takino, *Chem. Pharm. Bull.* **31** (1983) 1059–1066.
[47] M. Karikura et al., *Chem. Pharm. Bull.* **38** (1990) 2859–2861.
[48] I. T. Johnson et al., *J. Nutr.* **116** (1986) 2270.
[49] S. Hiai, H. Yokoyama, H. Oura, *Chem. Pharm. Bull.* **29** (1981) 490–494.
[50] US 4 755 504, 1988 (Y. Liu).
[51] H. Matsuda, K. Samakawa, M. Kubo, *Planta Med.* **56** (1990) 19–23.
[52] M. H. Khan, F. M. Sullivan in J. Robson, F. Sullivan (eds.): *Symposium on Carbenoxolone Sodium*, Butterworths, London 1968, p. 5.
[53] S. Shibata et al., *Chem. Pharm. Bull.* **35** (1987) 1910–1918.
[54] T. Kakuno, K. Yoshikawa, S. Arihara, *Tetrahedron Lett.* **32** (1991) 3535–3538.
[55] C. R. Sirtori et al., *Lancet* 1977, 275.
[56] D. G. Oakenful, D. L. Topping, R. J. Illman, D. E. Fenwick, *Nutr. Rep. Int.* **29** (1984) 1039.

[57]   H. Yoshimura, K. Watanabe, N. Ogawa, *Eur. J. Pharmacol.* **146** (1988) 291–297; *Eur. J. Pharmacol.* **150** (1988) 319–324.
[58]   Takeda Chem. Ind., GB 2 179 042, 1986 (H. Oshio, M. Kuwahara, T. Komiya).
[59]   S. Arichi, T. Hayashi, M. Kubo, DE-OS 2 828 851, 1980; T. Takamoto, S. Arichi, Y. Uchida, DE-OS 3 042 117, 1981.
[60]   M. Amoros, R. L. Girre, *Phytochemistry* **26** (1987) 787–791.
[61]   H. Tokuda, T. Konoshima, M. Kozuka, T. Kimura, *Cancer Lett. (Shannon, Irel.)* **40** (1988) 309–317.
[62]   K. Hirabayashi et al., *Chem. Pharm. Bull.* **39** (1991) 112–115.
[63]   X. H. Zhou et al., *Chem. Pharm. Bull.* **39** (1991) 1250–1252.
[64]   I. Maharaj, K. J. Froh, J. B. Campbell, *Can. J. Microbiol.* **32** (1986) 414.
[65]   Wakunaga Yakuhin K.K., DE-OS 3 207 841, 1982 (O. Tanaka, S. Hiroshima, N. Yata).
[66]   R. Tschesche, K. H. Richert, *Tetrahedron* **20** (1964) 387.
[67]   S. B. Mahato, A. N. Ganguly, N. P. Sahu, *Phytochemistry* **21** (1982) 959–978.
[68]   Y. Hirai, S. Sanada, Y. Ida, J. Shoji, *Chem. Pharm. Bull.* **32** (1984) 295–301.
[69]   D. C. Jain, *Phytochemistry* **26** (1987) 1789–1790.
[70]   Y. Hirai, S. Sanada, Y. Ida, J. Shoji, *Chem. Pharm. Bull.* **34** (1986) 82–87.
[71]   C. O. Okunji, C. N. Okeke, H. C. Gugnani, M. M. Iwu, *Int. J. Crude Drug Res.* **28** (1990) 193–199.
[72]   Y. Watanabe, S. Sanada, Y. Ida, J. Shoji, *Chem. Pharm. Bull.* **32** (1984) 3994–4002.
[73]   G. Wulff, T. Tschesche, *Chem. Ber.* **102** (1969) 1253.
[74]   S. Kiyosawa et al., *Chem. Pharm. Bull.* **16** (1968) 1162.
[75]   H. Matsura et al., *Chem. Pharm. Bull.* **36** (1988) 3659–3663.
[76]   S. Grünweller, E. Schröder, J. Kesselmeier, *Phytochemistry* **29** (1990) 2485–2490.
[77]   S. C. Sharma, R. Chand, B. S. Bhatti, O. P. Sati, *Planta Med.* **46** (1982) 48–51.
[78]   M. S. Kamel et al., *Chem. Pharm. Bull.* **39** (1991) 1229–1233.
[79]   R. K. Gupta, D. C. Jain, R. S. Thakur, *Phytochemistry* **25** (1986) 2205–2207.
[80]   R. Tschesche, P. Lauven, *Chem. Ber.* **104** (1971) 3549.
[81]   H. U. Lüning, E. Schlösser, *Z. Pflanzenkrankh. Pflanzenschutz* **82** (1975) 699.
[82]   T. Ikenaga et al., *Planta Med.* **54** (1988) 140–142.
[83]   I. Jizba et al., *Chem. Ber.* **104** (1971) 837.
[84]   J. A. Maga, *CRC Crit. Rev. Food Sci. Nutr.* **15** (1980) 371.
[85]   S. F. Osman in T. Swain, R. Klieman (eds.): *Recent Advances in Phytochemistry*, vol. **14**, Plenum Press, New York 1979, p. 75.
[86]   Y. Sato, N. Ikegawa, E. Mosettig, *J. Org. Chem.* **24** (1959) 893; *J. Org. Chem.* **25** (1960) 783.
[87]   J. G. Roddick, *Phytochemistry* **13** (1974) 9–25.
[88]   R. Kuhn, I. Löw, H. Trischmann, *Chem. Ber.* **88** (1955) 1492, 1690.
[89]   R. Kuhn, I. Löw, *Chem. Ber.* **94** (1961) 1088, 1096.
[90]   L. H. Briggs et al., *J. Chem. Soc.* 1950, 3013.
[91]   K. Schreiber, H. Ripperger, *Liebigs Ann. Chem.* **655** (1962) 114.
[92]   G. Adam, K. Schreiber, *Chem. Ber.* **99** (1966) 3173.
[93]   Y. Sato, H. B. Latham, *J. Am. Chem. Soc.* **78** (1956) 3146.
[94]   H. Ripperger, K. Schreiber, H. Budzikiewicz, *Chem. Ber.* **100** (1967) 1741.
[95]   A. Ehmke, *Planta Med.* **52** (1987) 507.
[96]   A. Uddin, H. C. Chaturvedi, *Planta Med.* **37**(1979) 90–92.
[97]   P. Adam, J. M. Trendel, P. Albrecht, *Tetrahedron Lett.* **32** (1991) 4179–4182.
[98]   N. Kawamura, H. Watanabe, H. Oshio, *Phytochemistry* **27** (1988) 3585–3591.

[99]   G. R. Fenwick, J. Lutomski, C. Nieman, *Food Chem.* **38** (1990) 119–143.
[100] H. Ishii et al., *Chem. Pharm. Bull.* **28** (1980) 2367–2383.
[101] G. Gosmann, E. P. Schenkel, *J. Nat. Prod.* **52** (1989) 1367–1370.
[102] S. Shimizu et al., *Chem. Pharm. Bull.* **36** (1988) 2466–2474.
[103] K. Hidaka et al., *Chem. Pharm. Bull.* **35** (1987) 524–529.
[104] O. Tanaka, R. Kasai, *Prog. Chem. Org. Nat. Prod.* **46** (1984) 1.
[105] D. Mandloi, P. G. Sant, *Phytochemistry* **20** (1981) 1687–1688.
[106] H. Wagner, H. Reger, *Dtsch. Apoth. Ztg.* **126** (1986) 1489–1493, 2613–2617.
[107] J. W. ApSimon, J. A. Buccini, S. Badripersaud, *Can. J. Chem.* **51** (1973) 850–855.
[108] J. W. ApSimon et al., *Can. J. Chem.* **58** (1980) 2703–2708.
[109] I. Kitagawa, M. Kobayashi, T. Sugawara, *Chem. Pharm. Bull.* **26** (1978) 1852–1863.
[110] R. Riccio et al., *J. Nat. Prod.* **48** (1985) 756–765.
[111] M. Iorizzi et al., *J. Nat. Prod.* **49** (1986) 67–78.
[112] N. Fusetani et al., *J. Nat. Prod.* **47** (1984) 997–1002.
[113] J. E. Gurst, Y. M. Sheikh, C. Djerassi, *J. Am. Chem. Soc.* **95** (1973) 628. J. W. ApSimon, J. Burnell, J. Eenkhoorn, *Synth. Commun.* **9** (1979) 215–217.
[114] I. Kitagawa et al., *Chem. Pharm. Bull.* **29** (1981) 1942–1950.
[115] I. Kitagawa, T. Nishino, M. Kobayashi, Y. Kyogoku, *Chem. Pharm. Bull.* **29** (1981) 1951–1956.
[116] I. Kitagawa, M. Kobayashi, M. Hori, Y. Kyogoku, *Chem. Pharm. Bull.* **37** (1989) 61–67.
[117] I. Kitagawa et al., *Chem. Pharm. Bull.* **26** (1978) 3722–3731.
[118] M. Girard et al., *Can. J. Chem.* **68** (1990) 11–18.
[119] T. Miyamoto et al., *Liebigs Ann. Chem.* 1990, 39–42.
[120] Y. Enomoto et al., *Br. J. Pharmacol.* **88** (1986) 259–267.

Silicon Compounds, Organic

LUTZ RÖSCH, Wacker Chemie GmbH, Burghausen, Federal Republic of Germany
PETER JOHN, Wacker Chemie GmbH, Burghausen, Federal Republic of Germany
RUDOLF REITMEIER, Wacker Chemie GmbH, Burghausen, Federal Republic of Germany

1.	Introduction	4306	4.3.3.	Other Nitrogen Compounds. 4337
2.	Fundamental Synthetic Routes for Organosilicon Compounds	4310	4.4.	**Organosulfur Compounds**. 4338
			4.4.1.	Mercapto and Sulfidic Organofunctions 4338
2.1.	Direct Synthesis of Organohalosilanes	4310	4.4.2.	Compounds Containing Sulfur–Oxygen Groups. 4339
2.2.	Grignard Synthesis	4314	4.5.	**Oxygen-Containing Compounds** 4339
2.3.	Syntheses with Alkali Metals	4315		
2.4.	Addition Reactions (Hydrosilylation)	4316	4.5.1.	Epoxy and Other Oxy Compounds. 4339
2.5.	Substitution Reactions of Si–H Bonds	4318	4.5.2.	Acrylates and Other Ester Functions 4340
2.6.	Coproportionation and Disproportionation	4319	4.5.3.	Acid Anhydrides and Other Carboxy Groups. 4341
3.	Silicon-Functional Organosilicon Compounds	4319	4.6.	**Other Organofunctions** 4343
			5.	**Other Organosilanes** 4343
3.1.	Halo- and Pseudohalosilanes	4321	5.1.	Tetraorganosilanes 4343
3.2.	Alkoxy- and Aryloxysilanes	4321	5.2.	Disilanes 4344
3.3.	Acyloxysilanes	4327	5.3.	Polysilanes 4345
3.4.	Oximino- and Aminoxysilanes	4327	6.	**Uses** 4345
3.5.	Aminosilanes, Amidosilanes, and Silazanes	4328	6.1.	Silylating Agents 4345
			6.2.	Physiologically Active Organosilicon Compounds 4347
3.6.	Hydrogensilanes	4329		
4.	Organofunctional Organosilicon Compounds	4330	6.3.	Silanes for Silicone Modification 4348
4.1.	Vinyl and Other Alkenyl Compounds	4330	6.4.	Silanes for Modification of Organic Polymers 4348
4.2.	Organohalogenated Compounds	4333	6.5.	Silanes in Catalysis 4351
			6.6.	Silanes for Use in Liquid Media 4352
4.3.	Nitrogen-Containing Compounds	4334	6.7.	Silane Coupling Agents 4352
4.3.1.	Cyanoalkyl Compounds	4335	6.8.	Silanes for Hydrophobic Surface Treatment 4354
4.3.2.	Organic Amino Compounds	4335		

| 6.9. | Analysis | 4355 | 8. | Economic Aspects | 4356 |
| 7. | Toxicology and Environmental Aspects | 4355 | 9. | References | 4357 |

1. Introduction

Organosilicon compounds are substances that contain at least one organic group directly bonded to silicon via a silicon–carbon bond. They can be subdivided into two main groups: monomeric compounds and polymers. Of the latter, the polymeric silicones, in which organosilicon units are linked by Si–O–Si bonds, are economically the most important. They are obtained from monomeric organohalosilanes by hydrolysis or methanolysis followed by polycondensation. Silicone production is thus the largest consumer of monomeric silanes. Polysilanes—compounds with Si–Si bonds—are increasing in importance. The monomeric organosilicon compounds are important components in a wide range of other applications apart from silicone chemistry. Their uses as auxiliary materials and as protecting groups in organic syntheses, as catalyst components in olefin polymerization, as protective agents for buildings, and as precursors for silicon carbide and silicon nitride coatings by chemical vapor deposition are important. An area with great potential is that of biologically active silicon compounds. Some organosilicon-based pharmaceuticals and crop protection agents are already on the market. Further developments in this area include silylated fragrance materials.

History. The first organosilicon compound, tetraethylsilane, was obtained in 1863 by FRIEDEL and CRAFTS by treatment of diethylzinc with tetrachlorosilane (Eq. 1) [15]. Two years later they prepared tetramethylsilane from dimethylzinc by an analogous method [16]. Subsequently, a series of new organosilicon compounds was prepared by treating tetrachlorosilane or tetraethoxysilane with organometallic reagents (Eq. 2). The work of PAPE, LADENBURG, and KIPPING was associated with this [17], [18]. The introduction of Grignard reagents into organosilicon chemistry by DILTHEY and KIPPING in the early 1900s was an important advance (Eq. 3) [19].

$$2\,Et_2Zn + SiCl_4 \longrightarrow Et_4Si + 2\,ZnCl_2 \quad (1)$$
$$Si(OEt)_4 + Et_2Zn \xrightarrow{Na} Et_xSi(OEt)_{4-x} \quad (2)$$
$$SiCl_4 + EtMgI \longrightarrow Et_xSiCl_{4-x} \quad (3)$$

Despite much research, organosilicon compounds remained a laboratory curiosity for a long time, and in 1936, KIPPING—in a summary of his investigations of nearly 40 years—still maintained that for this class of compound, important applications were neither known nor to be expected [18]. At the same time, however, Corning Glass and General Electric in the United States had already recognized that insulating materials resistant to high temperature could be produced from these compounds [20]. With the development of the direct synthesis—the production of methylchlorosilanes by the reaction of silicon with chloromethane in the presence of a catalyst—by ROCHOW [21] in the United States and MÜLLER in Germany [22], the basis for economical silicone production was finally established. Another milestone was the discovery in 1947 of the hydrosilylation reaction, in particular the oxidative addition of Si–H groups to carbon–carbon multiple bonds [23]. The discovery that this reaction can be controlled by using transition-metal catalysts [24] allowed economical access to a large number of new organosilicon compounds and their use in silicone chemistry and other areas.

Nomenclature. In nearly all organosilicon compounds, particularly the industrially important ones, the silicon atom is tetravalent. Most organosilicon compounds therefore contain tetracoordinate silicon. Compounds with higher coordination numbers occur as reactive intermediates in many reactions, while compounds with multiple bonds to silicon exhibit lower coordination numbers. The group of biologically active silatranes [25] with pentacoordinate silicon and the numerous investigations into hypervalent organosilicon compounds [26], whose chemistry may achieve industrial importance in the future, are also worth mentioning. In principle, organosilicon compounds can be regarded as derivatives of silane (SiH_4) in which the hydrogens have been replaced completely or partially by hydrocarbon groups:

$$SiH_4 \xrightarrow[-nH]{+nR} R_nSiH_{4-n} \qquad (4)$$

The remaining hydrogens can also be replaced by functional groups X. The resulting compounds are known as mono-, di-, or trifunctional organosilicon compounds, depending on the number of groups present in the molecule.

$$R_nSiH_{4-n} \xrightarrow[-mH]{+mX} R_nSiH_{4-n-m}X_m \qquad (5)$$

Almost all monomeric organosilicon compounds that contain tetracoordinate silicon can be derived from this general formula.

For a more precise classification the R groups are divided into "organofunctional" and "simple" organic groups. The latter are generally alkyl or aryl groups to which, however, a certain degree of functionality must be assigned, since it is possible to exchange alkyl groups for other alkyl or aryl groups, hydrogen, or other substituents in the presence of catalytically active Lewis acids such as $AlCl_3$. Alkyl groups can also be converted into organofunctional groups: e.g., chloromethylated silanes are formed in the chlorination of methylsilanes under UV irradiation. Examples:

Simple organic groups:	$-CH_3$, $-C_2H_5$, $-C_6H_5$
Organofunctional groups:	$-CH=CH_2$, $-C\equiv CH$,
	$-CH_2-CH=CH_2$,
	$-CH_2Cl$, $-\overset{\displaystyle \underset{\|}{O}}{C}-R$
	$-CH_2CH_2NR_2$,
	$-CH_2-CH-CH_2$ with O bridging
Silicon-functional groups:	halogens, $-NH_2$, $-NR_2$, $-S-C\equiv N$, $-OR$, $-OH$ (silanols), $-OM$ (metal silanolates)

The special IUPAC nomenclature rules for organosilicon compounds have not yet been issued. According to general IUPAC rules, tetravalent organosilicon compounds are called silanes and the ligands should be listed in alphabetical order, but hydrogen should not be named. The following order, which deviates from this, has become established in the technical literature: organofunctional ligand, simple organic ligand, alkoxy groups, halogen.

Compounds in which silicon is incorporated into a ring are known as silacyclo compounds (e.g., silacyclobutane). Silenes contain a Si–C double bond. Silicon chains and rings are named according to the number of silicon atoms: disilane, trisilane, cyclohexasilane. Certain functional groups have their own names: $\equiv$Si–OH = silanol, $\equiv$Si–SH = silanethiol, $\equiv$Si–O–Si$\equiv$ = siloxane. Some examples of nomenclature are:

Chemical formula	Name
Me$_4$Si	tetramethylsilane
Me(H)SiCl$_2$	dichloromethylsilane
Ph$_2$SiH$_2$	diphenylsilane
H$_2$NCH$_2$CH$_2$(Me)Si(OMe)$_2$	2-aminoethyldimethoxymethylsilane
Me$_3$SiOH	trimethylsilanol
(Me$_2$SiO)$_3$	hexamethylcyclotrisiloxane
Me$_2$(H)SiSiH$_2$Me	1,1,2-trimethyldisilane
(Me$_2$Si)$_6$	dodecamethylcyclohexasilane

Structure and Reactivity. Like carbon, silicon preferentially assumes a tetrahedral arrangement of its four valences in forming compounds. The larger atomic radius of silicon compared to carbon (C, 77 pm; Si, 177 pm) leads, for example, to lower barriers to rotation at silicon–element bonds than in the corresponding carbon compounds. Also, the completely different bond energies of silicon and carbon (for examples, see Table 1) and the different electronegativities of the two elements (Pauling electronegativities: Si = 1.8, C = 2.5) result in the organic chemistry of silicon being fundamentally different from that of carbon. The high affinity of silicon for oxygen is the reason why only compounds with Si–O bonds (SiO$_2$, silicates) occur naturally. Although addition reactions to unsaturated systems are important in synthetic organic chemistry, in silicon chemistry this type of reaction is primarily of academic interest because, thus far, species with multiple bonds to silicon have been difficult to prepare. If the problem of synthesizing compounds with Si–C or Si–Si double bonds economically is solved, the

Table 1. Examples of bond energies for silicon and carbon

Compound	Bond energy, kJ/mol	
	X = Si	X = C
$H_3X–H$	378	439
$H_3X–CH_3$	343	378
$H_3X–OH$	452	386
$H_3X–F$	565	460
$H_3X–Cl$	381	354
$H_3X–SiH_3$	310	343

door could be opened to a new sphere of industrial organosilicon chemistry. In organosilicon chemistry, S_N2 nucleophilic substitution at silicon plays a much more important role than in the case of carbon. Depending on the nucleophile, an Si–C bond can be formed or broken. It is particularly likely to be broken if a functional group is present in the β-position (β-effect). Nucleophilic substitution always occurs particularly readily if the bond strength of the new silicon–substituent bond is significantly higher and no steric hindrance occurs. Because of this, alkylhalosilanes react with water, for example, with the loss of hydrogen halide to form silanols in the first step (Eq. 6). The latter can then react further under alkaline (nucleophilic attack of OH^- at the silicon atom) or acid (electrophilic attack of H_3O^+ at the oxygen atom) conditions to give oligo- or polysiloxanes.

$$R_{4-n}SiX_n + nH_2O \longrightarrow R_{4-n}Si(OH)_n + nHX \quad (6)$$

The conversion of Si–X into Si–O is facilitated by the ability of the silicon atom to form pentacoordinate transition states.

The high reactivity of Si–X bonds (X = halogen, N, S), together with the high chemical and thermal stability of bonds between silicon and simple organic groups, is characteristic of organosilicon compounds. These properties have led to the broad range of applications of this class of compounds.

Because of the very large number of known organosilicon compounds, the most important fundamental synthesis routes for organosilanes are discussed first. Following this the classes of organosilicon compounds that are particularly important industrially and economically are discussed.

2. Fundamental Synthetic Routes for Organosilicon Compounds

2.1. Direct Synthesis of Organohalosilanes

The process described independently by ROCHOW [21] and MÜLLER [22], involving the direct reaction of haloalkanes with elemental silicon in the presence of a copper catalyst, is currently the standard synthesis used by all major silicone producers.

$$\text{Si} + \text{RX} \xrightarrow{250-320\,°C} \text{R}_3\text{SiX} + \text{R}_2\text{SiX}_2 + \text{RSiX}_3 + \ldots$$
R = CH$_3$: Cat. = Cu
R = C$_6$H$_5$: Cat. = Ag   X = F, Cl, etc.   (7)

Although the process is, in principle, applicable to any organic halide, it has been successful commercially only in the production of methylchloro- and phenylchlorosilanes. Longer-chain haloalkanes and unsaturated compounds such as vinyl chloride or allyl chloride generally give low yields of organosilanes, consisting of mixtures that are difficult to separate. Similarly poor results are obtained when polychlorinated compounds such as α,ω-dichloroalkanes or dichloromethane are used [28]–[30]. For this reason, only the reaction of silicon with chloromethane is described below.

Mechanism. Although the mechanism of the direct process has been investigated intensively since its discovery, it has not yet been fully elucidated. A detailed review of older work on this subject can be found in [31], while more recent studies are described in [32]. The reaction of chloromethane and silicon in the presence of copper is a heterogeneously catalyzed gas–solid reaction. In the first step the copper catalyst and silicon react to form Cu$_3$Si. Whether this η-phase is the actual reactive species has been disputed, but it does guarantee that silicon and copper are in close proximity. This permits the formation of active Si–Cl groups on the surface via the chlorination of silicon by CuCl. The latter is formed by the decomposition of a copper–chloromethane adduct, which also yields hydrogen, methane (and other hydrocarbons), and carbon as side products. Since the Si–Cl species are formed next to the copper–chloromethane adducts, the transfer of a methyl group with formation of a surface-bound Cl–Si–Me transition state is probable. Further chloromethane transfer finally leads to Me$_2$SiCl$_2$, which is liberated. CuCl remains behind at the catalyst surface and can initiate a new cycle. The chain reaction can be terminated by combination of two surface-bound Si–Cl groups or by reaction of Si–Cl with CuCl. The resulting SiCl$_2$ groups give MeSiCl$_3$ by further reaction with a Cu–MeCl adduct. The formation of Me$_3$SiCl and other side products can be explained similarly. Methyl radicals or silylenes in the gas phase [33] are not important for propagation of the reaction.

Table 2. Main monomeric silicon-containing products of direct chloromethylsilane synthesis

Compound	bp, °C	Proportion in the crude silane, wt%
Me$_2$SiCl$_2$ [75-78-5]	70	70–90
MeSiCl$_3$ [75-79-6]	66	5–15
Me$_3$SiCl [75-77-4]	57	2–4
MeHSiCl$_2$ [75-54-7]	41	1–4
Me$_2$HSiCl [1066-35-9]	35	0.1–0.5

Promoters. If copper alone is used as catalyst in the direct synthesis a slow reaction occurs with low degrees of silicon conversion and unsatisfactory selectivity for the target product dichlorodimethylsilane. By using metallic promoters, both the activity of the catalyst mixture and the selectivity of the synthesis can be improved. The addition of antimony, cadmium, aluminum, zinc, tin, or combinations thereof has been found to be effective. For example, a catalyst mixture with 94.49 wt% Si, 5 wt% Cu, 0.5 wt% Zn, and 0.01 wt% Sn gave a dichlorodimethylsilane selectivity of 80 mol% [34]. A selectivity of 90% has been reported with a similar catalyst system [35]. The promoter concentration is critical; for example, too high a tin content inhibits reaction. The action of promoters can be explained as follows: they facilitate the transfer of chlorine and methyl groups to silicon and the diffusion of silicon into copper. Further information is given in [34] and references cited therein.

Nonmetals such as phosphorus can also be used as promoters [36].

Product Distribution. The direct process never follows exactly the "ideal equation," according to which only dichlorodimethylsilane should be formed. Many other products are formed in varying quantities in temperature-, catalyst-, or raw material-dependent side reactions. Besides methyltrichlorosilane and chlorotrimethylsilane, incompletely substituted silanes, such as dichloromethylsilane, are also formed. Furthermore, the crude silane mixture contains saturated and unsaturated hydrocarbons and 3–8% disilanes of general formula Me$_n$Si$_2$Cl$_{6-n}$. Table 2 lists the distribution of the main monomeric silicon-containing products and their boiling points. The compounds Me$_4$Si [75-76-3], HSiCl$_3$ [10025-78-2], SiCl$_4$ [10026-04-7], and others are also formed in small amounts. All the organosilicon compounds formed can be used in silicone chemistry or other fields, the disilanes generally after catalytic or thermal cleavage to monosilanes. The product ratio causes problems in certain cases. The target is to obtain a crude silane mixture in which the individual silanes are obtained as closely as possible to the ratio in which they will be used in the production of functional silanes and silicones, by means of suitable reaction conditions. Fortunately, dichlorodimethylsilane, which is particularly important, is formed in ca. 80% yield. However, with chlorotrimethylsilane, chlorodimethylsilane, and dichloromethylsilane, shortages sometimes occur. These can be counteracted by altering the product ratio in the direct synthesis via addition of gases or metals.

Zinc, magnesium, or aluminum in quantities up to 10% or more has a strong alkylating effect via methylmetal halides formed as intermediates, so that the content of chlorotrimethylsilane in the crude silane can increase to 5–20%. The metals ultimately act as halogen scavengers, forming the halides $ZnCl_2$, $MgCl_2$, or $AlCl_3$ [37], [38].

Calcium silicide (10–80% Ca) reacts with chloromethane in the presence of zinc and copper as catalysts to give a mixture with a high proportion of more highly alkylated silanes (ca. 61% Me_3SiCl and ca. 8% Me_4Si) [39].

The proportion of hydrogen-containing organochlorosilanes can be increased considerably by adding H_2 [40] or HCl [41]. A $MeHSiCl_2$ + Me_2HSiCl selectivity of >80 mol% has been reported [42]. If silicon is reacted with HCl alone, $HSiCl_3$ and $SiCl_4$ are obtained.

Raw Materials. Formerly, ca. 97% pure silicon was used. After the significant effect of even small quantities of added metals on the selectivity of the reaction and the catalyst activity was recognized, a purity of >99% is now aimed for. The impurities are generally aluminum, calcium, iron, and titanium. The presence of certain elements (e.g., Pb), which act as inhibitors, is particularly critical. Even 50 ppm of lead in the catalyst mixture halves the silicon conversion, and the Me_2SiCl_2 selectivity decreases considerably. The reactivity of silicon depends on its morphology, which can be affected by the conditions under which liquid silicon is cooled in the production process.

The purity of other raw materials should also be as high as possible. Copper, produced by cementation or electrolysis, must therefore be subjected to further purification. Partially oxidized copper is used, whose average composition is approximately Cu_2O. The use of a Cu–Si alloy or other copper compounds (e.g., CuCl or copper oxalate) is also possible [43]. Promoter elements such as zinc are used either as the metal, the oxide, or the chloride. Most producers obtain chloromethane by treating methanol with hydrogen chloride. In this way the hydrogen chloride formed in the conversion of organochlorosilanes to silicones can be recycled. If methanolysis is carried out instead of hydrolysis (Eq. 8), chloromethane is formed directly and can be recycled into the reaction after purification.

$$2 \equiv SiCl + 2\ CH_3OH \longrightarrow\ \equiv Si-O-Si \equiv\ + 2\ CH_3Cl + H_2O \qquad (8)$$

Production and Composition of Catalyst Mixture. Silicon and the copper catalyst are either finely ground separately and then mixed with other catalyst components (promoters) or used in the form of a ground copper–silicon alloy. A particularly active catalyst mixture is obtained by atomizing silicon or Cu–Si alloys, which may already contain promoters [44]. During grinding the particle size must be adjusted so that homogeneous fluidization is obtained in the fluidized-bed reactor. For silicon the particle sizes are 30–350 µm, and for the copper catalyst, 5–30 µm. The catalyst mixture can be activated before use by tempering at 1000 °C in a stream of N_2 or preferably H_2. The direct synthesis, however, also activates the catalyst mixture. Usual

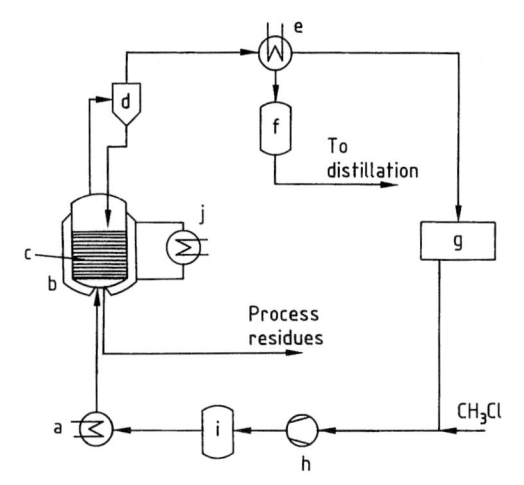

Figure 1. Production of chloromethylsilanes
a) Preheater; b) Reactor; c) Catalyst mixture; d) Cyclone; e) Heat exchanger; f) Crude silane container; g) MeCl purification; h) Compressor; i) MeCl container; j) Heating/cooling system

catalyst mixtures contain 2–6 % copper and 0.05–0.5 % zinc. Other promoters such as antimony or tin can also be added in small amounts [35], [45].

Industrial Process. The catalyst mixture is introduced into a fluidized-bed reactor and fluidized with a strong, tangentially introduced stream of chloromethane at ca. 300 °C and 2–5 bar (Fig. 1). The gas phase is partially freed from solid components in a cyclone at the head of the reactor, and separated solids are returned to the reactor. The fine portion passing through the cyclone enters the crude silane after condensation of the gas stream and is isolated during workup. Part of it can then be reprocessed. Condensation gives gaseous chloromethane and a crude liquid silane mixture. After purification, chloromethane is recycled to the reactor, and the crude silane mixture is distilled. Since silicon is consumed in the reaction and some of the catalyst is entrained in the gas stream, new catalyst mixture must be added continually. Its composition in the reactor must be kept as constant as possible. Silicon conversions of 90–98 % are achieved, and chloromethane conversions of 30–90 %, depending on process conditions (pressure, reactivity of the catalyst mixture, etc.). The strongly exothermic direct synthesis requires precise and effective temperature control because high time–space yields and high selectivity for dichlorodimethylsilane are possible only at ca. 300 °C. This is achieved by several measures, which generally supplement each other. Heat is removed by a reactor cooling jacket and sometimes by a heat exchanger in the reactor. The reaction can be slowed by pumping in an inert gas so that less heat is liberated. Thorough mixing of the catalyst mixture in the CH_3Cl stream is particularly important.

Distillation of Crude Silane. The crude silane mixture (for composition, see p. 4311) is separated into individual silanes in columns with very high separating capacity, connected in series. Since the organic chlorosilanes are building blocks for silicones, they must fulfill high purity requirements. Even a small amount of tri- or monofunctional silanes in the dichlorodimethylsilane can bring about chain branching ($MeSiCl_3$)

or chain termination (Me$_3$SiCl). The high purity required for Me$_2$SiCl$_2$ is achieved by distillation of the Me$_2$SiCl$_2$ obtained in the fractional distillation of the crude silane mixture.

Corrosion Problems. Steel is the preferred construction material for the direct process plant. Because methylchlorosilanes, particularly CH$_3$Cl, react with zinc, tin, magnesium, and aluminum at elevated temperature, these metals must not be used as alloy components. In some parts of the plant, enamel or Teflon must also be used.

Direct Synthesis with Halogen-Free Compounds. Many experiments have been described on the application of direct synthesis to halogen-free compounds. None of these routes have become important economically. A summary can be found in [46]. Treatment of silicon with dimethyl ether would be particularly interesting because it should give halogen-free methoxymethylsilanes. Such compounds could also serve as starting materials for the production of silicones. This would avoid the corrosion problems caused by HCl in the methods used up till now. Although a reaction between silicon and dimethyl ether can be achieved [47], the drastic reaction conditions required and the low yields of useful products make this route unlikely to become an economical alternative to conventional direct synthesis.

2.2. Grignard Synthesis

The Grignard synthesis of organosilanes, in particular organohalosilanes, was introduced by DILTHEY and KIPPING [19]. It still attracts much interest because of its universal preparative possibilities. For the synthesis of special silanes by the introduction of certain saturated or unsaturated hydrocarbon groups, this is frequently the best synthetic route.

Silanes with halogen or alkoxy groups can be used as starting materials. For example, SiCl$_4$ and HSiCl$_3$ can be treated with methyl Grignard reagents to give the silanes known from the direct synthesis (e.g., Eq. 9):

$$HSiCl_3 + RMgX \longrightarrow MgXCl + RSiHCl_2 + R_2SiHCl + R_3SiH \qquad (9)$$

Mono-, di-, and trichlorosilanes can act as starting materials for mixed alkyl–aryl-chlorosilanes:

$$RSiCl_3 + ArMgX \longrightarrow MgXCl + RSiArCl_2 + RSiAr_2Cl + RSiAr_3 \qquad (10)$$
Ar = Aryl

Ethers are the most frequently used solvents in Grignard synthesis (diethyl ether, dibutyl ether, THF, ethylene glycol ethers). The reactivity is particularly high in THF. In this solvent, vinyl groups have been introduced for the first time by using vinylmagnesium bromide [48], and Si–H groups can be substituted with alkyl groups [49].

The first step in Grignard synthesis is preparation of the Grignard solution. In some cases it can be prepared in situ in the presence of the halo- or alkoxysilane. This route also offers the possibility of a continuous Grignard synthesis [50]. The nature of the halogen or alkoxy group bonded to silicon, the number and nature of the organic groups already present in the starting silicon compound, and the size of the group to be introduced are some important factors for the result of the Grignard synthesis. Steric hindrance can make reactions proceed sluggishly or not at all, but it can also allow the formation of a desired, partially substituted product in high yield. Without steric hindrance the reaction is generally complete only after all halogens have been replaced. If fluorosilanes are used, particularly high degrees of substitution with bulky organic groups can be achieved because of the small size of the fluorine atom.

In some cases the reaction can be controlled by addition of catalysts. The reaction of dichlorodiphenylsilane [80-10-4] with tert-butyl Grignard in THF in the presence of CuCN or CuSCN gives tert-butylchlorodiphenylsilane [58479-61-1] in 75% yield (Eq. 11); without catalyst, no reaction occurs [51].

$$\text{tert-BuMgX} + \text{Ph}_2\text{SiCl}_2 \xrightarrow{\text{Cat.}} \text{MgXCl} + \text{tert-BuPh}_2\text{SiCl}$$
$$\text{Cat.: e.g., CuCN, CuSCN, Me}_3\text{SiCN} \qquad (11)$$

Even in the absence of steric hindrance, the yield of partially substituted intermediates can be increased by using an appropriate stoichiometry of starting materials. Generally, however, separation of the resulting mixtures is uneconomical.

Silicon-bonded alkoxy groups react with Grignard reagents more slowly than halogens; the reactions are thus easier to control. For example, diisobutyldimethoxysilane can be synthesized from tetramethoxysilane and isobutyl Grignard in 95% yield and >97% purity [52].

Disadvantages of Grignard syntheses are the high consumption of anhydrous, mostly volatile, and readily flammable solvents and the formation of large amounts of salts. They are also often accompanied by side reactions. Because of this they are used industrially only in exceptional cases (e.g., for the production of organosilicon compounds of particularly high value).

2.3. Syntheses with Alkali Metals

The reaction of halosilanes with organic compounds of alkali metals or in situ with the reaction products of haloalkanes and alkali metals is an alternative or a supplement to the Grignard synthesis and is also usually carried out only on a laboratory scale:

$$\text{RLi} + \text{ClSi}\equiv \longrightarrow \text{RSi}\equiv + \text{LiCl} \qquad (12)$$
$$\text{RCl} + \text{ClSi}\equiv + 2\,\text{Na} \longrightarrow \text{RSi}\equiv + 2\,\text{NaCl} \qquad (13)$$

Because of the high reactivity of organometallic compounds, they can also attack solvents. However, their advantage is that highly sterically demanding organic groups can often be introduced when this is not possible with a Grignard reagent [53].

With the smaller methyl or ethyl groups, all halogens are generally replaced. Even Si–OR and Si–H groups can react with organolithium compounds [54], [55].

Organozinc and organoaluminum compounds can also be used for alkylations [56].

2.4. Addition Reactions (Hydrosilylation)

Silanes and siloxanes that contain Si-bonded hydrogens can add to compounds with C–C multiple bonds (Eqs. 14 and 15). On addition to alkynes, alkene derivatives are formed initially, to which a second molecule of the Si–H compound can be added, provided no steric hindrance occurs.

$$\equiv SiH + R_2C=CR_2 \longrightarrow \equiv SiCR_2CR_2H \quad (14)$$
$$\equiv SiH + RC\equiv CR \longrightarrow \equiv SiCR=CRH \quad (15)$$

Hydrosilylation is used industrially for syn-thesizing alkylsilanes and functional silanes, cross-linking silicone polymers, binding silicone polymers to organic polymers, and removing impurities that contain Si–H groups from silane fractions of the direct synthesis.

Two different hydrosilylation routes are possible. Both make use of the high enthalpy of formation, estimated to be 160 kJ/mol for the addition of an Si–H group to a simple alkene, and the fact that the Si–H bond is readily activated. The first method developed is based on a chain mechanism involving silyl radicals [23]:

$$HSiCl_3 \longrightarrow H\cdot + \cdot SiCl_3 \quad (16)$$
$$\cdot SiCl_3 + CH_2=CHCH_3 \longrightarrow Cl_3SiCH_2\dot{C}HCH_3 \quad (17)$$
$$Cl_3SiCH_2\dot{C}HCH_3 + HSiCl_3 \longrightarrow Cl_3SiCH_2CH_2CH_3 + \cdot SiCl_3 \quad (18)$$

Silyl radicals can be generated thermally, by decomposition of radical initiators (e.g., acyl peroxides, azonitriles), by UV light, or by γ-radiation. With this method a large number of side products can be formed by decomposition of the initiators and polymerization of the alkene components. It is used industrially only in exceptional cases.

The route used almost exclusively in industry is the transition-metal-catalyzed addition of Si–H groups to multiple-bond systems.

In 1953, some supported transition metals were found to be very effective hydrosilylation catalysts. In particular, Pt–active charcoal is an highly active catalyst for the reaction of $HSiCl_3$ with acetylene, ethylene, butadiene, allyl chloride and $H_2C=CF_2$ at ca. 130 °C [57]. In 1957, SPEIER reported that chloroplatinic acid, $H_2PtCl_6 \cdot 6\,H_2O$, and its salts are catalysts that allow very rapid addition of a wide range of silanes to various alkenes, sometimes even below room temperature. This provided the possibility of hydrosilylating complex organic molecules. A summary of this work can be found in [58].

Mechanism. In the transition-metal-catalyzed addition reaction the ability of the transition metal to change its coordination number with a simultaneous change in oxidation state is exploited. This is shown in the example of platinum, where the reaction takes place at particularly active atoms on the metal surface or in the coordination sphere of the dissolved complex (Eq. 19). The Si–H bond adds oxidatively to platinum, and a π-bond to the olefin is formed.

$$X_3SiH + \underset{/}{\overset{\backslash}{C}}=\underset{\backslash}{\overset{/}{C}} + PtL_4 \rightleftharpoons \underset{\underset{L}{\overset{|}{C}}\underset{L}{\overset{|}{L}}}{\overset{\overset{\backslash C}{\overset{\|}{\underset{/C}{}}}{-}Pt-SiX_3}{\overset{H}{|}}}$$

$$\Big\updownarrow (L) \quad (19)$$

$$PtL_4 + H-\overset{|}{\underset{|}{C}}-\overset{|}{\underset{|}{C}}-SiX_3 \longleftarrow H-\overset{|}{\underset{|}{C}}-\overset{|}{\underset{|}{C}}-\underset{\underset{L}{\overset{|}{L}}\underset{L}{\overset{|}{L}}}{\overset{L}{Pt}}-SiX_3$$

L = ligand

Both steps are reversible. The reaction is completed by addition of hydrogen to the π-bonded molecule, which is also reversible. This is followed by reductive elimination of the silicon group with the σ-bonded organic ligand. When the final product is liberated the platinum catalyst is reformed and can reenter the cycle [59]. This mechanism explains most phenomena observed in the reaction. It is supported by the fact that in some cases, the intermediates can be characterized unambiguously. A more recent example is described in [60]. Colloidal solutions of platinum and other metals are also particularly active catalysts (see [61], [62], and references cited therein).

In general the β-addition product is preferentially formed in hydrosilylation:

$$X_3SiH + \underset{/}{\overset{R\backslash}{C}}=\underset{\backslash}{\overset{/}{C}} \longrightarrow \underset{\beta}{H-\overset{R}{\underset{|}{C}}-\overset{|}{\underset{|}{C}}-SiX_3} + \underset{\alpha}{X_3Si-\overset{R}{\underset{|}{C}}-\overset{|}{\underset{|}{C}}-H} \quad (20)$$

However, by using special catalysts (e.g., Rh or Pd compounds), the α-addition product can be isolated as the main component [63].

Olefins with terminal double bonds generally react faster than inner olefins. With the latter the terminal addition product is also formed because isomerization of the olefin occurs in the reversible first reaction step [64]. Silanes vary in reactivity depending on the nature of the substituents. Chlorosilanes are the most reactive, followed by alkoxysilanes; purely organic silanes and siloxanes frequently react sluggishly.

Catalysts. In industry, platinum catalysts such as Pt–activated charcoal, Pt–silica gel, $H_2PtCl_6 \cdot 6\,H_2O$–vinylsiloxane (Karstedt solution), $H_2PtCl_6 \cdot 6\,H_2O$ (Speier catalyst), and Pt olefin complexes are the most widely used. Palladium and rhodium compounds are also frequently used. For special applications a large number of catalysts based on other metals (main-group, transition, and rare-earth metals) have been developed [65]. In homogeneous catalysis the usual platinum concentrations are in the milligram-per-kilogram range, based on the silane used.

Hydrosilylation of Other Multiple-Bond Systems. Other multiple-bond systems such as

$$\overset{\backslash}{\underset{/}{C}}=O,\ \overset{\backslash}{\underset{/}{C}}=N-,\ -N=C=O,\ \text{and}\ -C\equiv N$$

can be hydrosilylated catalytically. These reactions and hydrosilylation in the presence of carbon monoxide are not described here; details can be found in [11], [58], [65].

2.5. Substitution Reactions of Si–H Bonds

Organic groups can also be bonded to silicon by substitution of Si–H bonds. The syntheses of PhSiCl$_3$ [98-13-5] (Eq. 21, an alternative to the direct synthesis) and MePhSiCl$_2$ [149-74-6] (Eq. 22) are important industrially.

$$HSiCl_3 + C_6H_5Cl \xrightarrow{\text{ca. }500\,°C} C_6H_5SiCl_3 + HCl \quad (21)$$

$$MeHSiCl_2 + C_6H_5Cl \xrightarrow{\text{ca. }500\,°C} C_6H_5MeSiCl_2 + HCl \quad (22)$$

The use of Friedel–Crafts catalysts, proposed earlier, is unnecessary because they favor side reactions and decomposition. The yields are 60–70 %; SiCl$_4$ or MeSiCl$_3$ and benzene are formed as side products [66].

Trichlorovinylsilane [75-94-5] can be obtained similarly from vinyl chloride and trichlorosilane [67].

In the presence of BCl$_3$ or AlCl$_3$, hydrocarbons can also undergo this type of reaction with the elimination of H$_2$ (Eq. 23).

$$\equiv SiH + RH \longrightarrow\ \equiv SiR + H_2 \quad (23)$$

A variant is shown in Equation (24):

$$\equiv SiH + ArH + Cl_2 \longrightarrow\ \equiv SiAr + 2\,HCl \quad (24)$$

For the mechanism of this reaction, see [68].

Other substitution reactions of Si–H bonds and other Si–X bonds are discussed in Chapter 3.

2.6. Coproportionation and Disproportionation

Co- and disproportionations are used for the conversion of silanes with undesired degrees of substitution into industrially utilizable products (Eq. 25):

$$\text{SiR}_4 + \text{SiR}^1_4 \underset{\text{Disproportionation}}{\overset{\text{Coproportionation}}{\rightleftharpoons}} 2\,\text{Si}(R, R^1)_4 \qquad (25)$$

The reaction is accelerated by Friedel–Crafts catalysts ($AlCl_3$, Na_3AlCl_6, CuCl), organic amines, and sodium alkoxides in the case of alkoxysilanes.

There are few limitations as to the nature of the substituents that can be exchanged because alkyl and aryl groups, alkoxy groups, halogen, and hydrogen atoms are all exchangeable [69], [70]. A simple example is the coproportionation shown in Equation (26), which proceeds at moderate temperature.

$$\text{MeSiCl}_3 + \text{Me}_3\text{SiCl} \xrightarrow{\text{Cat.}} 2\,\text{Me}_2\text{SiCl}_2 \qquad (26)$$

In practice, to increase the yield of Me_2SiCl_2 high- and low-methyl fractions from the direct synthesis are coproportionated catalytically in a similar way [71].

3. Silicon-Functional Organosilicon Compounds

This chapter deals mainly with organosilicon compounds of the general formula $R_n SiX_{4-n}$ having at least one Si–X group that can be cleaved hydrolytically (X = halogen, OR^1, NR^1_2; R, R^1 = H, alkyl, aryl). Fundamental aspects of these silanes are described in [8], [10]. An overview of the data for important individual compounds is given in [72].

Reactivity. Equation (27) describes the protolysis of silanes with cleavable groups X [73].

$$R_n SiX_{4-n} + (4-n)\,HY \longrightarrow R_n SiY_{4-n} + (4-n)\,HX \qquad (27)$$

Industrially this is carried out with many reactants HY and has led to various industrial processes, depending on the nature and number of R groups, and the nature of the leaving group X and the nucleophile Y (Scheme 1, with X = Cl). Some syntheses also exploit the interconversion of the silicon-functional derivatives [e.g., shifting the equilibrium by removing volatile components through distillation (Eqs. 28 and 29)].

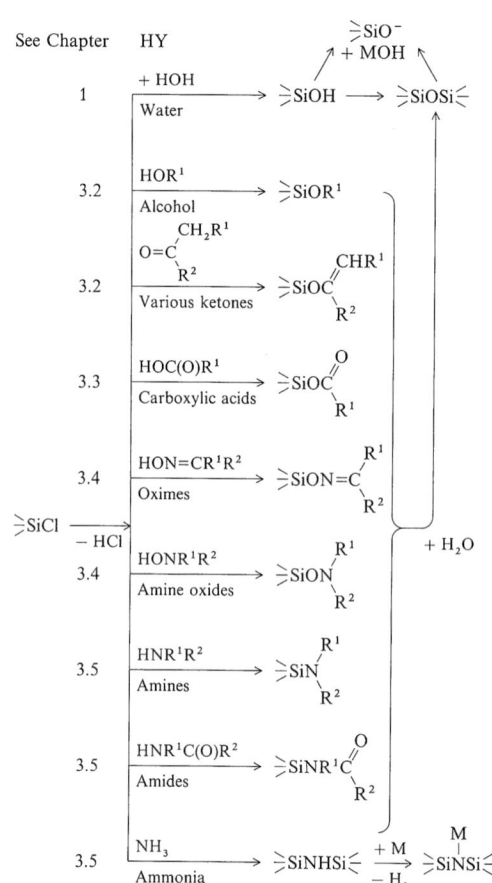

Scheme 1. Industrially important reactions of organochlorosilanes R_nSiCl_{4-n} with nucleophiles Y, according to Eq. (27) (protolysis)

See Chapter	HY		Product
1	+ HOH, Water	$\geq$SiCl $\xrightarrow{-HCl}$	$\geq$SiOH $\longrightarrow$ $\geq$SiOSi$\leq$ (via $\geq$SiO$^-$ + MOH)
3.2	HOR¹, Alcohol		$\geq$SiOR¹
3.2	O=C(CH₂R¹)(R²), Various ketones		$\geq$SiOC(CHR¹)(R²)
3.3	HOC(O)R¹, Carboxylic acids		$\geq$SiOC(O)R¹
3.4	HON=CR¹R², Oximes		$\geq$SiON=C(R¹)(R²)
3.4	HONR¹R², Amine oxides		$\geq$SiON(R¹)(R²)
3.5	HNR¹R², Amines		$\geq$SiN(R¹)(R²)
3.5	HNR¹C(O)R², Amides		$\geq$SiNR¹C(O)R²
3.5	NH₃, Ammonia		$\geq$SiNHSi$\leq$ $\xrightarrow{+M, -H_2}$ $\geq$SiNSi$\leq$ (M)

(+ H₂O returns to $\geq$SiOSi$\leq$)

R¹, R² = mostly H; alkyl, aryl; can also be different or organofunctional groups
M = (Alkali) metal

$$\equiv SiNHR + HON=CR^1R^2 \xrightarrow{Cat.} \equiv SiON=CR^1R^2 + H_2NR \quad (28)$$

$$\equiv SiOR^1 + (MeCO)_2O \xrightarrow{Cat.} \equiv SiOCOMe + MeCOOR^1 \quad (29)$$

If silane compounds are reacted with one another, the thermodynamic equilibrium (as opposed to a purely statistical distribution) can be shifted in the direction of a specific component (Eqs. 30 and 31).

$$2\,RSi(NHR^1)_3 + RSiCl_3 \xrightarrow{Cat.} 3\,RSiCl(NHR^1)_2 \qquad (30)$$

$$RR^1Si(OR^2)_2 + RR^1SiCl_2 \xrightarrow{Cat.} 2\,RR^1Si(OR^2)Cl \qquad (31)$$

This can be attributed to differences in the free enthalpy of reaction, which has been calculated for many combinations [74]. Other metal atoms can also be incorporated in such reactions:

$$TiCl_4 + n\,Me_3SiOR^1 \longrightarrow Cl_{4-n}Ti(OR^1)_n + n\,Me_3SiCl \qquad (32)$$
$$n = 1-3$$

3.1. Halo- and Pseudohalosilanes

Chlorosilanes are produced by the processes described in Chapter 2. They are generally liquids and are thermally quite stable. Their overall high reactivity in protolytic processes is intermediate in the series $SiF \ll SiCl < SiBr < SiI$ and decreases with increasing substitution at silicon in the order $SiCl_4 > RSiCl_3 > R_2SiCl_2 > R_3SiCl$ and with increasing steric hindrance in the group R in the series H, CH_3, C_2H_5, $CH(CH_3)_2$, $C(CH_3)_3$.

The interest in organic synthesis has concentrated for some time on trialkylsilyl derivatives with sterically demanding R groups and particular leaving groups X. The highly reactive [75], light-sensitive iodotrimethylsilane is being used increasingly in industry. It is produced by in situ cleavage of hexamethyldisilane (see Section 5.2) in solution with elemental iodine. The crystalline, volatile *tert*-butylchlorodimethylsilane (*mp* 89 °C, *bp* 125 °C) also has specific applications. Besides its synthesis from *tert*-butyllithium and dichlorodimethylsilane, new routes have also been proposed, analogous to Equation (11). The sterically demanding chlorodimethylhexylsilane [67373-56-2] is formed from chlorodimethylsilane and tetramethylethylene over aluminum chloride [76] according to Equation (14). For the uses of these compounds and information on pseudohalosilanes Me_3SiX (e.g., X = CN/NC, NCO, NCS, N_3) the literature cited in Section 6.1 should be consulted.

3.2. Alkoxy- and Aryloxysilanes

The alkoxylation of chlorosilanes to give alkoxysilanes (Eq. 33) is of major industrial importance.

$$R_nSiCl_{4-n} + (4-n)\,HOR^1 \longrightarrow R_nSi(OR^1)_{4-n} + (4-n)\,HCl \qquad (33)$$
R = H, alkyl, aryl; also organofunctional;
R^1 = Me, Et, etc.

Table 3. Important chloro- and ethoxymethylsilanes Me_nSiX_{4-n}

Compound	CAS no.	M_r	bp, °C	Flash point*, °C
$SiCl_4$**	[10026-04-7]	169.9	57	
$MeSiCl_3$	[75-79-6]	149.5	66	11
Me_2SiCl_2	[75-78-5]	129.1	70	−5
Me_3SiCl**	[75-77-4]	108.6	57	−20
$Si(OEt)_4$	[78-10-4]	208.3	168	50
$MeSi(OEt)_3$	[2031-67-6]	178.3	143	33
$Me_2Si(OEt)_2$	[78-62-6]	148.3	113	13
Me_3SiOEt	[1825-62-3]	118.3	76	−14
Me_4Si	[75-76-3]	88.2	26	<−20

* Measured on ≥ 98% pure silanes (DIN 51755).
** $SiCl_4$ and Me_3SiCl form an azeotrope.

Alkoxysilanes are preferred because of their advantages in further use. They do not produce any hydrogen chloride (TLV 7.5 mg/m^3 ; in comparison, methanol = 262 mg/m^3, ethanol = 1880 mg/m^3). Acid-sensitive substances (e.g., acetone) or basic catalysts can be used in processing. Higher boiling points and flash points compared to chloro compounds (Table 3) permit the use of open apparatus under certain circumstances, and corrosion-resistant materials need not be employed. Alkoxysilanes hydrolyze more slowly and in a more controlled manner than chlorosilanes and allow access to silanols [77], defined silicone resins, or sol–gel precursors. Mixed condensations of alkoxysilanes give better yields of the target product [78] (Eq. 34), compared to chlorosilanes.

$$RSiCl_3 + Me_3SiCl \xrightarrow[\text{2) Water}]{\text{1) Isopropanol}} RSi(OSiMe_3)_3 \quad 94\%$$
$$R = CH_2CH_2CH_2Cl \tag{34}$$

Alkoxysilanes also offer advantages in the Grignard synthesis (Section 2.2). Coupling with sodium and organic halogen compounds proceeds similarly [79].

Significant uses of silanes [e.g., as coupling agents or for polyethylene cross-linking (Section 6)] were able to be developed to their current major importance only with the introduction of neutral alkoxy compounds.

Synthesis. The formation of alkoxysilanes according to Equation (33) is often associated with undesired side reactions, such as elimination of haloalkanes (Eq. 8) or olefins, resulting in the simultaneous formation of siloxanes. The reactions proceed more cleanly if the hydrogen chloride liberated is removed by (tertiary) amines. The additional costs of these bases and the necessary solvents and the moderate yields led to the use of processes that do not require neutralization and allow recycling of the large quantities of HCl liberated. The best industrial process is continuous, solvent-free production in a corrosion-resistant distillation column [80]. Chlorosilane, prereacted with alcohol, is fed into the upper third of the column countercurrent to the alcohol vapor coming up from the circulation evaporator at the bottom. HCl gas with a low silane content escapes via the condenser. If pure silanes are used, the quality of the alkoxysilane removed from the base of the column depends on the boiling point and water content of the alcohol. It does not generally need to be distilled. Usually alcohol-sensitive groups such as SiH, $SiCH_2CH_2Cl$, or $SiCH_2CH_2CN$ remain unreacted, since the alcohol can be used in strictly stoichiometric quantities. The action of excess alcohol on SiH groups, especially in the presence of bases, leads to loss of hydrogen according to Equation (35), often with a dangerous pressure buildup.

$$\equiv SiH + HOR^1 \xrightarrow{Cat.} \equiv SiOR^1 + H_2 \quad (35)$$
$$R^1 = Alkyl \text{ (or also H)}$$

In the absence of free alcohol, $HSi(OEt)_3$ [998-30-1] and potassium ethoxide form the synthetically interesting pentacoordinate compound $[HSi(OEt)_4]K$ [81]. Reactions of the type shown in Equation (35) can also be used for the synthesis of silanols or alkoxysilanes (e.g., by using Pd catalysts; see also Section 3.6).

The reaction of simple alcohols with a hydrolyzable group other than SiCl is used only in exceptional cases for economic reasons. Silyl ethers of higher-boiling alcohols or phenols are formed from methoxy- or ethoxysilanes, the best results being obtained with alkaline catalysts (e.g., alkoxides) and by distillation of the volatile alcohol. The quite selective formation of diethoxymethylsilane [2031-62-1], which is difficult to obtain by the above route (see Eq. 33), from oligomeric methylhydrogensiloxane [82] (Eq. 36) is noteworthy:

$$\begin{bmatrix} H \\ | \\ SiO \\ | \\ Me \end{bmatrix}_n + n\, MeSi(OEt)_3 \xrightarrow{AlX_3} \begin{bmatrix} OEt \\ | \\ SiO \\ | \\ Me \end{bmatrix}_n + n\, MeSiH(OEt)_2 \quad (36)$$

Alkoxytrimethylsilanes are readily formed from the highly reactive nitrogen derivatives (see Section 3.5). Ethers of *tert*-butanol are obtained from the chlorosilane with amines or alcoholates, and phenol ethers are obtained without the addition of HCl-binding reagents. The phenoxysilanes have particularly high thermal stability: $Si(OC_6H_5)_4$ [1174-72-7] boils without decomposition at ca. 405 °C.

The most important reaction of the alkoxysilanes is with water. While earlier work was concerned with preparative aspects, more recent investigations on hydrolytic behavior have focussed on the kinetics in dilute, homogeneous, predominantly solvent-free solutions [73], [83]–[86]. Alkoxysilanes are generally insoluble or sparingly soluble in water. They undergo loss of alcohol (Eq. 37) to give reactive silanols. The first step is rate limiting.

Table 4. Industrially important methyloligosiloxanes

Structure *	CAS no.	mp, °C	bp, °C
HMe$_2$SiOSiMe$_2$H	[3277-26-7]	<−100	71
Me$_3$SiOSiMe$_3$	[107-46-0]	ca. −65	100
VMe$_2$SiOSiMe$_2$V	[2627-95-4]	ca. −103	140
(Me$_2$SiO)$_3$	[541-05-9]	62	135
(Me$_2$SiO)$_4$	[556-67-2]	17	170
(Me$_2$SiO)$_5$	[541-02-6]	−44	210
(HMeSiO)$_4$ **	[2370-88-9]	ca. −69	135
(VMeSiO)$_4$ **	[2554-06-5]	ca. −40	229

* V = CH$_2$=CH.
** Mixture of positional isomers.

$$R_n\text{Si}(OR^1)_{4-n} + H_2O \xrightarrow{\text{Cat.}} R_n\text{Si}(OR^1)_{3-n}(OH) + R^1OH$$
$$\downarrow + H_2O \quad (37)$$
$$R_n\text{Si}(OR^1)_{2-n}(OH)_2 + R^1OH$$
$$\downarrow \text{etc.}$$

The silanols are solvated by water because of their higher polarity and their capacity to form hydrogen bonds, sometimes resulting in complete miscibility. Condensation to give siloxanes, which are generally insoluble, occurs parallel to hydrolysis (Eq. 38).

$$2 \equiv \text{SiOH} \longrightarrow \equiv \text{SiOSi} \equiv + H_2O \quad (38)$$

Compared with halogens the alkoxy group is a poorer leaving group. Its reactivity toward nucleophiles is limited and decreases roughly in the order OR^1 = OCH$_3$ > OC$_2$H$_5$ > OCH(CH$_3$)$_2$ > OC$_6$H$_5$ > OC(CH$_3$)$_3$. The relationships among pH, rate of hydrolysis, and stability shown for 3-glycidoxypropyltrimethoxysilane [2530-83-8] (Fig. 2) hold approximately for the formation and condensation of all silanols. In general, precise pH control is important for the reproducibility of the reaction and the quality of the hydrolysis product. Even traces of chlorinated hydrocarbons, formed during the alkoxylation of industrial methylchlorosilanes from olefinic side products of Rochow silanes (Section 2.1), can disturb the reaction by uncontrolled liberation of HCl [87]. The stability of the silanols R_nSi(OH)$_{4-n}$ toward condensation increases with increasing n. Unlike the quite insensitive phenylsilanols, such as Ph$_2$Si(OH)$_2$ [947-42-2], compounds with the general formula R$_3$SiOH condense in the presence of traces of acid with the elimination of water to give disiloxanes. Under neutral conditions they can be stored almost indefinitely. Steric hindrance in the R group also leads to stabilization. Because of hydrogen bonding, Me$_3$SiOH [1066-40-6] (M_r 90) boils at ca. 100 °C, as does its "dimer" Me$_3$SiOSiMe$_3$ (M_r 162). The unstable dimethylsilanediol [1066-42-8] [77] may be an intermediate in the hydrolysis of dichlorodimethylsilane to form silicone fluids. In addition to linear polydimethylsiloxanes (PDMSs) the primary

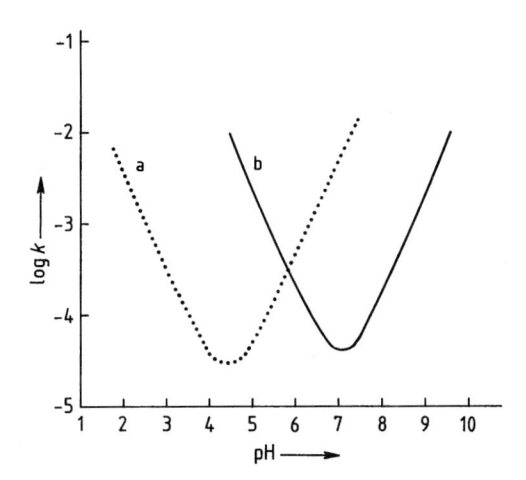

Figure 2. pH-dependent behavior of 3-glycidoxy-propyltrimethoxysilane in aqueous solution (ca. 0.25 mol/L) [83]
a) Silanetriol condensation; b) Silane hydrolysis; k = rate constant

hydrolysate contains cyclic compounds $(Me_2SiO)_n$ ($n \geq 3$; for CAS numbers and physical properties, see Table 4). The high ring strain of $(Me_2SiO)_3$ not only allows facile ring opening to give siloxane polymers (Eq. 39) but also reaction with silanols to give ambivalent siloxanes [88] (Eq. 40).

$$n\,[SiMe_2O]_3 \xrightarrow{\text{Cat. or heat}} [SiMe_2O]_{3n} \quad (39)$$

$$n\,[SiMe_2O]_3 + RMe_2SiOH \xrightarrow{\text{Cat.}} R[Me_2SiO]_{3n+1}H \quad (40)$$

R = (Functional) alkyl group

Cyclic siloxanes react particularly readily with chlorosilanes such as $MeRSiCl_2$ (R = H, Me, $CH=CH_2$) to form linear, short-chain, α,ω-halogenated siloxanes [89], which can be worked up hydrolytically to yield polymers. Siloxane cleavage also occurs with highly reactive silanols formed in situ from chloro- or alkoxysilanes, water, and a catalyst [90], often selectively (silanolysis, Eq. 41; X = Cl, OR^1).

$$2\,RSiX_3 + 3\,HMe_2SiOSiMe_2H \xrightarrow[H^+]{3\,H_2O} 2\,RSi(OSiMe_2H)_3 + 6\,HX \quad (41)$$

The reactivity of well-defined oligosiloxanes is reported in [91].

Water-soluble alkylsilanetriols can be stabilized only for a limited period in dilute solution under weakly acidic conditions. On warming, aeration of the solvent, or increasing the pH, they condense to resinous, generally water-insoluble polymers.

In strongly alkaline media, alkali-metal siliconates **1** are formed from $RSi(OR^1)_3$ (Eq. 42).

$$n\text{ MeSi(OMe)}_3 \text{ or } n\text{ MeSiCl}_3 \xrightarrow{\text{H}_2\text{O/OH}^-} \left[\begin{array}{c}|\\\text{MeSiO}^-\\|\\\text{O}\\|\end{array}\right]_n$$

$$\mathbf{1}$$

$$\xrightarrow[-\text{HCO}_3^-]{\text{H}_2\text{O/CO}_2} [\text{MeSiO}_{3/2}]_n \quad (42)$$

These are widely used to render mineral substrates hydrophobic. They react with atmospheric carbon dioxide to form insoluble, water-repellent methylsilicone resin, which is permeable to water vapor. It is used for impregnating mineral construction materials such as roofing tiles, or for protection against rising damp, particularly in old walls with low alkali content. More recently, silicone microemulsions have been used successfully (see Section 6.8).

A special category of alkoxysilanes is the silenol ethers, which can be obtained according to Equation (43) [92].

$$\text{CH}_3\text{SiCl}_3 + 3\,\text{O}=\underset{\underset{\text{R}^2}{|}}{\text{C}}-\text{CH}_2\text{R}^1 + 3\,\text{B} \longrightarrow$$

$$\text{CH}_3\text{Si}(\text{O}-\underset{\underset{\text{R}^2}{|}}{\text{C}}=\text{CHR}^1)_3 + 3\,\text{B}\cdot\text{HCl} \quad (43)$$

Moisture converts the enolized leaving group back into the starting neutral ketone [e.g., acetone ($R^1 = H$, $R^2 = Me$) or 5,5-dimethyl-1,3-cyclohexanedione (see Section 6.3)].

Esters of orthosilicic acid, $\text{Si}(\text{OR}^1)_4$, and their lower condensation stages are not organosilanes in the strictest sense. However, because of their industrial importance they are covered here. Unlike organo(organoxy)silanes, tetraalkoxysilanes are synthesized directly from silicon [93] or suitable natural silicates [94] and alcohols. Many side products are formed in this reaction, so that the overall energetic advantages of this direct synthetic route are offset by an uneconomical workup. Therefore products from the reaction of SiCl_4 with ethanol and water, which have a calculated SiO_2 content of 40 wt%, are used in practice [95]. These so-called ethyl silicates 40 [*26352-16-9*] are mixtures of linear, branched, and cyclic ethoxysiloxanes, which are stable on storage; their viscosity is ca. 4 mm^2/s. Unlike analogous condensation products of the toxic tetramethyl silicate, the cheap ethoxy compounds are used in large quantities. Water or atmospheric moisture causes loss of ethanol to give a strongly adhering silicic acid binding agent. Hydrolysis and cross-linking can be accelerated by catalysts. Examples of uses are:

1) Cross-linking of silicone rubber (see Table 6).
2) Inhibition of fiber pilling on polyester fabrics [as a mixture with alkoxyphosphorus compounds; 2-(diethylphosphono)ethyltriethoxysilane has been used as an alternative, see Section 4.6].
3) Consolidation of weathered sandstone for renovation of buildings (catalyst alkylacyltin compounds) [96], with retention of the open-pored character of the stone.

4) Protection of metallic construction components (bridges, containers, ships, pipelines): finely divided zinc pigments and other additives form impermeable coatings (cold galvanization) with silicic acid on steel, which protect against water and chloride ions and are resistant up to 420 °C. Even if the layer is damaged, the protective action is retained, since a local element is formed with zinc as the anode [97].
5) Making molds for precision casting: In many work cycles thin layers of ethyl silicate and fireproof pigments are deposited on top of each other on wax patterns and hardened in air or an ammonia atmosphere. After melting of the wax and firing at ca. 1000 °C the metal is cast into the porous mold.

3.3. Acyloxysilanes

Acyloxysilanes, $R_nSi(OCOR^1)_{4-n}$, are less readily formed than the alkoxy compounds. Acetoxysilanes ($R^1 = Me$) are very important industrially as cross-linking agents in one-component silicone sealants (RTV 1 systems; Section 6.3). They can only be produced economically from the reactive chlorosilanes:

$RSiCl_3 + 3\ Ac_2O \longrightarrow RSi(OAc)_3 + 3\ AcCl$ (44)
$RSiCl_3 + 3\ HOAc \longrightarrow RSi(OAc)_3 + 3\ HCl$ (45)
$R = CH_3, C_2H_5, HC=CH_2;\ Ac = COCH_3$

The first, older variant (Eq. 44) requires removal of acetyl chloride by distillation. Heating in the presence of traces of catalytically active metals can lead to reformation of acetic anhydride and oligomerization (Eq. 46).

$2 \equiv SiOAc \longrightarrow\ \equiv SiOSi \equiv\ + Ac_2O$ (46)

Continuous countercurrent reaction with acetic acid is more suitable (Eq. 45) [98]. The products are practically free of chloride, acid, and anhydride, and yields are close to 100 %. The compounds, known as acetic acid hardeners, are spontaneously hydrolyzable, corrosive, high-boiling liquids with melting points between 5 and 45 °C.

3.4. Oximino- and Aminoxysilanes

The trend toward RTV 1 systems, whose cross-linking agents eliminate neutral products, has made the long-known oxime and aminoxy cross-linking agents even more important industrially. Intensive work has been carried out on the production of alkyloximinosilanes; the oximino group (Ox) is currently almost exclusively – $ON=C(CH_3)C_2H_5$ [99]. The best process is the reaction shown in Equation (47) [100].

$$\text{MeSiCl}_3 + 6\,\text{HOx} \xrightarrow[<60\,°C]{\text{Solvent}} \text{MeSi(Ox)}_3 + 3\,\text{HOx} \cdot \text{HCl} \quad (47)$$

The amphoteric 2-butanone oxime acts as both a reactant and a "base." With hydrogen chloride it forms a liquid, readily separable phase that can be reacted with alkali to give 2-butanone oxime again. Careful dosage and monitoring lower the danger of vigorous decomposition of the acidic oxime hydrochloride.

Aminoxysilanes, $R_n\text{Si}(\text{ONR}^1R^2)_{4-n}$, are formed by reaction of hydroxylamines with alkylchlorosilanes, with removal of the liberated hydrogen chloride by bases [101]; another possible route is shown in Equation 53.

3.5. Aminosilanes, Amidosilanes, and Silazanes

(Alkylamino)alkylsilanes. Because of the high affinity of amines for HCl the industrial synthesis according to Equation (48) requires an excess of base. Triethylamine (A) can be used as co-base and solvent [102]. The granular, crystalline, readily separable hydrochloride formed can be worked up for reuse.

$$\equiv\text{SiCl} + \text{H}_2\text{NR}^1 + A \longrightarrow \equiv\text{SiNHR}^1 + A \cdot \text{HCl} \quad (48)$$
$$\text{Silylamine or (mono-)silazane}$$

When carbon dioxide is passed through, carbamatosilanes $\equiv\text{SiOCONHR}^1$ are formed from silylamines [103].

Amidosilanes. The amidosilanes $R_n\text{Si}-[\text{NR}^1\text{C(O)}R^2]_{4-n}$ ($R, R^1 = \text{Me}$; $R^2 = \text{Me}$ or $C_6H_5CH_2$), which are useful neutral cross-linking agents, are formed from chlorosilanes and organic amides by using strongly alkaline HCl scavengers (e.g., sodium metal or methoxide) or from Si–H compounds [104].

The silicon-monofunctional derivatives of urea and acetamide (see Table 5), used in large quantities for the synthesis of pharmaceuticals, also belong to this group of substances. N,N'-Bis(trimethylsilyl)urea (BSU) is produced industrially in quantitative yield from hexamethyldisilazane (HMDS) and urea, with, for example, a catalytic quantity of an ammonium salt or an organic sulfonamide [105] (Eq. 49).

$$(\text{Me}_3\text{Si})_2\text{NH} + \text{H}_2\text{NCONH}_2 \xrightarrow[120\,°C]{\text{Cat.}} (\text{Me}_3\text{SiNH})_2\text{CO} + \text{NH}_3 \quad (49)$$

The N,O-bis(trimethylsilyl)acetamides are among the most powerful trimethylsilyl-transfer agents. Production involves the conversion of acetamide into the liquid BSA by means of Me₃SiCl–trialkylamine via the solid monosilyl derivative (MSA) [106] (Eq. 50).

$$\underset{\text{Acetamide}}{\text{MeC}(=\text{O})\text{NH}_2} \longrightarrow \underset{\text{MSA}}{\text{MeC}(=\text{O})\text{NHSiMe}_3} \longrightarrow \underset{\text{BSA}}{\text{MeC}(\text{O-SiMe}_3)=\text{N-SiMe}_3} \quad (50)$$

On heating in the presence of Fe ions, for example, bis(trimethylsilyl)amides decompose to hexamethyldisiloxane and the corresponding nitrile, as indicated by the dashed lines in Equation (50).

Disilazanes. Many chlorosilanes react with ammonia to give not stable, simple silylamines $\equiv$SiNH$_2$, but disubstitution products with the SiNHSi building block. Hexamethyldisilazane has achieved major industrial importance. The exothermic reaction according to Equation (51) can be carried out in solvents such as hexane, toluene, or isododecane, but is best performed in warm HMDS itself [107].

$$2\,\text{Me}_3\text{SiCl} + 3\,\text{NH}_3 \longrightarrow (\text{Me}_3\text{Si})_2\text{NH} + 2\,\text{NH}_4\text{Cl} \quad (51)$$

The latter variant leads to the formation of coarse-grained ammonium chloride, which can be readily stirred and filtered and gives better turnover and yield. The suggestion has been made that the salt be separated by treatment with water. HMDS forms strong hydrogen bonds and is only weakly basic. Alkali and alkaline-earth metals form salts with it, e.g., LiN(SiMe₃)₂ [*4039-32-1*].

3.6. Hydrogensilanes

Organohydrogensilanes (H-silanes) are obtained by the general synthetic methods described in Chapter 2. The cheapest raw material for Si–H compounds, trichlorosilane [*10025-78-2*], is produced from silicon and hydrogen chloride. An Si–H synthesis from halosilane and metal hydride can only be considered for special, high-value sil(ox)anes [108]. Cleavage of Si–H bonds (and Si–Ph) is effected by halogens, giving halogen derivatives. H-Substitution also takes place on protolysis (e.g., Eq. 35). Other reactions of Si–H silanes involving loss of hydrogen are summarized in [109].

Many of the reactions shown in Scheme can be carried out with retention of the Si–H group. Insertion into Si–H bonds occurs on hydrosilylation (Section 2.4). In this way, the compounds R$_n$SiH(OR¹)$_{3-n}$, which can be produced from the corresponding chlorosilanes and alcohols, can be converted into functional alkoxyalkylsilanes (see Chapter 4). With amine or metal catalysts the alkoxy H-silanes sometimes undergo H-redistribution. In the absence of SiC-bonded groups, monosilane SiH₄ [*7803-62-5*], which is spontaneously flammable in air, can be formed [110]. The methylhydrogensiloxanes, which are used widely in industry, can react spontaneously with traces of alkali [111] according to Equation (52), to give gaseous methylsilane, MeSiH$_3$ [992-94-9].

$$3\,\overline{\{\mathrm{MeSi(H)O}\}} \xrightarrow{\text{Cat.}} \mathrm{MeSiH_3(g)} + 2\,\overline{\{\mathrm{MeSiO_{3/2}}\}} \quad (52)$$

Siloxanes with the structural elements **2**, **3**, or **4**, formed according to Equation (53), are effective condensation cross-linking agents.

$$\overline{\{\mathrm{MeSi(H)O}\}} + \mathrm{HY} \xrightarrow{\text{Cat.}} \overline{\{\mathrm{MeSiO}\atop|\atop\mathrm{Y}\}} + \mathrm{H_2} \quad (53)$$

$$\mathbf{2-4}$$

2: Y = NR^{1}C(O)R^2 [104];
3: Y = OCH$_2$CH$_2$NR$^1_2$ [112];
4: Y = ONR$^2_2$ [113]; R^1 = e. g., Me; R^2 = e. g., Me, Et

4. Organofunctional Organosilicon Compounds

Compared with the compounds described in Chapter 3, which react exclusively at the labile substituent X on the silicon, in this chapter derivatives with functionalized organic groups R^1 are considered. These silanes can be described by the general formula X$_n$R$_{3-n}$Si–R^1–Y, except for the alkenyl-substituted compounds. The alkyl substituents R and the leaving groups X correspond to those in Chapter 3. The Y groups, which are generally attached by stable bonds, are decisive for the activity of this class of compounds. These organic functional groups are used to bond quite different organic polymers to one another and to inorganic surfaces via siloxane bridges. The Y group generally participates itself in chemical bonding (e.g., in coupling agents, cross-linking agents) or at least increases the compatibility of different materials and surfaces.

The industrial development of this class of substances is closely associated with that of hydrosilylation. Vinylsilanes were first used, followed by aminopropyl- and mercaptosilanes [114]. In 1963 the methacrylic function, and later the epoxy and cationic styrene derivatives, came on the market.

4.1. Vinyl and Other Alkenyl Compounds

Industrial use of organofunctional silanes began in the early 1950s with the development of vinylsilanes and their reinforcing effect in glass-fiber–polyester plastics. Alkenyl compounds still dominate as strengthening additives in polymer applications and, in particular, as reactive components in many silicone formulations.

Today vinylsilanes are produced on a large scale mainly by continuous processes. The preferred route is catalytic addition of hydrogen-bearing chlorosilanes to acetylene.

$$HC \equiv CH + HSiRR^1 \overset{X}{\underset{}{|}} \xrightarrow{Pt}$$

$$X = Cl$$

$$H_2C=CH-\overset{X}{\underset{|}{Si}}RR^1 + R^1R\overset{X}{\underset{|}{Si}}CH_2CH_2\overset{X}{\underset{|}{Si}}RR^1 \quad (54)$$

In 1953 the use of platinum on solid carriers for this reaction was patented [57]. For example, trichlorosilane and acetylene react at ca. 130 °C in a tubular reactor to give a product mixture with 70% CH_2=$CHSiCl_3$ and 19% of the diadduct after a few minutes. Formation of $Cl_3SiCH_2CH_2SiCl_3$, derivatives of which are used to a limited extent as "bridging hardeners," can be suppressed by using a large acetylene excess and short residence time. Since heterogeneous catalysts such as Pt–active charcoal are often rapidly deactivated, soluble complexes such as hexachloroplatinic acid (Speier catalyst) [58] are now used. These are added in high-boiling solvents such as dichlorobenzene and lead to formation of the desired vinylsilanes in >90% yield without overpressure [115]. Through control of the temperature and reagent stoichiometry, the target product can also be removed continuously from the reactor [116]. In this way the subsequent reaction to give the diadduct, which acts simultaneously as a solvent, is suppressed still further. Addition reactions of alkylsilanes ($RHSiCl_2$) or H-substituted siloxanes proceed analogously. In particular, reactions with dichloromethyl- and chlorodimethylsilane are of industrial interest; those with the less reactive H_2SiCl_2 are somewhat less so.

Vinyl compounds can also be obtained from vinyl chloride by substitution reactions of Si–H bonds (see Section 2.5) [67].

The Grignard synthesis is worthwhile only for vinyl-rich intermediates that are virtually inaccessible by hydrosilylation.

Allyl compounds are also produced on a large scale. Tri- and dichloroallylsilanes can be prepared by replacing the halogen (e.g., in allyl chloride) with H-silanes. The resulting condensation with loss of HCl can be carried out at elevated temperature or with tertiary amines and CuCl as an effective catalyst [117], [118]. The allyl group can also be generated by subsequent HCl elimination from chloropropylsilanes, which are available on an industrial scale. Like the allyldimethylsilyl chain stoppers, these compounds also play an important role as reactive building blocks for polysiloxanes. The monochlorosilane precursor is produced better by cleavage of methyl-rich disilanes with allyl chloride, as is allyltrimethylsilane, used in pharmaceutical synthesis [64], [119]. The Grignard coupling of CH_2=$CHCH_2X$ with Si–X compounds (Section 2.2) is the most universal method, but because of the low time–space yield, it is limited to laboratory syntheses and the production of intermediates that cannot be produced directly by the above methods.

The lower stability of the Si–C bond toward strong acids and bases, compared with vinylsilanes, is also a sign of the special reactivity of the allyl group.

Other industrially interesting alkenyl compounds include adducts of H-sil(ox)anes with vinylcyclohexene, linear diolefins, or substituted alkynes.

The addition of trichlorosilane or dichloromethylsilane to vinylcyclohexene gives the desired cyclohexenylethyl compound with excellent selectivity because of the considerably higher reactivity of the terminal double bond. The cyclohexene structure is, moreover, suitable for the production of epoxysilanes and epoxysiloxanes (see Section 4.5.1) after replacement of the chlorine substituent. The process is more difficult to control when several double bonds of equal reactivity are present in the starting material. Diolefins such as hexadiene or octadiene must be present in at least two- to fourfold excess if good yields of the monoadduct are desired. Isomers derived from internal olefin structures may also be formed.

The resulting alkenylsilicon compounds are used, like the allyl derivatives, for particularly reactive silicones [e.g., LTC (low-temperature curing) systems for coating plastic films or paper] [120]. Such systems can also be produced by incorporating industrially relevant allyl ether precursors. Some of these newly developed siloxanes can even be cross-linked cationically after rearrangement of the allyl group to a propenyloxy group (cf. vinyl ethers) [121].

Butadiene gives mono adducts with high selectivity, but the resulting allylsilanes ($\equiv$SiCH$_2$CH=CHCH$_3$) are not suitable for addition reactions because of the terminal methyl group.

Sil(ox)anes with a butadienyl substituent are, like the conjugated diolefin itself, very reactive. Such compounds can easily be produced by the hydrosilylation of butynols, which are available on an industrial scale [122]. Monoaddition to the triple bond (Eq. 55) is highly selective and gives two isomeric adducts, both of which readily eliminate water.

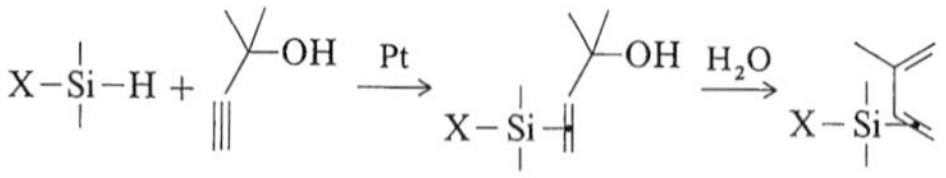

$$X-\overset{|}{\underset{|}{Si}}-H + \underset{|||}{\overset{\diagup OH}{\diagdown}} \xrightarrow{Pt} X-\overset{|}{\underset{|}{Si}}\diagup\diagdown\overset{OH}{\diagup} \xrightarrow{H_2O} X-\overset{|}{\underset{|}{Si}}\diagup\diagdown \qquad (55)$$

This elimination is generally carried out in situ, in the presence of the next reaction partner, which then becomes bonded to the isoprenylsilicon intermediate, for example, in a [2+4] cycloaddition (see Section 4.5.3).

Silylacetylenes can be produced relatively easily (e.g., from chlorosilanes and acetylene derivatives), but because of their low stability, particularly toward protic compounds, they have not achieved industrial importance thus far.

4.2. Organohalogenated Compounds

Silanes with halogen-containing side groups play a key role in silicone chemistry and particularly as reactive intermediates for the introduction of various Y functions. The downstream products of allyl chloride are the most important. 3- or γ-chloropropylsilanes are currently produced on a large scale. The production process starts with a catalytic addition, which is often carried out continuously (see Scheme 2). In the case of trichlorosilane it gives the desired chloropropylsilane with a selectivity up to ca. 83% [123]. The formation and addition of propene, occurring as a side reaction, can be interpreted as Cl–H exchange. It increases in the order $HSiCl_3$ < $MeHSiCl_2$ < Me_2HSiCl < Me_2HSiPh [124]. The resulting losses can, however, also be explained by the intermediate decomposition of a β-adduct, which is perhaps therefore not isolable. Accordingly, methallyl chloride gives fewer side products.

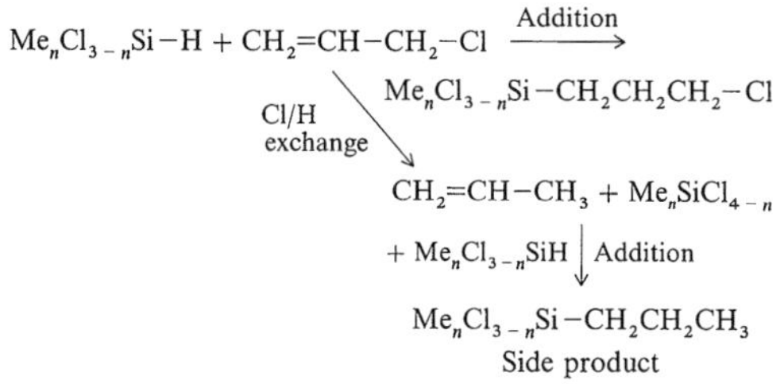

$$Me_nCl_{3-n}Si-H + CH_2=CH-CH_2-Cl \xrightarrow{Addition} Me_nCl_{3-n}Si-CH_2CH_2CH_2-Cl$$

$$\downarrow Cl/H \text{ exchange}$$

$$CH_2=CH-CH_3 + Me_nSiCl_{4-n}$$

$$+ Me_nCl_{3-n}SiH \downarrow Addition$$

$$Me_nCl_{3-n}Si-CH_2CH_2CH_3$$
Side product

Scheme 2. Addition of silanes to allyl chloride

The (chloropropyl)chlorosilanes are alkoxylated on a large scale by a reaction analogous to that shown in Equation (33) (Section 3.2). The methoxy and ethoxy derivatives are almost the only industrially important ones. They are additives for various formulations and, in particular, reactive intermediates for the production of other organofunctional silanes. They are also accessible directly by a new process starting from allyl chloride and H-alkoxysilanes with iridium catalysts [125].

The corresponding bromo- and iodopropyl compounds are of only minor importance, despite their higher reactivity in nucleophilic reactions, since the addition of iodine or phase-transfer catalysts is generally sufficient to activate the organochlorine group.

Chloromethyl compounds have achieved a certain importance (e.g., in the synthesis of biologically active substances) because of the many derivatives that can be formed from them (see Section 6.2). They are generally produced by radical chlorination of methylsilanes from the direct synthesis. Such processes generally exhibit modest selectivities and in the silicone industry are used only for methyltrichloro and chlorotrimethylsilane. The corresponding monochlorinated derivatives, $ClCH_2SiCl_3$ and $ClCH_2(CH_3)_2SiCl$, can be used in the same way as the chloropropylsilanes described above. Compounds with additional organic functional groups can be obtained from chloromethyllithium as the reactive intermediate [126].

Chlorophenylsilanes are produced by direct chlorination with Friedel–Crafts catalysts. Substitution in the aromatic ring leads to stabilization, making the compounds suitable for high-temperature applications.

β-Chloroethylsilanes show exceptionally high reactivity, which is further enhanced in the presence of bases. These compounds are readily produced by the reaction of gaseous HCl with chlorovinylsilanes, which undergo smooth anti-Markownikow addition in the presence of soluble aluminum catalysts [127]. The terminal carbon atom is rendered positively charged by the so-called β-effect, and the loss of the anionic leaving group is activated. In the presence of moisture the functional group decomposes according to Equation (56). The rate of ethylene formation depends on the pH and the nature of the groups bonded to silicon [128]. This has been used to control ripening processes.

$$-\mathrm{SiCH_2CH_2-X} \xrightarrow{H_2O} \underset{H_2C\overset{+}{-}CH_2}{\overset{\diagdown Si\diagup}{}} X^- \xrightarrow{H_2O}$$

$$-\mathrm{SiOH} + \mathrm{H_3O^+} + \mathrm{CH_2=CH_2} \quad (56)$$

Industrial fluoroorganosilanes differ fundamentally from the organohalogen compounds considered thus far. The systems that are of interest in industrial applications are strictly speaking not organofunctional since they are actually very inert. Because α- and β-C-fluorinated compounds are of comparatively low stability, fluoro substituents are incorporated only from the γ-position onward. The most important monomers are derived from 3,3,3-trifluoropropene. Trifluoropropylsilanes are produced by the catalytic addition of H-silanes.

Longer-chain highly fluorinated side groups lead to considerably lower surface tension. A monolayer of $CF_3(CF_2)_7CH_2CH_3SiO\equiv$ reaches values of < 8 mN/m [129]. In practical applications, such substituents, which can also be introduced by hydrosilylation, give particularly favorable effects with regard to solvent resistance, wettability, and adhesion/release. In addition, syntheses with SiH–fluoroaromatics (condensation, H_2 cleavage), Cl–F exchange, or Grignard precursors are known but have not achieved industrial importance.

4.3. Nitrogen-Containing Compounds

Parallel to the development of vinyl systems, nitrogen derivatives—in particular aminoorganosilanes—gained increasing industrial importance. These very reactive compounds currently play an important role in many silicone formulations, as well as in a large number of other applications involving compatibility, adhesion, or bonding of different substrates.

4.3.1. Cyanoalkyl Compounds

Cyanoethylsilanes were synthesized very early on by Si–H additions of chlorosilanes to acrylonitrile. Such processes have been carried out on a large scale since reduction of the alkoxylated downstream products to give the industrially interesting aminopropyl group became possible [130]. The acrylonitrile process is, however, carried out not with the usual transition-metal catalysts (which sometimes give the unstable α-adducts exclusively), but in the presence of tertiary amines or phosphines. Presumably, anionic intermediates are involved [131].

$$Cl_3SiH \xrightarrow{NR_3} \{Cl_3\bar{Si}-H-\overset{+}{N}R_3\} \xrightarrow{CH_2=CHCN} Cl_3SiCH_2CH_2CN \quad (57)$$

Addition of CuCl leads to more effective systems [132]. The corresponding trialkoxy compounds are produced by an alkoxylation process analogous to that described in Section 3.2 or by the addition of HCN to alkoxyvinylsilanes [133].

Longer-chain cyano compounds can be produced by noble-metal-catalyzed hydrosilylation of alkenyl nitriles such as allyl cyanide. The resulting cyanopropylsilanes are used, for example, in chromatography.

Organically bonded nitrile functions can be hydrolyzed to give free carboxylic acids. Hydrogenation is described in Section 4.3.2.

4.3.2. Organic Amino Compounds

Aminopropyl systems are the most important organofunctional sil(ox)anes. "Primary" aminosilanes are preferred because of their high reactivity.

Aminopropyltriethoxysilane [919-30-2], the most important individual compound, can be produced by three methods.

The acrylonitrile process (Section 4.3.1) is still regarded as very economical despite the necessary expenditures for safety. Cyanoethylsilane originating from the addition step (Eq. 57) is ethoxylated and then hydrogenated over Raney nickel at high hydrogen pressure [130].

$$Cl_3SiCH_2CH_2-CN \xrightarrow[2) H_2, Ni]{1) EtOH} (EtO)_3SiCH_2CH_2CH_2-NH_2 \quad (58)$$

The hydrosilylation of allylamine appears as a direct route to be actually simpler than the other, multistage processes, which in addition require pressurized reactors. The addition of triethoxysilane is, however, somewhat inhibited with the normal platinum catalysts and gives too much of the β-adduct (i.e., internal addition, 10–25%), which causes problems in many applications. Rhodium phosphine complexes increase the overall yield of the desired γ-product to > 80% so that the process now appears to be competitive [134].

$$(EtO)_3SiH + CH_2=CHCH_2NH_2 \xrightarrow{Rh}$$
(59)

$$(EtO)_3SiCH_2CH_2CH_2NH_2 + (EtO)_3Si\overset{Me}{\underset{|}{C}}HCH_2NH_2, Si(OEt)_4, H_2$$

This reaction is an example of the advantageous use of expensive rhodium compounds, which generally exhibit higher selectivity and lower sensitivity to inhibiting substances. The reaction mode and yield are now comparable to those of the more reactive (and with platinum compounds sufficiently selective) N-trimethylsilylallylamine [65].

The ammonia substitution process starts from chloropropyltriethoxysilane [5089-70-3], and requires a large excess and high pressure of NH_3. With a 75-fold molar quantity of NH_3 with respect to the chloro component, selectivities for the primary amine of up to 81% have been reported [135]. The correspondingly low space–time yields can only be improved if increased proportions of di- and even trisubstituted downstream products of the initially formed aminopropylsilane are acceptable.

$$(RO)_3SiCH_2CH_2CH_2Cl \xrightarrow{NH_3}$$
$$[(RO)_3SiCH_2CH_2CH_2]_nNH_{(3-n)}$$
(60)

Whether the NH_3 process is economical depends strongly on the possibilities for using the side products, particularly the bis(3-trialkoxysilylpropyl)amine (e.g., as coupling agents).

The substitution process with primary or even secondary amines or with ethylene-bridged polyamines as the nucleophiles is considered more advantageous. γ-(Ethylenediamino)propylalkoxysilanes are produced industrially by a reaction analogous to Equation (60). A three- to fivefold excess of diamine is generally sufficient to achieve yields of monomer > 80%. One equivalent of the strong base acts as an HCl acceptor and forms a readily separable liquid phase. The very reactive methoxy compounds are of particular industrial importance.

In many applications N-(2-aminoethyl)-3-aminopropyltrimethoxysilane [1760-24-3] competes with or supplements aminopropyltriethoxysilane, the most widespread organofunctional silane. The analogous dimethoxymethylaminosilane [3069-29-2] is the active building block of many amine fluids (for treating textiles, etc.).

Other important amino compounds are the N-methyl-, N,N-dimethyl-, and N-cyclohexylaminopropylsilanes. These are also produced by the substitution process. Here the removal of the amine hydrochloride, which is produced simultaneously, is problematic. Chloride-free pure products in good yields are obtained only after subsequent treatment with alkoxides or other strong bases [124]. Such secondary or tertiary amino groups impart advantages in certain applications (e.g., less thermal yellowing). They can therefore be considered alternatives to the standard silanes discussed previously. They are also used in the preparation of quaternary ammonium and betaine structures.

Organic amino functions at the end of siloxane chains are becoming increasingly important for the synthesis of block structures with organic polymer segments. Primary amino groups are preferred.

Bis(aminopropyl)tetramethyldisiloxane (aminopropyl chain stopper [*2469-55-8*]) is not sufficiently accessible either by substitution with NH_3 or by a route analogous to Equation (57). Since the direct addition of tetramethyldisiloxane (cf. Eq. 59) is also unsatisfactory, cyclic hydrosilylation of the silazane from chlorodimethylsilane and allylamine has been proposed [136].

$$CH_2=CH_2CH_2-NH-Si(Me)_2H \xrightarrow{Pt}$$

$$\overline{Me_2SiCH_2CH_2CH_2-NH} \quad (61)$$

The resulting Si–N ring is very reactive and can be cleaved hydrolytically to give the free aminopropyl group. The route involving phthalimide (Gabriel synthesis) or the treatment of the corresponding chloropropyldisiloxane with sodium diformylamide appears to be more straightforward [137].

As an alternative for certain applications the ethylenediaminopropyl stopper function may be considered, which is readily produced by direct substitution (cf. Eq. 60).

Aromatic amino compounds, which are used for the modification of particularly thermostable composites (e.g., in polyimides) can be produced by hydrogenation of nitrophenyl precursors (accessible according to Eq. 66 [138]).

4.3.3. Other Nitrogen Compounds

Other interesting nitrogen compounds can be produced by nucleophilic substitution of chloropropylsilanes, including isocyanate, urethane, ureido [123], and imidazole derivatives. Some of these compounds can also be produced by hydrosilylation of the corresponding allyl precursors and further derivatization. They are versatile intermediates in silicone chemistry.

Quaternary ammonium compounds (quats) are produced mainly by substitution reactions of chloropropyl precursors. The reaction with long-chain amines such as octadecyldimethylamine [139] is of industrial importance because it leads to surface-fixable biocides (Section 6.2). Methyl-substituted quats, which are also accessible from alkylaminosilanes, such as the side products shown in Equation (60) ($n = 2, 3$) act as ion exchangers and, if the chains are longer, also as phase-transfer catalysts [123].

The many compounds produced from aminosilanes include the cationic styrene compound $CH_2=CHC_6H_4CH_2N^+H_2CH_2CH_2NH–CH_2CH_2CH_2Si(OMe)_3$ [*34937-00-3*], which is produced by substitution of *p*-vinylbenzyl chloride and was introduced in 1972 as a copolymerizable coupling agent [114].

4.4. Organosulfur Compounds

4.4.1. Mercapto and Sulfidic Organofunctions

Mercaptofunctional silanes with spacers of more than one carbon between sulfur and silicon were reported as stable compounds in the early 1960s. These monomers, which are important in the rubber industry, are produced mainly from the corresponding chloropropyl precursors. Since alkali-metal hydrogensulfides give too much bis-(silylpropyl)sulfide in this reaction, the mercaptopropyl compound can, in practice, be obtained directly only with H_2S – amine combinations [140]. Pure monomers are better produced from protected S-nucleophiles such as thiourea. The isothiuronium salts formed initially are then cleaved with base (e.g., ammonia).

$$(RO)_{3-n}Me_nSi(CH_2)_3Cl \xrightarrow{+S=C(NH_2)_2} (RO)_{3-n}Me_n Si(CH_2)_3S^+C(NH_2)_2Cl^-$$

$$\xrightarrow[-(H_2N)_2C=NH_2{}^+Cl^-]{+NH_3} (RO)_{3-n}Me_nSi(CH_2)_3-SH \quad (62)$$

Since complete separation of the guanidinium salts is often difficult, addition of solvents or cleavage with alkylamines, which form liquid phases, has been proposed [123], [141]. The same mercapto compounds can also be produced by addition of H_2S to the corresponding cyanoethylsilanes or by their reductive cleavage with H_2–CoS_x [142]. β-Mercaptoethyl compounds can be produced from vinyl compounds and H_2S [143].

The thiol group is very versatile, bonding, for example, by radical addition to alkenyl systems under mild conditions. This predominantly anti-Markownikow addition is of considerable industrial importance in the cross-linking of various polymers.

Silanes with sulfidic sulfur bridges are of considerable industrial importance. They are produced mostly by direct substitution with sodium polysulfides Na_2S_x [123]. The average chain length of S_x is decisive for the reactivity of the compounds (e.g., in rubber vulcanization). The chain length can be regulated by addition of sulfur or by adjusting the pH during substitution with disodium tetrasulfide. This process is used industrially on a large scale. The most important compound, the tetrasulfane [(EtO)$_3$-SiCH$_2$CH$_2$CH$_2$]$_2$S$_4$ [*40372-72-3*], is used as a strengthening additive for the treatment of fillers and for the cross-linking (vulcanization) of polyolefin elastomers.

Disulfidic compounds, whose sulfur bridges can be readily cleaved by oxidation (see Section 4.4.2) or reversible reduction to thiol groups, can be obtained by the same route.

4.4.2. Compounds Containing Sulfur–Oxygen Groups

The ready oxidizability of sulfidic sulfur compounds (e.g., with H_2O_2) is used for the synthesis of sulfonate-functional siloxanes, which are used as ion exchangers, for example [144].

Such groups are preferably produced directly from the corresponding chloropropyl precursors. Substitution with Na_2SO_3 gives sulfopropylsilane–silanol intermediates, which are suitable for the production of hydrophilic to tensidic silicone oils [145]. Sulfoethylsilanols can be synthesized from vinylsilanes and bisulfite [146].

Thiosulfate groups are introduced by direct substitution with $S_2O_3^{2-}$ (Eq. 63) [147]. The organically bonded thiosulfate group (Bunte salt) undergoes hydrolytic cleavage to give mercaptopropylsiloxanes.

$$(RO)_{3-n}Me_nSi(CH_2)_3Cl \xrightarrow[-NaCl]{+ Na_2S_2O_3,\ H_2O} (RO)_{3-n}Me_nSi(CH_2)_3S_2O_3^-\ Na^+ \quad (63)$$

Compounds in which sulfur is bonded via oxygen are generally less stable to hydrolysis and are not dealt with here.

4.5. Oxygen-Containing Compounds

4.5.1. Epoxy and Other Oxy Compounds

In epoxy groups, oxygen is part of a highly activated ring structure. Since oxiranes can be opened even by traces of HCl, chlorosilanes are generally not suitable as starting materials. Industrial production of such silicon compounds starts mostly from alkoxy H-silanes or siloxanes and epoxidized alkenes such as allyl glycidyl ether or vinylcyclohexene oxide [124], [148]. Hydrosilylation with chlorine-free platinum catalysts or, for less pure reactants, with rhodium catalysts leads directly to monomers such as 3-glycidoxypropyltrimethoxysilane [2530-83-8] or 3,4-epoxycyclohexylethyltrimethoxysilane [3388-04-3], which are used as, e.g., coupling agents (Section 6.7).

The epoxidation of alkenylsil(ox)anes (Eq. 64) avoids toxic, potentially mutagenic, oxirane precursors [149], [150].

$$-O-\underset{|}{Si}-R^1-CH=C\diagup \xrightarrow{R-CO-OOH} -O-\underset{|}{Si}-R^1-CH-C\diagup \overset{O}{\diagdown} \quad (64)$$

This relatively mild oxidation is particularly suitable for polysiloxanes with epoxy groups in the chain. The degree of functionality can be controlled most readily by equilibration with the alkenylsilane precursor.

Whereas vinylic oxiranes, such as the α- or β-chloro or hydroxy compounds, rearrange relatively easily or eliminate olefins (Peterson reaction [151]), higher epoxy compounds behave like their organic analogues. Their high reactivity with acids and, in particular, with nucleophiles has led to a large number of individual applications.

Hydroxyalkyl groups are stable only if the OH group is in the γ-position or further from the silicon; they then react like normal alcohols. Direct hydrosilylation of allyl alcohol is now rarely used industrially because of considerable hydrogen formation in alcoholysis side reactions [124] and because of the toxicity of the alcohol. Since these problems are only partially solved by using silyl protecting groups, less critical precursors are used, such as allyl acetals or allyl esters, which can be hydrolyzed readily to hydroxy compounds after Si–H addition. Comparable intermediates are also obtained by the esterification of chloropropyl precursors with carboxylic acid salts. Similar hydroxy compounds are accessible by hydrosilylation of higher alkenols or glycol monoallyl ethers. Losses due to alcoholysis decrease with increasing chain length and with decreasing polarity of the components. This also holds for the addition to allylphenol.

Longer-chain poly(ethylene glycol) or poly(propylene glycol), which lead to systems with hydrophilic to tensidic properties, are added in the form of their monoallyl ethers, mostly directly to the oligomeric sil(ox)anes (siloxanes–oxyalkylenes, etc.).

Hydroxyl groups can also be formed by reaction of O-nucleophiles, in the simplest case water, with epoxyalkylsilanes.

4.5.2. Acrylates and Other Ester Functions

As early as 1963, methacrylic-functional silanes were introduced as particularly effective modifiers for organic polymers containing carboxyl groups. These are produced almost exclusively by hydrosilylation of allyl methacrylate (Eq. 65) [24].

$$X_3Si-H + H_2C=CHCH_2OCOC(Me)=CH_2 \xrightarrow{Pt} X_3Si(CH_2)_3OCOC(Me)=CH_2 \quad (65)$$

This reaction leads to the desired 3-methacryloxypropyl group with good selectivity because of the considerably higher reactivity of the allyl group. The most important of these compounds, the trimethoxy compound [*2530-85-0*] (X = OMe), is produced industrially either by direct addition of trimethoxysilane or preferably via the corresponding trichlorosilyl precursor [*7351-61-3*] (X = Cl) in the presence of radical scavengers [152]. The second method provides access to other alkoxy derivatives.

The reaction of γ-functional precursors with acrylates is an alternative. Chloropropylsil(ox)-anes can, for example, be esterified directly with salts of acrylic or methacrylic acid. Such processes, which are carried out in solution or heterogeneously with phase-transfer catalysts, are particularly suitable for products richer in siliconbound methyl groups, which are less readily accessible from allyl acrylates. The formation of esters from hydroxypropyl precursors and acrylic acid or the comparable transesterification with methyl methacrylate in the presence of tin catalysts [153] is preferably carried out on linear siloxanes. According to recent patents, similar systems can be produced by reaction of aminofunctional sil(ox)anes with acrylic acid chloride [154].

Other oxygen-bonded esters include the acyloxypropylsil(ox)anes discussed in Section 4.5.1.

Alkyl-bonded esters, such as the addition products of acrylates and vinylacetic acid, are dealt with in Section 4.5.3.

4.5.3. Acid Anhydrides and Other Carboxy Groups

Carboxy compounds play an important role in industrial polymer chemistry because of their dipolar nature. The modification of such systems can take place by radical grafting with acrylate groups or by nucleophilic coupling with activated carboxyl groups.

Anhydride-functional silicon compounds are particularly suitable for this. The introduction of such side groups is generally carried out by hydrosilylation of appropriate alkenyl precursors. Whereas the addition of sil(ox)anes to 5-norbornene-2,3-dicarboxylic acid anhydride [826-62-0] leads only to mixtures of isomers that are difficult to purify [155], the analogous reaction with allylsuccinic acid anhydride [7539-12-0] gives readily distillable monomers. The easy-to-handle 3-triethoxysilylpropylsuccinic acid anhydride [93042-68-3] is of particular interest [156].

Carboxyorgano groups can also be introduced into silanes by palladium-catalyzed reaction of disilanes with aromatic acid chlorides (e.g., from 1,2,4-benzenetricarboxylic anhydride [1204-28-0]) [138].

$$R\text{-}C_6H_3\text{-}COCl + (ClMe_2Si)_2 \xrightarrow[-CO, -Me_2SiCl_2]{Pd} R\text{-}C_6H_3\text{-}SiMe_2Cl \quad (66)$$

R = 3,4-(CO)$_2$O from benzene-1,2,4-tricarboxylic acid

4-(Chlorodimethylsilyl)phthalic anhydride [116088-82-5], produced by this route, can be used as a stopper function for bonding longer-chain siloxanes to OH- or NH-functional organic blocks. Comparable structures are synthesized better by the [2+4] cycloaddition of maleic anhydride to the isoprenyl (butadienyl) compounds described in Section 4.1 [157].

$$\text{OSi} \diagup\!\!\!\diagdown \text{R} + \begin{matrix} \text{CO} \\ \text{O} \\ \text{CO} \end{matrix} \xrightarrow{[2+4]} \text{OSi}\diagup\!\!\!\diagdown\text{R}\begin{matrix}\text{CO}\\\text{O}\\\text{CO}\end{matrix} \qquad (67)$$

The hydrosilylation of acrylates has attracted renewed interest as a result of the use of silylketene acetals in group-transfer polymerization (GTP, see Section 6.5). The resulting reactive intermediates can be regarded as neutral silyl-transfer agents like the α-silyl esters, which contain a labile Si–C bond. Such compounds, which are relatively unstable compared to silicones, are formed along with others in the hydrosilylation of acrylic esters [158].

$$CH_2=CH-COOR^1 + HSiR \xrightarrow{Cat.} \qquad (68)$$

$$\underset{\alpha}{R\overset{|}{Si}CHCOOR^1} + \underset{\beta}{R\overset{|}{Si}CH_2CH_2COOR^1}$$
$$\qquad \text{Me}$$

$$+ R\overset{|}{Si}OC(OR^1)=CH-Me$$

1,4-addition

The product ratio depends on the ester group, the catalyst used, and especially on the starting silane and the stoichiometry. The stable β-adducts are mainly of interest in silicone chemistry.

Methacrylic esters give terminal products with considerably higher selectivity, particularly when dichloromethylsilane is the H-component [24]. The resulting propionates can be equilibrated into organosiloxanes and subsequently saponified to the free acid, as is possible for cyanoalkyl groups under more forcing conditions.

Such carboxy-functional siloxanes have been produced from the corresponding silalactones [159]. The ester bond of the β-adducts is activated by the neighboring Si–Cl group, so that even catalytic quantities of tertiary or quaternary bases B are sufficient to liberate R^1Cl (usually chloromethane) according to Equation (69).

$$\underset{\text{Cl}}{\overset{\text{Me}}{Cl_2SiCH_2}}\underset{}{\overset{\text{Me}}{CHCOOR^1}} \xrightarrow{B} Cl(\overset{\text{Me}}{Si}CH_2\overset{\text{Me}}{CHCOO})_n\overset{\text{Me}}{Si}CH_2\overset{\text{Me}}{CH} \qquad (69)$$

Above 130 °C the lactones gradually rearrange into oligomeric acid chlorides [160], which are also suitable for the production of propionate-functional organosiloxanes.

Longer-chain carboxy functions are accessible by hydrosilylation of the trimethylsilyl esters of vinylacetic acid, undeceneoic acid, etc. [161].

The organometallic introduction of acid groups via haloalkyl precursors and carbonyl compounds with activated CH groups, or by using magnesium and CO_2, appears too costly for industrial applications.

Direct nucleophilic substitution of haloalkylsilanes with carboxylates plays a significant role. Besides the monoesters or amides of dicarboxylic acids, derivatives of chloroacetic acid should be mentioned in particular here.

Treatment of chloroacetic acid derivatives with dialkylaminopropylsil(ox)anes gives betaine-modified products, which are intended for cosmetic and tenside applications [162]. Similar carboxy-functional products are also accessible via mercaptoacetic acid [163].

4.6. Other Organofunctions

Carbon-bonded phosphorus groups are particularly stable in the β-position or further from silicon. The phosphonate esters $\equiv$SiCH$_2$CH$_2$–P(O)(OR)$_2$ can, for example, be saponified with strong acids without cleaving the ethylene bridge between the heteroatoms [164]. These compounds, which are of major industrial interest (see Section 6.4), are produced by radical addition of phosphite esters to vinylalkoxysilanes (hydrophosphorylation). The reaction with diethyl phosphite in the presence of peroxide catalyst to give 2-(diethylphosphono)ethyltriethoxysilane [757-44-8] is carried out on a large scale [165]. Similar phosphonate esters can be obtained by the Arbuzov reaction of trialkyl phosphites with haloalkylsilanes.

Phosphine groups can also be introduced by radical addition (e.g., of HPPh$_2$ to vinylsilanes) or by substitution with alkali-metal phosphides [166]. Such ligands are suitable for fixing noble metals, particularly rhodium catalysts.

Organometallic groups bound to organosilyl groups directly or via carbon bridges generally occur only as reactive intermediates in the synthesis of the compounds discussed here. Examples are ferrocene, titanium, or tin complexes, numerous Grignard reagents, and the π/σ-bonded active catalysts of various hydrosilylation reactions. They are not described in detail here.

5. Other Organosilanes

5.1. Tetraorganosilanes

In tetraorganosilanes, four organic groups are bonded to silicon via carbon. Symmetrical compounds have four identical organic substituents, and unsymmetrical ones up to four different organic groups on silicon. Carbosilanes, compounds in which silyl groups are bonded, for example, via CH$_2$ groups also belong to this class [30].

Production. Symmetrical tetraorganosilanes are generally produced by treatment of SiX$_4$ (X = halogen, OR, H) with organometallic reagents [167]. Tetramethylsilane is formed in small quantities in the direct synthesis. It is also produced as the target product by treatment of Me$_3$SiCl with methyl Grignard. For the synthesis of unsymmetrical products, numerous processes have been proposed in which SiCl$_4$ or R$_n$SiX$_{4-n}$

($n = 1-3$, X = mostly halogen) is treated with mixtures of organometallic reagents; for examples, see [168].

Properties and Uses. Tetraorganosilanes are relatively inert chemically except for compounds bearing organofunctional groups. Pyrolytic decomposition of tetramethylsilane occurs only above 600 °C. This compound is attacked only slowly by concentrated sulfuric acid (Eq. 70).

$$Me_4Si \xrightarrow{H_2SO_4} Me_3SiOSiMe_3 + CH_4 \qquad (70)$$

Tetraorganosilanes can, however, be used as alkylating agents [169] (Eq. 7).

$$R-\overset{O}{\underset{\|}{C}}-Cl + Me_3Si-C\equiv C-R^1 \xrightarrow{AlCl_3}$$

$$R-\overset{O}{\underset{\|}{C}}-C\equiv C-R^1 + Me_3SiCl \qquad (71)$$

Tetramethylsilane is used in coproportionation reactions [170] (see Section 2.6). It is also used as a reference in ^{1}H, ^{13}C, and ^{29}Si NMR spectroscopy and to fill ionization chambers in particle accelerators [171].

5.2. Disilanes

Chloromethyldisilanes $Me_nSi_2Cl_{6-n}$ ($n = 1-6$) are formed as side products in the direct synthesis (3–8% in the crude silane). Only hexamethyldisilane [1450-14-2] is produced specifically and on a large scale from Me_3SiCl by the Wurtz reaction (Eq. 72).

$$2\,Me_3SiCl + 2\,Na \longrightarrow Me_6Si_2 + 2\,NaCl \qquad (72)$$

To achieve complete conversion of sodium, many process variants have been proposed; see, for example, [172]. The electrochemical synthesis of Me_6Si_2 from Me_3SiCl, which has been further developed recently, is still not used in industry [173]. Most of the disilane fraction from the direct synthesis ($Me_nSi_2Cl_{6-n}$, $n = 1-6$, main products $n = 2-4$) is cleaved to give monosilanes. On acidic cleavage with hydrogen chloride, H-containing silanes are also obtained (Eq. 73).

$$Me_2ClSi-SiMeCl_2 + HCl \xrightarrow{Cat.} Me_2SiCl_2 + MeHSiCl_2 \qquad (73)$$

Cleavage with amine- or phosphorus-containing catalysts gives chlorine-containing polysilanes as well as monosilanes [174]. Chlorine-free polysilanes can be produced as the target product by disproportionation of methoxymethyldisilanes in the presence of NaOMe [175].

The synthetic potential of disilanes is noticeably gaining interest in industry. Besides the reactions described above, in recent years a range of methods has been developed in which new Si–C bonds can be formed by catalytic cleavage of the Si–Si bond. An example is shown in Equation (66). For further examples, see [176].

5.3. Polysilanes

Organopolysilanes are compounds in which organosilyl groups are linked by Si–Si bonds. They are increasingly being used as starting materials for silicon carbide ceramics, fibers [177], and composites derived therefrom [178]; as photoresists in microelectronics; and as photoinitiators. They also possess nonlinear optical properties. A review about the productioncan of silanes can be found in [179].

6. Uses

The rapid development of industrial silane syntheses created economical access to completely new substances, reaction principles, and surprising applications. The possible uses of mono- and oligomeric organosil(ox)anes, apart from the production of polymeric silicones, are given below (see also Chap. 4). Only some important uses are described here. Further information can be found in [123], [180].

6.1. Silylating Agents

Silanes have long been used in pharmaceutical research and production. One group of processes is characterized by the fact that silicon compounds enter only temporarily into the production cycle of the target product.

Silanes as Protecting Groups. In organic synthesis the presence of sensitive protic groups (OH, NH, COOH) can hinder reactions or make purification more difficult. These problems can be circumvented by temporarily substituting the protons using suitable silylating agents R_3SiX (X = halogen, amine, amide; see Table 5) [181]. The choice of the X group must take into consideration the possible pH sensitivity of the substrate molecule. The protecting group R_3Si can generally be introduced with high reactivity and selectivity. It can activate desired reactions or hinder undesired attack in another position. The derivatized molecule exhibits better solubility in organic solvents and increased volatility due to the absence of hydrogen bonds. After the desired

Table 5. Important silanes for introduction of protecting groups

Chemical name	Structure	CAS no.	Cleavage product
For standard applications			
Dichlorodimethylsilane	Me_2SiCl_2	[75-78-5]	acidic
Chlorotrimethylsilane *	Me_3SiCl	[75-77-4]	acidic
Hexamethyldisilazane *	$(Me_3Si)_2NH$	[999-97-3]	alkaline
N,N'-Bis(trimethylsilyl)urea *	$(Me_3SiNH)_2CO$	[18297-63-7]	neutral
For special applications			
N,O-Bis(trimethylsilyl)acetamide *	$CH_3C[=NSi(CH_3)_3]OSi(CH_3)_3$	[10416-59-8]	neutral
Iodotrimethylsilane *	Me_3SiI	[16029-98-4]	acidic
Hexamethyldisilane *	$Me_3SiSiMe_3$	[1450-14-2]	
tert-Butylchlorodimethylsilane	Me_3CSiMe_2Cl	[18162-48-6]	acidic

* Incorporated into industrial Me_3Si recycling.
** The highly reactive 1 : 1 molar mixture gives NH_4Cl.

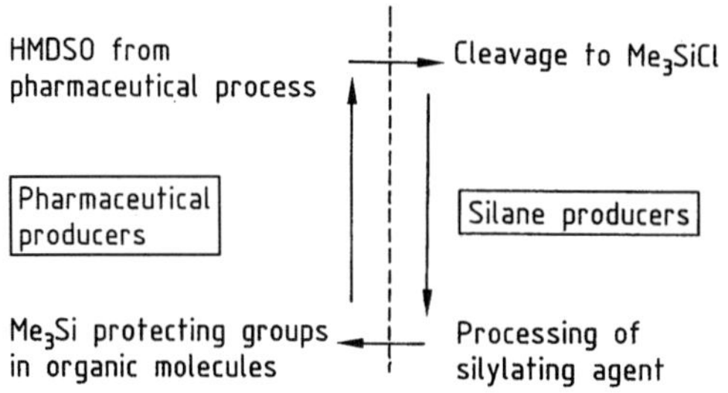

Figure 3. Recycling of Me_3Si protecting groups

reactions have been carried out, the protecting group is carefully removed by alcoholysis or hydrolysis, to finally form $Me_3SiOSiMe_3$ (HMDSO). To increase the overall economy and decrease environmental pollution, recycling of Me_3Si groups has been established since 1972 (Fig. 3). About 70% of the HMDSO wastes returned to the silicone producers can be recovered. The cleavage of HMDSO to give chlorotrimethylsilane can, for example, take place by continuous reaction with dichlorodimethylsilane [89] or hydrogen chloride [182]. The more selective, nitrogen-containing, silylating agents that can be produced from it (Section 3.5) are quite inert in the pure state toward the groups to be protected (and also to moisture). By addition of acids or metal catalysts their activity is drastically increased under certain conditions.

One of the most important uses of trimethylsilanes is in the production of completely or semisynthetic β-lactam antibiotics; for this and the use of more recent silylating agents, see [75], [181], [183].

The use of silanes as reagents in organic synthesis has also been investigated intensively [8], [184]. The stereochemical behavior of organosilanes is particularly interesting [187]. Relatively little is known of the extent to which this has been exploited industrially.

Silanes for Enzyme Immobilization. Many biotechnological processes [186] involve the covalent anchoring of enzymatically active macromolecules on inert carriers (e.g., porous glass) over which the reaction medium is (continuously) passed. One variant of this is covalent bonding by using silane coupling agents [123] (see Section 6.7).

6.2. Physiologically Active Organosilicon Compounds

In physiologically active organosilicon compounds at least one silicon atom is a component of the active molecule itself. Because of the differences in reactivity between carbon and silicon atoms, in some cases the synthesis can be carried out more specifically and in higher yield than with the carbon analogue [187].

In many cases the silicon compound has modified activity. Provided hydrolysis of the SiOC bonds can occur, they are degraded more rapidly in the organism and thus permit a higher dosage with fewer side effects under certain circumstances.

Organosilicon Pharmaceuticals. The above-mentioned organosilanes exhibit specific physiological activity only in exceptional cases (Chapter 7). Following the relatively late discovery and investigation of silatranes **5** (from ca. 1960 onward) [25], which have a unique, broad spectrum of activity depending on the structure of the SiC-bonded group, a range of potential organosilicon pharmaceuticals has been systematically researched [188].

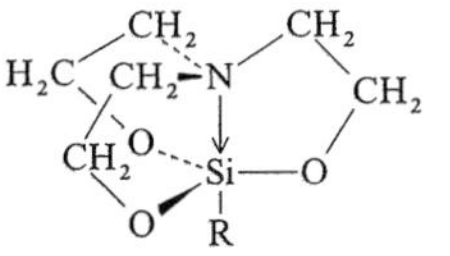

Only a few of these agents are used in the treatment of humans: Cisobitan (against prostate cancer) in Northern Europe; Migugen and Mival (hair growth and wound-healing agents) in the CIS.

Organosilicon Biocides. Recently, tetraorganosilanes with specific substitution in (hetero)aromatic ring systems have achieved importance as biocides. The first important market product was the broad-spectrum fungicide Flusilazol (**6**) [189], which is soon to be followed by the insecticide Silafluofen (**7**) [190].

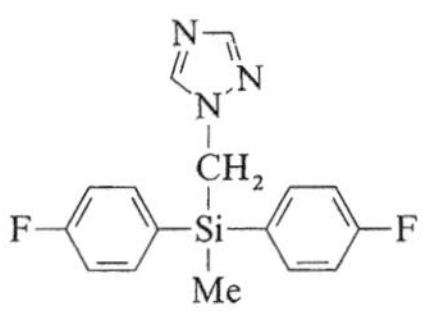

6 [85509-19-9]

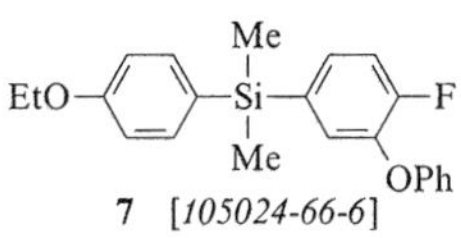

7 [105024-66-6]

The antimicrobial activity typical of ammonium compounds with longer-chain alkyl substituents is also exhibited by silanes such as $(RO)_3Si(CH_2)_3N^+MeR_1R_2Cl^-$. Here the Si function acts only as an anchor toward mineral substrates or organic fibers. The primary uses are in the treatment of textiles (carpets, sportswear) and the purification of water [191].

6.3. Silanes for Silicone Modification

A relatively small range of moisture-sensitive organosilanes is used as cross-linking agents in cold-cross-linking silicone sealants (RTV systems). The most important compounds are listed in Table 6. The quantities used are several thousand tonnes per year. Silane coupling agents can also be used for cross-linking (see Section 6.7) [192].

Organofunctional silicones are obtained by introducing the corresponding silanes (see Table 7) into polysiloxanes by condensation or equilibration reactions. The generally fluid products are used in reprography, in textile finishing, and in polish additives.

In coating technology, specific oligosiloxanes that undergo cross-linking by cationic ring-opening polymerization are now used [193]. In the production of siliconized release paper, the reaction of dimethylpolysiloxanes or methylcyclosiloxanes containing 2-(3-epoxycyclohexyl)-ethyl groups is initiated by strong Brönsted acids. The latter are formed, for example, from diaryliodonium salts on UV irradiation [194]. The reaction is not inhibited by oxygen.

6.4. Silanes for Modification of Organic Polymers

Organosil(ox)anes are being used increasingly for improving the production and properties of polymers.

Cross-Linking of Polyethylene. In some applications the SiOH condensation mechanism, which is so successful for silicones, is used for cross-linking and better adhesion. Two industrial processes were introduced in the 1970s under the names Sioplas and Monosil, in which trimethoxyvinylsilane [2768-02-7] is grafted onto polyethylene (PE, sometimes modified) by means of dicumyl peroxide (DICUP) [195]. Although the industrial processes differ (Fig. 4), the reaction principle is identical. After molding, the silyl function bonded to the polymer via an ethylene bridge is hydrolyzed by external moisture (hot water bath, steam). The resulting silanol groups then undergo cross-linking, which is catalyzed by premixed dibutyltin dilaurate (DBTL) [77-58-7] (Eq. 74).

Table 6. Important cross-linking agents for RTV silicone rubber *

Silane type	CAS no.	Cleavage product
Two-component systems		
Si(OEt)$_4$	[78-10-4]	Ethanol
Ethyl silicate 40	[26352-16-9]	Ethanol
One-component systems		
MeSi(OMe)$_3$	[1185-55-3]	Methanol
H$_2$C=HCSi[OC(Me)=CH$_2$]$_3$	[15332-99-7]	Acetone
Me/EtSi(OCOMe)$_3$	[4253-34-3]/[17689-77-9]	Acetic acid
MeSi[ONC(Me)Et]$_3$	[22984-54-9]	Butanone-2-oxime
Mixture of oligomers		Diethylhydroxyl amine + *N*-methylacetamide
MeSi(OEt)[N(CH$_3$)C(O)⟨⟩]$_2$	[16230-35-6]	*N*-methylbenzamide
MeSi(NHC$_6$H$_{11}$)$_3$	[15901-40-3]	Cyclohexylamine

* Difunctional analogues are chain extenders.

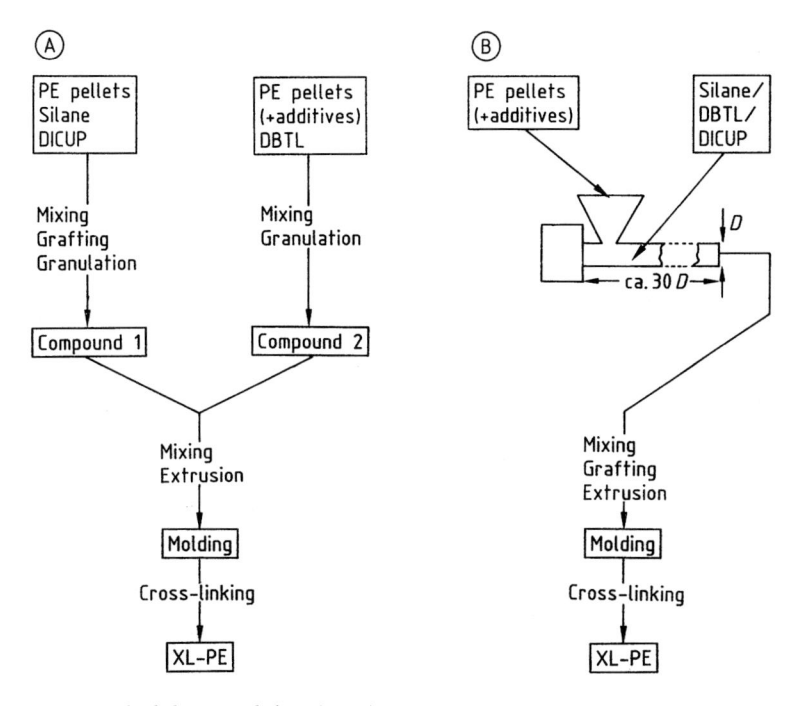

Figure 4. Polyethylene cross-linking (XL-PE)
A) Sioplas process (two stages); B) Monosil (one stage)

$$\underset{(OMe)_3}{\underset{|}{Si}}-\underset{|}{\overset{CH_2}{\underset{|}{CH_2}}}-CH_2-CH_3 \xrightarrow[-MeOH]{H_2O} \underset{(MeO)_2\ OH}{\underset{|}{Si}}-\underset{|}{\overset{CH_2}{\underset{|}{CH_2}}}-CH_2-CH_3 \xrightarrow[-H_2O]{DBTL}$$

$$\Big\rbrace-CH_2CH_2\underset{(OMe)_2}{\overset{(MeO)_2}{\underset{|}{\overset{|}{Si}}}}OSiCH_2CH_2-\Big\lbrace \qquad (74)$$

Repetition of reaction is possible

The Sioplas process separately produces two granulated polyethylene compounds in classical hot mixers. These are stored temporarily and can later be extruded as ca. 19 : 1 mixture with standard equipment. In a new process variant [196], vinylsilane is copolymerized with ethylene under high pressure. The resulting peroxide-free compound 1 (in Fig. 4) is very uniform and more stable to storage than grafted types. The cross-linked products are used in quantities of several hundred thousand tonnes per year as insulating materials, predominantly in low-voltage cables. They remain relatively stable dimensionally on thermal overloading. Their use in underfloor heating pipes or in self-cross-linking EPM or EDPM rubber articles is also common. The flameproofing of XL-PE with metal hydroxides is described in [197].

Cross-linking via silanol groups is also used in other polymers (e.g., vinyl acetate – acrylate dispersions [198], polyacrylates, or polyoxyalkylenes [199]) that are used as coatings, adhesives, or casting compounds.

Special Polymers. Unsaturated sil(ox)anes, in particular, have been proposed for copolymeric incorporation into standard polyolefins or polyacrylates. An important use is in raw materials for hard contact lenses, where a proportion of defined methacrylic-functional siloxanes [200] improves oxygen permeability. To enhance resistance to scratching, the surfaces of plastics are treated with silica-reinforced hydrolysates from, for example, $MeSi(OMe)_3$ (acrylic glass panes and consumer articles; polycarbonate spectacle lenses [201]). Polymers from short-chain or cyclic methylhydrosiloxanes, such as $[HMeSi(O)]_4$ [2370-88-9], with dicyclopentadiene and sometimes other diolefins, which are formed by platinum-catalyzed addition, are relatively new. They are recommended as nonstick, prehardened, impregnating resins for glass-fiber prepregs. The latter are laminated to form carrier plates for printed electrical circuits [202]. The field of sil(ox)ane-modified organic polymers is described in detail in [203].

The rubber additive polysulfidesilane (Section 4.4.1 and Table 7) is both a coupling agent in mixtures with silicate fillers and a source of sulfur for rubber cross-linking. This unique, low-toxicity substance lowers the tendency to scorch (to undergo cross-linking too early), so that, for example, thick-walled tires for heavy-goods vehicles can be vulcanized without difficulty. In addition, it gives rubber a high reversion stability. In terms of quantity used, this is one of the most industrially important silanes [123].

Stabilizers. Silane polymer additives have a wide range of structures and effects. Incorporation of $(CH_3O)_3SiCH_2CH_2CH_2Cl$ [*2530-87-2*] or polysulfidesilane (see above) increases the thermal stability of filled EPDM rubber and of EPDM covulcanized with silicone rubber [204]. Phenylsil(ox)anols are suitable for flameproofing high-temperature polymers [205]. Sterically hindered phenolic or amine-containing alkoxysilane antioxidants bind migration-free to polymers and fillers owing to hydrolysis and condensation and thus ensure long-term stability [206]. Addition of (chlorine-free) organosilanes can delay the aging of PE insulating materials in a damp environment (treeing) [207]. Cyanacrylates adhere to deactivating surfaces (e.g., wood) more reliably after the addition of sila crown ethers [208].

6.5. Silanes in Catalysis

Silanes themselves are generally not catalytically active. However, their ability to form complexes with active metal compounds, whereby the catalyst system displays specific action, and the fixing of catalytically active centers to sil(ox)ane structures are of interest.

Donor Silanes. In the production of polypropylene with classical Ziegler–Natta catalysts the desired isotactic material is formed in unsatisfactory proportions. By using alkoxysilanes as external donors [209] an increase to >96% isotactic material (heptane extraction) is achieved. The total catalyst requirement can be decreased to less than 1:20 000, so that separation from the polymer is unnecessary. Silanes $(CH_3O)_2SiRR^1$ with at least one sterically demanding group (e.g., R = cyclo-C_6H_{11}, R^1 = Me) [*17865-32-6*] have been shown to be the most effective donors. Bulk polymerization (Spheripol, Addipol, and Catalloy processes) and gas-phase processes (Unipol) are widely carried out by using these cocatalysts.

Siloxanes in Other Catalyst Systems. In high-temperature vulcanizing silicone rubbers, the Karstedt catalyst [60] is currently the most important platinum complex. It contains vinyl siloxane(s) as the complex-forming components, which give rise to the extremely high activity.

Mechanically tough, spherically uniform micro- and mesoporous particles can be produced by hydrolysis of suitable organofunctional silanes. They adsorb traces of metal or organic materials from (aqueous) solutions [210]. If they are coated with (noble) metal ions, highly active catalyst systems are formed [123]. Catalytically active groups can also be bonded to standard carriers via ambivalent silanes. For example, silane quats fixed on aluminosilicate (see Section 6.2) are excellent for the diproportionation of $MeHSiCl_2$ to MeH_2SiCl [*993-00-0*] or $MeSiH_3$ [*992-94-9*] [211].

In group-transfer polymerization [212] the reaction of α,β-unsaturated carbonyl monomers [e.g., (meth)acrylates] is initiated by *O*-silylketene acetals and traces of certain anions at room temperature. This relatively new process works quasi-anionically and is a living polymerization, whose end products have a very narrow molecular mass distribution. They are used as paint additives.

6.6. Silanes for Use in Liquid Media

As a result of their high thermal stability, low pour points, and relatively flat temperature–viscosity curves, silicic acid esters of higher alcohols and peralkylated silanes have been proposed as high-performance lubricants for air- and spacecraft [213]. Phenylmethyltrisiloxanes are used as diffusion pump oils, and methylsilane glycol esters have been developed as brake fluids for auto racing [214]. For durability, modern car engines (specifically those made from light alloys) require corrosion inhibition on the interior metal walls. To prevent gel formation, various organofunctional silanes are added to the water–glycol cooling medium, whose pH is adjusted to slightly alkaline [215]. The addition of water-soluble organofunctional silanes to detergents [216] improves dirt removal from hard surfaces (rinsing agents) and lowers surface corrosion in enameled dishwashers, washing machines, and dryers. Effective wetting additives are of interest in agriculture, e.g., the glycol-modified trisiloxane $Me_3SiOSi(Me)$ $[(CH_2)_3(OCH_2CH_2)_nOCH_3]$ $OSiMe_3$ (n = ca. 8) [87244-72-2 *]. It temporarily lowers the surface tension of water (at 0.1 % addition) to < 21 mN/m [217].

6.7. Silane Coupling Agents

Of all the organosilanes the silane coupling agents have resulted in the most important technological progress. From the early 1950s they were intensively developed, starting with the vinyl- and aminoalkoxysilanes, and they currently influence nearly all areas of technology, including the medical and analytical branches [114]. The field of silane coupling agents is covered by extensive literature [83], [84], [218], [219]. Characterization, synthesis, and some product-specific remarks on the use of coupling agents can be found in Chapter 4. Here, only the mode of action and the commercial products (Table 7) are mentioned and more recent directions of development are indicated.

Silane coupling agents are usually applied as prehydrolyzed, dilute (ideally 0.1–0.5 %, aqueous) solution, but also from the gas phase, by spraying, dipping, or direct mixing and (less often) in the form of primers (silane partial hydrolysates) on surfaces, fibers, or filler particles. Typical positively reacting inorganic [83] substrates are glass (fibers, hollow spheres), quartz (powder), (synthetic) silicas, wollastonite (calcium metasilicate), mica, kaolin, hydrated aluminum oxide, and among the metals, aluminum, steel, zinc, and copper (alloys) [220]. Additional complex-forming interactions of NH, SH, or COOH groups are believed to occur on metal surfaces.

Table 7. Organosilicon coupling agents in general use

Functionality	Chemical structure	CAS no.
Vinyl	$(MeO)_3SiCH=CH_2$	[2768-02-7]
	$(MeOC_2H_4O)_3SiCH=CH_2$	[1067-53-4]
Chloro	$(MeO)_3SiC_3H_6Cl$	[2530-87-2]
Monoamino	$(EtO)_3SiC_3H_6NH_2$	[919-30-2]
Diamino	$(MeO)_3SiC_3H_6NHC_2H_4NH_2$	[1760-24-3]
Mercapto	$(MeO)_3SiC_3H_6SH$	[4420-74-0]
Polysulfide	$[(EtO)_3SiC_3H_6]_2S_x$ (x = ca. 4)	[40372-72-3]
Epoxy	$(MeO)_3SiC_3H_6OCH_2CH\!-\!\!\underset{O}{\overset{}{\diagdown\!\!\diagup}}\!CH_2$	[2530-83-8]
	$(MeO)_3SiC_2H_4\text{-}\langle\text{cyclohexyl-epoxide}\rangle$	[3388-04-3]
Methacrylic	$(MeO)_3SiC_3H_6OCOC(CH_3)=CH_2$	[2530-85-0]
Styrylcationic	$(MeO)_3SiC_3H_6NHC_2H_4NHCH_2\langle\text{-}\rangle CH=CH_2 \cdot HCl$	[34937-00-3]

The function Y must generally be adapted to the polymer type. Epoxy and aminoorganosilanes can be used quite universally. Reinforced or filled thermosets, thermoplastics, and elastomers become industrially and economically more competitive by using silane coupling agents. Producers' leaflets give more detailed information on the areas of use and handling. The following is valid as a chemical basis for the coupling-promoting effect of organofunctional silanes with the structure $(RO)_3SiCH=CH_2$ or $(RO)_3SiR^1Y$ (Table 7) at interfaces between mineral or metallic substances and organic polymers. Silanols formed by optimal hydrolysis (see Section 3.1) of functional alkoxysilanes condense with the elimination of water, and strong MOSi bonds are formed with the hydroxyl containing substrate surface (M = Si, Al, Fe, etc.) and by the silane molecules with one another. A flexible siloxane intermediate phase, occasionally of varying thickness, must be formed, whose functional groups on the side facing the polymer can become bonded to the organic phase (covalent chemical bonding and interpenetration). The basic mechanisms and their effect on the solution of coupling problems are currently being investigated physicochemically [221]. Nevertheless, relationships between theory and activity are not always clear, so a test of practical application must be the deciding factor. Occasionally, silanes also act between different polymers. Compounds with fewer than three OR leaving groups can also be active but are rarely used in practice (to some extent for cost reasons, cf. Section 4.2), although their more hydrophobic intermediate phase increases their resistance to water.

Mode of Action. Reaction of the silanol groups of the intermediate phase with the OH groups of the inorganic substrate strongly alters the polarity and wettability of the latter, so that the polymeric binding agents, adhesives, or coatings, which are generally of low polarity, can spread and adhere better. As a result of the breakup of hydrogen bonds, penetration of the boundary layer by the diffusion of surrounding moisture is strongly hindered. Thus, consumer articles made of such fiber- or particle-reinforced composites (boats, tanks for liquids, sports articles) have a considerably increased lifetime. Often the mechanical properties can be clearly improved, even in the dry state, and the filler content increased (more economical at the same performance level), or areas of use and temperature ranges can be extended. The electrical values (dielectric strength, loss factor tan δ) are higher and less affected by the entry of water.

At the performance level attained today, improvements in activity can be achieved with new silane structures only in individual cases. Furthermore, restrictive laws in industrial nations are forcing deeper study of interaction phenomena in the combined use of known basic materials [219], [222]. Problems associated with the binding of mineral fibers and fillers in bulk plastics such as polyethylene or polypropylene [223], the incorporation of nonoxide fillers such as calcium carbonate (organotitanates, for example, are used here), and the reinforcement of high-temperature polymers, such as polycarbonates, polyimides, or polysulfones, have still not been solved satisfactorily. The standard silanes (Table 7) tend to yellow above 130–180 °C and bring about structural degradation. More recent developments include temperature-stable aromatic silanes containing carboxyl groups, for example [138], multifunctional silanes [224], stable aqueous silane solutions free of alcohol groups, or "polymeric silanes" [225]. If economic considerations are not taken into account, cationic, silanized, finely divided filler particles could become important as additives for washing powders, particularly for hard water [226].

6.8. Silanes for Hydrophobic Surface Treatment

The modification of (mineral) surfaces with nonfunctionalized silanes has been studied to a similar extent as the use of silane coupling agents [227]. The aim of treatment (generally with inexpensive alkylchloro- or alkylalkoxysilanes) is the elimination of polar HOM groups, the removal of adsorbed moisture, or improved wetting by organic media. Hydrophilic surfaces become hydrophobic and lipophilic. This is particularly effective for particle fillers, increasing the maximum degree of filling or improving mechanical, electrical, and chemical resistance. Silanized silicas attract much interest for the thixotropic modification of certain epoxide resins and especially for improving the mechanical properties of silicone rubber. A compound produced by cocondensation of $Si(OEt)_4$ and $Me_3SiOSiMe_3$ is claimed to exhibit particularly good properties [228]. Another example of this use of silicon compounds is in the protection of building materials of all types (masonry, plaster, natural stone, concrete) against weathering. Alkylalkoxysilanes [229] or silicone microemulsions [230] are used for this purpose, for example. The current state of the art is represented by the water-repellent silanization of glass instruments for chemical and medical use (e.g., ampules) or in the production of light bulbs. The solvent-, dirt-, and fat-repellent (soil release effect) action of alkylsilyl groups is surpassed by certain fluoroalkylsilanes (see Section 4.2).

Small quantities of volatile sil(ox)anes (e.g., $Me_3SiOSiMe_3$) are used in plasma chemical vapor deposition. The extremely thin polymeric silicon-containing layers formed, protect, for example, vapor-deposited aluminum headlight reflectors [231] and are used for the production of microstructures on electronic chips [232].

6.9. Analysis

The analysis of silanes and silicones is described in [233]. The use of organosil(ox)-anes in analysis and preparative substance separation is of growing importance (currently several thousand kilograms per year). In liquid chromatography (reversed phase HPLC) of mixtures of substances [234], stationary phases of defined polarity that are attached by stable bonds to the silica gel carrier are produced by using silanes (frequently octyl or octadecyl, but also organofunctional types). These phases are suitable for the separation of pesticides, protein components, or optical isomers and for ion chromatography. Information is provided by equipment suppliers. The methods for protective silylation described in Section 6.1 enable the analysis of sensitive organic (natural) products as well as short-lived silanols by GC or GC/MS (Lentz technique). Even mineral analyses are possible by using Me_3Si derivatization [235]. Traces of silicon in the environment can be characterized by $Me_3SiOSiMe_3$ – catalyst [233], and surfaces contaminated with silicones can be cleaned by using the same mixture.

7. Toxicology and Environmental Aspects

The physiological action of silanes is strongly structure dependent. An overall description is not possible, so in this chapter only some important aspects are discussed.

Tetraorganosilanes are nontoxic, provided they do not contain any toxic organofunctional substituents. However, for silicon-functional compounds, toxicity must always be a consideration, particularly when they contain readily hydrolyzable groups. This leads to attack on the mucous membranes and the eyes, particularly at high volatility. Organohalosilanes can also cause severe skin irritation or corrosion, due to the hydrogen halides liberated on exothermic hydrolysis and the associated localized dehydration. In the case of contamination, washing with copious amounts of water is indispensable as a first-aid measure.

Organoalkoxysilanes do not generally exhibit pronounced toxicity; however, the formation of alcohol during hydrolysis must be considered. The widely used tetraethoxysilane has a TLV of 10 ppm. In comparison, the volatile compounds $Si(OMe)_4$ (TLV = 1 ppm) and $HSi(OMe)_3$ [85], [236] are highly toxic. On inhalation, severe headache and visual disorders occur, which can lead to permanent blindness. Liver and kidney damage has also been reported [237].

Organofunctional silanes are generally considered of low toxicity. Depending on the substituents, special effects can also occur [188]. In the case of physiologically active silicon compounds (Section 6.2) these properties are specifically exploited.

Widely varying toxicity can be found in series of silanes that contain both organofunctional and silicon-functional groups. The combination of amino and alkoxy groups in particular, present in silatranes, for example [25], [238], leads to quite pronounced toxicity in some cases.

A considerable proportion of physiologically active organosilicon compounds is processed inhouse to polymers that are of low activity or harmless. During production and transport, attention must be given to safe handling of the monomers. Depending on the vapor pressure, flammability, corrosivity, etc., suitable precautions must be taken, which are set out in safety data sheets. In industry, halosilanes should be stored in double-walled steel tanks under an inert gas. Laboratory quantities are kept in glass vessels; plastic containers are generally unsuitable.

8. Economic Aspects

The main source of monomeric organosilicon compounds is the crude silane mixture produced by the direct synthesis. Its production has increased greatly during the last decade. In 1977 the quantity of crude silane was 380 000 t; in 1985, 600 000 t; and in 1991, ca. 10^6 t. About 95% of this is used for the production of silicones. This explains why the largest silicone producers are also the most important producers of monomeric silanes. In the United States, these companies are Dow Corning, General Electric, and Union Carbide; in Europe, Bayer (Germany), Hüls (Germany), Rhône-Poulenc (France), and Wacker-Chemie (Germany); in Japan, Shin-Etsu, Toray (joint enterprise with Dow Corning), and Toshiba (joint enterprise with General Electric). Most of these companies operate worldwide; the assignment by country is made by head office. There are also production units in the CIS and in China.

According to estimates, ca. 40% of the world market is accounted for by the United States, 30% by Western Europe, 15% by Japan, and 15% by the rest of the world. The total turnover in silicones was ca. 7.5×10^9 DM in 1991 [239]. Besides the crude silane mixture, essentially only trichlorosilane is used as a starting material for other silanes, especially organofunctional silanes; ca. 30 000 t of $HSiCl_3$ is estimated to have been used for this purpose in 1991. The high-value silanes obtained by this route (price per kilogram between 10 and 60 DM; for special silanes, even higher) are used, among other things, as coupling agents, cross-linking agents, and for surface treatment. The world market for these applications was estimated to be 10^9 DM in 1991 (United States 46%, Europe 46%, Japan 8%) [240]. These types of silanes are produced also by other companies besides silicone producers, such as PCR Chemicals (United States), Degussa (Germany), and Chisso (Japan).

Other applications of monomeric organosilanes are of comparatively minor economic importance. Silylating agents, which are mainly used as reagents in the pharmaceutical industry, had a market volume of 75×10^6 DM (ca. 7000 t) in 1991. Organosilanes used as catalyst components in polypropylene production had a turnover of ca. 15×10^6 DM (400 t) in 1991.

9. References

General References

[1] A. R. Bassindale, P. P. Gaspar (eds.): *Front. Organosilicon Chemistry*, R. Soc. Chem., Cambridge, U.K., 1991.
[2] S. Patai, Z. Rappoport (eds.): *The Silicon-Heteroatom Bond*, Wiley-Interscience, Chichester 1991.
[3] S. Patai, Z. Rappaport (eds.): *The Chemistry of Organic Silicon Compounds*, **2** vols., Wiley-Interscience, Chichester 1989.
[4] J. M. Zeigler, F. W. G. Fearon (eds): "Silicon-Based Polymer Science," *Adv. Chem. Ser.* **224** (1990).
[5] E. Lukevics, O. Pudova, R. Sturkovich: *Molecular Structure of Organosilicon Compounds*, Ellis Horwood, Chichester 1989.
[6] J. Y. Corey, E. R. Corey, P. P. Gaspar (eds.): *Silicon Chemistry*, Ellis Horwood, Sussex 1988.
[7] E. G. Rochow: *Silicon and Silicones*, Springer Verlag, Berlin 1987.
[8] S. Pawlenko: *Organosilicon Chemistry*, W. de Gruyter, Berlin 1986.
[9] H. Sakurai (ed.): *Organosilicon and Bioorganosilicon Compounds*, Ellis Horwood, Chichester 1985.
[10] D. A. Armitage: "Organosilanes," in G. Wilkinson, F. G. A. Stone, E. W. Abel (eds.): *Comprehensive Organometallic Chemistry*, vol. **2**, Pergamon Press, Oxford 1982, pp. 3–204.
[11] F. O. Stark, J. R. Falender, A. P. Wright: "Silicones," in G. Wilkinson, F. G. A. Stone, E. W. Abel (eds.): *Comprehensive Organometallic Chemistry*, vol. **2**, Pergamon Press, Oxford 1982, pp. 306–363.
[12] *Winnacker-Küchler*, 4th ed., **6,** 816–852.
[13] *Houben-Weyl*, 4th ed., **13/5,** 1–502.
[14] R. J. Voorhoeve: *Organosilanes, Precursors to Silicones*, Elsevier, Amsterdam 1967.

Specific References

[15] C. Friedel, J. M. Crafts, *Justus Liebigs Ann. Chem.* **127** (1863) 28–32.
[16] C. Friedel, J. M. Crafts, *Justus Liebigs Ann. Chem.* **136** (1865) 203–211.
[17] A. Ladenburg, *Justus Liebigs Ann. Chem.* **173** (1874) 143–166; C. Pape, *Justus Liebigs Ann. Chem.* **222** (1884) 354–374.
[18] F. S. Kipping, *Proc. R. Soc. London Ser. A* **159** (1937) 139–147.
[19] W. Dilthey, F. Eduardoff, *Ber. Dtsch. Chem. Ges.* **37** (1904) 1139–1142; F. S. Kipping, *Proc. Chem. Soc. London* **20** (1904) 15–16.
[20] In [7] pp. 64–80.
[21] General Electric, US 2 380 995, 1941 (E. G. Rochow).
[22] VEB Silikonchemie, DD 5448, 1942 (R. Müller).

[23] L. H. Sommer, E. W. Pietrusza, F. C. Whitmore, *J. Am. Chem. Soc.* **69** (1947) 1881; C. A. Burkhard, R. H. Krieble, *J. Am. Chem. Soc.* **69** (1947) 2687–2689.
[24] J. L. Speier, J. A. Webster, G. H. Barnes, *J. Am. Chem. Soc.* **79** (1957) 974–979.
[25] M. G. Voronkov, V. M. Dyakov, S. V. Kirpichenko, *J. Organomet. Chem.* **233** (1982) 1–147.
[26] R. J. P. Corriu, J. C. Young in [2] pp. 1–66.
[27] J. Y. Corey in [3] pp. 1–56.
[28] In [14] pp. 186–219.
[29] R. Müller, *Z. Chem.* **25** (1985) 309–318.
[30] G. Fritz, *Angew. Chem.* **99** (1987) 1150–1171; *Angew. Chem. Int. Ed. Engl.* **26** (1987) 1111–1132.
[31] In [14] pp. 244–282.
[32] M. P. Clarke, *J. Organomet. Chem.* **376** (1989) 165–222.
[33] M. P. Clarke, I. M. T. Davidson, *J. Organomet. Chem.* **408** (1991) 149–156.
[34] L. D. Gasper-Galvin, D. M. Sevenich, H. B. Friedrich, D. G. Rethwisch, *J. Catal.* **128** (1991) 468–478.
[35] W. J. Ward, A. Ritzer, K. M. Carroll, J. W. Flock, *J. Catal.* **100** (1986) 240–249.
[36] Dow Corning, US 4898960, 1986 (V. D. Dosaj, R. L. Halm, O. K. Wilding).
[37] General Electric, US 2464033, 1949 (W. F. Gilliam).
[38] Wacker Chemie, US 2877254, 1959 (E. Enk, S. Nitzsche).
[39] General Electric, US 2887501, 1959 (B. A. Bluestein).
[40] Union Carbide, US 4973725, 1989 (K. M. Lewis, R. A. Cameron, J. M. Larnerd).
[41] Dow Corning, US 4966986, 1989 (R. L. Halm, R. H. Zapp).
[42] M. G. R. T. de Cooker, J. H. N. de Bruyn, P. J. van den Berg, *J. Organometal. Chem.* **99** (1975) 371–377.
[43] VEB Nünchritz, DD 250536, 1986 (W. Walkow et al.).
[44] Bayer, DE 3841417, 1988 (K. Feldner, B. Degen, G. Wagner, M. Schulze).
[45] R. Schliebs, J. Ackermann, *Chem. Unserer Zeit* **21** (1987) 121–127.
[46] J. Y. Corey in [3] p. 24.
[47] Union Carbide, US 4593114, 1985 (M. L. Kenrick, B. Kanner).
[48] H. Normant, C. R. Hebd. Seances Acad. Sci. **239** (1954) 1510–1512.
[49] H. Gilman, E. A. Zuech, *J. Am. Chem. Soc.* **79** (1957) 4560–4561.
[50] V. I. Zhun, M. K. Ten, *Khim. Promst. (Moscow)* 1989, 15–18; *Chem. Abstr.* **111** (1989) 153903 g.
[51] Dow Corning Toray, EP-A 405560, 1990 (A. Shirahata). A. Shirahata, *Tetrahedr. Lett.* **30** (1989) 6393–6394.
[52] Schering, EP-A 348693, 1989 (J. Graefe, W. Uzick, U. Weinberg).
[53] M. Weidenbruch, K. Kramer, *J. Organomet. Chem.* **291** (1985) 159–163.
[54] H. Gilman, S. P. Massie, *J. Am. Chem. Soc.* **68** (1946) 1128.
[55] M. G. Voronkov, N. G. Romanova, L. G. Smirnova, *Collect. Czech. Chem. Commun.* **23** (1959) 1013.
[56] Dow Corning, US 4888435, 1989 (K. M. Chadwick, R. L. Halm, B. R. Keyes).
[57] Union Carbide, US 2637738, 1953 (G. H. Wagner).
[58] J. L. Speier, *Adv. Organometal. Chem.* **17** (1979) 407–447. B. Marciniec (ed.): *Comprehensive Handbook on Hydrosilylation*, Pergamon Press, New York 1992.
[59] A. J. Chalk, J. F. Harrod, *J. Am. Chem. Soc.* **87** (1965) 16–21.
[60] P. B. Hitchcock, M. F. Lappert, N. J. W. Warhurst, *Angew. Chem.* **103** (1991) 439–441; *Angew. Chem. Int. Ed. Engl.* **30** (1991) 438–440.
[61] L. N. Lewis, R. J. Uriarte, N. Lewis, *J. Catal.* **127** (1991) 67–74.

[62] L. N. Lewis, K. G. Sy, G. L. Bryant, P. E. Donahue, *Organometallics* **10** (1991) 3750–3759.
[63] B. Marciniec, E. Máckowska, J. Guliúski, W. Urbaniak, *Z. Anorg. Allg. Chem.* **529** (1985) 222–228.
[64] In [11] pp. 310–313.
[65] I. Ojima in [3] pp. 1479–1526.
[66] In [12] p. 827.
[67] Hüls, EP-A 438 666, 1990 (W. Hange, H. Dietsche, C.-D. Seiler).
[68] In [11] pp. 317, 318.
[69] In [11] pp. 307–309.
[70] V. D. Sheludrakov, V. I. Zhun', M. K. Ten, *Zh. Obshch. Khim.* **57** (1987) 567–571; *J. General Chem. USSR (Engl. Transl.)* **57** (1987) 495–499.
[71] Bayer, DE-OS 3 410 644, 1984 (K. Feldner, W. Grape).
[72] O. L. Flaningham in A. L. Smith (ed.): *The Analytical Chemistry of Silicones*, Wiley & Sons, New York 1991, pp. 523–541.
[73] A. L. Bassindale, P. G. Taylor in [3]pp. 839–892.
[74] D. R. Weyenberg, L. G. Mahone, W. H. Atwell, *Ann. N. Y. Acad. Sci.* **159** (1969) 38–55.
[75] G. A. Olah, G. K. Prakash, R. Krishnamurti in G. L. Larson (ed.): *Advances in Silicon Chemistry*, vol. **1**, JAI Press, Greenwich 1991, pp. 1–64.
[76] Ciba-Geigy, EP-A 177 454, 1985 (K. Oertle, H. Wetter).
[77] General Electric, US 4 395 563, 1981 (S. E. Hayes).
[78] Dow Corning, BP 914 460, 1961 (J. L. Speier).
[79] J. L. Speier, D. L. Kleyer, *Organometallics* **10** (1991) 3046–3049.
[80] Wacker Chemie, US 4 298 753, 1980 (A. Schinabeck et al.).
[81] R. J. P. Corriu, C. Guerin, B. J. L. Henner, Q. Wang, *Organometallics* **10** (1991) 3200–3205.
[82] Bayer, EP-A 4 310, 1979 (O. Schlak, H.-H. Moretto).
[83] E. P. Plueddemann: *Silane Coupling Agents*, 2nd ed., Plenum Press, New York 1991 (and references cited therein).
[84] F. D. Blum, W. Meesiri, H.-J. Kang, J. E. Gambogi, *J. Adhes. Sci. Technol.* **5** (1991) 479–496. F. D. Osterholtz, E. R. Pohl, *J. Adhes. Sci. Technol.* **6** (1992) 127–149.
[85] G. J. Kallos, J. C. Tou, R. M. Malczewski, W. F. Boley, *Am. Ind. Hyg. Assoc. J.* **52** (1991) 259–262.
[86] D. E. Leyden, J. B. Atwater, *J. Adhes. Sci. Technol.* **5** (1991) 815–829.
[87] R. H. Chung, S. E. Hayes, *J. Organomet. Chem.* **265** (1984) 135–139; Dow Corning, US 4 732 996, 1987 (K. W. Moorhead, K. L. Reading, D. J. Rengering, A. P. Wright).
[88] Toray, US 5 045 621, 1989 (T. Suzuki).
[89] Wacker Chemie, US 4 113 760, 1977 (V. Frey et al.).
[90] Wacker Chemie, DE-OS 3 716 372, 1987 (C. Trieschmann, J. Müller, K. H. Wegehaupt); Shin-Etsu, EP-A 435 654, 1990 (Y. Yamamoto, T. Matsuda).
[91] P. V. Wright in K. J. Ivin, T. Saegusa (eds.): *Ring-Opening Polymerisation*, vol. **2**, Elsevier, London/New York 1984, pp. 1055–1133.C. J. C. Edwards, R. F. T. Stepto in J. A. Semlyen (ed.): *Cyclic Polymers*, Elsevier, London/New York 1986, pp. 85–166.
[92] Shin-Etsu, US 3 819 563, 1974 (T. Takago, T. Sato, H. Aoki).C. Rochin, O. Babot, R. Duboudin, *J. Organomet. Chem.* **281** (1985) C 24–C 28; General Electric, US 4 210 596, 1978 (J. A. Cella).
[93] R. J. Ayen, J. H. Burk, *Mater. Res. Soc. Symp. Proc.* **73** (1986) 801–808.
[94] G. B. Goodwin, M. E. Kenney, *Inorg. Chem.* **29** (1990) 1216–1222.
[95] Wacker Chemie, US 4 209 454, 1978 (W. Graf, V. Frey, P. John, N. Zeller).
[96] M. Roth, *Bautenschutz + Bausanierung* **2** (1979) 12–15.

[97] E. le Coz, *DEFAZET Dtsch. Farben Z.* **32** (1978) no. 12, 56–60.
[98] Wacker Chemie, US 4 176 130, 1978 (P. John, W. Feichtner, W. Graf, V. Frey). General Electric, US 4 329 484, 1981 (L. P. Petersen).
[99] Rhone-Poulenc, US 4 918 209, 1988 (P. Baule, F. Chizat). Hüls, EP-A 381 840, 1989 (G. Zoche).Wacker Chemie, DE 4 104 725, 1991 (P. John, C. Braunsperger).
[100] Allied, US 4 400 527, 1981 (C. T. Mathew, H. E. Ulmer).
[101] K. C. Pande, R. E. Ridenour, *Chem. Ind.* (*London*) **1970**, 56. Shin-Etsu, JP 51 (76)–19 728, 1974 (M. Takamizawa, Y. Inone, H. Yoshioka); *Chem. Abstr.* **85** (1976) 33181 t.
[102] D. A. Armitage in [2] pp. 367–394 and 448–466.
[103] Union Carbide, US 4 400 526, 1982 (B. Kanner, C. L. Schilling, S. P. Hopper). General Electric, US 4 631 346, 1985 (J. L. Webb, C. E. Olsen).
[104] General Electric, US 4 602 094, 1985 (T. D. Mitchell), and references cited therein. Bayer, US 4 739 088, 1986 (R. Endres, A. de Montigny).
[105] Wacker Chemie, US 3 992 428, 1976 (H. Müller, I. Bauer, E. Schmidt, R. Riedle). Gist-Brocades, EP 43 630, 1981 (C. A. Bruynes, T. K. Jurriens).
[106] Dynamit Nobel, DE-OS 3 443 960; DE 3 443 961, 1984 (H.-J. Kötzsch, H.-J. Vahlensieck).
[107] Th. Goldschmidt, DE-AS 2 645 703, 1976 (G. Körner, H.-J. Patzke).
[108] Th. Goldschmidt, EP-A 266 633, 1987 (G. Körner, G. Weitemeier, D. Wewers). Dow Corning, US 5 015 624, 1990 (W. J. Schulz).
[109] J. Y. Corey in G. L. Larson (ed.): *Advances in Silicon Chemistry*, vol. **1**, JAI Press, Greenwich 1991, pp. 327–387.
[110] Dynamit Nobel, US 4 016 188, 1975 (H.-J. Kötzsch, H.-J. Vahlensieck).
[111] E. L. Zichy, *J. Organomet. Chem.* **4** (1965) 411–412.
[112] Imperial Chemical Ind. GB 1 152 251, 1966 (R. M. Gibbon, E. K. Pierpoint).
[113] General Electric, US 3 441 583, 1965 (R. A. Murphy).
[114] E. P. Plueddemann, *J. Adhes. Sci. Techn.* **5** (1991) 261–277.
[115] Rhone-Poulenc, US 3 404 169, 1964 (M. Gaignon, M. Lefort).
[116] Wacker Chemie, US 3 793 358, 1972 (S. Nitzsche, I. Bauer, W. Graf, N. Zeller).
[117] T. K. Sarkar, *Synthesis* 1990, 969–983.
[118] N. Furuya, T. Sukawa, *J. Organomet. Chem.* **96** (1975) C 1–C 3.
[119] H. Matsumoto et al., *J. Organomet. Chem.* **148** (1978) 97–106.
[120] Dow Corning, US 4 609 574, 1985 (J. R. Keryk, P. Y. K. Lo, L. E. Thayer).
[121] Wacker Chemie, EP-A 396 130, 1990 (C. Herzig, D. Gilch).
[122] Wacker Chemie, US 5 041 594, 1991 (C. Herzig).
[123] U. Deschler, P. Kleinschmit, P. Panster, *Angew. Chem.* **98** (1986) 237–253; *Angew. Chem. Int. Ed. Engl.* **25** (1986) 236–252.
[124] In [11] pp. 314–316.
[125] Union Carbide, US 4 658 050, 1986 (J. M. Quirk, B. Kanner).
[126] T. Kobayashi, K. H. Pannell, *Organometallics* **10** (1991) 1960–1964.
[127] Wacker Chemie, DE 3 144 020, 1981 (R. Artes, V. Frey, P. John, M. Scherer).
[128] M. J. Gregory, *J. Chem. Soc., Perkin Trans. II* 1973, 1699–1702.
[129] M. J. Owen, D. E. Williams, *J. Adhes. Sci. Techn.* **5** (1991) 307–320.
[130] Union Carbide, US 2 930 809, 1956 (V. B. Jex, D. L. Bailey).
[131] R. A. Benkeser, K. M. Foley, J. B. Grutzner, W. E. Smith, *J. Am. Chem. Soc.* **92** (1970) 697–698.
[132] A. B. Rajkumar, P. Boudjouk, *Organometallics* **8** (1989) 549–550.
[133] Union Carbide, US 3 595 897, 1968 (E. S. Brown, E. A. Rick, F. D. Mendicino).

[134] Union Carbide, US 4 556 722, 1985 (J. M. Quirk, S. Turner).
[135] Dynamit Nobel, DE-AS 2 749 316, 1977 (F.-R. Kappler, C.-D. Seiler, H.-J. Vahlensieck).
[136] General Electric, DE-OS 3 546 376, 1985 (J. L. Webb, E. Cathryn).
[137] H. Yinglin, H. Hongwen, *Synthesis* 1990, 122–124.
[138] J. D. Rich, *J. Am. Chem. Soc.* **111** (1989) 5886–5893.
[139] Dow Corning, US 3 560 385, 1968 (C. A. Roth).
[140] J. E. Bittell, J. L. Speier, *J. Org. Chem.* **43** (1978) 1687–1688.
[141] Dynamit Nobel, DE 3 346 910, 1983 (C.-D. Seiler, H.-J. Vahlensieck).
[142] Union Carbide, DE-OS 2 300 912, 1973 (J. Y. Pui Mui).
[143] Phillips Petroleum, US 3 890 213, 1973 (R. P. Louthan).
[144] Degussa, DE-OS 3 226 093, 1982 (P. Panster, H. Grethe, P. Kleinschmit).
[145] Wacker Chemie, DE-OS 4 135 170, 1991 (R. Hager, B. Deubzer, J. Wolferseder).
[146] Minnesota Mining, US 4 267 213, 1979 (B. R. Beck, F. T. Sher, G. V. D. Tiers).
[147] Wacker Chemie, DE-OS 4 135 142, 1991 (R. Hager, B. Deubzer).
[148] Union Carbide, EP-A 262 642, 1987 (J. M. Quirk, B. Kanner).
[149] Degussa, DE-OS 3 528 006, 1985 (U. Deschler, A. Grund, G. Prescher).
[150] Wacker Chemie, DE 4 128 894, 1991 (C. Herzig, D. Gilch, J. Bindl).
[151] P. R. Hudrlik, E. L. O. Agwaramgbo in [6] pp. 95–104.
[152] Dow Corning, US 4 780 555, 1988 (H. M. Bank).
[153] Rhône-Poulenc, US 4 940 766, 1987 (M. Gay, E. Canivenc).
[154] Dow Corning, US 4 861 907, 1989 (A. P. Wright, D. J. Bunge).
[155] S. A. Swint, M. A. Buese, *J. Organomet. Chem.* **402** (1991) 145–153.
[156] Wacker Chemie, DE-OS 3 301 807, 1983 (T. Lindner, P. John, N. Zeller, R. Riedle).
[157] Wacker Chemie, US 5 015 700, 1990 (C. Herzig, J. Esterbauer).
[158] K. Takeshita et al., *J. Org. Chem.* **52** (1987) 4864–4868.
[159] Dow Corning, US 4 788 313, 1983 (G. Chandra, D. R. Juen).
[160] Dow Corning, US 4 329 483, 1981 (J. Speier).
[161] Dow Corning, EP-A 196 169, 1986 (K. J. Woodward, R. M. Edmund).
[162] Dow Corning, US 4 847 397, 1988 (F. Sawaragi, H. Taniguchi).
[163] Dow Corning, US 4 599 438, 1983 (J. W. White, S. Westall, B. J. Griffiths).
[164] G. H. Barnes, M. P. David, *J. Org. Chem.* **25** (1960) 1191–1194.
[165] Wacker Chemie, DE-AS 2 219 983, 1972 (G. Künstle, H. Liberda, H. Spes).
[166] M. Czakova, M. Capka, *J. Mol. Catal.* **11** (1981) 313–322.
[167] In [13] pp. 31–78.
[168] Ethyl Corp., US 4 711 965; 4 711 966, 1987 (G. E. Nelson).
[169] In [8] pp. 124–127. J. Ipaktschi, A. Heydari, *Angew. Chem.* **104** (1992) 335–336; *Angew. Chem. Int. Ed. Engl.* **31** (1992) 331.
[170] M. Bordeau, S. M. Djamei, R. Calas, J. Dunogues, *J. Organomet. Chem.* **288** (1985) 131–138.
[171] S. Ochsenbein, D. Schinzel, A. Gonidec, W. F. Schmidt, *Nucl. Instrum. Methods Phys. Res., Sect. A* **273** (1988), 654–656; *Chem. Abstr.* **110** (1989) 65467 s.
[172] Rhône-Poulenc, EP-A 255 453, 1988 (J. S. Ferlet).
[173] A. Kunai, T. Kawakami, E. Toyoda, M. Ishikawa, *Organometallics* **10** (1991) 2001–2003.
[174] Dow Corning, US 4 534 948, 1981 (R. H. Baney).
[175] Wacker Chemie, EP 214 664, 1986 (V. Frey, B. Pachaly, N. Zeller).
[176] Y. Ito in [1] pp. 391–398.
[177] J. Lipowitz, *Am. Ceram. Soc. Bull.* **70** (1991) 1888–1894.

[178] K. S. Mazdiyasni (ed.): *Fiber Reinforced Ceramic Composites-Materials*, Processing and Technology, Park Ridge (USA), 1990.

[179] R. West in [3] pp. 1207–1240.

[180] A. Tomanek: *Silicone und Technik: Ein Kompendium für Praxis, Lehre und Selbststudium*, Hanser Verlag, München 1990. A. Tomanek:
Silicones and Industry: A Compendium for Practical Use, Instruction and Reference, Hanser Verlag, München 1992.

[181] H. Menzel: *Pharmaceutical Manufacturing International*, Sterling Publication, London 1989, pp. 129–134. P. Kochs, *Chem. Ztg.* **113** (1989) 225–238.

[182] Dynamit Nobel, DE-OS 3 151 677, 1981 (H.-J.Kötzsch, H.-J. Vahlensieck).

[183] M. Lalonde, T. H. Chan, *Synthesis* 1985, 817–845; H. Wetter, K. Oertle, *Tetrahedron Lett.* **26** (1985) 5515–5518.

[184] W. P. Weber: *Silicon Reagents for Organic Synthesis*, Springer Verlag, Berlin 1983. R. Anderson, *Synthesis* 1985, 717–734. E. W. Colvin:
Silicon in Organic Synthesis, R. E. Krieger Publishing, Malabar 1985. E. W. Colvin:
Silicon Reagents in Organic Synthesis, Academic Press, London 1988. G. L. Larson in [2] pp. 763–808. J. K. Rasmussen in G. L. Larson (ed.): *Advances in Silicon Chemistry*, vol. **1**, JAI Press, Greenwich (1991) pp. 65–187.

[185] I. Fleming in [9] pp. 197–211. R. J. P. Corriu, C. Guerin, J. J. E. Moreau in [2] pp. 305–370. I. Fleming, *Pure Appl. Chem.* **62** (1990) 1879–1886.

[186] A. Rosevear, J. F. Kennedy, J. M. S. Cabral (eds.): *Immobilized Enzymes and Cells*, Adam Hilger, Bristol 1987.

[187] S. Barcza in [6] pp. 135–144.

[188] R. Tacke, H. Linoh in [3] pp. 1143–1206. R. Tacke et al., *J. Organomet. Chem.* **417** (1991) 339–353.

[189] Du Pont, US 4 510 136, 1983 (W. K. Moberg). W. K. Moberg, D. R. Baker, J. G. Fenyes, *ACS Symp. Ser.* **443** (1991) 1–14.

[190] Hoechst, EP-A 224 024, 1986 (H. H. Schubert et al.). Dainippon, JP-Kokai 03 (91)–99 003, 1989 (K. Sugamoto, Y. Namite); *Chem. Abstr.* **115** (1991) 201146 u.

[191] W. C. White, R. L. Gettings in D. E. Leyden (ed.): *Chemically Modified Surfaces*, vol. **1**, Silanes, Surfaces, and Interfaces, Gordon and Breach, London 1986, pp. 107–140. K. J. Hüttinger, M. F. Jung, M. C. Schnell, *Chem.-Ztg.* **114** (1990) 161–165.

[192] Wacker Chemie, US 4 801 673, 1987 (E. Bosch, F. Neuhauser, A. Schiller, O. Sommer).

[193] J. Stein, R. P. Eckberg, *J. Coated Fabr.* **20** (1990) 24–42.

[194] J. V. Crivello, J. L. Lee, *J. Polym. Sci., Polym. Chem. Ed.* **28** (1990) 479–503. General Electric, EP-A 412 430, 1990 (J. V. Crivello, J. L. Lee).

[195] H.-G. Fritz, S. Ultsch in Verein Dt. Ingenieure (ed.): *Polymerreaktionen und reaktives Aufbereiten in kontinuierlichen Maschinen*, VDI-Verlag, Düsseldorf 1988, pp. 243–295. S. Ultsch, H.-G. Fritz, *Plast. Rubber Process Appl.* **13** (1990) 81–91.

[196] Mitsubishi, EP 193 317, 1986 (I. Ishino, A. Ohno, T. Isaka). Nippon Oil, US 4 412 042, 1982 (K. Matsuura, N. Noboru, M. Miyoshi).

[197] Nippon Unicar, EP-A 370 518, 1979 (K. Horita, S. Hayashi, T. Koshijama).

[198] Wacker Chemie, DE-AS 2 148 456–2 148 458, 1971 (E. Bergmeister, P.-G. Kirst, H. Wiest).

[199] Kanegafuchi, US 4 837 401, 1985 (R. Hirose, S. Yukimoto, K. Isayama). Kanegafuchi, US 5 011 900, 1989 (S. Yukimoto, T. Hirose, H. Wakabayashi, K. Isayama).

[200] B. Arkles, *CHEMTECH* **13** (1983) 542–555.

[201] Seiko Epson, US 5 015 523, 1987 (H. Kawashima, M. Nakashima, T. Mogami).

[202] Hercules, US 5 008 360, 1989 (J. K. Bard, J. S. Burnier).Hercules, US 5 013 809, 1989 (R. T. Leibfried).
[203] I. Yilgör, J. E. McGrath, *Adv. Polym. Sci.* **86** (1988) 1–86. W. Gardiner, J. W. White: "High Value Polym.," *Spec. Publ. R. Soc. Chem.* **87** (1991) 98–108.
[204] Shin-Etsu, US 4 201 698, 1978 (K. Itoh, T. Oshima).Mitsui, EP-A 314 396, 1988 (T. Tojo, K. Okamoto, A. Matsuda, E. Louis).
[205] General Electric, EP-A 415 072 and A 415 073, 1990 (L. N. Lewis et al.).
[206] Enichem Sintesi, EP-A 162 523 and A 162 524, 1985 (A. Greco, L. Cassar, C. Neri et al.).
[207] Licentia, DE-OS 3 628 554, 1986 (J. Wartusch). Licentia, DE-OS 3 702 209, 1987 (J. Wartusch, H. Andreß, W. Gölz).
[208] Loctite, US 4 906 317, 1985 (J.-C. Lin).
[209] M. Härkönen, J. V. Seppälä, T. Väänänen, *Makromol. Chem.* **192** (1991) 721–734.
[210] Degussa, EP-A 416 271 and A 416 272, 1990 (P. Panster et al.).
[211] Wacker Chemie, US 4 870 200, 1988 (R. Ottlinger, A. Rengstl, R. Jira).
[212] D. Y. Sogah, W. B. Farnham in [9] pp. 219–230.
[213] V. K. Gupta et al., *Lubr. Eng.* **46** (1990) 706–711. K. J. L. Paciorek et al., *Ind. Eng. Chem. Prod. Res. Dev.* **30** (1991) 2191–2194.
[214] Castrol, US 3 994 948, 1974 and US 4 141 851, 1978 (H. F. Askew, C. J. Harrington, G. J. J. Jayne).
[215] Dow Chemical, EP-A 111 013, 1982 (R. T. Jernigan). Korea advanced Institute (KAIST), US 4 873 011, 1988 (I. N. Jung, S. Y. Hwang, C. S. Lee).
[216] Procter and Gamble, US 4 005 024/25/28/30, 1975 (D. C. Heckert, D. M. Watt et al.); EP 75 986-90, 1982 (C. R. Barrat, J. R. Walker, J. Wevers, H. Ernst).
[217] M. Knoche, H. Tamura, M. J. Bukovac, *J. Agric. Food Chem.* **39** (1991) 202–206.
[218] V. Chvalovsky in V. Chvalovsky, J. M. Belama (eds.): *Organosilicon Compounds,* Plenum Press, New York 1984, pp. 1–33.
[219] G. Tesoro, Y. Wu, *J. Adhes. Sci. Technol.* **5** (1991) 771–784.
[220] P. Walker, *J. Adhes. Sci. Technol.* **5** (1991) 279–305.
[221] K. P. Hoh, H. Ishida, J. L. Koenig, *Polym. Compos.* **9** (1988) 151–157; D. J. Ondrus, F. J. Boerio, K. J. Grannen, *J. Adhes.* **29** (1989) 27–42. W. J. van Ooij, A. Sabata, *J. Adhes. Sci. Technol.* **5** (1991) 843–863. E. Nishio, N. Ikuta, H. Okabayashi, *J. Anal. Appl. Pyrolysis* **18** (1991) 261–268.
[222] P. G. Pape, E. P. Plueddemann, *J. Adhes. Sci. Technol.* **5** (1991) 831–842.
[223] PCR, US 4 975 509, 1988 (W. G. Joslyn, A. D. Ulrich, M. E. Wilson).
[224] Toshiba, DE 4 010 128, 1990 (H. Motegi, T. Sunaga, M. Zanbayashi).Toshiba, JP-Kokai 03 (91)–188 085, 1989 (H. Mogi, T. Sunaga, M. Zenbayashi); *Chem. Abstr.* **116** (1992) 83922 q.
[225] B. Arkles, J. Steinmetz, J. Zazyczny, P. Mehta, *J. Adhes. Sci. Technol.* **6** (1992) 193–206.
[226] E. P. Plueddemann in D. E. Leyden, W. T. Collins (eds.): *Chemically Modified Surfaces,* vol. **3**, "Chemically Modified Oxide Surfaces," Gordon and Breach, London 1990, pp. 281–294.
[227] D. E. Leyden, W. T. Collins (eds.): *Chemically Modified Surfaces,* vol. **2**, "Chemically Modified Surfaces in Science and Industry," Gordon and Breach, London 1988; H. A. Mottola, J. R. Steinmetz (eds.), *Chemically Modified Surfaces,* vol. **4**, Elsevier, Amsterdam 1992, supplements the series of monographs on silanes on surfaces cited in [196], [226]
[228] K. E. Polmanteer, H. L. Chapman, M. A. Lutz, *Rubber Chem. Technol.* **58** (1985) 939–974.
[229] K. M. Roedder., *Bautenschutz + Bausanierung* **10** (1987) 143–148.
[230] H. Mayer, M. Roth, *Bautenschutz + Bausanierung* **13** (1990) pp. 1–4.

[231] G. Benz, *Bosch Tech. Ber.* **8** (1986/87) 219–226.
[232] J. N. Helbert, N. Saha, *J. Adhes. Sci. Technol.* **5** (1991) 905–925.E. Babich et al., *Microelectron. Eng.* **13** (1991) 47–50.
[233] A. L. Smith (ed.): *The Analytical Chemistry of Silicones,* J. Wiley & Sons, New York 1991.
[234] W. Cheng, M. McCown, *J. Chromatogr.* **318** (1985) 173–185.K. K. Unger (ed.): *Handbuch der HPLC,* GIT Verlag, Darmstadt 1989. K. Kimata, N. Tanaka, T. Araki, *J. Chromatogr.* **594** (1992) 87–96.
[235] A. M. Dunster, J. R. Parsonage, E. A. Vidgeon, *Mat. Sci. Technol.* **5** (1989) 708–713.
[236] G. B. Kolesar et al., *Fundam. Appl. Toxicol.* **13** (1989) 285–295.
[237] In [8] pp. 130–134 and references cited therein.
[238] M. A. Horsham, C. J. Palmer, L. M. Cole, J. E. Casida, *J. Agric. Food Chem.* **38** (1990) 1734–1738.
[239] D. Hunter, *Chem. Week,* Feb. 19 (1992) 24–25.
[240] *Compatibilizers,*Hewin International Inc., Amsterdam 1991.

Sorbic Acid

Erich Lück, Hoechst Aktiengesellschaft, Frankfurt/Main, Federal Republic of Germany
Martin Jager, Hoechst Aktiengesellschaft, Frankfurt/Main, Federal Republic of Germany
Nico Raczek, Hoechst Aktiengesellschaft, Frankfurt/Main, Federal Republic of Germany

1. Introduction 4365
2. Physical Properties 4366
3. Chemical Properties and Derivatives 4366
4. Production 4368
5. Environmental Protection ... 4369
6. Quality Specifications....... 4369
7. Analysis 4370
8. Storage and Transportation .. 4370
9. Legal Aspects............. 4370
10. Mode of Action and Uses.... 4371
11. Economic Aspects 4373
12. Toxicology............... 4373
13. References............... 4374

1. Introduction

The *trans,trans* isomer of 2,4-hexadienoic acid is known as sorbic acid [110-44-1].

$$H_3C-CH=CH-CH=CH-COOH$$

The name is derived from the scientific name of the mountain ash, *Sorbus aucuparia* Linnaeus. In 1859 Hofmann obtained an oil with a characteristic odor from the juice of unripe rowan berries by distillation. He named the most important component of this oil parasorbic acid. Parasorbic acid [10048-32-5] is the δ-lactone of sorbic acid. Strong acids or alkalis convert it into the isomeric sorbic acid [3]. The parasorbic acid content of unripe rowan berries is ca. 0.1% [4].

Sorbic acid occurs in the form of 2-sorboyl-1,3-dimyristin [7175-63-5] in the fatty deposits of certain aphids [5].

Doebner established the structure of sorbic acid in 1890 [6], and in 1900 also described the first total synthesis [7].

In 1939/40 Müller [8] in Germany and Gooding in the United States [9] independently discovered the antimicrobial activity of sorbic acid. Industrial supplies of sorbic acid and potassium sorbate became available during the 1950s, first in the United States, shortly thereafter in Germany, and later in Japan. This opened the way to a broad range of applications in the preservation of food, animal feeds, pharmaceuticals, cosmetics, and other non-food applications. Because it is harmless from a physiological standpoint and exhibits favorable sensory characteristics sorbic acid has in the meantime become the leading preservative for foods.

2. Physical Properties

Sorbic acid crystallizes in needles or plates with a weak characteristic odor and a slightly acidic taste; mp 132–135 °C; bp 228 °C (decomp.); n_D^{20} 1.4248, density 1.204. Sorbic acid is soluble in 100 mL of water to the extent of 0.16 g at 20 °C, 0.58 g at 50 °C, and 3.9 g at 100 °C. The solubility in anhydrous lower molecular mass alcohols and anhydrous acetic acid is ca. 11–12 g per 100 mL. Liquid fats dissolve 0.5–1 g of sorbic acid per 100 mL.

Sorbic acid begins to sublime above 60 °C. The vapor pressure at 20 °C is < 0.001 kPa; at 100 °C it is 0.25 kPa, and at 120 °C 1.3 kPa. Sorbic acid is volatile in steam without decomposition. This property is important for its isolation from foods for analytical purposes.

The dissociation constant of sorbic acid is 1.73×10^{-5} at 25 °C, and the pK_a is 4.76. Its pH-dependent dissociation behavior is important in the use of sorbic acid as a preservative. Only the undissociated form has antimicrobial activity.

3. Chemical Properties and Derivatives

The chemical behavior of sorbic acid is determined by the carboxyl group and the conjugated double-bond system. Reactions at the carboxyl group correspond to those of other carboxylic acids; salts, esters, and other acid derivatives are easily prepared by the usual methods. The carboxyl group can be reduced selectively with lithium aluminum hydride to give sorbyl alcohol, 2,4-hexadien-1-ol [17102-64-6] [10]. Unlike sorbic acid itself, the alkali salts are readily water soluble, and they are used preferentially for the preservation of systems containing water.

Potassium sorbate [24634-61-5], prepared by dissolving sorbic acid in aqueous potassium hydroxide, is of great practical importance. It is sold commercially as a white powder and in granulated form. The stability of potassium sorbate is strongly dependent on its water content, which must be kept well below 0.5%. At room temperature ca. 140 g potassium sorbate can be dissolved in 100 mL of water. Saturated potassium sorbate solutions contain ca. 58 wt% potassium sorbate.

Calcium sorbate [7492-55-9], with a very low water solubility, can be produced by the reaction of sorbic acid with calcium hydroxide in water. Because of its particularly high resistance to oxidation, calcium sorbate is suitable for the production of packaging materials with a preserving effect [11].

The *lower alkyl esters* of sorbic acid also exhibit antimicrobial properties [12]. Unlike sorbic acid, these are also active in neutral and weakly alkaline media. In contrast to sorbic acid and its salts, the sorbate esters have a pronounced odor. The esters of sorbic acid with long chain fatty acids are described as suitable for bread preservation because, in contrast to normal sorbic acid they do not inhibit the growth of the bakery yeasts. However, nowadays a special type of sorbic acid has been developed that can also be used for this purpose.

Reactions of sorbic acid involving the double-bond system are often complex, frequently resulting in mixtures of products. This is due to the differing reactivities of the two double bonds and to side reactions and such subsequent transformations as isomerizations, rearrangements, double-bond migration, and polymerization.

In the case of addition reactions and partial hydrogenation, preferential attack is usually observed at the 4,5 double bond. Even with long reaction times chlorine addition is incomplete, leading to mixtures of various chlorohexenoic acids with an average chlorine content of 38–48 wt%. Addition of bromide in the presence of organic solvents leads to 2,3,4,5-tetrabromohexanoic acid [62284-99-5]. The major product of partial bromination in aqueous medium is 4,5-dibromo-2-hexenoic acid [19147-46-7] [13].

Pure crystalline sorbic acid is surprisingly resistant to air oxidation given that it is doubly unsaturated. At room temperature it remains unchanged for years, although its stability in the solid state depends very much on purity. The shelf life is reduced considerably by traces of solvents, heavy metals, or isomeric hexadienoic acids, which can form as side products during sorbic acid synthesis. The same applies to potassium sorbate. Sodium sorbate [7757-81-5] is unstable in the solid state and is therefore not produced industrially. In air, but not in solution, sodium sorbate is converted very rapidly into the sodium salt of 4,5-epoxy-2-hexenoic acid [14]. The oxidation of aqueous solutions of sorbates or solutions of sorbic acid in organic solvents gives rise to a large number of carbonyl compounds, although this reaction depends upon the presence of a large quantity of oxygen is strongly increased by exposure to sunlight [15], [16]. In solutions, the presence of oxygen causes oxidative degradation which may result in brown discoloration [83]. Oxygen-free solutions of sorbic acid remain unchanged even upon exposure to daylight [17]. In intermediate-moisture foods slow degradation of sorbic acid becomes apparent after storage for several months in the presence of air [18]. Potassium dichromate oxidation of acidic, aqueous sorbic acid solutions leads to reproducible yields of malonaldehyde [542-78-9]. The analytical determination of sorbic acid acid is based on a color reaction of malonaldehyde with 2-thiobarbituric acid (see Chap. 7).

Sorbic acid that has been incorporated into foods is at least as stable as the important constituents of the foods themselves (e.g., vitamins, flavors, and aromatic substances). The theoretical potential for sorbic acid to undergo autoxidation thus poses no problems in practice with respect to the preservation of foods.

4. Production

The first described synthesis of sorbic acid consisted of a condensation of *trans*-2-butenal [123-73-9] with malonic acid [141-82-2] in pyridine [19]. It is named Doebner synthesis after its inventor. Many other syntheses are based on similar principles [94] – [100]. Most are of no commercial interest because of low yields, expensive starting materials, or high production cost. This also applies, for example, to a process practiced temporarily on an industrial scale in the 1950s and 1960s in the United States by Union Carbide, which involved the oxidation of 2,4-hexadienal [142-83-6] to sorbic acid in the presence of catalysts [20] – [22]. This process was abandoned largely because it gave up to 20% of the isomeric hexadienoic acids as well, which are much less stable to storage than sorbic acid and could be removed only through an expensive purification process. 2,4-Hexadienal can also be oxidized to sorbic acid by microorganisms [23], [101].

An alternative preparation of sorbic acid involves isomerization of 2,5-hexadienoic acid [38867-16-2] by boiling with aqueous alkali. 2,5-Hexadienoic acid is in turn available by treatment of allyl chloride [107-05-1] with acetylene [74-86-2], carbon monoxide, and water in the presence of tetracarbonylnickel as catalyst [24] – [26]. The acidic cleavage of 5-vinyl-γ-butyrolactone [21963-38-2], obtained from 1,3-butadiene [106-99-0] and acetic acid in the presence of redox catalysts, is of no industrial importance [27] – [31]. Production of sorbic acid by the addition of carbon dioxide to 1,3-pentadiene [2004-70-8] in the presence of nickel complexes has also been described [32].

Another common process starts with ketene [463-51-4] and 2-butenal [123-73-9] [33]. In the presence of salts of divalent transition metals as catalysts, ketene and 2-butenal react at 20 – 80 °C to give a polymeric ester of 3-hydroxy-4-hexenoic acid [26811-78-9] with a molecular mass ≥ca. 2000. This polyester can be cleaved to give sorbic acid in good yield [34] by either bases or acids (e.g., hydrochloric acid [35]);

alternatively, cleavage can be effected by metal-complex catalysts, [36].

The polyester is also suitable for the production of various sorbic acid derivatives. For example, in the presence of acidic esterification catalysts it can be reacted with alcohols to give sorbate esters [38].

The crude sorbic acid must be purified, generally by recrystallization from an aqueous solution [39], from a mixture of water and an alcohol or acetone [40], or from organic solvents that are immiscible with water, such as methyl acetate [41]. Treatment with organic solvents [42] and purification by steam distillation [43] have also been described.

5. Environmental Protection

Sorbic acid shows antimicrobial activity against many fungi, yeasts, and bacteria (see Chap. 8). However, this is only the case if the population of the target organisms is small. Many microorganisms are capable of degrading and metabolizing sorbic acid when present in large numbers [44], including gram-negative sulfate-reducing bacteria [45]. It is for this reason that sorbic acid is readily degradable in soil. Sorbic acid also shows high degradability (95% within 6 d) in the Zahn–Wellens test (OECD 302 B). Sorbic acid and the sorbates are not hazardous with respect to water and are not subject to any hazardous materials classification.

The acute fish toxicity (LC_{50} for the zebra barbel) is very low (>1000 mg/L after 48–96 h). Thus, sorbic acid is incorporated in the lowest German water hazard class (WGK"0") [148].

At the normal application concentrations of 0.05–0.2%, sorbic acid displays very little toxicity to plants. Neither the roots nor the above-ground portions are damaged, from which it can be concluded that the ecotoxicity of the substance generally must be very low.

In the absence of local regulations to the contrary, sorbic acid and sorbates can be disposed of or incinerated along with household waste in the same way as foods.

6. Quality Specifications

Sorbic acid and sorbates are used almost exclusively in foods, animal feeds, cosmetics, and pharmaceuticals, and quality specifications are therefore adapted to the legal requirements applicable to these areas. In the EC the regulation of 2. December 1996, concerned specifically with additives is valid, which is based on an EC directive. The purity standards of the third edition of the Food Chemicals Codex are applicable in the United States. JECFA, a joint committee of experts from the United Nations Food and Agriculture Organization (FAO) and the World Health Organization (WHO) has also developed a set of purity specifications. Furthermore, sorbic acid and potassium sorbate are cited in various pharmacopeias.

All these requirements specify a purity of at least 99% and limit the content of aldehydes, arsenic, and toxicologically relevant heavy metals. Neither sorbic acid nor its salts should show signs of discoloring during 90 min of heating at 105 °C.

7. Analysis

Steam distillation is a common method for the quantitative isolation of sorbic acid from the investigated material. For qualitative detection and quantitative determination the red coloration is used that sorbic acid produces with 2-thiobarbituric acid after oxidation with potassium dichromate to malonaldehyde. As a polyunsaturated compound, sorbic acid displays a pronounced absorption maximum at ca. 260 nm (depending on the pH of the solution), which can be likewise used for quantitative determination [84]. Various methods are described and published that utilize HPLC for analysis. For analyses in special foods see [85]–[88], for determination of sorbic acid in foods and cosmetics in general see [89]–[91].

Standardized methods of detecting sorbates (GC, TLC, and HPLC) have been published in the revised edition of the Swiss Foodstuffs Manual (1992). A method exists for detecting sorbates in liquid sweeteners in accordance with § 35 of the Federal German food law (L. 57.22.99). Rather unconventional techniques of detecting sorbic acid such as ion chromatography or capillary isotachophoresis [92] have not become established so far in routine use. X-ray structural analysis of sorbic acid has also been described [93].

8. Storage and Transportation

Sorbic acid and its salts can be stored for several months. They should be stored in a cool, dry environment with protection from light. This applies especially to finely powdered sorbates.

Sorbic acid and sorbates are not subject to any hazardous goods classification. They can be stored and transported in the same way as ordinary foods.

9. Legal Aspects

Sorbic acid is permitted worldwide as a preservative for foods. Approval lists differ considerably from country to country, however, because of differences in eating habits. In the United States sorbic acid and its salts (which includes the sodium, potassium, and calcium salts) are classified as "generally recognized as safe" (GRAS). (Sodium sorbate is not produced industrially and is generally not used because of its instability).

In the EC sorbic acid is included as number E 200 in the list of approved preservatives, which also recognizes potassium sorbate (E 202) and calcium sorbate (E 203).

Sorbic acid is also permitted as an additive for animal feeds and cosmetics. Sorbic acid and potassium sorbate are listed in various pharmacopeias and are thus recognized as acceptable preservatives for pharmaceutical products as well.

10. Mode of Action and Uses

Certain potential uses for sorbic acid described in the patent literature (e.g., as a raw material for chemical reactions or polymerizations) have remained unimportant in practice because of the relatively high cost of the substance. Sorbic acid and the sorbates are used almost exclusively as preservatives. Because of their lack of toxicity the main areas of application are in foods, animal feeds, tobacco, cosmetics, and pharmaceuticals, as well as in packing materials for these substances and in other products that come into contact with human or animal skin in some way.Sorbic acid, potassium sorbate, and especially calcium sorbate are also used in some technical applications[46]

Some details of the antimicrobial activity of sorbic acid remain unexplained [47], [48]. Sorbic acid inhibits many enzymes in the microbial cell, namely enolase and lactate dehydrogenase, which are involved in the carbohydrate metabolism [105]. Sorbic acid interferes also, though not very specifically, with enzymes of the citric acid cycle including succinate dehydrogenase, fumarase, aspartase, malate dehydrogenase, isocitrate dehydrogenase, and α-ketoglutanate dehydrogenase[106]–[108]. Sorbic acid also impairs the catalase/peroxidase system [109], [110] and enzymes containing a sulfhydryl group [111]. The primary centers of attack in the cell may differ in bacteria, yeasts, and fungi. In other reports, it has been increasingly assumed that sorbic acid is also active against the cell wall [49]. Thus, sorbic acid inhibits the absorption of amino acids in bacteria. Proton-flow into the cell also increases as a result of partial destruction of the cell membrane, requiring the cell to expend additional energy to compensate for the resulting potential differences [50]. Resistance in the true sense of the word (i.e., an increase in the limiting inhibiting concentration under the influence of sub-threshold amounts of sorbic acid) is not observed in either *Echerichia coli* [51] or fungi [52]. Primarily the undissociated sorbic acid enters the cell wall of microorganisms, i.e., for food preservation the undissociated proportion of sorbic acid is most effective. Owing to its low dissociation constant sorbic acid, unlike other preservative acids, can also be employed for preserving weakly acid foods with a high pH. There are reasons to believe that also dissociated sorbic acid has an antimicrobial action. However this action is about 100 times weaker than that of the undissociated acid [112], [113].

Sorbic acid is a microbiostatic and not a microbiocidal agent. Sorbic acid activity is directed mainly toward yeasts and molds, including organisms responsible for aflatoxin formation [53]. Among bacteria, those that are catalase positive are inhibited more strongly than those that are catalase negative, with aerobic bacteria affected the most. However, bacteria can also be suppressed by a combined action of sorbic acid and other inhibiting factors [114]–[125]

Some microorganisms are capable of incorporating sorbic acid into their metabolism and thereby degrading it [44]. This is only possible, however, if sorbic acid is present below the threshold concentration and the microorganism population is very high. This means that sorbic acid is suitable for the preservation only of hygienically clean ingredients with low microorganism counts.

An important incentive for the use of sorbic acid and sorbates in foods is their virtual neutrality with respect to taste and odor, together with the fact that sorbic acid does not react with food components because of its chemical inertness.

The application concentration of sorbic acid is usually 500–2500 ppm. Higher concentrations are required the higher the pH of the product to be preserved. Low concentrations suffice if the system to be preserved is subject to other protective factors, such as low pH. In the case of wine ca. 50–200 ppm sorbic acid or a corresponding amount of potassium sorbate is adequate because the effectiveness is enhanced by the wine's low pH and the sulfur dioxide and ethanol present.

The favorable partition coefficient of sorbic acid between oil and water is important in the preservation of fat emulsions such as margarine or mayonnaise and cosmetic lotions. Thus, a relatively high proportion of the sorbic acid in an emulsion remains in the aqueous phase, which is alone susceptible to microbiological attack [54]. Sorbic acid is the agent of choice for the preservation of butter and margarine with a low fat content.

Sorbic acid and its salts are very important for preventing the development of mold in cheeses of all types. The activity of sorbic acid against agents responsible for mycotoxin formation plays a special role in this context [55]. Depending on the type of cheese, sorbic acid can be added to the cheese itself (cottage cheese, processed cheese), potassium sorbate can be added to the salt bath, or the cheese surface can be treated with a solution of suspension of sorbic acid, potassium sorbate, or calcium sorbate (hard cheese). The role of fungistatic packaging materials and coating materials containing sorbic acid or its salts is diminishing [56].

Undesirable mold growth on hard sausages can be suppressed by treatment with 10–20% potassium sorbate solution [57]. Replacement of the nitrite used in meat processing by sorbic acid has been widely discussed. While sorbic acid is inactive towards *Clostridium* species in neutral medium, (i.e., under optimal growth conditions) it does suppress the formation of toxins very effectively at lower pH (i.e., pH ca. 5.5) in concentrations of 0.1–0.2%. Sorbic acid can be considered as a good alternative to nitrite from a microbiological viewpoint provided sodium chloride is present as well. However, the red coloration and the pickling aroma preferred in cured meats do not develop in the absence of nitrite, which has prevented the successful use of sorbic acid in this application [58]–[61]. The use of potassium sorbate in combination with ascorbic or others acids gives a good and long curing [126]–[129].

The addition of potassium sorbate to fermenting vegetables leads to a particularly pure fermentation, because sorbic acid inhibits the undesirable proliferation of molds more than it does the desired lactic acid fermentation [62], [63]. Sorbic acid is also used for preserving all types of fruit products, including prunes, fruit pulps, marmelades and jams, fruit juices, jellies, and candies. Potassium sorbate in combination with sulfur dioxide is very important for stabilizing wine against undesired after-fermentation.

Baked goods can be protected against mold formation by the addition of sorbic acid to the dough. In yeast products (e.g., bread) sorbic acid may disrupt the action of the yeast, but a special form of the acid, characterized by a coarser particle size, shows virtually no detrimental effect on yeast activity [64].

Sorbic acid and potassium sorbate are acquiring increasing importance in the preservation of cosmetic products [130] (especially baby care products). Because of their good efficacy, low toxicity, and good skin compatibility they are favorably estimated in the literature [131]–[136].

In accordance with the "hurdle theory" [137] a combination of sorbates with modern physical preservation methods [138], [139] or other preservatives can be advantageous [140].

11. Economic Aspects

Sorbic acid and its salts, particularly potassium sorbate, have been produced industrially since the mid-1950s, primarily in the United States, Japan, and the Federal Republic of Germany. Current world demand exceeds 30 000 t/a.

12. Toxicology

The LD_{50} value of sorbic acid is ca. 10 g per kilogram body weight for rats [65]. Other authors give an LD_{50} of 7.4 g [141], 8.7 g [142], or 7.5 g [143] per kilogram body weight. As an acid, sorbic acid causes irritation to the mucous membranes, but skin irritation is observed only with very sensitive individuals.

Sorbic acid was first introduced as a food preservative at a time when extensive toxicological testing for new food additives was already mandatory. For this reason numerous studies have been published on acute, subacute, subchronic, and chronic tolerance (for a summary see [1, vol. II, pp. 13–31], [2, pp. 205–224] and [66], [67]). Sorbic acid has probably undergone the most thorough toxicological investigations of any preservative.

The allergenic potential of sorbic acid is extremely low, since, as a low molecular substance, it cannot cause an antibody response. Moreover, formation of covalent bonds with proteins, which might result in immediate hypersensitivity, is not known

[144]. Pseudo-allergic reactions to sorbic acid as a food additive are relatively rare [145]–[147].

The presence of 10% sorbic acid in rat feed causes increased growth and an increase in liver weight in the subchronic test, but no other damage [68]. The increase in liver weight observed at these high dose levels is interpreted as hypertrophy caused by excess work [69]. The increase in weight can be traced back to sorbic acid's caloric value.

Rats tolerate the lifelong addition of 5% sorbic acid to their feed [69]. Carcinogenesis studies (rats, oral exposure, two years) have shown that low concentrations of sorbic acid (1.5–10% in the feed) produce no abnormalities compared with control animals with respect to growth, blood count, and the state and function of 12 internal organs. Adding 10% sorbic acid to the feed leads to somewhat less weight gain and an increase in the size of the thyroid, liver, and kidneys [70]. Similar results have been obtained with mice [71].

Sorbic acid is not teratogenic [72]. All mutagenicity and genotoxicity investigations have also produced negative results [67], [73]–[77]. Sorbic acid is also not carcinogenic [70], [71].

Sorbic acid is metabolized like other fatty acids; degradation occurs via the typical β-oxidation pathway [69], [78]–[79].

The allergy-producing potentials and phototoxicities of both sorbic acid and potassium sorbate are extraordinarily low. Just one case is of allergic reaction has been reported (a type 4 allergic reaction with a low impact on the patient) [147], although sorbic acid has been used for more than 40 years all over the world. The CIR (Cosmetic Ingredient Review) Expert Panel of the Cosmetic, Toiletry, and Fragrance Association (CTFA) has classified both substances as "safe" at concentrations and under conditions typical for cosmetic agents [80].

In the case of sodium sorbate, which is not utilized in foods because of its tendency to undergo oxidation, recent studies have demonstrated a low level of genetic toxicity potential in vitro (e.g., micronuclei induction, sister chromatid exchanges). No analogous effects are observed with sorbic acid or potassium sorbate [81], [82], [135], [149].

13. References

General References

[1] E. Lück, M. Jager: *Antimicrobial Food Additives - Characteristics, Uses, Effects* 2nd ed., Springer Verlag, Berlin 1996.
[2] J. N. Sofos: *Sorbate Food Preservatives*, CRC Press, Boca Raton, Fla., 1989.

Specific References

[3] A. W. Hofmann, *Justus Liebigs Ann. Chem.* **110** (1859) 129–140.
[4] E. Letzig, W. Handschack, *Nahrung* **7** (1963) 591–605.
[5] J. H. Bowie, D. W. Cameron, *J. Chem. Soc.* **1965**, 5651–5657.
[6] O. Doebner, *Ber. Dtsch. Chem. Ges.* **23** (1890) 2372–2377.

[7] O. Doebner, *Ber. Dtsch. Chem. Ges.* **33** (1900) 2140–2142.
[8] BASF, DE 881 299, 1939 (E. Müller).
[9] The Best Foods, US 2 379 294, 1940 (C. M. Gooding).
[10] R. F. Nystrom, W. G. Brown, *J. Am. Chem. Soc.* **69** (1947) 2548.
[11] E. Lück, *Dtsch. Lebensm. Rundsch.* **50** (1962) 353–357.
[12] Dragoco, DE 1 139 610, 1961 (G. Nowak).
[13] R. Fittig, J. B. Barringer, *Ann. Chem. Pharm.* **161** (1872) 307–328.
[14] E. H. Farmer, A. T. Healey, *J. Chem. Soc. (London)* 1927, 1064–1067.
[15] T. Sabalitschka, H. Marx, *Naturwissenschaften* **46** (1959) 648.
[16] L. Pekkarinen, P. Rissanen, *Suom. Kemistil. B* **39** (1966) no. 3, 50–56.
[17] L. Pekkarinen, *Z. Lebensm. Unters. Forsch.* **139** (1968) 23–29.
[18] D. A. Ledward, *Food Addit. Contam.* **7** (1990) 677–683.
[19] O. Doebner, *Ber. Dtsch. Chem. Ges.* **33** (1900) 2140–2142.
[20] Union Carbide, DE 1 043 314, 1953 (H. C. Chitwood, B. T. Freure).
[21] Union Carbide, US 2 930 801, 1953 (A. E. Montagna, L. V. McQuillen).
[22] Hoechst, DE 1 244 768, 1963 (H. Fernholz, G. Jacobsen).
[23] Nippon Gosei, US 4 997 756, 1988 (M. Hasegawa, Y. Honda).
[24] Montecatini, BE 571 889, 1957 (G. P. Chiusoli).
[25] Montecatini, BE 586 166, 1958 (G. P. Chiusoli).
[26] Montecatini, BE 590 945, 1959 (G. P. Chiusoli).
[27] Nippon Gosei, US 4 022 822, 1974 (Y. Tsujino, M. Miyashita, T. Hashimoto).
[28] Monsanto, EP 55 936, 1981 (J. P. Coleman, R. C. Hallcher, D. E. McMackins).
[29] Monsanto, EP 83 914, 1981 (J. P. Coleman, R. C. Hallcher).
[30] Monsanto, EP 83 304, 1981 (R. C. Hallcher).
[31] J. P. Coleman et al., *Tetrahedron* **47** (1991) 809–829.
[32] H. Hoberg, D. Schaefer, *J. Organomet. Chem.* **255** (1983) C 15–C 17.
[33] Hoechst, DE 1 042 573, 1956 (H. Fernholz, E. Mundlos).
[34] Hoechst, DE 1 049 852, 1956 (H. Fernholz, E. Mundlos).
[35] Nippon Gosei, DE 2 526 716, 1975 (K. Sekiyama, Y. Taga, S. Fujita).
[36] Hoechst, DE 2 165 219, 1971 (H. Hey, H.-J. Arpe).
[37] Hoechst, DE 1 059 899, 1956 (H. Fernholz).
[38] Stamicarbon, US 3 056 830, 1961 (S. Koopal, U. Verstrijden, W. Pesch, J. J. M. Deumens).
[39] Hoechst, DE 2 331 668, 1973 (H. Fernholz, H.-J. Schmidt, F. Wunder).
[40] Hoechst, DE 1 618 357, 1967 (O. Probst, G. Roscher, H. Oehme).
[41] Wacker, DE 2 103 051, 1971 (H. Liberda, G. Künstle).
[42] Monsanto, EP 153 292, 1984 (H. C. Brown, R. P. Crowley, D. N. Heintz, J. R. Ryland).
[43] Chemcell, CA 879 082, 1969 (R. S. Smith, E. L. Jeans).
[44] J. L. Kinderlerer, P. V. Hatton, *Food Addit. Contam.* **7** (1990) 657–669.
[45] S. Schnell, C. Wondrak, G. Wahl, B. Schink, *Biodegradation* **2** (1991) 33–41.
[46] E. Lück, *Food Addit. Contam.* **7** (1990) 711–715.
[47] J. N. Sofos, F. F. Busta, *J. Food Prot.* **44** (1981) 614–622.
[48] J. N. Sofos, M. D. Pierson, J. C. Blocher, F. F. Busta, *Int. J. Food Microbiol.* **3** (1986) 1–17.
[49] T. Eklund, *J. Appl. Bacteriol.* **48** (1980) 423–432.
[50] T. Eklund, *J. Gen. Microbiol.* **131** (1985) 73–76.
[51] H. Lück, E. Rickerl, *Z. Lebensm. Unters. Forsch.* **109** (1959) 322–329.
[52] E.-M. Lukas, *Zentralbl. Bakteriol. Parasitenkd. Infektionskrankh. Hyg., Abt. 2* **117** (1964) 485–509.
[53] K. H. Wallhäußer, E. Lück, *Dtsch. Lebensm. Rundsch.* **66** (1970) 88–92.

[54] E. Becker, I. Roeder, *Fette Seifen Anstrichm.* **59** (1957) 321–328.
[55] K. H. Wallhäusser, E. Lück, *Dtsch. Lebensm. Rundsch.* **66** (1970) 88–92.
[56] E. Lück, *Dtsch. Lebensm. Rundsch.* **50** (1962) 353–357.
[57] L. Leistner, I. Y. Maing, E. Bergmann, *Fleischwirtschaft* **55** (1975) 559–561.
[58] M. C. Robach, J. N. Sofos, *J. Food. Prot.* **45** (1982) 374–383.
[59] J. N. Sofos, F. F. Busta, *Food Technol.* **34** (1980) no. 5, 244–251.
[60] J. N. Sofos, F. F. Busta, C. E. Allen, *J. Food Prot.* **42** (1979) 739–770, 1099.
[61] E. Lück, *Fleischwirtschaft* **64** (1984) 727–733.
[62] R. N. Costilow, W. E. Ferguson, S. Ray, *Appl. Microbiol.* **3** (1955) 341–345.
[63] E. Lück, *Ind. Obst Gemüseverwert.* **51** (1966) 410–414.
[64] Hoechst, EP 75 289, 1981 (E. Lück, K. H. Remmert).
[65] H. J. Deuel, R. Alfin-Slater, C. S. Weil, H. F. Smyth, *Food. Res.* **19** (1954) 1–12.
[66] R. Walker, *Food Addit. Contam.* **7** (1990) 671–676.
[67] British Industrial Biological Research Association: *Sorbic Acid and its Common Salts, Toxicity, Profile*, 1987.
[68] G. E. Demaree, D. W. Sjogren, B. W. McCashland, F. P. Cosgrove, *J. Am. Pharm. Assoc. Sci. Ed.* **44** (1955) 619–621.
[69] K. Lang, *Arzneim. Forsch.* **10** (1960) 997–999.
[70] I. F. Gaunt, K. R. Butterworth, J. Hardy, S. D. Gangolli, *Food Cosmet. Toxicol.* **13** (1975) 31–45.
[71] R. J. Hendy et al., *Food Cosmet. Toxicol.* **14** (1976) 381–386.
[72] Food and Drug Research Laboratories: *Teratologic Evaluation of FDA 73-4, Potassium Sorbate; Sorbistat in Mice and Rats*, PB-245 520, National Technical Information Service, US Department of Commerce, Springfield, Mass., 1975.
[73] Litton Bionetics: *Mutagenic Evaluation of Compound FDA 73-4, Potassium Sorbate*, PB-245 434, National Technical Information Service, US Department of Commerce, Springfield, Mass., 1974.
[74] R. Münzner, C. Guigas, H. W. Renner, *Food Chem. Toxicol.* **28** (1990) 397–401.
[75] R. Jung, C. Cojocel, W. Müller, D. Böttger, E. Lück, *Food Chem. Toxicol.* **30** (1992) 1–7.
[76] D. Schiffmann, J. Schlatter, *Food Chem. Toxicol.* **30** (1992) 669–672.
[77] J. Schlatter et al., *Food Chem. Toxicol.* **30** (1992) 843–851.
[78] H. J. Deuel et al., *Food. Res.* **19** (1954) 13–19.
[79] M. Fingerhut, B. Schmidt, K. Lang, *Biochem. Z.* **336** (1962) 118–125.
[80] Cosmetic, Toiletry and Fragrance Association (CTFA), *J. Am. Coll. Toxicol.* **7** (1988) 837–880.
[81] J. Schlatter et al., *Food Chem. Toxicol.* **30** (1992) 843–851.
[82] D. Schiffmann, J. Schlatter, *Food Chem. Toxicol.* **30** (1992) 669–672.
[83] B. R. Thakur, R. K. Singh, S. S. Arya, *Food Rev. Int.* **10** (1994) 71–91.
[84] F H. Lückmann, D. Melnick, *Food Res.* **20** (1955) 649–654.
[85] W. Flak, R. Schaber, *Mitt. Klosterneuburg* **38** (1988) 10–16.
[86] J. Kantasubrata, Imamkhasani, *ASEAN Food J.* **6** (1991) 155–158.
[87] S. Küppers, *J. Assoc. Off. Anal. Chem.* **71** (1988) 1068–1071.
[88] F. Olea Serrano, I. Lopes, N. Revilla., *J. Liqu. Chromatogr.* **14** (1991) 709–717.
[89] C. Reifschneider, C. Klug, M. Jager, *Seifen Öle Fette Wachse* **120** (1994) 650–654.
[90] U. Hagenauer-Hener, C. Frank, U. Hener, A. Mosandl, *Dtsch. Lebensm. Rundsch.* **83** (1990) 348–351.
[91] L. Bui, C. Kooper, *J. Assoc. Off. Anal. Chem.* **70** (1987) 892–896.
[92] J. Karovicova, J. Polonski, P. Simko, *Nahrung* **35** (1991) 543–544.
[93] P. Cox, *Acta Crystallogr.* **C50** (1994) 1620–1622.

[94]   R. Joly, C. Amiard, *Bull. Soc. Chim. France* (1947) 139.
[95]   E. Philippi, *Monatsh. Chem.* **51** (1929) 277.
[96]   N. G. Poljanskij, *Chim. Prom.* (1963) no. 1 20.
[97]   G. M. Robinson, R. Robinson, *Nature (London)* **151** (1943) 195.
[98]   J. Klein, E. D. Bergmann, *J. Amer. Chem. Soc.* **79** (1957) 3452.
[99]   L. Canonica, T. Bacchetti, *Gazz. Chim. Ital.* **83** (1953) 1043.
[100] Tarchorminskie Zalkady Farmaceutiycne "Polfa", PL 47 632 (I. Nagrodzka, B. Chechelska, B. Morawski, L. Brzechffa).
[101] Nippon Gosei, DE 3 930 338 A1, 1990 (M. Hasegawa, Y. Honda).
[102] Nippon Gosei, JP 05 032 583 A2, 1993 (I. Kakimoto, Y. Takehiko).
[103] Nantong, CN 1 112 915 A, 1995 (Ding, Caifeng, Yao, Jungsheng).
[104] Traian Vuia, RO 102 186 B1, 1991 (S. Tiberiu Mircea, S. Tiberiu Ladislau, S. Zlatimir).
[105] J. J. Azukas, PhD Thesis Michigan State Univ. (1962).
[106] H. J. Rehm, *Zentralblatt Bakteriol. Parasiten Infektion Hyg. II. Abt.. 2* **121** (1967) 492–502.
[107] G. K. York, *Food Res.* **20** (1955) 60–65.
[108] H. J. Rehm, P. Wallhöfer, *Naturwissenschaften* **51** (1964) 13–14.
[109] H. Lück, *Biochem Z.* **328** (1957) 411–419.
[110] H. Lück, *Z. Lebensm. Unters. Forsch.* **108** (1958) 1–9.
[111] W. Martoadiprowito, J. R. Whitaker, *Biochem. Biophys. Acta* **77** (1963) 536–544.
[112] H. J. Rehm, E. M. Lukas, *Zentralblatt Bakteriol. Parasiten Infektion Hyg. II. Abt.. 2* **117** (1963) 306–318.
[113] T. Eklund, *J. Appl. Bakteriol.* **54** (1983) 383–389.
[114] G. K. York: " *Studies on the Inhibition of Microbes by Sorbic Acid*" PhD Thesis University of California, Davis 1960.
[115] L. O. Emard, R. H. Vaughn, *J. Bacteriol.* **63** (1952) 487.
[116] T .A. Bell, J. L. Etchells, A. F. Borg, *J. Bacteriol.* **63** (1959) 573.
[117] H. .J. Rehm, *Z. Lebensmittel Unters. Forsch.* **115** (1961) 293.
[118] M. Lerche, G. Reuter, *Zentralbl. Bakteriol. Parasitenkd. Infektionskrankh. Hyg. I. Abt.,Orig* **179** (1960) 354.
[119] M. L. Couceiro, B. Regueiro, *An. Bromatol. (Madrid)* **14** (1962) 263.
[120] A. A. Lentzner, M. A. Toom, M. N. Woronina, M. E. Mikelsaar, *Lab Delo* (1967) 301.
[121] R. H. Vaughn, , L. O. Emard, *Bacteriol. Proc.* **54** (1954) 25.
[122] G .K. York, R. H. Vaughn, *Food Res.* **20** (1955) 60.
[123] H. Raj, *Canad. J. Microbiol.* **12** (1966) 191.
[124] S. M. Finegold, E. E. Sweeny in :*Antimicrobial Agents Chemotherapy,* 1961, p. 911.
[125] F. W. Beech, J. G. Carr, *J. Sci. Food Agric.* **11** (1960) 38.
[126] Rhenus Rheinische Getränke Industrie Binz & Binz GmbH, DE 1 692 110, (P. Flesch, G. Bauer).
[127] Armour Co., US 4 277 508, (U. Sato, A. Miller, L. J. Zimont.)
[128] Rhenus Rheinische Getränke Industrie Binz & Binz GmbH, DE 1 119 640, (P. Flesch, J. Hader).US 3 099 566, (P. Flesch, J. Hader).
[129] Rhenus Rheinische Getränke Industrie Binz & Binz GmbH, DE 1 155 313, (P. Flesch, J. Hader).
[130] C. Reifschneider, C. Klug, M. Jager : "Sorbic acid - The gentle alternative," in : *Proceedings of the Preservatec Conference 1995,* Ziolkowsky Verlag, Augsburg pp. 1–15.
[131] K. Schrader, *Dragoco Report* **17** (1970) 3.
[132] E. Birggal, *Kosmetik Parfum Drogen Rdsch.* **17** (1970) 1/2, 5.
[133] R. Goldschmidt, *Seifen Öle Fette Wachse* **96** (1970) A2, 7.

[134] R. Woodford, E. Adams, *Amer. Perfumer Cosmetics* **85** (1970) no. 3, 25.
[135] M. A. Kassem, A. G. Mattha, *Pharmac. Acta Helvetiae* **45** (1970) 345.
[136] A. Schunk, *Seifen Öle Fette Wachse* **122** (1996) 136–144.
[137] L. Leistner, *Fleischwirtschaft* **66** (1986) 10–15.
[138] T. Vardag, M. Jager, *Food Marketing Technology* **10** (1996) 58–67.
[139] T. Vardag, M. Jager, in: *Proceedings of the Preservatec Conference 1996,* Ziolkowsky Verlag, Augsburg pp. 33–40.
[140] T. Augustin et al., *Getränkeindustrie* **6** (1996) 407.
[141] H. F. Smyth, C .P. Carpenter, *J. Ind. Hyg. Tox.* **30** (1948) 63–68.
[142] I. Sado, *Nippon Fisagaku Zasshi* **28** (1973) 463–476.
[143] J. A. Troller, R. A. Olsen, *J. Food Sci.* **32** (1967) 228–232.
[144] S. Vieths, K. Fischer, L. I. Dehne, K. W. Bögl, *Ernährungs Umschau* **41** (1994) 140–143, 186–190.
[145] L. Rosenhall, *Eur. J. Respir. Dis.* **63** (1982) 410–419.
[146] M. Hannuksela, T. Haahtela, *Allergy* **42** (1987) 56–575.1
[147] M. Häberle, *Ernährungs Umschau* **36** (1989) 8–16.
[148] C. S. Koch, *Seifen Öle Fette Wachse* **120** (1994) 655–660.
[149] F. Würgler, J. Schlatter, P. Maier, *Mutat. Res.* **283** (1992) 107–111.

Starch

JAMES R. DANIEL, Department of Foods and Nutrition, Purdue University, West Lafayette, Indiana 47907, United States

ROY L. WHISTLER, Whistler Center for Carbohydrate Research, Purdue University, West Lafayette, Indiana 47907, United States

HARALD RÖPER, Cerestar, Eridania Béghin-Say, Vilvoorde, Belgium Chap. 3

1.	Starch	4379
1.1.	Raw Materials	4380
1.2.	Molecular Structure and Composition	4381
1.3.	Biosynthesis and Structure	4381
1.3.1.	Biosynthesis of Amylose, Amylopectin and Involved Enzymes	4381
1.3.2.	Structure of the Starch Granule	4382
1.4.	Physicochemical Properties	4382
1.4.1.	X-Ray Patterns	4384
1.4.2.	Rheological Properties	4385
1.4.3.	Water Activity	4387
1.4.4.	Amylose Complexes	4388
1.5.	Composition of Native Starches	4388
1.6.	Industrial Starch Production Processes	4389
1.6.1.	Corn and Sorghum Starch	4389
1.6.2.	Wheat Starch	4391
1.6.3.	Potato Starch	4392
1.6.4.	Rice Starch	4394
1.6.5.	Tapioca Starch	4394
1.6.6.	Arrowroot Starch	4395
1.6.7.	Sago Starch	4395
1.7.	Uses of Native Starches	4395
2.	Modified Starches	4396
2.1.	Physically Modified Starches	4396
2.2.	Chemically Modified Starches	4397
2.2.1.	Acid-Modified Starch	4397
2.2.2.	Oxidized Starch	4399
2.2.3.	Cross-Linked Starch	4402
2.2.4.	Starch Esters	4404
2.2.5.	Starch Ethers	4407
2.2.6.	Cationic Starch	4408
3.	Economic Aspects	4411
4.	References	4412

1. Starch

Starch [*9005-25-8*], the principal source of dietary calories to the world's human population, has many chemical and physical characteristics that set it apart from other food components and give it numerous applications. Food processors and other industrial users should be familiar with starch's structural and behavioral characteristics so as to make use of its properties. The chemistry and technology of starch has been reviewed in detail [1].

Unlike other carbohydrates and edible polymers, starch occurs as discrete particles called starch granules. These particles are unique to the plant source and, therefore, starch from every plant type is different in appearance, properties, and particle size

Figure 1. Representative structure of linear amylose

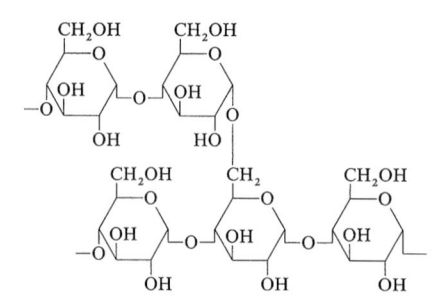

Figure 2. Representative structure of amylopectin, including (1,6)-α-D branch point

distribution. The size distribution of wheat starch granules during endosperm development has been reported [2]. Microscopists can readily identify the plant source of most commercial starches.

Starch in granular form is generally composed of two types of molecules, amylose and amylopectin. *Amylose* [9005-82-7] is a linear (1,4)-α-D-glucan (Fig. 1), although there is now some evidence of a few 1,6 branches in some amyloses. *Amylopectin* [9037-22-3] (Fig. 2) is a branched, bushlike structure containing both (1,4)-α-D linkages between D-glucose residues and (1,6)-α-D branch points. Normal starches contain ca. 75% amylopectin molecules. Amylopectin is a much larger molecule than amylose as shown in Table 1. It is these characteristic molecular sizes, amylose–amylopectin ratios, and granular structures which give each type of starch its unique properties.

1.1. Raw Materials [1], [3]–[5]

Plants grown for starch production include corn, sorghum, wheat, rice, tapioca, arrowroot, sago, and potato. Other noncommercial sources of starch include quinoa and amaranth, whose starches have been isolated and examined with respect to chemical and physical characteristics and possible uses [6]–[9].

Table 1. Physicochemical properties of amylose and amylopectin

Property	Amylose	Amylopectin
Molecular mass	50 000 – 200 000	one to several million
Glycosidic linkages	mainly (1,4)-α-D-	(1,4)-α-D-, (1,6)-α-D-
Susceptibility to retrogradation	high	low
Products of action of β-amylase	maltose	maltose, β-limit dextrin
Products of action of glucoamylase	D-glucose	D-glucose
Molecular shape	essentially linear	bush-shaped

Table 2. Diameter and gelatinization temperature of starch granules

Source	Mean diameter, μm	Gelatinization temperature, °C
Corn	15	62 – 71
Wheat	20 – 22	53 – 64
Rice	5	65 – 73
White potato	33	62 – 68
Sweet potato	25 – 50	82 – 83
Tapioca	20	59 – 70

1.2. Molecular Structure and Composition

The molecular structure and composition of amylose, amylopectin, and starch granules is largely described in Sections 1, 1.3.2, 1.4.1, 1.4.4, and 1.5 as well as Tables 1, 2, 3, 4 [49]. Purified amylose forms strong, transparent films and fibers, having properties similar to cellulose [9004-34-6], except for the greater swellability of amylose and its alkaline solubility. Amylose molecules have degrees of polymerization (DP) of 350 – 1000 compared to amylopectin which has a DP of several thousand.

1.3. Biosynthesis and Structure

1.3.1. Biosynthesis of Amylose, Amylopectin and Involved Enzymes

Normal starch, containing both amylose and amylopectin, is biosynthesized by two enzymes, a *chain-lengthening enzyme*, which is apparently identical for both amylose and the linear portions of amylopectin chains, and a *branching enzyme* that gives rise to amylopectin. Evidence suggests that a branching enzyme requires an amylose chain of ca. 35 – 40 units before it can transfer a portion of that chain to form a branch on another starch molecule.

When starch granules are examined under normal light or by scanning electron microscopy, growth rings are visible around the hilum (see below) which may become wider in successive layers as the granule grows. Some evidence points to the beginning of amylopectin molecules at an inner lammella edge and extending to the outer edge.

1.3.2. Structure of the Starch Granule

Individual starch molecules that are formed by the action of starch synthetase nucleate at a point called the *hilum* (the botanical center of the granule). These starch molecules are produced and deposited in a radial, spherocrystalline arrangement. Synthetase enzymes cling to the surface of starch granules although there is no membrane. The surface consists of end units of starch molecular chains deposited side-by-side during granular growth. Linear amylose molecules occur interspersed among the branched amylopectin molecules. Linear amylose molecules tend to form helices which are often double. Although the structure of amylopectin is thought to be branched and perhaps bushlike (Fig. 3 A) there are an increasing number of investigators who believe—as NIKUNI [10], [11] originally proposed and FRENCH [12], [13] asserted—that the molecules have what may be called a tassel-on-a-string structure (see Fig. 3 B). In this structure short chains of 12–15 D-glucopyranosyl units occur about every 25 units of the main chain, which may consist of more than a thousand D-glucopyranosyl units.

The spherocrystalline radial arrangement of starch molecules in the granule is inferred from the Maltese cross that is seen on microscopic examination of starch granules under polarized light [1] (see Fig. 4). The center of the cross is at the hilum. X-Ray patterns of starch granules show that cereal starches possess an *A pattern*, which indicates chains of amylose or outer chains of amylopectin in antiparallel double helices separated by interstitial water. Tuber and root starches produce *B patterns*, with up to 30% water present in sheets and columns. These patterns result primarily from the otherwise uncomplexed amylose molecules or long outer chains of amylopectin. Amylopectin molecules seem to be oriented with their reducing ends interior to the granule, as would be expected from outward biosynthetic growth. In addition to X-ray evidence, crystallinity is also evident from measurements by differential scanning calorimetry. This method measures the energy for melting crystalline regions when heated [14].

1.4. Physicochemical Properties

Undamaged starch granules are insoluble in cold water, but can reversibly absorb water and swell when their aqueous dispersion is heated. In water at 25 °C, the percentage increase in granule diameter ranges from 9.1% for normal (native) corn starch to 22.7% for waxy (100% amylopectin) corn starch. As temperature is increased,

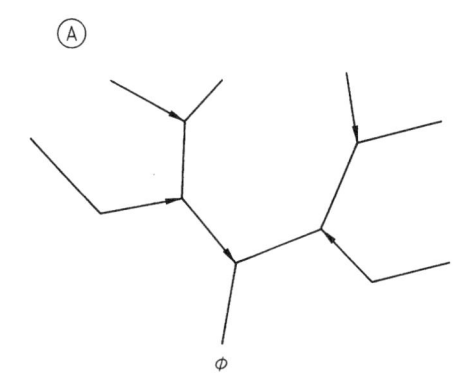

Figure 3. A) Bushlike branched model of amylopectin (ϕ = reducing end group); B) Tassel-on-a-string model of amylopectin (ϕ = reducing end group)

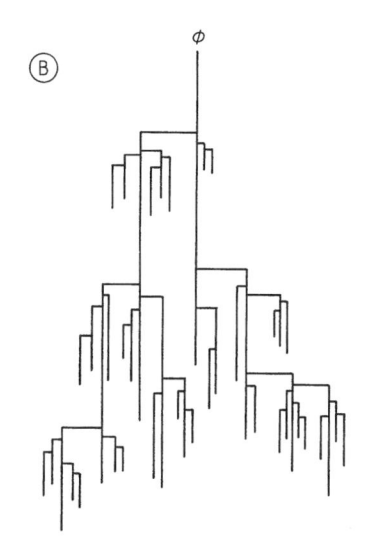

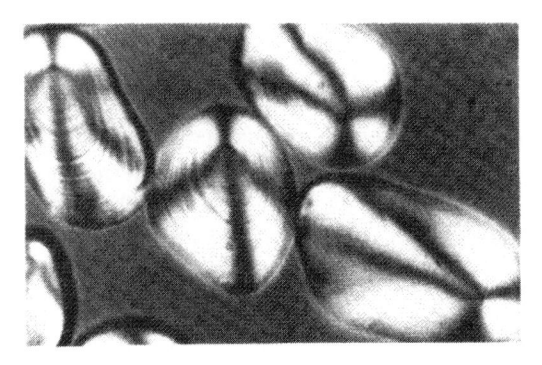

Figure 4. Micrograph of potato starch in polarized light (magnification ca. ×250)

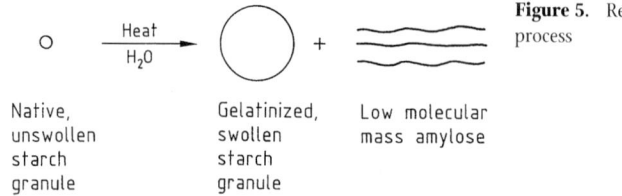

Figure 5. Representation of starch gelatinization process

the molecules in the starch granule vibrate vigorously, breaking intermolecular bonds, thus allowing increased interaction with water molecules. This penetration of water combined with the increased separation of the starch chains increases randomness and decreases crystallinity in the granule. Continued heating in the presence of free water results in complete loss of crystallinity, as evident by loss of birefringence of the starch granules and changes in the X-ray diffraction pattern. The temperature range over which birefringence disappears is called the *gelatinization temperature range* (see Table 2). Gelatinization usually occurs over a small temperature range with larger starch granules generally gelatinizing first and smaller granules later. During gelatinization, granules swell extensively (Fig. 5). Thus, a 1% starch slurry in cold water has low viscosity, but on heating a thick paste is produced wherein almost all the water has entered the granules and combined with them. This causes the granules to press tightly against each other. The viscosity of the paste results from resistance to the flow of the swollen granules that now occupy nearly the entire volume. Highly swollen granules are fragile and disintegrate on mild stirring. This causes a significant decrease in paste viscosity.

The swelling and gelatinization of wheat, barley, and maize starch granules have been studied with respect to the effect of amylose, amylopectin, and lipid content [15]. A similar investigation has been performed with rice starch [16]. In the case of wheat, barley, and maize starch it was concluded that granular swelling during gelatinization was mainly a function of amylopectin content. Gelatinization has been studied by scanning electron microscopy and a honeycomb-like structure has been noted on gelatinization [17]. Visual characteristics of starch pastes have been investigated, concentrating specifically on clarity and whiteness [18]. Potato starch paste exhibited the greatest clarity, wheat and corn were intermediate in clarity, and high amylose corn paste was of very low clarity.

Starch swelling rates are greatly increased at pH 10.0 and above. At low pH, the peak viscosity (see Section 1.4.2) of starch pastes is markedly reduced, along with a rapid loss of viscosity on cooking. This is due to extensive hydrolysis of starch to form non-thickening dextrins (low molecular mass oligosaccharides derived from starch).

1.4.1. X-Ray Patterns

The linear nature of corn [19] and potato [20] amylose was proposed in 1940 and confirmed [21] three years later. However, evidence now exists that some amyloses may contain a few very long branches.

X-ray diffraction has been useful in examining the structure of both intact starch and amylose. Starch naturally occurs in three crystalline modifications designated A (cereal), B (tuber), and C (smooth pea and various beans). Starch which is precipitated from solution, or complexed with organic molecules, adopts the so-called *V structure* (Verkleisterung). Amylose also exists in A, B, C, and V structures. MARCHESSAULT and coworkers [22] studied the chain conformation of B-amylose and proposed that *B-amylose* is helical with an integral number of α-D-glucopyranosyl residues per turn. The density of B-amylose suggests a helical structure with six α-D-glucopyranosyl residues per unit cell and 3–4 molecules of water of hydration. Sixfold helices are energetically favored and have a hydrogen bond between the C-2 hydroxyl group of one α-D-glucopyranosyl unit and the C-3 hydroxyl group of the succeeding sugar unit. Left-handed helices are slightly favored over right-handed ones. It was thus concluded that solid-state B-amylose exists as a left-handed sixfold helix. Amylose triacetate, which exists in a left-handed helix, forms B-amylose upon saponification [23]. Calculations show that *V-amylose*, having a fiber repeat of about 80 pm, is more stable as a left-handed helix than as a right-handed one [24]. The easy conversion of V-amylose to B-amylose [25] implies that no extensive molecular reorganization accompanies the reversal of chain chirality. It seems likely that amylose, in the solid state, exists as a left-handed helical polymer with six α-D-glucopyranosyl units per turn.

A model proposed by KREGER [26] in which amylose exists as a threefold helix has been examined [27]. Although this model is consistent with some of the amylose X-ray data, it has not received wide acceptance. It should be noted that the model of the starch granule of FREY–WYSSLING [27] employs radially oriented low-molecular-mass amylose as a three-fold helix. This explains the granule's positive birefringence and proposes a higher-molecular-mass amylose (DP > 280) as a sixfold helix running tangentially to the granule surface.

1.4.2. Rheological Properties

For thickening foods, sizing and coating papers, textile sizing, use in drilling muds, adhesive formulations, and many other applications, starch gelatinization is required. When starch is cooked, flow behavior of the slurry changes significantly as the suspension becomes a mixture of swollen granules, partially disintegrated granules, and molecularly dispersed granule contents (mostly amylose). The cooked product is termed a starch paste and transition from a granular suspension to a cooked paste is accompanied by an increase in apparent viscosity. Although the term viscosity is generally used, starch pastes are non-Newtonian and a better description is that pastes develop resistance to deformation or flow.

On cooling, paste consistency increases because side-by-side molecular associations form junction zones that increase paste resistance to deformation or flow. Therefore, when cooled, pastes may remain fluid or form gels with considerable strength. Starch manufacturers and users both require techniques for judging starch quality and flow

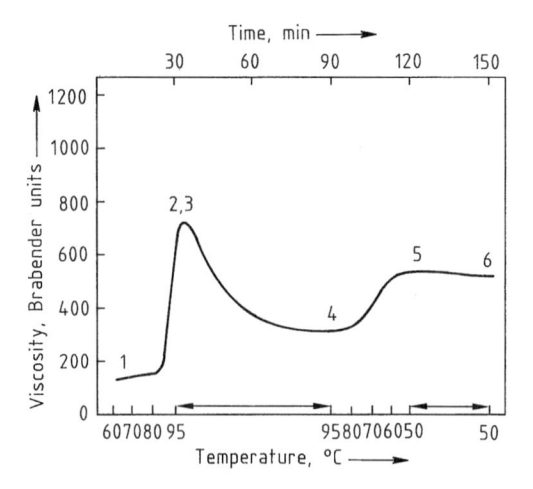

Figure 6. Representative Brabender amylograph curve
For explanation of points 1–6 see above.

behavior in its applications. For these reasons measurements of viscosity, gel rigidity, and gel strength have been devised.

A *Brabender viscoamylograph* provides information on gelatinization as well as on the properties of the cooled paste. This instrument records the torque needed to counteract the viscosity that develops when a starch suspension is subjected to heating and cooling regimes. The temperature at which the first significant rise in viscosity occurs depends on starch concentration. It is usually higher than the gelatinization temperature that is determined by loss of birefringence. Viscosity is measured in arbitrary units (Brabender units) that reflect paste consistency; for this reason, Brabender results are described in terms of paste properties and pasting temperatures at specific times. The Amylograph Handbook describes the theory, construction, applications, sources of error, troubleshooting, and adjustment of Brabender units [28].

A typical Brabender curve is shown in Figure 6. Six important points on the Brabender curve are generally recognized:

1) *Pasting Temperature.* Indicates onset of paste formation; varies with starch type or modification and with additives in the starch slurry.
2) *Peak Viscosity.* Cited regardless of the temperature at which the peak is attained.
3) *Viscosity at 95 °C.* Related to ease of cooking the starch.
4) *Viscosity after 1 h at 95 °C.* Shows cooked paste stability, or lack of stability under relatively low shear.
5) *Viscosity at 50 °C.* Measures the setback (increase in viscosity) that occurs when the hot paste is cooled.
6) *Viscosity after 30 min at 50 °C.* Related to stability of cooked paste under simulated use conditions.

Other chemicals in the starch dispersion may have a significant effect on gelatinization and gelation. This is especially true in the use of starch in foods. High *sugar* concentrations lower the rate of starch gelatinization, the peak viscosity, and the gel strength.

Disaccharides are more effective in delaying gelatinization and reducing peak viscosity than monosaccharides [29]. Sugars may decrease gel strength by exerting a plasticizing action and interfering with formation of starch junction zones. *Lipids,* such as the triacylglycerol fats and oils, and lipid-related materials, such as mono- and diglycerides may also affect starch gelatinization (Section 1.4.4). Wheat starch contains up to 1% lipids consisting mainly of lysophosphatidylcholines of palmitic and linoleic acids which are principally complexed with amylose. Such complexes retard starch hydration and lower peak viscosity on gelatinization. These complexes also are more slowly hydrolyzed by starch amylases.

Because of the essentially neutral character of starch, low concentrations of *salts* have little effect on gelatinization and gelation. Exceptions include potato amylopectin, which contains some phosphate groups, and manufactured ionic starches. With these starches salts may either increase or decrease swelling, depending on conditions [29].

Starch-thickened *acidic* foods (pie fillings, salad dressings) usually have pH values in the range 4–7 which has little effect on starch gelatinization. At lower pH values starch hydrolysis occurs, yielding nonthickening dextrins (starch breakdown to oligosaccharides). To avoid thinning in acidic starch-thickened foods, a *cross-linked starch* (see Section 2.2.3) may be selected. Since cross-linked molecules are quite large, extensive hydrolysis is required to decrease viscosity significantly.

1.4.3. Water Activity

In foods and other systems, water is not just a medium for reaction, but is also an active ingredient used to control reactions, texture, and general physical and biological behavior. However, total water content is not the important factor, but rather the availability of water (water activity, a_w). Water activity may be defined as the ratio of water vapor pressure over a sample p to the water vapor pressure over pure water p_0 under the same conditions

$$a_w = \frac{p}{p_0}$$

Alternatively, it may be expressed as the equilibrium relative humidity *ERH* of the atmosphere surrounding a sample divided by 100

$$a_w = \frac{ERH}{100}$$

Water activity is affected by strong water-binding agents or hydrophilic parts of other molecules. If water-binding agents are present in starch preparations, water activity is lower and gelatinization is inhibited to varying extents. Strong water-binding constituents retard starch gelatinization by binding and holding the water needed for gelatinization.

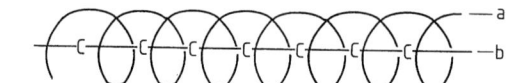

Figure 7. Complexation of a fatty acid-like chain with helical amylose segment (a = helical amylose; b = fatty acid chain)

1.4.4. Amylose Complexes

Helical amylose molecules can entrap other organic molecules, such as fatty acids as well as inorganic molecules such as iodine. Such complexes are called inclusion compounds and the amount of iodine binding is a measure of amylose content. Amylose can be separated from amylopectin by addition of n-butanol [71-36-3] to a hot dispersion and allowing the mixture to cool slowly. On cooling, crystals of butanol–amylose separate and can be removed by filtration or centrifugation [30].

Lipids and emulsifiers, such as mono- and diacylglycerol, affect starch gelatinization. Fats that complex with amylose retard granule swelling. Fat, in the absence of emulsifiers, decreases the temperature at which maximum viscosity occurs but does not affect the maximum viscosity. Corn starch normally gelatinizes to produce a maximum viscosity at 92 °C, but in the presence of 9–12 % fat, maximum viscosity occurs at 82 °C.

Monoacylglycerols with acyl chain length of 16–18 carbon atoms increase the gelatinization temperature. They also decrease the temperature of maximum viscosity, the temperature of gel formation, and the gel strength. Fatty acids or the fatty acid chains of monoacylglycerols form inclusion complexes with amylose, and possibly with longer outer chains of amylopectin (see Fig. 7). Such complexes are not easily leached from the granule and inhibit entry of water into the starch granule. The lipid–amylose complex also limits retrogradation by interfering with the formation of junction zones between starch molecules.

1.5. Composition of Native Starches

Normal starches contain about 25 % linear amylose (Table 3). Particular varieties of corn, called *high-amylose corn*, produce starches with amylose contents up to 85 %, although most commercial varieties contain ca. 65 % amylose. High-amylose starches are difficult to gelatinize, often requiring temperatures in excess of 100 °C at > 1 bar. At the other extreme, some starches consist only of branched *amylopectin* (Table 3). These include waxy corn, waxy barley, and waxy or glutinous rice. They produce clear cooked pastes, resembling those of root or tuber starches. These starch pastes are fairly stable to retrogradation due to their lack of amylose.

The term *retrogradation* or setback denotes spontaneous changes during the aging of starch solutions, pastes, or gels that lead to an increase in the degree of order. Retrogradation is crystallization due to the formation of hydrogen bonds between hydroxyl groups of neighboring starch molecules. Retrogradation kinetics of numerous starches

Table 3. Comparison of composition and properties of some commercial starches

	Starch content, wt%	H$_2$O content, wt%	Protein content, wt%	pH	Amylose content, wt%	Amylopectin content, wt%	Specific gravity
Corn							
Normal	88	11	0.35	5	28	72	1.5
Waxy	88	11	0.28	5	0	100	1.5
High-amylose					65–85	15–35	
Sorghum	88	11	0.37	5	28	72	1.5
Tapioca		12		6.3–6.5	16	84	
Arrowroot					21	79	
Sago					26	74	
Potato	80	17–18	trace		20	80	
Wheat					30	70	
Rice							
Normal		12	0.37		20–30	70–80	1.5
Waxy		12	0.13		0	100	1.5

have been studied by differential scanning calorimetry and rheology [31], [32]. Potato and pea starches are prone to retrogradation while waxy rice and modified waxy corn are least susceptible.

1.6. Industrial Starch Production Processes

1.6.1. Corn and Sorghum Starch

Corn (maize, *Zea mays* L.) has been grown for 5000–7000 years and was probably initially cultivated in Central Mexico. Early corn was quite different from modern varieties, which were derived by both random and scientific selection. Corn types available include popcorn, sweet corn, dent corn, flint corn, and flour corn. Corn's availability at low prices, its storability, and 70% starch content has greatly expanded its use in commercial production of starch.

Grain sorghum (*Sorghum bicolor* M.), was probably initially cultivated 5000–7000 years ago in Eastern Africa. Although sorghum resembles corn in chemical composition, it is generally considered to be inferior for food, feed, and industrial uses. Sorghum can be cultivated effectively in semi-arid regions and is milled similarly to corn.

A single kernel of corn is a caryopsis (a berry) borne on an ear. This berry matures about 60 days after pollination and corn is typically harvested in late summer or early fall. In sorghum, the seeds are borne on a terminal, bisexual head, or rachis.

Corn (and sorghum) wet milling consists of several major steps (Fig. 8): (1) cleaning the grain; (2) steeping; (3) milling and fraction separation; (4) starch processing; and (5) starch drying.

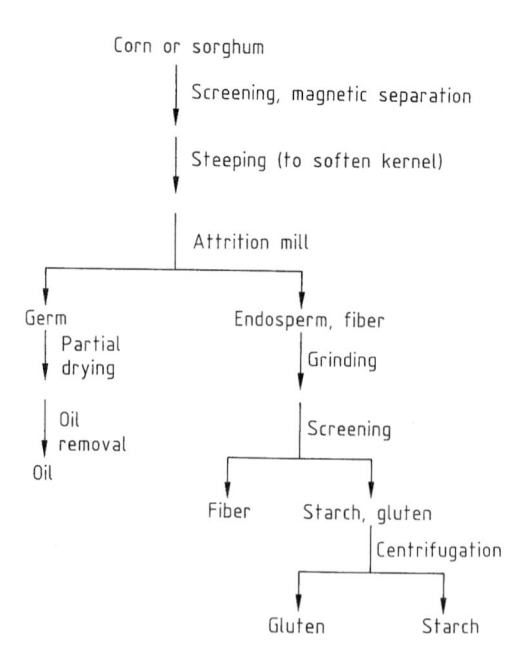

```
Corn or sorghum
    │
    ▼
Screening, magnetic separation
    │
    ▼
Steeping (to soften kernel)
    │
    ▼
Attrition mill
    │
    ├──────────────┐
    ▼              ▼
  Germ        Endosperm, fiber
    │              │
Partial         Grinding
drying             │
    │              ▼
   Oil          Screening
 removal           │
    │         ┌────┴────┐
   Oil        ▼         ▼
            Fiber   Starch, gluten
                       │
                  Centrifugation
                       │
                  ┌────┴────┐
                  ▼         ▼
                Gluten    Starch
```

Figure 8. Flow sheet for corn (sorghum) starch production

Cleaning. Grain is first cleaned by screening and by magnetic means to remove foreign material including metal.

Steeping. The grain is steeped to soften the kernel and facilitate its separation into components. This process involves steeping in countercurrent water containing 0.10% sulfur dioxide at 48–52 °C for 30–40 h. Sulfur dioxide combats growth of spoilage organisms and maximizes the starch yield. Sulfur dioxide also reacts with the protein matrix to disperse the glutelin by breaking disulfide bonds and releasing extra starch. Proper steeping temperature is critical to promote the growth of lactic acid bacteria. Growth of these bacteria lowers the steep pH, restricting the growth of other organisms. Steep water pH is typically ca. 4.0, and is highly buffered. Steep water is concentrated, dried by absorption on corn fiber, and sold as cattle food. Alternatively, it is further processed to remove phytic acid or sold as nutrient for commercial fermentation to produce, for example, antibiotics.

Milling and Fraction Separation. After steeping, grain is milled to obtain separation of components. Steeped grain is first put through an *attrition mill* (Bauer type) to release the oil-containing germ which is removed by centrifugation in hydroclones. Next, the residual grain is *ground* between stones to release starch granules from their containment in proteinaceous cells. The slurry is *screened* to separate the corn hull or seed coat which is recovered as a fibrous mass. The remaining starch slurry is more strongly centrifuged to separate starch granules that are washed free of protein and salts and are dried in continuous driers. The isolated germ is partially dried prior to hexane extraction or mechanical removal of corn oil.

Centrifugation. Usually the removal of starch is performed in centrifuges or hydrocyclone batteries that can separate not only the starch but also the insoluble gluten fraction of protein.

Gluten is less dense (1.1 g/cm^3) than starch (1.5 g/cm^3) and is thus easily separated by centrifugal forces. Starch thus produced contains 1–2% protein, which may be removed by a second starch washing with water and centrifugation to give a starch with a final protein content of < 0.38%. In the case of *sorghum*, starch and gluten are more difficult to separate because the sorghum pericarp is fragile compared to corn. Separation is improved if the kernels are dehulled before milling. This produces a cleaner separation of starch from gluten, giving a starch of low color and protein content.

The final starch slurry can be dried and sold as unmodified starch, or the starch can be modified (see later sections) and dried, or gelatinized and dried. It also can be acid or enzyme hydrolyzed to a syrup which may be used directly or fermented to produce ethanol [64-17-5] or other chemical feedstocks.

Drying. Starch is usually dried by flash drying. In this method starch slurry is centrifuged or filtered and the moist starch is introduced at the bottom of a stream of rapidly moving hot air (93–127 °C). Drying is rapid and the starch is collected in cyclones. Drying parameters can be used to control both bulk density and particle size. Evaporation of the water requires so much energy that the wet-milling industry is the second most energy intensive food industry.

1.6.2. Wheat Starch

Wheat has been utilized by humans for 8000–9000 years. Its cultivation probably began in what is now Iraq. The first instance of commercial manufacture of wheat starch was in England in the 1500s. Wheat is mainly used in human food, while on the contrary much of the corn grown is used in animal feed, and the world annually produces more wheat than corn. Wheat is harvested by mechanical means and wheat starch is separated by one of several processing methods such as the Martin, Batter, Fesca, ammonia, or the acid process, by wet wheat milling, whole wheat fractionation, the Rasio, or the hydroclone process. The first two processes are most common.

Martin Process. The Martin (dough ball) process, invented in 1835, uses wheat flour as its raw material (Fig. 9). Wheat flour and water (2:1) are blended and the dough is allowed to rest, hydrate, and strengthen the gluten matrix. The dough is then kneaded while washing with water to remove the starch. The residual gluten is partially dried by roller compression and then flash dried. Starch slurry from the dough washer is screened to remove any remaining gluten and then fine screened to remove bran. The starch slurry is centrifuged to produce a high solids starch cake which is flash dried. Starch is dried to 10–12% moisture and contains ca. 0.3% protein.

Figure 9. Flow sheet of the Martin process

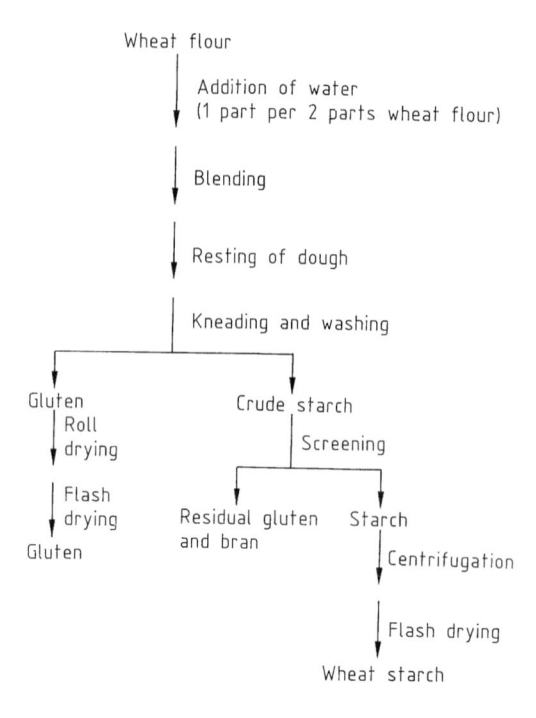

```
Wheat flour
    │
    │ Addition of water
    │ (1 part per 2 parts wheat flour)
    ▼
  Blending
    │
    ▼
  Resting of dough
    │
    ▼
  Kneading and washing
    │
    ├──────────────┐
    ▼              ▼
  Gluten       Crude starch
    │              │
  Roll         Screening
  drying           │
    │         ┌────┴────┐
  Flash       ▼         ▼
  drying   Residual   Starch
    │      gluten       │
  Gluten   and bran   Centrifugation
                        │
                        ▼
                     Flash drying
                        │
                        ▼
                   Wheat starch
```

Batter Process. The Batter process is derived from the Martin process but differs in the way the dough is formed and the way it is subsequently treated (Fig. 10). Wheat flour and water (ca. a 1:1 ratio) are mixed and the produced dough is allowed to rest for 30 min. The dough is then vigorously mixed with twice its weight of water. Gluten is separated from this mixture by screening and bran is removed from the starch by additional screenings. Centrifugal separation of the starch slurry separates the large starch granules (*type A starch*) from the small granules, damaged granules, and pentosan complexes (*type B starch*). Type A starch is essentially pure and may be dewatered and flash dried. Type B starch is processed again in a similar way but is of lower purity. Large wheat starch granules are extensively used in making carbonless copy paper where the granules serve as agents to prevent premature bursting of the ink microcapsules.

1.6.3. Potato Starch

Potato starch is produced in large quantities in Europe, especially in the Netherlands, Poland, and Germany. It once was produced in more than 150 plants in the United States alone, but is no longer produced there. Potato starch is usually isolated from cull potatoes, surplus potatoes, or waste streams from potato processing plants (Fig. 11).

Cull or surplus potatoes are washed with water to remove dirt and other foreign matter. They are then disintegrated by a saw blade rasp or hammermill and the mashed

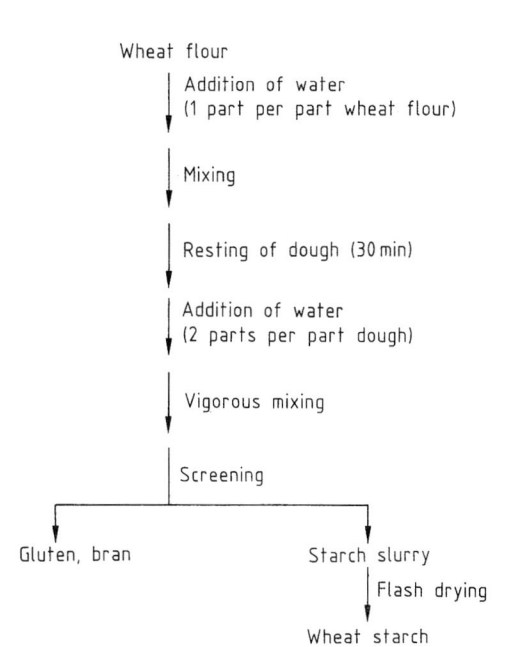

Figure 10. Flow sheet of the Batter process

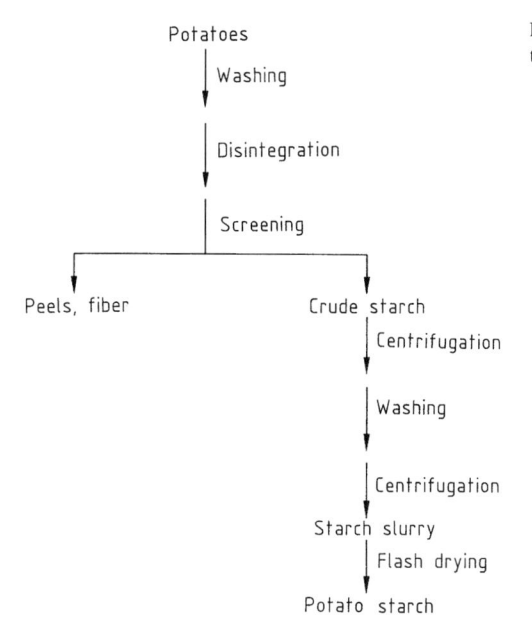

```
Wheat flour
   │
   │ Addition of water
   ▼ (1 part per part wheat flour)
Mixing
   │
   ▼
Resting of dough (30 min)
   │
   │ Addition of water
   ▼ (2 parts per part dough)
Vigorous mixing
   │
   ▼
Screening
   ├──────────────┐
   ▼              ▼
Gluten, bran   Starch slurry
                  │
                  ▼ Flash drying
               Wheat starch
```

Figure 11. Flow sheet for potato starch production

```
Potatoes
   │
   ▼ Washing
Disintegration
   │
   ▼
Screening
   ├──────────────┐
   ▼              ▼
Peels, fiber   Crude starch
                  │
                  ▼ Centrifugation
               Washing
                  │
                  ▼ Centrifugation
               Starch slurry
                  │
                  ▼ Flash drying
               Potato starch
```

product is screened to remove peels and fiber. During this process water containing sulfur dioxide is used to preserve color and inhibit enzymatic browning. Screening separates starch from pulp. The pulp can be reground and a second extraction performed to obtain a total starch yield of 12–19%, based on raw potatoes.

Starch isolated as described above is reslurried to remove water solubles and then partially dried in a continuous centrifuge. This is followed by another washing and centrifugation to obtain a purified slurry of potato starch. The slurry obtained from these operations is filtered and flash dried (< 175 °C inlet temperature). The potato starch, dried to 17 – 18 % moisture, is subsequently screened and packaged.

1.6.4. Rice Starch

Rice (*Oryza sativa* L.) cultivation probably originated in Asia where the crop is still very important. Rice is an important commercial crop in India, China, the United States, and Italy. In the United States more than 90 % of the rice crop is produced in Arkansas, Louisiana, Texas, and California. Rice grain consists of a hull surrounding an edible berry, or caryopsis. The outer fibrous layers of the caryopsis conceal a starchy endosperm and the germ. The endosperm consists principally of starch granules and protein. Little matrix protein is evident in rice endosperm.

Broken rice is reacted with 0.3 – 0.5 % sodium hydroxide solution to soften the grain and assist in protein removal. The alkali treated grain is hammermilled and the broken cell walls are removed by screening. The rice starch slurry is purified by centrifugation and dried.

1.6.5. Tapioca Starch

Tapioca starch is produced by processing the tuberous roots of the cassava plant (*Manihot utilissima* P., *Manihot esculenta*), a native of many equatorial regions. Cassava roots may be "sweet" (containing ≤ 50 mg of HCN per kilogram fresh root) or "bitter" (containing ≥ 250 mg of HCN per kilogram fresh root). Sweet root varieties are grown for food starch production while bitter varieties are used for other industrial purposes. In both cases, hydrogen cyanide content is eliminated or reduced during processing.

Root harvesting is usually manual and yields vary from 11 200 – 44 800 kg/ha. Typical root composition is 70 % water, 24 % starch, 2 % fiber, 1 % protein, and 3 % fats, minerals, and sugars. Tubers need to be processed within 48 h of harvest. They are first washed to remove dirt and then the outer skin is removed. The root is chopped into small pieces and passed through a rasp disintegrator. In this process hydrogen cyanide is released by natural enzymes and removed by washing.

Cassava root pulp is washed on screens. The starch passes through but the fibers are retained. The fiber fraction is used in fertilizer or cattle feed. After screening the starch slurry is centrifuged to separate the starch from fiber fines and other soluble material. Starch collected in this manner can be reslurried and recentrifuged to obtain greater purity. Sulfur dioxide (0.05 %) is usually added to process water to inhibit microbial growth. Slurried starch from the purification process is dewatered by centrifugation or filtration followed by drying on a drum, belt, or tunnel, or by flash drying. Flash drying is most common. Tapioca starch is mainly used unmodified but can be modified as is done with corn starch. The viscosity of tapioca starch dispersions is a function of plant variety, geographical conditions, harvest time, root age, soil fertility, rainfall, and manufacturing practices during starch production.

1.6.6. Arrowroot Starch

Arrowroot starch comes from the root of the tropical perennial *Maranta arundinacea*. Roots are harvested after 6–12 months growth and contain 20% or more starch, extracted as described for tapioca starch. The outer skin of arrowroot root must be completely removed to prevent off-color and off-taste in the starch. Arrowroot starch is mainly a product of China, Brazil, and St. Vincent in the West Indies.

1.6.7. Sago Starch

Sago starch is obtained from stems of several species of palms, principally *Metroxylon, Arenga,* and *Mauritia sp.* Trees of eight or more years old are used. Principal production occurs in Sarawak and New Guinea. Cut palm trunks are split and the pith removed and kneaded in water to release the starch which is screened to remove fiber, filtered, and dried. A single palm trunk can yield 100–200 kg sago starch. Sago starch granules are 20–60 μm in diameter.

Some comparative properties of commercial starches are given in Table 3.

1.7. Uses of Native Starches

Uses in Foods. Underivatized (native) starch is widely used in foods, both as an isolated, pure product and as a constituent of cereal grains, where it thickens and forms gels. It performs this function by undergoing gelatinization (see Section 1.4.). In principle, the function of starch is to act as a "water sink", absorbing much of the available free water and providing desirable structure and texture in many foods. Starch may also be cooked and dried to provide a cold water dispersible, pregelatinized product which is the basis of many instant gel and pudding products.

Pea starches have been proposed for use in preparation of gels requiring ca. 50% less starch than comparable corn starch gels, in production of extruded products, in production of pulpy products via freeze–thaw technology, and in production of instant starches with cold water swelling and gelling properties [33].

The viscoelastic behavior of starch gels as a function of time has been reported [34].

Starch may be hydrolyzed and/or isomerized to produce corn syrup sweeteners, corn syrup solids, or high fructose corn syrup (HFCS). Alternatively, it may be fermented to produce ethanol and other industrial chemicals (→ Ethanol).

Table 4. Properties of various corn starch derivatives

Type	Gelatinization temperature range, °C	Distinguishing properties
Normal	62–72	poor freeze–thaw stability
Waxy	63–72	little retrogradation
High-amylose	66–92	granules less birefringent than those of normal starches
Acid-modified	69–79	decreased hot-paste viscosity compared to unmodified starches
Cross-linked	higher than unmodified, dependent on degree of cross-linking	reduced peak viscosity, increased paste stability
Acetylated	55–65	excellent paste clarity and stability, good freeze–thaw stability
Phosphate, monoesters	56–66	reduced gelatinization temperature, reduced retrogradation
Hydroxypropyl	58–68 (DS 0.04)*	increased paste clarity, reduced retrogradation, good freeze–thaw stability

* DS = degree of substitution.

Other Uses. Unmodified starch also finds applications as textile sizings and adhesives, and is useful as a wet-end adhesive in the paper industry. It is used in oil field applications (in its pregelatinized form) where it functions as a suspending agent and maintains drilling fluid viscosity and decreases fluid loss.

In the pharmaceutical industry starch is used as a tablet binder, and in cosmetics rice starch in particular is employed as a dusting powder.

2. Modified Starches

Starch may be modified in numerous ways, both physically and chemically. These modifications are discussed in the following sections. Some properties of various corn starch derivatives are shown in Table 4.

2.1. Physically Modified Starches

Pregelatinized starch is a common ingredient in foods as well as other industrial applications. It is commercially prepared by putting a starch slurry in a trough between two counterrotating rolls internally heated by steam. The starch slurry is heated above the gelatinization temperature and is then drawn downward between the nip of the hot rolls to form a thin film. The film dries as the rolls turn and is scraped off and ground to produce a fine powder. Pregelatinized starch can also be prepared by spray drying a gelatinized starch paste. Pregelatinized starch quickly rehydrates in cold water. In foods it can be uniformly incorporated without heating to provide thickening, binding, and other properties. Pregelatinized starch is an essential component in foods in which cooking is not employed, such as in instant puddings, pie fillings, and cake frostings.

Pregelatinized starches have been incorporated into extruded oat-based snack foods where they provide well expanded, crisp products and lead to reduced product rancidity. Incorporation of starch in these foods also reduces wear and tear on the cooker extruder [35].

Pregelatinization of starch typically increases ease of digestibility, although not necessarily in a linear fashion [36]. Although pregelatinized starches are rapidly digested, they do not differ in their calorific value from other starches except for the case of high-amylose starch, which is so difficult to solubilize that any ungelatinized granules may be unavailable for digestion. Some precooked starches give low food value to some animals because the high viscosity produced on eating may induce satiety so that the animal eats less [37], [38].

2.2. Chemically Modified Starches

Starch molecules, both in free and granular form, are subject to chemical modification. Thus, starch is modified in various ways to produce acid-modified (thin-boiling), oxidized, cross-linked (e.g., by formation of distarch phosphates or adipates), partially esterified (either as carboxylate esters or phosphate esters), or partially etherified starch, or converted to cationic derivatives. These modifications, normally present at very low concentrations in starch preparations, produce dramatic differences in the physical and chemical properties of starch and lead to a multitude of uses in both food and nonfood applications.

2.2.1. Acid-Modified Starch

Treatment of starches such as dent corn starch with acid at a temperature below the gelatinization temperature produces virtually no change in granular appearance [39], and only a small change is noted in granular birefringence patterns [40], [41]. However, changes are produced that affect the gelation of acid-modified starch such as a lower hot paste viscosity [42]–[49], lower intrinsic viscosity [50], lower iodine affinity [50], and an increased solubility in water at temperatures just above the gelatinization temperature [44], [51], [52].

Acid treatments of gelatinized starch date back to 1811, when KIRCHOFF produced D-glucose [*50-99-7*] from starch by acid hydrolysis. Treatment of ungelatinized starch was reported by NAEGELI in 1874 [53] who found that such treatments dissolved much of the starch granule and left a residue of short starch molecules (DP ca. 25) (Naegeli dextrins). Commercial production of acid-modified starch dates to between 1897 and 1901 [54], [55].

Table 5. Relation of fluidity of starch dispersions to alkali number

Relative fluidity	Alkali number*
20	14.5
40	15
90	41.5

* Alkali number = milliequivalents of alkali consumed per 10 g of calculated dry starch during a standard 1 h digestion in 0.1 M sodium hydroxide at 100 °C.

Acid-modified, or thin-boiling starch has a lower molecular mass, or degree of polymerization than native starch [56], [57]. For example, the DP of potato starch, which is initially 1630, drops to 990 after 4 h treatment with 0.2 N HCl at 45 °C.

Acid-modified starch is prepared today by heating a starch slurry with 36–40% solids content to 40–60 °C (below the gelatinization temperature of the starch) with hydrochloric acid for one to several hours. When the desired degree of hydrolysis is achieved, as judged by the viscosity of the cooked modified starch, the acid is neutralized, and the granular modified starch is filtered, washed, and dried. Sometimes other acids are used such as sulfuric acid, with different reaction conditions.

The change in iodine binding capacity (see Section 1.4.4) or iodine affinity (which reflects the amount of amylose or linear starch material) after acid treatment varies with the type of starch. With corn starch only a little change in iodine affinity occurs for moderate acid conversions (relative product fluidity in the range 40; fluidity = 1/viscosity). This result is taken as an indication of preferential acid attack on the amylopectin component of the starch.

As the molecular size of starch decreases due to acid hydrolysis, the number of reducing end groups increases. This is measured commercially by the alkali number which increases as the extent of hydrolysis increases (there is an increase in fluidity) [58] (see Table 5). As acid hydrolysis proceeds, starch solubility in hot water increases [52].

Acid treatment causes hydrolysis of glycosidic bonds in the starch molecules

For *dispersed starch* the (1,4)-α-D linkages are more susceptible to acid cleavage than the (1,6)-α-D branch points [59]. However, in *granular starch*, portions containing many of the (1,4)-α-D linkages (amylose, long outer chains of amylopectin) are present in the crystalline regions of the granule. Hence, the (1,6)-α-D linkages become more accessible and susceptible to acid hydrolysis. For hydrolysis of potato starch, moisture conditioning without swelling may facilitate hydrolysis [57]. This, with other data, suggests that acid conversion occurs in two steps:

1) An initial attack on the amorphous regions, rich in amylopectin, especially the vulnerable (1,6)-α-D branch points

2) A slower attack on the more highly organized and crystalline regions of amylose and amylopectin

Uses. Acid-modified starches are employed in *textile manufacture* where they function as warp sizes [60], [61] to increase yarn strength and abrasion resistance during weaving. These uses have somewhat declined due to increased use of poly(vinyl alcohol). Starches with different fluidities are used for specific types of yarns. Thus, low fluidity starch is used for heavy yarns and high fluidity starch is used for light yarns. Starch concentrations of 10–12% are usual for such sizing applications.

Acid-modified starch is also used in the manufacture of *gypsum board* for dry wall construction where it serves to bond plaster and paper together.

Acid-modified starches find use in *starch gum confectionery*, an example being candy orange slices. To produce proper texture, the acid-modified starch is boiled together with sugar, corn syrup, and water to form highly concentrated pastes which form very firm gels on cooling. Not surprisingly, the granule structure is almost completely disintegrated in the cooking process. The composition of a typical starch gum confectionery is given below:

Sucrose	45.4 kg
Corn syrup (dextrose equivalent 63)	68.1 kg
Acid-modified starch (fluidity 70)	31.8 kg
Water	26.5 L

Production of such confectioneries begins with heating of corn syrup, water, and acid-modified starch to just below boiling in a steam kettle. When the starch has been dispersed, sugar is added and the mixture is heated to 103 °C and the heat removed. The water content of the cooked paste is adjusted to 22–23% and the mixture is put through an injection cooker at 140 °C. Flavor and color are added and the paste poured into molds and allowed to set to a firm gel. As a final step, the molded gel is coated with sucrose using a sugar sander.

Thin-boiling starch is also applied in *paper manufacture* as a paper size where it increases the strength of Kraft linerboard and improves bleached board printability. It is typically applied at 66–71 °C and pH 7.5–8.0. A high solid, acid-modified carrier starch is used to increase the production rate of corrugated board [62].

2.2.2. Oxidized Starch

Oxidation of starch has a long history beginning with LIEBIG [63] who used chlorine to oxidize starch in 1829. Somewhat later LIEBEN and REICHARDT [64] used bromine. These oxidations by halogens, and the hypohalites produced by their reaction with water, may proceed in four different ways, producing distinct products.

1) *Oxidation of reducing end aldehyde to carboxyl groups*, producing aldonic acid end units; namely, D-gluconic acid end units.

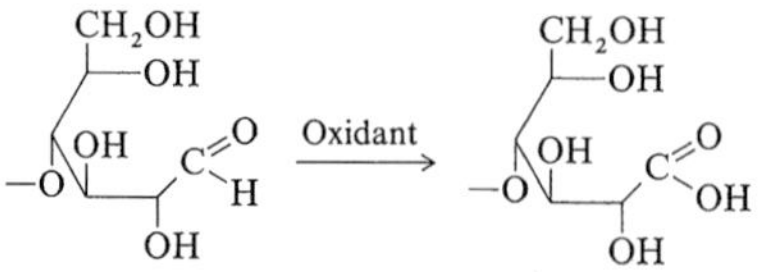

2) *Oxidation of the C-6 Methylol Group to a Carboxyl Group.* D-Glucuronic acid was isolated and identified from starch oxidized by bromine [65]. The oxidation probably proceeds by way of the 6-aldehyde derivative.

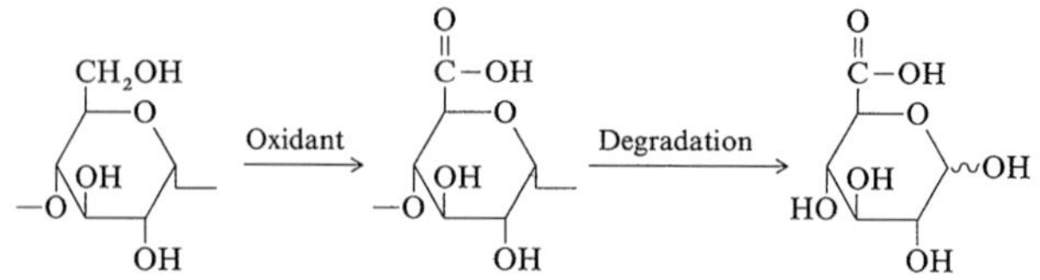

3) *Oxidation of Starch Secondary Hydroxyl to Ketone Groups.* This reaction is shown below for oxidation of the 3-OH to a carbonyl group. Bromine oxidation of starch [66] resulted in a product from which an oxime could be prepared, thus indicating the presence of ketone groups.

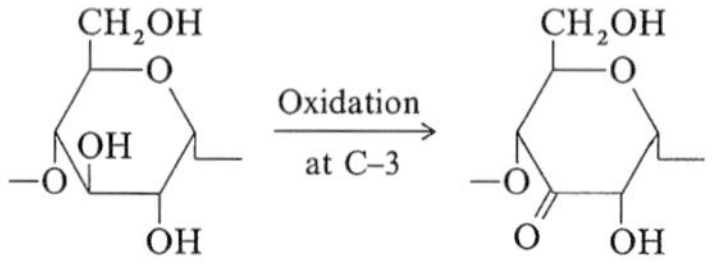

4) *Oxidation of 2,3-Glycol Units to Dialdehyde and Dicarboxylic Acid Units (Glycol Cleavage).* Evidence for this reaction was first obtained by FARLEY and HIXON [65]. Oxidative attack varies with the location of hydroxyl groups and with the nature of the oxidant. Oxidation at C-1, C-2, C-3, and C-6 determines the properties of the resulting starch product.

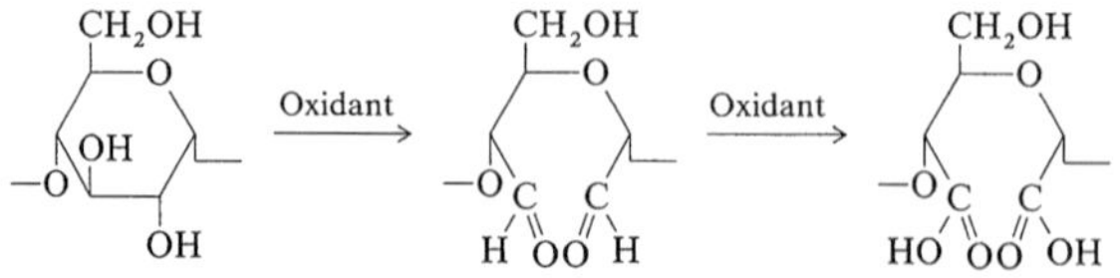

Common sodium hypochlorite-oxidized starch is granular in form with few surface changes noted by scanning electron microscopy. It retains its birefringence and granular appearance [67]; it is whiter because of the bleaching action of sodium hypochlorite. Oxidation occurs mainly in the amorphous regions of the granule. Oxidized starch is less stable to heat than native starch, tending to turn yellow or brown upon heating due to further reaction of carboxyl or aldehyde groups produced by the oxidation [68].

Heated aqueous dispersions of oxidized starches have lower intrinsic viscosities which depend on the extent of oxidation [67] and have a lower gelatinization temperature range than native starches [69], [70]. The aqueous dispersions have greater clarity than those of unmodified starch and exhibit a lower tendency to gel or setback [71].

Production. Commercially a controlled amount of chlorine gas is bubbled into dilute aqueous alkali to produce hypochlorite ions. The amounts of available sodium hydroxide and chlorine determine which of the four types of oxidation will principally occur. Reaction is performed on a rigorously stirred starch slurry of about 33–44% solids content. The pH of the slurry is adjusted to 8–10 with sodium hydroxide and chlorine added to produce hypochlorite with 5–10% available chlorine. Temperature is maintained at 21–38 °C by controlling the rate of hypochlorite addition or by external cooling. When the desired degree of oxidation is obtained (usually determined by viscosity analysis) the pH is decreased to 5–7 and excess chlorine is destroyed by reaction with reducing agents such as sodium bisulfite or sulfur dioxide. The modified starch is isolated by centrifugation or filtration and is washed and dried.

Uses. Oxidation introduces an increasing number of carbonyl and carboxyl groups into the starch molecules, which in turn cause changes in the chemical and physical properties [72], [73]. Most commercial oxidized starches contain about 1.1% carboxyl groups. These groups play a predominant role in determining the properties of the starch derivative which is used mostly in the *paper industry*. 80–85% of oxidized starch is used in paper production where it acts as a coating binder for high-solid pigment coating colors. In this application, pigment compatibility is important and the starch should not affect the rheology or the water binding characteristics of the coating color [69], [74].

Oxidized starch is used also as a surface size in paper and paperboard manufacture. Here it seals pores, improves surface strength, and provides ink holdout. As a wet-end additive in paper manufacture, oxidized starch reduces pigment retention by acting as a pigment dispersant [75].

Oxidized starches have a long history of use in *textile manufacture* where they function primarily as warp sizes but also in finishing and printing. As warp sizes they provide abrasion resistance for yarn but are now being replaced by synthetic polymers.

Oxidized starch has been used in *laundry finishing* and in making *construction materials* such as wallboard and ceiling tiles. In these applications, starch provides sizing and adhesive characteristics.

Lightly oxidized starches have application in *breading batters* for deep fried foods where they give good adhesion of the breading to the food [76].

Periodate oxidized starch is a good substitute for formaldehyde or glutaraldehyde as *hardening agents* in gelatin-immobilization of yeast cells [77].

2.2.3. Cross-Linked Starch

Cross-linking of starch reinforces the intermolecular binding by introducing covalent bonds to supplement natural intermolecular hydrogen bonds. Cross-linking restricts granule swelling, decreases peak viscosity on cooking, and increases the stability of the gelatinized granule. Most cross-linked starches are quite lightly derivatized, containing ca. one cross-link for every 100 to 3000 anhydro-D-glucosyl units (degree of substitution, DS 0.01 – 0.003).

Cross-linked starch granules also appear unchanged by microscopic investigation but they are more resistant to swelling than unmodified starch. This is especially evident when their dispersions are placed in a Brabender viscoamylograph or similar instrument. Cooked cross-linked starch granules are less fragile than native starch granules and more resistant to high shear, elevated temperatures, or low pH [78]. At low degree of substitution an increase in peak viscosity is observed when measured by a Brabender amylograph but with high DS no viscosity peak occurs when aqueous dispersions are gelatinized. Loss of peak viscosity is caused by constrained swelling of the granule due to extensive cross-linking of the starch molecules. If the degree of cross-linking is sufficient, gelatinization will not take place at 100 °C or even in an autoclave. Cross-linked starches of low DS ($<$ 1) have improved textures; they are less rubbery and stringy than pastes and gels made from unmodified native starches. Cross-linking is sometimes performed in combination with oxidation [79], phosphorylation [80], hydroxyalkylation [80], or esterification [81].

Production. Cross-linking reactions are usually performed on starch slurries which are then reacted with specific bi- or polyfunctional reagents such as phosphorus oxychloride, sodium trimetaphosphate, epichlorohydrin, or mixed anhydrides such as those of acetic and adipic acid. Slurry reaction temperatures may range from 25 to 50 °C. After cross-linking is complete, the cross-linked starch is isolated by filtration, washed to remove excess reagents and salts, and dried.

The reaction of starch with *phosphorus oxychloride* to produce a distarch phosphate is shown below:

$$2\,\text{Starch}-\text{OH} + \text{OPCl}_3 \xrightarrow[-\text{NaCl}]{\text{NaOH}} \text{Starch}-\text{O}-\overset{\overset{\displaystyle O}{\|}}{\underset{\underset{\displaystyle O^-\text{Na}^+}{|}}{\text{P}}}-\text{O}-\text{Starch}$$

Typically, the reaction is carried out at pH 8 to 12 and with 0.005 to 0.25% phosphorus oxychloride. Frequently, sodium chloride or sodium sulfate is added to the slurry to inhibit granule swelling and starch leaching from the granule [82]. When the concentration of phosphorus oxychloride is increased from 0.005 – 0.25% to 1% or more, the resulting starch phosphates are highly cross-linked.

Epichlorohydrin is commonly used [79] to cross-link starch under alkaline conditions, as shown below:

Table 6. Solubility of waxy and cross-linked waxy sorghum in dimethylsulfoxide at 25 °C

Starch	Time, h	Solubility, %
Waxy	20	100
Cross-linked waxy	45	20

$$\text{Starch}-\text{OH} + \text{H}_2\text{C}\underset{\text{O}}{-}\text{CH}-\text{CH}_2-\text{Cl} \xrightarrow{\text{OH}^-} \text{Starch}-\text{O}-\text{CH}_2-\underset{\text{OH}}{\text{CH}}-\text{CH}_2\text{Cl}$$

$$\downarrow \text{OH}^-$$

$$\text{Starch}-\text{O}-\text{CH}_2-\underset{\text{OH}}{\text{CH}}-\text{CH}_2-\text{O}-\text{Starch} \xleftarrow{\underset{\text{Starch-OH}}{\text{OH}^-}} \text{Starch}-\text{O}-\text{CH}_2-\text{CH}\underset{\text{O}}{-}\text{CH}_2$$

Common *mixed anhydride* cross-linking agents are those of acetic and dicarboxylic acids, such as adipic acid [81]. Other reagents used to cross-link starch include (but are not limited to) vinyl sulfone, diepoxides, cyanuric chloride, toluene-2,4-diisocyanate, *N,N*-bismethylene-bisacrylamide, phosgene, imidazolides of polybasic carboxylic acids, and aldehydes such as formaldehyde, acetaldehyde, or acrolein [83].

Properties. Cross-linked starches are less soluble in hot water than native starches even though the degree of molecular cross-linking is very low [84]. Material extractable by dimethyl sulfoxide (DMSO) is also reduced by cross-linking (see Table 6) [51].

Cross-linking of starch at low degrees has little effect on in vitro digestibility. The calorific value of the cross-linked derivative is the same as that of the unmodified, native starch [85].

Cross-linked starch provides a stable, high viscosity paste that may be subject to high temperature, high shear, or low pH.

Uses. Cross-linked starches with a low DS are extensively used in *foods*. In extrusion cooking, high degrees of cross-linking are necessary to prevent complete granular disintegration. When a short, salve-like consistency is desired, cross-linked waxy corn, potato, or tapioca starch, which is usually also phosphorylated, acetylated, or hydroxypropylated, is used. Cross-linked starch provides texture in salad dressings due to its stability to both low pH and the high shear experienced during homogenization. These derivatized starches are also used in batter mixes for deep-fried foods, fruit pie fillings, puddings, cream style corn, soups, gravies, sauces, and baby foods [78], [86], [87]. In some food systems the starch imparts a desirable pulpy texture [88].

Cross-linked waxy starches are constituents of cake mixes where they improve cake volume, crumb texture, and increase shelf-life, due to decreased rate of becoming stale [89]. Acid-converted cross-linked starches are dispersible in cold water and form gels [90], they are thus used in instant puddings.

Cross-linked starches are used also in antiperspirants [91] and textile printing pastes where they produce high viscosity and a short, cohesive texture [92]. They have been employed in oil-well drilling muds, printing inks, charcoal briquette binders, fiberglass sizing [93], and textile sizing [94].

2.2.4. Starch Esters

Carboxylate Esters. Starch acetates [9045-28-7] have been known for more than 100 years [95]. Those of high DS are soluble in organic solvents with a low polarity but because of the branched amylopectin molecules they are not economically competitive with similar cellulose derivatives. Material with a lower degree of substitution (0.3 – 1.0) is water soluble, whereas most commercial products are even more lightly substituted (DS 0.01 – 0.2). In anhydrous media, the acetylation reaction is very sluggish and, in addition, the granules require prior swelling to allow penetration by the acetylation reagents. Starch acetates can be produced in many ways. Commercial processes typically employ acetic anhydride, with or without catalysts.

Reaction with Acetic Anhydride [108-24-7] *in Pyridine* [83]. Starch granules treated with acetic anhydride alone at 20 °C for 5 months showed no reaction. Pyridine treatment [96] rendered the starch granule reactive [97]:

Reaction with Acetic Anhydride. Starch does not react with acetic anhydride at ambient tem-perature and only slightly at 90 – 140 °C. Cooking and disruption of starch granules followed by ethanol precipitation increases reactivity but acidic catalysts cause starch degradation [98].

Reaction with Acetic Anhydride and Glacial Acetic Acid. Reaction requires the addition of an acidic catalyst such as sulfuric acid, *p*-toluenesulfonic acid, perchloric acid, or phosphoric acid. The yield of acetylated (and degraded) starch acetate increases at elevated temperature.

Reaction with Acid Anhydride in Dimethyl Sulfoxide (DMSO) [99]. Starch derivatives of acetic, propanoic, and butanoic anhydrides have been prepared up to a DS of 0.08. Triethylamine serves as a catalyst and acid scavenger.

Reaction with Carboxylic Acids. Starch does not react with aqueous acetic acid but with *glacial acetic acid* at 100 °C for 5 – 13 h [100], to give a product with 3 – 6 % content of acetyl groups. Treatment of starch with *concentrated formic acid* leads to gelatinization and simultaneous esterification [101].

Acetylation with Ketene. This reaction [102] can produce starch with an acetyl content of 2.2 – 9.4 %. The reaction is typically performed in acetic acid, diethyl ether, or acetone with an acid catalyst, the most effective of which are sulfuric acid and *p*-toluenesulfonic acid.

These reactions typically lead to a considerable degree of acetylation in the starch granule. More lightly substituted starches (low DS starch acetates) can be produced, e.g., by reaction of starch with *acetic anhydride* in the presence of a base such as 3% sodium hydroxide [103]. The reaction is carried out at pH 7–11 and alternative bases such as other alkali metal hydroxides, calcium hydroxide, or sodium carbonate can be employed. Low DS starch acetates can be produced by reaction of granular starch with *vinyl acetate* [108-05-4] [104]:

$$\text{Starch}-\text{OH} + \text{CH}_3\overset{\overset{\text{O}}{\|}}{\text{C}}-\text{O}-\text{CH}=\text{CH}_2 \xrightarrow{\text{Na}_2\text{CO}_3} \text{Starch}-\text{O}-\overset{\overset{\text{O}}{\|}}{\text{C}}-\text{CH}_3 + \text{CH}_3\text{CHO}$$

This reaction requires basic catalysts, such as alkali metal hydroxides, amines, or sodium carbonate. Acetylation is sometimes combined with hydroxypropylation to produce starches having excellent freeze–thaw stability [105].

Properties. Starch acetates with degrees of substitution of ca. 0.5 have roughly the same solubility as normal starch. Starch esters are readily cleaved by alkali, for example, acetylated granular corn starch of 1.8% acetyl content can be deacetylated in 4 h at 25 °C at pH 11. At low DS (< 1), microscopic examination shows no difference between the granule of starch ester and the native starch granule. Low DS starch acetates have a lower gelatinization temperature, an increased hot cooked paste viscosity, easier dispersion on cooking, and a decreased setback or final gel strength.

Higher DS starch acetates, with up to 15% acetyl content, are soluble in water at 50–100 °C [106], [107]. When the acetyl content exceeds 40%, they are only soluble in organic solvents with lower polarity (aromatic hydrocarbons, ketones, nitroalkanes, etc.). Such highly substituted derivatives are insoluble in water, diethyl ether, aliphatic alcohols, and aliphatic hydrocarbons [107].

Starch acetates with a high degree of substitution (DS 2–3) have increased densities, specific rotations, and melting temperatures.

Uses. **Low DS starch acetates** have major use in *foods* where they provide thickness, body, and texture. Certain food applications require a starch derivative with a bland taste, stability to low pH, stability to high shear and temperature, and freeze–thaw stability. Acetylation and/or cross-linking provide such qualities.

Starches containing *0.5–2.5% acetyl content* (the FDA permits up to 2.5% acetyl content) provide gels of exceptional brilliance and clarity. Such starch acetates are used in canned, baked, and dry foods such as instant gravies and pie fillings. They are used in baby foods and cream pie filling to provide texture. Products which are frozen and thawed (perhaps several times) before use also benefit from a content of starch acetates. Such products include frozen pies and frozen gravies.

Low DS starch acetates are also used in *nonfood applications* such as warp sizing in textiles to provide good yarn adhesion, tensile strength, and flexibility. Starch acetate is easily removed due to its solubility in water [108]. Starch acetates have also been used as fiberglass forming sizes [109] and in the paper industry as a surface size where it improves printability and gives uniform surface strength, porosity, solvent resistance, and abrasion resistance [108]. Such starch esters have also found use in gummed tape formulations.

High DS starch acetate derivatives have fewer uses but have been employed in thermoplastic molding [110] and in films as plasticizers [111].

Phosphate Esters. Because of their ionic character, cooked pastes of starch phosphate monoesters are "thinned" or decreased in viscosity when salts are added. Paste viscosity is also affected by pH [112].

Starch phosphate esters react with aluminum, titanium, and zirconium salts to form insoluble precipitates [113]. Starch phosphates also react with cationic dyes such as methylene blue [114], [115]. Starch phosphate esters are chemically compatible with gelatin, vegetable gums, poly(vinyl alcohol), and polyacrylates [116], [117].

Introduction of a few *distarch phosphate groups* which cause chain cross-linking greatly increases the acid stability of the phosphorylated starch derivative. In starch monophosphate esters, 28% of the phosphate ester groups are present at C-2, 9% are at C-3, and 63% at C-6 of the corresponding D-glucose units. The pK_1 ranges between 1 and 2, and pK_2 between 6 and 7.

Production. Starch phosphate esters can be prepared by reaction of starch with inorganic phosphates such as sodium tripolyphosphate [7758-29-4] [116]. The reagent produces phosphate esters with a DS up to 0.01:

$$\text{Starch-OH} + \text{Na}_5\text{P}_3\text{O}_{10} \longrightarrow \text{Starch-O-}\underset{\underset{\text{O}^-\text{Na}^+}{|}}{\overset{\overset{\text{O}}{\|}}{\text{P}}}\text{-O}^-\text{Na}^+ + \text{Na}_3\text{HP}_2\text{O}_7$$

To do so, starch is impregnated with phosphate salts in an aqueous slurry. Starch and phosphate are mixed together for 10–30 min, then filtered. Typically 40–60% of the phosphate salt is retained by the starch. The pH is carefully adjusted to 5–8.5, if the pH is too high, distarch phosphate cross-links will be formed. The final starch phosphate salt is heated at lower temperature (60 °C) to remove moisture, then heated at 120–170 °C to effect phosphorylation. This drying–reaction process can be performed by belt or spray drying [118].

Starch phosphates with a *low degree of substitution* can be produced by extrusion of starch with sodium tripolyphosphate [119]. Maximum degree of substitution (DS = 0.0053) is obtained at 200 °C, tripolyphosphate concentration ≥ 1.4 g/100 mL, and pH 8.5. Pastes of these starch phosphates have exceptional clarity.

Properties. Cooked aqueous dispersions of starch phosphates compared to natural starch have increased gel clarity, increased viscosity, a more cohesive texture, and gels that are stable to retrogradation [116].

Uses. Starch phosphate monoesters find application in paper, textiles, adhesives, as scale inhibitors, in flocculation, and numerous other applications. In the *paper industry*, starch phosphates are used as wet-end additives which improve paper strength and filler retention [120], while in *textiles* these esters act as sizes for polyester–cotton, polyester–rayon, and pure cotton yarns [117]. They are also used as thickeners in textile printing inks beause they improve ink penetration into cotton.

In *adhesives,* these esters improve storage stability and promote rapid bond strength. Use of 10 mg/L of starch phosphates prevents or inhibits *scale-forming deposits* [121]. *Coal washery tailings* can be flocculated by starch phosphates, in *food* they function as effective oil-in-water emulsifiers [122] as well as good thickeners and stabilizers for puddings [123].

Starch phosphates have also been used in *medicinal films* for the treatment of skin wounds and burns [124]. The film promotes rapid healing and reduces incidence of infection.

2.2.5. Starch Ethers

The two major types of commercially important starch ethers are hydroxyethyl and hydroxypropyl ethers.

Production. The *hydroxyethyl derivatives* are typically produced by reaction of ethylene oxide [75-21-8] with starch under alkaline conditions:

$$\text{Starch}-\text{OH} + \text{CH}_2\underset{\text{O}}{-}\text{CH}_2 \xrightarrow{\text{NaOH}} \text{Starch}-\text{O}-\text{CH}_2\text{CH}_2-\text{OH}$$

A starch slurry of 35–45% solids content is made alkaline with sodium hydroxide, purged with nitrogen gas, and then ethylene oxide is added. Depending on the DS desired, the reaction is conducted for a while at 25–50 °C and the hydroxyethylated product is filtered, washed to remove excess reagent and salts, and dried. The mechanism apparently involves abstraction of a hydroxyl proton from a starch molecule by alkali, followed by $S_N 2$ attack of the starchate anion on the highly strained oxirane ring. Commercial hydroxyethyl starch has a DS of about 0.2. Most of the newly formed ether groups (76–85%) are formed at the oxygen bound to C-2, because the order of hydroxyl reactivity is 2-OH > 3-OH ≈ 6-OH [125].

Hydroxypropyl derivatives are prepared in a similar manner as hydroxyethyl derivatives, following the same mechanism with most substitution occurring at 2-OH. To an aqueous starch slurry of 32–38 wt%, sodium hydroxide is added to a concentration of 0.5–1.0% (based on the dry weight of the starch). Sodium sulfate (5–15%) is sometimes added to prevent starch swelling. Propylene oxide [75-56-9] is added in the amount of 5–10% of the dry weight of starch and the reaction carried out at 52 °C for ca. 24 h. The hydroxypropyl starch is filtered, washed, and dried.

Properties. Hydroxyethyl [9005-27-0] and hydroxypropyl [9049-76-7] starch are the commercially most important starch ethers. At low degree of substitution the physical properties of these ethers are similar to those of low DS starch acetates. As the DS increases the gelatinization temperature range decreases [126], but rate of granule swelling, clarity of the dispersion [127], and dispersion cohesiveness increases. The tendency to gel or setback is lowered, and solubility in methanol or ethanol is increased [128].

It is possible to prepare films from hydroxypropyl high-amylose corn starch which are water-soluble, transparent, and oxygen impermeable [129]. Hydroxyalkylation results in decreasing digestibility at increasing degree of substitution.

Lightly substituted (low DS) hydroxyethyl or hydroxypropyl starch is similar in properties and appearance to unmodified starch. An increasing DS of either derivative decreases inter-starch hydrogen bonding which lowers pasting temperature and inhibits retrogradation. Hydroxyethyl starch pastes are more stable to added salts including those present in hard water than are the cationic or anionic starches. Hydroxyethyl ether linkages are stable over a wide pH range.

Uses. *Hydroxyethyl starch* is used in the *paper industry* as a surface sizing and coating [108] where it provides strength, stiffness, and ink holdout. It is also applied in coatings for pigmented paper [130] to achieve good printing quality. Hydroxyethyl starch is also a wet-end additive in papermaking [120] and has been used in many *adhesives* such as bag pastes, case sealing, and label and envelope adhesives because of its excellent water holding and film forming properties [128].

In *textiles,* low DS starch ethers are employed as warp sizing. In *foods,* starch ethers (which are also usually cross-linked) provide thickness and texture and the ability to hold water at low temperature. They are used in milk-based products [131] and salad dressings [132]. The good freeze–thaw stability of the starch ethers permits their use in frozen puddings, sauces, gravies, and fruit pie fillings [133]. Hydroxypropyl, high-amylose starches produce edible, water-soluble food films [134].

2.2.6. Cationic Starch

Cationic starches include tertiary aminoalkyl ethers, quaternary ammonium ethers, aminoethylated starches, cyanamide derivatives, starch anthranilates, and cationic dialdehyde starch (although the last three are commercially less important).

Tertiary Aminoalkyl Derivatives. These derivatives are produced by reaction of the appropriate chloroalkane tertiary amine with starch under alkaline conditions, followed by acidification to produce the desired cationic derivative:

$$\text{Starch} - \text{OH} + \text{Cl} - \text{CH}_2\text{CH}_2 - \text{N}(\text{CH}_2\text{CH}_3)_2$$

$$\downarrow \text{1) OH}^- \quad \text{2) H}^+$$

$$\text{Starch} - \text{O} - \text{CH}_2\text{CH}_2 - \overset{\text{H}}{\underset{|}{\text{N}}}{}^+(\text{CH}_2\text{CH}_3)_2 \; \text{Cl}^-$$

Reactions are conducted at pH 10.5–12 and 25–50 °C. The mechanism involves the reaction of starch with a nitrogen heterocycle to produce the desired product. Substitution mainly occurs at C-2 [135].

Quaternary Ammonium Starch Derivatives. These derivatives are obtained by reaction of starch with 2,3-epoxypropyltrimethylammonium chloride [136]:

$$Cl-CH_2-\underset{OH}{CH}-CH_2-N^+(CH_3)_2 \ Cl^- \quad \text{Chlorohydrin form}$$

$$\downarrow OH^-$$

$$\underset{O}{CH_2-CH}-CH_2-N^+(CH_3)_3 \ Cl^- \quad \text{Epoxide form}$$

$$\downarrow \text{Starch-OH}$$

$$\text{Starch}-O-CH_2-\underset{OH}{CH}-CH_2-N^+(CH_3)_3 \ Cl^-$$

Starch is reacted with this reagent in a 40–46% slurry at pH 11.0–12.0. The ratio of alkali to epichlorohydrin is ca. 2.8:1 and excess reagent is removed by vacuum distillation or solvent extraction.

Aminoethylated Starches. These derivatives are produced by reaction of starch with ethyleneimine [151-56-4] [137] followed by acidification:

$$\text{Starch}-OH + \underset{\underset{H}{N}}{CH_2-CH_2} \longrightarrow \text{Starch}-O-CH_2CH_2-NH_2$$

$$\downarrow HX$$

$$\text{Starch}-O-CH_2CH_2-N^+H_3 \ X^-$$

This reaction is performed with a starch slurry and typically carried out at 75–120 °C [138].

Cyanamide Derivatives. Cyanamide [420-04-2] and its dialkyl derivatives react with granular starch at pH 10–12 to produce iminocarbamate derivatives which can be acidified to give cationic starch derivatives [139]. This reaction is conducted under alkaline conditions

$$\text{Starch}-OH + R_2N-C\equiv N \xrightarrow{OH^-} \text{Starch}-O-\underset{\parallel}{C}-NR_2$$
$$\overset{H}{\underset{}{N}}$$

$$\downarrow HX$$

$$\underset{\text{Starch}-O-\underset{\parallel}{C}-NR_2}{N^+H_2 \ X^-}$$

Starch Anthranilates. Starch can be reacted with isatoic anhydride at pH 7.5–9.0 [140]

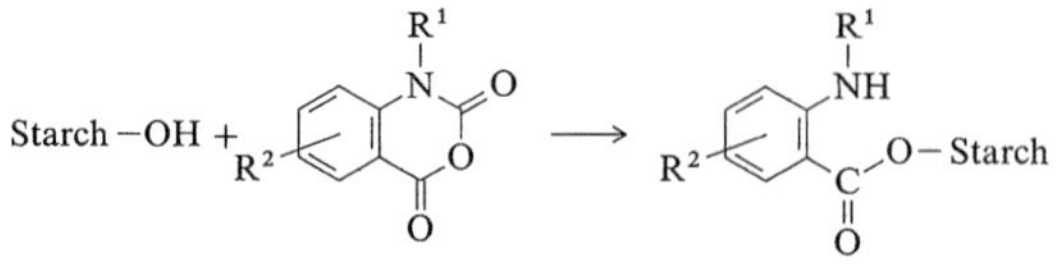

Diazotization of the resultant amine with nitrous acid produces the desired cationic starch derivative

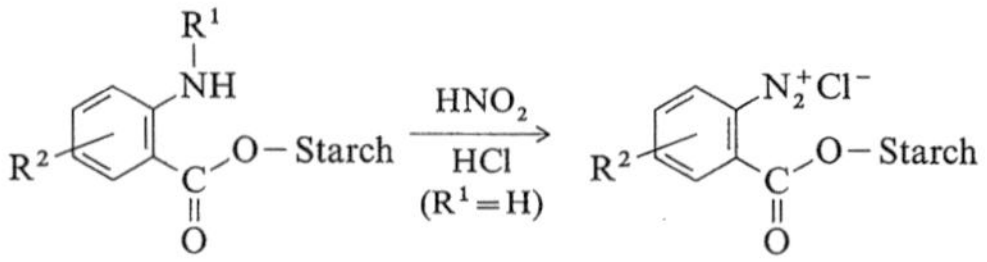

Cationic Dialdehyde Starch. Starch can be converted into a cationic form by initial oxidation to dialdehyde starch followed by reaction with hydrazine [*302-01-2*] or hydrazides (the reaction with betaine hydrazide hydrochloride is shown below) [141]

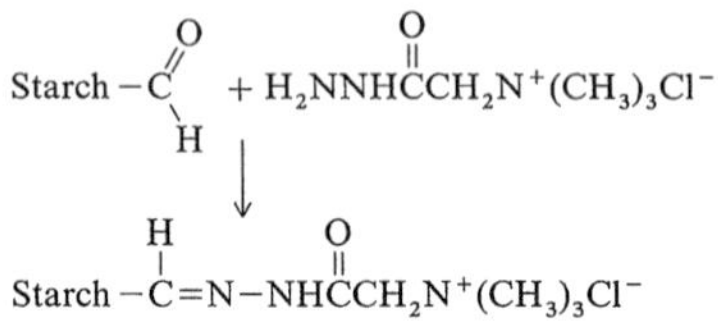

$$\text{Starch}-\underset{H}{\overset{O}{\underset{\|}{C}}} + H_2NNH\overset{O}{\overset{\|}{C}}CH_2N^+(CH_3)_3Cl^-$$

$$\downarrow$$

$$\text{Starch}-\overset{H}{\underset{|}{C}}=N-NH\overset{O}{\overset{\|}{C}}CH_2N^+(CH_3)_3Cl^-$$

Granular starch is dispersed in an aqueous slurry (15% solids) and reacted with a hydrazine or hydrazide derivative (3–5% concentration) at an initial pH of 4.5. The pH drops to between 2.5 and 3.2 as the reaction proceeds at 90–95 °C for 2–3 h.

Properties. The gelatinization temperature range of cationic derivatives decreases as the DS increases [142]. Such cationic starch dispersions have increased stability and clarity compared to dispersions of the native starches from which they are derived. Also, zeta potential measurements show that these cationic starch derivatives have a net positive charge at a pH of between 4 and 9 [143].

Cationic starches react chemically with anionic dyes such as Light Green SF Yellowish or Acid Fuchsin [114]. On electrophoresis these derivatives migrate toward the cathode [144]. Commercial cationic starches typically have nitrogen contents of 0.2–0.4% [145].

Uses. Cationic starches are used in the paper industry as wet-end additives and sizings, as coating binders for clay, as warp sizing agents in textiles, as flocculation agents for clay, titanium dioxide, coal, iron ore, silt, anionic starch, and cellulose, and for many other uses.

As a wet-end additive in *paper*, cationic starches increase the retention of fines and improve paper strength while also increasing drainage [146]. Cationic starches may also be used as paper sizing agents [143].

Table 7. Starch production worldwide and in the EU 1995

	World	EU*
Raw material input, 10^6 t		
Maize	44.22 (53%)	5.8 (37%)
Potato	13.65 (16%)	6.6 (42%)
Wheat	5.27 (6%)	3.3 (21%)
Tapioca	18.25 (22%)	
Others	2.6 (3%)	
Total	83.99	15.7
Starch produced, 10^6 t		
Maize starch	27.6 (74%)	3.6 (55%)
Potato starch	2.7 (7%)	1.3 (20%)
Wheat starch	2.92 (8%)	1.7 (25%)
Tapioca starch	3.65 (10%)	
Others	0.35 (1%)	
Total	37.15	6.6

*Source: Association des Amidonneries de Cereales de la C.E.E. (AAC), 1996.

As a *coating binder*, cationic starch promotes the binding of clay to fiber [147]. In *textiles* it is a warp sizing agent [148] and provides good lubrication and yarn abrasion resistance. The ability of cationic starch to flocculate fines is seen in its use in treating raw primary sludge [149]. Its ionic nature makes it a useful *emulsion breaker* for either water-in-oil or oil-in-water emulsions [150].

3. Economic Aspects

Global Starch Production [151] (Table 7). 99% of the global starch production of $\approx 37 \times 10^6$ t is from maize, manioc/cassava, wheat and potatoes, with 27.6×10^6 t (74%) maize starch, 3.7×10^6 t (10%) tapioca starch, 2.9×10^6 t (8%) wheat starch and 2.7×10^6 t (7%) potato starch. Starch is predominantly produced in highly industrialized countries such as the United States, the EU and Japan. A breakdown of starch production by continent is given below:

America	18.95×10^6 t	(51%)
Asia	9.66×10^6 t	(28%)
Europe	7.43×10^6 t	(20%)
Africa	0.74×10^6 t	(2%)
Australia/Oceania	0.37×10^6 t	(1%)

In small production units, mainly in Asia, starches are separated also from other starch-containing materials. These materials include: amaranth, arrowroot, banana, canna, cow cockle, faba/mung beans, kouzou, lentils, lotus roots, quinoa, sago palm, sorghum, sweet potatoes, taro, water chestnut, wild rice, and yam.

Table 8. Consumption of starch and starch derivatives by sector in the EU in 1995*

Sector	Consumption	
	10^6 t	%
Food	3.3	55
Confectionery and drinks	1.98	33
Processed food	1.32	22
Nonfood	2.7	45
Paper and corrugating	1.68	28
Chemicals fermentation products	0.78	13
Other industrial products	0.18	3
Feeds	0.06	1
Total	6.0	

*Source: AAC 1996.

Starch Production in the EU (Table 7). The starch industry in Western Europe comprises 20 different companies, which produced 6.6×10^6 t of starch in 1995 with a share of 55 % maize starch (3.6×10^6 t), 25 % of wheat starch (1.7×10^6 t) and 20 % of potato starch (1.3×10^6 t).

EU Consumption of Starch and Starch Derivatives (Table 8). In 1995, 6.0×10^6 t of starch products were consumed in the EU, split into 3.1×10^6 t of starch hydrolysates (starch sweeteners), including isoglucose and polyols (52 %); 1.7×10^6 t of native starches (28 %) and 1.2×10^6 t of modified starches (20 %). 55 % of these 6.0×10^6 t were consumed in the food sector and 45 % in the nonfood sector.

The breakdown by application areas/market sectors was as follows: sweets and drinks (confectionery, drinks, fruit processing) 33 %, processed food (convenience food, bakery, food ingredients and food preparations, dairy products and ice cream) 22 %, paper and corrugating (papermaking, corrugating and paper processing) 28 %, chemical and fermentation products 13 %, other industrial products 3 %, and feeds, 1 – 2 %.

Starch Production in the United States [152]. In 1994/1995, 2.6×10^6 t of starch products (corn starch, modified starches, dextrins) and 12.8×10^6 t of refinery products (glucose syrups, high fructose corn syrups, dextrose, corn syrup solids, maltodextrins, and fructose) were produced in the United States. High fructose corn syrup (HFCS) 42 accounted for 3.7×10^6 t, HFCS 55 for 5.1×10^6 t of these 12.8×10^6 t. From the total of 15.3×10^6 t of starches and starch products, 0.44×10^6 t were exported.

4. References

[1] R. Whistler, J. BeMiller, E. Paschall (eds.): *Starch: Chemistry and Technology,* 2nd ed., Academic Press, New York 1984.
[2] D. Bechtel, I. Zayas, L. Kaleikau, Y. Pomeranz, *Cereal Chem.* **67** (1990) 59.
[3] G. Inglett (ed.): *Corn: Culture, Processing, Products,* AVI Publishing, Westport 1970.
[4] Y. Pomeranz, L. Munck (eds.): *Cereals: A Renewable Resource,* Amer. Assn. Cereal Chemists, St. Paul 1981.

[5] R. Whistler, J. Daniel in O. Fennema (ed.): *Food Chemistry*, Marcel Dekker, New York 1985, pp. 69–137.
[6] R. Singhal, P. Kulkarni, *Starch/Staerke* **42** (1990) 5.
[7] R. Singhal, P. Kulkarni, *Starch/Staerke* **42** (1990) 102.
[8] O. Paredes-Lopez, M. Schevenin, D. Hernandez-Lopez, A. Carabez-Trejo, *Starch/Staerke* **41** (1989) 205.
[9] A. Cortella, M. Pochettino, *Starch/Staerke* **42** (1990) 251.
[10] Z. Nikuni et al., *Mem. Inst. Sci. Ind. Res., Osaka Univ.* **26** (1969) 1.
[11] Z. Nikuni, *Staerke* **30** (1978) 105.
[12] D. French, *J. Jpn. Soc. Starch Sci.* **19** (1972) 8.
[13] D. French in R. Whistler, J. BeMiller, E. Paschall (eds.): *Starch: Chemistry and Technology*, 2nd ed., Academic Press, New York 1984, pp. 183–247.
[14] L. Slade, H. Levine in R. Millane, J. BeMiller, R. Chandrasekaran (eds.): *Frontiers in Carbohydrate Research*, Elsevier, New York 1989, 215.
[15] R. Tester, W. Morrison, *Cereal Chem.* **67** (1990) 551.
[16] R. Tester, W. Morrison, *Cereal Chem.* **67** (1990) 558.
[17] L. Jing-ming, Z. Sen-lin, *Starch* **42** (1990) 96.
[18] S. Craig, C. Maningat, P. Seib, R. Hoseney, *Cereal Chem.* **66** (1989) 173.
[19] K. Meyer, M. Wertheim, P. Bernfeld, *Helv. Chim. Acta* **23** (1940) 865.
[20] K. Meyer, P. Bernfeld, W. Hohenemeser, *Helv. Chim. Acta* **23** (1940) 885.
[21] W. Hassid, R. McCready, *J. Am. Chem. Soc.* **65** (1943) 1157.
[22] J. Blackwell, A. Sarko, R. Marchessault, *J. Mol. Biol.* **42** (1969) 379.
[23] A. Sarko, R. Marchessault, *J. Am. Chem. Soc.* **89** (1967) 6454.
[24] V. Rao, P. Sundararajan, C. Ramakrishnan, G. Ramachandran in G. Ramachandran (ed.): *Conformation of Biopolymers*, vol. **2**, Academic Press, London 1967, p. 721.
[25] F. Senti, L. Witnauer, *J. Am. Chem. Soc.* **70** (1948) 1438.
[26] D. Kreger, *Nature* **158** (1946) 199.
[27] A. Frey-Wyssling, *Am. J. Bot.* **56** (1969) 696.
[28] W. Shuey, K. Tipples: *The Amylograph Handbook*, Am. Assoc. Cereal Chem., St. Paul 1980.
[29] E. Osman in R. Whistler, E. Paschall (eds.): *Starch: Chemistry and Technology*, vol. **II**, Academic Press, New York 1967, pp. 163–215.
[30] M. Glicksman, R. Sand in R. Whistler, J. BeMiller (eds.): *Industrial Gums, Polysaccharides and Their Derivatives*, Academic Press, New York 1973, pp. 197–263.
[31] P. White, I. Abbas, L. Johnson, *Starch/Staerke* **41** (1989) 176.
[32] P. Roulet, W. MacInnes, D. Gumy, P. Würsch, *Starch/Staerke* **42** (1990) 99.
[33] R. Stute, *Starch/Staerke* **42** (1990) 178.
[34] C. Biliaderis, J. Zawistowski, *Cereal Chem.* **67** (1990) 240.
[35] *Food Manufacture*, Sept. 1990, p. 21.
[36] M. Wooton, M. Chaudry, *J. Food Sci.* **45** (1980) 1783.
[37] S. Fleming, *J. Food Sci.* **47** (1981) 1.
[38] S. Fleming, J. Vose, *J. Nutr.* **109** (1979) 2067.
[39] R. Walton: *Comprehensive Survey of Starch Chemistry*, Chemical Catalog Co., New York 1928.
[40] H. Brown, G. Morris, *J. Chem. Soc.* **55** (1889) 449.
[41] D. French in R. W. Kerr (ed.): *Chemistry and Industry of Starch*, 2nd ed., Academic Press New York 1950, p. 158.
[42] G. V. Caeser, E. E. Moore, *Ind. Eng. Chem.* **27** (1935) 1447.
[43] W. Gallay, A. C. Bell, *Can. J. Res. Sect. B* **14** (1936) 360.

[44] W. Gallay, A. C. Bell, *Can. J. Res., Sect. B* **14** (1936) 381.
[45] J. Katz, *Textile Res. J.* **9** (1939) 146.
[46] H. Schopmeyer, G. E. Felton, US 2 319 637, 1943.
[47] W. Bechtel, *Cereal Chem.* **24** (1947) 200.
[48] W. Bechtel, *J. Colloid Sci.* **5** (1950) 260.
[49] E. Mazurs, T. Schoch, F. Kite, *Cereal Chem.* **34** (1957) 141.
[50] S. Lansky, M. Kooi, T. Schoch, *J. Am. Chem. Soc.* **71** (1949) 4066.
[51] H. Leach, T. Schoch, *Cereal Chem.* **39** (1962) 318.
[52] H. Leach, L. McCowen, T. Schoch, *Cereal Chem.* **36** (1959) 534.
[53] C. Naegeli, *Ann. Chem.* **173** (1874) 218.
[54] B. Bellmas, DE 110 957, 1897.
[55] C. Duryea, US 675 822, 1901.
[56] R. Kerr, *Staerke* **4** (1952) 39.
[57] J. Cowie, C. Greenwood, *J. Chem. Soc.* (1957) 2658.
[58] W. Gallay, *Can. J. Res. Sect. B* **14** (1936) 391.
[59] M. Wolfrom, A. Thompson, C. Timberlake, *Cereal Chem.* **40** (1963) 82.
[60] *Posselts Textile J.* **21** (1917) 9.
[61] W. Cathcart, *Textile World* **59** (1921) 2896.
[62] C. Musselmann, E. Bovier, US 4 014 727, 1977.
[63] J. Liebig, *Ann. Phys. Chem.* **15** (1829) 541.
[64] A. Lieben, E. Reichardt, *Dtsch. Chem. Ges.* **8** (1875) 1020.
[65] F. Farley, R. Hixon, *Ind. Eng. Chem.* **34** (1942) 677.
[66] G. Felton, F. Farley, R. Hixon, *Cereal Chem.* **15** (1938) 678.
[67] J. Schmorak, M. Lewin, *J. Polymer Sci., Part A* **1** (1963) 2601.
[68] V. Prey, S. Fischer, *Staerke* **28** (1976) 125.
[69] D. Lucas, C. Fletcher, *Paper Ind.* **40** (1959) 810.
[70] V. Prey, S. Fischer, S. Klinger, *Staerke* **28** (1976) 259.
[71] R. Mellies, C. Mehltretter, F. Senti, *J. Chem. Eng. Data* **5** (1960) 169.
[72] J. Schmorak, D. Mejzler, M. Lewin, *Staerke* **14** (1962) 278.
[73] J. Potze, P. Hiemstra, *Staerke* **15** (1963) 217.
[74] C. Cairns, *TAPPI* **57** (1974) 85.
[75] H. Brill, *TAPPI* **38** (1955) 522.
[76] Anheuser-Busch Inc., US 3 767 826, 1973 (J. Fruin).
[77] E. Alteriis, P. Parascandola, V. Scardi, *Starch* **42** (1990) 57.
[78] O. Wurzburg, C. Szymanski, *Agr. Food Chem.* **18** (1970) 997.
[79] U.S. Department of Agriculture, US 2 989 521, 1961 (F. Senti, R. Mellies, C. Mehltretter).
[80] Corn Products Co., US 2 801 242, 1957 (R. Kerr, F. Cleveland, Jr.).
[81] National Starch and Chemical Corp., US 2 935 510, 1960 (O. Wurzburg).
[82] American Maize Products Co., US 2 328 537, 1943 (G. Felton, H. Schopmeyer).
[83] M. Rutenberg, D. Solarek in R. Whistler, J. BeMiller, E. Paschall (eds.): *Starch: Chemistry and Technology*, 2nd ed., Academic Press, New York 1984, pp. 324–326.
[84] F. Kite, T. Schoch, H. Leach, *Bakers Dig.* **31** (1957) 42.
[85] World Health Organization, *Food Additive Series* 1974, no. 5, 329, 345.
[86] A. E. Staley, DE 2 541 513, 1976 (R. Van Schanefelt, J. Eastman, M. Campbell).
[87] National Starch and Chemical Corp., US 3 052 545, 1962 (J. Ducharme, H. Black, Jr., S. Leith).
[88] National Starch and Chemical Corp., US 3 579 341, 1971 (N. Marotta, P. Trubiano).
[89] American Maize Products, US 3 346 387, 1967 (J. Evans, C. MacWilliams).

[90]   National Starch and Chemical Corp., US 4 229 489, 1980 (C. Chin, M. Rutenberg).
[91]   Unilever, DE 2 837 088, 1979 (D. Chaudhuri, M. Stebles).
[92]   A. E. Staley, US 3 069 410, 1962 (C. Smith, J. Tuschoff).
[93]   PPG Industries, US 3 887 389, 1975 (J. Hedden).
[94]   Penick and Ford, US 3 438 913, 1969 (E. Hjermstad).
[95]   W. Jarowenko in N. Bicales (ed.): *Encyclopedia of Polymer Science and Technology*, vol. **12**, Interscience, New York 1970, p. 787.
[96]   R. Lohmar, C. Rist, *J. Am. Chem. Soc.* **72** (1950) 4298.
[97]   J. Mullen, E. Pacsu, *Ind. Eng. Chem.* **34** (1942) 1209.
[98]   R. Whistler, *Adv. Carbohyd. Chem.* **1** (1945) 279.
[99]   M. Rutenberg, W. Jarowenku, L. Ross, US 3 038 895, 1962.
[100] C. Cross, E. Bevan, J. Tranquair, *Chem. Ztg.* **29** (1932) 2083.
[101] H. Roberts in R. Whistler, E. Paschall (eds.): *Starch: Chemistry and Technology*, vol. **II**, Academic Press, New York 1967, chap. 13.
[102] E. Middleton, US 1 682 220, 1928.
[103] O. Wurzburg in R. Whistler (ed.): *Methods in Carbohydrate Chemistry*, vol. **4**, Academic Press, New York 1964, p. 286.
[104] C. Smith, J. Tuschoff, US 2 928 828, 1960.
[105] S. Takahashi, C. Maningat, P. Seib, *Cereal Chem.* **66** (1989) 499.
[106] E. Degering in R. Kerr (ed.): *Chemistry and Industry of Starch*, 2nd ed., Academic Press, New York 1950, chap. 10.
[107] C. Burkhard, E. Degering, *Rayon Text. Mon.* **23** (1942) 416.
[108] A. Harsveldt, *Chem. Ind. (London)* 1961, 2062.
[109] National Starch and Chemical Corp., US 3 481 771, 1969 (A. Doering).
[110] J. Mullen, E. Pacsu, *Ind. Eng. Chem.* **35** (1943) 381.
[111] R. Whistler, G. Hilbert, *Ind. Eng. Chem.* **43** (1944) 911.
[112] R. Kerr, F. Cleveland, US 2 884 412, 1959.
[113] R. Whistler, J. Daniel in R. Whistler, J. BeMiller, E. Paschall (eds.): *Starch: Chemistry and Technology*, 2nd ed., Academic Press, New York 1984, chap. 6.
[114] E. Snyder in R. Whistler, J. BeMiller, E. Paschall (eds.): *Starch: Chemistry and Technology*, 2nd ed., Academic Press, New York 1984, pp. 661–674.
[115] D. Christianson et al., *Cereal Chem.* **46** (1969) 372.
[116] R. Hamilton, E. Paschall in R. Whistler, E. Paschall (eds.): *Starch: Chemistry and Technology*, vol. **II**, Academic Press, New York 1967, chap. 14.
[117] Benckiser-Knapsack, Höchst, DE 2 426 404, 1975 (A. Kling, W. Traud, W. Hansi, H. Jalke).
[118] R. Kerr, US 2 884 413, 1947.
[119] E. Salay, C. Ciacco, *Starch*/Staerke 42 (1990) 15.
[120] B. Hofreiter in J. Casey (ed.): *Pulp and Paper Chemistry and Chemical Technology*, 3rd ed., Wiley, New York 1981, p. 1475.
[121] J. Benckiser, GB 1 233 637, 1971.
[122] National Dairy Products Corp., GB 938 717, 1963.
[123] M. Tessler, US 3 719 662, 1973.
[124] American-Maize Co., US 3 238 100, 1966 (H. Meyer, R. Milloch, V. Shreeram, T. Tsuzuki).
[125] H. Merkus, J. Mourits, L. deGalan, W. deJong, *Stärke* **29** (1977) 406.
[126] S. El-Hinnaway et al., *Starch/Staerke* **34** (1982) 112.
[127] T. Schoch, *TAPPI* **35** (1952) no. 7, 22 A.

[128] E. Hjermstad in R. Whistler, J. BeMiller (eds.): *Industrial Gums*, Academic Press, New York 1973, pp. 601–615.
[129] W. Roth, C. Mehltretter, *Food Technol.* **21** (1967) 72.
[130] A. Harsveldt, *TAPPI* **45** (1962) 85.
[131] National Starch and Chemical Corp., US 3 628 969, 1971 (R. Vilim, H. Bell).
[132] Merck & Co., US 4 105 461, 1978 (J. Racciato).
[133] General Foods Corp., US 3 669 687, 1972 (A. D'Ercole).
[134] American-Maize Products Co., US 3 427 951, 1969 (F. Mitan, L. Jokay).
[135] T. Shiroza, K. Furihata, *Agric. Biol. Chem.* **46** (1982) 1425.
[136] E. Tasset, US 4 464 528, 1984.
[137] R. Kerr, H. Neukom, *Staerke* **4** (1952) 255.
[138] Dow Chemical, US 3 846 405, 1974 (J. McClendon).
[139] H. Prietzel, *Staerke* **22** (1970) 424.
[140] S. Parmerter, US 3 620 913, 1971.
[141] USDA, US 3 087 852, 1963 (B. Hofreiter, G. Hamerstrand, C. Mehltretter).
[142] E. Paschall in R. Whistler, E. Paschall (eds.): *Starch: Chemistry and Technology*, vol. **II**, Academic Press, New York 1967, pp. 403–422.
[143] D. Greif, L. Gaspar in W. Reynolds (ed.): *Dry Strength Additives*, TAPPI Press, Atlanta 1980, pp. 95–117.
[144] D. Greif, L. Gaspar in W. Reynolds (ed.): *Dry Strength Additives*, TAPPI Press, Atlanta 1980, chap. 4.
[145] J. Marton, T. Marton, *TAPPI* **59** (1976) 121.
[146] J. Marton, *TAPPI* **63** (1980) 87.
[147] E. Mazzarella, L. Hickey, *TAPPI* **49** (1966) 926.
[148] Hubinger, US 3 793 310, 1974 (L. Elizer).
[149] D. Halabisky, *TAPPI* **60** (1977) 125.
[150] Chemed Corp., US 4 088 600, 1978 (T. Tutein, A. Harrington, J. Jacob).
[151] Fachverband der Stärke-Industrie e.V., "Stärke Fortschritt durch Tradition 1946–1996", Bonn 1996, pp. 48–49.
[152] F. Schierbaum, "U.S. Corn Refining Industry 1994/95, A Survey on Member Companies, Product Lines and Shipments", *Starch/Staerke* **48** (1996) no. 5, 201.

Styrene

DENIS H. JAMES, Dow Chemical, Freeport, Texas 77 541, United States

WILLIAM M. CASTOR, Dow Chemical, Freeport, Texas 77 541, United States

1.	Introduction	4417
2.	Physical Properties	4418
3.	Chemical Properties	4419
4.	Production	4422
4.1.	Catalytic Dehydrogenation of Ethylbenzene	4422
4.1.1.	Reaction Mechanisms	4422
4.1.2.	Adiabatic Dehydrogenation	4424
4.1.3.	Isothermal Dehydrogenation	4425
4.1.4.	Distillation of Crude Styrene	4426
4.1.5.	Oxidative Hydrogen Removal	4427
4.2.	Styrene – Propylene Oxide Process	4428
4.3.	Styrene from Butadiene	4429
4.4.	Styrene from Toluene	4430
4.5.	Styrene from Pyrolysis Gasoline	4430
5.	Quality and Testing	4431
6.	Storage and Transportation	4431
7.	Uses and Economic Aspects	4432
8.	Related Monomers	4434
8.1.	Vinyltoluene	4434
8.2.	Divinylbenzene	4435
8.3.	α-Methylstyrene	4436
8.4.	Chlorostyrene	4437
8.5.	Vinylbenzyl Chloride	4438
9.	Toxicology and Occupational Health	4439
10.	References	4440

1. Introduction

Styrene [100-42-5], also known as phenylethylene, vinylbenzene, styrol, or cinnamene, C_6H_5–$CH=CH_2$, is an important industrial unsaturated aromatic monomer. It occurs naturally in small quantities in some plants and foods. In the nineteenth century, styrene was isolated by distillation of the natural balsam storax [1]. It has been identified in cinnamon, coffee beans, and peanuts [2], and it is also found in coal tar.

The development of commercial processes for the manufacture of styrene based on the dehydrogenation of ethylbenzene [100-41-4] (→ Ethylbenzene) was achieved in the 1930s. The need for synthetic styrene – butadiene rubber during World War II provided the impetus for large-scale production. After 1946, this capacity became available for the manufacture of a high-purity monomer that could be polymerized to a stable, clear, colorless, and cheap plastic. Peacetime uses of styrene-based plastics expanded rapidly, and polystyrene is now one of the least expensive thermoplastics on a cost-per-volume basis.

Styrene itself is a liquid that can be handled easily and safely. The activity of the vinyl group makes styrene easy to polymerize and copolymerize. When the appropriate technology became available through licensors styrene was quickly transformed into a bulk-commodity chemical, growing to a world-wide capacity estimated at 17×10^6 t/a in 1993. For reviews of this important industrial history see [3], [4].

2. Physical Properties

Styrene is a colorless liquid with a distinctive, sweetish odor. The most important physical properties of styrene are summarized below. Data appearing in the literature are not always consistent, varying as a consequence both of the method of measurement and the purity of the monomer. Most manufacturers make available the physical properties they regard as valid [5]. Important general physical properties of styrene monomer are as follows:

M_r	104.153
bp	145.15 °C
fp	−30.6 °C
Critical density, D_c	0.297 g/mL
Critical pressure, P_c	3.83 MPa
Critical temperature, T_c	362.1 °C
Critical volume, V_c	3.37 mL/g
Flammable limits in air	1.1 – 6.1 vol%
Flash point, Tag Closed Cup (TCC)	31.1 °C
Autoignition point	490 °C
Heat of combustion, ΔH_c, constant pressure (25 °C)*	−4.263 MJ/mol
Heat of formation, ΔH_f	
gas (25 °C)	147.4 kJ/mol
liquid (25 °C)	103.4 kJ/mol
Heat of fusion, ΔH_m	−11.0 kJ/mol
Heat of polymerization, ΔH_p(25 °C)	−69.8 kJ/mol
Heat of vaporization, ΔH_v	
(25 °C)	421.7 J/g
(145 °C)	356.7 J/g
Volume expansion coefficient	
(20 °C)	9.783×10^{-4} °C^{-1}
(40 °C)	9.978×10^{-4} °C^{-1}
Q value	1.0
e value	0.8
Volume shrinkage on polymerization, typical	17.0 %
Solubility of oxygen (from air)	
(15 °C)	53 mg/kg
(25 °C)	50 mg/kg
(35 °C)	45 mg/kg

* All reactants and products are gases.

Vapor pressure is a key property in the design of styrene distillation equipment. It is essential that the same data set be used for both column design and column assessment. Vapor-pressure data we believe to be accurate are as follows:

Temperature, °C	Vapor pressure, kPa
20	0.6
50	3.2
80	12.2
100	25.7
145.2	101.3

Antoine equation:

$$\log_{10} P = [6.08201 - 1445.58/(209.43 + T)]$$

Styrene is miscible with most organic solvents in any ratio. It is a good solvent for synthetic rubber, polystyrene, and other non-cross-linked high polymers. Styrene is sparingly soluble in polyvalent hydroxy compounds such as glycol and diglycol monoethers. Styrene and water are sparingly soluble in each other (Table 1).

The composition of the azeotropic mixture at standard pressure is ca. 66 wt % styrene and 34 wt % water, and the minimum boiling point is 94.8 °C.

Important parameters for monitoring the production and use of styrene are the refractive index and the density (Table 2). Table 3 lists supplementary physical properties of styrene.

3. Chemical Properties

The most important reaction of styrene is its polymerization to polystyrene, but it also copolymerizes with other monomers. The copolymerization with butadiene to give Buna S synthetic rubber (emulsion styrene–butadiene rubber, E-SBR) was the reaction that led initially to the development of the styrene industry. Other characteristic reactions are described below. The literature has been well summarized [3], [6], [7].

Oxidation of styrene in air is of special importance. The reaction leads to high molecular mass peroxides by way of free radicals. Styrene is also oxidized to various other compounds, including benzaldehyde, formaldehyde, and formic acid. Other typical alkene reactions are observed with stronger oxidizing agents:

Table 1. Mutual solubilities, styrene/water, wt%

Temperature, °C	Water in styrene	Styrene in water
0	0.02	0.018
10	0.04	0.023
25	0.07	0.032
50	0.12	0.045

Table 2. Refractive index and density of styrene

Temperature, °C	Refractive index, n_D	Density, g/mL
0		0.9223
20	1.54682	0.9050
25	1.54395	0.9007
30	1.54108	0.8964
60		0.8702
100		0.8355

$n_D = 1.55830 - 0.000574\, T$

Table 3. Other physical properties of styrene

Temperature, °C	Viscosity, mPa s*	Surface tension, mN/m	Specific heat capacity, C_p, $J\,g^{-1}\,K^{-1}$
0	1.039	34.5	1.6367
20	0.762	32.3	1.6907
40	0.588	30.0	1.7489
60	0.469	27.8	
80	0.385	25.6	
100	0.324	23.5	

* ln viscosity (mPa s) = $-4.2488 + 1170.8/T$

Styrene can undergo addition to the double bond of the side chain as well as substitution in the ring. Treatment with bromine in the cold gives addition, leading to styrene dibromide. This crystallizes well, and has been used both to characterize styrene and to determine it quantitatively in solution [8]. Splitting off hydrogen bromide from styrene dibromide gives α-bromostyrene, while phenylacetylene can be formed by heating with calcium oxide. Addition of chlorine forms styrene dichloride, which can be converted to α-chlorostyrene and β-chlorostyrene by elimination of hydrogen chloride.

The halohydrins are important intermediates in preparative chemistry. They react with alkali to form styrene oxide, while further hydrolysis leads to phenyl glycol. Iodohydrin is formed from styrene in the presence of iodine, mercury(II) oxide, and water. 1-Phenylethanol is an important intermediate product in the perfume industry. It is formed by the hydration of styrene via quantitative addition of mercury(II) acetate:

$$\text{Ph-CH=CH}_2 \xrightarrow[\text{H}_2\text{O, Tetrahydrofuran}]{(\text{CH}_3\text{COO})_2\text{Hg}} \text{Ph-CH(OH)CH}_2\text{HgOOCCH}_3 \xrightarrow{\text{NaBH}_4} \text{Ph-CH(OH)CH}_3$$

The corresponding methyl ether is obtained by addition of methanol (e.g., at 135–150 °C in the presence of sulfuric acid):

$$\text{Ph-CH=CH}_2 + \text{CH}_3\text{OH} \xrightarrow{\text{H}^+} \text{Ph-CH(OCH}_3)\text{CH}_3$$

Numerous reactions of styrene with sulfur and nitrogen compounds have been reported [3]. Thus, styrene reacts with sulfur dioxide, sulfur monochloride, sodium or ammonium dithionite, mercaptans, aniline, amines, diazomethane, and sodium hydrazide to give cleavage of the C=C double bond. Heating styrene with sulfur at high temperature gives hydrogen sulfide, styrene sulfide, and diphenylthiophenes.

Styrene sulfide

Styrene forms solid complexes with copper and silver salts at low temperature. These compounds are suitable for the purification of styrene, or for separating styrene from mixtures with other hydrocarbons. The color of the copper complexes makes this metal and its alloys unsuitable for use in contact with styrene.

Styrene also undergoes many cyclization reactions [7]; e.g.:

$$\text{Ph-CH=CH}_2 + \text{CH}_2\text{=CH-CH=CH}_2 \longrightarrow$$

Lithium alkyls, such as C_2H_5Li, initiate a polymerization of styrene. In the presence of ethers, styrene is polymerized almost completely by sodium. Controlled polymerization (telomerization) with olefins has generated a great deal of industrial interest, leading to new polymers that show promise as plasticizers, lubricants, and textile auxiliaries. Aromatic hydrocarbons such as benzene can be added to styrene in the presence of aluminum chloride, just as they can to other olefins.

$$C_6H_5-CH=CH_2 + C_6H_6 \xrightarrow{AlCl_3} (C_6H_5)_2CHCH_3$$

The high reactivity of the double bond, which is due to the resonance-stabilized aromatic ring, usually means that reactions must be carried out in several steps, with the double bond protected until the final step.

4. Production

Production of styrene is world-wide, and dates back more than 50 years. However, new variations on established processes are continually being developed, and new companies are building plants or buying out older producers. The feedstock for all commercial styrene manufacture is still ethylbenzene (→ Ethylbenzene). This is converted to a crude styrene that requires finishing to separate out the pure product.

4.1. Catalytic Dehydrogenation of Ethylbenzene

Direct dehydrogenation of ethylbenzene to styrene accounts for 85 % of commercial production. The reaction is carried out in the vapor phase with steam over a catalyst consisting primarily of iron oxide. The reaction is endothermic, and can be accomplished either adiabatically or isothermally. Both methods are used in practice.

4.1.1. Reaction Mechanisms

The major reaction is the reversible, endothermic conversion of ethylbenzene to styrene and hydrogen:

$$C_6H_5CH_2CH_3 \rightleftharpoons C_6H_5CH=CH_2 + H_2 \quad \Delta H(600\,°C) = 124.9 \text{ kJ/mol}$$

This reaction proceeds thermally with low yield and catalytically with high yield. As it is a reversible gas-phase reaction producing 2 mol of product from 1 mol of starting material, low pressure favors the forward reaction.

Competing thermal reactions degrade ethylbenzene to benzene, and also to carbon:

$C_6H_5CH_2CH_3 \longrightarrow C_6H_6 + C_2H_4 \qquad \Delta H = 101.8 \text{ kJ/mol}$
$C_6H_5CH_2CH_3 \longrightarrow 8\,C + 5\,H_2 \qquad \Delta H = 1.72 \text{ kJ/mol}$

Styrene also reacts catalytically to toluene:

$C_6H_5CH{=}CH_2 + 2\,H_2 \longrightarrow C_6H_5CH_3 + CH_4$

The problem with carbon production is that carbon is a catalyst poison. When potassium is incorporated into the iron oxide catalyst, the catalyst becomes selfcleaning (through enhancement of the reaction of carbon with steam to give carbon dioxide, which is removed in the reactor vent gas).

$C + 2\,H_2O \longrightarrow CO_2 + 2\,H_2 \qquad \Delta H = 99.6 \text{ kJ/mol}$

Typical operating conditions in commercial reactors are ca. 620 °C and as low a pressure as practicable. The overall yield depends on the relative amounts of catalytic conversion to styrene and thermal cracking to byproducts. At equilibrium under typical conditions, the reversible reaction results in about 80% conversion of ethylbenzene. However, the time and temperature necessary to achieve equilibrium give rise to excessive thermal cracking and reduced yield, so most commercial units operate at conversion levels of 50–70 wt%, with yields of 88–95 mol%.

Dehydrogenation of ethylbenzene is carried out in the presence of steam, which has a threefold role:

1) It lowers the partial pressure of ethylbenzene, shifting the equilibrium toward styrene and minimizing the loss to thermal cracking
2) It supplies the necessary heat of reaction
3) It cleans the catalyst by reacting with carbon to produce carbon dioxide and hydrogen.

Many catalysts have been described for this reaction [9]. One catalyst, Shell 105, dominated the market for many years, and was the first to include potassium as a promoter for the water-gas reaction. This catalyst is typically 84.3% iron as Fe_2O_3, 2.4% chromium as Cr_2O_3, and 13.3% potassium as K_2CO_3. It has good physical properties and good activity, and it gives fair yields [10].

In recent years, the situation has become more complex. The market has become more competitive, causing manufacturers to seek new catalysts that produce higher yields without compromising activity or physical properties, or catalysts that meet specific requirements. The Süd-Chemie Group, which includes Nissan Girdler Catalyst in Japan, Süd-Chemie in Germany, and United Catalysts (UCI) in the United States, now has the major share of the catalyst market with its G-64 and G-84 types [11]. Shell also remains active through a joint partnership with American Cyanamid called Criterion

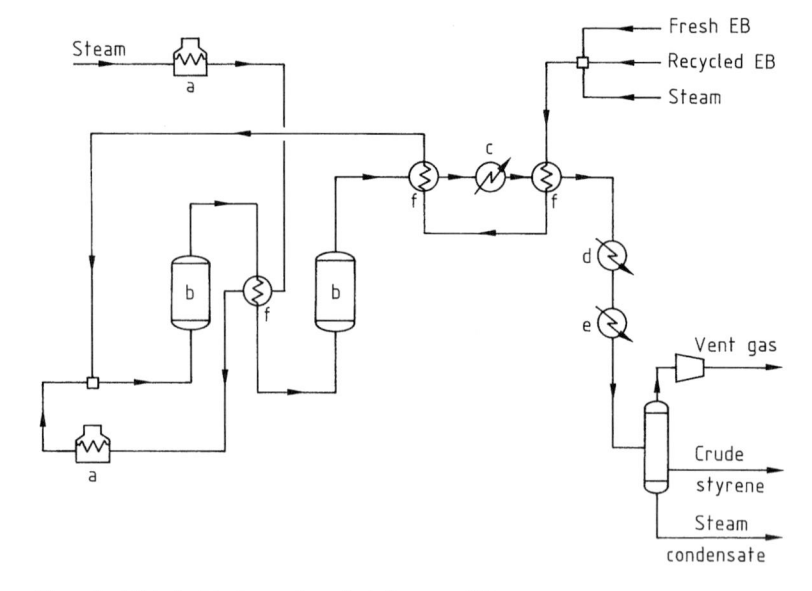

Figure 1. Adiabatic dehydrogenation of ethylbenzene (EB)
a) Steam superheater; b) Reactor; c) High-pressure steam; d) Low-pressure steam; e) Condenser; f) Heat exchanger

Catalyst. In addition to Criterion 105, a series of new Criterion catalysts is available, including C-115 and C-025 HA. Dow Chemical and BASF manufacture their own catalysts to suit their specific needs, and there are other small producers as well. A catalyst life of ca. two years is claimed.

4.1.2. Adiabatic Dehydrogenation

Over 75 % of all operating styrene plants carry out the dehydrogenation reaction adiabatically in multiple reactors or reactor beds operated in series (Fig. 1). The necessary heat of reaction is applied at the inlet to each stage, either by injection of superheated steam or by indirect heat transfer.

Fresh ethylbenzene feed is mixed with recycled ethylbenzene and vaporized. Dilution steam must be added to prevent the ethylbenzene from forming coke. This stream is further heated by heat exchange, superheated steam is added to bring the system up to reaction temperature (ca. 640 °C), and the stream is passed through catalyst in the first reactor. Adiabatic reaction drops the temperature, so the outlet stream is reheated prior to passage through the second reactor. Conversion of ethylbenzene can vary with the system, but is often about 35 % in the first reactor and 65 % overall. The reactors are run at the lowest pressure that is safe and practicable. Some units operate under vacuum, while others operate at a low positive pressure. The steam : ethylbenzene ratio fed to the reactors is chosen to give optimum yield with minimum utility cost. The reactor effluent is fed through an efficient heat recovery system to minimize energy consump-

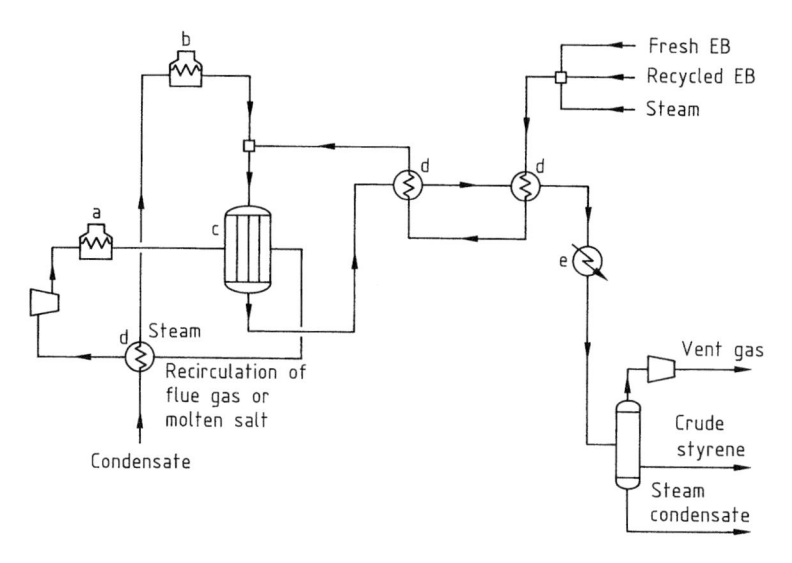

Figure 2. Isothermal dehydrogenation of ethylbenzene (EB)
a) Heater; b) Steam superheater; c) Reactor; d) Heat exchanger; e) Condenser

tion, condensed, and separated into vent gas, a crude styrene hydrocarbon stream, and a steam condensate stream. The crude styrene goes to a distillation system. The steam condensate is steam-stripped, treated, and reused. The vent gas, mainly hydrogen and carbon dioxide, is treated to recover aromatics, after which it can be used as a fuel or a feed stream for chemical hydrogen. The complete technology is for sale from various licensors (Chap. 7).

4.1.3. Isothermal Dehydrogenation

Isothermal dehydrogenation (Fig. 2) was pioneered by BASF and has been used by them for many years. The reactor is built like a shell-and-tube heat exchanger. Ethylbenzene and steam flow through the tubes, which are packed with catalyst. The heat of reaction is supplied by hot flue gas on the shell side of the reactor–exchanger [12]. The steam:oil mass ratio can be lowered to about 1:1, and steam temperatures are lower than in the adiabatic process. A disadvantage is the practical size limitation on a reactor–exchanger, which restricts the size of a single-train plant to about 150×10^3 t/a, translating into increased capital for large plants.

Lurgi GmbH operates an isothermal reactor system that uses a molten salt mixture of sodium, lithium, and potassium carbonates as the heating medium. The multitubular reactor is operated at ca. 600 °C under vacuum and a steam:ethylbenzene ratio of 0.6–0.9. High conversion and selectivity are claimed [13]. A demonstration plant has been operated since 1985 at Mantova, Italy, by Montedison. The technology is offered for license by Lurgi, Montedison, and Deggendorfer, but so far no more units have been built.

4.1.4. Distillation of Crude Styrene

A typical crude styrene from the dehydrogenation process consists of:

Benzene (*bp* 80 °C)	1 %
Toluene (*bp* 110 °C)	2 %
Ethylbenzene (*bp* 136 °C)	32 %
Styrene (*bp* 145 °C)	64 %
Others	1 %

The separation of these components is reasonably straightforward, but residence time at elevated temperature needs to be minimized to reduce styrene polymerization. At least three steps are involved. Benzene and toluene are removed first, and either sent to a toluene dehydrogenation plant or further separated into benzene for recycling and toluene for sale. Ethylbenzene is then separated and recycled to the reactors. Finally, styrene is distilled away from the tars and polymers under vacuum to keep the temperatureas low as possible. Figure 3 A shows a typical distillation train. The variant shown in Figure 3 B is the Monsanto approach, where the major split is accomplished first, followed by separation of the benzene – toluene mixture.

Ethylbenzene and styrene, having similar boiling points, require 70 – 100 trays for their separation depending on the desired ethylbenzene content of the finished styrene. If bubble-cap trays are used, as in old plants, a large pressure drop over the trays means that two columns in series are necessary to keep reboiler temperatures low. Low-pressure drop trays, such as the Linde UCC sieve tray [14], permit this separation to be achieved in one column [15]. The most modern plants use packing in place of trays [16], [17]. This results in less pressure drop, giving a lower bottom temperature, shorter residence time, and hence less polymer. Sulzer has done pioneering work in the field of packings for distillation [18]. Koch, Norton, and Glitch also produce packings for this purpose.

A polymerization inhibitor (distillation inhibitor) is needed throughout the distillation train. Sulfur was originally used, but environmental constraints make sulfur tar unacceptable as a fuel. Many new inhibitors have been marketed, usually aromatic compounds with amino, nitro, or hydroxy groups (e.g., phenylenediamines, dinitrophenols, and dinitrocresols). Uniroyal is especially active in this field. The distillation inhibitor tends to be colored and is thus unacceptable in the final product; finished monomer is usually inhibited instead with *tert*-butylcatechol (TBC) (10 – 50 mg/kg) during storage and transportation.

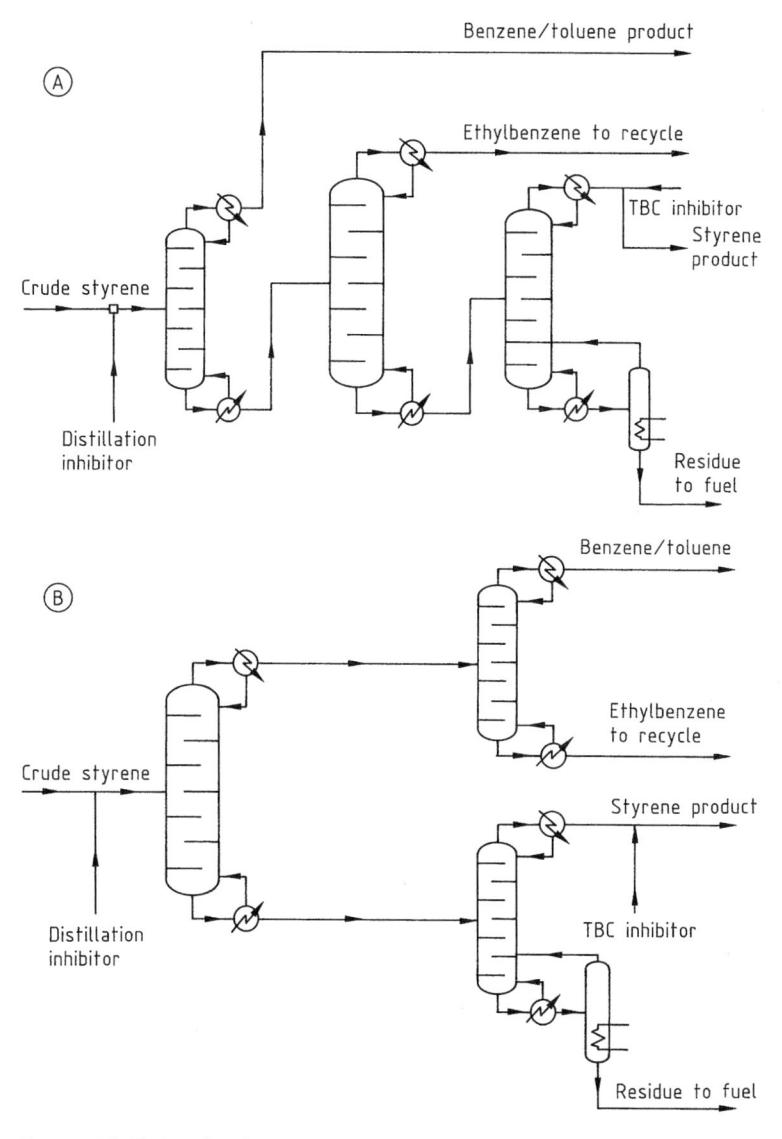

Figure 3. Distillation of crude styrene
A) Standard approach; B) Monsanto approach

4.1.5. Oxidative Hydrogen Removal

Many efforts have been made to remove hydrogen from the process to favor the forward reaction. A new process is now being offered for licensing by Lummus, based on technology of UOP [19]. It employs a noble metal catalyst that selectively oxidizes hydrogen, allowing the ethylbenzene conversion to be increased to over 80% while

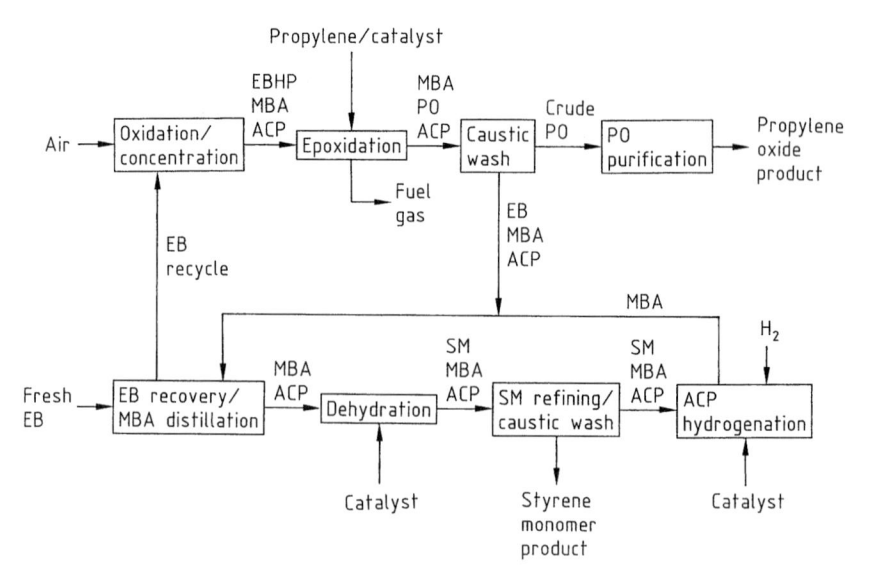

Figure 4. Styrene–propylene oxide process
ACP acetophenone; EB ethylbenzene; EBHP ethylbenzene hydroperoxide; MBA methylbenzyl alcohol; PO propylene oxide; SM styrene monomer

maintaining good styrene selectivity [20]. The process was originally named Styro-Plus, but is now called the SMART SM Process. It has performed well in a pilot unit (Mitsubishi Petrochemical, Japan) since 1985. It is mainly offered as a retrofit to existing units as a way of gaining extra capacity. It is claimed to have higher variable costs, but a lower capital cost than a conventional plant. Several licenses have been sold, but the corresponding plants are not yet operating.

4.2. Styrene–Propylene Oxide Process

The only other route to commercial production of styrene involves coproduction of propylene oxide, illustrated in simplified form in Fig. 4 [21]. The first step is direct air oxidation of ethylbenzene at ca. 130 °C and 0.2 MPa. This gives ethylbenzene hydroperoxide (EBHP); α-methylbenzyl alcohol (MBA) and acetophenone (ACP) are also formed. Conversion is held to about 13% in this step to minimize byproducts. The selectivity for ethylbenzene to EBHP is approximately 90%, and the selectivity to MBA and ACP is 5–7%.

$$C_6H_5C_2H_5 + O_2 \longrightarrow C_6H_5CH(OOH)CH_3 \quad \text{EBHP}$$
$$C_6H_5C_2H_5 + 1/2\,O_2 \longrightarrow C_6H_5CH(CH_3)OH \quad \text{MBA}$$
$$C_6H_5C_2H_5 + xO_2 \longrightarrow C_6H_5CO(CH_3) + \text{acids} \quad \text{ACP}$$

Ethylbenzene hydroperoxide is then reacted with propylene in the presence of a metallic catalyst to form propylene oxide (PO) and more MBA. Liquid-phase molybdenum or heterogeneous titanium catalysts are used at ca. 110 °C and 4 MPa. The conversion of EBHP is nearly complete, with 70–85% selectivity to PO and > 70% selectivity to MBA.

$$C_6H_5CH(OOH)CH_3 + C_3H_6 \longrightarrow C_3H_6O + C_6H_5CH(CH_3)OH \quad PO$$

To improve yields, ACP is hydrogenated to MBA in the liquid phase at 90–150 °C and ca. 8 MPa. The catalyst is a mixture of ZnO and CuO. Approximately 90% of the ACP is converted, with 92% selectivity to MBA.

$$C_6H_5CO(CH_3) + H_2 \longrightarrow C_6H_5CH(CH_3)OH$$

Finally, the MBA is dehydrated to styrene at 250 °C and low pressure over a suitable metal oxide catalyst (often Al_2O_3).

$$C_6H_5CH(CH_3)OH \longrightarrow C_6H_5CH=CH_2 + H_2O$$

The patents issued to Halcon, Arco, and Shell on this process are extensive. The resulting ratio of styrene to PO is about 2.5. The process requires clean-up and purification steps to generate specification products.

The first commercial development of the process was by a joint venture company, Oxirane, formed by Halcon and Atlantic Richfield (ARCO); ARCO became the sole owner in 1980. Independently, Shell developed its own process based on similar chemistry. This route offers a way to manufacture propylene oxide without the need for chlorine as in the chlorohydrin process, but it is subject to market fluctuations in styrene and PO demand that do not match the stoichiometry of the process. The process requires a large capital investment and produces styrene at a higher cost than the conventional process, but credits for the coproduct PO can make the overall operation profitable. About 15% of the world's supply of styrene is now made by this process.

4.3. Styrene from Butadiene

Another route to styrene that is being heavily researched starts with the Diels–Alder dimerization of 1,3-butadiene to 4-vinylcyclohexene-1 (VCH) [100-40-3] (→ Butadiene):

$$2 \, C_4H_6 \longrightarrow C_6H_9CH=CH_2$$

This reaction is exothermic, and can be accomplished either thermally or catalytically. Thermal processes [22], [23] require a temperature of ca. 140 °C and a pressure of ca. 4 MPa. The thermal approach is most suitable for use with purified butadiene. Yields of VCH are ca. 90%. The catalytic process is based on nitrosyl halide–iron complexes, and operates at 0–80 °C and 0.1–1.30 MPa [24], [25]. The yield is almost quantitative, and purification is not difficult. This route can accept either the raw C_4 stream from a naphtha or gas-oil steam cracker, or purified butadiene.

VCH is then dehydrogenated to ethylbenzene [26], [27] or, under more severe conditions, oxidatively dehydrogenated directly to styrene [28], [29]:

$$C_6H_9CH=CH_2 \longrightarrow C_6H_5CH_2CH_3 \longrightarrow C_6H_5CH=CH_2$$

This route to styrene is not yet economically attractive, but the availability and price of butadiene in the future could make it so.

4.4. Styrene from Toluene

There have been many attempts to find a route to styrene starting from toluene (→ Toluene). Toluene is readily available, is usually at least 15% cheaper than benzene, and is not as toxic. However, no process has become commercially competitive. Monsanto worked extensively on a process for styrene starting with air oxidation of toluene to give stilbene [30]. This used a fluidized bed of supported lead oxide catalyst:

$$2\,C_6H_5CH_3 + O_2 \longrightarrow C_6H_5CH=CHC_6H_5 + H_2O \quad \Delta H = -77.26 \text{ KJ g}^{-1} \text{ mol}^{-1}$$

Stilbene is then reacted with ethylene over a molybdenum catalyst to give styrene:

$$C_6H_5CH=CHC_6H_5 + C_2H_4 \longrightarrow C_6H_5CH=CH_2 \quad \Delta H = -15.99 \text{ KJ g}^{-1} \text{ mol}^{-1}$$

The idea attracted interest, but the project has been formally abandoned.

Another route being researched is the alkylation of toluene with methanol over zeolite catalysts [31]–[33]. The selectivity with respect to toluene is claimed to be high, but it is only about 50% based on methanol. No commercial plants have been announced.

4.5. Styrene from Pyrolysis Gasoline

An aromatic mixture ("pyrolysis gasoline;" → Benzene) including styrene is obtained from the thermal cracking of naptha or gas-oil. Recovery of the styrene has been proposed on the basis of extractive distillation with dimethylformamide or dimethylacetamide, by adsorption, by complex formation, or by membrane separation. So far none of these processes has been commercially exploited.

Table 4. ASTM styrene specifications (1993)

	ASTM D 2827-92	Typical analysis	ASTM method
Purity, %	99.7	99.8	D 3799*
Color, Pt–Co, max.	10	7	D 1209
Aldehydes, mg/kg	200	30	D 2119
Peroxides, mg/kg	100	5	D 2340
Polymer, mg/kg	10	0	D 2121
Inhibitor, mg/kg	10–15	12	D 4590
Impurities			D 5135**

* By freezing point.
** By gas chromatography.

5. Quality and Testing

The specifications and methods of analysis for styrene have changed through the years. Almost all manufacturers use ASTM D 2827-92 in their sales specifications, which calls for a minimum purity of 99.7 %, but much of the styrene in today's competitive market is of much higher purity. Historically, styrene purity has been determined by freezing point, the method referred to in ASTM D 2827-92. Gas chromatography (GC) is used to determine specific impurities, and in practice, most manufacturers now use GC to determine overall purity by subtracting the total impurities from 100 %. Table 4 lists ASTM specifications and typical analysis for styrene monomer.

The major impurities in styrene monomer are a function of the process variables. Ethylbenzene content varies depending on the effort put into the main ethylbenzene–styrene distillation column. α-Methylstyrene, isopropylbenzene (cumene), n-propylbenzene, and minor amounts of ethyltoluene and vinyltoluene in finished styrene are a function of the separation power of the styrene still. Xylene content is influenced by the purity of the feed ethylbenzene. Phenylacetylene is produced by the dehydrogenation process in amounts depending on the catalyst used, typically 50–150 mg/kg. Phenylacetylene must be removed by a hydrogenation step; it is absent from material made by the styrene–propylene oxide process.

6. Storage and Transportation

Styrene is a flammable, reactive monomer. It has a flash point of 31 °C and a flammable range of 1.1–6.1 vol % in air. It undergoes exothermic polymerization quite readily (0.02 %/h at 25 °C) liberating 69.8 kJ per mole of reacted monomer. Failure to remove this heat from an enclosed container could theoretically produce a temperature of 300 °C; uncontrolled polymerization may therefore lead to pressure build-up in a closed container, or a potentially explosive vapor cloud. However, styrene has been used on a large scale industrially for more than 50 years, and extensive experience is available with respect to preventive measures against fire and explosion, permitting problem-free storage and safe shipping of the monomer. Data are available from all styrene producers [5], [34].

Transportation of styrene monomer is subject to regulation in all countries (e.g., DOT in the United States and the VbF guidelines in Germany). The DOT identification number is UN 2055, and the reportable quantity for a spill is 454 kg.

Leakage can easily be detected owing to the characteristic styrene odor and a low odor threshold (0.005 mg/kg in air).

To increase its shelf life, styrene monomer is inhibited, typically with 4-*tert*-butylcatechol (TBC). Hydroquinone has also been used, but it is not as effective. At a TBC level of 12 mg/kg, a shelf life of 6 months is predicted at 20 °C; this falls to 3 months at 30 °C. This inhibitor requires trace amounts of oxygen to render it effective. To ensure a margin of safety, the recommended minimum oxygen level in the monomer is 15 mg/kg. The solubility of oxygen from the air in styrene monomer is 50 mg/kg. To prevent the occurrence of hazardous polymerization, styrene should always be kept cool, and appropriate inhibitor and oxygen levels should be maintained.

Styrene in storage is frequently padded with inert gases, which are partially soluble in the monomer. The consequences of desorption of these gases must be considered in subsequent handling. When the oxygen level in the inert gas pad is < 8 vol%, the possibility of fire or explosion is eliminated.

Storage and shipping containers for styrene may be of standard steel or aluminum. Rust acts as a catalyst for the polymerization of styrene, so inorganic zinc linings are recommended for storage tanks. Copper and brass fittings must be avoided because they can lead to discoloration of the styrene. Contamination with bases must also be avoided, as these react with the inhibitor, making it ineffective. Loading on a common carrier in a compartment next to a heated product is an unacceptable practice.

Foam is the preferred medium for fighting styrene fires. Water fog dissipates vapor clouds and provides cooling for structural supports, but rarely extinguishes the fire. Styrene monomer, with its high volume resistivity, can acquire and hold a static electric charge during transfer. Effective grounding measures must therefore be taken to eliminate uncontrolled electric discharge.

7. Uses and Economic Aspects

Styrene monomer is used as a feedstock in a variety of polymer products: thermoplastics, elastomers, dispersions, and thermoset plastics. The following breakdown represents an estimate of the distribution of the monomer, as well as some of the uses of its derivatives [35].

Approximately 65% of the styrene produced goes into polystyrene. This is used to make a wide range of products, from toys, housings for room air conditioners, and television cabinets, to cassettes, combs, and furniture parts. Polystyrene can also be foamed to produce insulation board, loose-fill packaging, and disposable food containers.

Approximately 6% still goes into styrene–butadiene rubber elastomers (SBR) for such uses as passenger car tires, industrial hoses, and footwear.

Approximately 7% goes to styrene–butadiene latexes, used in tufted carpet, paper coatings, and as a component in latex paints.

Approximately 9% goes into styrene–acrylonitrile copolymer (SAN) and terpolymers of acrylonitrile, butadiene, and styrene (ABS); SAN is used for drinking tumblers and battery cases, ABS for piping, automotive components, refrigerator doorliners, and shower stalls.

A further 7% is combined with unsaturated polyester resins in fiberglass-reinforced boats, storage tanks, shower units, and simulated marble products.

The remainder goes to miscellaneous uses, especially involving blends with other thermoplastics for a constantly growing line of products, including ion-exchange resins and adhesives.

World capacity for styrene in 1993 was estimated at ca. 17×10^6 t/a, divided among the most important markets as follows:

North America	35%
Western Europe	27%
Japan	16%
Korea	7%
Far East (remainder)	5%
Eastern Europe	5%
South America	4%
Middle East	1%

Over 80 companies are now producing styrene monomer, but the pace of announcements of new plants, shutdown of old plants, and business mergers makes this a constantly changing number.

The majority of the producing companies have licensed their technology. A very rough estimate of capacities for the various technologies is as follows:

Badger	40%
ABB Lummus Crest	24%
PO/Styrene	13%
Dow Chemical	11%
BASF	7%
Miscellaneous	5%

Badger, a Raytheon Company, entered the styrene business in 1955. They acquired and developed portions of the Cosden, Union Carbide, and Fina styrene technology. Lummus Crest is part of the ABB group, and has acquired and developed the Monsanto styrene process, also merging technologies and resources with UOP and Unocal. ARCO and Shell own their own styrene–PO technologies. Dow and BASF have their own dehydrogenation technologies, and have both been making styrene for over 50 years. The size and age of the industry means that it is well reported and analyzed [34]–[36].

World-scale single-train units have a capacity of 450×10^3 t/a, but units down to 20×10^3 t/a still find their niche.

Benzene [*71-43-2*] (→ Benzene), used for making ethylbenzene, accounts for ca. 65 % of the cost of styrene. Large swings in the price of benzene are reflected in that of styrene, which also responds to variation in supply and demand. The price of styrene has fluctuated from a high of $ 1100/t in the late 1980s to $ 450/t in 1992 [37]. High profit margins tempt new producers into the business, and because they can buy their way in quickly via the licensors, overcapacity is easily reached (as in 1993), giving rise to a classic business cycle.

8. Related Monomers

Styrene can be substituted in the side chain or in the ring to give many other compounds; only the following have attracted commercial interest.

8.1. Vinyltoluene

Vinyltoluene, [*25013-15-4*] *p*-methylstyrene, $CH_3C_6H_4CH=CH_2$, is a specialty monomer, with properties similar to those of styrene [38]. It is available commercially in two different isomer mixes. For many years it has been supplied as a 68/32 mixture of the *meta* and *para* isomers, reflecting the composition of the feedstock ethyltoluene (ethylmethylbenzene), from the Friedel–Crafts synthesis. This is usually called vinyltoluene (VT). The advent of a *para*-specific route to ethyltoluene using zeolite technology has made available a 97 % *para*-isomer product called PMS [*627-97-9*] to differentiate it from the original mix. The VT mix was originally developed to compete with styrene, with the idea that its feedstock (toluene) would give it a competitive advantage over a benzene-based material. However, benzene remained cheap and available, and VT proved to be more costly to make than styrene, so it has not gained a large market share. It has nevertheless maintained a specialty position because of its unique properties. Table 5 lists the physical properties of VT and PMS.

The reactivity of VT is similar to that of styrene. Thus, it undergoes both homopolymerization and copolymerization. Polymerization can be initiated by exposure to strong acids, peroxides, perchlorates, or hypochlorites. It also reacts similarly to styrene with oxygen and halogens, although the presence of an extra side chain makes the chemistry more complex.

The dehydrogenation of ethyltoluene to vinyltoluene is similar to that described for ethylbenzene to styrene. It is a vapor-phase process, employing superheated steam and a suitable catalyst. The amount of *o*-ethyltoluene in the feed should be kept at a minimum because it can form indene in the cracker, which impairs the properties of the final polymer. Distillation of the crude product must be performed carefully

Table 5. Physical properties of VT and PMS

	VT	PMS
M_r	118.18	118.18
bp, °C	171.2	170.0
fp, °C	−77	−34
Flammable limits, vol% in air	1.1–5.2	
Flash point, TCC, °C	52.8	
Autoignition point, °C	575	
Heat of polymerization, ΔH_p, kJ/mol (25 °C)	−70	−70
Heat of vaporization, ΔH_v, J/g (25 °C)	426.1	410.9
Density, g/mL (25 °C)	0.8930	0.8920
Refractive index, n_D^{25}	1.5395	1.5408
Specific heat capacity (liquid), $J\,g^{-1}\,K^{-1}$ (20 °C)	1.715	
Viscosity, mPa s (25 °C)	0.79	0.79
Surface tension, mN/m (25 °C)	31.0	34.0
Solubility in water, mg/kg (25 °C)	90	

under vacuum to limit the temperature and thus the rate of polymerization. A suitable polymerization inhibitor is also necessary. The complexity of the crude reactor product requires elaborate separation by distillation to give a specification product. The monomer is stored under refrigeration at 10 °C, and it is typically inhibited with TBC (12 mg/kg) during storage and transportation.

Toxicologically, VT has been considered to be very similar to styrene, and normal precautions should be taken when handling it (see Chap. 9). The ACGHI TLV is 50 mg/kg. VT is flammable, but its vapor pressure does not produce a flammable mixture in air below 38 °C. It polymerizes more rapidly than styrene, and polymerization can be initiated by exposure to strong acids, peroxides, perchlorates, or hypochlorites. Contact between vinyltoluene and a base will remove the TBC inhibitor.

Vinyltoluene is used in copolymers and as a specialty monomer in paint, varnish, and polyester formulations.

8.2. Divinylbenzene

Divinylbenzene (DVB) [1321-74-0], $C_6H_4(CH=CH_2)_2$, is a cross-linking agent that improves polymer properties. The commercial monomer is mainly a mixture of m- and p-DVB (typical ratio 2.25:1), diluted with ethylvinylbenzenes. Physical properties of the 55% commercial mixture are listed below:

M_r	130.191
bp	195 °C
fp	−45 °C
Pseudocritical pressure	2.462 MPa
Pseudocritical temperature	369 °C
Flammable limits in air	0.8–5.5 vol%

Flash point, TCC	65.6 °C
Autoignition point	505 °C
Heat of vaporization, ΔH_v (195 °C)	350.6 J/g
Vapor pressure	
(100 °C)	4.4 kPa
(140 °C)	20.0 kPa
(180 °C)	67.0 kPa
Density (25 °C)	0.9084 g/mL
Refractive index, n_D^{25}	1.5585
Surface tension (25 °C)	32.1 mN/m
Viscosity (25 °C)	1.007 mPa s
Solubility	
DVB in water (25 °C)	0.0052 %
Water in DVB (25 °C)	0.054 %

The method of manufacture is analogous to that for styrene: endothermic dehydrogenation of an isomer mix of diethylbenzenes obtained as a side-stream of ethylbenzene production. Diethylbenzene is vaporized, diluted with superheated steam, and then passed over a catalyst at ca. 600 °C. The reactor effluent is mainly *m*- and *p*-DVB, the corresponding ethylvinylbenzenes, and unreacted diethylbenzene. Most of the *o*-diethylbenzene undergoes ring closure to naphthalene. Separation of this mixture must compete with the ready polymerization of DVB. Separation can be achieved by vacuum distillation with suitable in-process polymerization inhibitors.

Divinylbenzene is more reactive than styrene, and can homopolymerize or copolymerize with many other monomers. Because of the two vinyl groups, the resulting polymers are cross-linked. Cross-linking improves solvent resistance, heat distortion, impact resistance, tensile strength, and hardness.

DVB is very similar to styrene monomer in its toxicological properties (Chap. 9). The ACGIH TLV for divinylbenzene is 10 mg/kg (53 mg/m^3). Its single-dose oral toxicity is low. Because DVB polymerizes so readily, additional precautions must be taken to prevent runaway reactions in storage. The monomer is inhibited with TBC (ca. 1000 mg/kg), and refrigerated storage at 10 °C is recommended.

Most DVB is used in ion-exchange resins based on suspension-polymerized beads of styrene–divinylbenzene. It is also used as a cross-linking agent in specialty rubber and other plastic applications.

8.3. α-Methylstyrene

α-Methylstyrene (AMS) [*98-83-9*], $C_6H_5C(CH_3)=CH_2$, is a monomer with a polymerization rate much lower than that of styrene. It can be produced commercially by the dehydrogenation of isopropylbenzene (cumene), but also occurs as a byproduct in the manufacture of phenol and acetone via the cumene oxidation process, now the major source of AMS (→ Acetone). Physical properties of AMS are listed below:

M_r	118.18
bp	165 °C
fp	− 23.2 °C
Critical density	0.29 g/mL
Critical pressure	4.36 MPa
Critical temperature	384 °C
Critical volume	3.26 mL/g
Flammable limits in air	0.7 – 3.4 vol%
Flash point, Cleveland Open Cup (COC)	57.8 °C
Heat of polymerization	−39.75 kJ/mol
Heat of vaporization, ΔH_v	
(25 °C)	404.6 J/g
(165 °C)	326.4 J/g
Specific heat capacity (liquid)	
(20 °C)	2.047 J g^{-1} K^{-1}
(100 °C)	2.176 J g^{-1} K^{-1}
Specific heat capacity (vapor) (25 °C)	1.2357 J g^{-1} K^{-1},
Solubility	
AMS in water (25 °C)	0.056 %
water in AMS (25 °C)	0.010 %
Density (20 °C)	0.9106 g/mL
Viscosity (20 °C)	0.94 mPa s

AMS undergoes reactions and polymerizations similar to those of styrene and vinyltoluene. However, the addition of an α-methyl group to the side chain changes its chemical reactivity. AMS monomer tends to be more stable than other vinyl aromatics. Homopolymers are best prepared by ionic catalysis, copolymers by radical polymerization.

Toxicologically, AMS is similar to styrene (Chap. 9). The ACGIH TLV is 50 mg/kg, and the STEL 100 mg/kg. AMS has a higher flash point than styrene, and it is less likely to polymerize in storage. It forms low molecular mass polymers slowly after the depletion of oxygen and/or inhibitor. The polymer thus formed is usually a viscous liquid. Uninhibited AMS in storage oxidizes slowly to acetophenone, aldehydes, and peroxides, which may affect the polymerization rate.

Homopolymers of AMS are used as plasticizers in paints, waxes, and adhesives. The monomer is also used to form a copolymer with methyl methacrylate. The copolymer has a high heat-distortion temperature, and has been approved for use in food applications. Its light color makes it useful in modified polyester and alkyd resin formulations.

8.4. Chlorostyrene

The two compounds *ortho-* and *para-* chlorostyrene [2039-87-4], ClC$_6$H$_4$CH=CH$_2$, are reactive monomers produced only in small quantities. The *para* isomer has a boiling point of 192 °C and a flash point of 60 °C. It is combustible. Chlorostyrene is more reactive than styrene, and is easily polymerized. It is typically inhibited with 3,5-di-*tert*-butylcatechol (250 mg/kg) for storage and transportation.

The monomer can be manufactured from either *o*- or *p*-chloroethylbenzene by oxidation to the corresponding alcohol, followed by dehydration. It is possible to produce pure monomeric *o*- or *p*-chlorostyrene as well as a mixture of the two.

Extensive toxicological data for chlorostyrene are not available, but it is assumed to be similar to styrene (Chap. 9), with an ACGIH TWA of 50 mg/kg.

Chlorostyrene is used to provide shorter cure times, better heat-distortion properties, and flame resistance in polyester resins.

8.5. Vinylbenzyl Chloride

Vinylbenzyl chloride (VBC) [*30030-25-2*], $CH_2=CHC_6H_4CH_2Cl$, is also called (chloromethyl)ethenylbenzene or α-chloromethylstyrene. It is a difunctional monomer with both a polymerizable double bond and a benzylic chlorine, and is an isomeric mixture made by the chlorination of VT. VBC has a high boiling point (98.9 °C at 1.3 kPa) and low flammability. Other physical properties are as follows:

M_r	152.62
bp (1.33 kPa)	98.9 °C
(101.325 kPa)	229 °C*
fp	−26 to −42 °C
Flash point, COC	104.4 °C
Autoignition point	610 – 620 °C
Heat of fusion, ΔH_m	−10.04 kJ/mol
Heat of vaporization, ΔH_v (157.2 °C)	334.0 J/g
Heat of polymerization, ΔH_p (195 °C)	−0.422 kJ/mol
Solubility in water (25 °C)	730 mg/kg

* Extrapolated value.

The reactivity of VBC is similar to that of VT. VBC is stored under refrigeration, and inhibited with 75 mg/kg of TBC and 800 mg/kg of nitromethane [38]. Its acute oral toxicity is claimed to be low, but the monomer vapor is a strong lachrymator. VBC can be transformed into a high molecular mass homopolymer, will copolymerize with many other monomers, and is used as comonomer in a variety of specialty plastics. It is also possible to carry out reactions on the chloromethyl group, either before or after polymerization. The chemistry of VBC has been covered in an excellent review [39].

9. Toxicology and Occupational Health

In view of the large-volume production of styrene and the widespread use of styrene-based polymers in general, and especially in food- contact use, there is much research and a huge literature on the toxicology of styrene [40], [41].

Styrene is slightly toxic. In rats, the acute oral LD_{50} is 500 – 5000 mg/kg, with an inhalation $LC_{50} > 3000$ mg/kg. Death from acute dermal exposure has not been reported. Styrene is a moderate skin irritant, especially if trapped against the skin by contaminated clothing. Styrene vapor at high concentration has an irritant effect on the eyes and mucous membranes of humans and animals. Controlled observations on humans have shown that no irritation occurs at a concentration of ca. 100 mg/kg [42], [43], but trace amounts of styrene in the workplace together with trace amounts of halogen can produce a lachrymator in the presence of bright sunlight.

Styrene can affect the central nervous system, producing fatigue and headache at a certain level of exposure, and narcosis if the level is sufficiently high. Signs of transient impairment of central nervous system function have been described in voluntary test subjects [44] and in workers [45] exposed to styrene in concentrations of 375 – 800 mg/kg. Impaired hearing and visual coordination have been detected after exposure to styrene at ca. 100 mg/kg. At 50 mg/kg, however, any negative effects of this type were considered doubtful or barely discernible [46]. Prenarcotic symptoms have been shown in workers exposed to a TWA of 50 mg/kg. They are clearly established at a TLV of 100 mg/kg [47].

A review of eight cohort studies involving nearly 50 000 employees over 46 years [48] showed no indication of human carcinogenicity related to styrene exposure. A comprehensive review of published data [49] concluded that there is little indication that styrene has any specific developmental or reproductive toxicity.

Long-term animal studies on styrene are more difficult to analyze. Limitations in the studies often preclude definitive conclusions. However, a review by BOYD et al. [50] concludes that there is no clear evidence of a carcinogenic response related to styrene exposure. It continues to be a problem to relate high-dose results in different animals to each other, and to project these to human response.

The metabolism of styrene proceeds through styrene-7,8-oxide to mandelic acid and phenylglyoxalic acid, which are then excreted in the urine [51]. There is a correlation in humans between urinary elimination of mandelic acid and the styrene concentration in the respiratory air if the latter is ≤ 100 mg/kg (420 mg/m^3). However, the styrene concentration in the blood seems to be a more reliable exposure index than either the mandelic acid content of urine or the styrene content in exhaled air.

The mouse seems to be a very sensitive species in styrene exposure studies. Mice apparently have a greater capacity than rats or humans to form styrene-7,8-oxide, and a lesser capacity to metabolize this chemical [52].

Nevertheless, the International Agency for Cancer Research classified styrene in 1987 as a possible human carcinogen, at the same time concluding that the evidence for carcinogenicity was limited, and in humans inadequate. This classification has provoked legislation, causing some countries and several states in the US to list styrene as a possible carcinogen. OSHA, on the other hand, decided in 1989 that styrene should not be regulated as a workplace carcinogen.

For styrene monomer, as for many other reactive substances, the toxic effect clearly depends on dose. Health risks can be expected at very high concentrations, but the no-observed-effect levels (NOEL) are considerably higher than those found in a well-managed operating facility, or in the background to which users of styrene-based products are exposed. The ACGIH TLVs for styrene are 50 mg/kg (213 mg/m^3) for an 8-h time-weighted average (TWA) exposure, with a 100 mg/kg (426 mg/m^3) short-term exposure limit (STEL), defined as a 15-min TWA [53]. The MAK value is 20 mL/m^3 (85 mg/m^3).

10. References

[1] E. Simon, *Justus Liebigs Ann. Chem.* **31** (1839) 265.
[2] D. H. Steele, MRI Project No. 6450, Midwest Research Institute, Kansas City, Mo., 1992.
[3] R. H. Boundy, R. F. Boyer, (eds.): *Styrene, Its Polymers, Copolymers and Derivatives,* Reinhold Publ. Co., New York 1952; reprinted by Hafner Publishing Corp., Darien, Conn., 1970.
[4] R. F. Boyer: "Anecdotal History of Styrene and Polystyrene," in R. Semour (ed.): *History of Polymer Science and Technology,* M. Dekker, New York 1982.
[5] Dow Chemical, The Dow Family of Styrenic Monomers, Form No. 115-00 608-1289 X-SAI, Midland, Mich. 1989.
[6] W. S. Emerson, *Chem. Rev.* **45** (1949) 183 ff.
[7] K. E. Coulter, H. Kehde, B. F. Hiscock in E. C. Leonard (ed.): *High Polymers,* Wiley-Interscience, New York, vol. XXIV, "Vinyl and Diene Monomers," part 2: "Styrene and Related Monomers," 1971.
[8] K. Meinel, *Justus Liebigs Ann. Chem.* **510** (1934) 129.
[9] D. L. Williams, "Styrene Catalysts, Past Present and Future," AICHE Spring Meeting, New Orleans, March 1988.
[10] Shell Chemical, Technical Bulletin on Shell 005, Shell 105 Dehydrogenation Catalysts, 78:14 Houston, Texas 1978.
[11] United Catalysts Inc., G-64 & G-84 Product Bulletin, 2nd ed. Louisville, Ky. 1987.
[12] *Ullmann,* 4th ed., **22**, 298.
[13] Lurgi, Low Energy Concept for Styrene Production, Technical Brochure 2173 e/4.89.
[14] Union Carbide Corp., US 3 282 576, 1966.
[15] J. C. Frank, G. R. Geyer, H. Kehde, *Chem. Eng. Prog.* **65** (1969) no. 2, 79.
[16] "Facelift for Distillation," *Chem. Eng.* 1987, March 2, 14.
[17] D. B. McMullen et al., *Chem. Eng. Prog.* **87** (1991) no. 7, 187.
[18] Sulzer Bros. Inc., GB 1 020 190, 1966.
[19] UOP Inc., US 4 435 607, 1984 (T. Imai).

[20] K. Egawa et al., *Aromatics* **43** (1991) 5–6.
[21] ARCO Chemical, Construction Permit Application for Propylene Oxide/Styrene Monomer Facility, Channelview, Texas, Texas Air Control Board Account No. HG-1575 W, July 1989.
[22] Phillips, US 4 117 025, 1978 (T. C. Liebert, W. A. McClintock).
[23] Exxon, US 2 943 117, 1960 (A. H. Gleason).
[24] Phillips, US 3 377 397, 1968 (P. L. Maxfiel).
[25] Phillips, US 4 144 278, 1979 (D. J. Strope).
[26] BASF, US 3 903 185, 1975 (H.-H. Vogel, H.-M. Weitz, E. Lorenz, R. Platz).
[27] ARCO, US 4 029 715, 1977 (R. W. Rieve, H. Shalit).
[28] Maruzen Oil, US 3 502 736, 1970 (M. Sato, K. Tawara).
[29] Montedison, DE 2 612 082, 1976.
[30] Monsanto, US 3 965 206, 1976 (H. W. Scheeline, J. J. L. Ma).
[31] Monsanto, US 4 115 424, 1978 (M. L. Unland, G. E. Barker).
[32] Monsanto, US 4 140 726, 1979 (M. L. Unland, G. E. Barker).
[33] Shell Oil, US 5 015 796, 1991 (L. H. Slaugh, T. F. Brownscombe).
[34] R. Kuhn, K. Birett: *Merkblätter gefährliche Arbeitsstoffe*, Blatt Nr. S 25, Verlag Moderne Industrie, München 1978.
[35] EPA-450/4-91-029, 11–12, Office of Air Quality, Research Triangle Park, N. C.T. Wett, *Chemical Business*, March 1993, 21.
[36] SRI International, Styrene, Report 33 C, Menlo Park, Ca., 1993.
[37] Chem Systems, Styrene/Ethylbenzene 91-9, Tarrytown, New York, 1992.
[38] Dow Chemical, Speciality Monomers Product Stewardship Manual. Form No. 505-0007-1290 JB, Midland, Mich. 1990.
[39] M. Camps, M. Chatzopoulos, J. P. Montheard, *J. Macromol. Sci., Macromol. Chem. Phys.*, Part C **22** (1982–83) no. 3, 343.
[40] J. Santodonato et al., "Monograph on Human Exposure to Chemicals in the Workplace; Styrene," PB 86-155 132, Syracuse Research Corp., Syracuse, N.Y., July 1985.
[41] *The SIRC Review*, vol. **1**, no. 1 and vol. **1**, no. 2. 1275 K St., N.W., Suite 400, Washington, DC 20 005.
[42] R. D. Stewart, H. C. Dodd, E. D. Baretta, A. W. Schaffer, *Arch. Environ. Health* **16** (1968) 656.
[43] R. L. Zielhuis et al.: *14th Int. Congr. Occup. Health*, Madrid, 3, 1092.
[44] C. P. Carpenter et al., *J. Ind. Hyg. Toxicol.* **26** (1944) no. 3, 68.
[45] H. Harkonen, *Int. Arch. Occup. Environ. Health* **40** (1977) 231.
[46] M. Oltramare et al.: *Editions Medecine et Hygiene*, Geneva 1974, p. 100.
[47] Y. Alarie, *Toxicol. Appl. Pharmacol.* **24** (1973) 279.
[48] G. Bond, K. Bodner, R. Cook, *SIRC Review* **1** (1991) no. 1, 43–55.
[49] N. A. Brown, *Reproductive Toxicol.* **5** (1991) 3.
[50] D. P. Boyd et al., *SIRC Review* **1** (1990) 5–23.
[51] Z. Bardodej, E. Bardodejova, *Amer. Ind. Hyg. Assoc. J.* **31** (1970) 206.
[52] D. H. Steele et al., *J. Agric. Food Chem.* **42** (1994) 1661.
[53] 1992–1993 Threshold Limit Values, American Conference of Governmental Industrial Hygienists, Cincinnati, Ohio 1992.

Sulfamic Acid

ADOLF METZGER, Hoechst Aktiengesellschaft, Frankfurt/Main, Federal Republic of Germany

1.	Introduction 4443	5.	Uses 4446
2.	Physical Properties 4443	6.	Toxicology............. 4446
3.	Chemical Properties....... 4444	7.	Sulfamates 4446
4.	Production 4445	8.	References............ 4447

1. Introduction

Sulfamic acid (IUPAC recommendation: amidosulfuric acid) has the formula NH_2SO_3H.

Sulfamic acid is a crystalline, strong inorganic acid. It has been produced industrially for ca. 50 years. In expectation of good sales, production plants were set up in a number of industrial countries during the 1950s to 1980s. However, as a result of a decrease in demand, production difficulties, and problems with the disposal of by-products, all the plants in Europe and the United States have been closed down. Plants are still operating in Japan (Nissan) and Taiwan (several producers).

2. Physical Properties

Pure sulfamic acid is a white, odorless, crystalline, nonhygroscopic compound. It is not volatile at room temperature. Some important data are given in [1], [2].

Density, 25 °C	2.126 g/cm^3
Specific heat	1.1467 J/g
Melting point	205 °C
Vapor pressure	
20 °C	0.8 Pa
100 °C	0.25 Pa
Dissociation constant	1.10×10^{-1}

Sulfamic acid is strongly dissociated in aqueous solution. The pH of a 1 % solution is 1.18. The variation in water solubility with temperature is shown in Table 1. The

Table 1. Solubility of sulfamic acid in water

Temperature, °C	Solubility, g/100 g	Concentration, wt%
0	14.7	12.8
10	18.5	15.6
20	21.3	17.5
30	26.1	20.7
40	29.5	22.8
50	32.8	24.7
60	37.1	27.0
70	41.9	29.5
80	47.0	32.0

solubility in water can be increased by the addition of other acids or salts. Sulfamic acid is only sparingly soluble in concentrated inorganic acids, and most organic solvents.

3. Chemical Properties

Sulfamic acid decomposes on heating above 209 °C. Sulfur trioxide, sulfur dioxide, water, ammonia, and nitrogen have been detected as decomposition products [3].

Aqueous solutions of sulfamic acid hydrolyze to ammonium hydrogen sulfate:

$$NH_2SO_3H + H_2O \longrightarrow NH_4HSO_4$$

The rate of hydrolysis depends on concentration, pH, and temperature. Dilute, aqueous sulfamic acid solutions are stable at room temperature. A 10% solution is 50% hydrolyzed at 80 °C in 10 h [4].

Chlorine, bromine, and chlorates oxidize sulfamic acid to sulfuric acid:

$$2\,NH_2SO_3H + KClO_3 \longrightarrow 2\,H_2SO_4 + N_2 + KCl + H_2O$$

Concentrated nitric acid reacts to give dinitrogen oxide on warming:

$$NH_2SO_3H + HNO_3 \longrightarrow H_2SO_4 + N_2O + H_2O$$

Nitrous acid converts sulfamic acid completely to nitrogen and sulfuric acid, so this reaction can be used for quantitative analysis:

$$NH_2SO_3H + HNO_2 \longrightarrow H_2SO_4 + H_2O + N_2$$

Reaction with thionyl chloride forms sulfamyl chloride:

$$NH_2SO_3H + SOCl_2 \longrightarrow ClSO_2NH_2 + SO_2 + HCl$$

Metal hydroxides, oxides, and carbonates are dissolved by sulfamic acid.

The reactions of sulfamic acid with organic compounds have been described in detail [5]. Sulfamic acid reacts with primary and secondary alcohols to alkylammonium sulfates:

$ROH + NH_2SO_3H \longrightarrow ROSO_2ONH_4$

Secondary alcohols react only in the presence of amines, which act as catalysts. Tertiary alcohols do not react. Aromatic alcohols, such as phenol, give sulfonates with sulfamic acid, phenylammonium sulfate being an intermediate:

$C_6H_5OH + NH_2SO_3H \longrightarrow C_6H_5OSO_2ONH_4 \longrightarrow HO-C_6H_4-SO_2-ONH_4$

Aldehydes form addition products with salts of sulfamic acid. Aromatics with unsaturated side-chains, e.g., styrene, are sulfonated by sulfamic acid, with retention of the side-chain double bond:

$C_6H_5CH=CH_2 + NH_2SO_3H \longrightarrow C_6H_5-CH=CH-SO_2ONH_4$

4. Production

Sulfamic acid is now produced solely by the urea process; the plants using ammonia and sulfur trioxide [6] have been closed down.

Following a suggestion by BAUMGARTEN, equimolar quantities of urea, sulfur trioxide, and sulfuric acid are reacted directly to sulfamic acid [7]:

$NH_2CONH_2 + SO_3 + H_2SO_4 \longrightarrow 2NH_2SO_3H + CO_2$

This is a strongly exothermic reaction. The process is carried out in two stages, based on the following reactions:

$NH_2CONH_2 + SO_3 \longrightarrow NH_2CONHSO_3H$
$NH_2CONHSO_3H + H_2SO_4 \longrightarrow 2\ NH_2SO_3H + CO_2$

Urea is stirred in excess sulfuric acid and sulfur trioxide in the first stage. The temperature should be kept below 40 °C to hinder the formation of carbon dioxide. In the second stage, the reaction product from the first stage reacts to form sulfamic acid and carbon dioxide at 50–80 °C in the presence of excess sulfur trioxide. After removal of the excess sulfur trioxide, sulfamic acid of purity > 95% can be obtained [8]. High-purity sulfamic acid is obtained by recrystallization.

According to a Nissan patent, sulfamic acid is produced by the reaction of urea and oleum ($x\,H_2SO_4 \cdot y\,SO_3$). The reaction product is recrystallized wet, and the crystallized acid is separated and dried. By grinding under high pressure, preferably 600–1500 kg/m^2, a fine crystalline product with average particle diameter ≤ 500 µm is obtained [9].

5. Uses

It is estimated that world annual production is ca. 60 000 t. Most of this is used in cyclamate production. The use of sulfamic acid in cleaning agents for carbonate- and phosphate-containing deposits, e.g., boiler scale, is based on its ability to form readily soluble salts and its relatively low corrosive effect on metals. Sulfamic acid is widely used for cleaning machines and instruments in the paper, sugar, dairy, and brewing industries, and for removing deposits in evaporation plants, heat exchangers, and cooling systems.

In some countries, a process involving the treatment of fatty or ethoxylated alcohols with sulfamic acid to produce raw materials for waxes is used on an industrial scale [10].

6. Toxicology

Sulfamic acid is not specifically toxic, and its physiological properties are typical of a strong mineral acid. Sulfamic acid dust has an irritant effect on the mucous membranes of the nose and pharynx, and the conjuctiva of the eyes. Oral administration of 1.6 g/kg to rats is lethal.

7. Sulfamates

The solubility of the alkaline-earth sulfamates is important for the utilization of sulfamic acid in industry [11].

Ammonium sulfamate can be obtained by neutralization of sulfamic acid with ammonia. Technical grade ammonium sulfamate was formerly produced by reaction of ammonia with sulfur trioxide under pressure.

Ammonium sulfamate is an effective softener for paper. Impregnated paper is not readily flammable [12]. Ammonium sulfamate has also been used for the removal of lime salts without leaving a residue, which would interfere with the tanning process in leather production.

8. References

[1] E. S. Taylor, R. P. Desch, A. J. Catotti, *J. Am. Chem. Soc.* **72** (1950) 74–77.
[2] E. J. King, G. W. King, *J. Am. Chem. Soc.* **84** (1962) 1212–1215.
[3] E. Divers, T. Haga, *J. Chem. Soc.* **69** (1896) 1634.
[4] J. M. Notley, *Trans. Inst. Met. Finish.* **52** (1974) 78.
[5] L. F. Audrieth, M. Sveda, H. Sisler, L. J. Butler, *Chem. Rev.* **26** (1940) 49–94.
[6] Ullmann, 4th ed., **22,** pp. 311–313.
[7] P. Baumgarten, *Ber. Dtsch. Chem. Ges. B* **69** (1936) 1929–1936.
[8] Du Pont, US 2 880 064, 1959.
[9] *Chem. Abstr.* **109** (1988) 22, 193174.
[10] General Anilin and Film Corp., US 2 758 977, 1958.
[11] G. B. King, I. F. Hooper, *J. Phys. Chem.* **45** (1941) 938.
[12] Du Pont, US 2 723 212, 1958. F. T. Blakemore, *Pap. Film Foil Converter* **23** (1949) 29.

Sulfinic Acids and Derivatives

RÜDIGER SCHUBART, Bayer AG, Leverkusen, Federal Republic of Germany

1.	Introduction 4449	5.1.	Sulfinyl Chlorides 4458
2.	Properties 4450	5.1.1.	Preparation................ 4458
3.	Preparation of Sulfinic Acids and their Salts............ 4452	5.1.2.	Reactions 4461
3.1.	Reduction of Sulfonic Acid Derivatives 4452	5.2.	Sulfinic Acid Esters 4462
		5.2.1.	Preparation................ 4462
		5.2.2.	Reactions 4463
3.2.	Sulfination with Sulfur Dioxide................. 4453	5.3.	Thiosulfinic Acid Esters..... 4464
		5.4.	Sulfinic Acid Amides 4464
3.3.	Oxidation of Thiols 4454	6.	1-Hydroxyalkanesulfinates ... 4465
3.4.	Nucleophilic Cleavage of Sulfones and Related Compounds 4454	6.1.	Properties................. 4465
		6.2.	Production 4466
		6.3.	Quality Specifications, Analysis 4467
4.	Reactions of Sulfinic Acids and their Derivatives 4457	7.	Formamidinesulfinic Acid.... 4468
4.1.	Alkylation of the Sulfinate Anion 4457	7.1.	Properties................. 4468
		7.2.	Production 4468
4.2.	Addition of Sulfinic Acids to Multiple Bonds 4458	8.	Industrial Uses of Sulfinic Acids and their Derivatives .. 4469
4.3.	Redox Reactions of Sulfinic Acids.................... 4458	9.	Toxicology................ 4470
5.	Sulfinic Acid Derivatives 4458	10.	References................ 4471

1. Introduction

Sulfinic acids are synthetic chemical building blocks that also occur in living organisms. For example, allicin, S-2-propenyl 2-propene-1-sulfinothioate [539-86-6] (**1**) is a component of extract of garlic [1]. It can be prepared by oxidation of diallyl disulfide with *m*-chloroperbenzoic acid [2]:

$$\text{\textasciitilde}S-S\text{\textasciitilde} \xrightarrow{[O]} \text{allicin (1)}$$

1

Allicin has antibiotic properties [3], and it has also been found to be an effective antithrombotic and an inhibitor of blood platelet aggregation [3], [4]. In inflammatory processes (e.g., allergies, asthma, rheumatism, etc.) allicin causes 100% inhibition of lipoxygenases [5] as well as inhibiting the formation of arachidonic acid [6], which plays an important role in such processes. Animal experiments have shown that cholesterol formation is also significantly lowered by allicin [7]. The biosynthesis of taurine, an important neurotransmitter in living organisms, proceeds from cysteine via cysteine-sulfinic acid and then hypotaurine [8]. More information in [9].

2. Properties

Physical Properties. Table 1 lists a number of aliphatic and aromatic sulfinic acids together with their CAS registry numbers, melting points, and methods of synthesis.

Chemical Properties. Sulfinic acids are compounds with the following general formula [33], [34]:

$$R-S(=O)-OH$$

Sulfur has a coordination number of 3 in these compounds, and is in the +4 oxidation state. Sulfinic acids have an asymmetric, pyramidal structure in which the sulfur atom constitutes a chiral center. Free sulfinic acids are relatively unstable compounds. On storage or heating they tend to disproportionate (via intermediates) with the loss of water to give sulfonic acids and S-esters of thiosulfonic acids [15], [16], [35]–[38]. Some sulfinic acids decompose as soon as they are formed. For example, benzo-1,3-thiazole-2-sulfinic acid desulfinates immediately upon acidification of the sodium salt, giving benzo-1,3-thiazole (**2**) [39]:

$$\text{benzothiazole-SH} \xrightarrow[\text{NaOH}]{[O]} [\text{benzothiazole-S(=O)-O}^-\text{Na}^+] \xrightarrow{H^+} \text{benzothiazole (2)} + SO_2$$

Aromatic sulfinic acids are generally more stable than long-chain aliphatic sulfinic acids, which are in turn more stable than their short-chain counterparts. Sulfinic acids

Table 1. Physical data for sulfinic acids whose methods of synthesis are described in the text (all compounds are colorless in the pure state)

Sulfinic acid	Molecular formula (molecular mass)	CAS registry no.	mp, °C	Method of synthesis	Reference
$C_2H_5-SO_2H$	$C_2H_6O_2S$ (94.13)	[598-59-4]	liquid	oxidation of ethanethiol with m-chloroperbenzoic acid reduction of ethanesulfonyl chloride with zinc cleavage of sulfones with thiolate	[10] [11] [12]
$H_3C-(CH_2)_3-SO_2H$	$C_4H_{10}O_2S$ (122.19)	[55109-28-9]	liquid	oxidation of butanethiol with m-chloroperbenzoic acid reduction of butanesulfonyl chloride with triethylaluminum cleavage of sulfones with sodium cyanide from butyllithium and sulfur dioxide	[10] [13] [14] [15]
$H_3C-(CH_2)_7-SO_2H$	$C_8H_{18}O_2S$ (178.30)	[3944-71-6]	5–8	reduction of octanesulfonyl chloride with triethylaluminum	[13]
$H_3C-(CH)_{11}-SO_2H$	$C_{12}H_{26}O_2S$ (234.40)	[26535-63-7]	35–36	from dodecylmagnesium chloride and sulfur dioxide cleavage of sulfones with thiolate	[16] [12]
$HO_2S-(CH_2)_4-SO_2H$	$C_4H_{10}O_4S_2$ (186.25)	[6340-77-8]	122–126	reduction of 1,4-butanebis(sulfonyl chloride) with sodium sulfite	[17], [18]
Cl_3C-SO_2H	$CHCl_3O_2S$ (183.44)	[7430-24-2]	liquid	reduction of trichloromethanesulfonyl chloride with hydrogen sulfide	[19]
F_3C-SO_2H	CHF_3O_2S (134.08)	[34642-42-7]	liquid	reduction of trifluoromethanesulfonic acid with hydrazine reduction of trifluoromethanesulfonic acid with zinc reduction of trifluoromethanesulfonic acid with potassium iodide in acetone	[20] [21] [22]
$4-CH_3-C_6H_4-SO_2H$	$C_7H_8O_2S$ (156.20)	[536-57-2]	84	reduction of 4-toluenesulfonyl chloride with triethylaluminum reduction of 4-toluenesulfonyl chloride with zinc reduction of 4-toluenesulfonyl chloride with lithium aluminum hydride from toluene and sulfur dioxide/aluminum trichloride from the diazonium salt of toluene and sulfur dioxide–copper powder	[13] [23] [24] [25] [26]
$4-(H_3C-CO-NH)-C_6H_4-SO_2H$	$C_8H_9NO_3S$ (199.23)	[1126-81-4]	155	reduction of p-acetamidobenzenesulfonyl chloride	[27], [28]
$C_6H_5-CH_2-C_6H_4-4-SO_2H$	$C_{13}H_{12}O_2S$ (232.30)	[58661-47-5]	70–72	intramolecular rearrangement of diarylsulfones with butyllithium	[29]
$H_3C-CO-CH_2-CH_2-SO_2H$	$C_4H_8O_3S$ (136.17)		94–95	cleavage of sulfones with sodium cyanide or thiolate	[30]
⟨S⟩-CO-CH_2-CH_2-SO_2H	$C_7H_8O_3S_2$ (204.27)	[65373-90-2]	91–92	cleavage of sulfones with sodium cyanide or thiolate	[30]
$H_2N-C(NH)-SO_2H$	$CH_4N_2O_2S$ (108.12)	[1758-73-2]	144 (decomp.)	oxidation of thiourea with hydrogen peroxide	[31]

Properties

are best stored as the alkali metal salts, from which they can be liberated with mineral acid [40].

Sulfinic acids are somewhat weaker acids than the corresponding sulfonic acids (→ Sulfonic Acids, Aliphatic), with pK_a values between 1.2 (aromatic sulfinic acids) and 2.2 (aliphatic sulfinic acids). Aromatic sulfinic acids generally crystallize readily as colorless compounds that are sparingly soluble in water but readily soluble in diethyl ether. Aliphatic sulfinic acids are often obtained as oily or low-melting compounds that are soluble in many organic solvents. Further information in [41].

3. Preparation of Sulfinic Acids and their Salts

Sulfinic acids can be synthesized by reduction of compounds containing sulfur in the +6 oxidation state, transformation of other sulfur +4 compounds, oxidation of sulfur +2 compounds, and nucleophilic cleavage of sulfone-like species; review in [42].

3.1. Reduction of Sulfonic Acid Derivatives

The preparation of sulfinic acids and their salts by reduction of sulfonic acid derivatives is straightforward, especially from the readily accessible sulfonyl chlorides.

$$R-SO_2Cl + 2e^- \longrightarrow R-SO_2^- + Cl^-$$

Sodium sulfite, zinc, iron, magnesium, and stannous chloride are frequently used as reducing agents [11], [18], [23], [28], [36], [40], [43]–[45]. The use of lithium aluminum hydride [24], alkylaluminum compounds [13], sodium in liquid ammonia [46], and electrochemical methods [47], has also been investigated. These alternative procedures often lead to high yields, but they have not been adopted on an industrial scale because of technical complications or cost. For additional synthetic information see [48].

Synthesis of 1,4-Butanedisulfinic Acid [17], [18]. A solution of 65.5 g (0.52 mol) anhydrous sodium sulfite in 310 mL of water is mixed with 128 g (0.52 mol) of magnesium sulfate heptahydrate to give a thick precipitate. This mixture is heated to 40–50 °C and 51 g (0.2 mol) of 1,4-butanebis(sulfonyl chloride) is added. The mixture is kept alkaline by the slow addition of ca. 23 g of magnesium oxide. The sulfinic acid precipitates as a magnesium salt. Acidification with dilute sulfuric acid, extraction with diethyl ether, and evaporation gives 1,4-butanedisulfinic acid, *mp* 122–126 °C.

Sodium Ethanesulfinate [11]. 436 g (3.4 mol) of ethanesulfonyl chloride is added dropwise with stirring over 3–4 h to a mixture of 500 mL of water, 500 g of ice, and 340 g (5.2 mol) of zinc dust. The temperature is maintained at +5 °C by continuous addition of 500 g of ice. The reaction mixture is stirred for 1 h and then a solution of 350 g of sodium carbonate decahydrate in 1.5 L of water at 80–90 °C is added over 1.5 h. The reaction mixture should remain distinctly alkaline. It is subsequently filtered, and the residue is washed with warm water. The filtrate and washings are evaporated to give 232 g of the sodium salt, from which the free sulfinic acid can be obtained.

3.2. Sulfination with Sulfur Dioxide

Sulfur dioxide sulfinates both aliphatic and aromatic organometallic compounds, (e.g., Grignard reagents [45], organoaluminums [49]), and particularly organolithiums [50], to give the corresponding sulfinic acid salts:

$$R\text{-}M + SO_2 \longrightarrow R\text{-}SO_2^- M^+$$

p-Dodecylbenzenesulfinic Acid [50]. A solution of 29.2 g (0.09 mol) *p*-dodecylbromobenzene in 120 mL of diethyl ether is added dropwise to 1.5 g (0.22 mol) of finely cut lithium. The reaction mixture is stirred under reflux for 1h and then cooled in an ice bath, after which sulfur dioxide is passed through the cooled solution for ca. 30 min. A further 100 mL of diethyl ether is added during the sulfination. The reaction mixture is centrifuged and the residue removed from the centrifuge vessel with dilute mineral acid. The resulting suspension is extracted with diethyl ether, the solution is evaporated, and the *p*-dodecylbenzenesulfinic acid residue in 53% yield is dried, *mp* 54–55 °C.

1-Dodecanesulfinic Acid [49]. A solution of 70 g (0.13 mol) tridodecylaluminum in 78 mL of toluene is added dropwise over 15 min to a mixture of 60 g (0.95 mol) of sulfur dioxide and 240 g of toluene at −45 °C. The reaction mixture is allowed to warm to room temperature and the solvent is removed by distillation in vacuo. The residue is the aluminum salt of dodecylsulfinic acid, from which the sulfinic acid is liberated in 98% yield, *mp* 35 °C.

Sulfur dioxide also reacts with aromatic hydrocarbons and aluminum trichloride to give sulfinic acids by a Friedel–Crafts-type mechanism [25], [40]. Alkanes react as well under high pressure to give the corresponding alkyl sulfinic acids [51], and alkenes undergo the same reaction at atmospheric pressure. Photochemical sulfinations of alcohols in liquid sulfur dioxide [52] and of alkenes and alkanes in the gas phase are also known [53]. Sulfination of aromatic diazonium salts with sulfur dioxide in the presence of copper salts or copper powder is often used for the synthesis of sulfinic acids [26], [43]. For the preparation of sodium 2,2,2-trifluoro-1,1-dichloroethanesulfinate see [54].

3.3. Oxidation of Thiols

Thiols can be easily oxidized to sulfinic acids, but it is often difficult to limit the oxidation of sulfur to the +4 state [55]. Halogens [56], hydrogen peroxide, and peracids [40] have been used as oxidizing agents. *m*-Chloroperbenzoic acid is particularly applicable because it does not result in further oxidation to sulfonic acids or sulfones [10]:

$$R-SH + 2\ \text{(3-Cl-C}_6\text{H}_4\text{-CO}_3\text{H)} \longrightarrow R-SO_2H + 2\ \text{(3-Cl-C}_6\text{H}_4\text{-CO}_2\text{H)}$$

Ethanesulfinic Acid [10]. A solution of 17.2 g (0.1 mol) *m*-chloroperbenzoic acid in 200 mL of dichloromethane at $-30\,°C$ is added slowly to a solution of 3.1 g (0.05 mol) of ethanethiol in 10 mL of dichloromethane (also at $-30\,°C$). The reaction mixture is allowed to stand overnight at $-30\,°C$ and the precipitated *m*-chlorobenzoic acid is removed by filtration. The filtrate is cooled to $-80\,°C$ and refiltered to remove residual *m*-chlorobenzoic acid, after which the solvent is evaporated in a stream of nitrogen. Analytically pure ethanesulfinic acid is obtained in 80–85% yield.

The oxidation of 2-mercaptobenzo-1,3-thiazole with hydrogen peroxide in alkaline medium gives the sodium salt of benzo-1,3-thiazole-2-sulfinic acid, which (see Chap. 2) is converted immediately upon acidification into benzo-1,3-thiazole (**2**, 70%) [39].

3.4. Nucleophilic Cleavage of Sulfones and Related Compounds (→ Sulfones and Sulfoxides)

Cleavage of the Sulfonyl Group at the α-Carbon Atom. The cleavage of aromatic sulfones carrying electron-withdrawing groups on the aromatic ring constitutes a general method for the synthesis of aromatic sulfinic acids [57]:

$$\text{(2,4-(O}_2\text{N)}_2\text{C}_6\text{H}_3\text{-SO}_2\text{-C}_6\text{H}_5) \xrightarrow{\text{NaOCH}_3} \text{C}_6\text{H}_5\text{-SO}_2\text{Na} + \text{2,4-(O}_2\text{N)}_2\text{C}_6\text{H}_3\text{-OCH}_3$$

Intramolecular rearrangement (Smiles rearrangement) of the lithium salt of 2-hydroxyphenyl phenyl sulfone gives lithium 2-phenoxyphenylsulfinate (**3**) [58]:

[Structure diagram: 2-(lithiooxy)phenyl phenyl sulfone → diphenyl ether with SO₂Li, labeled **3**]

The reaction of 2-methylphenyl phenyl sulfone (**4**) with butyllithium is similar to the Smiles rearrangement [29]:

[Structure diagram: 2-(lithiomethyl)phenyl phenyl sulfone → diphenylmethane derivative with SO₂Li, labeled **4**]

For further information see [59]; for electrochemical cleavage of sulfones see [60].

The facile cleavage of symmetrical aliphatic sulfones with sodium cyanide when both β-positions are activated by electron-withdrawing substituents is of greater interest [30]:

$$CH_3-\underset{O}{\overset{O}{C}}-CH_2-CH_2-SO_2-CH_2-CH_2-\underset{O}{\overset{O}{C}}-CH_3 \xrightarrow{NaCN}$$

$$CH_3-\underset{O}{\overset{O}{C}}-(CH_2)_2SO_2Na \;+\; CH_3-\underset{O}{\overset{O}{C}}-(CH_2)_2-CN$$

Similarly, α-disulfones are readily cleaved with sodium cyanide [14], and β-cyanosulfones with sodium thiolate (see below) [61]. Furthermore, the cleavage of phthalimidomethyl sulfones with alkoxide or thiolate proceeds smoothly to give sulfinic acids in high yield [12]:

[Structure diagram: phthalimide-N-CH₂-SO₂-R + R'-SNa → RSO₂Na + phthalimide-N-CH₂-S-R']

If the cleavage is carried out using thiolate, a phthalimidomethyl sulfide is formed as a byproduct. This sulfide can be reoxidized and itself cycled through the cleavage process, a reaction that permits the high-yield conversion of any thiol to the corresponding sulfinic acid.

General Example [12]. An alkylsulfonylmethylphthalimide (0.02 mol) is added to a solution of 0.51 g (0.022 mol) of sodium ethoxide in 65 mL of ethanol under nitrogen. The reaction mixture is stirred under reflux until a homogeneous solution is formed. After evaporation to dryness the residue is extracted several times with warm benzene. The sulfinate remains as an insoluble powder in a 92–100% yield, purity 94–98%.

Cleavage of Thiosulfonic Acid Esters. The cleavage of thiosulfonates with thiolate [62], sulfinate [63], and hydroxyl ions [64] can also be used for the preparation of sulfinic acids:

$$RSO_2-S-R' + R''SNa \longrightarrow R\underset{\underset{O}{\|}}{S}-ONa + R'SSR''$$

Cleavage at the β-Carbon Atom of a Sulfonyl Group. Both sulfones with β-hydrogens [65] and β-disulfones [66] undergo fragmentation with base to give the corresponding sulfinic acid salts:

$$R-CH_2-CH_2-SO_2R \xrightarrow{NaOH} R-CH=CH_2 + R-SO_2Na + H_2O$$

The following reaction sequence also gives high yields of sulfinic acids [67]:

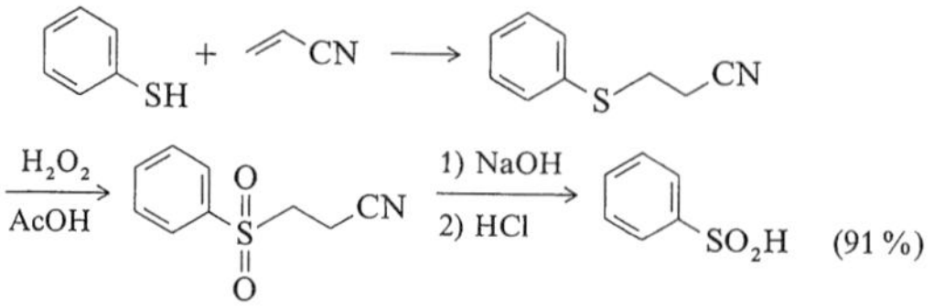

In addition, sulfinic acids can be prepared by alkylation of sulfonyl carbanions with subsequent elimination of sulfinate [68], [69]:

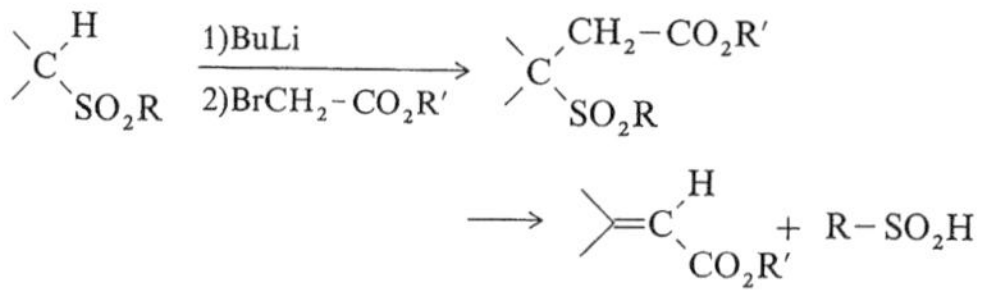

Cleavage of Sulfolenes. The ring-opening of a cyclic sulfone with a Grignard reagent represents another synthetic possibility. For example, in the case of 2-methylsulfolene (**5**) the product is a doubly unsaturated sulfinic acid [70]:

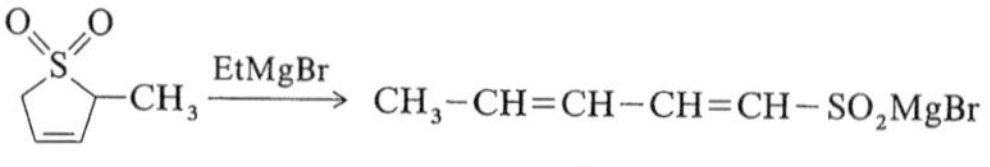

Addition of Thionyl Chloride to an Olefin with Subsequent Hydrolysis. Thionyl chloride adds to olefins in the presence of aluminum trichloride forming a complex that can be hydrolyzed with water to a sulfinic acid [71]:

$$H_2C=CH_2 + SOCl_2 \xrightarrow{AlCl_3} Cl\diagdown\diagup\overset{\overset{O}{\|}}{\underset{+}{S}}AlCl_4^- \xrightarrow[-HCl]{H_2O} Cl\diagdown\diagup SO_2H$$
$$ -AlCl_3$$

For further syntheses see [72].

4. Reactions of Sulfinic Acids and their Derivatives

4.1. Alkylation of the Sulfinate Anion

The sulfinate anion can be alkylated either at sulfur or oxygen depending on the reaction conditions, forming a sulfone or a sulfinic acid ester, respectively. Reaction with a soft alkylating agent (in the sense of the Pearson HSAB principle [73]–[75]) such as an alkyl halide, gives a sulfone [76]:

$$R-SO_2Na + R-X \longrightarrow R-SO_2-R + NaX$$

Hard alkylating agents, such as triethyloxonium tetrafluoroborate [77] or diazomethane [78], give sulfinic acid esters:

$$R-SO_2Na + (C_2H_5)_3O^+BF_4^- \longrightarrow R-\underset{\underset{O}{\|}}{S}-OEt$$

The solvent may also have a significant effect on the observed ratio of sulfone to sulfinic acid ester [79].

The alkylation reaction is an effective approach to preparing dialkyl sulfones, aralkyl sulfones, and β-hydroxysulfones [75], [80]–[82]. Reaction of sulfinates with epoxides leads to β-hydroxysulfones [83]; with lactones, sulfonylcarboxylic acids are formed [84]. Sulfones are also formed by the reaction of sulfinates with systems containing activated double bonds [85]–[87]. Formaldehyde reacts with sulfinate to give hydroxymethyl sulfone [88]. Simultaneous addition of an amine leads to an N-methylsulfonylamine in a reaction analogous to the Mannich reaction [89]. On the other hand, diaryl sulfones are obtained by nucleophilic substitution of haloaromatics containing electron-withdrawing substituents (e.g., 2-nitrochlorobenzene) [90]:

$$Ar-SO_2Na + Ar'X \longrightarrow Ar-SO_2-Ar'$$

[1-(Phenylsulfonyl)-aceto]-4-nitrobenzene [82]. 244 g (1 mol) of (1-bromoaceto)-4-nitrobenzene and 164 g (1 mol) of sodium benzenesulfinate are suspended in 1 L of ethanol and boiled for 4 h. After cooling, the precipitate is separated by filtration, boiled with acetone, and filtered again. The filtrates are combined, evaporated, and cooled. The crystalline sulfone is isolated by filtration in 76% yield, *mp* 138 °C.

For the synthesis of thioacyclic sulfones see [91].

4.2. Addition of Sulfinic Acids to Multiple Bonds

Polarized double-bond systems react with aliphatic and aromatic sulfinic acids in a process similar to the Michael reaction. Acrylic acids and their derivatives [92], vinyl ketones [93], nitroolefins [94], vinyl sulfones [95], quinones [96]–[98], and quinone imines [98] all undergo this reaction with sulfinic acids to form sulfones; e.g.:

$$R-SO_2H + \text{\hspace{1mm}}{=}\hspace{-1mm}\text{CN} \longrightarrow R-SO_2-CH_2-CH_2-CN$$

The addition of a sulfinic acid to an α-ethynyl ketone sulfone gives an α-oxo-β-sulfonylalkene in good yield [99]. A *cis*-1,2-disulfonylethylene is formed with 1-phenylsulfonylpropyne [100].

4.3. Redox Reactions of Sulfinic Acids

Sulfinic acids are readily oxidized to sulfonic acids by, for example, peroxides [101], hypochlorites [102], and oxygen [103]. α-Disulfones are formed with permanganate or cobalt(III) salts [104]. The reduction of sulfinic acids with lithium aluminum hydride gives disulfides [105]. As they are relatively unstable compounds, sulfinic acids undergo acid-catalyzed disproportionation to sulfonic acids and thiosulfonates [15], [16], [35]–[37].

5. Sulfinic Acid Derivatives

5.1. Sulfinyl Chlorides

5.1.1. Preparation

From Disulfides. Both aliphatic and aromatic sulfinic acids form sulfinyl chlorides. Low yields are generally obtained from the direct reaction of a sulfinic acid with thionyl chloride [106], but chlorination of a disulfide has proven to be an effective method of synthesis. The chlorination is carried out at 0–10 °C in acetic anhydride [107], [108]:

$$R-S-S-R + 2\,Ac_2O \xrightarrow{3\,Cl_2} 2\,RS\overset{O}{\underset{Cl}{\diagup\!\!\!\diagdown}} + 4\,AcCl$$

The course of the reaction is clearly indicated by the disappearance of the orange-red color of the intermediate sulfenyl chloride (RSCl).

Methanesulfinyl Chloride [107]. A mixture of 23.5 g (0.25 mol) of freshly distilled dimethyl disulfide and 51 g (0.5 mol) of acetic anhydride is chlorinated at −10 to 0 °C. During the course of the reaction the color changes from yellow to red and then disappears, at which point the addition of chlorine is suspended, leading to 126 g of reaction mixture. This is distilled in a 45 cm Vigreux column at 2 kPa with cooling. Acetyl chloride distills at 0 °C, and at 47–48 °C the almost colorless methanesulfinyl chloride is obtained in 83–86 % yield.

Many sulfinyl chlorides decompose to varying degrees in the course of distillation, and in aromatic cases there is even a danger of explosion. Sulfinyl chlorides may also decompose slowly during storage, causing pressure to build up in a closed container.

From Trithianes. Chloromethanesulfinyl chloride (**7**) can be prepared by the chlorination of trithiane (**6**) in acetic anhydride at 0 °C [109]:

$$\underset{6}{\text{(trithiane)}} + 6\,Cl_2 + 3\,Ac_2O \longrightarrow 3\,Cl-CH_2-\underset{7}{\overset{O}{\underset{\|}{S}}}-Cl + 6\,AcCl$$

Chloromethanesulfinyl Chloride. A mixture of 46 g (0.33 mol) of *sym*-trithiane (**6**) and 102 g (1 mol) of acetic anhydride is chlorinated at 0 °C. Residual chlorine and resulting acetyl chloride are evaporated in vacuo and the chloromethanesulfinyl chloride (**7**) is distilled in 55 % yield (*bp* 42–62 °C at 2 kPa).

For α-heteroatom-substituted methanesulfinic acids and their derivatives see [110]; for the synthesis of trimethylammoniomethanesulfinate see [111].

From Sulfuranes (RSHal₃). The selective hydrolysis of sulfurane also gives a sulfinyl halide [112]:

$$CF_3-SF_3 \xrightarrow{H_2O} CF_3-\overset{O}{\underset{\|}{S}}-F$$

$$Ph-SF_3 + Ph-CHO \longrightarrow Ph-\overset{O}{\underset{\|}{S}}-F + Ph-CHF_2$$

For the photolytic addition of sulfur dioxide to perfluoroalkenes giving sulfinyl fluorides see [113].

Trichlorosulfuranes [114], [115] are converted into sulfinyl chlorides in high yield in alcohol [116], carboxylic acids [117], or acetic anhydride [118]:

$$R-SCl_3 + (CH_3CO)_2O \longrightarrow R-\overset{O}{\underset{\|}{S}}-Cl + 2CH_3-\overset{O}{\underset{Cl}{C}}$$

Because of the instability of sulfuranes, the compounds are usually synthesized immediately prior to use by further chlorination of the corresponding sulfenyl chloride.

From Thioesters. Another method for producing sulfinyl chlorides involves the chlorolytic cleavage of thioesters in acetic anhydride, leading to a mixture of the desired sulfinyl chloride and acetyl chloride [119]:

$$Ph-S-C(O)CH_3 \xrightarrow[Ac_2O]{Cl_2} Ph-S(O)-Cl + AcCl$$

8

Phenylsulfinyl Chloride (**8**) [119]. A mixture of 30.4 g (0.2 mol) of phenyl thioacetate and 20.4 g (0.2 mol) of acetic anhydride is chlorinated at −10 to 0 °C. The color changes to yellow, then red, and then back to yellow. The reaction mixture is heated to 80 °C under reduced pressure (3.5 kPa), until the residue begins to boil, after which it is cooled; yield of **8**: 31.5–31.8 g (98.1–99.1%).

By the Addition of Thionyl Chloride to Alkenes. Addition of ethylene to thionyl chloride in the presence of aluminum trichloride gives the corresponding acid chloride complex, which upon hydrolysis forms 2-chloroethanesulfinic acid (**9**) [71]:

$$CH_2=CH_2 + SOCl_2 \xrightarrow{AlCl_3} Cl-CH_2CH_2-\overset{O}{\underset{\|}{S}}^+ AlCl_4^-$$

$$\xrightarrow[-HCl]{H_2O} Cl-CH_2CH_2-\overset{O}{\underset{\|}{S}}-OH$$
$$-AlCl_3 \qquad \mathbf{9}$$

2-Chloroethanesulfinic Acid (**9**) [71]. A mixture of 32 g (0.27 mol) of thionyl chloride and 32 g (0.24 mol) of aluminum trichloride is saturated with 6 L (0.27 mol) of ethylene. The reaction mixture is hydrolyzed with water and the resulting byproducts are extracted with benzene. The residue is extracted with diethyl ether, which is then dried over magnesium sulfate and evaporated in vacuo. The yield of **9** is 98%.

By Treatment of CH-Acidic Compounds with Thionyl Chloride. CH-acidic compounds react readily with thionyl chloride to give sulfinyl chlorides. For example, formylsulfinyl chloride can be obtained from isobutyraldehyde (2-methylpropanal) in 81% yield [120]:

1-Formyl-1-methylethanesulfinyl Chloride (**10**) [120]. 72 g (1 mol) of 2-methylpropanal is added dropwise to 360 g (3 mol) of thionyl chloride at 50 °C with cooling. As soon as the vigorous production of hydrogen chloride has subsided, the temperature is raised to 80 °C. When hydrogen chloride formation ceases completely excess thionyl chloride is removed by distillation at atmospheric pressure and the residue is distilled under vacuum, leading to 124 g, of **10**, bp 83–84 °C at 1.6 kPa.

For an analogous synthesis of adamantanesulfinyl chloride in 72 % yield see [121].

5.1.2. Reactions

Sulfinyl chlorides are important starting materials for the preparation of sulfinic acid esters (see Section 5.2) and amides (Section 5.4), as well as thiosulfinic acid esters (Section 5.3). For a transformation of sulfinyl chlorides into sulfines ($R_2C=S=O$) see [122]. Phenylsulfinyl chloride can be converted into a diarylthiosulfonate ester in dichloromethane in the presence of pyridine and 1-hydroxypyridine-2-thione [123].

Asymmetric aromatic sulfoxides can be prepared by the Friedel–Crafts reaction of an aromatic sulfinyl chloride with a benzene derivative [124], [125].

5.2. Sulfinic Acid Esters

5.2.1. Preparation

From Sulfinyl Chlorides and Alcohols. Sulfinic acid esters are readily prepared from sulfinyl chlorides and alcohols in an inert solvent [126]:

$$R-\underset{Cl}{\overset{O}{\overset{\|}{S}}} + R'-OH \xrightarrow{-HCl} R-\underset{O-R'}{\overset{O}{\overset{\|}{S}}}$$

The hydrogen chloride formed can be removed by applying a vacuum [126], or by the addition of a base such as potassium carbonate [127] or pyridine [128]. For a one-pot synthesis of sulfinic acid esters (50–88%) see [129].

Methyl Methanesulfinate [126]. A mixture of 47 g (0.5 mol) of dimethyl disulfide and 102 g (1 mol) of acetic anhydride is chlorinated to give methanesulfinyl chloride. The latter is then treated with 35 g (1.1 mol) of cold anhydrous methanol at −30 °C. Hydrogen chloride is removed by applying a vacuum and the reaction mixture is heated to 30 °C until gas formation has ceased. *p*-Toluidine can be added to assist in the removal of methanesulfinyl chloride. The reaction mixture is filtered and the filtrate is distilled using an 18 cm Vigreux column in yield 71%.

From Disulfides by Chlorination in Alcohol. If chlorination of a disulfide to give the corresponding sulfinyl chloride is carried out in an alcohol, the sulfinic acid ester is obtained directly [130]:

$$R-S-S-R + 4R'OH \xrightarrow{Cl_2} 2R-\underset{O-R'}{\overset{O}{\overset{\|}{S}}} + 2R'Cl + 4HCl$$

The oxidation of disulfides with lead tetraacetate in alcohol also leads to sulfinic acid esters in 30–90% yield [131]. For an analogous reaction with diphenyl disulfide see [132]. Electron-withdrawing substituents on the aromatic ring inhibit this oxidation. *N*-Bromsuccinimide also acts as an oxidizing agent in the synthesis of sulfinic acid esters (5–78% yield) [133]. For cyclic sulfinic acid esters see [134].

Other Methods. The preparation of sulfinic acid esters from sulfinyl chlorides and diazoalkanes in polar solvents is described in [135] and from sulfinic acids and diazomethane in [78].

The reaction of benzenesulfinyl chloride with diazomethane is described in [109]. 4-Methylbenzenesulfinic acid reacts with diphenyldiazomethane in dichloromethane to give a sulfone as the major product, but in dimethylsulfoxide the sulfinic acid ester is formed in 100% yield [135]. Another route involves the reaction of a sulfinate with a hard alkylating agent [78] (see Section 4.1).

Unlike carboxylic acids, sulfinic acids cannot be esterified by acid catalysis because under these conditions they disproportionate to sulfonic acids and thiosulfonates [15], [16], [35]–[37] (Section 4.3). However, esterification can be accomplished in a large excess of alcohol in the presence of dicyclohexyl carbodiimide [136]:

$$R-S(=O)-OH + R'OH \xrightarrow{C_6H_{11}-N=C=N-C_6H_{11}} R-S(=O)-OR'$$

Sulfinic acid esters can also be obtained in good yield by the alcoholysis of N-sulfinylphthalimides [137]:

$$\text{Phthalimide-N-S(=O)-R} + R'OH \longrightarrow R-S(=O)-OR' + \text{phthalimide-NH}$$

Synthesis of sulfinic acid esters by the alcoholysis of N-sulfinylsulfonamides is described in [138].

5.2.2. Reactions

Rearrangements. Alkane-, alkene-, and alkynesulfinic acid esters undergo acidcatalyzed rearrangement to sulfones [139]; for example:

$$R-S(=O)-OR' \longrightarrow R-S(=O)_2-R'$$

For further information see [140].

Conversion into Sulfoxides. Sulfinic acid esters react with Grignard reagents to give optically active sulfoxides [141]:

$$Ph-S(=O)-OR + R'MgBr \longrightarrow Ph-S(=O)-R'$$

This reaction is highly stereospecific, and is, therefore, frequently used for the determination of absolute configurations.

Claisen Condensation. The Claisen condensation (→ Esters, Organic) of sulfinic acid esters with enolates leads to β-oxosulfoxides [142], [143], which can also be obtained by the reaction of sulfinyl carbanions with carboxylic acid esters:

$$R-\overset{O}{\underset{\|}{C}}-CH_2-R' + CH_3-O-\overset{O}{\underset{\|}{S}}-C_6H_5$$

$$\xrightarrow{\text{Base}} R-\overset{O}{\underset{\|}{C}}-\overset{H}{\underset{R'}{C}}-\overset{O}{\underset{\|}{S}}-C_6H_5$$

For further reactions (e.g., the transformation of sulfinic acid esters into disulfides) see [144]. Information about MS-Research of sulfinic acid esters in [145].

5.3. Thiosulfinic Acid Esters

Thiosulfinic acid esters are readily prepared from the corresponding disulfides by careful oxidation with, for example, *m*-chloroperbenzoic acid [2]. Allicin (**1**) can be obtained in good yield by this method.

The disproportionation of sulfinyl chlorides in water also gives thiosulfinic acid esters [146]. For further references see [147].

5.4. Sulfinic Acid Amides

Sulfinic acid amides can be readily produced from sulfinic acid chlorides and amines:

$$R-\overset{O}{\underset{Cl}{S}} + HN\overset{R'}{\underset{R''}{}} \longrightarrow R-\overset{O}{\underset{N(R'')}{S}}R' + HCl$$

The reaction is carried out in inert, dry solvents in the presence of base or excess amine [148]–[150].

General Method of Preparation [149]. Sulfinyl chloride (0.2 mol) is added dropwise to a solution of 0.4 mol of amine in 150 mL of dichloromethane under nitrogen in the course of 1 h. The temperature of the reaction mixture is maintained at −20 to −40 °C. It is stirred for a further hour at room temperature, after which ammonium chloride is filtered off. The filtrate is concentrated by evaporation, and the product is distilled in vacuo or crystallized.

Sulfinic acid amides can also be prepared by the reaction of sulfinylamines with Grignard reagents [72], [151], [152]:

$$R'-N=S=O \xrightarrow[\text{2) } H_2O]{\text{1) RMgX}} R-\overset{\overset{O}{\|}}{\underset{\underset{H}{N}}{S}}-R' + Mg(OH, X)$$

Good yields are obtained with aromatic sulfinylamines. The Grignard complex should be decomposed carefully with 10% cold ammonium chloride solution.

For a discussion of the preparation of N-sulfinylnonafluorobutanesulfonamide via an ene reaction and its subsequent conversion into the sulfine–sulfone imide see [138].

Sulfinic acid amides can also be prepared directly by the oxidation of sulfenic acid amides. Thus, 2-benzo-1,3-thiazolesulfenamides can be transformed into 2-benzo-1,3-thiazolesulfinamides by careful oxidation. The latter can be used as vulcanization accelerators for natural and synthetic rubbers [150]. Sodium hypochlorite is generally used as the oxidizing agent [150]:

Percarboxylic acids and hydrogen peroxide are also suitable.

N,N-Dicyclohexyl-2-benzo-1,3-thiazolesulfinic Acid Amide [153]. A 15% aqueous sodium hypochlorite solution (150 g) is added dropwise over 10 minutes to a suspension of 48.3 g of N,N-dicyclohexyl-2-benzo-1,3-thiazole-sulfenamide in 300 mL of methanol. The reaction mixture is boiled for 1 h and then cooled to 0 °C. The product is collected by filtration, washed, and dried; *mp* 135–137 °C.

Additional references regarding the preparation of benzo-1,3-thiazolesulfinic acid amides in yields of ca. 90% can be found in [154].

6. 1-Hydroxyalkanesulfinates

The sodium salt of a 1-hydroxyalkanesulfinic acid was first prepared by BAZLEN in 1904 [155], [156].

6.1. Properties

Sodium hydroxymethanesulfinate [*149-44-0*] is commercially available as the dihydrate $HOCH_2SO_2Na \cdot 2\,H_2O$, which crystallizes as white needles, *mp* 63 °C. Its solubility in water at room temperature is 60 g/100 g. The salt is sparingly soluble in alcohol, and can be easily obtained pure from 60% methanol. In moist air it decomposes slowly to products with noxious odors. The water of crystallization is lost upon heating to 120 °C, and then from 125 °C the compound undergoes a very exothermic decomposition to methanethiol, hydrogen sulfide, some formaldehyde, and sulfur dioxide, during which the temperature rises to 160 °C. The residue consists of sodium hydroxymethanesulfonate, sodium thiosulfate, and sodium sulfite. In solution the decomposition is pH and temperature dependent. Sodium hydroxymethanesulfinate reacts with sodium hydrogen sulfite to give sodium dithionite ($Na_2S_2O_4$) and sodium hydroxymethanesulfonate. The hydroxyl group is acidic and susceptible to salt formation.

The monozinc salt of hydroxymethanesulfinic acid $(HOCH_2SO_2)_2Zn$ is readily soluble in water, but zinc oxidomethanesulfinate $(OCH_2SO_2)Zn$ is only sparingly soluble. The former crystallizes from water at 80 °C in anhydrous form, at 60 °C as the trihydrate, and at 20 °C as the tetrahydrate (rhombohedric leaflets). By contrast, the sparingly soluble dizinc salt precipitates from cold solution as a trihydrate and from warm solution as a monohydrate. The amorphous and crystalline calcium salts of hydroxymethanesulfinic acid and also zinc 1-oxidoethanesulfinate have been described as well. The latter is readily cleaved, and acts as a reducing agent at 50 °C. Derivatives of other aldehydes, such as benzaldehyde and butyraldehyde, also decompose readily, reducing vat dyes even at low temperature (see Chap. 8). The nature of sodium hydroxymethane sulfinate in aqueous solution is described in detail in [157].

6.2. Production

Sodium hydroxymethanesulfinate can be prepared by reacting sodium dithionite with formaldehyde in the presence of alkali, or by catalytic reduction of sodium hydroxymethanesulfonate with hydrogen. It also results from the zinc dust reduction of sodium hydroxymethanesulfonate, which is in turn prepared in situ from zinc oxide, sulfurous acid, and formaldehyde. A variant of this process involves the reduction of aqueous sulfur dioxide (sulfurous acid) with zinc dust to zinc dithionite, which is then converted with formaldehyde into the zinc salts of hydroxymethanesulfonic and hydroxymethanesulfinic acids. The sulfonic acid portion is reduced to the desired sulfinic acid salt in the presence of zinc and sodium hydroxide:

$$Zn + 2SO_2 \xrightarrow{H_2O} ZnS_2O_4$$

$$ZnS_2O_4 + 2H_2O + 4CH_2O \longrightarrow (HO-CH_2-SO_3)_2Zn + (HOCH_2-SO_2)_2Zn$$

$$(HOCH_2-SO_3)_2Zn + 2Zn \longrightarrow 2(-OCH_2-SO_2)Zn + ZnO + H_2O$$

$$(-OCH_2SO_2)Zn \xrightarrow{NaOH} HOCH_2-SO_2Na + ZnO$$

General Synthesis. Zinc dust is suspended in 5 times its volume of water and sulfur dioxide is passed in at 30–35 °C until all the zinc has dissolved. Aqueous formaldehyde (30–40 %) is then added at 50–60 °C until the solution no longer reduces indigo in the cold. After addition of the calculated quantity of zinc dust, the reaction mixture is heated to 90 °C, and this temperature is maintained until virtually no more zinc compounds dissolve. Sodium hydroxide is subsequently added and the mixture is filtered. The clarified solution of sodium hydroxymethanesulfinate is evaporated in vacuo until it acquires the consistency of honey; $HOCH_2SO_2Na \cdot H_2O$ solidifies within a few hours, after which it can be ground.

The synthesis also produces zinc hydroxymethanesulfinate in soluble form, together with the zinc salt of the sulfonic acid. The former can be separated by evaporation in vacuo and subsequent drying. Zinc oxidomethanesulfinate can be obtained from zinc hydroxymethanesulfinate by the addition of zinc salts and sodium hydroxide. Other salts, such as the calcium salt, are precipitated by treating sodium hydroxymethanesulfinate with the appropriate chloride (e.g., $CaCl_2$). Other aldehydes can be used in place of formaldehyde to synthesize alternative hydroxyalkanesulfinic acids. Many aldehydes have been shown to be effective [156]. Hydroxyalkanesulfinic acids are intermediates in the reduction of aldehydes and ketones with dithionite [158].

6.3. Quality Specifications, Analysis

The industrially important salts of hydroxymethanesulfinic acid decompose in moist air. Their packaging must, therefore, provide protection against air and moisture. The products present neither a fire hazard nor a tendency to self-ignite.

Sodium hydroxymethanesulfinate is commercially available in 98–100% purity; sodium hydroxymethanesulfonate and sodium sulfite are present as impurities. The commercial zinc salt contains 88–90% sulfinate based on anhydrous product. The main impurities in this case are sodium sulfate and zinc hydroxymethanesulfonate. The proportion of sulfinate in the calcium salt is 70% based on anhydrous product, with the corresponding reactants present as impurities. Other salts of industrially produced alkanesulfinic acids (e.g., sodium hydroxy-1-ethanesulfinate) are avoided in pure form because of their odors.

The content of alkanehydroxysulfinate can be determined iodometrically in the absence of air. The method is analogous to determination of the content of dithionite, in which titration is carried out with acidic indigo carmine solution. The required temperature for the titration depends on the nature of the aldehyde.

7. Formamidinesulfinic Acid

Formamidinesulfinic acid (**11**) [*1758-73-2*], $CH_4N_2O_2S$, is used in certain branches of the textile industry [159].

$$\underset{H_2N}{\overset{HN}{\diagdown}}\!\!\!\!\!\!\!\!>\!\!-SO_2H$$
11

7.1. Properties

Formamidinesulfinic acid is a white crystalline compound, *mp* 128 °C (decomp.), in [159] *mp* 144 °C. It is sparingly soluble in water and other common solvents. The compound dissolves in alkaline solutions with decomposition, but it can be heated to 100 °C in concentrated sulfuric acid without decomposition. It reacts with formaldehyde and sodium hydroxide to form sodium hydroxymethanesulfinate, and with sodium hydrogensulfite to give sodium dithionite. Further information in the references given in [160]; for spectrochemical study see [161].

7.2. Production

Formamidinesulfinic acid was first obtained by BARNETT, in 1910 by the oxidation of thiourea with hydrogen peroxide:

$$\underset{H_2N}{\overset{H_2N}{\diagdown}}\!\!\!C\!=\!S \xrightarrow{H_2O_2} \mathbf{11}$$

Since then a number of syntheses have been described in the literature, involving the use of hydrogen peroxide in a variety of solvents. Methanol:water (40:60) leads to formamidinesulfinic acid in 96.5 % yield and 98 % purity [162].

Formamidinesulfinic Acid (**11**) [162]. Thiourea (80 g) is dissolved in 800 g of methanol:water (40:60) at 20 °C and oxidized with 210 g of 35 % hydrogen peroxide at −20 to −10 °C over the ourse of 30 min. The reaction mixture is then stirred for a further 10 min. The product is obtained in 96.5 % yield and 98 % purity. For further information see [163].

Other synthetic procedures can be found in the literature, including oxidation of thiourea with chlorine dioxide [164], photochemical oxidation with singlet oxygen [165], and oxidation with hydrogen peroxide in water [166] or carbon tetrachloride [167]. Oxidation of substituted thioureas to the corresponding formamidinesulfinic acids has also been described [168].

Oxidimetric methods and titration with indigo in the presence of sodium hydrogencarbonate are suitable means for determining purity.

8. Industrial Uses of Sulfinic Acids and their Derivatives

Sulfinic acids and their derivatives are used in a wide range of industrial applications, serving as:

1) Catalysts for the polymerization of styrene, butadiene, and methacrylic acid esters at low temperature [10], [72], as well as for polymerization of acrylonitrile [169]
2) Catalysts for rapid and mild isomerization of disubstituted double bonds [170]
3) Surface-active agents for the pretreatment of raw cotton as a way to avoid bleaching [171]
4) Antistatic additives to facilitate dyeing of polyamides [172]
5) Agents to improve the dyeing of linear polyesters [173]
6) Additives for polyacrylonitrile solutions subject to spinning [174]
7) Plant-growth regulators [175], crop-protection agents [176], [177], and fungicides [178]
8) Fragrances and flavors for foods, perfumes, and tobacco [179], [180]
9) Activators for nitrogen-containing propellants [181] (e.g., azodicarbonamide [182])
10) Emulsifiers and dispersing agents for oil–water emulsions [183], [184], agents for floating ores [185]
11) Intermediates in the production of crease-resistant finishes [186]
12) Intermediates in the production of pharmaceuticals [187]
13) Antifogs in silver halide photography [188]
14) Agents for stimulating mucous membranes of the mouth and nose providing a cooling sensation similar to that of menthol [189]
15) Vulcanization accelerators [190], [191]
16) Reducing agents for dyes [192], as well as elsewhere in the dye industry [193]
17) Reducing agents for aldehydes and ketones [158], [194].

Several hydroxyalkanesulfinic acid salts are available commercially. For example, sodium hydroxymethanesulfinate is sold under the trade names Rongalite C [195] (BASF) and Brüggolit (Brüggemann, Germany), and as various Hydrosulfite formulations (Ciba-Geigy and Rohner). It is used mainly in direct and discharge printing for converting vat dyes into their leuco forms and as reducing catalyst for synthetic rubber thermoplastics and for bleaching polyesters. A redox potential corresponding to that of dithionite is attained only at 75–80 °C, so the reduction is carried out in a steam bath.

Zinc hydroxymethanesulfinate is sold under the trade name Decrolin (BASF) and as various Hydrosulfite preparations (Ciba-Geigy, Rohner, and Pechiney, France). The zinc salt is used in the wool industry and also as a bleaching agent for such materials as soaps, fatty acids, glycerol, sizing agents, and for corrosion color printing on polyamide, wool, silk etc.

Calcium hydroxymethanesulfinate is commercially available under the trade name Rongalite H (BASF). It is used for vat-dye printing on acetate, polyamide, silk, and wool fabrics. Such prints display exceptional stability toward storage.

Hydroxyalkanesulfinic acids derived from other aldehydes, such as acetaldehyde, benzaldehyde, or butyraldehyde, as well as products condensed with ammonia, are available under such trade names as Rongalite 2 PH – A and 2 PH – B_{fl} and Rongalite FD_{fl}(BASF). These vary in their redox potentials. For example, the redox potential of Rongalite 2 PH lies between those of Hydrosulfite and Rongalite C (sodium hydroxymethanesulfinate), and that of Rongalite FD is above Rongalite C. These products have proven useful in two-phase and fill printing as well as in the textile industry.

For the use of sulfinic acids as fungicides see [178], for other applications in the dye industry see [192], [193], and as reducing agents for aldehydes and ketones see [158], [194].

Formamidinesulfinic acid is sold under the trade names Reduktionsmittel F (Degussa) [196] GLO Lite TD (Glo-Tex Chemicals, United States), Arolite TD Concentrate (Arol Chemical Products Company, United States), Thio Urea Dioxide and tec light (Tohai Denka Kogya Co., Japan). Formamidinesulfinic acid can be used for the reduction of disulfides [197] and the hydrogenation of aldehydes [198], as a processing aid in noble-metal extraction [199], as a dye stabilizer for pyrazoline brighteners [200], as a bleaching agent in the paper industry [201], as an agent for improving silver halide photography [202], for anaerobic textile wastewater treatment [203], and for catalyzing the reactions of olefins with dibromodifluoromethane [204]. Further sulfinic acid salts for synthetic intermediates are available: 4-chlorobenzenesulfinic acid sodium salt (Rohner, Bayer) [205]; benzenesulfinic acid sodium salt (Bayer) [206], the analogous zinc salt (Rohner) [207]; 4-methylbenzenesulfinic acid sodium salt (Rohner) [208]; 4-nitrobenzenesulfinic acid sodium salt (Bayer) [209]; 4-chloro-3-nitrobenzenesulfinic acid sodium salt [210].

9. Toxicology

Published toxicological data regarding these compounds are sparse. The reported LD_{50} values are in the region of a few grams per kilogram. Benzenesulfinic acid is said to cause slight skin irritation [211], [112].

The LD_{50} value of sodium hydroxymethanesulfinate is ca. 6400 mg/kg (oral), that of zinc hydroxymethanesulfinate 1280 mg/kg oral, rat [213]. The former is slightly irritating to the mucous membranes, but not to the skin (Safety Data Sheet, Brüggolit

C) [214]. The latter causes skin irritation and moderate mucous membrane irritation [213]. Further LD$_{50}$ values (oral, rat) for some sulfinic acid salts are listed below:

Formamidinesulfinic acid	1120 mg/kg skin irritant [196]
4-Chlorobenzenesulfinic acid sodium salt	> 5000 mg/kg [205]
Benzenesulfinic acid sodium salt	> 5000 mg/kg [206]
Benzenesulfinic acid zinc salt	> 5000 mg/kg [207]
4-Methylbenzenesulfinic acid sodium salt	3200 mg/kg [208]
4-Nitrobenzenesulfinic acid sodium salt	> 5000 mg/kg [209]
4-Chloro-3-nitrobenzenesulfinic acid sodium salt	> 5000 mg/kg [210]

In case of an emergency, the appropriate safety data sheets should be consulted for all these products.

10. References

[1] E. Block, S. Naganathan, D. Putman, Shu Hai Zhao, *Pure Appl. Chem.* **65** (1993) 625. X. Yan, Zh. Wang, Ph. Barlow, *Food Chem.* **47** (1993) 289. H. Jansen, B. Müller, K. Knobloch, *Planta Med.* **53** (1987) 559.

[2] F. Freeman, Xiao Bo Ma, R. J. San Lin, *Sulfur Lett.* **15** (1993) 253.

[3] R. J. Apitz-Castro, M. K. Jain, US 4 665 088, 1987.

[4] E. Block et al., *J. Am. Chem. Soc.* **108** (1993) 7045. S. F. Mohammad, St. C. Woodward, *Thromp. Res.* **44** (1986) 793.

[5] H. Wagner, W. Dorsch, EP 299 424, 1989. E. Block et al., *J. Am. Chem. Soc.* **110** (1988) 7813.

[6] K. K. Nippon Kokan, JP 02 204 487, 1990, *Chem. Abstr.* **114** (1991) 6523.

[7] R. Gebhardt, *Lipides* **28** (1993) 613.

[8] M. Tappaz, K. Almarghini, F. Legay, A. Remy, *Neurochem. Res.* **17** (1992) 849. A. J. Khodair, A. A. Swelin, A. A. Abdel-Wehab, *Phosphorus Sulfur* 2 (1976) 173.

[9] N. Kharash, A. S. Arora, *Phosphorus and Sulfur* **2** (1976) 1–50.

[10] W. G. Filby, K. Günther, R. D. Penzhorn, *J. Org. Chem.* **38** (1973) 4070. Gesellschaft für Kernforschung, DE-OS 2 322 199, 1973.

[11] Fiat Final Rep. 1313 I (1948) 357.

[12] M. Uchino, K. Suzuki, M. Sekiya, *Synthesis* **11** (1977) 794.

[13] H. Reinheckel, D. Jahnke, *Chem. Ber.* **99** (1966) 1718. Dt. Akad. Wiss., DE-AS 1 248 656, 1965.

[14] W. M. Ziegler, R. Connor, *J. Am. Chem. Soc.* **62** (1940) 2596.

[15] E. Wellisch, E. Gipstein, O. J. Sweeting, *J. Org. Chem.* **27** (1962) 1810.

[16] C. S. Marvel, R. S. Johnson, *J. Org. Chem.* **13** (1948) 822.

[17] American Cyanamid, US 2 917 540, 1958. Du Pont, US 2 993 932, 1961; US 2 315 514, 1938.

[18] M. T. Beachem et al., *J. Am. Chem. Soc.* **81** (1959) 5430.

[19] U. Schöllkopf, P. Hilbert, *Justus Liebigs Ann. Chem.* 1973, 1061.

[20] H. W. Roesky, *Angew. Chem.* **83** (1971) 890.

[21] R. N. Haszeldine, J. M. Kidd, *J. Chem. Soc.* 1955, 2901.

[22] J. B. Hendrickson, A. Giga, J. Wareing, *J. Am. Chem. Soc.* **96** (1974) 2275.

[23] F. C. Whitmore, F. H. Hamilton, *Org. Synth. Coll.* **I** (1941) 492.

[24] L. Field, F. A. Grunwald, *J. Org. Chem.* **16** (1951) 946.

[25] E. Knoevenagel, J. Kenner, *Chem. Ber.* **41** (1908) 3315.

[26] L. Gattermann, *Chem. Ber.* **32** (1899) 1136.
[27] J. Thomas, *J. Chem. Soc.* **95** (1909) 342.
[28] S. Smiles, C. M. Bere, *Org. Synth. Coll.* **1** (1948) 7.
[29] W. E. Truce, W. J. Ray, *J. Am. Chem. Soc.* **81** (1959) 481.
[30] P. Messinger, H. Greve, *Justus Liebigs Ann. Chem.* **1977**, 1457. Hoechst, DE-AS 1 129 477, 1960.
[31] J. Boeseken, *Recl. Trav. Chim. Pays-Bas* **55** (1936) 1040. E. W. Tillitson, US 2 493 471, 1950.
[32] BASF, DE 222 195, 1910.
[33] H. Bredereck, G. Brod, G. Höschele, *Chem. Ber.* **88** (1955) 438.
[34] S. Detoni, D. Hadzi, *J. Chem. Soc.* 1955, 3163.
[35] H. Bredereck et al., *Angew. Chem.* **70** (1958) 268.
[36] J. L. Kice, K. W. Bowers, *J. Am. Chem. Soc.* **84** (1962) 605.
[37] J. L. Kice, N. E. Pawlowski, *J. Org. Chem.* **28** (1963) 1162.
[38] *Houben Weyl*, 4th ed., **E 11**, part 1, pp. 64, 88, suppl.
[39] J. Metzger, H. Plank, *Bull. Soc. Chim. Fr.* 1956, 1701.
[40] W. E. Truce, A. Murphy, *Chem. Rev.* **48** (1951) 69.
[41] C. J. Stirling, *Int. J. Sulfur Chem, B* **6** (1971) nr. 4, 277.
[42] L. Field, *Synthesis* 1978, 713; 1972, 101.
[43] *Houben-Weyl*, 4th. ed., **IX,** pp. 289, 303.
[44] B. Lindberg, *Acta Chem. Scand.* **17** (1963) 377.
[45] P. Allen, Jr., *J. Org. Chem.* **7** (1942) 23.
[46] D. B. Hope, C. D. Morgan, M. Wälti, *J. Chem. Soc. C.* 1970, 270. W. E. Truce, D. P. Tate, D. N. Burdge, *J. Am. Chem. Soc.* **82** (1960) 2872.
[47] L. Horner, H. Neumann, *Chem. Ber.* **98** (1965) 3462. Hoechst, DE-OS 1 493 664, 1964.
[48] C. J. M. Stirling, *Int. J. Sulfur. Chem. Part. B* **6** (1971) 280. L. Field, *Synthesis* 1972, 120. L. Field, *Synthesis* 1978, 713. K. K. Anderson, *Compr. Org. Chem.* **3** (1979) 317. E. Wendschuh, K. Dölling, *Z. Chem.* **20** (1980) 122.
[49] Continental Oil, BE 603 176, 1961. K. Ziegler, DE-AS 1 050 762, 1957.
[50] W. E. Truce, J. F. Lyons, *J. Am. Chem. Soc.* **73** (1951) 126.
[51] IG Farbenindustrie, GB 321 843, 1929.
[52] J. R. Nool, P. C. van der Hoeven, W. P. Haslinghuis, *Tetrahedron Lett.* 1970, 2531.
[53] F. S. Dainton, K. I. Irvin, *Trans. Faraday Soc.* **46** (1950) 374.
[54] W. Huang, *Youje Huaxue* **12** (1992) 12; *Chem. Abstr.* **117** (1992) 191 273.
[55] H. Berger, *Red. Trav. Chim. Pays-Bas* **82** (1963) 773.
[56] I. Scheinfeld, J. C. Parham, S. Murphy, G. B. Brown, *J. Org. Chem.* **34** (1969) 2153.
[57] N. Kharash, R. Swidler, *J. Org. Chem.* **19** (1954) 1704.
[58] J. F. Bennet, R. E. Zahler, *Chem. Rev.* **49** (1951) 273. W. E. Truce, E. M. Kreider, W. W. Brand, *Org. React. (N.Y.)* **18** (1970) 100.
[59] *Houben Weyl* **E 11**, part 1, p. 614, suppl.
[60] L. Horner, H. Neumann, *Chem. Ber.* **98** (1965) 1715.
[61] W. E. Truce, F. E. Roberts, Jr., *J. Org. Chem.* **28** (1963) 593.
[62] A. J. Parker, N. Kharash, *Chem. Rev.* **59** (1959) 583. S. Smiles, D. T. Gibson, *J. Chem. Soc.* **125** (1924) 176.
[63] D. T. Gibson, J. D. Loudon, *J. Chem. Soc.* 1937, 487.
[64] G. Leandri, A. Tundo, *Ann. Chim. (Rome)* **44** (1954) 340.
[65] G. W. Fenton, C. K. Ingold, *J. Chem. Soc.* 1928, 3127; 1929, 2338.
[66] W. M. Ziegler, R. Connor, *J. Am. Chem. Soc.* **62** (1940) 2596.
[67] W. E. Truce, F. E. Roberts, *J. Org. Chem.* **28** (1963) 593.

[68]   J. B. Hendrickson, A. Giga, J. Wareing, *J. Am. Chem. Soc.* **96** (1974) 2275.
[69]   M. Julia, B. Badet, *Bull. Soc. Chim. Fr.* 1976, 525.
[70]   R. C. Krug, J. A. Rigney, G. R. Tidulaar, *J. Org. Chem.* **27** (1962) 1305.
[71]   A. J. Titov, A. N. Baryshnikova, *Dokl. Akad. Nauk SSSR* **157** (1964) 139. *Chem. Abstr.* **61** (1964) 9396.
[72]   Bayer, DE 975 943, 1943.
[73]   G. W. Fenton, C. K. Ingold, *J. Chem. Soc.* 1929, 2338. J. Büchi, H. R. Füeg, A. Aebi, *Helv. Chim. Acta* **42** (1959) 1368.
[74]   C. M. Suter in: *Tetravalent Organic Sulfur Compounds*, J. Wiley, New York 1944, p. 660. C. M. Suter in: *The Organic Chemistry of Sulfur*, J. Wiley, New York 1948, p. 658.
[75]   G. E. Vennstra, B. Zwanenburg, *Synthesis* 1975, 519. L. Field, R. D. Clark, *Org. Synth. Coll.* **4** (1963) 674.
[76]   G. E. Vennstra, B. Zwanenburg, *Synthesis* 1975, 519.
[77]   M. Kobayashi, *Bull. Chem. Soc. Jpn.* **39** (1966) 1296.
[78]   F. Arndt, A. Scholz, *Justus Liebigs Ann. Chem.* **510** (1934) 62.
[79]   J. C. Meek, J. S. Fowler, *J. Org. Chem.* **33** (1968) 3422.
[80]   P. Allen, Jr., L. S. Karger, J. D. Haygood, J. Shrensel, *J. Org. Chem.* **16** (1951) 767.
[81]   I. B. Douglass, J. F. Ward, R. V. Norton, *J. Org. Chem.* **32** (1967) 324.
[82]   J. Tröger, O. Beck, *J. Prakt. Chem.* **87** (1913) 289.
[83]   C. C. J. Culvenor, W. Davies, N. S. Heath, *J. Chem. Soc.* 1949, 278.
[84]   T. L. Gresham et al., *J. Am. Chem. Soc.* **74** (1952) 1323.
[85]   O. Bayer, *Angew. Chem.* **61** (1949) 229.
[86]   J. E. Herweh, R. M. Fantazier, *J. Org. Chem.* **41** (1976) 116. H. Gilman, L. F. Cason, *J. Am. Chem. Soc.* **72** (1950) 3469.
[87]   W. B. Price, S. Smiles, *J. Chem. Soc.* 1928, 3154.
[88]   L. Field, P. H. Settlage, *J. Am. Chem. Soc.* **73** (1951) 5870.
[89]   H. Brederick, E. Bäder, *Chem. Ber.* **87** (1954) 129. E. Bäder, H. D. Hermann, *Chem. Ber.* **88** (1955) 41.
[90]   F. Ullmann, G. Pasdermadian, *Chem. Ber.* **34** (1901) 1150. C. W. Ferry, J. S. Buck, R. Baltzly, *Org. Synth.* **22** (1942) 31.
[91]   N. H. Nilson, C. Jacobsen, A. Senning, *Chem. Commun.* 1971, 314.
[92]   Hoechst, DE-AS 1 097 434, 1961. R. Kerber, J. Starnick, *Chem. Ber.* **104** (1971) 2035.
[93]   H. Gilman, L. F. Cason, *J. Am. Chem. Soc.* **72** (1950) 3469. IG Farbenindustrie DE 676 013, 1938.
[94]   L. F. Cason, C. C. Wanser, *J. Am. Chem. Soc.* **73** (1951) 142.
[95]   W. E. Truce, E. Wellisch, *J. Am. Chem. Soc.* **74** (1952) 5177. IG Farbenindustrie, DE 663 992, 1938.
[96]   H. Burton, E. Hoggarth, *J. Chem. Soc.* 1945, 468.
[97]   J. Walker, *J. Chem. Soc.* 1945, 630.
[98]   S. Pickholz, *J. Chem. Soc.* 1946, 685.
[99]   K. Bowden, E. A. Braude, E. R. H. Jones, *J. Chem. Soc.* 1946, 945. E. P. Kohler, G. R. Barrett, *J. Am. Chem. Soc.* **46** (1924) 747.
[100] C. J. M. Stirling, *J. Chem. Soc.* 1964, 5856.
[101] A. G. Davies, R. Feld, *J. Chem. Soc.* 1956, 665.
[102] S. Atkin, *Anal. Chem.* **19** (1947) 816.
[103] L. Horner, O. H. Basedow, *Justus Liebigs Ann. Chem.* **612** (1958) 108.

[104] P. Allen, Jr., L. S. Karger, J. D. Haygood, J. Shrensel, *J. Org. Chem.* **16** (1951) 767. G. C. Denzer, Jr., P. Allen, Jr., P. Conway, J. M. van der Veen, *J. Org. Chem.* **31** (1966) 3418.
[105] J. Strating, H. J. Backer, *Recl. Trav. Chim. Pays-Bas* **69** (1950) 638.
[106] F. Kurzer, *Org. Synth.* **4** (1963) 937.
[107] I. B. Douglass, R. V. Norton, *J. Org. Chem.* **33** (1968) 2104.
[108] M. L. Kee, I. B. Douglass, *Org. Prep. Proced.* **2** (1970) no. 3, 235.
[109] G. G. Venier, H.-H. Hsich, H. J. Barager (III), *J. Org. Chem.* **38** (1973) 17.
[110] E. Wenschuh, *Z. Chem.* **24** (1984) 126.
[111] J. F. King, S. Skanieczny, *Phosphorus Sulfur* **25** (1985) 11–20.
[112] J. M. Shreeve, *Acc. Chem. Res.* **6** (1973) 387.
[113] D. Sianesi, G. C. Bernadi, G. Moggi, *Tetrahedron Lett.* 1970, 1313.
[114] K. R. Brown, J. B. Douglass, *J. Am. Chem. Soc.* **73** (1951) 57 87.
[115] J. B. Douglass, K. R. Brown, F. T. Martin, *J. Am. Chem. Soc.* **74** (1952) 5770.
[116] J. B. Douglass, D. R. Poole, *J. Org. Chem.* **22** (1957) 536. US 3 253 028, 1957/1962. *Chem. Abstr.* **65** (1966) 3751.
[117] J. B. Douglass, B. S. Farah, E. G. Thomas, *J. Org. Chem.* **26** (1961) 1996.
[118] J. B. Douglass, R. V. Norton, *J. Org. Chem.* **33** (1968) 2104.
[119] Mi Lo Kee, J. B. Douglass, *Org. Prep. Proced.* **2** (1970) 235.
[120] Bayer AG, DE 1 167 326, 1962.*Chem. Abstr.* **61** (1964) 1758.
[121] H. Stetter, M. Krause, W.-D. Last, *Chem. Ber.* **102** (1969) 3357.
[122] B. Zwanenburg, *Phosphorus, Sulfur and Silicon,* **43** (1989) 1–2, 1–24.
[123] W. Sas, *J. Chem. Res. Synop.* 1993, 160.
[124] Goodrich Co., US 4 032 505, 1977.
[125] Goodrich Co., US 4 055 540, 1977.
[126] I. B. Douglass, *J. Org. Chem.* **30** (1965) 633.
[127] C. J. M. Stirling, *J. Chem. Soc.* 1963, 5741.
[128] A. Heesing, M. Jaspers, I. Schwermann, *Chem. Ber.* **112** (1979) 2903.
[129] M. Mikolajcyk, J. Drabowicz, *Synthesis* 1974, 124.
[130] I. B. Douglass, *J. Org. Chem.* **39** (1974) 563.
[131] L. Field, C. B. Hockel, J. M. Locke, J. E. Lawson, *J. Am. Chem. Soc.* **83** (1961) 1256.
[132] L. Field, C. B. Hockel, J. M. Locke, *J. Am. Chem. Soc.* **84** (1962) 847.
[133] P. Brownbridge, J. C. Jowett, *Synthesis* **3** (1988) 252.
[134] R. St. Henion, *Org. Chem. Bull.* **41** (1969) 4.
[135] J. S. Meek, J. S. Fowler, *J. Org. Chem.* **33** (1968) 3422. M. Kobayashi, H. Minato, H. Fukuda, *Bull. Chem. Soc. Jpn.* **46** (1973) 1266.
[136] Y. Miyaji, H. Minato, M. Kobayashi, *Bull. Chem. Soc. Jpn.* **44** (1971) 862.
[137] D. N. Harpp, T. G. Back, *J. Org. Chem.* **38** (1973) 4328.
[138] A. J. Ruming, G. Kresze, *Phosphorus Sulfur* **29** (1986) 49.
[139] D. Darwish, R. McLaren, *Tetrahedron Lett.* 1962,1231. A. C. Cope, D. E. Morrison, L. Field, *J. Am. Chem. Soc.* **72** (1950) 59. C. J. M. Stirling, *Chem. Commun.* 1967, 131.
[140] C. J. M. Stirling, *Chem. Commun.* 1967, 131. S. Braverman, *Int. J. Sulfur Chem. Part C* **6** (1971) 149.
[141] K. K. Andersen et al., *J. Am. Chem. Soc.* **86** (1965) 5637.
[142] R. M. Coates, H. D. Pigott, *Synthesis* 1975, 319.
[143] M. Axelrod et al., *J. Am. Chem. Soc.* **90** (1968) 4835.
[144] M. Koboyaski, A. Yamamoto, *Bull. Chem. Soc. Jpn.* **39** (1966) 2736.
[145] A. J. Khodair, A. A. Swelin, A. A. Abdel–Wehab, *Phosphorus Sulfur* **2** (1976) 173.

[146] K. Kondo, A. Negishi, I. Ojima, *J. Am. Chem. Soc.* **94** (1972) 5786.
[147] F. Freeman, *Chem. Rev.* **84** (1984) 117–135.
[148] I. B. Douglass, B. S. Farah, *J. Org. Chem.* **23** (1958) 805.
[149] Y. H. Chiang, J. S. Luloff, E. Schipper, *J. Org. Chem.* **34** (1969) 2397.
[150] Monsanto, FR 1 529 050, 1967; GB 1 139 349, 1966.
[151] A. Sonn, E. Schmidt, *Chem. Ber.* **57** (1924) 1355.
[152] D. Klamann, C. Sass, M. Zelenka, *Chem. Ber.* **92** (1959) 1910. Fuji Film Co., US 3 498 792, 1965.
[153] Monsanto, US 3 541 060, 1967/1970.
[154] Ugine Kuhlman, BE 738 304, 1968; FR 2 041 585, 1969.
[155] M. Bazlen, BASF, US 855 566, 1904. M. Bazlen, *Chem. Ber.* **42** (1900) 4634.
[156] M. Mulliez, C. Naudy, *Tetrahedron* **49** (1993) 2469.
[157] J. S. Edgar, *Phosphorus Sulfur* **2** (1976) 181.
[158] J. G. de Vries, R. M. Kellogg, *J. Org. Chem.* **45** (1980) 4126.
[159] *Römpp*, 9th ed., Thieme-Verlag, 1990, p. 1426.
[160] Mitsubishi Gas Chemical, EP 607 448, 1993. S. Iwasaki, T. Oka, JA 49 040 451, 1974. A. Kojima, M. Okawara, *Chem. Lett.* 1984, 2125. Bayer, DE 4 240 708, 1992. Bayer, EP 600 339, 1993.
[161] D. De Filipo, G. Pouticelli, E. F. Tragu, A. Lai, *J. Chem. Soc., Perkin Trans.* **2** (1972) 1500.
[162] Kuroda (Nihon) Chemicals KK, JP 50 062 934, 1975; *Chem. Abstr.* **83** (1975) 96 423.
[163] M. Drifoglu, W. N. Marmer, R. L. Dudley, *Text. Res. J.* **62** (1992) 94. G. Wank, A. Mursyidi, *J. Photochem. Photobiol. A* **64** (1992) 263. G. Rabai, R. T. Wang, K. Kustin, *Int. J. Chem. Kinet.* **25** (1993) 53.
[164] G. Rabai, R. T. Wang, K. Kustin, *Int. J. Chem. Kinet.* **25** (1993) 53.
[165] G. Crank, A. Mursyidi, *J. Photochem. Photobiol. A* **64** (1992) 263.
[166] Japan Oils and Fat Co., JP 45 017 665, 1970; *Chem. Abstr.* **73** (1970) 98 387.
[167] E. Jourdan-Laforte, FR 2 040 797, 1971; *Chem. Abstr.* **75** (1971) 48 427.
[168] I. I. Havel, R. Q. Kluttz, *Synth. Commun.* **4** (1974) 389.
[169] Teijen Ltd, JP 47 025 464, 1972; *Chem. Abstr.* **78** (1973) 17 055.
[170] T. W. Gibson, P. Strassburger, *J. Org. Chem.* **41** (1976) 791.
[171] Agrotex, DL 129 567, 1977.
[172] Phillips Petroleum, US 4 059 653, 1976.
[173] Du Pont, GB 856 917, 1958.
[174] Bayer, US 3 511 800, 1970.
[175] Bayer, DE-OS 2 110 773, 1971.
[176] R. Wegler: Chemie der Pflanzenschutz- und Schädlingsbekämpfungsmittel, vol. 5, Springer, Heidelberg 1977, pp. 233, 503.
[177] Phillips Petroleum, US 2 955 980, 1958.
[178] M. Kling, US 5 270 058, 1993; *Chem. Abstr.* **120** (1994) 99 425.
[179] Int. Flavors and Fragrances Inc., US 3 906 119, 1973.
[180] Int. Flavors and Fragrances Inc., US 3 966 989, 1975.
[181] Bayer, DE-OS 2 102 177, 1971.
[182] Uniroyal Inc., EP 101 198, 1984; *Chem. Abstr.* **100** (1984) 211 426. Armstrong Co., US 4 104 301, 1978; *Chem. Abstr.* **90** (1979) 71 929.
[183] Dt. Akad. Wiss., DE-AS 1 248 656, 1965.
[184] Du Pont, US 2 315 514, 1938.
[185] V. Petrovich, US 3 890 222, 1974.

[186]   American Cyanamid, US 2 917 540, 1958.
[187]   Hoechst, DE-AS 1 129 477, 1960.
[188]   Fuji Film Co., US 3 498 792, 1965.
[189]   Wilkinson Sword Ltd., BE 802 469, 1972.
[190]   Monsanto, US 3 532 693, 1967. Ugine Kuhlmann, FR 2 037 001, 1969.
[191]   Uniroyal Inc., FR 1 581 710, 1969; *Chem. Abstr.* **73** (1970) 16 057.
[192]   Sandoz Products, GB 1 591 616, 1981; *Chem. Abstr.* **96** (1982) 53 748.
[193]   Yu. U. Polenov et al., *Zh. Prikl. Khim. (Leningrad)* **63** (1990) 1622; *Chem. Abstr.* **114** (1991) 64 230.
[194]   S. K. Chung, *J. Org. Chem.* **46** (1981) 5457.
[195]   *Römpp*, 9th ed., Thieme Verlag, 1992, p. 3916.
[196]   Degussa Safety Data Sheet, 00 122/1 580 006 000/SDB/D001, 1994.
[197]   A. Sazasz et al., HU 55 763, 1991; *Chem. Abstr.* **116** (1992) 6562.
[198]   S. L. Huang, T.-Y. Chen, *J. Chin. Chem. Soc. (Taipei)* **21** (1974) 235; *Chem. Abstr.* **82** (1975) 139 916.
[199]   Degussa AG, DE 4 028 239, 1992; *Chem. Abstr.* **117** (1992) 94 231. Degussa AG, DE 4 028 240, 1992; *Chem. Abstr.* **117** (1992) 94 230. Sumitomo Metal Mining Co., JP 1 136 911, 1987; *Chem. Abstr.* **111** (1989) 185 579.
[200]   Ciba Geigy AG, EP 396 503, 1990; *Chem. Abstr.* **114** (1991) 188 036.
[201]   J. Maier, *Wochenbl. Papierfabr.* **12** (1993) 111; M. Hammann, WO 9 221 814, 1992; *Chem. Abstr.* **119** (1993) 74 932. Degussa AG, DE 3 923 728, 1991; *Chem. Abstr.* **114** (1991) 209 447. Interox International S.A., WO 9 207 139, 1992; *Chem. Abstr.* **117** (1992) 92 512. P. Pettit, *Apprita J.* **45** (1992) 385.
[202]   Konica Co., EP 552 650, 1993; *Chem. Abstr.* **120** (1994) 231 815. Y. Fu, S. Ji, B. Zhon, X. Ren, *Ganguang Kexue Yu Kuang Huaxue* **10** (1992) 103; *Chem. Abstr.* **120** (1994) 177 952. Fuji Photo Film Co., JP 5 165 132, 1993; *Chem. Abstr.* **120** (1994) 204 455. Konishiroku Photo Ind., JP 5 224 335, 1993; *Chem. Abstr.* **120** (1994) 120 622.
[203]   D. L. Michelsen, M. Padaki, *Hazard. Ind. Wastes* **25** (1993) 218.
[204]   H. F. Wu, B. N. Huang, *Chim. Chem. Lett.* **4** (1993) 683; *Chem. Abstr.* **120** (1994) 162 932.
[205]   Bayer, Safety Data Sheet, 016 962/01, 1993.
[206]   Bayer, Safety Data Sheet, 018 604/05, 1993.
[207]   Bayer, Safety Data Sheet, 323 520/04, 1993.
[208]   Bayer, Safety Data Sheet, 378 945/02, 1994.
[209]   Bayer, Safety Data Sheet, 692 100/01, 1993.
[210]   Bayer, Safety Data Sheet, 321 668/02, 1993.
[211]   Bayer AG, Leverkusen, personal communication.
[212]   *Ullmann*, 3rd. ed., **16,** 490.
[213]   BASF, Ludwigshafen, Safety Data Sheets ET 00 338.
[214]   Brüggemann, Safety Data Sheet, Bruggolit C, SD 00/01 A. DOC.

Sulfones and Sulfoxides

KATHRIN-MARIA ROY, Langenfeld, Federal Republic of Germany

1.	Sulfones	4477	2.2.2.	Reaction of Sulfur Dioxide with Aromatics. 4489
1.1.	Physical and Chemical Properties	4477	2.2.3.	Sulfoxides from Sulfinates 4490
1.2.	Synthesis	4479	2.2.4.	α,β-Unsaturated Sulfoxides 4490
1.2.1.	Oxidation of Thioethers (Sulfides)	4479	2.2.5.	Optically Active Sulfoxides. 4490
1.2.2.	Alkylation of Sulfinic Acid Derivatives	4480	2.3.	Reactions 4491
1.2.3.	Addition of Sulfinic Acid Derivatives to Multiple Bonds . .	4481	2.3.1.	Reactions of α-Sulfinyl Carbanions 4491
1.2.4.	Rearrangement of Sulfinates . . .	4482	2.3.2.	Reactions with Grignard Reagents 4492
1.2.5.	Sulfones from Sulfonyl Carbanions	4482	2.3.3.	Pummerer Rearrangement. 4492
1.2.6.	Addition of Sulfur Dioxide	4483	2.3.4.	Oxidation and Reduction. 4493
1.3.	Reactions	4484	2.4.	Uses 4493
1.4.	Industrial Uses	4485	2.5.	Dimethyl Sulfoxide 4494
2.	Sulfoxides	4486	2.5.1.	Production 4495
2.1.	Physical and Chemical Properties	4486	2.5.2.	Uses. 4495
2.2.	Synthesis.	4487	2.5.3.	Reactions 4495
2.2.1.	Oxidation of Sulfides	4487	3.	Toxicology 4496
			3.1.	Sulfones 4496
			3.2.	Dimethyl Sulfoxide 4496
			4.	References. 4497

1. Sulfones

1.1. Physical and Chemical Properties

Sulfones are described by the following general structural formula, in which the groups R^1 and R^2 can be hydrocarbon substituents, including cyclic ones, of any combination:

$$R^1-\overset{O}{\underset{O}{\overset{\|}{S}}}-R^2$$

Table 1. Physical data and molecular formulae for some industrially important sulfones

Sulfone	mp, °C (bp, °C/kPa)	Molecular formula (molecular mass)
(tetrahydrothiophene 1,1-dioxide, sulfolane structure)	(285/101.3)	$C_4H_8O_2S$ (120.16)
$4\text{-}CH_3\text{-}C_6H_4\text{-}SO_2\text{-}C_6H_4\text{-}4\text{-}CH_3$	158	$C_{14}H_{14}O_2S$ (246.33)
$4\text{-}Br\text{-}C_6H_4\text{-}SO_2\text{-}C_6H_4\text{-}4\text{-}Br$	172	$C_{12}H_8Br_2O_2S$ (376.08)
$CH_3\text{-}(CH_2)_3\text{-}SO_2\text{-}(CH_2)_3\text{-}CH_3$	46	$C_8H_{18}O_2S$ (178.30)
$HO\text{-}CH_2\text{-}CH_2\text{-}SO_2\text{-}CH_2\text{-}CH_2\text{-}OH$	58	$C_4H_{10}O_4S$ (154.19)
$C_6H_5\text{-}CH_2\text{-}SO_2\text{-}CH_2\text{-}C_6H_5$	149–153	$C_{14}H_{14}O_2S$ (246.33)
$C_2H_5\text{-}SO_2\text{-}CH_2\text{-}CONH_2$	99	$C_4H_9NO_3S$ (151.19)
$CH_3\text{-}(CH_2)_2\text{-}SO_2\text{-}CH_2\text{-}CONH_2$	104	$C_5H_{11}NO_3S$ (165.21)
$C_6H_5\text{-}SO_2\text{-}CH_2\text{-}CH_2\text{-}OH$	(184/19)	$C_8H_{10}O_3S$ (186.23)
$CO_2H\text{-}CH_2\text{-}CH_2\text{-}SO_2\text{-}CH_2\text{-}CH_2\text{-}CO_2H$	222–224	$C_6H_{10}O_6S$ (210.21)
$C_2H_5\text{-}SO_2\text{-}(CH_2)_2\text{-}CH_3$	25	$C_5H_{12}O_2S$ (136.21)
$(CH_3)_2CH\text{-}SO_2\text{-}CH(CH_3)_2$	35–36	$C_6H_{14}O_2S$ (150.24)
$CH_3\text{-}SO_2\text{-}CH_3$	109	$C_2H_6O_2S$ (94.13)
$CH_3\text{-}(CH_2)_2\text{-}SO_2\text{-}(CH_2)_2\text{-}CH_3$	26	$C_6H_{14}O_2S$ (150.24)
$CH_3\text{-}(CH_2)_7\text{-}SO_2\text{-}(CH_2)_7\text{-}CH_3$	73	$C_{16}H_{34}O_2S$ (290.51)
$CH_3\text{-}SO_2\text{-}CH_2\text{-}C_6H_5$	126–127	$C_8H_{10}O_2S$ (170.23)
$CH_3\text{-}SO_2\text{-}(CH_2)_3\text{-}CH_3$	28–30	$C_5H_{12}O_2S$ (136.21)
$CH_3\text{-}SO_2\text{-}C_6H_4\text{-}4\text{-}CH_3$	88–89	$C_8H_{10}O_2S$ (170.23)
$4\text{-}Cl\text{-}C_6H_4\text{-}CH_2\text{-}SO_2\text{-}C_6H_4\text{-}4\text{-}CH_3$	167–168	$C_{14}H_{13}ClO_2S$ (280.78)
$CH_2\!=\!CH\text{-}CH_2\text{-}SO_2\text{-}C_6H_4\text{-}4\text{-}CH_3$	50–52	$C_{10}H_{12}O_2S$ (196.27)
$C_2H_5CO_2\text{-}CH_2\text{-}SO_2\text{-}C_6H_4\text{-}4\text{-}CH_3$	91–93	$C_{11}H_{14}O_4S$ (242.30)
$4\text{-}CH_3\text{-}C_6H_4\text{-}SO_2\text{-}C_6H_5$	125–126	$C_{13}H_{12}O_2S$ (232.30)
$2,2',4\text{-}(CH_3)_3\text{-}C_6H_2\text{-}SO_2\text{-}C_6H_4\text{-}4\text{-}OCH_3$	136	$C_{16}H_{18}O_3S$ (290.38)
$4\text{-}CH_3\text{-}C_6H_4\text{-}SO_2\text{-}C_6H_3\text{-}2,4\text{-}(CH_3)_2$	51–52	$C_{15}H_{16}O_2S$ (260.36)
$C_6H_5\text{-}SO_2\text{-}C_6H_4\text{-}4\text{-}C_2H_5$	92–93	$C_{14}H_{14}O_2S$ (246.33)
$4\text{-}CH_3O\text{-}C_6H_4\text{-}SO_2\text{-}CH_3$	120	$C_8H_{10}O_3S$ (186.23)
$2,4\text{-}(CH_3)_2\text{-}C_6H_3\text{-}SO_2\text{-}CH_3$	56	$C_9H_{12}O_2S$ (184.26)
$C_6H_5\text{-}SO_2\text{-}C_6H_5$	121	$C_{12}H_{10}O_2S$ (218.28)
$4\text{-}CH_3\text{-}C_6H_4\text{-}SO_2\text{-}C_6H_5$		$C_{13}H_{12}O_2S$ (232.30)
$4\text{-}CH_3\text{-}C_6H_4\text{-}CH\text{-}SO_2\text{-}C_6H_4\text{-}4\text{-}NHCOCH_3$ $\quad\quad\quad\quad\; \vert$ $\quad\quad\quad\; CH_2\text{-}NO_2$	185–187	$C_{17}H_{18}N_2O_5S$ (362.41)
$C_6H_5\text{-}CH\text{-}SO_2\text{-}C_6H_5$ $\quad\; \vert$ $CH_2\text{-}NO_2$	186–187	$C_{14}H_{13}NO_4S$ (291.33)
$4\text{-}CH_3O\text{-}C_6H_4\text{-}CH\text{-}SO_2\text{-}C_6H_4\text{-}4\text{-}Cl$ $\quad\quad\quad\quad\; \vert$ $\quad\quad\; CH_2\text{-}CO\text{-}C_6H_4\text{-}4\text{-}CH_3$	158	$C_{23}H_{21}ClO_4S$ (428.94)
$C_6H_5\text{-}CH\text{-}SO_2\text{-}C_6H_4\text{-}4\text{-}N(CH_3)_2$ $\quad\; \vert$ $CH_2\text{-}CO\text{-}C_6H_5$	191–192	$C_{23}H_{23}NO_3S$ (393.51)
$C_6H_5\text{-}SO_2\text{-}CH_2\text{-}OH$	57–60	$C_7H_8O_3S$ (172.20)
$4\text{-}CH_3\text{-}C_6H_4\text{-}SO_2\text{-}CH(OH)\text{-}(CH_2)_2\text{-}CH_3$	78	$C_{11}H_{16}O_3S$ (228.31)
$CH_3\text{-}(CH_2)_{11}\text{-}SO_2\text{-}CH_2\text{-}OH$	64–66	$C_{13}H_{28}O_3S$ (264.43)

Typical sulfones are colorless and odorless in the pure state and are very stable both thermally and chemically. They are protonated only by the strongest acid systems. Unlike sulfoxides, sulfones form relatively weak hydrogen bonds.

The presence of a sulfonyl group in the molecule is shown by strong absorptions at 1300–1320 cm^{-1} and 1140–1160 cm^{-1} in the IR spectrum. Physical properties of industrially important sulfones are listed in Table 1.

1.2. Synthesis [1, pp. 223, 1129], [2], [3, p. 171]

1.2.1. Oxidation of Thioethers (Sulfides)

Sulfones are generally synthesized by oxidation of sulfides via sulfoxides as intermediates. A common oxidizing agent in the laboratory is hydrogen peroxide in acetic acid. It is also often used in industry, but requires reaction temperatures of 70 °C up to reflux in many cases [4], [5]. In the presence of various metal catalysts, oxidation with hydrogen peroxide proceeds under milder conditions. Thus, β-hydroxy sulfides can be oxidized selectively and in good yield to the corresponding sulfones by using titanium, vanadium, molybdenum, or tungsten salts [6], as well as selenium dioxide [7], as catalysts. A very effective oxidizing agent, which gives high-purity products in high yield, is *m*-chloroperbenzoic acid (MCPBA). This reagent can often be used if the molecule contains other sensitive groups, as in many pharmaceutically active substances [8]. Alkyl thiocyanates are converted to sulfonyl cyanides by MCPBA [9]

$$R-SCN \xrightarrow{MCPBA} R-SO_2-CN$$

Other oxidizing agents used include potassium permanganate [10], sodium perborate [11], and dimethyldioxirane [12]. An extremely simple, rapid process is sulfide oxidation with periodic acid in the presence of ruthenium tetroxide as catalyst. This reaction is carried out in a two-phase solvent mixture and is, therefore, also suitable for the synthesis of water-soluble sulfones [13]. A related method using osmium tetroxide in the presence of a tertiary amine *N*-oxide gives good yields [14]. In industrial processes, gaseous chlorine is also used as the oxidizing agent (see, for example [15]). Recently, the potassium peroxomonosulfate complex Oxone has gained importance [16]. With this reagent, allyl sulfoxides, for example, can be converted into the corresponding sulfones [17]. In the presence of wet clays such as kaolin or montmorillonite, the oxidation of various sulfides in dichloromethane gives the sulfone almost quantitatively [18].

Oxidation of Phenyl Vinyl Sulfide with H_2O_2 in Glacial Acetic Acid [4]. To a solution of 19.7 g (0.145 mol) of phenyl vinyl sulfide in 70 mL of glacial acetic acid, 56 mL (0.5 mol) of 30% hydrogen peroxide is added dropwise in such a way that the reaction temperature becomes 70 °C. The reaction mixture is subsequently heated for 20 min under reflux. After cooling, 150 mL of diethylether and 200 mL of water are added, and the organic phase is separated. The organic phase is subsequently washed with water (50 mL) and sodium chloride solution (50 mL) and evaporated, giving 18–19 g (74–78%) of phenyl vinyl sulfone as a colorless solid, *mp* 64–65 °C.

Oxidation of Sulfides with Oxone in the Presence of Wet Montmorillonite [18]. Commercial clay (montmorillonite or kaolin; 10 g) is treated with 2 g of distilled water in eight portions. After each addition the mixture is shaken vigorously for several minutes, until a loose powder is formed. Then 1.2 g of this catalyst is suspended in 6 mL of dichloromethane (7 mL for the kaolin catalyst) and treated with Oxone (2.5 equivalents based on the sulfide). After addition of 1 mmol sulfide the reaction vessel is purged with argon, closed, and stirred for 2 h at room temperature. After filtration of the catalyst on a glass frit, washing with ca. 100 mL of dichloromethane, and evaporation of the solvent, the pure sulfone (>98% purity) is obtained almost quantitatively.

A particularly simple and rapid process allows the synthesis of sulfonyl-substituted propanones on a large scale [19]. Here, chloroacetone is reacted with various sulfides. The intermediate β-keto sulfide is not isolated but is oxidized directly with Oxone to the sulfone.

$$CH_3-CO-CH_2Cl + R-SH \xrightarrow{\text{NaOH}}_{\text{MeOH}}$$
$$CH_3-CO-CH_2-S-R \xrightarrow{\text{Oxone}}_{\text{H}_2\text{O}}$$
$$CH_3-CO-CH_2-SO_2-R$$

1.2.2. Alkylation of Sulfinic Acid Derivatives

Alkylation of the ambidentate sulfinate ion with soft alkylating agents, such as alkyl halides, is a widely used method for synthesis of sulfones:

$$R^1-SO_2^-Na^+ + R^2-X \longrightarrow R^1-SO_2-R^2$$

Aryl sulfones, in particular, in which the second sulfonyl substituent can be an alkyl [20], an alkynyl [21], a β-keto [22], or an allyl [23] group, can be produced by this process. These reactions are catalyzed, for example, by copper(I) bromide [24], phase-transfer catalysts [20], [21], [25], or ultrasound [26].

A related reaction is that of arylsulfonyl chlorides with alkyl iodides or benzyl chlorides in the presence of telluride. The reaction is based on a reactivity inversion, where the original electrophilic sulfonyl group is converted to the nucleophilic sulfinate ion by electron transfer from the telluride ion [27]:

$$Ar-SO_2Cl \xrightarrow{Te^{2-}} [Ar-SO_2^-] \xrightarrow{R-X} Ar-SO_2-R$$

Substituted vinyl and allyl sulfones are obtained in good yield by palladium-catalyzed coupling of arylsulfonyl chlorides with organotin compounds [28].

$$Ar-SO_2Cl + Bu_3Sn-CH=CH-R \xrightarrow{(Ph_3P)_4Pd}$$
$$Ar-SO_2-CH=CH-R$$

The sulfonylation of various organometallic reagents with arylsulfonyl fluorides is a one-step sulfone synthesis, which gives alkyl, trimethylsilylmethyl, and allyl sulfones in good yield [29]. Cycloalkenes can be converted particularly easily into the corresponding vinyl sulfones by reaction with tetrabutylammonium sulfinate [30].

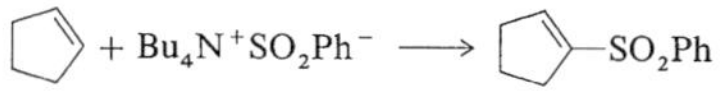

[diagram: cyclopentene + Bu₄N⁺SO₂Ph⁻ → cyclopentyl-SO₂Ph]

1.2.3. Addition of Sulfinic Acid Derivatives to Multiple Bonds

Isolated double bonds do not generally react with sulfinic acids or their salts to give sulfones. Iodosulfonylation and sulfonylmercuration, which both take place regioselectively [31], are exceptions. The adducts formed initially are converted by dehydrohalogenation or demercuration into the corresponding vinyl sulfones (see Scheme 1).

Conjugated dienes [32] and α,β-unsaturated carbonyl compounds [33] can also be converted regioselectively into unsaturated sulfones by using iodosulfonylation.

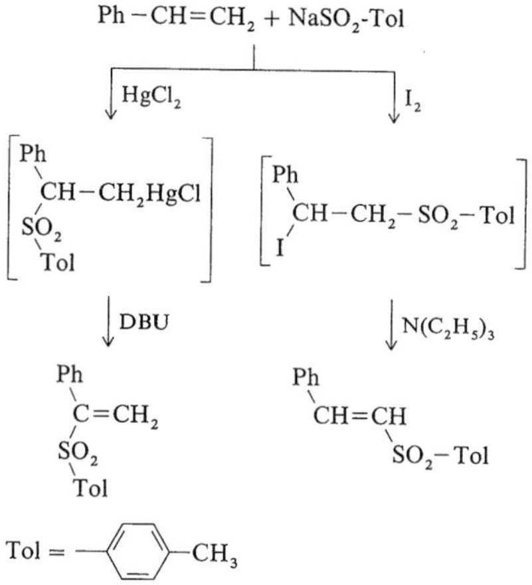

[Scheme 1 diagram:

Ph–CH=CH₂ + NaSO₂-Tol

↓ HgCl₂   ↓ I₂

[Ph–CH(SO₂Tol)–CH₂HgCl]   [Ph–CH(I)–CH₂–SO₂–Tol]

↓ DBU   ↓ N(C₂H₅)₃

Ph–C(SO₂Tol)=CH₂   Ph–CH=CH–SO₂–Tol

Tol = 4-methylphenyl

DBU = 1,5-Diazabicyclo[4.3.0]non-5-ene]

Scheme 1.

The addition of sulfinic acids to polarized double or triple bonds is an excellent method of synthesizing sulfones [34]. A large number of olefins, such as vinyl aldehydes [35] and ketones [36], nitroolefins [37], acrylic acid derivatives [38], vinyl sulfones [39], quinones [40], and quinonimines [41] have been used successfully as acceptor compounds.

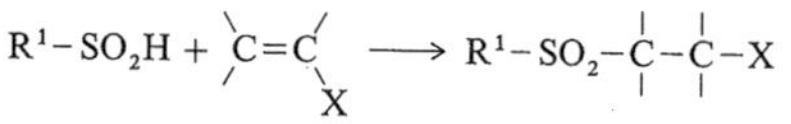

$$R^1-SO_2H + \,\,{}^{\backslash}_{/}C=C^{/}_{\backslash X} \longrightarrow R^1-SO_2-\overset{|}{\underset{|}{C}}-\overset{|}{\underset{|}{C}}-X$$

$X = -CHO, -CO-CH_3, -CO-N(C_2H_5)_2, -CO-O-R^2, -NO_2, -SO_2-Ph$

Sulfinic acid also adds readily to sulfonyl-substituted allenes and acetylenes to give disulfonyl derivatives [42].

$$Ph-SO_2H + H_3C-C\equiv C-SO_2Ph$$

$$\longrightarrow \begin{array}{c} Ph-SO_2 \\ \diagdown \\ C=C \\ \diagup \diagdown \\ CH_3 H \end{array} \begin{array}{c} SO_2-Ph \end{array}$$

$$Ph-SO_2H + H_2C=C=CH-SO_2-Ph$$

$$\longrightarrow H_2C=C \begin{array}{c} SO_2Ph \\ \diagdown \\ CH_2-SO_2Ph \end{array}$$

Unlike sulfinic acids, the addition of sulfonyl halides to unsaturated C–C systems is a radical reaction in which mixtures of isomers can be formed. To suppress competing polymerization reactions, the addition is carried out in the presence of copper(I) or copper(II) salts. The β-halo sulfones formed initially can be converted into vinyl sulfones by elimination of hydrogen halide [43].

$$R-SO_2X + \begin{array}{c} \diagdown \diagup \\ C=C \\ \diagup \diagdown \\ H \end{array} \xrightarrow{CuCl \text{ or } CuCl_2}$$

$$R-SO_2-\underset{H}{\overset{}{C}}-\underset{}{\overset{X}{C}}- \xrightarrow{-HX} R-SO_2-\overset{}{C}=\overset{}{C}$$

1.2.4. Rearrangement of Sulfinates

Sulfinates rearrange to sulfones under various conditions. Acids [44], tertiary amines [45], or palladium(0) complexes [46] are examples of suitable catalysts. Allyl-substituted sulfinic acid esters can also be isomerized thermally [46], [47].

$$R^1-O-\underset{}{\overset{O}{\underset{\|}{S}}}-R^2 \xrightarrow{\Delta} R^1-\underset{O}{\overset{O}{\underset{\|}{\overset{\|}{S}}}}-R^2$$

1.2.5. Sulfones from Sulfonyl Carbanions

α-Sulfonyl carbanions are formed from sulfones containing α-hydrogens by the action of bases, such as Grignard reagents, alkali-metal amides, sodium hydride, or butyllithium. These carbanions can be converted into a variety of new sulfones by alkylation, acylation, or addition to unsaturated systems.

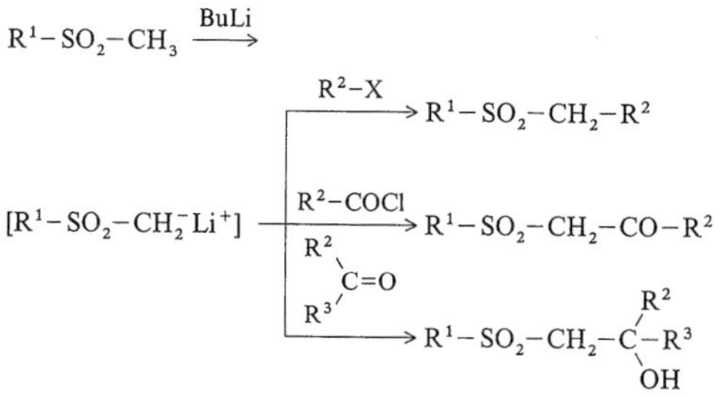

$$R^1-SO_2-CH_3 \xrightarrow{BuLi} [R^1-SO_2-CH_2^- Li^+]$$

$$\begin{array}{l} \xrightarrow{R^2-X} R^1-SO_2-CH_2-R^2 \\ \xrightarrow[R^2]{R^2-COCl} R^1-SO_2-CH_2-CO-R^2 \\ \xrightarrow[R^3]{C=O} R^1-SO_2-CH_2-\underset{OH}{\overset{R^2}{\underset{|}{\overset{|}{C}}}}-R^3 \end{array}$$

Treatment of α-sulfonyl carbanions with acid chlorides [48] or esters [49] gives β-keto sulfones, which can rearrange with allyl bromide to give α-allyl-substituted β-keto sulfones [50].

The Knoevenagel condensation of aryl sulfones with aldehydes or ketones is also frequently used. The intermediate products eliminate water to form vinyl sulfones [51] (see Scheme 2).

The Knoevenagel condensation, with subsequent allyl sulfoxide–sulfenic acid ester rearrangement, is of synthetic importance in the production of γ-hydroxy-α,β-unsaturated sulfones. Here piperidine is used as the base [52] (see Scheme 3).

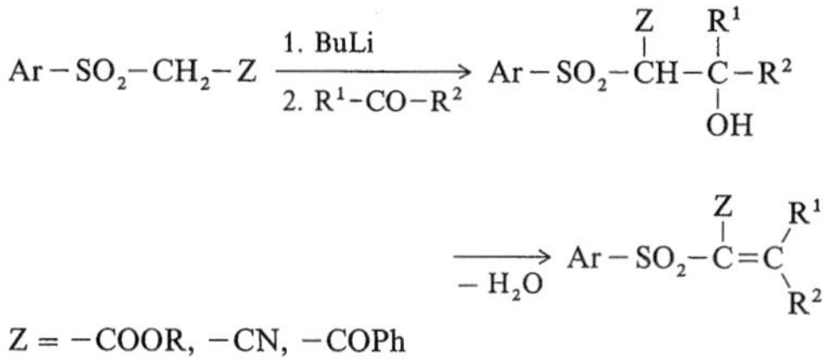

$$Ar-SO_2-CH_2-Z \xrightarrow[2.\ R^1-CO-R^2]{1.\ BuLi} Ar-SO_2-\overset{Z}{\underset{|}{C}}H-\overset{R^1}{\underset{OH}{\overset{|}{\underset{|}{C}}}}-R^2$$

$$\xrightarrow{-H_2O} Ar-SO_2-\overset{Z}{\underset{|}{C}}=\overset{R^1}{\underset{R^2}{C}}$$

Scheme 2. $Z = -COOR, -CN, -COPh$

$$Ph-SO_2-CH_2-SO-Ar$$

$$\xrightarrow[2.\ R^1\text{CH-CHO}\ /\ R^2]{1.\ \text{piperidine NH}} Ph-SO_2\diagup\!\!\diagdown\overset{OH}{\underset{R^1}{\overset{|}{\underset{|}{C}}}}R^2$$

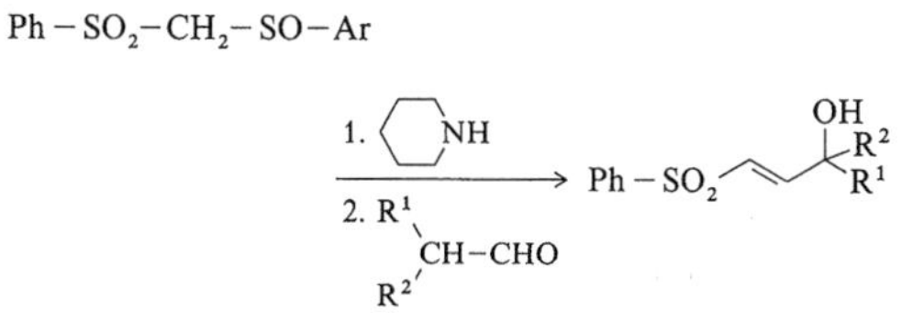

Scheme 3.

1.2.6. Addition of Sulfur Dioxide

By addition of sulfur dioxide to aryldiazonium salts, aryl sulfonyl radicals are formed, which can react with olefins to give alkyl aryl sulfones [53].

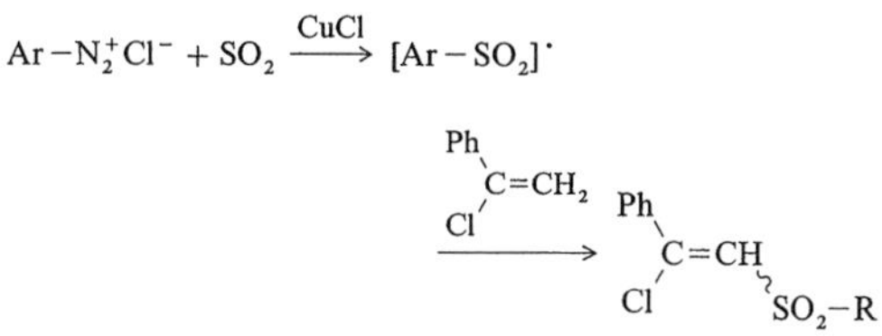

$$Ar-N_2^+Cl^- + SO_2 \xrightarrow{CuCl} [Ar-SO_2]^\bullet$$

$$\xrightarrow[Cl]{\overset{Ph}{\diagdown}C=CH_2} \overset{Ph}{\underset{Cl}{\diagdown}}C=CH\diagdown SO_2-R$$

1,3-Dienes undergo stereoselective cycloaddition with sulfur dioxide to give sulfolenes. This reaction is reversible [54].

$$\text{diene} + SO_2 \underset{\Delta}{\rightleftharpoons} \text{sulfolene}$$

1.3. Reactions

Reductive Elimination of the Sulfonyl Group. Reductive elimination of the sulfonyl group is of great synthetic importance because sulfones are used as intermediates in many syntheses. Alkali metals in ammonia or low molar mass amines [55], sodium amalgam [56], and aluminum amalgam [57] have been described as reducing agents. Sulfur dioxide is eliminated from unsaturated, cyclic sulfones by potassium dispersed with ultrasound [58].

$$\text{cyclic sulfone} \xrightarrow[\text{ultrasound}]{K} R-CH=CH-CH=C\begin{smallmatrix}R\\R^1\end{smallmatrix}$$

The simultaneous reduction and desulfonylation of α-alkylidene β-keto sulfones is effected under mild conditions by using sodium hydrogentelluride [59].

$$R-C_6H_4-CH=C(SO_2Ph)(CO-Ph) \xrightarrow{NaHTe} R-C_6H_4-CH_2-CH_2-CO-Ph$$

The sulfonyl group can be eliminated photochemically from β-keto sulfones by using a β-keto ester in the presence of a ruthenium(II) salt, so that many other functional groups are not attacked [60].

Elimination of Sulfinate. The base-catalyzed elimination of organic sulfinate groups from sulfones is often used preparatively. Arylsulfonyl [61] and, in particular, trifluoromethanesulfonyl groups are good leaving groups [62]:

$$R-CH_2-CH_2-SO_2-Ar \xrightarrow{\text{Base}} R-CH=CH_2 + Ar-SO_2H$$

Ramberg – Bäcklund Reaction. Bases effect the elimination of sulfur dioxide from α-halo sulfones with α-hydrogens via an episulfone intermediate, to give olefins [63].

Smiles and Truce–Smiles Rearrangement. Diaryl and aralkyl sulfones undergo base-catalyzed rearrangement to form sulfinic acids [64]. In the *Smiles rearrangement* the migrating aryl group must be activated by an electron-withdrawing group.

In the related *Truce–Smiles rearrangement* a nonactivated aryl or tertiary alkyl group migrates to a carbanion formed by lithiation of an *o*-methyl group [65].

1.4. Industrial Uses

Unsaturated sulfones can be used as intermediates for pharmaceuticals and pesticides and for modification of polymers [66]. Many reactive dyes also contain unsaturated sulfonyl groups, (e.g., see [67]). The possible uses of sulfones described below are taken from the recent extensive patent literature. For uses of sulfolane, see → Benzene

Uses in Crop Protection. 1,1-Diiodomethyl sulfone [68], heterocyclic sulfones [69], and α-sulfonyloximes [70] have been described as *plant fungicides*. Certain sulfones are effective *herbicides* [71]. Others act as *insecticides* and *acaricides* [72].

Pharmaceutical Uses. Dihydropyridine sulfone derivatives are *calcium antagonists*, which are said to have a longer period of activity and fewer side effects than nifedipine [73]. Alkylsulfonyl-substituted steroid pyrazoles and thiazoles exhibit *antiandrogenic activity* [74].

Sulfonylnitromethane derivatives are aldose reductase inhibitors and can be used for the treatment of certain secondary effects of *diabetes* and *galactosemia* [75]. A range of sulfonyl-substituted benzazepines and imidazoles have been patented as agents for the treatment of *gastrointestinal disorders* [76]. Fluoro sulfones are said to be very effective leukotriene antagonists with *antiallergic* and *anti-inflammatory activity* [77]. Aryl- or heteroaryl-substituted sulfones exhibit activity against *picornaviruses* [5] and *retroviruses* [78]. Sulfonylcyclopentenones are *antitumor agents* and promote *bone formation* [79].

Others. Piperidinyl sulfones are used as light stabilizers in polymers, particularly polyolefins [80]. Certain heterocyclic sulfones can be used as *corrosion protection agents* for metals [81]. Sulfones are used as hardeners [82] and couplers [83] in *photographic materials*. Diazodisulfones can be used in the production of *semiconductors* and as photosensitive reagents for *printing plate materials* [84]. Certain fluorinated sulfones have *nonlinear optical properties* [85]. Sulfone percarboxylic acids are *bleaching agents* with little damaging effect on colors and good storage stability [86]. *Fabric softeners* contain esters or ethers of thiodiglycol sulfones [87].

2. Sulfoxides

2.1. Physical and Chemical Properties

Sulfoxides are colorless, hygroscopic liquids or solids. They dissolve readily in protic solvents to form strong hydrogen bonds. Because of the polar nature of the S–O bond, sulfoxides form adducts with salts of most metals.

Because of the pyramidal structure of sulfoxides, unsymmetrically substituted compounds of this series form enantiomers, with the sulfur atom as the chiral center.

$$R^1 \overset{S}{\underset{R^2}{\diagdown}} O$$

In sulfoxides, sulfur has the intermediate oxidation state of +4. Sulfoxides can therefore be reduced to form sulfides (thioethers) and oxidized to form sulfones by reversible redox reactions.

$$R^1-S-R^2 \underset{Red.}{\overset{Ox.}{\rightleftharpoons}} R^1-\underset{\underset{O}{\parallel}}{S}-R^2 \rightleftharpoons R^1-\underset{\underset{O}{\parallel}}{\overset{\overset{O}{\parallel}}{S}}-R^2$$

Physical properties of important sulfoxides are listed in Table 2.

Table 2. Physical data and molecular formulas for some industrially important sulfoxides

Sulfoxide	mp, °C (bp, °C/kPa)	Molecular formula (molecular mass)
$C_6H_5-CH_2-SO-CH_2-C_6H_5$	134–135	$C_{14}H_{14}OS$ (230.33)
$C_6H_5-SO-CH_3$	33–34	C_7H_8OS (140.21)
$C_6H_5-SO-C_6H_5$	69–71	$C_{12}H_{10}OS$ (202.28)
$C_6H_5-SO-CH_2-COOH$	118–119	$C_8H_8O_3S$ (184.21)
$CH_3-CH_2-SO-CH_2-CH_3$	(45–47/0.015)	$C_4H_{10}OS$ (106.19)
tetrahydrothiophene-S-oxide (cyclic structure with S=O)	67–68	$C_5H_{10}OS$ (118.20)
cyclic ketone-sulfoxide structure	109–110	$C_5H_8O_2S$ (132.18)
cyclic structure	91–92	$C_7H_{12}O_2S$ (160.24)
$C_6H_5-SO-CH_2-CH_3$	(68–70/0.005)	$C_8H_{10}OS$ (154.23)
$CH_3-(CH_2)_3-SO-(CH_2)_3-CH_3$	(105/0.3)	$C_8H_{18}OS$ (162.30)
4-$H_3C-C_6H_4-SO-CH_3$	38–39	$C_8H_{10}OS$ (154.23)
$C_6H_5-CH_2-SO-CH_2-CH_3$	51–52	$C_9H_{12}OS$ (168.26)
$C_6H_5-SO-CH_2-C_6H_5$	122–123	$C_{13}H_{12}OS$ (216.30)
$CH_3-(CH_2)_2-SO-(CH_2)_2-CH_3$	(73–75/0.1)	$C_6H_{14}OS$ (134.24)
$C_6H_5-CH_2-SO-CH_3$	53–54	$C_8H_{10}OS$ (154.23)
$C_6H_5-SO-CH=CH_2$	(105–110/0.15)	C_8H_8OS (152.22)
4-$H_3CO-C_6H_4-SO-CH_3$		$C_8H_{10}O_2S$ (170.23)
$C_6H_5-SO-CH_2-CH=CH_2$		$C_9H_{10}OS$ (166.24)
$C_6H_5-CO-CH_2-SO-C_6H_4-4-CH_3$	83–84	$C_{15}H_{14}O_2S$ (258.34)
4-$CH_3-C_6H_4-SO_2-CH_2-SO-C_6H_4-4-CH_3$	117–118	$C_{15}H_{16}O_3S_2$ (308.42)

2.2. Synthesis

2.2.1. Oxidation of Sulfides [1, pp. 207, 665], [2], [3, p. 121], [88], [89]

The oldest and most frequently used method for the production of sulfoxides is by oxidation of sulfides. Since further oxidation to the corresponding sulfones can occur, a range of selective reagents and methods has recently been developed. Oxidation with hydrogen peroxide, which has long been known, can be carried out in the presence of various catalysts. Acids, such as acetic [90], [91], perchloric [92], and sulfuric acids [93], [94] are suitable. A more recent industrial process uses 50% sulfuric acid and an organic acid, such as formic or acetic acid, in a 1:1 ratio as the catalyst [93]. Similarly, sterically hindered sulfides can be oxidized selectively and in high yield with hydrogen peroxide in the presence of a catalyst mixture of sulfuric acid and a secondary or tertiary alcohol [94]. Other suitable catalysts are titanium trichloride [95], selenium dioxide [96], tellurium dioxide [97], and vanadium pentoxide [98], which can act simultaneously as an indicator of excess hydrogen peroxide. Good yields of sulfoxide are obtained by using the complex from urea and hydrogen peroxide (UHP complex) together with half an equivalent of phthalic anhydride [99]. The use of organic peroxides, such as cyclohexyl or *tert*-butyl hydroperoxide has also been known for a long time [100]. Sulfide oxidation takes place quantitatively at −78 °C using 2-methoxy-2-hydroperoxypropane produced in situ from 2,3-dimethyl-2-butene and ozone [101]. Among the organic peroxycarboxylic acids, peracetic [102], perbenzoic [103], and monoperoxyphthalic acids [104], in particular, are effective oxidizing agents. One of the best selective oxidizing agents is *m*-chloroperoxybenzoic acid, which not only gives sulfone-free sulfoxides [105], but also reacts with a broad range of substrates [106] including unstable unsaturated thioethers [4], multifunctional heterocyclic thioesters [107], thioacetals [108], disulfides [109], and penicillins [110].

The longest-known oxidizing agent, used for sulfide oxidation as early as 1865, is nitric acid [111]. In the presence of gold (III) salts and by phase-transfer catalysis, oxidation with aqueous nitric acid in nitromethane is selective; oxidizable groups such as vinyl, tertiary amino, hydroxy, and diol groups are not attacked [112]. Very good yields of sulfoxide are also obtained with acetyl and benzoyl nitrates [113], and with the inorganic oxidants thallium(III) nitrate [114] and ceric ammonium nitrate (CAN). Oxidation with CAN alone is limited to diaryl sulfides, since in the presence of an α-hydrogen in the substrate molecule, the sulfoxide formed undergoes a Pummerer rearrangement under reaction conditions [115]. When CAN oxidation is carried out with sodium bromate as a co-oxidizing agent [116] or by phase-transfer catalysis [117], dialkyl and aryl alkyl sulfides can also be oxidized selectively. One of the most frequently used methods for effecting the sulfide – sulfoxide transformation is oxidation with sodium metaperiodate in aqueous methanol at 0 °C [118]. Modified processes employ sodium metaperiodate on an alumina carrier [119] or water-insoluble tetrabutylammonium metaperiodate in boiling chloroform [120]. Iodosobenzene [121] and *o*-iodosylbenzoic acid [122] are also effective oxidizing agents. Selective oxidation with halogen compounds, such as sodium bromite [123], benzyltrimethylammonium tribromide [124], and particularly sodium hypochlorite has been described. The latter can be used in alkaline media at pH > 10 for the production of pharmacologically active compounds [125].

In addition to sulfones, sulfoxides are also obtained by reaction of sulfides with Oxone [16]. In the presence of wet alumina, phenyl alkyl sulfides are converted to either sulfone-free sulfoxides or sulfones, depending on the quantity of Oxone used [126]. Diastereoselective oxidation with Oxone has also been described [127].

Synthesis of 4-Chlorobenzyl-N,N-Diethylcarbamoyl Sulfoxide [94]

$$Cl-\underset{}{\langle\underline{}\rangle}-CH_2-\overset{O}{\underset{\|}{S}}-\overset{O}{\underset{\|}{C}}\underset{\underset{C_2H_5}{N}}{\diagdown}C_2H_5$$

10 mol (2.575 kg) 4-chlorobenzyl-N,N-diethylthiocarbamate are dissolved in 1.5 L of toluene and treated with 0.4 L of formic acid and 0.4 L of 50% sulfuric acid. Then 1.07 kg of 35% hydrogen peroxide (11 mol) is added dropwise with stirring and ice-cooling in such a way that the temperature of the reaction mixture does not exceed that of the surroundings. After the mixture has been stirred for 4 h at room temperature, the organic phase is separated and neutralized with 20% aqueous NaOH. The toluene phase is washed with water and evaporated to give the sulfoxide in 95% yield (purity 95%).

Synthesis of Phenyl Vinyl Sulfoxide [4]

$$Ph-\overset{O}{\underset{\|}{S}}-CH=CH_2$$

A solution of 20 g (0.147 mol) phenyl vinyl sulfide in 250 mL of dichloromethane, is cooled to $-78\,°C$, and a solution of 25.4 g (1.0 equivalent) of m-chloroperoxybenzoic acid in 200 mL of dichloromethane is added dropwise with stirring over 30 min. The reaction mixture is then warmed for 1 h in a water bath at $30\,°C$ and poured into 300 mL of saturated hydrogencarbonate solution. The organic phase is separated, and the aqueous phase is extracted with 3250 mL of dichloromethane. The combined organic phases are dried over magnesium sulfate, and after evaporation of the solvent, the residue is distilled. About 15–16 g (68–70%) of the sulfoxide is obtained as a colorless liquid, *bp* $98\,°C$ (80 Pa).

2.2.2. Reaction of Sulfur Dioxide with Aromatics

Simple aromatic compounds can be converted into the corresponding sulfoxides in good yield with sulfur dioxide and an equimolar quantity of Magic Acid ($FSO_3H \cdot SbF_5$). Unsymmetrical sulfoxides are also accessible by this route [128].

$$2\,Ar-H + SO_2 \xrightarrow{FSO_3H \cdot SbF_5} Ar-\overset{O}{\underset{\|}{S}}-Ar$$

2.2.3. Sulfoxides from Sulfinates

Reaction of sulfinates with organometallic compounds, such as Grignard reagents and organocopper or organolithium reagents, is important for the production of optically active sulfoxides.

$$R^1-S(=O)-O-R + R^2-MgX \longrightarrow R^1-S(=O)-R^2$$

An extension of this reaction is the Claisen condensation of enolate anions with methyl arylsulfinates, giving α-keto sulfoxides [129].

$$R^1-C(O^-Na^+)=CH-R^2 + Ar-S(=O)-O-CH_3 \longrightarrow R^1-C(=O)-CH(R^2)-S(=O)-Ar$$

2.2.4. α,β-Unsaturated Sulfoxides

A range of vinyl sulfoxides is accessible through trapping reactions of thermally produced sulfenic acids with alkynes [130].

$$R^1-SOH + R^2-C{\equiv}CH \longrightarrow R^1-S(=O)-CH=CH-R^2$$

2.2.5. Optically Active Sulfoxides

The production of chiral sulfoxides, which play an important role in asymmetric synthesis, is described extensively in the recent literature. The following methods are used: asymmetric synthesis, optical and kinetic separation of enantiomers, and stereospecific synthesis; for reviews, see [3], [131].

2.3. Reactions

2.3.1. Reactions of α-Sulfinyl Carbanions

Strong bases, such as butyllithium, sodium hydride, or lithium dialkylamide, deprotonate sulfoxides with α-hydrogens to give α-sulfinyl carbanions. The latter react stereoselectively with organic electrophiles to form C–C bonds [1], [2, p. 583].

$$H_3C-\overset{O}{\underset{\|}{S}}-CH_2-H \xrightarrow{BuLi} H_3C-\overset{O}{\underset{\|}{S}}-CH_2^- Li^+$$

$$\xrightarrow{R-CH_2-Br} H_3C-\overset{O}{\underset{\|}{S}}-CH_2-CH_2-R$$

$$\xrightarrow{R-CO-O-R'} H_3C-\overset{O}{\underset{\|}{S}}-CH_2-\overset{O}{\underset{\|}{C}}-R$$

$$\xrightarrow{R^1\diagdown_{C=O}^{R^2}} H_3C-\overset{O}{\underset{\|}{S}}-CH_2-\overset{R^1}{\underset{OH}{C}}-R^2$$

Subsequent reactions consist of reductive cleavage of the sulfinyl group to a ketone and elimination with the formation of a double bond. These processes have been used for the synthesis of various ketones and carboxylic acids [132].

$$Ph-\overset{O}{\underset{\|}{C}}-\underset{R}{CH}-\overset{O}{\underset{\|}{S}}-CH_3 \xrightarrow{Al-Hg} Ph-\overset{O}{\underset{\|}{C}}-CH_2-R$$

$$Ph-\overset{O}{\underset{\|}{C}}-\underset{CO-O-C_2H_5}{CH}-\overset{O}{\underset{\|}{S}}-CH_3 \xrightarrow{\Delta}$$

$$Ph-\overset{O}{\underset{\|}{C}}-CH=CH-CO-O-C_2H_5$$

For α-halogenation of sulfoxides, which also proceeds via sulfinyl carbanions, various reagents have been described [1], [3].

$$R^1-\underset{\underset{O}{\|}}{S}-\overset{R^2}{\underset{R^3}{CH}} \xrightarrow{X^+} R^1-\underset{\underset{O}{\|}}{S}-\overset{R^2}{\underset{X}{C}}-R^3$$

Halogenation with N-halosuccinimides [133] and chlorination with N-,N-dichloro-p-toluenesulfonamide [134] give good yields.

2.3.2. Reactions with Grignard Reagents

Ligand exchange reactions between heterocyclic sulfoxides and Grignard reagents are used for production of the corresponding heterocyclic Grignard reagents. Subsequent reaction of these organometallic derivatives with carbonyl compounds provides a route to new heterocyclic compounds [135], [136].

$$\text{Py-SO-Ph} \xrightarrow{\text{Ph-MgBr}} \text{Py-MgBr} \xrightarrow{\text{Ph-CHO}} \text{Py-CH(OH)Ph}$$

A variant of this reaction is ligand coupling. The heterocyclic Grignard compound (**2**) formed initially reacts with the educt sulfoxide (**1**) to give the coupling product (**3**) [137].

$$\underset{\mathbf{1}}{\text{Py-SO-CH}_2\text{Ph}} \xrightarrow{\text{R-MgX}} \underset{\mathbf{2}}{\text{Py-MgX}} \xrightarrow{\mathbf{1}} \underset{\mathbf{3}}{\text{Py-CH}_2\text{Ph}}$$

This reaction is used mainly for the synthesis of bisaryls [138], [139].

2.3.3. Pummerer Rearrangement

In the Pummerer rearrangement, sulfoxides having at least one α-hydrogen are treated with acids or strong electrophiles such as carboxylic acid anhydrides or chlorides to form α-substituted sulfides as the primary products.

$$\underset{R^2}{\overset{R^1}{\diagdown}}\text{CH-S(=O)-R}^3 + (R^4\text{-CO-})_2\text{O} \longrightarrow R^2\text{-C}(R^1)(\text{O-CO-R}^4)\text{-S-R}^3$$

The Pummerer rearrangement is used widely in synthesis [1], [3]. For example, α-dicarbonyl compounds are formed from β-keto sulfoxides on acidic workup.

$$\text{R-CO-CH}_2\text{-S(=O)-CH}_3 \xrightarrow{H^+} \text{R-CO-CHO} + \text{H}_3\text{C-SH}$$

Cycloalkyl phenyl sulfoxides are converted into vinyl sulfides under conditions of the Pummerer rearrangement [140].

2.3.4. Oxidation and Reduction

Oxidation of sulfoxides to sulfones not only is a frequently occurring side reaction in the production of sulfoxides, but is also used for the synthesis of sulfones (see Section 2.2.1).

Reduction of sulfoxides to sulfides is achieved with hydrogen in ethanol on Pd–C catalysts [141] or selectively with dichloroborane [142] even in the presence of acid chloride, ester, nitrile, or nitro groups. Deoxygenation of sulfoxides with zinc–dichlorodimethylsilane under very mild conditions [143] and with polystyryldiphenylphosphine [144] has also been described. These reactions have no effect on a wide variety of functional groups and permit a cleaner product isolation.

2.4. Uses

Many sulfoxides are used as intermediates in the production of pharmaceuticals, plant protection agents, and other physiologically active substances.

Sulfoxides also play an important role in the synthesis of optically active substances. A selection of recent publications in the patent literature on the use of sulfoxides is given below.

Crop Protection and Veterinary Drugs
  Insecticides and acaricides [145].
  Bactericides and fungicides [146].
  Plant growth regulators [147].
  Anthelmintics in veterinary medicine [148].

Pharmaceuticals
  Agents for treating gastrointestinal disorders [149].
  Enantiomerically pure sulfoxides as cardiovascular medicines [150].
  Sulfoxide derivatives of 5-thiaprostaglandin E_1 act as vasodilators and inhibit thrombocyte aggregation [151].
  Thiocarbamoylsulfoxide derivatives exhibit antihypertensive action [152].
  Imidazolyl sulfoxides are anti-inflammatory and are also said to be suitable for treating asthma and skin diseases such as psoriasis [153].
  Trichlororuthenium complexes with a sulfoxide group have antitumor activity [154].
  Agents for treating allergic reactions, particularly asthma [155].
  Methionine sulfoxide lowers the liver toxicity of acetaminophen in humans [156].

Table 3. Comparison of physical properties of dimethyl sulfide and dimethyl sulfoxide

	Dimethyl sulfide	Dimethyl sulfoxide
mp, °C	−98	18.5
bp, °C	38	189
d, g/mL	0.846	1.101

Tetrahydrothiophen-1-oxide and its derivatives are inhibitors of alcohol dehydrogenase and therefore act as antidotes for alcohol poisoning [157].
Various sulfoxides are enzyme inhibitors [158].
Thiocarbamate sulfoxides have been described as agents for combating alcohol addiction [159].
Agents for treating sun-damaged skin [160].
Sulfoxides support the percutaneous absorption of a range of pharmaceuticals [161].

Others
Bis(2-hydroxyethyl) sulfoxide is used to treat oily hair [162].
Sulfoxides act as stabilizers for polymers and lubricating agents against heat, light, and oxygen [163].
Polyalkoxylated thioglycol sulfoxides are used as fabric softeners [164].
Use in refrigerating liquids for refrigerators and air-conditioning equipment [165].
Coupling agents in silver halide color films [166].

2.5. Dimethyl Sulfoxide

$$H_3C-S^{\delta+}(-O^{\delta-})-CH_3$$

Dimethyl sulfoxide (DMSO) is a colorless, odorless, strongly hygroscopic liquid. It is readily miscible with water and other protic solvents such as alcohols. This property can be attributed to the formation of strong hydrogen bonds between the polarized S–O bond and water molecules. The strong polarity of the S–O bond, with a partial negative charge on oxygen (see formula) is shown by the high dipole moment (4.3 ± 0.1 D at 20 °C) and basicity ($pK_B = -2.7$) of dimethyl sulfoxide. Pure DMSO forms chain-like polymeric association complexes at room temperature. This gives rise to considerably higher melting and boiling points compared to dimethyl sulfide (see Table 3).

$$\cdots S(CH_3)(CH_3)-O\cdots S(CH_3)(CH_3)-O\cdots S(CH_3)(CH_3)-O\cdots$$

2.5.1. Production

In principle, DMSO can be produced by oxidation of dimethyl sulfide as described in Section 2.3.1. These methods are, however, often suitable only for small quantities. In industry, DMSO is produced by catalytic oxidation of dimethyl sulfide with oxygen or by oxidation with nitrogen dioxide [167]. The oxidation of dimethyl sulfide with oxygen at 7.2 MPa and 105 °C gives good yields of DMSO [168].

2.5.2. Uses

Dimethyl sulfoxide is an excellent solvent [169] that is widely used for reactions in modern chemistry. Because of its high polarity and good solvation of cations, DMSO dissolves not only a large number of organic compounds, but also many metal salts, particularly those of alkali and alkaline-earth metals. DMSO is also a good solvent for many polymers.

A range of different pharmacological properties and effects have been described for DMSO [170]. For example, it increases the capacity of the skin to absorb drugs, and exhibits both analgesic and bacteriostatic activity. It can be used for the treatment of infection by retroviruses [171] and as an anti-inflammatory agent. Furthermore, DMSO has been described as a preservative and cryo-protective agent for organ and tissue transplants [172].

In plant protection, DMSO can be used to ward off mammals (e.g., deer [173]) and as a growth stimulant [174].

2.5.3. Reactions

The so called dimsyl anion, formed on treatment of dimethyl sulfoxide with base, can react further like other sulfinyl carbanions (see Section 2.3.1).

$$H_3C-SO-CH_3 \xrightarrow{NaH} H_3C-SO-CH_2^-$$

Cyclic ketones react with DMSO in the presence of KOH to give 2-methylene-cycloalkanols [175].

The methylation of various aromatic systems by the dimsyl anion is also of interest [176].

3. Toxicology

3.1. Sulfones [177]–[181]

The toxicity of sulfones is generally low. The sulfonyl group is relatively stable metabolically. However, for some compounds, enzymatic reduction to the sulfide has been discussed. For the solvent *sulfolane* (tetramethylene sulfone), on oral administration to rats, an average lethal dose of 1.8–2.5 g/kg has been found. Various mammals show neurotoxic symptoms after oral administration, injection, inhalation, or dermal absorption of sulfolane. No information is available on the carcinogenic or teratogenic activity of sulfolane. *p-Chlorophenyl phenyl sulfone*, used as an insecticide, is a skin irritant; in animal experiments, liver and kidney damage occurs. For *2,4,5,4′-tetrachlorodiphenyl sulfone* (tetradifon, an acaricide), very low toxicity has been found in animal experiments.

3.2. Dimethyl Sulfoxide [178], [181]–[184]

Dimethyl sulfoxide has very low *acute toxicity*, and chronic effects are observable only at high doses. The lethal dose for humans is estimated to be ca. 10–12 mL/kg [177]. The toxic effects of DMSO, such as skin rash and itching after dermal application, hemolysis after intravenous infusion, and gastrointestinal disorders after swallowing, depend strongly on the concentration. DMSO–water mixtures have an irritant effect only at concentrations > 70 % DMSO.

In the mammalian organism, sulfoxides can be reduced to sulfides by tissue or microbial enzyme systems. Enzymatic oxidation to sulfones has also been described, therefore the toxic effect of sulfoxides can arise from both of these redox states.

Teratogenic and *mutagenic activity* has been described for DMSO. DMSO strengthens the action of other toxic substances due to absorption through the skin.

In handling DMSO, attention should be paid to the fact that hazardous chemical reactions can occur in some cases. Explosions have been reported as a result of thermal decomposition and on reaction with organic acid chlorides, nonmetal halides, perchloric acid, perchlorates, nitrates, and sodium hydride.

Only limited data can be found in the literature on the toxicity of other sulfoxides.

4. References

[1] *Houben-Weyl,* 4th ed., **9**, 223, **E 11**, 1132.
[2] S. Patai, Z. Rappoport, C. J. M. Stirling (eds.): *The Chemistry of Sulphones and Sulphoxides,* J. Wiley & Sons, New York 1988.
[3] T. Durst in D. N. Jones (ed.): *Comprehensive Organic Chemistry 3,* Pergamon Press, Oxford 1979.
[4] L. A. Paquette, R. V. C. Carr, *Org. Synth.* **64** (1986) 157.
[5] Synphar, EP 335 646, 1989 (M. Daneshtala et al.).
[6] C. S. Giam, K. Kikukawa, D. A. Trujillo, *Org. Prep. Proced. Int.* **13** (1981) 137. Sumitomo Pharmaceuticals, EP 322 800, 1989 (N. Ohashi, K. Fujimoto).
[7] J. Drabowicz, P. Lyzwa, M. Mikolajczyk, *Phosphorus Sulfur* **17** (1983) 169.
[8] J. B. Doherty et al., *J. Med. Chem.* **33** (1990) 2513.
[9] J. M. Blanco et al., *Tetrahedron: Asymmetry* **3** (1992) 749.
[10] K. Peseke, U. Schoenhusen, *J. Prakt. Chem.* **332** (1990) 679. S. T. Purrington, A. G. Glenn, *Org. Prep. Proced. Int.* **17** (1985) 227.
[11] A. McKillop, J. A. Tarbin, *Tetrahedron Lett.* **24** (1983) 505. W. Zhou, L. Zhang, X. Xu, Z. Zhang, *Wuji Huaxue Xuebao* **8** (1992) 88; *Chem. Abstr.* **117** (1992) 191 428 n.
[12] W. Adam, L. Hadjiarapoglou, *Tetrahedron Lett.* **33** (1992) 469.
[13] C. M. Rodriguez, J. M. Ode, J. M. Palazon, V. S. Martin, *Tetrahedron* **48** (1992) 3571.
[14] S. W. Kaldor, M. Hammond, *Tetrahedron Lett.* **32** (1991) 5043. W. Priebe, G. Grynkiewicz, *Tetrahedron Lett.* **32** (1991) 7353.
[15] Monsanto, US 4 966 731, 1990 (Y. Chou). Monsanto, EP 480 900, 1992 (Y. Chou).
[16] B. M. Trost, D. P. Curran, *Tetrahedron Lett.* **22** (1981) 1287.
[17] J. Holoch, W. Sundermeyer, *Chem. Ber.* **119** (1986) 269.
[18] M. Hirano, J. Tomaru, T. Morimoto, *Bull. Chem. Soc. Jpn.* **64** (1991) 3752.
[19] R. Davis, *Synth. Commun.* **17** (1987) 823.
[20] J. K. Crandall, C. Pradat, *J. Org. Chem.* **50** (1985) 1327. J. Wildeman, A. M. van Leusen, *Synthesis* 1979, 733.
[21] Z.-D. Liu, Z.-C. Chen, *Synth. Commun.* **22** (1992) 1997.
[22] A. Borchardt, A. Kopkowski, W. Wasilewski, A. Zakrzewski, *Chem. Abstr.* **118** (1992) 38 529 x.
[23] A. Borchardt, H. Janota, A. Zakrzewski, *Chem. Abstr.* **114** (1987) 81 139 z.
[24] Z.-Y. Yang, D. J. Burton, *J. Chem. Soc. Perkin Trans. 1,* 1991, 2058.
[25] G. Bram, A. Loupy, M. C. Roux-Schmitt, J. Sansoulet, T. Strzalko, J. Seyden-Penne, *Synth.* 1987, 56.
[26] G. K. Biswas, S. S. Jash, P. Bhattacharyya, *Indian J. Chem. Sect. B* **29 B** (1990) 491.
[27] H. Suzuki, Y. Nishioka, S. J. Padmanabhan, T. Ogawa, *Chem. Lett.* 1988, 727.
[28] S. S. Labadie, *J. Org. Chem.* **54** (1989) 2496. J.-L. Parrain, A. Duchene, J.-P. Quintard, *Tetrahedron Lett.* **31** (1990) 1857.
[29] L. L. Frye, E. L. Sullivan, K. P. Cusack, J. M. Funaro, *J. Org. Chem.* **57** (1992) 697.
[30] K. S. Kim, T. K. Kim, C. S. Hahn, *Chem. Abstr.* **113** (1989) 5812 c.
[31] K. Inomata et al., *Chem. Lett.* 1986, 289. K. Inomata et al., *Bull. Chem. Soc. Jpn.* **60** (1987) 1767.
[32] J. Barluenga et al., *J. Chem. Soc. Perkin Trans. 1,* 1987, 2605.
[33] C. Najera, B. Baldo, M. Yus, *J. Chem. Soc. Perkin Trans. 1,* 1988, 1029.
[34] C. J. M. Stirling, *Int. J. Sulfur Chem. Part B* **6** (1971) 277.
[35] G. K. Cooper, L. J. Dolby, *J. Org. Chem.* **44** (1979) 3414.

[36] J. Fayos, J. Clardy, L. J. Dolby, T. Farnham, *J. Org. Chem.* **42** (1977) 1349. P. Messinger, K. Treudler, *Arch. Pharm. (Weinheim Ger.)* **321** (1988) 441.
[37] P. Messinger, *Arch. Pharm. (Weinheim Ger.)* **306** (1973) 458.
[38] R. Kerber, J. Starnick, *Chem. Ber.* **104** (1971) 2035.
[39] J. M. Bazavova, V. M. Neplyuev, M. O. Lozinskii, *Zh. Org. Khim.* **18** (1982) 865; *Chem. Abstr.* **97** (1982) 91851.
[40] H. Maruyama, T. Hiraoka, *J. Org. Chem.* **51** (1986) 399.
[41] K. Bailey, B. R. Brown, B. Chalmers, *J. Chem. Soc. Chem. Commun.* 1967, 618.
[42] C. J. M. Stirling, *J. Chem. Soc.* 1964, 5856.
[43] J. Sinnreich, M. Asscher, *J. Chem. Soc. Perkin Trans. 1,*1972, 1543. Y. Amiel, *J. Org. Chem.* **36** (1971) 3697.
[44] A. C. Cope, D. E. Morrison, L. Field, *J. Am. Chem. Soc.* **72** (1950) 59. R. W. Hoffmann, W. Sieber, *Justus Liebigs Ann. Chem.* **703** (1967) 96.
[45] S. Braverman, T. Globerman, *Tetrahedron* **30** (1974) 3873.
[46] K. Hiroi, M. Yamamoto, Y. Kurihara, H. Yonezawa, *Tetrahedron Lett.* **31** (1990) 2619.
[47] G. Büchi, R. M. Freidinger, *J. Am. Chem. Soc.* **96** (1974) 3332.
[48] M. W. Thomsen, B. M. Handwerker, S. A. Katz, R. B. Belser, *J. Org. Chem.* **53** (1988) 906.
[49] K. Kondo, D. Tunemoto, *Tetrahedron Lett.* 1975, 1007.
[50] J. W. Lee, D. Y. Oh, *Bull. Korean Chem. Soc.* **12** (1991) 347; *Chem. Abstr.* **115** (1991) 114 077 d.
[51] J. W. Lee, D. Y. Oh, *Bull. Korean Chem. Soc.* **10** (1989) 392; *Chem. Abstr.* **112** (1989) 157 769 m. D. B. Reddy et al., *Phosphorus Sulfur Silicon Relat. Elem.* **70** (1992) 325. D. Villemin, A. Ben Alloum, *Synth. Commun.* **21** (1991) 63.
[52] E. Dominguez, J. C. Carretero, *Tetrahedron Lett.* **31** (1990) 2487. B. M. Trost, T. A. Grese, *J. Org. Chem.* **56** (1991) 3189.
[53] V. M. Naidan, G. D. Naidan, *Zh. Obshch. Khim.* **50** (1980) 2611; *Chem. Abstr.* **94** (1981) 191 834. N. D. Obushak, E. E. Bilaya, N. J. Ganushchak, *Zh. Org. Khim.* **27** (1991) 2372; *Chem. Abstr.* **116** (1991) 235 172 k.
[54] P. Robson, P. R. H. Speakman, D. G. Stewart, *J. Chem. Soc. C* 1968, 2180.
[55] W. E. Truce, D. P. Tate, D. N. Burdge, *J. Am. Chem. Soc.* **82** (1960) 2872.
[56] B. M. Trost, H. C. Arndt, P. E. Strege, T. R. Verhoeven, *Tetrahedron Lett.* 1976, 3477.
[57] E. J. Corey, M. Chaykovsky, *J. Am. Chem. Soc.* **87** (1965) 1345.
[58] T. S. Chou, M. L. You, *J. Org. Chem.* **52** (1987) 2224.
[59] X. Huang, H. Zhang, *Synth. Commun.* **19** (1989) 97.
[60] M. Fujii et al., *Bull. Chem. Soc. Jpn.* **61** (1988) 495.
[61] T. Mandai et al., *Tetrahedron Lett.* **24** (1983) 4993. C. Herve Du Penhoat, M. Julia, *Tetrahedron* **42** (1986) 4807.
[62] M. Julia, B. Badet, *Bull. Chem. Soc. Chim. Fr.* 1976, 525.
[63] L. A. Paquette, *Org. React. (N. Y.)* **25** (1977) 1. D. Scarpetti, P. L. Fuchs, *J. Am. Chem. Soc.* **112** (1990) 8084.
[64] R. D. G. Cooper, *J. Am. Chem. Soc.* **94** (1974) 1018.
[65] W. E. Truce, E. J. Madaj, Jr., *Sulfur Rep.* **3** (1983) 259.
[66] Nissan Chem. Ind., JP 63 216 860, 1987 (K. Ogura, T. Fujimo).Kurakay, EP 282 915, 1988 (T. Onishi et al.). Hoffmann-La Roche, EP 298 404, 1988 (K. Bernhard, S. Jaggli, P. Kreienbuhl, U. Schwieter). Roussel Uclaf, EP 348 254, 1989 (D. Babin, J. P. Demoute, J. Tessier).
[67] Bayer, EP 418 664, 1990 (M. Hoppe, K. Herd, H. Henk, F. Stoehr).BASF, EP 492 236, 1991 (C. Marschner et al.). Hoechst, EP 489 360, 1991 (J. Dannheim).
[68] Abbott Laboratories, EP 218 095, 1986 (D. S. Kenney, J. C. Kane, B. N. Devisetty).

[69]   Bayer, EP 486 798, 1991 (W. Brandes et al.). Duphar Int. Res., EP 301 613, 1989 (H. Dolman, J. Kuipers).
[70]   Hoechst, WO 92/11 237, 1991 (P. Braun, H. Mildenberger, B. Sachse, F. Zurmühlen). Bayer, EP 205 076, 1986 (C. Fest, W. Brandes, G. Hänssler, P. Reinecke).
[71]   Rhône-Poulenc, EP 351 332, 1989 (P. Desbordes, M. Euvrard).Sumitomo, JP 63 208 564, 1987 (O. Kirino et al.).
[72]   Abbott Laboratories, EP 258 878, 1987 (A. J. Crovetti, R. A. Smith, B. E. Melin, F. M. H. Casati).ICI, EP 273 549, 1987 (R. A. E. Carr, M. J. Bushell).
[73]   Nisshin Flour Milling, DE 3 620 632, 1986 (T. Takahashi, K. Hagihara, Y. Suzuki).
[74]   Sterling, EP 207 375, 1986 (R. G. Christians, M. R. Bell, J. L. Herrmann, C. J. Opalka). Sterling, US 5 053 405, 1990 (G. M. Pilling, J. P. Mallamo).
[75]   ICI, WO 90/08 761, 1990 (D. R. Brittain et al.).
[76]   Smithkline Beecham, WO 88/07 858, 1988 (W. E. Bondinell, H. S. Ormsbee). Tanabe Seyaku, EP 270 091, 1987 (Y. Honma et al.).
[77]   Merck Frosst Canada, EP 288 202, 1988 (H. W. R. Williams, R. N. Young).
[78]   Research Corporation Technologies, WO 92/06 683, 1992 (F. W. Wassmundt).
[79]   Teijin, EP 338 796, 1989 (A. Hazata et al.).
[80]   Pennwaldt, EP 225 981, 1985 (L. K. Huber, J. L. Reilly).
[81]   Bayer, DE 3 921 691, 1989 (W. Broda et al.).
[82]   Konica, JP 02 110 544, 1988 (M. Nishizeki, N. Tachibana, N. Kagawa).
[83]   Kodak, WO 90/13 852, 1990 (M. Crawley).
[84]   Wako Pure, EP 440 375, 1991 (F. Urano, M. Nakahata, H. Fujie, K. Oono).
[85]   Du Pont, WO 91/08 198, 1990 (L. T. Cheng, A. E. Feiring, W. Tam).
[86]   Monsanto, EP 267 175, 1987 (D. R. Dyrott, D. P. Getman, J. K. Glascock).
[87]   Henkel, WO 91/06 534, 1989 (A. Meffert et al.).
[88]   J. Drabowicz, M. Mikolajczyk, *Org. Prep. Proced. Int.* **14** (1982) 45.
[89]   M. Madesclaire, *Tetrahedron* **42** (1986) 5459.
[90]   O. Hinsberg, *Ber. Dtsch. Chem. Ges.* **43** (1910) 289.
[91]   G. Schill, P. R. Jones, *Synthesis* 1974, 117.
[92]   A. Ceruiani, G. Modena, P. E. Todesca, *Gazz. Chim. Ital.* **90** (1970) 383.
[93]   Nihon Tokushu Noyaku Seizo K. K., EP 91 052, 1983 (J. Saito, K. Shiokawa, T. Takemoto).
[94]   J. Drabowicz, P. Lyzwa, M. Popielarczyk, M. Mikolajczyk, *Synthesis* 1990, 937.
[95]   Y. Watanabe, T. Numata, S. Oae, *Synthesis*1981, 204.
[96]   J. Drabowicz, M. Mikolajczyk, *Synthesis* 1978, 758.
[97]   K. S. Kim, H. J. Hwang, C. S. Cheong, C. S. Hahn, *Tetrahedron Lett.* **31** (1990) 2893.
[98]   F. E. Hardy, R. P. H. Speakman, P. Robson, *J. Chem. Soc. C* 1969, 2334.
[99]   R. Balicki, L. Kaczmarek, P. Nantka-Namirski, *Liebigs Ann. Chem.* 1992, 883.
[100] L. Bateman, K. R. Hargrave, *Proc. Roy. Soc. London Ser. A* **224** (1954) 389.
[101] P. Leriverend, M.-L. Leriverend, *Synthesis* 1987, 587.
[102] Taiho Pharmaceutical Co., JP-Kokai 58 222 064, 1983 (S. Yasumoto et al.); *Chem. Abstr.* **100** (1984) 209 381 c.
[103] P. Laur in A. Senning (ed.): *Sulfur in Organic and Inorganic Chemistry*, vol. **3**, Marcel Dekker, New York 1971, p. 203.
[104] B. Rajanikanth, B. Ravindranath, *Indian J. Chem. Sect. B* **23 B** (1984) 877.
[105] A. L. Ternay, Jr., L. Ens, J. Herrmann, S. Evans, *J. Org. Chem.* **34** (1969) 940.
[106] S. Florio, J. L. Leng, C. J. M. Stirling, *J. Heterocyclic Chem.* **19** (1982) 237. D. E. Beattie, R. Crossley, K. H. Dickinson, G. M. Dover, *Eur. J. Med. Chem. Chim. Ther.* **18** (1983) 277.

[107] F. H. Walker, US 4304-916, 1981; *Chem. Abstr.* **96** (1982) 122 654 u.
[108] R. Kaya, N. R. Beller, *J. Org. Chem.* **46** (1981) 196.
[109] F. Freeman, C. N. Angeletakis, *J. Org. Chem.* **50** (1985) 793.
[110] J. A. Webber, E. M. van Heyningen, R. T. Vasileff, *J. Am. Chem. Soc.* **91** (1969) 5674.
[111] C. Märcker, *Justus Liebigs Ann. Chem.* **136** (1865) 891.
[112] F. Gasparini et al., *Tetrahedron* **39** (1983) 3181. F. Gasparini et al., *J. Org. Chem.* **55** (1990) 1323.
[113] R. Low, H. P. W. Vermeeren, J. J. A. van Asten, W. J. Ultee, *J. Chem. Soc. Chem. Commun.* 1976, 496.
[114] Y. Nagano et al., *Tetrahedron Lett.* 1977, 1345.
[115] T. L. Ho, C. M. Wong, *Synthesis* 1972, 562.
[116] T. L. Ho, *Synth. Commun.* **9** (1979) 237.
[117] E. Baciocchi, A. Piermattei, R. Ruzziconi, *Synth. Commun.* **18** (1988) 2167.
[118] C. R. Johnson, J. E. Keiser, *Org. Synth.* **46** (1966) 78.
[119] K. T. Liu, Y. C. Tong, *J. Org. Chem.* **43** (1978) 2717.
[120] E. Santaniello, A. Manzocchi, C. Farachi, *Synthesis* 1980, 563.
[121] K. R. Roh, K. S. Kim, Y. H. Kim, *Tetrahedron Lett.* **32** (1991) 793.
[122] H. E. Folsom, J. Castrillion, *Synth. Commun.* **22** (1992) 1799.
[123] T. Kageyama, Y. Ueno, M. Okawara, *Synthesis* **1983,** 815. Nippon Silica Industrial Co., JP-Kokai 2 040 354, 1990 (T. Morimoto); *Chem. Abstr.* **113** (1990) 39 528 e.
[124] S. Kajigaeshi, K. Murakawa, S. Fujisaki, T. Kakinami, *Bull. Chem. Soc. Jpn.* **62** (1989) 3376. Ube Industries, JP-Kokai 2 268 150, 1990 (S. Kajisori); *Chem. Abstr.* **114** (1990) 184 990 z.
[125] Zambon S.p.A., EP 125 654, 1984 (M. Meneghin); *Chem. Abstr.* **102** (1985) 95 392 r.
[126] R. P. Greenhalgh, *Synlett* 1992, 235.
[127] G. J. Quallich, J. W. Lackey, *Tetrahedron Lett.* **31** (1990) 3685.
[128] K. K. Laali, D. S. Nagvekar, *J. Org. Chem.* **56** (1991) 1867.
[129] R. M. Coates, H. D. Pigott, *Synthesis* 1975, 319. H. J. Monteiro, J. P. de Souza, *Tetrahedron Lett.* 1975, 921.
[130] E. Block, J. O'Connor, *J. Am. Chem. Soc.* **96** (1974) 3929. D. N. Jones, P. D. Cottam, J. Davis, *Tetrahedron Lett.* 1979, 4977.
[131] H. B. Kagan, *Phosphorus Sulfur* **27** (1986) 127.
[132] M. Trost, K. K. Leung, *Tetrahedron Lett.* 1975, 4197. G. A. Russell, L. A. Ochrymowycz, *J. Org. Chem.* **34** (1969) 3624.
[133] S. Iriuchijima, G. Tsuchihashi, *Synthesis* 1970, 588.
[134] Y. H. Kim, S. C. Lim, H. R. Kim, D. C. Yoon, *Chem. Lett.* 1990,79.
[135] N. Furukawa et al., *Tetrahedron Lett.* **27** (1986) 3899. T. Shibutani, H. Fujihara, N. Furukawa, S. Oae, *Heteroat. Chem.* **2** (1991) 521.
[136] Wako Pure Chem. Ind., JP 62 212 370, 1986 (N. Furukawa, T. Fujiwara, N. Shibuya); JP 64 003 169, 1987 (N. Furukawa, T. Fujiwara, N. Shibuya).
[137] T. Kawai et al., *Phosphorus Sulfur* **34** (1987) 139.
[138] S. Oae, T. Kawai, N. Furukawa, *Phosphorus Sulfur* **34** (1987) 123.
[139] Wako Pure Chem. Ind., JP 03 024 058, 1989 (S. Wakabayashi, J. Uenishi, S. Daikyo).
[140] P. Bakuzis, M. L. F. Bakuzis, *J. Org. Chem.* **50** (1985) 2569.
[141] K. Ogura, M. Yamashita, G. Tsuchihashi, *Synthesis* 1975, 385.
[142] H. C. Brown, N. Ravindran, *Synthesis* 1973, 506.
[143] K. Nagasawa, A. Yoneta, T. Umezawa, K. Ito, *Heterocycles* **26** (1987) 2607.
[144] R. A. Amos, *J. Org. Chem.* **50** (1985) 1311.

[145] Chem. AG Bitterfeld, DE 4 032 414, 1990 (B. Biber et al.). ICI, EP 273 549, 1987 (R. A. E. Carr, M. J. Bushell).
[146] Suntory, JP 3 197 483, 1989 (M. Ishiguro, R. Tanaka). ICI, EP 390 394, 1990 (F. F. Morpeth, M. Greenhalgh). BASF, DE 3 328 770, 1983 (H. Graf et al.).
[147] Bayer, EP 119 457, 1984 (H. G. Schmitt, K. Lürssen, K. Wedemeyer).
[148] Elf Aquitaine, EP 209 462, 1986 (J. C. Gautier, J. Komornicki, J. Foix, G. Pastor). Richter Gedeon, DE 3 506 998, 1985 (J. Kreidl et al.).
[149] Hoechst, EP 298 440, 1988 (H. J. Lang, K. Weidmann, A. W. Herling).
[150] Merck, DE 3 621 112, 1986 (R. Jonas, H. Wurziger, J. Pinlats, M. Klockow).
[151] Teijin, JP 59 175 465, 1983 (A. Hasato, T. Tanaka, S. Kurozumi).
[152] Mitsubishi, JP 3 232 853, 1990 (H. Okujima et al.).
[153] Hoffmann-La Roche, US 4 973 599, 1989 (N. W. Gilman, W. Y. Chen).
[154] Boehringer Biochemia Robin, EP 471 709, 1990 (E. Alessio et al.).
[155] Boehringer, EP 232 820, 1987 (E. Reinholz et al.).
[156] G. M. Rosen, US 4 314 989, 1982.
[157] Univ. Iowa Res. Foundation, US 4 482 568, 1984 (B. V. Plapp, V. K. Chadha).
[158] Merck, US 4 617 301, 1984 (A. A. Patchett, M. J. Wyvratt). Merck, US 4 670 470, 1985 (R. A. Firestone).
[159] Faiman, US 5 153 219, 1991 (M. D. Faiman, B. W. Hart, A. Madan).
[160] Hoffmann-La Roche, US 5 061 733, 1990 (G. F. Bryce, S. S. Shapiro).
[161] Sagami Chem. Res. Centre, JP 2 036 131, 1988 (K. Matsui, T. Aoyanagi, K. Suzuki, M. Yamamura).
[162] L'Oreal, US 4 888 164, 1986 (J. F. Grollier, J. Maignan).
[163] Ciba Geigy, EP 473 549, 1991 (H. R. Meier, P. Dubs).
[164] Henkel, DE 4 021 694, 1990 (G. Stoll, P. Daute, J. Wegener, F. Berger).
[165] Lubrizol, EP 404 903, 1989 (S. T. Jolley, M. F. Salomon).
[166] Fuji Photo Film, JP 59 180 559, 1983 (T. Kamio, K. Aoki, J. Arakawa).
[167] L. Field, *Synthesis* 1972, 101.
[168] P. E. Correa, D. P. Riley, *J. Org. Chem.* **50** (1985) 1787.
[169] D. Martin, A. Weise, H.-J. Niclas, *Angew. Chem.* **79** (1967) 340. W. O. Ranky, D. C. Nelson in N. Kharash (ed.): *Organic Sulfur Compounds*, vol. **1**, Pergamon Press, New York 1961, p. 170. A. J. Parker, *Q. Rep. Sulfur Chem.* **3** (1968) 185.
[170] S. W. Jacob, R. Herschler, *Cryobiology* **23** (1986) 14. C. F. Brayton, *Cornell Vet.* **76** (1986) 61. B. N. Swanson, *Rev. Clin. Basic Pharmacol.* **5** (1985) 1.
[171] Rhoderton, EP 320 271, 1988 (J. M. Vichipascu).
[172] N. G. Leveskis, US 4 512 337, 1981.
[173] Boehringer, DE 1 642 289, 1967 (H. Hildebrand).
[174] Kemer Agric. Res. Inst., SU 1 386 071, 1985 (N. L. Voronova, V. J. Kozyakov, A. M. Yufferrov).
[175] B. A. Trofimov, O. V. Petrova, A. M. Vasil'tsov, A. J. Mikhaleva, *Izv. Akad. Nauk SSSR Ser. Khim.* 1990, 1601.
[176] G. A. Russell, S. A. Weiner, *J. Org. Chem.* **31** (1966) 248. J. K. Stamos, *Tetrahedron Lett.* **26** (1985) 2787.
[177] W. Braun, A. Dönhardt: *Vergiftungsregister*, Thieme Verlag, Stuttgart 1982.
[178] A. G. Renwick in L. A. Damani (ed.): *Sulphur-Containing Drugs and Related Organic Compounds*, vol. **1**, Part B, Ellis Horwood Limited, Chichester 1989, p. 133.
[179] Industrieverband Pflanzenschutz e.V. (ed.): *Wirkstoffe in Pflanzenschutz- und Schädlingsbekämpfungsmitteln*, Pressehaus Bintz-Verlag, Offenbach 1982.

[180] M. V. Cone: *Chemical Hazard Information Profile-Draft Report. Sulfolane. Office of Toxic Substances,* US EPA, Washington D.C., 1984.
[181] *Ullmann,* 4th ed., **22,** 334.
[182] C. C. Willhite, P. J. Katz, *JAT J. Appl. Toxicol.* **4** (1984) 155.
[183] L. F. Rubin, *Ann. New York Acad. Sci.* **411** (1983) 6.
[184] L. Roth, U. Weller: *Gefährliche chemische Reaktionen,* 13. + 14. Ergänzung, ecomed 1992.